Handbook of Research on Driving Transformational Change in the Digital Built Environment

Jason Underwood
University of Salford, UK

Mark Shelbourn
University of Salford, UK

IGI Global
PUBLISHER of TIMELY KNOWLEDGE

A volume in the Advances in Civil and Industrial Engineering (ACIE) Book Series

Published in the United States of America by
IGI Global
Engineering Science Reference (an imprint of IGI Global)
701 E. Chocolate Avenue
Hershey PA, USA 17033
Tel: 717-533-8845
Fax: 717-533-8661
E-mail: cust@igi-global.com
Web site: http://www.igi-global.com

Library of Congress Cataloging-in-Publication Data

Names: Underwood, Jason, 1969- editor. | Shelbourn, Mark, 1973- editor.
Title: Handbook of research on driving transformational change in the digital built environment / Jason Underwood and Mark Shelbourn, editors.

Description: Hershey, PA : Engineering Science Reference, 2021. | Includes bibliographical references and index. | Summary: "This book focuses on some of the current developments in practice and education within the construction industry towards facilitating the transformation in the digitally-built environment. In particular, from a practice perspective, developments are presented to enhance the client's understanding of digitally-enabled collaboration, interoperability and open standards, and maturity/capability and offers approaches to embedding digital construction within education"-- Provided by publisher.
Identifiers: LCCN 2020025671 (print) | LCCN 2020025672 (ebook) | ISBN 9781799866008 (h/c) | ISBN 9781799866022 (eISBN)
Subjects: LCSH: Building information modeling. | Building information modeling--Study and teaching..
Classification: LCC TH438.13 .H35 2021 (print) | LCC TH438.13 (ebook) | DDC 624.0285--dc23
LC record available at https://lccn.loc.gov/2020025671
LC ebook record available at https://lccn.loc.gov/2020025672

This book is published in the IGI Global book series Advances in Civil and Industrial Engineering (ACIE) (ISSN: 2326-6139; eISSN: 2326-6155)

British Cataloguing in Publication Data
A Cataloguing in Publication record for this book is available from the British Library.

For electronic access to this publication, please contact: eresources@igi-global.com.

Advances in Civil and Industrial Engineering (ACIE) Book Series

Ioan Constantin Dima
University Valahia of Târgovişte, Romania

ISSN:2326-6139
EISSN:2326-6155

Mission

Private and public sector infrastructures begin to age, or require change in the face of developing technologies, the fields of civil and industrial engineering have become increasingly important as a method to mitigate and manage these changes. As governments and the public at large begin to grapple with climate change and growing populations, civil engineering has become more interdisciplinary and the need for publications that discuss the rapid changes and advancements in the field have become more in-demand. Additionally, private corporations and companies are facing similar changes and challenges, with the pressure for new and innovative methods being placed on those involved in industrial engineering.

The **Advances in Civil and Industrial Engineering (ACIE) Book Series** aims to present research and methodology that will provide solutions and discussions to meet such needs. The latest methodologies, applications, tools, and analysis will be published through the books included in **ACIE** in order to keep the available research in civil and industrial engineering as current and timely as possible.

Coverage

- Ergonomics
- Earthquake engineering
- Quality Engineering
- Construction Engineering
- Engineering Economics
- Coastal Engineering
- Optimization Techniques
- Materials Management
- Operations Research
- Hydraulic Engineering

IGI Global is currently accepting manuscripts for publication within this series. To submit a proposal for a volume in this series, please contact our Acquisition Editors at Acquisitions@igi-global.com or visit: http://www.igi-global.com/publish/.

The Advances in Civil and Industrial Engineering (ACIE) Book Series (ISSN 2326-6139) is published by IGI Global, 701 E. Chocolate Avenue, Hershey, PA 17033-1240, USA, www.igi-global.com. This series is composed of titles available for purchase individually; each title is edited to be contextually exclusive from any other title within the series. For pricing and ordering information please visit http://www.igi-global.com/book-series/advances-civil-industrial-engineering/73673. Postmaster: Send all address changes to above address.

Titles in this Series

For a list of additional titles in this series, please visit: http://www.igi-global.com/book-series/advances-civil-industrial-engineering/73673

Structural Dynamics and Static Nonlinear Analysis From Theory to Application
Mourad Belgasmia (Setif 1 University, Algeria)
Engineering Science Reference • © 2021 • 347pp • H/C (ISBN: 9781799843993) • US $195.00

Handbook of Research on Developing Smart Cities Based on Digital Twins
Matteo Del Giudice (Politecnico di Torino, Italy) and Anna Osello (Politecnico di Torino, Italy)
Engineering Science Reference • © 2021 • 674pp • H/C (ISBN: 9781799870913) • US $295.00

Data-Driven Optimization of Manufacturing Processes
Kanak Kalita (Vel Tech Rangarajan Dr. Sagunthala R&D Institute of Science and Technology, India) Ranjan Kumar Ghadai (Sikkim Manipal Institute of Technology, Sikkim Manipal University, Majhitar, India) and Xiao-Zhi Gao (University of Eastern Finland, Finland)
Engineering Science Reference • © 2021 • 298pp • H/C (ISBN: 9781799872061) • US $225.00

Developing and Monitoring Smart Environments for Intelligent Cities
Zaigham Mahmood (University of Northampton, UK & Shijiazhuang Tiedao University, China)
Engineering Science Reference • © 2021 • 367pp • H/C (ISBN: 9781799850625) • US $215.00

Applications and Challenges of Maintenance and Safety Engineering in Industry 4.0
Alberto Martinetti (University of Twente, The Netherlands) Micaela Demichela (Politecnico di Torino, Italy) and Sarbjeet Singh (Luleå University of Technology, Sweden)
Engineering Science Reference • © 2020 • 321pp • H/C (ISBN: 9781799839040) • US $265.00

Reconstructing Urban Ambiance in Smart Public Places
Hisham Abusaada (Housing and Building National Research Center (HBRC), Egypt) Ashraf M. Salama (University of Strathclyde, Glasgow, UK) and Abeer Elshater (Ain Shams University, Egypt)
Engineering Science Reference • © 2020 • 280pp • H/C (ISBN: 9781799838562) • US $225.00

Additive Manufacturing Applications for Metals and Composites
K.R. Balasubramanian (National Institute of Technology, Tiruchirappalli, India) and V. Senthilkumar (National Institute of Technology, Trichy, India)
Engineering Science Reference • © 2020 • 348pp • H/C (ISBN: 9781799840541) • US $205.00

IGI Global
PUBLISHER of TIMELY KNOWLEDGE

701 East Chocolate Avenue, Hershey, PA 17033, USA
Tel: 717-533-8845 x100 • Fax: 717-533-8661
E-Mail: cust@igi-global.com • www.igi-global.com

List of Contributors

Table of Contents

Detailed Table of Contents

Calls for the digital transformation of the construction sector in part revolve around a need for productivity improvements, with a focus upon project time and cost enhancements. The purpose of this work is to provide a state-of-the-art analysis of design management (DM) usually employed to oversee design quality by coordinating design information, typically on behalf of a construction contractor. DM methods, activities, and processes with respect to the potential and underutilisation of building information modelling (BIM) are discussed. A synthesis of recent research efforts is provided identifying further emerging, disruptive, but underutilised digital tools and technologies, which when integrated with BIM, are capable of supporting DM processes. This chapter will aid practitioners and researchers in the design, implementation, and management of digital tools, and provide greater support to the DM function on modern construction projects. It will also be of use to students for a grounding in BIM and BIM-related technologies.

The high rates of fatalities and injuries in the construction industry highlight the importance of work health and safety (WHS). Building information modelling (BIM) could significantly improve WHS management in construction, but integration has been limited. The chapter explores the application of BIM for WHS management and the rationale for market leaders' role to catalyse a BIM-enabled WHS management ecosystem through procurement strategies and tendering proficiency. An international comparative analysis of BIM WHS is presented along with a BIM drivers and barriers critique, including a clustering into seven areas of market, perceptions and attitudes, leadership, supply chain integration and collaborative practice, procurement strategies and tendering proficiency, decision tools, and technical functionality. A focussed decision framework is outlined. The findings provide insights for researchers and practitioners to understand how a BIM-enabled WHS management system can be facilitated and supported through the project lifecycle.

water accumulation have also been determined which may help for the construction of water storage or rainwater harvesting facilities. The chapter shows the role of GIS-based techniques for the hydrological analysis of difficult terrains which can be applied for planning and management of water storage facilities in underdeveloped areas.

This chapter describes blockchains and illustrates this explanation using the results of a prototype project for an industrial application for a construction project. The chapter describes the application and how modular software components can be used to assemble a blockchain solution. The chapter concludes with a design of the system architecture. The background to blockchain technology includes a description of the evolving nature due to communal, open software consortia and an accelerated prototyping of systems. Four recommendations are made in the chapter. These include the need to form consortia for prototyping applications, encouraging government involvement, the need for engagement with the open software development community, and the suggestion that systems should be designed to support Lean production. A final section offers a range of discussion topics on the current state of the technology and where to expect area of increased interest. These are summarized in three areas: Lean management, Industry 4.0 and smart cities, and topics around privacy and security.

The AEC industry is facing a digital transformation that is improving the efficiency of services involved in designing, building, and operating assets and the users' well-being. Such a transformation towards sustainable smart cities is underpinned by two disruptive technologies: building information modelling (BIM) and internet of things (IoT). In this chapter, the authors present a review of studies that have focused on the identification of BIM and IoT technologies applicable to AEC industry processes, describing suitable strategies for data collection, storage and sharing, and fields of application. They also present data from two case studies that help us to understand how BIM can support better facilities management. From the literature, it was found that process improvements are a predominant research focus, reinforcing that successful BIM and IoT adoption goes beyond the acquisition of technology. The case studies revealed that a framework for prioritising areas is still needed for giving direction to researchers and practitioners concerning where to start the digital transition.

The fourth industrial revolution transforms the built environment through several highly intelligence systems such as drones, 3D printers, robotics, as well as building information modelling (BIM) and geographic information systems (GIS). This transformation has widely been applied in buildings. However, to establish environmental sustainability in the built environment, this transformation needs to be expanded in other sectors that impact climate change such as construction and demolition waste materials. The aim of this chapter is to introduce a new conceptual model that can measure construction and demolition waste in real time and optimise their carbon footprint spatially. A quantitative methodology that embraces a measuring protocol and geospatial research method is proposed for this purpose. The proposed model is capable of measuring the recycling potential result in construction sites to support the circular economy as well as to mitigate the associated carbon dioxide emissions (CO2) with materials' embodied energy, transportation, and treatment.

Digital infrastructure comes up with an enormous opening and possibility to deepen the knowledge horizon of those who have an instinct for teaching and learning. Likewise, digital infrastructure benefits to reduce information management. Managing the educational system can be categorized into two important parts, namely managing knowledge and managing information. Digital infrastructure facilitates huge opportunities to expand the knowledge and ease of information management, to examine the issues of the infrastructure of ICT and lack of awareness about the new technology about the training and support in education, and to discuss smart education with the implementation of digital infrastructure to the development of education classrooms and institutions. The chapter also presents a framework of ICT infrastructure scenario and smart infrastructure strategies for education. This chapter examines the development of new emerging technologies to build digital infrastructure applications to enhance the education system.

The recent publication of the Roadmap to Digital Transition for Ireland's Construction Industry 2018-2021 clearly identified training and education as key priorities in the development of core BIM competences to stimulate the transition towards a more collaborative digital working environment. It identified an urgent need for a consistent and coherent digital experience for students in Irish education and industry to help grow capacity and maturity in the use of BIM and other innovative techniques. The higher education sector has a vital role to play. Key to this will be ensuring appropriate graduate knowledge, skills and competences, ongoing professional development and upskilling of higher education staff, and significant collaboration with industry. This chapter will present a best practice example of academic-industry collaboration, which resulted in the delivery of a flexible Level 8 programme for industry, the implementation of a BIM strategy within an academic context, and the establishment of a focused research group, which is currently engaged in ongoing applied research.

Lack of BIM skilled professionals and lack of education and BIM training are key challenges to BIM implementation. The provision of BIM education in higher education institutions is seen as the main solution to these challenges. This chapter is part of an on-going Ph.D. research about the implementation of BIM in the Dominican Republic, a country with interest in implementing BIM but suffering from many challenges including lack of BIM education. This study aims to present the development of a framework to guide the integration of BIM in university curricula in the D.R., which is part of a comprehensive framework, outcome of the Ph.D. research. The framework was developed from a thorough literature on BIM education as well as an analysis of existing BIM Education frameworks. Moreover, findings of the research about BIM education in the D.R. are presented to describe the presence of BIM education in the country.

The construction and project management graduates are entering an ever-changing workforce that will require a smarter way of working. Creation, use, and management of building information modelling (BIM) models is a critical part of this smarter world. The aim of this research is to develop a threshold capability framework within the context of a broader digital construction project management curriculum to enable the global integration of BIM into a construction management curriculum. This chapter reports the evaluation of the framework through an analysis of 21 interviews with key stakeholder groups. Much effort is required to guide Australian construction practitioners to embrace a greater use of BIM in practice. This research identifies that the concept of graduate resilience skills in students is critical for the success of such a transition. Infusing construction management digital literacy is a long term and evolving exercise, and confidence in delivery capability must be simultaneously built.

Preface

The last decade has witnessed a global trend in the drive for digital transformational change in the construction sector and wider built environment. While the construction sector is an integral part of the global economy (GDP) and a significant contributor to employment, in contrast to the productivity gains made in other sectors, its productivity has been on a downward trajectory or flat lining at best; being widely criticised for not being as productive as it should be (Deloitte, 2021; WEF, 2018; Wilen & Gadsby, 2020; McKinsey & Company, 2020). Such low productivity and inefficiencies, which have implications for the wider growth of national economies, have been attributed to the traditional nature of the sector in relation to as fragmented siloed structures, lowest-cost mentality, lack of a collaborative and team working culture, adversarial behaviours, and a lack of investment in innovation and technology. In an attempt to address these innate sector issues, national Government strategies have responded by mandating Building Information Modelling (BIM) on central government public procured built asset projects with the aim of transitioning towards driving whole-life value and better outcomes from their portfolios of built assets through the procurement of better quality digital information. This 'Golden Thread' (of digital information) means having appropriate information at hand for the right people to make informed decisions within the expected timeframes throughout the life cycle of built assets. The authors in Chapter 1 explore the perceptions, awareness, and operations on Industry 4.0's innovation and technological growth, from the context of public sector construction clients in Malaysia.

While the digital transformation in many developed nations has been driven through central government industrial strategies and BIM mandates, studies indicate a year-on-year increase in the awareness and utilisation of BIM, filtering through to BIM being used more on private sector projects. This suggests an industry developing its own momentum for driving BIM and its digital transformation forward with private sector clients taking the lead in their adoption of BIM and facilitating the drive for digital transformational change in the construction sector and wider built environment. In addition, it is evident that BIM is beginning to become business as usual as it becomes utilised on all or the majority of projects, with the recognition that the digitalisation of the construction sector will transform the way it operates and the industry as a whole; improving the sector's productivity, speed of delivery, health and safety, and ability to meet sustainability challenges. The authors of Chapters 4 and 5 explore the improvement of construction health and safety through BIM and digitalisation, while the authors of Chapter 10 introduce a new conceptual model to measure construction and demolition waste in real-time and spatially optimise the carbon footprint. On the other hand, studies, prior to the COVID pandemic, also suggested there was still a perception that the sector is not moving quickly enough into the adoption of supporting digital processes (ICE, 2020; NBS, 2020; WEF, 2018). However, the impact of the COVID-19 crisis is expected to accelerate the transformation of the industry (McKinsey & Company, 2020).

The drive for transformational change in the digital built environment that has been facilitated through BIM as a first step towards the digitalization of the industry, involves the application of people, processes, and enabling technologies and tools to generate and manage information about built environment assets during their whole life (Underwood et al., 2018; WEF, 2018). Fostering better outcomes from the built environment through the procurement and delivery of better digital information requires standardised information management processes through which the right people, with the required digital competencies are able to deliver the right information, at the appropriate time in order to make better informed decisions, which is enabled through the appropriate enabling digital technology and tools.

PEOPLE

While a key focus for businesses in the short-term is on improving digital skills to drive productivity and efficiencies, the key barriers to the utilisation of BIM and digital transformation continues to be attributed to a lack of in-house digital skills/expertise and training, and talent shortages (NBS, 2020; The Construction Manager, 2020; ICE, 2020; WEF, 2018). Furthermore, the most influential barrier to the uptake of BIM continues to be consistently attributed to the lack of individual digital competency pertaining to the dimensions of knowledge, skills, abilities, experience, behaviours, and attitude (Ozorhon & Karahan, 2017; Santos *et al.*, 2017; Liao *et al.*, 2018; Antwi-Afari *et al.*, 2018). Moreover, such studies collectively acknowledge the need for education and training as a resolution, but from distinct and limited perspectives, mainly focused on integrating BIM education and training within Higher Education (HE), with little emphasis on the role that organisations play in upskilling/reskilling their in-house culture to drive effective adoption. Furthermore, such low-levels of digitalisation could impact on talent retention and recruitment (ICE, 2020; McKinsey & Company, 2020). The authors of Chapter 16 present an approach for strengthening collaborative capacity in the construction industry, while the emergence of an effective digitally-enabled competency management ecosystem approach is proffered by the authors of Chapter 17 to address the upskilling/reskilling issue.

The silo nature of built environment HE, in direct response to meeting the accreditation requirements of their courses by specific professional intuitions, has significantly contributed to reinforcing the traditional innate silo fragmented mentality of the construction sector. In addition, the rate of digital transformational change within the sector raises the issue as to whether HEIs are able to keep pace by providing graduates that are fit-for-purpose. Therefore, education needs to reform in order to be able to provide prospective employees with the required digital skills, including the interdisciplinary skills required for digital collaboration (WEF, 2018). However, in parallel with the increasing momentum of digital transformational change within the industry, BIM/digital construction has also continued to become more widespread and be embedded within curricula across various levels of education. However, the approach has tended to be ad hoc and without consistency, and driven bottom-up by individual academics or schools/departments, rather than top-down as a strategic objective set at a school/department or institutional level. The authors of Chapters 12 and 13 introduce an approach in Ireland and the Dominican Republic, respectively, to implement BIM/digital construction into curricula using industry collaborations and frameworks. Initiatives such as the development of learning outcome frameworks and establishing graduate digital competencies aim to facilitate consistency in the development and delivery of digitally-enabled built environment training and education (see Chapter 16 for a discussion on how this can be achieved). However, as is similarly evident in the industry, the understanding, acceptance, and importance of digital transformation amongst HE academics within the built environment, engineering, architecture, etc. is still considerably low (NATSPEC, 2020). Therefore, a significant challenge to the transformation of built environment education from one that reinforces a disciplinary silo mentality is changing the culture and mindset that exists among many academics in relation to the digital transformation agenda. Chapter 14 demonstrates how such silos can be broken down in a Construction Management programme from Australia, and Chapter 15 uses case studies from the USA, UK, and Ireland, to signpost positive practices that educational institutions might embed in their program development of digital education.

In summary …

"...digital technologies are transforming every aspect of the built environment. Helping BE education take advantage of this new and connected world is important to the toolkit for the future BE professionals..."

PROCESS

While many still perceive BIM as 3D models or simply technology/tools, more of the industry are identifying with BIM and the digital transformation of the industry as a collaborative process, rather than the adoption and utilisation of a particular type of technology. The facilitation of BIM-enabled collaboration as a transformative approach in the built environment is discussed by the authors of Chapter 2, while the authors of Chapter 3 focus on how Design Management could aid practitioners and researchers in the design, implementation and management of digital tools, and provide greater support to the DM function on modern construction projects. The UK has played an active role in driving forward BIM adoption at a national level: through government initiatives, the strong support network that has been established, and its central role in developing standards. The success of the PAS 1192 series of standards has resulted in establishing a standardised (digital) information management process over the whole life cycle of a built asset, using BIM developed through the UK Government's commitment to drive digital transformational change through the mandate of BIM on central government public procured built asset projects. In addition, the success of the wider adoption of the PAS 1192 series of standards beyond the UK has subsequently has led to its internationalisation through its translation into the BS EN ISO 19650 standard (BSi, 2020). The indication of this success is that the level of utilisation of the BS EN ISO 19650 standard outside of the UK is similar to that of the UK.

TECHNOLOGY

In contrast to other sectors, construction has traditionally tended not to be innovative and has been slow in their adoption of technology and digitization (ICE, 2020; McKinsey, 2016). Fragmented industry structures/supply chains in construction complicate the value creation potential of sectors in that the average organisation in them is relatively small, 90+% of the industry is represented by SMEs and the average size of those organisations is 5 or fewer people. However, the construction sector can increase the value creation through innovation as the industry supply chains and clients and owner/operators evolve through the adoption of BIM and their digital transformational change.

This serves to facilitate establishing the foundational fundamentals for digital transformation enabled through a greater degree of collaborative and integrated structures, culture, processes, etc. along with improved and efficient information management and a shift from a narrowed focus on 'lowest/first-cost' capital delivery procurement to 'value-driven' life-cycle management and digital procurement of assets, and a data-driven approach. Thereby, such transformation facilitates the ability to capture value from data; increasing the sector's potential for value creation through utilising such innovation as Big Data, AI, Machine Learning, etc. by developing the sector's capabilities to capitalise on data together with strong market incentives to do so. Developing the industry's capabilities to capitalise on data through the adoption of a data-driven approach enables fostering better societal, economic, business, and environmental outcomes from the built environment through embracing digital twins and an ecosystem of connected digital twins (i.e. National Digital Twin) (CCSIC, 2020; National Infrastructure Commission, 2017; WEF, 2018; McKinsey & Company, 2020).

Technologies such as cloud computing, immersive technologies (e.g. virtual, augmented and mixed reality), and data capturing (e.g. drones, laser scanning, etc.) are now maturing and their wider utilisation is evident within the construction sector (McKinsey & Company, 2018; NBS, 2020). Examples of strategies for implementation of such technologies are discussed by the authors of Chapters 6 to 9. 3D printing, modularization, and robotics; digital twin models; and artificial intelligence and analytics are poised to be transformational for the construction industry and wider built environment. These are currently at an early adopter stage but are expected to be utilised within the next five years as an enabler to the evolving drive for digital transformational change within the sector. The authors of Chapter 18 provide a fitting conclusion by providing further understanding and clarity on smart cities with the inclusion of several initiatives and examples already applied in practice within the UK and across other countries.

This book pulls together leading authors in many of the key fields to showcase innovative research that is focused on addressing aspects of the issues outlined above.

ORGANISATION OF THE BOOK

The book is organised into 18 chapters. A brief description of each of the chapters is provided below:

Chapter 1 explores people's perceptions, awareness and operations on Industry 4.0's innovation and technological growth, which constitutes a new scenario for the future construction industry. The authors provide an overview of the knowledge, attitude, and practices (KAP) amongst public sector construction clients in Malaysia.

Chapter 2 focuses on the current improvements and approaches in facilitating the BIM-enabled collaboration as a transformative approach in the built environment.

Chapter 3 discusses Design Management (DM) methods, activities, and processes with respect to the potential and underutilisation of Building Information Modelling (BIM). The authors provide a synthesis of recent research efforts and identify further emerging, disruptive but underutilised digital tools and technologies, which when integrated with BIM, is capable of supporting DM processes. The authors propose the chapter will aid practitioners and researchers in the design, implementation and management of digital tools, and provide greater support to the DM function on modern construction projects, while also being useful to students for a grounding in BIM and BIM-related technologies.

Chapter 4 presents an international comparative analysis of BIM Work Health and Safety and a critique of the BIM drivers and barriers, clustering into seven areas, and outlines a focused decision framework. The authors also provides insights for researchers and practitioners to understand how a BIM-enabled Work Health and Safety management system can be facilitated and supported through the project lifecycle.

Chapter 5 focuses on the perception of professionals in the Hong Kong construction industry of the application of BIM for site safety. The authors suggest there being strong support for using BIM to improve safety outcomes and the potential use in planning, visualization and communication as the main reasons for supporting BIM for safety. They further suggest issues such as data accuracy, BIM competence, and resource implications as some of the concerns with BIM for safety.

Chapter 6 introduces a new conceptual model to measure construction and demolition waste in real-time and optimise their carbon footprint spatially. The authors contend that the proposed model is capable of measuring the recycling potential result in construction sites to support the circular economy as well as to mitigate the associated carbon dioxide (CO2) emissions with materials' embodied energy, transportation, and treatment phase.

Chapter 7 explores the application of modern innovative digital technologies in construction including virtual reality (VR)/augmented reality (AR), block chain, 3D printing, building information modelling (BIM), and off-site manufacturing. The chapter also includes recent case studies and examples from the New Zealand construction industry.

Chapter 8 explores the role of GIS-based techniques for the hydrological analysis of difficult terrains which can be applied for planning and management of water storage facilities in underdeveloped areas. The authors discuss the creation of Digital Elevation Models (DEMs) to study the topography of the area by using Point Krigging method of gridding.

Chapter 9 provides a description to blockchain technology and illustrates this explanation using the results of a prototype project for an industrial application for a construction project. The chapter also discusses the application and how modular software components can be used to assemble a blockchain solution. The chapter concludes with a design of the system architecture and a set of recommendations.

Chapter 10 present a review of studies that have focused on the identification of BIM and IoT technologies applicable to AEC industry processes, describing suitable strategies for data collection, storage and sharing, and fields of application. The authors also discuss data from two case studies that help to understand how BIM can support better facilities management.

Chapter 11 presents a framework of ICT infrastructure scenario and smart infrastructure strategies for education. The authors also examine the development of new emerging technologies to build digital infrastructure applications to enhance the education system.

Chapter 12 presents a best practice example of academic-industry collaboration that has resulted in the delivery of a flexible Level 8 programme for industry in Ireland. The chapter discusses the implementation of a BIM strategy within an academic context and the establishment of a focused research group, which is currently engaged in ongoing applied research.

Chapter 13 presents the development of a framework to guide the integration of BIM in university curricula in the Dominican Republic. The authors discuss the findings of the research in BIM education in the Dominican Republic to provide an insight into the presence of BIM education in the country.

Chapter 14 reports on the evaluation of a Threshold Capability framework within the context of a broader Digital Construction Project Management Curriculum to enable the global integration of BIM into a construction management curriculum. The authors suggest that the concept of Graduate Resilience skills in students is critical for the success for the transition of Australian construction practitioners embracing a greater use of BIM in practice. Furthermore, the authors argue that infusing construction management digital literacy is a long term and evolving exercise, and confidence in delivery capability must be simultaneously built.

Chapter 15 discusses how higher education is challenged to prepare design, engineering, and construction students who are ready to enter the workforce. This challenge is complex as educators strive to meet the criteria set forth by their institution, policy decisions/regulation, accreditation standards, and industry expectations. While workforce requirements typically respond to the national economy, digital transformation in the construction industry is global. Therefore, to gauge the levels of the challenges faced in academia, four case studies are used to compare different approaches to the mandated criteria. The authors summarise the opportunities and limitations construction educators face in the United States, Great Britain, and Ireland. They also signpost some of the positive practices that educational institutions might embed in their program development of digital (transformation) education.

Chapter 16 presents an approach for strengthening collaborative capacity in the construction industry. The authors propose 133 competencies based on a Collaborative Practice Model developed from six qualitative case studies. In addition, the authors discuss the mobilization of Simulation-Based Training (SBT) grounded in detailed scenarios linked to their model to address the challenge of development and an assessment instrument and a novel Collaborative Practice Index for the challenge of assessment. The authors also discuss digitizing the SBT as part of future work.

Chapter 17 explores how we create and support a digitally enabled, agile, competent, and ultimately, productive workforce and determine the key research questions that need to be addressed if Digital Built Britain is to provide return on investment, and succeed as the catalyst for evolving the manner in which we conceive, plan, design, construct, operate, and interact with the built environment. The authors propose a vision of a digital competency management ecosystem where interdependent stakeholders are incentivised to work together in coopetition to create, capture, infer, interpret, specify, integrate, accredit, apply, use, monitor, and evolve competence as a working (data) asset.

Chapter 18 concludes and provides a further understanding and clarity on smart cities with inclusion of several initiatives and examples already applied in practice within the UK and across other countries. The authors provide key recommendations for policy developers, local authorities, academics and students interested in the field.

Jason Underwood
University of Salford, UK

Mark Shelbourn
Birmingham City University, UK

REFERENCES

Antwi-Afari, M. F., Li, H., Pärn, E. A. & Edwards, D. J. (2018). Critical success factors for implementing building information modelling (BIM): A longitudinal review. *Automation in Construction, 91*(November), 100–110. doi:10.1016/j.autcon.2018.03.010

BSi. (2021). *BIM - Building Information Modelling - ISO 19650: Continually improve your standards in built environment.* Available at https://www.bsigroup.com/en-GB/iso-19650-BIM/

CCSIC. (2020). *Flourishing Systems: Re-envisioning infrastructure as a platform for human flourishing.* Cambridge Centre for Digital Built Britain.

Construction Manager. (2020). *Annual BIM survey: Poor digital skills hold back adoption*. Charted Institute of Building (CIOB). Available at https://www.constructionmanagermagazine.com/annual-bim-survey-poor-digital-skills-hold-back-adoption/

Deloitte. (2021). *2021 engineering and construction industry outlook: Explore digital transformation and business model trends*. Available from: https://www2.deloitte.com/us/en/pages/energy-and-resources/articles/engineering-and-construction-industry-trends.html

ICE. (2020). *Digital Transformation: Overcoming cultural and behavioural barriers. Institution of Civil Engineers*. ICE.

Liao, L., Ai, E., & Teo, L. (2018). Organizational Change Perspective on People Management in BIM Implementation in Building Projects. *Journal of Management Engineering*, *34*(3), 1–13. doi:10.1061/(ASCE)ME.1943-5479.0000604

McKinsey & Company. (2016). *Imagining construction's digital future*. Available at https://www.mckinsey.com/business-functions/operations/our-insights/imagining-constructions-digital-future

McKinsey & Company. (2018). *Seizing opportunity in today's construction technology ecosystem*. McKinsey & Company. Available at https://www.mckinsey.com/business-functions/operations/our-insights/seizing-opportunity-in-todays-construction-technology-ecosystem

McKinsey & Company. (2020). *The next normal in construction How disruption is reshaping the world's largest ecosystem*. Executive summary. Author.

National Infrastructure Commission. (2017). *Data for the public good*. National Infrastructure Commission.

NATSPEC. (2020). *ICIS - International BIM Education Report*. NATSPEC.

NBS. (2020). *10th Annual BIM Report 2020*. NBS Enterprises Ltd.

Ozorhon, B., & Karahan, U. (2017). Critical Success Factors of Building Information Modeling Implementation. *Journal of Management Engineering*, *33*(3), 04016054. Advance online publication. doi:10.1061/(ASCE)ME.1943-5479.0000505

Santos, R., Costa, A. A., & Grilo, A. (2017). Bibliometric analysis and review of Building Information Modelling literature published between 2005 and 2015. *Automation in Construction*, *80*, 118–136. Advance online publication. doi:10.1016/j.autcon.2017.03.005

Underwood, J., Shelbourn, M., & Fleming, A. (2018). *Streamlining a Design, Manufacture and Fitting Workflow within a UK Fit-out SME: A BIM Implementation Case Study. Contemporary Strategies and Approaches in 3D Information Modelling* (B. Kumar, Ed.). IGI-Global. doi:10.4018/978-1-5225-5625-1.ch010

WEF. (2018). *Shaping the Future of Construction: An Action Plan to Accelerate Building Information Modeling (BIM) Adoption*. World Economic Forum.

Wilen, A., & Gadsby, R. (2020). *UK Construction Industry Forecast 2021-2022*. Available from: https://www.glenigan.com/market_analysis/construction-industry-forecast/?ppc_keyword=&LeadSource=Web%20-%20PPC%20campaign&gclid=Cj0KCQjwyN-DBhCDARIsAFOELTkLBUYFaQbdhR77qTuh-1k6N2tR_zpe-3AviyvEiLbE6QU8_pJqq8oQaAuKDEALw_wcB

Section 1

Strategy

Chapter 1
A Knowledge, Attitude, and Practices (KAP) Study on Industry 4.0 Among the Public Sector Construction Client in Malaysia

Nadia Safura Zabidin
Universiti Teknologi MARA, Malaysia

Sheila Belayutham
https://orcid.org/0000-0003-2009-2603
Universiti Teknologi MARA, Malaysia

Che Khairil Izam Che Ibrahim
https://orcid.org/0000-0002-7728-3702
Universiti Teknologi MARA, Malaysia

ABSTRACT

The impact of Industry 4.0 will amplify globalization, automation, virtualization, integration, and agility towards the construction industry's development in light of different stakeholders in the industry. In this chapter, people's perceptions, awarenesses, and operations on Industry 4.0's innovation and technological growth, which constitutes a new scenario for future construction industry, will be explored. The outcome of this Industry 4.0 study provides an overview of the knowledge, attitude, and practices (KAP) among the public sector construction client in Malaysia. This would enable the establishment of the baseline (reference value) for future intervention strategies by having construction public sector, particularly the client, as the driver for change towards construction digitalisation.

DOI: 10.4018/978-1-7998-6600-8.ch001

INTRODUCTION

The new technological globalisation has revolutionized the transportation, communication and operation that changes human civilization, with new invention of machineries for minimum workforce practices. In 1784, the water and steam power engine were introduced as the first Industrial Revolution's production, followed by the second industrial revolution, where the electricity has led to labour and mass production in the beginning of the twentieth century. In the beginning of 1970s, the industry became automated through the use of electronic and information technology (IT) system. The fourth industrial revolution that engages with the term Industry 4.0 is defined as a design, fabrication, usage and operational transformation of products or services by blurring the distinction between the physical, digital and biological aspects. The latest industrial revolution builds and extends the impact of digitalisation in unprecedented ways, throughout the different economic sectors across the world.

The digital transformation through Industry 4.0 has been embraced by various sectors, such as manufacturing, finance, information and communication technology, as well as healthcare. In regards to that, construction, being one of the nation's building sector has been challenged by the current paradigm shift towards digitalisation in the industry. The construction industry is known to be slow in adopting technologies due to several well-identified factors such as the reluctance to change, fragmentation, lack of talented workforce, as well as complex projects with temporal organisations, multiple stakeholders and uncertain working environment. In addition, issues such as disputes on the lack of standards, higher cost in terms of equipment, regulatory compliance, legal and contractual uncertainties have proved to be a challenge for construction to embrace Industry 4.0. In contrary, previous studies have also shown that certain big players in the construction industry have well rode the wave of digital transformation and shown potential benefits by using digital machineries to predict failures, enhance process efficiencies, safety enhancement and reduction in construction cost and time. Therefore, one of the important initiatives in enhancing digitisation in construction is by building the culture of digital transformation across the industry and should be led by example by major players in the industry, particularly the public sector in construction.

The public sector in construction is an important entity to drive the economy of a nation. Public sector projects are generally driven by the central or local government. Big scale public funded infrastructure projects would deliver job opportunities to private sectors, such as contractors and consultants, as well as being the enabler for other sectors, such as manufacturing and finance. Given the highly influential nature of the entity, it is reasonable to perceive that the change in culture, particularly on digitalisation of the industry should begin with the construction public sector, particularly the client. Therefore, the government and public sector construction clients should be in a comfortable setting with Industry 4.0. Previous studies on Industry 4.0 in construction have investigated the insights among different stakeholders in order to strive towards effective transformation of construction life cycle and value chain through proper regulatory and legal framework. Studies have often focused on the practical and application aspect of Industry 4.0, such as specific technologies of Big Data, Building Information Modelling and Internet of Things. However, most of the studies have focused on the designers, contractors and suppliers, with little thought given to the client, particularly of the public sector. Exploring the current state of knowledge, attitude and practices among the public sector construction client is essential to understand the current state and further acts as the benchmark for improvement.

Therefore, this study aims to explore the current knowledge, attitude and practices (KAP) of Industry 4.0 among the Public Sector Construction Client. The outcome of this exploration could be set as a baseline for further improvements or interventions. The objectives of this study are as follows:

1. To identify the knowledge on Industry 4.0 among public sector construction client.
2. To recognise the feeling and attitude of public sector construction client towards Industry 4.0.
3. To identify the applications of Industry 4.0, practiced by the public sector construction clients.

Outcomes from the KAP study (knowledge gap, perception and practices) could be used as a baseline towards crafting intervention programmes to fill the current gap towards Industry 4.0 technological adoptions.

BACKGROUND

The construction industry, being one of the nation's building sector, has been challenged by the current paradigm shift towards digitalisation that impacts the economy, environment and people. Despite the vast transformation in digitisation, the construction industry is still facing the lack of Industry 4.0 ready talent, insufficient investments and other challenges that impede the movement towards adopting new technologies. However, the need to adopt and adapt technological transformation strongly influences the urgency to improve and transform the current traditional construction methods into more adequate and sustainable project value chain. Hence, this section explores the overview of Industry 4.0 in the construction industry, along with the insights on construction public sector clients' role towards the industry's transformation. The outcome of this literature exploration sets the scene for this study.

Industry 4.0

In the past few years, Industry 4.0 has become one of the most frequently discussed topics worldwide, as new changes take place in production, networking, data processing and people, towards establishing a comprehensive set of smart system. Industry 4.0, which was originally introduced during the Hannover Fair, Germany in year 2011 has heavily influenced the academia, as a steep upward trend could be observed in the number of papers published in this subject matter. In reference to the Scopus online database, the number of research papers on Industry 4.0 have drastically increased from year 2015 (208 publications) to 2019 (4054 publications), with most of the publications observed in the area of engineering, followed by computer science. In specific, the engineering subject area could be narrowed down to industrial and manufacturing engineering, electrical engineering and electronic engineering (Liao *et al.*, 2017). Various nature of research were discussed across the Industry 4.0 domain, such as concepts and theories, key technologies and interaction, and sustainability, through methodologies such as conceptual review, case study, simulation, experimentation, survey and prototype research approach (Kamble *et al.*, 2018).

There were various definitions of Industry 4.0, as proposed by previous researchers from the manufacturing perspective, such as "Industry 4.0 is the combination between Internet of Things and Services, Cyber Physical Systems, Cloud Computing and Big Data by self-optimization devices and machineries to enhance industry performance, inventing novel products and start a new business model" (Raja Sreedharan & Unnikrishnan, 2017). The key technological enabler for Industry 4.0 is a complex system that integrates all fundamental elements into value-added systems to embrace digital, communications, computer and automation technology by eliminating the digital and physical world boundaries (Pereira & Romero, 2017). As most of the authors have outlined, the definition of Industry 4.0 consists of high technology driven and intelligent services to control, organize and transform the entire value chain, where the epitome of Industry 4.0 affects the whole product lifecycle through the emergence in digital technologies.

The vision of digitisation has been powered by the nine pillars of Industry 4.0 that shapes the future by creating interactions between human and machine. The foundation of Industry 4.0 used in the manufacturing industry transforms the production lifecycle as a fully integrated, automated and optimized flows, which leads to better efficiencies and relationship among different stakeholders along the process (Vaidya *et al.,* 2018). Furthermore, it will be easier to assemble and analyse data across the machineries, faster and flexible configuration towards producing higher-quality products with minimal cost of processes by shifting the manufacturing productivity, growth, economics as well as changing the workforce talents (Rüßmann *et al.,* 2015). The nine technologies which are driving the industries forward are autonomous robots, simulation, system integration, internet of things, cybersecurity, the cloud, additive manufacturing, augmented analysis and big data analytics.

The technological advancements have commonly been dominated by the manufacturing industry, which has been adopting autonomous robots to solve complex production, logistic, activities and to control machineries in factories (Erboz, 2018). The big data analytic capabilities with variety of algorithm and data, from various channels in the production flow were used to monitor production, supporting predictive analysis, uncovering hidden patterns and unknown correlations (Zhong *et al.,* 2017). Besides that, the use of smart sensor-based application that combines Internet of Things and the clouds allows real-time data collection, effective analysis and synchronization of data management, which have been adopted in the manufacturing industries, healthcare and transportations (Kamble *et al.,* 2018). Although the advancements of the Industry 4.0 have been widely adopted to provide significant benefits in the aforementioned industries, the question still remains of whether similar trends could be observed in the construction engineering field.

INDUSTRY 4.0 IN CONSTRUCTION

The construction industry is one of the major economic contributors, which provides basic amenities and socioeconomic infrastructures that benefit a country's development, as well as to improve the living standards of the society. Although the construction industry's contribution is relatively small, as compared to the other anchored sectors, such as manufacturing and services, the industry still provides a sustainable economic growth in the domestic, as well as international markets. The construction industry is able to transform a country into competitive, sturdy and expanding nation by delivering complex mega infrastructure projects that generate huge scale of employment to the society. This has resulted in the need to adopt and ride along the wave of Industrial Revolution 4.0, in order to attain more smart materials, intelligent machineries, as well as digitalised technologies. Construction is mostly interrelated through the manufacturing and transportation sectors which provides machineries, materials, chemicals and more, with multiple stakeholders' involvement in projects. However, the construction industry is still known to be slow in adopting technologies due to several well-identified factors, such as the reluctance to change, lack of talented workforce, peculiar characteristics of projects with temporal organisations, legal and contractual uncertainties that have proved to be a challenge for construction to embrace Industry 4.0.

The construction industry is unique with complex characteristics in each one of the project designs and built phases, in meeting the specific needs of stakeholders, from developer, designer, engineer, contractor, supplier and the others (Tougwa, 2018). Industry 4.0 in the construction industry sector as the combination of various interdisciplinary technologies and concept that enables the digitization. automation and integration to transform the construction value chain (Oesterreich & Teuteberg., 2016). Furthermore, Oesterreich & Teuteberg (2016) have identified three key elements (smart factory, simulation and modelling, and digitisation and virtualisation) and 17 concepts adapted in the construction industry, such as cyber-physical system, radio-frequency identification, automation, modularization, building information modelling and the others. Even though the characteristics of the construction industry are not similar to the manufacturing industry, most of the concepts of Industry 4.0 in construction are still in parallel with the manufacturing counterpart.

In regards to the three key elements (Oesterreich & Teuteberg, 2016) of Industry 4.0 in construction, a wide range of smart factory integration in the construction industry deals with the integration of cyber-physical system (CPS) concepts between virtual worlds and physical construction, and to automate construction processes, such as real-time monitoring of the use of personal protective equipment (PPE). The other element introduces the essence of Building Information Modelling (BIM) as a 3-dimensional (3D) model presentation through the use of information and communication technology (ICT) in coordinating construction lifecycle and project management (Husain *et al.*, 2018). The last element, digitisation and virtualization, such as cloud-based solution through the use of any communication devices that can easily be accessed by the internet as a data sharing platform and digital collaboration in real-time project development. Most of the previous studies have portrayed specific Industry 4.0 applications and there is still a gap to be explored on the capability of the various construction stakeholders in embracing Industry 4.0, particularly the client.

PUBLIC SECTOR CONSTRUCTION CLIENT

In certain countries, particularly in Malaysia, the notion of Industry 4.0 has only recently been officially introduced in the construction industry, and the grey or unclear governmental initiative towards construction industry is apparent, as compared to the other sectors such as manufacturing. It is imperative that the initiative should first look into the ability and capability of stakeholders in the industry, such as the client, contractors and designers, as the transition of construction industry alongside technological advancements relies very much on the uptake of the people in the industry. The nature of different construction stakeholders and its organizations would suggest different kinds of challenges along the way of adopting and implementing Industry 4.0 in the project lifecycle. Given the globalised nature of Industry 4.0, government agencies should take the initial step to establish structured measures to create awareness and to promote the adoptions of Industry 4.0 in the construction industry. As a governmental agency that deals largely with construction, they often take the role as a client, where projects are commonly awarded to the bid-winning consultant or contractors to complete a construction project, in accordance to the clients' desire and needs.

Generally, the government agency under the Ministry of Works in Malaysia, namely the Public Works Department (PWD) remains the largest client in the construction public sector and is responsible for the commissioning of public projects. Big scale public funded infrastructure projects would deliver job opportunities to private sectors, such as contractors and consultants, as well as being the enabler for other sectors, such as manufacturing and finance. The construction project development authorized by agencies such as the PWD emphasizes on basic infrastructures, such as roads, houses, schools, hospitals, irrigation, drainage, and physical maintenance projects that could benefit and increases the living standards of the society. Nonetheless, the public construction projects are also faced with common construction issues such as project delays, cost overhead and lack of quality standards, which affects the state of a country's economy and eventually the reputation of PWD. Therefore, the drive in adopting new technological advances as one of the strategies to enhance the productivity in construction has been proposed in the Construction Industry Transformation Programme 2016-2020 (CIDB, 2016). However, it is essential to make known the ability and capability of a governmental agency, before embarking on a journey towards technology, as essentially, organizational transformation requires the people and process to be in proper alignment before deciding to move into the technological aspect.

The commitment of public sector construction client, such as PWD towards expanding the construction industry's performance has resulted in positive uptake on the adoption of innovation and technologies. As reported by the Australian construction industry, there is significant capabilities in PWD, which makes such agencies appear to be innovative, competent and influential in promoting innovation throughout the construction sector (Manley, 2006). The Building Information Modelling (BIM) (one of the construction-related Industry 4.0 elements mentioned in Oesterreich & Teuteberg, 2016) has been widely used by developed countries, such as the United States of America (USA), Australia and United Kingdom that creates a paradigm shift in the construction industry (Latiffi *et al.,* 2017). In Malaysia, BIM still poses a low maturity level since there is a lack of BIM demand by the PWD (Porwal & Hewage, 2013). Hence, a BIM-partnering concept based on PWD construction projects has been introduced, as a way forward with BIM, by enforcing the use of BIM in the contract of projects. In general, one of the most important factors in enhancing digitisation in construction is by building a culture of digital transformation across the industry and this should be led by key players in the industry, particularly public sector construction client.

CASE STUDY

Methodology

A Knowledge, Attitude and Practice (KAP) study has been adopted as the research tool, as this study intents to explore what is known, believed and done by the public sector client in regards to Industry 4.0 and its technological advancements. A KAP study provides avenues to quantitative and qualitative data information that intends to reveal the characteristic, behaviour, factors and idea of the study (Gumucio, 2011).The KAP study has its roots from the field of medicine and made its way into other fields, including construction, where the method has been adopted into subject areas such as construction safety (Che Ibrahim & Belayutham, 2020; Goh & Chua, 2015; Toh *et al.,* 2016), water resource (Fapohunda *et al.,* 2014), sustainable development (Ahamad & Ariffin, 2018), health and safety management (Chileshe & Dzisi, 2010), and engineering education (Mohamad *et al.,* 2019). However, limited study was found on the subject of Industry 4.0, particularly from the context of construction. Therefore, it is essential for a KAP study to be conducted in this subject matter, as a future reference of the current state.

This study adopts and adapts the comparative case study approach, where data was collected from three case projects managed by the PWD, which is one of the biggest public sector construction clients in Malaysia. The comparative case study approach was established to analyse and synthesize the similarities, difference and patterns across three case projects in order to enrich the generalization of the findings in collecting the information on what is perceived, expected and executed on Industry 4.0. The three case projects, from hereon named as Project A, Project B and Project C have been selected based on the scale of the projects, whereby all projects have contract values above MYR 100 million. Besides that, these projects are on-going projects under the jurisdiction of the Malaysian PWD. Project A involves the extension of a public hospital building, Project B is the construction of a new municipal council office building and Project C is the development of a transportation project. At the time of data collection, these three projects have been of the highest value under the management of the PWD and given the high value, these projects should have been in compliance with the current technological advancements, as they possess the capability to adopt such advancements.

The main data collection method used in this KAP study was face-to-face data collection. The structured survey was divided into two main sections, whereby the first section involves attaining the respondents' demographic information, such as organisation type, age, gender, education level, years of practice and position. The second section involves the three parts of KAP, whereby the first part of the KAP survey gathers information on the general idea/ overview of Industry 4.0, exposure, definitions and familiarity of the Industry 4.0 components that highlights the Knowledge (K) section of the study. The importance of Industry 4.0 adoptions, benefits and effectiveness were measured to evaluate the stakeholders' Attitude (A) pertaining to Industry 4.0. The final part intents to explore the accessibility of technological practices and the current education, in light of respondents' real practices, situation and responses, in order to address the Practices (P) aspect. All data will be further analysed through statistical descriptive and inferential analysis, as well as to compare and contrast between the different case studies. In order to complement the structured approach, semi-structured interviews have also been conducted among the key personnel from the case projects.

The KAP Study on Industry 4.0

Findings for the KAP study on Industry 4.0 among public sector construction clients will be reported as follows, where first, the demographic information of the respondents will be delineated. This is followed by the different sections that addresses the Knowledge, Attitude and Practices (KAP) of the respondents. The questionnaires were analysed using the Statistical Packages for the Social Sciences (SPSS) package version 23 and comprises of descriptive and inferential analysis in order to achieve the objectives of this study.

DISTRIBUTION OF RESPONDENTS

The selection of the targeted respondents begun with a list of on-going projects that was recommended by the Head of Department of the Public Works Department (PWD). Further discussion was conducted that led to the selection of three projects as study cases, after taking into consideration of factors, such as accessibility to project stakeholders, value of project and the potential use of technological advancements in the project. A total of 42 responses have been collected from the three case studies, identified as Project A, B and C, where Project A consists of 18 respondents (42.9%), 15 respondents from Project B (35.7%) and 9 respondents from Project C (21.4%). Figure 1 shows the distribution of respondents for this study.

Figure 1. Distribution of respondents by the projects

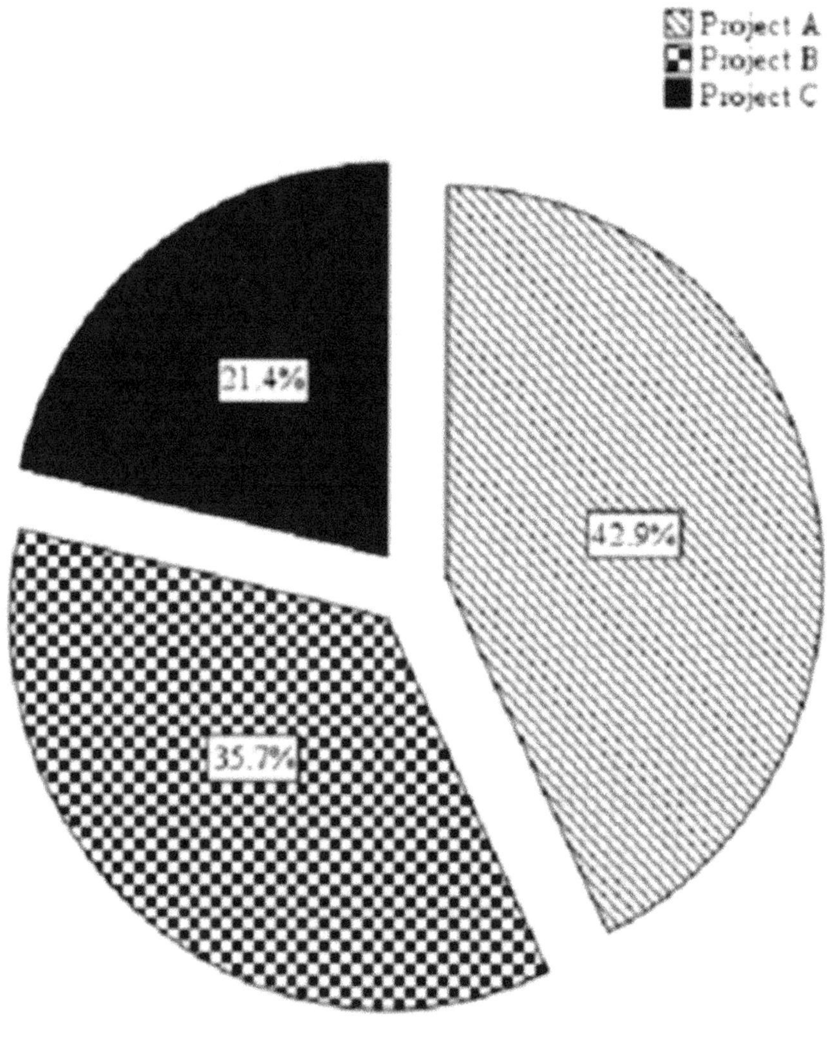

From the respondents, 69% were male, while another 31% were female. In reference to Table 1, most of the respondents are aged between 20 to 29 years old (38.1%), followed by 30 to 39 years old (33.3%), 40 to 49 years old (16.7%) and more than 50 years old (11.9%). The education level among the respondents were mostly bachelor's degree (66.7%), followed by diploma (21.4%), masters (9.5%) and PhD (2.4%). With respect to their years of practice in the construction industry, most of the respondents were having 1 to 5 years (35.7%) experience, followed by 5 to 10 years (28.6%), 10 to 15 years (14.3%), 15 to

20 years (2.4%) and 19.0% of the respondents were having more than 20 years of working experience. Most of the respondents were civil or project engineers (35.7%), while the others (26.2%) consisted of architect, mechanical and electrical engineer, quantity surveyor, site engineer (9.5%), project manager (9.5%), design team (7.1%), project planner (4.8%), project supervisor (2.4%), assistant manager (2.4%) and director (2.4%). The different ranges of age, position, education level and experience between the respondents will eliminate the bias analysis and provides a more genuine feedback on the current subject of technological advancements in Industry 4.0.

Table 1. Demographic profiles of the respondents

	Descriptions	Frequency	Percentage (%)
Age	20-29	16	38.1
	30-39	14	33.3
	40-49	7	16.7
	> 50	5	11.9
Years of practice	1-5	15	35.7
	6-10	12	28.6
	11-15	6	14.3
	16-20	1	2.4
	> 20	8	19.0
Designation	Director	1	2.4
	Project Manager	4	9.5
	Assistant Manager	1	2.4
	Design team	3	7.1
	Civil/ Project Engineer	15	35.7
	Site Engineer	4	9.5
	Project Planner	2	4.8
	Project Supervisor	1	2.4
	Others	11	26.2

Knowledge (K) on Industry 4.0

Figure 2 presents the respondents' period of first knowing about Industry 4.0. 45.2% of the respondents have only heard of Industry 4.0 less than one year ago, presumably due to the fact that the Malaysian government has only initiated the National Policy on Industry 4.0 in year 2018. On the other hand, 19.05% do not have any knowledge on Industry 4.0. Following that, Table 2 portrays the sources of knowledge on Industry 4.0, where 73.8% of the respondents have heard about Industry 4.0, mostly through online media (25.8%), such as websites and online publications, due to the ease of accessibility for new information and knowledge. 24.2% of the respondents have known about this term from attending conference, seminar and workshops, as the PWD frequently held roadshows to expand the publics' and staffs' knowledge on new transformation in the construction industry. The other sources of information were obtained through print media (newspaper and magazines) (22.6%), information from within the organization itself (17.7%), educational institutions (8.1%) and others (1.6%).

Figure 2. Period first heard of Industry 4.0

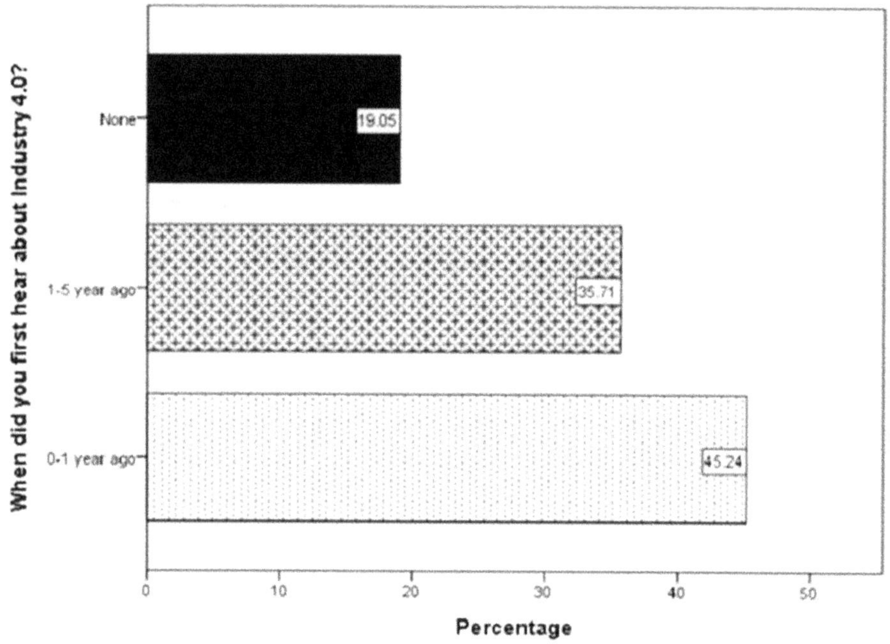

Table 2. Sources of knowledge on Industry 4.0

Knowledge sources	Frequency	Percentage (%)
Company	11	17.7
Tertiary institutions (University/ College)	5	8.1
Conference/ Seminar/ Workshop	15	24.2
Print Media (Newspaper or magazines)	14	22.6
Online media (Website or publication)	16	25.8
Others	1	1.6

The respondents were subsequently asked, if they have attended any Industry 4.0 training courses and their level of understanding on the concept, as well as the understanding on the applications of Industry 4.0. From Figure 3, only eight (8) respondents have attended Industry 4.0 training courses, with 50% (4) respondents from Project A, 37.5% (3) from Project B and 12.5% (1) from Project C. The measure of understanding was queried and identified using Likert scale, which ranges between very poor to having excellent understanding. Based on the respondents who have attended training courses, Table 3 shows that only two (2) respondents, each from Project B and C were having good level of understanding regarding the concept of Industry 4.0. Meanwhile, only one respondent who has attended training course is having good level of understanding on the application of Industry 4.0. The remaining respondents who have attended training courses admitted that they were having between fair (4) and poor (1) level of understanding on the concept and applications of Industry 4.0. Meanwhile, most of the respondents who have not attended any Industry 4.0 training courses claimed to have poor to very poor level of understanding, with only one responded (from Project B) to have excellent understanding in regards to Industry 4.0 knowledge. This particular respondent from Project B is a senior assistant director, who have been heavily involved in BIM Projects in the PWD Malaysia and was also responsible for developing the BIM standard guideline.

Figure 3. Respondents attended Industry 4.0 training course

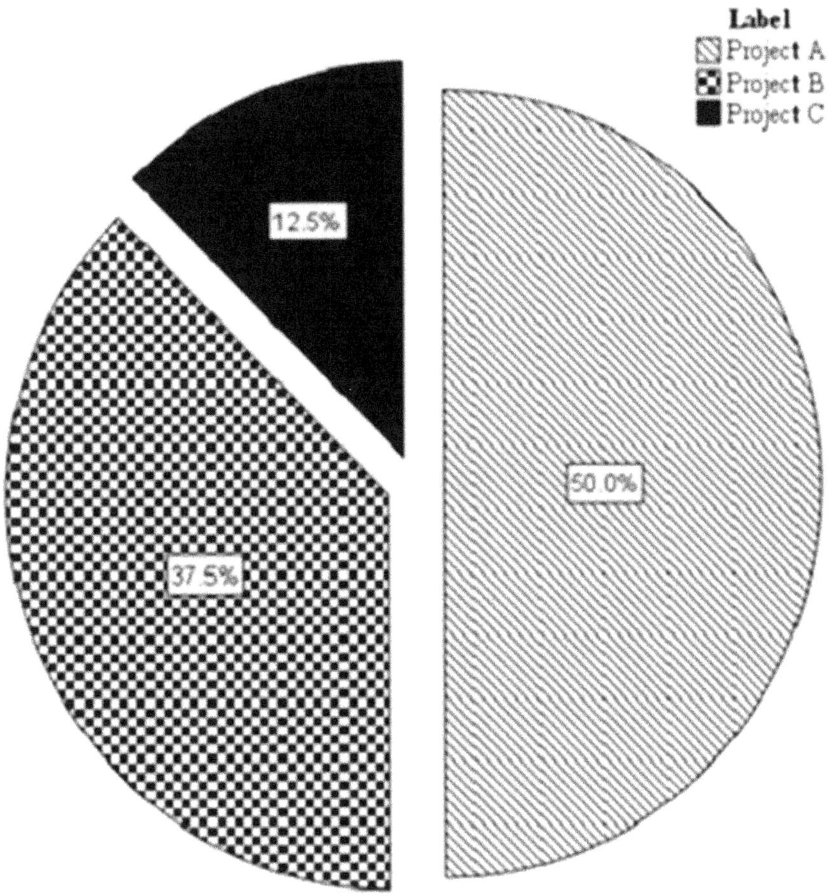

A Chi-Square test has further been conducted to analyse the respondents' level of understanding on Industry 4.0 concept and applications with regards to attending Industry 4.0 training courses. A non-parametric test (Chi-Square) were conducted to evaluate whether there is an association between two or more variables in a table and to test how likely the data fits with the expected distribution (Singhal & Rana, 2015).The results in Table 3 shows that there is an association between Project A respondents who have attended training courses, with their level of understanding on the concept of Industry 4.0, since the alpha value is less than 0.005. However, the other two projects (B and C) do not show statistical significance and there is no strong evidence to suggest a relation between training and level of understanding. For the level of understanding on applications, only Project C showed statistical significance between training and the understanding on applications. The other two, Project A and B did not show any association between training and the understanding on the applications of Industry 4.0. In summary, there is a relationship between training and the understanding of Industry 4.0 concept in Project A, while there is also relationship between training and the understanding of Industry 4.0 applications in Project C. Respondents in Project A finds that training increases their understanding on the concept, while Respondents in Project C feels that training increases their understanding on the application.

Table 3. Crosstabulation on the effect of attending training course with the level of understanding on industry 4.0 concept and applications

	Project		**Level of understanding on the concept of Industry 4.0**					**Total**	**Chi-Square** ($p < 0.05$)
			Very poor	**Poor**	**Fair**	**Good**	**Excellent**		
Have you attended any Industry 4.0 training course?	Project A	Yes	0	0	4	0	0	4	0.004
		No	3	7	1	3	0	14	
	Project B	Yes	0	1	1	1	0	3	0.740
		No	4	2	3	2	1	12	
	Project C	Yes	0	0	0	1	0	1	0.145
		No	2	2	4	0	0	8	
	Project		**Level of understanding on the application in Industry 4.0**					Total	Chi-Square ($p > 0.05$)
			Very poor	Poor	Fair	Good	Excellent		
Have you attended any Industry 4.0 training course?	Project A	Yes	0	0	4	0	0	4	0.160
		No	4	6	2	2	0	14	
	Project B	Yes	0	1	0	2	0	3	0.358
		No	4	1	3	3	1	12	
	Project C	Yes	0	0	0	1	0	1	0.029
		No	3	2	3	0	0	8	

Further on, the respondents were requested to select the definition of Industry 4.0, based on their understanding. Four options were given, where each respondent is allowed to pick more than one answer. The most selected (agreed upon) definition of Industry 4.0 is 1) "*The introduction of internet of things (IoT), services (IoS) and data (IoD) technology to allow machines to manage themselves and the use of automation to computer, communication and internet*", followed by 2) "*A new, emerging structure and integrated communications network for a widely automated exchange of information between production and processes*", 3) "*A combination between real world, virtual world and traditional operations based on information and communication technology (ICT) adoption*" and 4) *A new level of organization and control over the entire value chain of the life cycle of products, geared towards increasingly individualized customer requirements*". Based on the selection, the respondents felt that the definition that best suited Industry 4.0 is the definition with various technological terms embedded in the definition.

Table 4. Familiarity with the Industry 4.0 concepts and applications

	Project A		Project B		Project C		Total	
	Mean	**Std. Dev.**	**Mean**	**Std. Dev.**	**Mean**	**Std. Dev.**	**Mean**	**Std. Dev.**
Autonomous Robots	2.28	1.07	2.93	1.34	2.44	1.33	2.55	1.23
Artificial Intelligence	2.56	1.38	2.93	1.44	2.78	1.20	2.74	1.34
Advanced Materials	2.72	1.41	2.93	1.44	2.33	1.00	2.71	1.33
Big Data Analytics	2.61	1.38	3.00	1.46	2.67	1.23	2.76	1.36
Cloud Computing	2.78	1.35	3.13	1.51	3.22	1.30	3.00	1.38
Additive Manufacturing	2.50	1.04	3.40	1.45	3.00	1.32	2.93	1.30
System Integration	2.83	1.10	2.80	1.42	3.00	1.80	2.86	1.35
Cybersecurity	2.61	1.24	2.87	1.46	3.00	1.50	2.79	1.35
Augmented Reality	2.56	1.34	2.87	1.36	2.56	1.24	2.67	1.30
Simulation.	3.00	1.28	3.13	1.46	3.22	1.20	3.10	1.30
Building Information Modelling	3.78	1.22	3.47	1.36	4.00	1.23	3.71	1.25
Internet of Things/ Services	2.94	1.26	3.20	1.21	3.33	1.00	3.12	1.17
Product-Lifecycle-Management	2.44	1.04	2.87	1.41	3.22	1.09	2.76	1.21
Mobile Computing	2.72	1.27	2.87	1.30	3.44	1.33	2.93	1.30
Augmented/ Virtual /Mixed Reality	2.50	1.25	2.93	1.39	3.22	1.39	2.81	1.33
Radio-Frequency Identification	2.61	1.24	2.67	1.54	3.33	1.00	2.79	1.32
Smart Factory	2.28	0.96	2.53	1.25	2.56	1.13	2.43	1.09
Human-Computer-Interaction	2.28	1.02	2.53	1.30	2.67	1.32	2.45	1.17
Modularisation	2.50	1.20	2.67	1.29	2.78	1.39	2.62	1.25
Cyber-Physical/ Embedded systems	2.17	1.04	2.47	1.51	2.44	1.24	2.33	1.24

Given that most of the respondents have heard of Industry 4.0 and some of them have even attended training courses on the subject matter, the level of familiarity on Industry 4.0 applications/ components were asked. The survey has listed ten (10) main Industry 4.0 components and another ten (10) Industry 4.0 construction related components, where respondents were asked to rate their familiarity with those elements, from one, being (least familiar) to five (extremely familiar). The Likert-scale items is an ordinal measurement scale with an appropriates descriptive analysis recommended were mode or median of central tendency and frequencies for variability standard deviation Likert-scale data analysis procedure (Harry N. Boone & Boone, 2012). From the findings, as shown in Table 4, the most familiar Industry 4.0 component among respondents from the three projects is Building Information Modelling (BIM), while the least familiar component is cyber-physical system/ embedded system. The trend is the same across all projects.

Attitude Towards Industry 4.0

In this section, the perception towards Industry 4.0 was asked to measure the acceptance, agreement and the importance perceived by respondents of the three construction projects. There were 10 questions concerning the attitude of respondents in the KAP survey. The first cluster in the attitude section contains four (4) basic questions that were to be answered by selecting either yes, no or uncertain. The results are shown in Figure 4. Firstly, the respondents were asked on the necessity for Industry 4.0 adoption or implementation. Sixteen (16) respondents from Project A opinionated that it was necessary to adopt or implement Industry 4.0 in the construction industry, followed by thirteen (13) respondents from Project B and seven (7) from Project C. In total, 85.7% of the respondents admitted that it is necessary to implement Industry 4.0 advancements in construction. Meanwhile, ten (10) of the respondents from Project A, nine (9) from Project B and four (4) respondents from project C felt they were obliged to adopt Industry 4.0 as it is part of their job. Asked whether it is possible for the industry to adopt Industry 4.0 practices, thirteen (13) respondents from Project A and B, as well as seven (7) respondents from Project C replied that it is possible. Fourteen (14) respondents each from Project A and B, and eight (8) respondents from Project C that made up 85.7% of the respondents agreed that the usage of Industry 4.0 will be very useful to them. In general, at least more than 80% of the respondents have shown positive feedback and attitude on the shift towards Industry 4.0.

Figure 4. Perception towards Industry 4.0

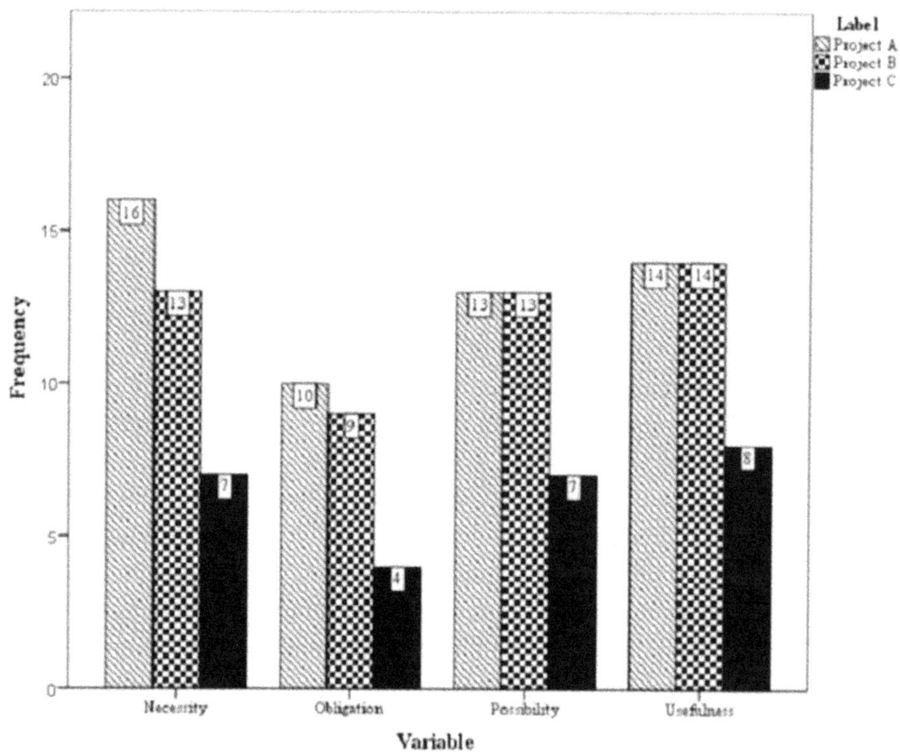

Table 5. Attitudes towards Industry 4.0

	Project A		Project B		Project C		ANOVA	
	Mean	Std. Dev	Mean	Std. Dev	Mean	Std. Dev	F	Sig. (*p*)
Importance of Industry 4.0	3.89	1.08	3.93	1.44	4.00	1.32	0.02	0.98
Improvement to current practice	3.83	1.15	3.93	1.39	4.00	1.22	0.06	0.94
Encourage positive attitudes	3.78	0.94	3.93	1.16	3.67	1.12	0.19	0.83
Importance for the involvement of top management	4.11	1.02	3.80	1.37	4.00	1.32	0.27	0.77
Involvement of stakeholder	4.11	1.02	3.93	1.28	3.56	1.33	0.66	0.52
External collaboration	4.00	1.03	3.53	1.46	3.78	1.48	0.53	0.59

In the second cluster of attitudes towards Industry 4.0, the questions raised consisted of the importance for the industry to adopt Industry 4.0 practices, as well as whether the adoption of Industry 4.0 will positively improve staffs' attitude (motivated, committed). In addition, the respondents were also asked on the importance of getting the involvement of all parties in Industry 4.0, as well as the need to collaborate for external sources of knowledge. The respondents were required to choose a response for the above queries based on a 5-point Likert scale, where one (1) being the least (lower end of the scale) and five (5) being extremely high (higher end of the scale). From Table 5, the mean scores and standard deviation between the three (3) project cases were analysed. The mean scores were used to measure central tendency or average of the Likert responses and the standard deviation is to measure the spread of the observations between the project cases. The mean scores across all projects are very close, on the perceived importance in adopting Industry 4.0, where Project A scored (3.89), Project B (3.93) and Project C (4.00), which represents moderate to important. Besides that, most of the respondents from Project A (3.83), Project B (3.93) and Project C (4.00) moderately agreed that the Industry 4.0 adoption will improve the current industry's practices and performance with a relatively close standard deviations between the different projects.

Within Project A (3.78), Project B (3.93) and Project C (3.67), the respondents have moderately agreed that the adoption of Industry 4.0 will increase staffs' motivation, commitment and performance. Project A (4.11) and Project B (4.00) respondents felt that it is important to get the top management's involvement in the adoption of Industry 4.0. Meanwhile, Project B (3.80) respondents felt less strongly on this matter. Most of the respondents from Project A (4.11) opinionated that it is important to get the stakeholders' involvement to adopt the Industry 4.0 concept. However, Project B (3.93) and Project C (3.56) respondents found it to be somewhat to moderately important. For the question on external collaboration, there was a huge gap between Project A's standard deviation with Project B and Project C because most of the Project A's (4.00) respondents strongly agree that it is necessary to have collaborations with other resource of knowledge. In regards to that, Project B (3.53) and Project C (3.78) respondents were found to be less convinced on the need to have external collaborations. In general, the standard deviations for Project A is smaller, as compared to Project B and C, which shows least variability among the respondents in Project A, as compared to Project B and C.

Upon further analysis, an omnibus statistic test of one-way analysis of variance (ANOVA) was adopted to determine whether there are any statistically significant differences between the means of the three (3) projects. In order to obtain a valid result, there were several key assumptions involved; there is no significant outliers and the data was normally distributed, based on the boxplot and Shapiro-Wilk test ($p < 0.05$), and the homogeneity of variances according to the Levene's test for equality of variances, $p = 0.0576$ were tested and checked. From the analysis, it was found that there were no significant differences ($p > 0.05$) between the three project cases on the attitude towards Industry 4.0 adoption in the construction industry development. The results show that most of the respondents have positive attitude towards the adoption and implementation of Industry 4.0 in the construction project.

Practices on Industry 4.0

In this section, the respondents were asked about the current practices of Industry 4.0 in or related to their project. The respondents were first asked if they have practiced or implemented any Industry 4.0 tools or concepts. Slightly more than half of the respondents (with 24 out of 42) claimed to have implemented Industry 4.0 in their current work environment. As presented in Figure 5, ten (10) respondents, each from Project A and B have practiced Industry 4.0, whereas only four (4) respondents from Project C were involved in the practice. However, eighteen (18) respondents claimed that they did not apply any of the Industry 4.0 concepts or applications in their tasks.

Following that, the respondents were asked to rate the different supporting practices/ facilities of Industry 4.0, using 5-point Likert scale from, 1-very poor, 2-poor, 3-fair, 4- good and 5-excellent. As shown in Table 6, the respondents from all projects opinionated that there were insufficient guidelines, research and development, as well as investments to expand the adoption of Industry 4.0 in the construction industry. There were insufficient guidelines provided by the government or authority towards the adoption of Industry 4.0 in construction, which places the variable at the lower end of the scale (poor to fair). Similarly, the respondents claimed that insufficient research and development (R&D) has been done to facilitate the implementation of Industry 4.0 in the construction industry, where Project A (2.44) and Project C (2.56) were providing poor values for R&D with Project B (3.07) respondents providing slightly better. Most of the respondents in Project B (3.27) opinionated that the investment on Industry 4.0 technologies is sufficient, unlike Project A (2.61) and Project B (2.44), which considered the investment to be at the lower end of the scale. In summary, respondents of Project B provided the most positive feedback, as compared to the other projects, on the supporting elements that could drive Industry 4.0, albeit the value being at the fair range.

Further on, the ANOVA test was done for this section, in order to discover any significant differences on the perception of Industry 4.0 practices between Project A, Project B and Project C. Based on the key assumptions of the data after being tested and checked, there is no significant outliers and the data between the three projects were normally distributed (Shapiro-Wilk test, $p < 0.05$). The homogeneity of variances (Levene's test) results are $p = 0.396$. According to the results, there is no significant differences ($p > 0.05$) between Project A (0.08), Project B (0.32) and Project C (0.22). Even though government agencies have initiated the Industry 4.0 national policy, the framework has focused more on the manufacturing sector. Due to the rather slow movement of Industry 4.0 in the Malaysian construction sector, the R&D works were rather slow, as well but has been picking-up in recent times, particularly by construction industry research and development centers, such as the Construction Research Institutes of Malaysia (CREAM).As for the investment on Industry 4.0 technologies, this variable heavily relies on the project cost and the organization's size that highlights the capability and capacity for the adoption and the provision for training of staffs.

Figure 5. The adoption of Industry 4.0 in current practice

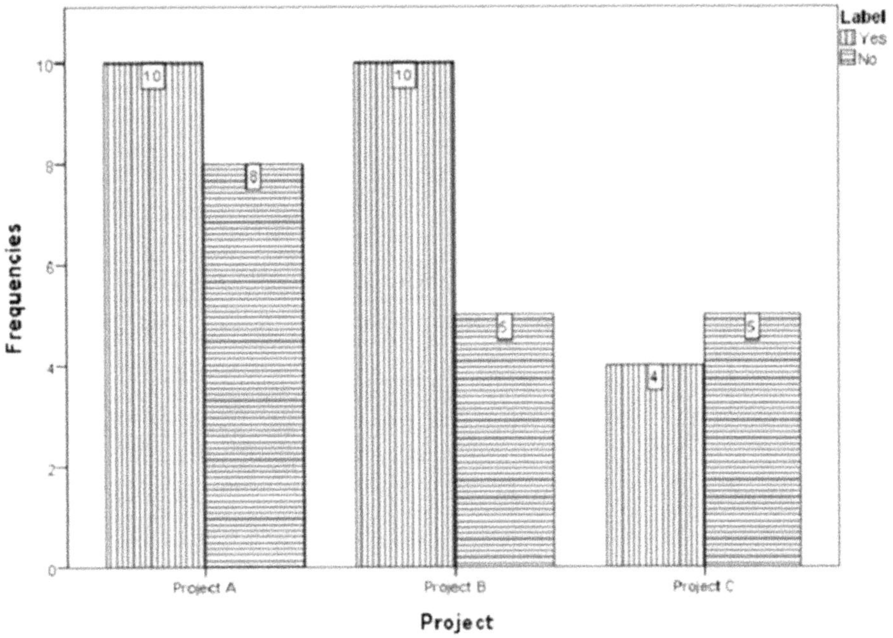

Table 6. The perception on supporting facilities for Industry 4.0 practices

		Mean	Std. Deviation	ANOVA	
				F	Sig,
Guidelines sufficiency	Project A	2.22	1.00	2.72	0.08
	Project B	3.13	1.30		
	Project C	2.56	1.01		
Research and Development (R & D) facilities sufficiency	Project A	2.44	1.15	1.18	0.32
	Project B	3.07	1.34		
	Project C	2.56	1.01		
Investment Sufficiency	Project A	2.61	1.29	1.56	0.22
	Project B	3.27	1.39		
	Project C	2.44	1.01		

The respondents were further requested to indicate their level of involvement with Industry 4.0 components, through 5-points percentage Likert scale from "0% – not involved" to "100% – extremely involved" and the findings are shown in Table 7. The analysis is presented in terms of comparison between the three different projects and Industry 4.0 applications used most by the respondents from all three projects is Building Information Modelling (BIM). This might be due to the fact that the PWD has mandated on the rules and regulations, where it is compulsory for all government projects that costs over MYR 100 million to use BIM during the design stage for authority approval on the submissions. Meanwhile, the least used applications are autonomous robots and cyber physical system as there is a lack of instruments and machineries in the market, in comparison to the manufacturing sector that has widely adopted autonomous robots in the manufacturing process. Even so, the responses from all respondents on the current applications are at the lower end of the scale, which indicates minimal involvement with Industry 4.0 technologies in their projects.

Table 7. Industry 4.0 practices

	Project A		Project B		Project C		Total	
	Mean	**Std. Dev.**	**Mean**	**Std. Dev.**	**Mean**	**Std. Dev.**	**Mean**	**Std. Dev.**
Autonomous Robots	1.33	0.97	1.33	0.62	1.67	1.12	1.40	0.89
Artificial Intelligence	1.39	0.78	1.47	0.74	1.89	1.17	1.52	0.86
Advanced Materials	1.50	1.10	1.47	0.74	1.89	1.36	1.57	1.04
Big Data Analytics	1.83	1.04	1.87	0.92	2.11	1.36	1.90	1.05
Cloud Computing	2.56	1.42	2.13	1.06	2.67	1.66	2.43	1.35
Additive Manufacturing	1.56	0.86	1.93	0.96	1.89	1.17	1.76	0.96
System Integration	2.33	1.33	2.13	0.99	2.11	1.17	2.21	1.16
Cybersecurity	1.72	0.89	1.53	0.74	2.00	1.00	1.71	.86
Augmented Reality	1.89	1.18	1.60	0.83	1.89	1.05	1.79	1.02
Simulation.	2.56	1.29	2.07	1.10	2.44	1.24	2.36	1.21
Building Information Modelling	4.11	1.49	3.53	1.36	3.78	1.39	3.83	1.41
Internet of Things/ Services	2.22	1.26	2.40	1.12	2.56	1.33	2.36	1.21
Product-Lifecycle-Management	1.78	1.06	1.93	1.03	2.22	1.20	1.93	1.07
Mobile Computing	2.17	1.20	1.87	.99	2.44	1.42	2.12	1.17
Augmented/ Virtual/ Mixed Reality	1.94	1.16	2.27	1.33	2.11	1.17	2.10	1.21
Radio-Frequency Identification	1.72	1.07	1.73	1.03	2.00	1.12	1.79	1.05
Smart Factory	1.44	0.98	1.60	0.83	1.89	1.27	1.60	0.99
Human-Computer-Interaction	1.33	0.69	1.67	0.98	2.33	1.41	1.67	1.03
Modularisation	1.61	1.04	1.60	0.83	2.22	1.56	1.74	1.11
Cyber-Physical/ Embedded systems	1.39	0.92	1.60	0.99	1.56	0.73	1.50	0.89

In addition to that, the respondents were also asked about the readiness of future graduates to adopt and implement Industry 4.0 in the construction engineering discipline. From the yes/no question, most of the respondents believe that, the future graduates were not Industry 4.0 ready as the current education system at universities do not fulfil the needs and requirements of Industry 4.0. As presented in Figure

6, only four (4) respondents from Project A, five (5) from Project B and two (2) respondents from Project C opinionated that the future graduates are Industry 4.0 ready. Furthermore, there is only one (1) respondent from Project A and six (6) respondents from Project B who thought that the current Higher Education Institutions are fulfilling the needs for Industry 4.0. In order to eradicate the gap between the education and industry's future needs, the respondents have suggested to introduce Industry 4.0 as early as secondary school (40.5%) and more on at the undergraduate level of education (45.2%) at universities. Figure 7 shows the breakdown of opinions on introducing Industry 4.0 in the education system. However, in order to change and improve the current teaching and learning processes, intervention from multiple stakeholders from the education and practicing industry is required.

Figure 6. Respondent perceptions towards Industry 4.0 in education

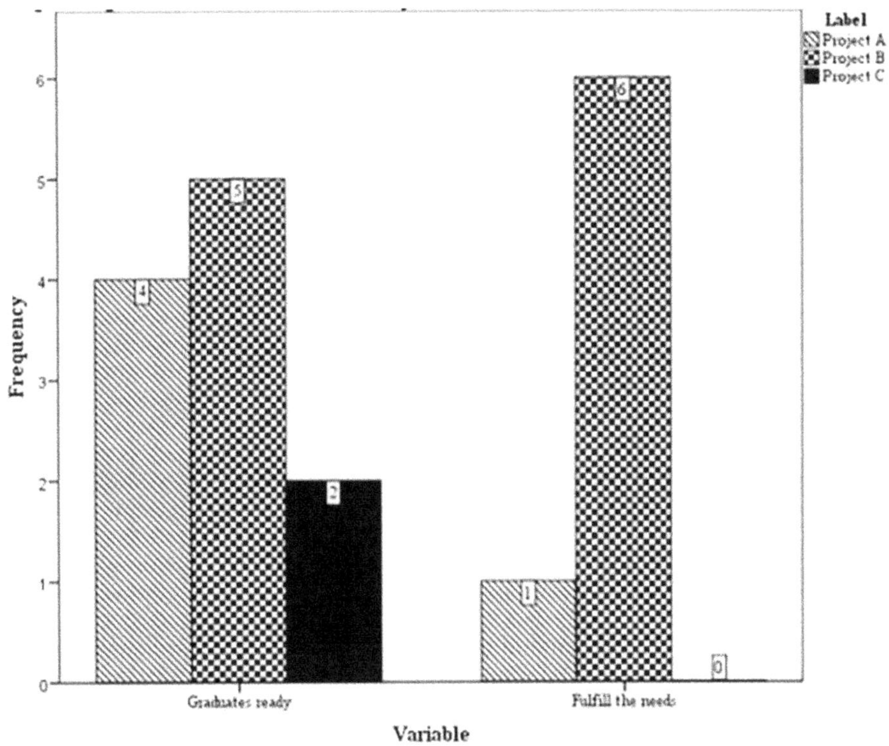

Figure 7. The level of education that needs to be prioritized in introducing and adopting the Industry 4.0 concepts

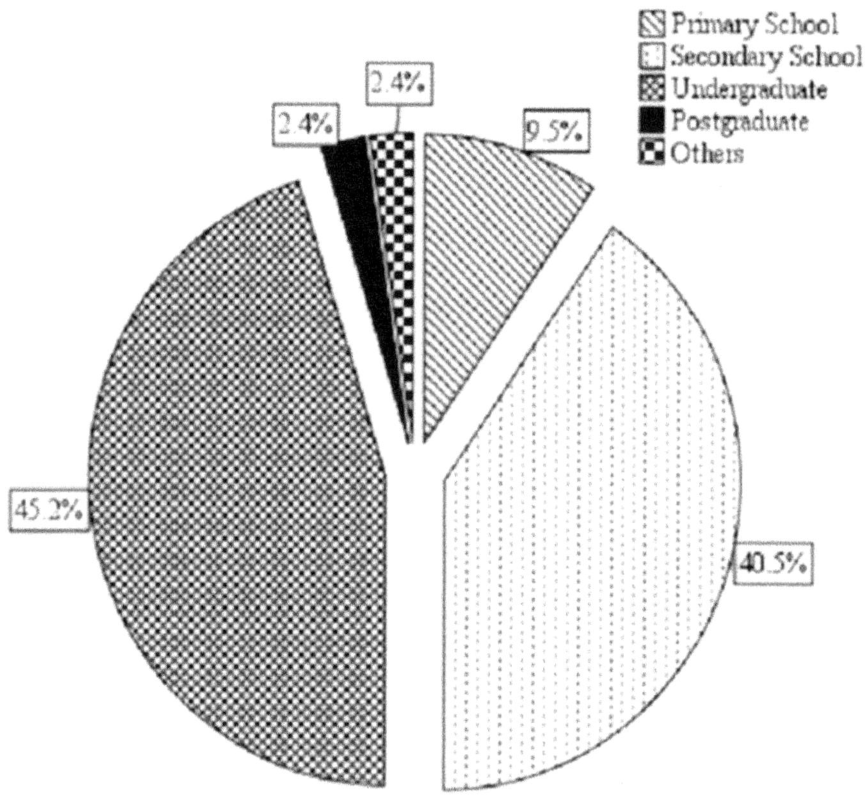

FINDINGS FROM THE SEMI-STRUCTURED INTERVIEW

For further clarifications in regards to this Industry 4.0 KAP study, a semi-structured interview using open-ended questions through face-to-face interviews was conducted by selecting one (1) key representative from each of the projects. The interviewees' selection has followed the following criteria: (1) have experience of more than 10 years in the industry; and (2) have knowledge on Industry 4.0. General questions were asked about the current state of Industry 4.0 practices and adoption in the projects, as well as potential barriers faced by the different stakeholders in each of the project. Furthermore, they were also asked on how the implementation of Industry 4.0 has benefited their project and the applications used in the projects.

Project A interviewee believed that the current construction industry is heading towards the vision of Industry 4.0. The importance of adopting new technological advancements would comprehensively boost a company's return on investment, create visibility and compatibility. The use of Industry 4.0 application, such as BIM and related software enables the prediction of future situations, reduces construction waste and creates better communication channel between project stakeholders. However, there is problem in the adoption, particularly the incompatibilities due to different software used by different stakeholders. The productivity of Industry 4.0 in Project A is based mostly on the company's income, and the interviewee has proposed that it should be based on the cost of project. In addition, there should be an extensive collaboration between the public and private sector, to create better awareness on Industry 4.0.

Project B interviewee was extremely positive towards the adoption of Industry 4.0 in the construction industry, as most of the government agencies have a special department named technology innovation that investigates and executes practices on Industry 4.0. The respondent also mentioned that Industry 4.0 adoption will save construction cost by incorporating detailed construction material information from all project stakeholders, such as architecture, civil work, mechanical, electrical and plumbing. Nonetheless, there are observed problems occurring at the construction stage, such as software misinterpretation, software incompatibility, expensive database system and data security. A sustainable construction development can be successfully achieved by having appropriate Industry 4.0 guidelines, framework and acts that could be provided by the relevant government agencies.

Lastly, representative for Project C feels that Industry 4.0 is a good start and way forward to improve the current construction industry. Industry 4.0 could offer transparent data information and eases the communication gap between construction project stakeholders. Furthermore, the engineer can foresee the future work and detect any discrepancies before the construction work being executed on site. According to the third interviewee, there is lack of Industry 4.0 expert involved in the projects, and this has affected the project time during the design stage. Besides that, there is also some issue during the construction stage due to disparities between workmanship on-site and software model output. Hence, comprehensive staff training courses should be provided to increase the Industry 4.0 skills throughout the construction project and industry at large.

The key findings from the semi-structured interviews indicate that all three interviewees agree that the Industry 4.0 adoption and implementation will bring good prospect and benefit to construction project development because of the advantages that could be offered by the Industry 4.0 concept. Nonetheless, the challenges may present in terms of the people, software and working environment, which are heavily influenced by the common construction performance measures, such as cost, time and quality. The respondents have suggested that the challenges could be eliminated by having discussion among Industry 4.0 professional and stakeholders involved in projects. Hence, the adoption of Industry 4.0 shall be widely introduced and implemented by Malaysian construction companies, beyond public sector clients.

DISCUSSION

Digitization of the construction industry, alongside Industry 4.0 practices acts as a paradigm shift towards creating a good culture for ensuring the future of construction. The main reason for developing an Industry 4.0 culture is to decrease the remoteness between people, technology and the construction industry. Findings from this study suggest that the construction public sector clients were generally aware (73.8%) of Industry 4.0, and mostly heard of the term within the last 5 years. Nonetheless, there are still a small percentage that has not heard about the current trend. It was also obvious that minimal trainings have been given to the respondents and it was found that trainings are beneficial to increase the understanding of respondents on Industry 4.0.

In regards to the familiarity of respondents across the different projects, the number does not vary much as the respondents have claimed to have moderate level of familiarity on the Industry 4.0 applications. In summary, even though the nature of the projects were different (hospital, office building, transportation infrastructure), the knowledge among the public sector clients are very similar to each other. Given the limited knowledge recognised from the previous sector, the positive attitude shown across respondents of the different projects provides a promising premise for construction development, towards the incorporation of Industry 4.0, as they have acknowledged the importance, usefulness and necessity of Industry 4.0 related technologies.

The attitude of players between different projects also did not show any significant differences, where most of the respondents have reflected positive attitudes in this subject matter. In regards to the practices section of this KAP study, the findings do co-relate with the Knowledge section. The limited applications (Refer to Table 7) of Industry 4.0 elements in these projects justifies the least familiarity (Refer to Table 4) among the respondents. Respondents across the different projects have echoed on the insufficient supporting facilities such as guidelines, research and investments, which hampers the notion of Industry 4.0. In addition, there is a need to educate future generations from as early as secondary school, up to universities, in order to have a good start in promoting digitisation culture. This will then generate future Industry 4.0 ready graduates and cultivate a deeper interest in the use of new technologies.

In general, the adoption of Industry 4.0 brings more benefits than disadvantages (as perceived by the respondents) and has developed greater motivation to the public sector clients in embracing Industry 4.0 practices. Even though the tools have not been fully utilized, there are some technologies that have been applied in the projects, which are BIM and cloud computing. This achievement can be traced back to the Malaysian government's mandate to public sector clients in using BIM as a new technological platform in public projects that exceed certain project value. In order to shape and drive the implementation of Industry 4.0 in construction, best practices among the public sector clients should be highlighted.

Given the current state of KAP among clients of public project, further efforts should be made to enhance the KAP through trainings and the involvement of supporting facilities, as the positive attitude among the respondents reflects the readiness of the people in embracing Industry 4.0. As the largest construction client, the readiness of the public sector will set the tune for other players, particularly the private sector to follow-suit. Hence, the best practices, good motivation and culture is the core built towards the notion of Industry 4.0, for the future of the construction industry.

FUTURE RESEARCH DIRECTIONS

This study has focused only on the public sector construction client (from the top management and site engineer position), with projects above MYR 100 million. Even though the study has provided a holistic view of Industry 4.0, among the different positions of the client's organization, across different nature of projects, the scope of study could be further enhanced in future studies by considering other stakeholders in construction. Furthermore, the effect of new technological advancement in the construction industry development should be investigated to discover the challenges and drivers of Industry 4.0 adoptions. The KAP survey and semi-structured interview approach have been complementary in enriching the data collected in this study. However, due to the limited number of respondents, the KAP study cannot be off-set beyond the scope of this study. Hence, future studies should focus to increase

e the number of case studies, as well as to conduct similar study among the private construction organisations. Further qualitative methodologies such as focus group could be used to facilitate in-depth studies on the understanding of construction industry players on Industry 4.0.

CONCLUSION

The findings of this KAP study on Industry 4.0 among the public sector construction client in Malaysia has shed some light on the current state of Industry 4.0. The findings suggest that at least 80% of the respondents are aware of the notion, but with limited familiarity of this subject matter. This aligns with the Industry 4.0 applications that have been implemented in the case projects, where similar application was observed across the different projects, which are Building Information Modelling (BIM) and Cloud Computing. The respondents have also echoed on the same least familiar elements, which is the cyber-physical system. Moving forward, it is suggested that trainings (on-the-job/ off-the-job/ distance education) should be provided to stakeholders as the trainings could enhance the knowledge of the people. In addition, supporting facilities such as guidelines and research, as well as early education on Industry 4.0 could create the digitalisation culture among the current and future generation, towards future-ready construction.

REFERENCES

Ahamad, N. R., & Ariffin, M. (2018). Assessment of knowledge, attitude and practice towards sustainable consumption among university students in Selangor, Malaysia. *Sustainable Production and Consumption, 16*, 88–98. doi:10.1016/j.spc.2018.06.006

Che Ibrahim, C. K. I., & Belayutham, S. (2020). A knowledge, attitude and practices (KAP) study on prevention through design: A dynamic insight into civil and structural engineers in Malaysia. *Architectural Engineering and Design Management, 16*(2), 131–149. doi:10.1080/17452007.2019.1628001

Chileshe, N., & Dzisi, E. (2010). Benefits and barriers of construction health and safety management (HSM). *Perceptions of practitioners within design organisations*, (Unit 07), 1–5.

CIDB. (2016). *Construction Industry Transformation Programme (CITP) 2016-2020.* Construction Industry Development Board Malaysiafile. doi:10.11113/jt.v78.4056

Erboz, G. (2018, July). How to define Industry 4.0. *The Main Pillars of Industry 4.0.*

Fapohunda, J. A., & Stephenson, P. P. (2014). *Knowledge, Attitude, and Perception (KAP) of Construction Participants towards Resource Waste in the Construction Industry*. Academic Press.

Goh, Y. M., & Chua, S. (2015). Knowledge, attitude and practices for design for safety: A study on civil & structural engineers. *Accident; Analysis and Prevention*. Advance online publication. doi:10.1016/j.aap.2015.09.023 PMID:26477455

Gumucio, S. (2011). *Data collection Quantitative methods The KAP survey model*. Knowledge, Attitude & Practices.

Harry, N., Boone, J., & Boone, D. A. (2012). Analyzing Likert Data. *Journal of Extension, 50*(2), 1456–1466. doi:10.100711172-017-1908-3

Husain, A. H., Razali, M. N., & Eni, S. (2018). Stakeholders' expectations on building information modelling (BIM) concept in Malaysia. *Property Management, 36*(4), 400–422. doi:10.1108/PM-02-2017-0013

Kamble, S. S., Gunasekaran, A., & Gawankar, S. A. (2018). Sustainable Industry 4.0 framework: A systematic literature review identifying the current trends and future perspectives. *Process Safety and Environmental Protection, 117*, 408–425. doi:10.1016/j.psep.2018.05.009

Latiffi, A. A., Brahim, J., & Fathi, M. S. (2017). Building information modelling (BIM) after ten years: Malaysian construction players' perception of BIM. *IOP Conference Series. Earth and Environmental Science, 81*(1), 012147. Advance online publication. doi:10.1088/1755-1315/81/1/012147

Liao, Y., Deschamps, F., Loures, E. F. R., & Ramos, L. F. P. (2017). Past, present and future of Industry 4.0 - a systematic literature review and research agenda proposal. *International Journal of Production Research, 55*(12), 3609–3629. doi:10.1080/00207543.2017.1308576

Manley, K. (2006). The innovation competence of repeat public sector clients in the Australian construction industry. *Construction Management and Economics, 24*(12), 1295–1304. doi:10.1080/01446190600934953

Mohamad, S. F., Abidin, U. F. U. Z., Saari, N., Zakaria, M. H., Ismail, A., & Man, N. A. (2019). Knowledge, attitude and practice towards the implementation of outcome-based education among the academics in Universiti Putra Malaysia. *International Journal of Engineering and Advanced Technology, 8*(5C), 1044–1053. doi:10.35940/ijeat.E1148.0585C19

Oesterreich, T. D., & Teuteberg, F. (2016). Understanding the implications of digitisation and automation in the context of Industry 4.0: A triangulation approach and elements of a research agenda for the construction industry. *Computers in Industry, 83*, 121–139. doi:10.1016/j.compind.2016.09.006

Pereira, A. C., & Romero, F. (2017). A review of the meanings and the implications of the Industry 4.0 concept. *Procedia Manufacturing, 13*, 1206–1214. doi:10.1016/j.promfg.2017.09.032

Porwal, A., & Hewage, K. N. (2013). Building Information Modeling (BIM) partnering framework for public construction projects. *Automation in Construction, 31*, 204–214. doi:10.1016/j.autcon.2012.12.004

Raja Sreedharan, V., & Unnikrishnan, A. (2017). Moving towards industry 4.0: A systematic review. *International Journal of Pure and Applied Mathematics, 117*(20), 929–936.

Rüßmann, M., Lorenz, M., Gerbert, P., Waldner, M., Justus, J., Engel, P., & Harnisch, M. (2015). Industry 4.0: The Future of Productivity and Growth in Manufacturing. *The Boston Consulting Group, 20*. Advance online publication. doi:10.100712599-014-0334-4

Singhal, R., & Rana, R. (2015). Chi-square test and its application in hypothesis testing. *Journal of the Practice of Cardiovascular Sciences, 1*(1), 69. doi:10.4103/2395-5414.157577

Toh, Y. Z., Goh, Y. M., & Guo, B. H. W. (2016). Knowledge, Attitude, and Practice of Design for Safety. Multiple Stakeholders in the Singapore Construction Industry, 1–11. doi:10.1061/(ASCE)CO.1943-7862.0001279

Tougwa, F. N. (2018). *The Construction Industry and Its Research Rationale in the Construction Industry and Its Research*. Academic Press.

Vaidya, S., Ambad, P., & Bhosle, S. (2018). Industry 4.0 - A Glimpse. *Procedia Manufacturing, 20*, 233–238. doi:10.1016/j.promfg.2018.02.034

Zhong, R. Y., Xu, X., Klotz, E., & Newman, S. T. (2017). Intelligent Manufacturing in the Context of Industry 4.0: A Review. *Engineering*, *3*(5), 616–630. doi:10.1016/J.ENG.2017.05.015

Section 2

Process: Digitally Enabled Collaboration

Chapter 2
The Role of Collaboration in the Implementation of BIM-Enabled Projects

Behzad Abbasnejad
https://orcid.org/0000-0002-0415-239X
RMIT University, Australia

Soroush Moeinzadeh
University Sains Malaysia, Malaysia

Alireza Ahankoob
RMIT University, Australia

Peter S. P. Wong
https://orcid.org/0000-0002-8429-2551
RMIT University, Australia

ABSTRACT

The effective undertaking of BIM as a systemic innovation requires tighter collaboration among multiple stakeholders and supply chain partners and entails a culture change. However, obtaining a high degree of collaboration is difficult given the nature of construction projects with a diverse supply chain and stakeholders involved in a project. Therefore, this study focuses on the current improvements and approaches in facilitating the BIM-enabled collaboration as a transformative approach in the built environment.

INTRODUCTION

The implications of digitalisation and automation, by utilising various technologies and processes, such as the Internet of Things (IoT), cloud computing and artificial intelligence (AI) have been the main concern of experts in the manufacturing (Lu, 2017) and Architecture, Engineering, Construction, and Opera-

DOI: 10.4018/978-1-7998-6600-8.ch002

tions (AECO) industries in terms of Industry 4.0 principles. Industry 4.0 embraces a preponderance of technologies, principles, and methods for increasing the autonomy, dynamism, flexibility, and accuracy of production systems and supply chains (Tortorella & Fettermann, 2018). Information and digital technologies of Industry 4.0 can be used for integration of processes at inter and intra-organisational levels which aims to introduce solutions for requirements of a firm's digitalisation and automatisation (Xu et al., 2018) within a smart factory paradigm, including interactions between machines (Strozzi et al. 2017).

Moreover, it allows the integration of information throughout the supply chain and the real-time transmission and processing of information for decision making objectives (Novais et al., 2019) in order to maximise value and reduce waste (Deng et al., 2019; Meng, 2020; Bortolini et al., 2019) and to contribute in the lean construction supply chain management. For instance, the use of Computer Aided Design (CAD), Computer Aided Engineering (CAE), Advanced Manufacturing Technologies (AMT and Geographical Information Systems (GIS) in lean construction supply chain (CSC) management processes help the decision making process for supplier selection, determining number of deliveries, and allocation of diverse and scattered consolidation centers (Deng et al., 2019).

Building Information Modelling (BIM) has been contemplated as one of the key technology-centric innovations for the digitalisation of the construction manufacturing context in recent years. Moreover, BIM has been established as a central platform to integrate both the traditional and recent technological advancements in the AECO industries. For example, through integration in the BIM environment, Virtual Reality (VR) and Radio-frequency Identification (RFID) can be used in prefabricated component construction projects (Li et al., 2018). Integrating BIM and VR facilitates production processes to reduce uncertainty and constraints, improve information sharing, and eliminate prefabrication housing production associated reworks (Li et al., 2018).

Additionally, BIM implementation has emerged to improve the collaboration level of the CSC (Deng et al., 2019; Meng, 2020). Concurrently, collaborative efforts of various stakeholders involved in BIM-enabled projects reinforced the BIM implementation process. By the application of BIM in a project, a new collaborative culture should be established in the project environment, which in turn, fosters supply chain and lean-based management in construction practices (Meng, 2020). The synergy in utilization of BIM and SCM decreases the number of and complexities of project tasks to increase overall productivity in construction processes in order to achieve earlier completion dates with the expected quality (Bortolini et al., 2019). However, the fragmented and complexity nature of the construction industry has been frequently cited as one of the most inhibiting factors in improving collaboration.

Nature and Characteristics of AECO Projects and Supply Chain

There is consensus among practitioners that the management of existing social problems linked to physical assets should not be addressed without considering the interactions between assets and the economic, social, and environmental effects of human behavior. This interrelation is followed by substantial challenges owing to fundamental complexities among the elements containing physical built assets and organisational systems used in their diverse phases, such as procurement, delivery, utilization, and disposal.

Complexity is the core characteristic of construction projects through their life cycle from pre-design to recycling and disposal phases. Building projects have diverse interacting and interdependent subsystems such as the network of various actors, the communication networks, and the external environment. To further illustrate this complexity, the procurement of assets contains complicated interactions among diverse stakeholders and organisations operating in supply networks with mutually exclusive perspectives and expectations about their configuration and exploitation. These networks include myriads of interactions which contain varying degrees of cooperation, competition, and conflict, e.g., social, economic, and legal, to name a few, that influence the management of a projects' processes.

Additionally, the fragmented nature of the construction industry and inefficient work practices interrupts the reliability and agility of the information and material flows among the involved parties. This literally illustrates the complexity of construction as a multidisciplinary field that needs integration of expertise and input from numerous constituencies including the exchange of different sources of information through complex networks. Thus, the combination of present complexities associated with technology implementation and inherent complexities of the construction processes bring about noteworthy challenges to decision makers.

The current conventional problem-solving approaches are no longer sufficient for dealing with these levels of complexities. The industry, accordingly, must continuously improve and evolve to be able to respond to the above-mentioned complexities and fulfil the continuously changing societal demands. One example of such demands is the inclination to use state-of-the-art technologies (e.g., VR and BIM) and innovations in the new phase in the industrial revolution (i.e. Industry 4.0) that focuses heavily on digitisation, automation, machine learning, real-time data, and interconnectivity (Liao et al. 2017).

Digitally-Enabled Construction Supply Chain (DCSC)

A supply chain consists of a variety of stakeholders/players/actors to perform a set of interlinked activities concerning the planning, coordination, collaboration and control of goods and services from raw materials to final products delivery to customers/clients (Ivanov et al., 2018). In particular, the SCM concept has been utilised for planning and controlling flows (e.g., material, service, information, and monetary flows) and activities (e.g., internal and external supply chain activities), along with addressing the relationships developed and the processes shared with customers and suppliers (Chen & Paulraj, 2004).

Nevertheless, supply chains constantly face diverse risks and challenges, such as complexity, uncertainty, and vulnerability. The traditional supply chain consists of scattered and isolated steps and does not have certain required attributes for business requirements. Furthermore, supply chain operations involve some conflicts. For example, their related information is stored locally creating functional, organisational, and cultural barriers. This not only decreases the efficacy and efficiency of collaboration among diverse members and functions but also increases the possibility of complications evolving at later stages which inevitably limits response options (Raj & Sharma, 2014).

These issues are expected to be resolved through utilising technological infrastructure and information systems that enable information flow to be integrated into supply chain processes (Abdel-Basset, Manogaran, and Mohamed, 2018). The new transformed digital supply chain (DSC or SC 4.0) has been introduced as a flawless integrated system. Regardless of products or services attributes (physical or digital), DSC is the main enabler for managing a supply chain (Büyüközkan & Göçer, 2018). The main goal of the Digitally-enabled Construction Supply Chain (DCSC) is to improve operational communication, coordination and collaboration among participants through a project's life cycle (Holweg et al., 2005) in order to increase the integration level throughout the chain.

DCSC, as a smart, value-driven process, has transformed supply chains through a novel innovation path. There are many innovations within the DCSC such as Augmented Reality (AR), Big Data, Cloud Computing, Robotics, Sensor Technology, Internet of Things (IoT), and BIM. Benefits of DCSC are basically derived from integrating myriads of various technology-centric innovations. However, most of these benefits are untouched since management of organisational transformations and project relationships aspects are usually overlooked and/or postponed.

Building Information Modelling (BIM)

BIM has been considered as a digital revolution pursuing Construction 4.0 (Dallasega et al., 2018). BIM encourages integration of the roles among all stakeholders on a project (Eastman et al., 2011) In addition, BIM supports the concept of Integrated Project Delivery (IPD), which is a novel project delivery approach to integrate people, systems, and business structures and practices into a collaborative process to reduce waste and optimize efficiency through all phases of the project life cycle (Azhar et al. 2011). In the BIM context, the ultimate aim is to develop a centralized platform in which all available information about a building can be stored; moreover, it enables all practitioners in different project phases (plan, design, execution, and operation and maintenance) to share necessary information through this platform.

BIM is a technology-based innovation that enables design and management of construction projects through simulating a virtual building model in the AECO industry (Eastman et al., 2011; Tah, 2012). BIM is not just software, but more a methodology that represents a new paradigm within the AECO industry (Azhar et al. 2011). BIM is defined as “a set of interacting policies, processes and technologies generating a methodology to manage the essential building design and project data in digital format throughout the building’s life-cycle” (Succar, 2009, p.357).

By evaluating various meanings of BIM, a specific shift from the isolated interpretation of BIM as a mere technology (Eastman et al., 2011; Azhar et al. 2011) towards the socio-technical perspective of BIM must be taken into consideration in the implementation process (Papadonikolaki et al., 2019; Oraee et al., 2019). BIM implementation requires a set of transformative improvement changes to be initiated within the AECO industry. Oesterreich and Teuteberg (2016), based on the socio-technical system theory (STST), argue that “BIM gets people and information working together effectively and efficiently through defined processes and technology” (RICS, 2014). The major key components of BIM implementation therefore are people, information, processes, policies, and technology. In fact, implementation of BIM-enabled processes, policies and technologies require effective and efficient management of human resources, information, business processes, structures, and collaborative efforts to name a few.

In the context of the DCSC, collaboration as the main socio-centric aspects of BIM implementation has been received significant attention in the extant literature (Papadonikolaki et al., 2019; Oraee et al., 2019). The relation between BIM and collaboration can be viewed from two different perspectives: (1) the role of collaboration in BIM implementation; and (2) the role of BIM in collaboration improvement. Collaboration is more than development of close information exchange based on relationships at an operational level; its implementation in tactical and strategic levels is also required across the supply chain (Barratt, 2004). Organisations are able to integrate processes at an operational level (Kahn & Mentzer, 1996), but if tactical and strategic level processes are unintegrated, the related performance benefits of this integration will be limited (Barratt, 2004). Moreover, operational, and tactical level integration can deliver far-reaching benefits despite the vague impact of gaps in the strategic levels of integration. The operational level of SCM focuses on information and material flows. SCM at strategic level, however, is relationship-based and focuses on partnership development of stakeholders to benefit from stronger collaboration.

BIM Implementation and Diffusion Process

The implementation of BIM is still considered as a complex and risky exercise and still several problems such as budgets, quality, and adhering to the time schedule have been experienced in several BIM projects. Multiple stakeholders who may previously have been operating in isolation need to complement each other to combine as a "system" in order to obtain a superior level of performance by BIM (Murphy, 2014). Succar (2009) has defined BIM as "a set of interacting policies, processes and technologies generating a methodology to manage the essential building design and project data in digital format throughout the building's life-cycle". Therefore, based on the Slaughter (1998) innovation models, BIM is classified as a system innovation (Murphy, 2014; Succar & Kassem, 2015) which requires integration of multiple dependent innovations. The deployment of BIM entails IT-based changes to occur within multiple organisations involved in the construction projects. These changes, in turn, create complicated interoperability issues. Effective undertaking of this interoperability requires tighter collaboration among multiple stakeholders and culture change. Effective undertaking of BIM as a systemic innovation requires tighter collaboration among multiple stakeholders and supply chain partners and entails a culture change. However, obtaining a high degree of collaboration is difficult given the nature of construction projects with diverse supply chains and stakeholders involved in a project. Supply chain collaboration can take place in various forms, including inter- and intra-organisational collaboration. Inter-organisational collaboration depicts relationships between two or several organisations in which the participating parties agree to invest resources, mutually achieve goals, share information, resources, rewards and responsibilities, as well as jointly make decisions and solve problems (Ma, Pal, & Gustafsson, 2019).

There are several key requirements for successful implementation of BIM and enabling a collaborative platform. This includes identification of key implementation phases/stages, enablers (e.g. critical success factors, tools, techniques, and decision support systems) and barriers. For successful BIM implementation, as a systemic innovation, there is a need to follow several interrelated stages. Slaughter (2000) found six stages for construction innovation implementation. These phases include identification, evaluation, commitment, preparation, use and post-use evaluation which are described as the following:

- Evaluation stage: the identified benefits should be evaluated based on the project objectives to measure the performance of the alternatives for key criteria.
- Commitment: company should commit to the selected innovation through allocating sufficient internal resources including financial, personnel, equipment, and material resources to the implementation of the innovation. In addition, long-term resources should be allotted to sustain the operation of the innovation (Slaughter, 2000).
- Preparation stage: people within the organisation responsible for innovation implementation, the project team and the construction firms as a whole should be prepared at this stage. Two major activities during this stage are prescribed: (1) providing the resources and (2) educating and training the involved personnel. The role of "project leader" at this stage is critical for coordinating and negotiating among the team members to make the key decisions during the actual use of the innovation (Nam & Tatum, 1997). Successful implementation of system innovation also requires a "system expertise and control" during the early phases to integrate the individual innovations (Hutcheson, Pearson, & Ball, 1996). At the preparation stage, special resources should be allocated, and a high degree of trials and testing are required to reduce the complexity and uncertainty associated with system innovation (Slaughter, 2000).

- Actual use: several adjustments and changes such as changes to processes and systems and also changes to the innovation are required at this stage to fit the intricate set of systems and to maximise the benefits of the innovation implementation (Hutcheson et al., 1996). During the actual use, there is a need for coordinating and adjusting the complementary innovations to obtain the overall objectives. The critical role at this stage is the presence of the decision makers with sufficient level of competency to control the resources and monitor the involved parties (Nam & Tatum, 1997) and to guide the modifications and changes (Slaughter, 2000).
- Post-evaluation stage: after the end of the project and before the project teams disperse - the expected benefits of use should be compared with the actual outcome. The lessons learned and benefits obtained from the project should be documented. This documented data should include technical, project, organisation, strategy and society related aspects (Slaughter, 2000). Since a system innovation consists of complementary innovations, the benefits will be realised when the evaluation of innovation is done across the combination of innovations and not only by considering individual innovation (Slaughter, 2000). Benefits assessment should consider both tangible and intangible benefits by considering a balanced scorecard (BSC) approach inclusive of immediate short-term and medium-term benefits such as rework reduction, and long-term benefits such as business performance improvement, and overall productivity improvement in different levels of project, organisation and industry.

Subsequently, the lessons learned, and the perceived benefits can be a useful resource for assessing the business processes (Buijs & Silvester, 1996) and to use for further projects. Moreover, the performance measurement should be continuously accomplished along all major implementation stages and essential feedback should be provided to utilise for further areas of improvement and next projects. The performance management can be facilitated by adopting an input-process-output model.

Additionally, for BIM tools and concepts to be implemented effectively they should be diffused successfully throughout a construction project and among various supply chain partners. Taylor and Levitt (2004) identified organisational variety, degree of interdependence, boundary strength and span to impact the diffusion of systemic innovations. For example, organisational variety is considered "high" if there is a tendency to use a different set of subcontractors for each trade classification from project to project. A long-term relationship with a particular set of subcontractors across projects would constitute "low" organisational variety.

Barriers of BIM Implementation and Diffusion

Several barriers inhibit wider collaboration in the context of BIM. These barriers can be categorised into a few major groups including financial and economic barriers, social-organisational and behavioural barriers, and technical issues. The influence of these barriers varies in different stages of BIM implementation. The first group are related to *financial and economic barriers* and the high cost of investment in BIM (Cheung, Rihan, Tah, Duce, & Kurul, 2012). Another group of barriers are associated with organisational issues. These include problems such as lack of top management support, problems with alignment of organisational structure with the new systems, information communication and dissemination as well as organisational culture.

The other group of barriers are *social-organisational and behavioural barriers* among individuals who are affected by BIM implementation. At the individual level resistance to change, misinterpretation and misuse of shared data and information, conflicting individual goals and expectations have been found to inhibit the effective implementation and diffusion of BIM tools and concepts. Organisational and project wide challenges include aspects pertaining to culture and structure issues, misaligned vision and strategy, tight timeframes, solid project management issues, changes in business processes, lack of rigorous implementation standards and guidelines, technical challenges include BIM software and hardware capability issues, compatibility issues, BIM technical standards, interoperability issues and the emergence of new BIM software and technologies and their associated complexities and the required learning curve to complete the tasks at an outstanding level. Among different barriers, people's resistance to change is perhaps the most critical barrier to the successful implementation of an innovation (Nisbet & Collins, 1978). Orlikowski, Yates, Okamura, and Fujimoto (1995) found that the use of an innovation is considerably shaped by activities of individuals that interact with the technology, modify it, and change the environment of its use.

Moreover, several problems happen as a result of *technical issues* and incompatibility in BIM. Organisations use different software packages and create different modellings which bring challenges to normal processes integration (Fox & Hietanen, 2007). Data interoperability is considered a major obstacle for BIM deployment (Grilo & Jardim-Gonçalves, 2010). Interoperability occurs when sharing of design information happens. The level of detail (LOD) in BIM models is another technical issue critical for BIM implementation (Gu et al., 2010) as it must correspond to the needs of different parties and hence is a key for maintaining interactivity. LOD determines the level of required information about a model element at a given time. Another set of difficulties are related to how BIM data are structured, nature of available data, data quality, data accuracy, and data integration matters. However, it has been widely promised that technical issues are likely to be resolved through the advancement in the BIM packages by vendors over time (Bryde, Broquetas, & Volm, 2013). In sum, different groups of barriers influence the process of BIM adoption, implementation, and diffusion which include economic barriers, management barriers, technical barriers, external environmental barriers (macro-level barriers) and socio-organisational and behavioural barriers.

Enablers of BIM Implementation and Diffusion

Enablers of BIM-enabled collaboration are categorised into different categories including (1) strategic and contextual factors including culture and support (management and leadership), (2) tasks and job specific factors in BIM-enabled processes such as team work, interaction and interfaces, (3) human resource (people) factors including individual competency assessment and training, (4) policy factors including BIM standards, guidelines, protocols and mandates (prescriptive/descriptive and mandate), and (5) business processes and performance management enablers such as maturity assessment tools, benefit assessment tools and business process mapping. Some of the most significant enablers will be discussed in the following.

Information sharing and communication, for instance, is found as a key enabler. Won et al. (2013) claimed that willingness to share information among project participants was considered as one of the most critical factors to adopt and implement BIM. BIM expands the work scope and blurs the boundary of stakeholders, which requires effective information and knowledge exchange, and efficient inter-organisational communication. In a BIM context, collaboration and synchrony communications still rely on organisational features, such as leadership. However, a high degree of collaboration among the involved parties is required for a system innovation such as BIM (Slaughter, 2000).

Another key enabler that is highly cited in the literature is trust. To date, the lack of trust in completeness and accuracy of 3D models has remained a major concern for the practitioners involved (Gu et al., 2010). At the inter-firm level, researchers believe that trust is a key element for cooperative relationships. The full benefits of BIM as a collaboration platform cannot be achieved without trust and respect between different stakeholders (Singh et al., 2011).

BIM policies, standards, protocols, guidelines, and manuals (prescriptive, descriptive, and/or mandated) along with a structured performance management system foster the maturity of organisational and project business processes as work is repeated, evaluated, and improved. Integration of human resource processes inclusive of collaborative and coordinative structures and efforts will generally increase with process maturity. A systematic individual competency assessment and development program which identifies and matches different individual skills and competencies such as technical skills, and leadership and change behaviour skills is greatly imperative if potential BIM maximum benefits are to be achieved. As maturity is improving to higher levels, processes are more likely to become more integrated, and a rigorous performance management system and a structured platform needs to be in place to enable and facilitate workflow and information sharing among organisational units, project members and other stakeholders as they collectively progress toward established vision and goals. Such a system should be supported through strategic planning systems, vertical and horizontal information channels, and a reporting structure that encourages managerial reviews. The aforementioned stages/phases, barriers and enablers are summarised in a framework (Figure 1) which forms a basis for the rest of this book chapter.

Theoretical View of BIM Implementation and Diffusion

Even though implementing BIM tools and concepts, if properly executed, is reported to bring about more benefits, BIM implementation especially in large construction projects is contemplated to be more challenging. Large projects involved more interfaces and more complicated processes. Additionally, different categories of conflicts are normally encountered on large construction projects such as internal conflicts and conflicts with external entities. Internal conflicts are experienced among the project participants, whereas conflicts with external entities are between the construction project and groups outside the project. Many large construction projects often experience interface conflicts because of the conflicting requirements or priorities from different stakeholders. Accordingly, the human and non-human interactions and interfaces in the implementation process of BIM-enabled projects are discussed by drawing upon the Socio-Technical System Theory (STST), Diffusion of Innovation Theory (DOIT), and Actor Network Theory (ANT) (see Figure 2).

According to STST, BIM consists of a technical subsystem (including all processes, tasks, and technology components necessary for the system) and a social subsystem (concerned with the structure of the work-system, the employees and their attitudes, knowledge, skills, values and interrelationships (Oesterreich & Teuteberg, 2016). Lyytinen and Newman (2008, p. 596) argued the STST posits that both subsystems interact, and such mutual-interaction view is essential for identifying and addressing issues related to the implementation of information systems. The components of both subsystems are as follow (Lyytinen & Newman, 2008):

Figure 1. Process of implementation and diffusion of systemic innovation

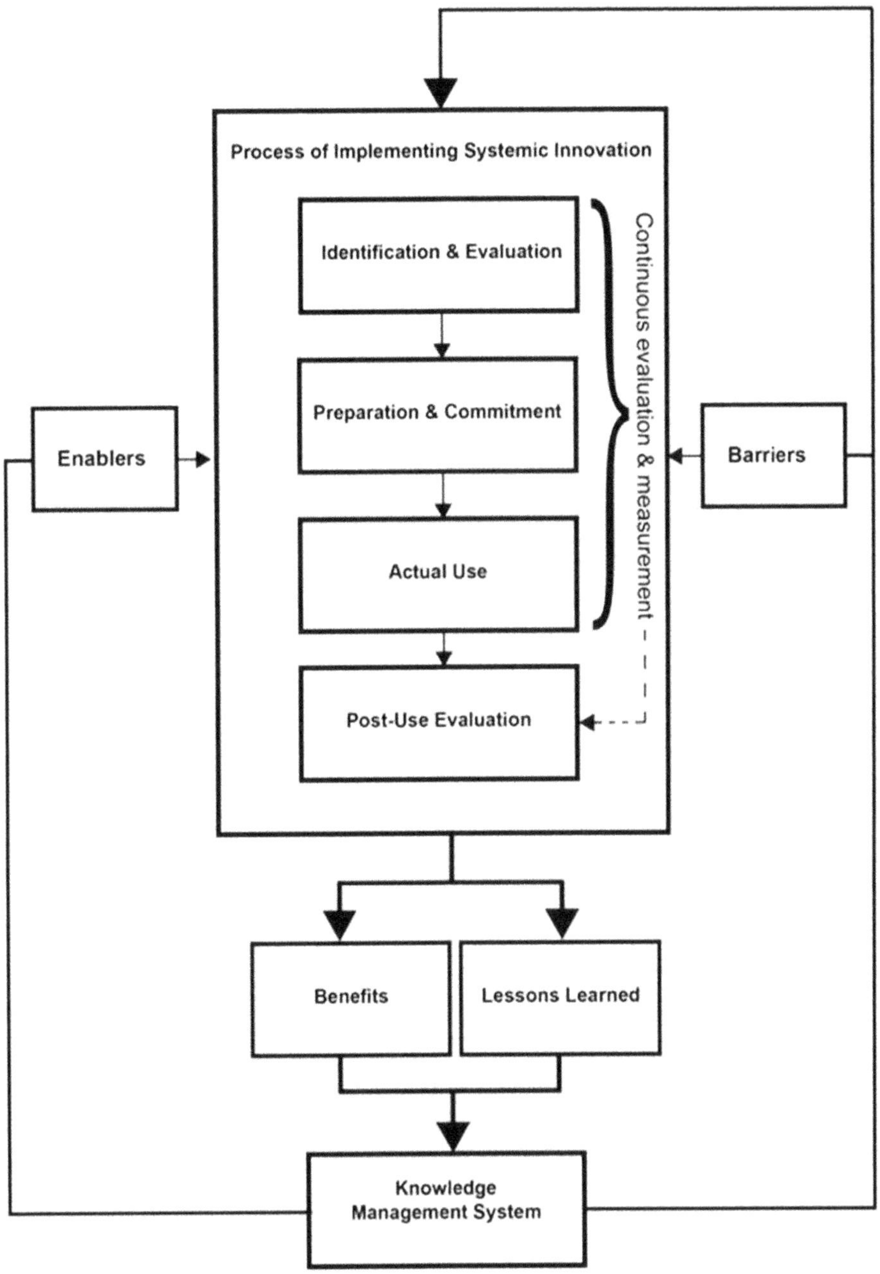

- The first aspect of the social subsystem are individuals or entities in a corporate context who "have a stake or can set up a requirement towards the organisation", which involves employees, managers, clients, clients, subcontractors, and suppliers.
- The second aspect of the social subsystem is the structure, characterized by institutional hierarchy, such as the formal work organisation, communication and governing structure, comprising values, beliefs, norms, common role requirements and behavioral traits.

- The first facet of the technical subsystem refers to the technology involving the software and hardware, the infrastructure together with methods and tools required for the implementation process.
- The second facet of the technological subsystem is the activities within which organisational goals, objectives and stakeholder criteria are met.

DOIT addresses patterns of how innovation spread throughout a group of adopters. In other words, DOIT explains the process wherein a new concept, idea or technology has been spread into the social system over a period (Rogers, 2010). Innovation diffusion plays an important role in theories describing information technology implementation. Rogers (2010) identifies five characteristics in the decision of innovations' implementation which can facilitate or hinder the success, including relative advantage over existing practice; compatibility with existing values of the adopter; overall complexity; observability, the degree that results can be observed; and trialability, the limited experiment of the innovation by the potential adopters (Burritt et al., 2019).

The discussion on interactions herein are also analysed through the lens of ANT which contemplates that both human and non-human actors influence the performance and outcomes of diverse interactions. ANT acknowledges that diverse people, ideas, and things join within networks, as a whole, and become centralized and can become disassociated or reconstituted. The contexts of BIM implementation are dynamic, which means a more detailed understanding of the implementation process can be developed; this eases the understanding of contextual realities.

According to ANT, the role of standards in the diffusion process of inventions is critical, because it enables various actors (1) to coordinate their competing interests in creating a successful actor network, (2) to integrate and produce the technological expertise necessary for implementation, and (3) to enhance the growth and regulation of the domestic and global market (Oesterreich & Teuteberg, 2016). In formation system literature claims that innovations in "complex technology systems" are joint contributions of various entities such as private firms, government agencies and individual project members rather than independent entities (Yoo et al., 2005).

Technologies and actors, by themselves, cannot assemble unilateral driving forces of implementation; a complicated series of negotiation might cause unexpected outcomes. Taking various influences into account, the concept of "relative boundedness" (Harty, 2008) is important in understanding human and non-human actors. This concept challenges the assumptions of coherent implementation processes. This helps to recognize negotiations and alignments, used for innovation purposes, not just for positioning mediators for competing concerns in practice. By following these dynamic interactions, actors, and artefacts, innovation is better understood. Detailed description of activities which establish implementation and processes, used for reducing or increasing relative boundedness, can be used for solving convoluted controversies about the nature of innovation in other complex landscapes (Harty, 2008). Negotiation protocols such as BIM protocol and policy documents constrain the agents' interactions, though these are useful and essential for governance aspects of BIM context.

BIM implementation is a dynamic process and the influential factors are subject to continuous changes along the process. Nevertheless, the initial step (see step 1) is the establishment of a BIM project vision and strategy to provide a direction for the project stakeholders. The socio-technical view of the BIM implementation process starts from a small-scale deployment which diffuses towards a large-scale deployment on a continuous improvement basis which is subjected to dynamic adaptations (see step 2). With the increase in interactions among both human and non-human actors and the learning and experimentation process (see step 3), the project vision might be subject to continual adjustments (see step 1) (Figure 2).

Figure 2. Socio-technical system view of BIM implementation and diffusion (Adapted from Ceschin, 2013)

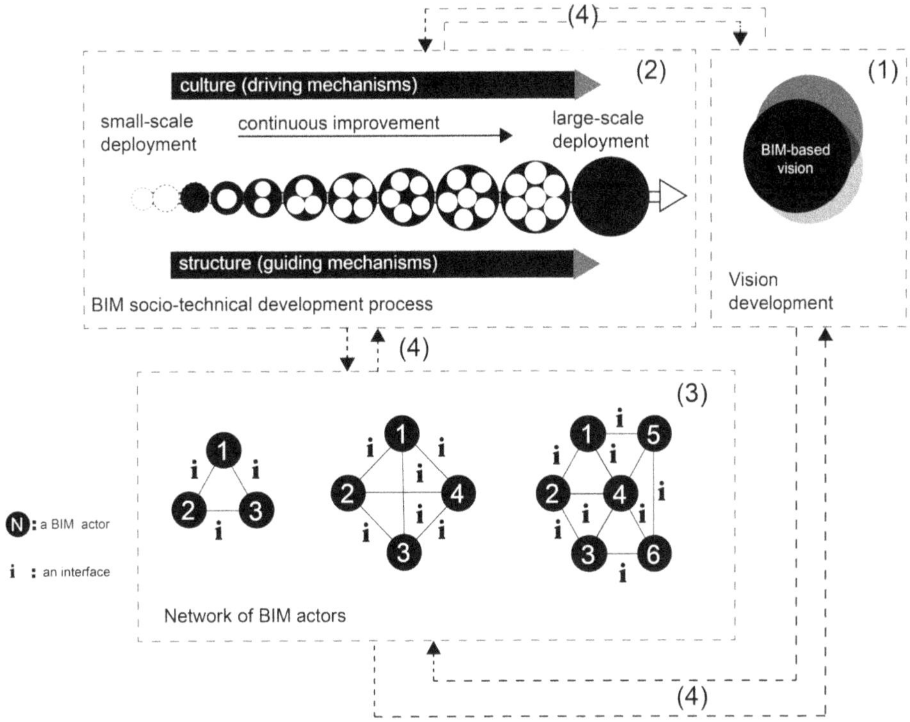

Relational Management in the BIM-enabled Construction Supply Chain Context

The construction industry is heavily reliant on small firms that often find work as subcontractors under a main contract. This prevalence of subcontracting is because of the nature of construction work which requires specialists (Yik et al., 2006) and the strategic choice of main contractors to offer flexibility as their major competitive advantage (Winch, 1998). Manu et al. (2013) also added other reasons including such as the ability to fine-tune labour flexibility, encourage quicker completion of tasks, transfer financial risk and dangerous activities, avoid workers' compensation cost, and align with changing product market demands. Although subcontractors offer organisational and managerial flexibility and provide professional services, the management of contractors and subcontractors relationships can be demanding and sophisticated.

The relationship management approach can be viewed through the following continual maturity model, i.e., C4 continuous improvement model (Figure 3):

- Communication: Communication is the establishment of interactions among organisations through specific methods (Wankmüller & Reiner, 2020). In the concept of coordination, after stakeholders take specific roles, communication channels will be created which is more formal comparing with cooperation; in cooperation concept, communication is for inter-organisational relationships and is more formal than collaboration; in collaboration concept, organisations, together, make decisions and communication channels are already established, standardized, and formalized in which they have shared rules and norms (Wankmüller & Reiner, 2020).

- Cooperation: In support of each other's goals rather than a shared goal.
- Coordination: This relationship approach is mutual and well-defined which organisations use to achieve their goals. Coordination encompasses formal relationships, mutual and common objectives and actions, and not necessarily a centralized authority.
- Collaboration: Occurs when two or more organisations are committed to work together toward mutual goals and operational benefits with openness and trust among each other.

Figure 3. C4 continuous improvement model (Adapted from Kotzab et al., 2019)

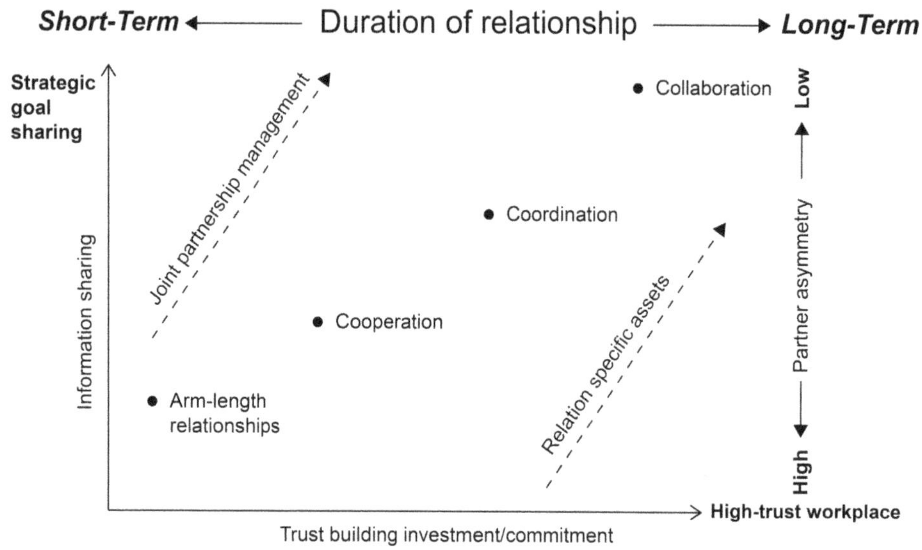

As Figure 3 depicts, collaboration is the climax relationship among involved organisations. This target is achieved when organisations continue their relationships with sharing a strategic goal and building trust amongst each other in the long term. Specific relationship assets and mutual partnership management will alleviate the difficulties of achieving this goal. Furthermore, the highly trustful interrelation among organisations creates integration. Integration encompasses collaboration and coordination between diverse BIM actors in a project which is necessary for achieving a unified version of work units. Therefore, in this chapter we mainly focus on coordination and collaboration as the major requirements for BIM integration.

Coordination

The dependencies among activities are managed through coordination. This relationship approach is mutual and well-defined which involved organisations use to achieve their goals. Coordination encompasses formal relationships, mutual and common objectives, and actions, and not necessarily a centralized authority. For example, Zailani et al. (2016) concluded that coordination is the main issue faced by Malaysian construction companies, among the other causes of project delay (e.g., resources, and business environmental issues). They identified that construction contractors can minimize and control project delays caused by coordination issues through sharing knowledge and working together with channel partners to monitor the key project processes, informing channel partners about upcoming needs of customers, constantly communicating with channel partners about future strategic needs, and looking for innovative ways of fully integrating processes (Zailani et al., 2016).

Collaboration

Collaboration, like "partnering", is a diffuse term and can be ascribed to any non-adversarial relationship (Bresnen & Marshall, 2000). Definitions of collaboration cannot be limited since it emerges in various forms in diverse cases. Collaboration occurs when two or more organisations are committed to work together toward mutual goals and operational benefits with openness and trust among each other (Chakkol et al., 2018). Collaboration has been touted as a facilitator of the BIM implementation process, as an aid to behavioural change within this process, and as an important component of CSC relationships. On the other side, BIM implementation is considered as a promising approach to improve collaborative efforts within the AECO industry.

However, there are some challenges for effective collaboration in the context of BIM projects. For example, each organisation has its own BIM vision, strategy, and plan. Furthermore, they may have uneven maturity levels of BIM implementation. In the BIM-enabled project context, vague definitions and establishment of various terminologies and ontologies can cause different interpretations and elaborations between researchers, clients, contractors, and the rest of involved stakeholders (Osipova, 2015). This misunderstanding might confine best practices' transfer across different contexts because the knowledge transferring process cannot be planned effectively (Bresnen & Marshall, 2000). In short, various stakeholders' objectives, procedures, and practices highly influence coordination and collaboration goals. Thus, the main challenge is to develop a proper, specific strategy and conceptualization approach which is yet broad enough to encompass a variety of arrangements and expectations and is distinguishable from other types of relationships.

Collaborative aspects broadly can be categorised into intra-organisational and inter-organisational collaboration (Figure 4). An intra-organisational collaboration system occurs among the internal working groups and individuals within an organisation and is partly reliant upon an information sharing platform that collects, processes, stores, and shares information between individuals and working groups at an organisational level. More importantly, an organisational culture and structure should be established to encourage organisational employees to come together to communicate, share and access documents, and keep track of BIM-enabled tasks. Inter-organisational and interfunctional supply chains are more effective with the coordinated and collaborative efforts between participants (Soosay & Hyland, 2015). Moreover, collaboration can be defined as a relationship among independent firms which have trust and openness in times when risks, rewards, and costs are shared among them (Olorunniwo & Li, 2010). Collaboration is the commitment of parties to work together to achieve mutual goals to create value and operational benefits through a trustful and open environment (Chakkol et al., 2018).

Figure 4. Model of horizontal and vertical collaboration

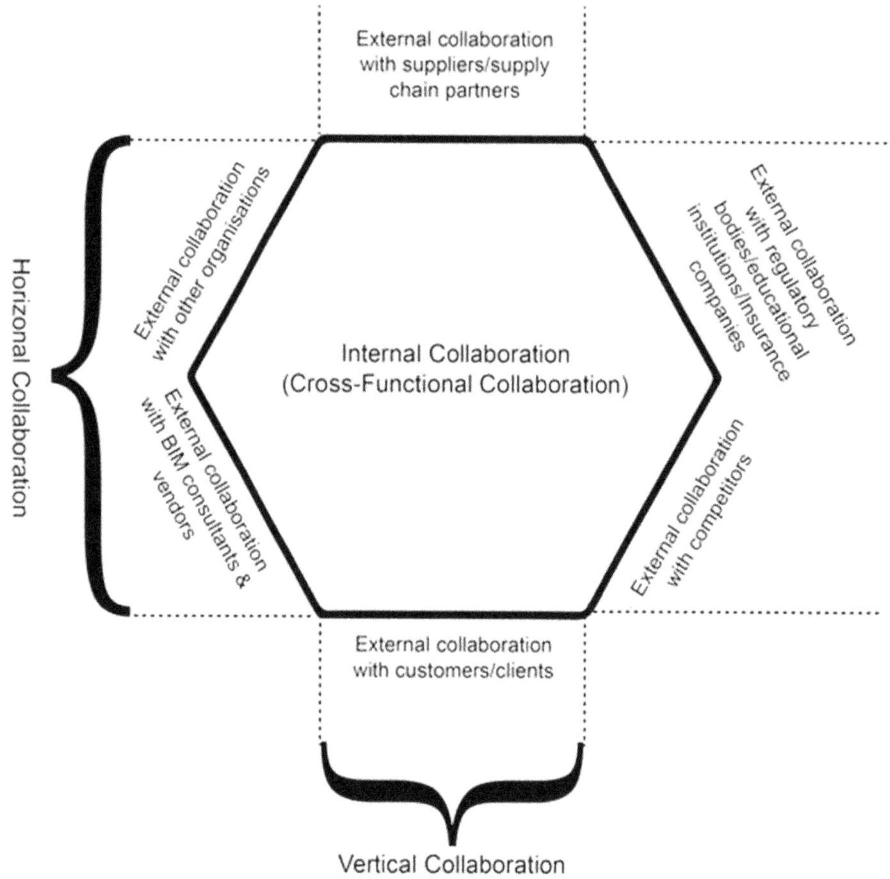

Elements of collaboration are (London & Pablo, 2017): Collaboration drivers (who and what drives collaboration), collaboration participants, collaborative goals, temporality of collaboration (To what extent is collaboration persistent/stable), and collaborative context (environment). There are many other elements of collaboration that have been identified in the various literature in and around SCM (Figure 5).

Different characteristics of collaboration will be discussed in the following section:

- Collaborative culture

  Collaborative culture is one of the most critical elements of collaboration. However, most existing corporate cultures cannot support collaboration either internally or externally (Ireland & Bruce, 2000). The use of functional thinking is frequent, and instead of supply chain processes, it is supported by organisational structures and performance measures which are aligned to functional activities. Several elements together form collaborative culture including trust, mutuality, information exchange, and openness and communication (Barratt, 2004).
  - Trust

Figure 5. Various elements of collaboration in the context of BIM implementation

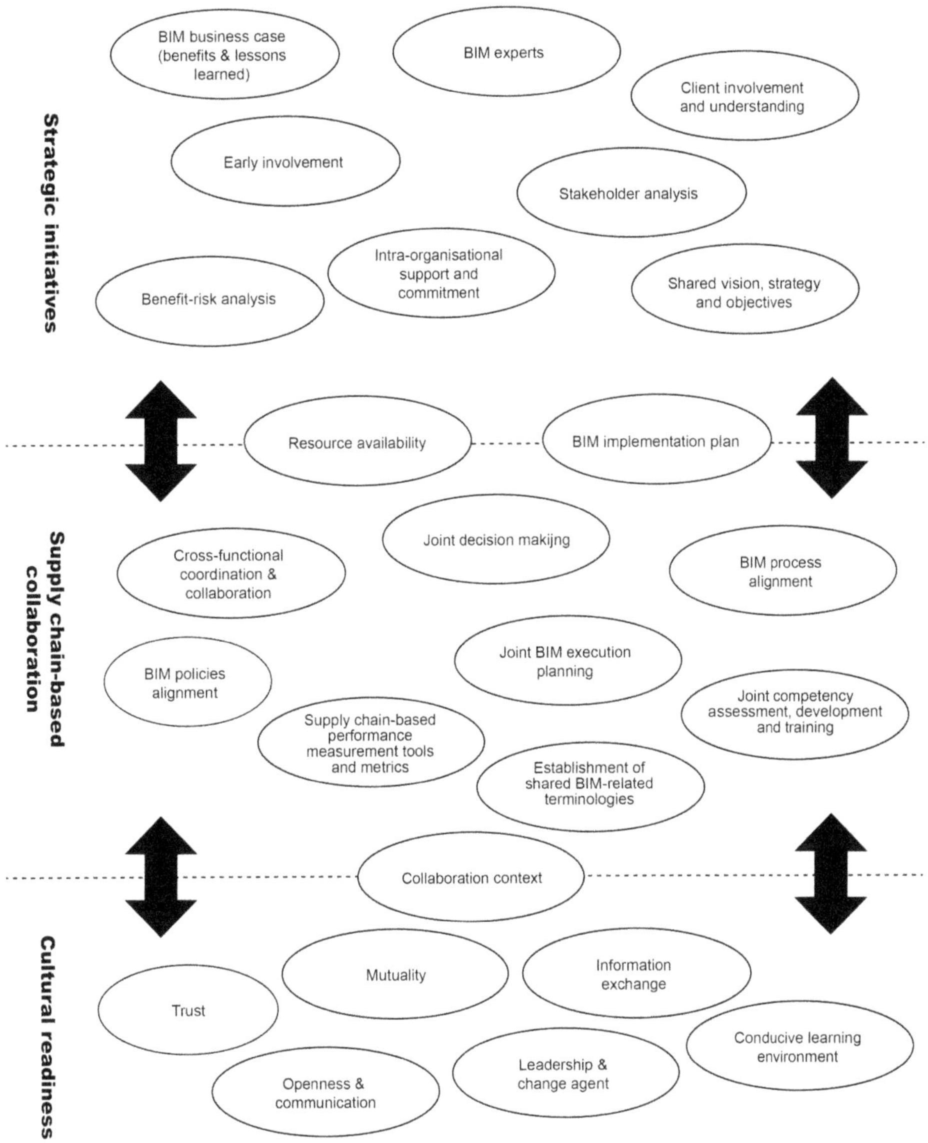

Trust is essential in maximizing the benefits of CSC integration and collaboration. Trust is an important aspect of relationships management. It is seen as a major component of supply chain relationships and can enhance project performance in construction (Lu et al., 2015). Due to high uncertainties within the project-based contexts, a trust-based collaborative culture is required to minimize delay and enhance project quality (Wu et al., 2017) to increase flexibility, promote high knowledge sharing rates and ensure supply chain commitments from the very early phases of a project life-cycle (Manu et al., 2015), which results in significant time and cost savings. Particularly, lack of trust between project partners is highlighted as one of the most influential barriers for collaboration and without effective collaboration the full benefits of the partnership cannot be realized (Chen et al., 2019). Both industry practitioners and scholars have criticised that the construction industry is much too reliant on competitive tendering and adversarial attitudes rather than trust-based collaborative approaches.

The concept of trust-based collaboration within supply chain literature has been viewed at different organisational levels. For instance, Barratt (2004) stresses the significance of 'external trust' development as well as 'internal trust' that can be even harder to develop. More concisely, Fang et al. (2008) explained that any task collaboration forms either a structured or unstructured 'co-entity', i.e., or collaborative entity is "an identified set of employees from each collaborating firm tasked to work together to achieve collaborative outcomes", relies on trust at three distinct levels: (1) "inter-organisational trust" or reciprocal trust of the collaborating firms, (2) "agency trust" or trust of each firm in its own employees appointed to the collaborative entity, and (3) "intra-entity trust" or interpersonal trust between the members of the collaborating firms within the collaborative entity. Thus, it is important to distinguish between interpersonal and inter-organisational trust (Roehrich & Lewis, 2010) since the maturity of the supply chain requires a degree of trust between all members in all levels.

- Mutuality

Mutuality refers to the condition or quality of being mutual, reciprocity, or mutual dependence. In the context of CSC, individuals and organisations must work together towards a common aim; mutuality of benefit, rewards, and risk sharing along with information sharing are the collaboration foundations (Barrat, 2004). Mutuality together with integrity and openness, as codes of conduct, help to maintain trust and create a sense of security by establishing a common sense of values that governs the actions of both parties. (Xu, 2019). Based on the Social Exchange Theory (Lambe et al, 2001), trust is strengthened by the formation of mutual and reciprocal rules in the relationship resulting in positive outcomes for both parties involved. In addition, mutuality is considered as one of the fundamental predictors of collaborative supply chain relationships, since together with other behavioural attributes such as trust and commitment, it can consolidate internal control and reinforce beneficial behaviour of partners (Tsanos & Zografos, 2016).

- Information Exchange/Sharing in the Supply chain

The exchange of information as part of the separate task distributions plays a key role in achieving the best possible joint performance (Lauer & Franke, 2020). For instance, if members share market demand data in real time for the creation of joint forecasting, they can implement a collaborative supply chain structure (Cannella et al., 2015).

Information exchange/sharing is one of the key components of supply chain integration (Tsanos & Zografos, 2016). Particularly, information clarity and consistency of information flows, plays a significant role in several accounts of changes in the supply chain and each of the following premises (Ahankoob et al., 2020). Firstly, intermediation is a possible obstacle to greater openness in the supply chain since it serves as a source of knowledge asymmetry and information impactedness. Secondly, intermediation inevitably drives up costs and is often a non-value-added activity (Barrat, 2004).

Enrichment of information, i.e. the rapid exchange of business data across the chain is beneficial and necessary. Information enrichment can be realized through a process integration scenario as industry 4.0 is moving towards the seamless supply chain, wherein all players think and act simultaneously as one entity. Through the fostering of knowledge exchange and joint decision-making in the BIM setting of the CSC, it is assumed that coordination and contingency adaptation arrangements can promote connections between parties. For instance, Dowsett and Harty (2019) argued that the substantial benefits of exchanging knowledge through the introduction of BIM are neglected, as the essential process enhancement required to do so are hindered by the broader organisational and cultural context influencing the project.

- Communication and understanding

In its current customer-centric configuration, the supply chain relies heavily on flows of communication to promote engagement between actors and to enable decision-making across all phases connecting upstream and downstream stakeholders (Gambetti et al., 2018). Information and communications technology (ICT) play a key role in the management of the supply chain, and the intensity of information technology usage. Communication improves strategic intangible assets across the chain, such as supply chain culture, image, and identity. Reciprocal communication, in particular, has enabled diverse organisations to establish mutual understanding and sense of ownership of interactions (Xu, 2019).

- Openness and honesty/integrity

From both intra- and inter-organisational views, a culture of openness and honesty is necessary (Barrat, 2004). Openness and integrity tend to build trust and commitment, contributing to increased trustworthiness in the reciprocal relationship; therefore, in a sense, each of these four components strengthens strong collaborative links (Prasanna & Haavisto, 2018). In order to sustain trust-based relationships along all projects, therefore, behavioral programs should be implemented by construction firms and supply chain members to institutionalize a relational code of conduct, e.g., openness, fairness and reciprocity, into inter-organisational interactions (Xu, 2019) to cope with supply chain uncertainties and fluctuations (Manu et al., 2015).

- Managing change

The introduction of the BIM system should not take place until a positive attitude is formed and sustained among the potential users (Abbasnejad et al., 2016). For instance, the introduction of the BIM system should not occur when most of the members feel threatened or forced. The sources and types of resistance are many, and they need to be properly identified, analysed and managed. Change management, in the construction context, often relies on identifying and enforcing processes and/or technologies to accommodate the improvements that could occur from the implementation of BIM in order to ensure higher efficiencies are resulted. Matthews et al. (2018) argued that this approach can be viewed as a continual cycle of renewal.

- Cross Functional Activities

In the BIM setting, a systematic and integrated strategy to performance-based design is required to include performance assessments derived from different domain simulations with a life-cycle view in the initial design phase. Such a holistic analysis of project performance requires the integration of multi-domain stakeholders such as clients, architects, engineers, facility managers and consultants. The use of integration tools enables collaborative work designed to allow any participant to exchange information with other experts, regardless of the software wherein the information is first created (Arayici et al., 2018).

- Process Alignment

  Process alignment represents the configuration of various functions of an organisation or project to operate together in a harmonious fashion in order to achieve shared goals, maximise efficiency and maintain competitive advantage (Hung et al., 2010). Successful implementation of BIM relies more on how effectively BIM-enabled processes are managed and integrated within all work processes both within and across organisational boundaries (Won et al., 2013).
- Joint Decision Making

  Most AECO decision-making activities depend on a multi-organisational and multidisciplinary structure that relies on a continuous feedback loop between various project functional units. An automated and rapid data sharing network can promote more open and collaborative decision-making in AECO applications (Du et al., 2018). For instance, BIM enhances the coordination of mechanical, engineering and plumbing (MEP) engineering under circumstances such as previous expertise, initial joint decision-making, and shared planning and operations (Papadonikolaki & Wamelink, 2017).
- Ontology (establishment of shared BIM-related terminologies)

  Furthermore, a set of declarative knowledge is called "ontology" which represents every critical concept in an application domain. Gruber (1995) defines ontology as "a formal, explicit specification of a shared conceptualisation". Ontology is a way of representing knowledge to facilitate joint understanding and thereby information sharing and communication. Providing that parties become aware of a meaningful interaction, its concept can be easily assessed and be modeled afterwards. Ontology consists of major concepts in a specific application domain, their attributes, types, and any probable restrictions. Standardised BIM-related definitions and terminologies should be established, and team members should reach a consensus on which terms they will use when they are collaborating.
- Supply Chain Metrics (Performance Measures)

  A performance metric is a collection of measures that seeks to calculate the efficiency and/or effectiveness of an action. The metric not only is a formula to calculate the measure but also is a description that denotes the measurement tile, how effectively it can be calculated, and where the data can be collected (Mishra et al., 2018). The nature of SCM shows that the overall performance of the integrated supply chain is of greatest priority. This is because various organisations may use different performance metrics.

  From a systematic perspective, for example, the Kaplan and Norton's Balanced Scorecard Approach, related to supply chain performance measurements, combines both financial and operational metrics that are applied at various supply chain, managerial, functional and team levels (Chen & Paulraj, 2004). In the context of the BIM project, organisations first must define success stories with respect to critical success factors (CSFs) and then sets clear measures to evaluate whether all the CSFs are being fulfilled (Chegu Badrinath & Hsieh, 2019). An organisation can acquire information from the supplier about BIM implementation process constraints and capabilities. Given such information about a supplier's knowledge and experience, the organisation may have an opportunity to adjust its own BIM implementation in order to avoid implementation failures. Similarly, an organisation may involve supply chain partners in their BIM implementation efforts to enrich their understanding about the BIM implementation process and to foster an environment for joint decision making and performance management.

It is critical to develop a BIM-based supply chain maturity assessment tool which envisions the expectations, the competencies and capabilities of various supply chain parties involved in the BIM implementation efforts. This also helps to create visibility in regard to the strategic objective of the supply chain to maintain a compatibility relationship among upstream and downstream of the BIM network actors. Zailani et al. (2016) discussed that such visibility and flexibility minimised the negative effect of resource issues and coordination issues on the project performance. Such a BIM-based supply chain maturity helps for managing the evolved processes and positive progress towards more joint decision making and relational behaviours and achieve both organisational and project goals which had been previously deemed to be conflicting in some instances.

- Resources Commitment

Enterprise-related problems such as resource management would have an impact on the pace and success of technological adoption. The implementation of BIM is expected to be more troublesome for small and medium-sized companies (SMEs) lacking the capital and ability to invest in technology (Hosseini et al., 2018). However, the long-term vision and commitment of managers to promote higher BIM activities are required to guarantee the digital transformation of the AEC industry. organisations in the construction industry should establish a BIM knowledge network with professional associations and universities so that external sources of advice and assistance can be established (Le et al., 2020).

- Intra-organisational Support

Limited resources within SMEs impact their ability to implement BIM in projects. Intra-organisational factors like incentive for adoption, the intra-company structure, incorporation of BIM-SCM vision into business model, intra-organisational BIM-related functions, and BIM used by the organisation can impact the BIM-enabled supply chain relationship (BIM-SCM) (Papadonikolaki & Wamelink, 2017). Intra-organisation BIM networks should be built under the governance of core BIM diffusion's drivers such as top general contractors and developers, in order to bring the latest BIM implementation knowledge back to the BIM project (Le et al., 2020).

- The Corporate Focus (vision & mission)

A concrete framework should be built to prevent risks in a large-scale drastic transition, such as a shift in the nature of the building project due to BIM implementation. An illustration of such a vision is that service providers by full implementation of BIM can win contracts in the future market as a long-term benefit. Thus, the vision and mission are achieving competitive advantages through full BIM utilization that reinforce the project team's change towards full BIM implementation (Liao & Teo, 2019).

- Demonstrating the Business Case (Lesson learn, Cost-benefit Analysis)

Some business factors are related to BIM, such as improved profitability, increased efficiency, enhanced information processing and better coordination and visualization (Amin & Abanda, 2017). In developing a business case, different analysis methods, e.g., the SWOT analysis, can be used to clarify the ability of the contractor in seeing precisely where BIM fits into their business at a certain time and in visualizing whether the use of BIM can maximize positive outcomes and reduce risks (Whitlock & Abanda, 2017). For instance, if the conversion of BIM data to existing facilities management systems is compatible or strengthens new operation modes and the importance of BIM is shown, otherwise it is doubtful that facility managers will be able to make the business case for the implementation of BIM (Love et al., 2015).

- The Role of Technology (for communication and information sharing)
  Innovation practices such as new product development (e.g., digital technology creation) and collaborations to promote design and process innovations are part of the technology development. Another aspect of this development is information sharing that builds effective relationships and fosters integration between suppliers and manufacturers in the supply chain, which results in better performance (Kim & Chai, 2017). Although the technical dimension of information sharing is important, the pace, quantity and quality of information that is exchanged really counts. (Prajogo & Olhager, 2012). The role of technology to share information allows organisations to access strategic supply chain information and not just transactional details, such as materials or purchase orders.

Collaboration is a satisfactory means for delivering sophisticated projects which involve diverse parties, suppliers, and customers (Davies & Hobday, 2005). Partnering firms must act collaboratively in order to manage mega projects successfully (Davies et al., 2016). In project-based settings, this action can be challenging in the short-term, especially when organisations are inexperienced in working collaboratively; this is a great risk for large-scale infrastructure projects. Developing collaborative culture, norms, and practices are time-consuming which is a challenge in time-limited projects (DeFillippi & Sydow, 2016).

Regardless of the project types, collaboration is a great challenge which requires effective governance and management aspects. Project practitioners' main goal is to obtain competitive advantage based on their abilities to collaborate, which can improve their governance in specific projects (Vangen et al., 2015). Project governance is a method of controlling a project and authorizing reliable and repeatable practices (Müller et al., 2015), one of which is collaborative practices. While flexibility of responding to unique customer's needs is necessary for delivering complex projects successfully (Davies & Hobday, 2005), the governance of these projects can be repeated across various projects improving efficiency in future ventures (Müller et al., 2015).

Formalising experiences on interorganisational collaboration is beneficial as they can be used in subsequent projects and tasks (e.g. Davies et al., 2016). This benefits organisations from both the required flexibility for output customisation and standardisation of processes which drive operational efficiency (Davies & Brady, 2000). However, the understanding of how to govern this kind of collaboration, and the use of experiential learning across diverse projects are still limited (Maylor et al., 2015). It is important to note that collaboration of time-limited, project-based operations also requires governance.

Construction project teams have recognized the urge to have better integration, cooperation, and coordination. One of the major concerns in this industry is project team integration, and practitioners must overcome this issue or at least provide a context to deal with cultural inconsistencies, distrustful relationships, and paradoxes in the "design-construction divide" which are the main obstacles to successful project completion. The management of construction projects with robust attitudes and loyalties (Walker, 2002), prohibit the process of change by having great coordination and cooperation among project's participants.

Despite the complexity and low innovation indices of the construction industry, important changes are happening in it through higher levels of dematerialization, online collaboration, and management efficiency which are positively affecting performance and introducing even more advanced systems and tools. Through the support of an integrated strategy via BIM, supply chain management can evolve as a strategic enabler in e-business systems to enhance collaboration and create stronger networks with a special emphasis on digital/virtual collaborative projects (Costa & Tavares, 2012). In this network, information and knowledge are deemed as competitive advantages, so businesses try to focus more on alliances, partnerships, and collaborations.

It is widely discussed that for IT-based collaboration to be customized and best fit the organisational culture either structured or unstructured collaboration can be stipulated. The structured collaboration entails establishment of more rules, guidelines and restrictions, however, unstructured collaboration relatively leaves members unregulated in terms of how they can collaborate and use internal collaborative tools and platforms, though some guidelines, and best practices have to be in place to guide the collaborative efforts of members. Different organisational cultures require either unstructured collaboration or structured collaboration. For instance, in companies where more innovative ideas need to come up, more unstructured types of collaboration are required. On the other hand, in most of the manufacturing firms where constant workflows need to be followed, a more structured type of collaboration is advisable. A well-organized structure would fit well with a structured collaboration model because it is hierarchy based and needs leadership to keep the flow moving. In addition to corporate culture, the structure of the firm needs to be evaluated first to get a better understanding of which collaboration model would best fit. For example, if a firm has an organic structure with no managerial hierarchy, then an unstructured collaboration would be best because this model fosters innovation in a group setting without any power struggle.

It can be argued that as AECO companies move from adopting innovative technologies (lower maturity levels) towards effective implementation and utilization of them (higher maturity levels) some form of both structured and unstructured collaboration is required, though the degree and significance of them varies in different adoption and implementation stage and based on the organisational context (culture and structure). For example, in the early stage of BIM adoption and implementation process, a company might need more innovative ideas to implement BIM tools and concepts more effectively thereby a more unstructured type of collaboration would fit better. As companies move towards higher maturity levels (more repeatability and quality in processes) more standardize and constant workflows and processes may evolve thus a higher degree of structured collaboration would fit best though yet a degree of unstructured collaboration might be required to foster innovation.

Inter-dependencies in Construction Supply Chain (CSC)

According to Fellows and Liu (2012), there are two broad models of interdependencies: the "competing forces model" (or differentiation or specialization) and the "integration, coordination, collaboration" model, in which the forces strive for equilibrium in dynamic environments. The competing forces models have been criticized to stifle collaborative efforts. The CSC requires more frequent and direct interaction to enable mutual adjustment. It also entails continuous adjustment of the working plans. However, as the plans will be continuously adjusted, it affects all the up-stream activities. CSCs, unlike other industries, are subject to both sequential and the two other interdependencies (pooled and reciprocal) which are used according to synchronising requirements of construction sites (Thompson, 2003):

- Sequential interdependencies: there is a direct interdependence among activities as an output of one activity is the input of the next one.
- Pooled interdependencies: each part has its own contribution to the whole and all parts are supported by the whole.

- Reciprocal interdependencies: the output of each activity is the input of other activities and affects it, which means that each work unit is reliant upon the other unit/s. Reciprocal interdependencies also involve pooled and sequential aspects.

Intra- and Inter-organisational Interfaces, Interactions, and Boundaries

It is critical to distinguish the difference between the three key terms of interface, interactions, and boundary. Interface is the point of interconnection between entities while interaction is the situation or occurrence in which two or more objects or events act upon one another to produce a new effect, and boundary is the dividing line or location between two domains, and/or supply chain parties. Interfaces can be categorized into different taxonomies such as organisational interfaces, contractual interfaces, and physical interfaces. organisational interfaces are the interactions of construction projects' participants. They include the relationships among diverse individuals and teams throughout the whole project life cycle. Thus, they contain a variety of companies and their subdivisions, especially within design and build practices. Contractual interfaces occur when grouped work elements are divided into distinct work packages in order to be suitable for design information availability or align with the principal contractor's program. Due to these work packages, interfaces are created, which are usually associated with specialist contractors. Physical interfaces are the actual, physical connections between two or more building elements or components. In construction projects, these interfaces happen inevitably, but their amount and complexity depend on the building's detailed design. If not managed properly, this will result in constructability issues i.e. "the ability of an element to be constructed with relative ease". Though constructability issues highly reliant upon competencies of individuals who are responsible to install the building elements.

Certain forms of procurement and delivery methods, especially the management forms, predominantly construction management, and management contracting, place a notable emphasis on interface management. A contract may provide an institutional framework guiding the course of cooperation rather than coordination and collaboration. Relational management and governance should overcome the adaptive limits of contracts (Luo, 2002) enabling the BIM actors to focus on collaboration. With the application of these methods, the responsibility will be appointed to the expert contractor. Lack of organisational boundaries arises potential issues as overlapping with other organisations on the project which can cause conflicts. By undertaking a project, the location of the formal (legal) organisational boundary of the firm does not change, but the informal boundary, at least the firm's role in the project, might be extended to include the whole project.

There are myriad systems of boundaries in construction projects: (1) the formal organisations' boundaries; (2) the boundaries around each organisation's activity groups operating on the project; (3) the boundaries around each within-an-organisation group on the project; and (4) the boundaries around informal groups which include participants of diverse organisations. Ease of movement across a boundary is referred to permeability. The boundary permeability is highly reliant upon cultural and behavioural aspects. One condition that causes conflict is the incompatibility of organisational cultures/ climates because of inappropriate behaviour, especially if the participants are not familiar with these differences.

Additionally, establishing and managing external linkages (boundary spanning, scouting, and coordinating, for instance) can happen within an organisation and/or across organisational boundaries. In boundary spanning, teams connect discrete parties and become critical channels for information sharing and innovation (e.g. Hargadon, 1998); this action underpins knowledge sharing and social networking.

Boundary spanning is appointed to specific individuals, called *designated boundary spanners*, and is more likely performed informally. Using social network analysis, this action can be analysed and depicted (Di Marco et al., 2010). This action usually occurs among subunits in an organisation, but it can also occur between separate organisations with more extensive relationship patterns, such as stakeholders' representations, coordinating groups, and searching information from external parties (Ancona & Caldwell, 1992). These activities are substantial for performance (Gladstein, 1984) and innovation (Hargadon, 1998).

In order to understand organisational functioning, team-based project organisations should focus on both inter-team activities and intra-team ones. Inter-team boundary activities depend on the inter-team goal and task interdependence, and team heterogeneity, organisational culture, and favourable external environment are critical moderators of the relationships in an inter-team, boundary activity, team and organisational performance (Drach-Zahavy & Somech, 2010).

Participants' performances are important for underpinning commitment and cooperation. For this issue, utilisation, and development of boundary management practices through boundary spanning can positively affect interaction and coordination among participants even with the ones who want to hold their personal goals.

Integration

Generally, integration encompasses collaboration and coordination among various BIM stakeholders in a project which is necessary for achieving a unified version of work units. Collaboration, as a component of integration, focuses on socialization, group work, and interaction among the project team (O'Reilly, Caldwell, & Barnett, 1989). It is obtained through structural mechanisms like job sharing, joint decision making, standardized practices, plans, policies, and other arrangements that increase flexibility, cooperation, and mutual goal achievement. Collaborative processes are followed by some possible errors of collective decision making like groupthink (Janis, 1972). However, collaborated well-organized work units and tasks may end in the wrong destination, so inevitably produce useless outputs which cannot help the involved actors to reach their strategic goals.

By achieving effective performance outcomes, another dimension of integration, i.e. coordination, is suggested. The general definition of coordination is the skillful and balanced simultaneous movement of diverse parts. Coordination helps the identification and arrangement of diverse activities that must be linked, (Fredrickson, 1986), and objectivise collaborative efforts which do not let the progress lose track toward effective outcomes (Schweiger, Sandberg, & Rechner, 1989). Structural mechanisms support coordination by harmonizing interprocess activities and controlling them toward common directions (Sherman & Keller, 2011).

Through the lens of process maturity, integration is achieved by both collaborative and coordinative characteristics. It is the state of association among organisational processes that boosts unified, harmonious efforts toward efficient achievement of organisational goals. The aim of an integrated supply chain is to remove all boundaries to ease the flow of material, cash, resources, and information to obtain an optimised supply chain afterwards.

In the context of construction, fragmentation occurs both horizontally and vertically. Horizontal fragmentation is about the multiplicity of actors, such as individuals, organisations, and business units, whose functions are virtually at the same stage (e.g. schematic design of a building). In vertical fragmentation, a process splits up into different components/stages which are allocated to separated functional actors, such as general contractor with subcontractors and/or structural engineering design with reinforcement supply. Therefore, to achieve a thorough integration approach, both vertical and horizontal integrations should be considered. Horizontal integration must be addressed to foster effective and efficient selection of participants for each stage (vertical) of a project (Figure 6). Additionally, Vrijhoef and Kosekela (2000) identified the four roles of construction supply chain (CSC) management. Respectively, the roles focus on:

1. The interface between the CSC and construction site.
2. The collaboration and integration aspects of the CSC itself.
3. Transforming activities from the construction site to the CSC.
4. Integrated management of the CSC and the construction site.

In sum, to obtain the full integration (a high degree of communication, coordination, collaboration, etc.) on a construction project, both vertical integration and horizontal integration must be addressed, fostered and managed through contemplating key relational aspects such as boundaries, interdependencies, interactions and interfaces among not only both the human and non-human actors but also by focusing on the interfaces between different construction phases (Figure 6) as well as between supply chain and construction site.

Figure 6. Model of construction projects horizontal and vertical integration

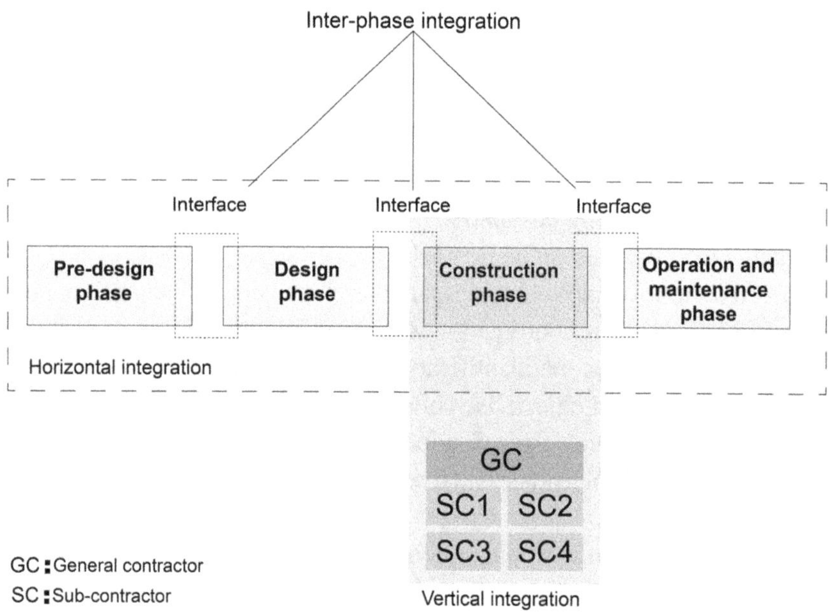

Broadly speaking, integration may occur within the following modes (Lahti et al., 2009):

- Function integration: Functional divisions/departments within an organisation are focusing on improving their own internal processes, procedures and use of internal resources and conducting performance measurement at functional level. It requires little business process change and may rely more on technology-enabled tools to support the integration of different functions such as scheduling, estimation, etc.

- Internal integration: Various organisational divisions/functions and organisational-wide processes are now defined; this allows individual functions to understand their roles in complex organisational wide and supply chain processes. Process-specific information is collected and shared within the entire organisation by using integrated systems and internal databases.
- External integration: Strategic stakeholders and supply chain partners should be identified, as well as key information the organisation requires from them to support its BIM-enabled business processes. Also, effective collaboration processes with key supply chain partners are implemented, and information is collected and shared with the value chain parties. External integration in regard to BIM-enabled projects may fall within the following two categories:
    - Multi-project integration: Requires not only integration of BIM technical and software tools, but also process and organisational alignment to work across multiple BIM-enabled projects.
    - Multi-firm integration: Requires seamless communication across CSC firms.

Additionally, in the context of BIM, integration should be established within the following pillars:

- ❖ Process integration: Integration in regard to interfaces, interactions, boundaries and interdependencies (relational integration),
- ❖ Policy integration: Integration of BIM protocols, plans, standards, guidelines for information sharing,
- ❖ Technological integration: Integration of different BIM models and functions (e.g. 3D model and scheduling and cost estimation), and
- ❖ People integration: Cultural integration (behavioural integration).

What is central to these types of integrations is the significance of information integration and distinction of targeted, expected and actual information to identify where the information gaps exist and how further improvement should be conducted to facilitate collaboration and integration efforts (Succar & Poiri, 2020). It is crucial to distinguish between information sharing and information integration. Information sharing refers to sharing both tacit and explicit knowledge via networkings and communities of practice, document management, use of information technologies, and the like. Information integration entails not only information sharing but also inter-organisational collaboration and coordination mechanisms. Technical integration entails interoperability and integration of BIM data through focusing on various software platforms and BIM data sharing standards. In other words, data integration entails not only the integration of data elements but also integration of related BIM standards and policies. All these components may be influenced by the macro-level and meso-level interventions such as institutional arrangements and industry norms and settings (such as government mandates), and organisational specific norms and characteristics (such as top management support).

CONCLUSION

AECO companies and their supply chains can take advantage of information sharing platforms and digital technology in the context of Industry 4.0. Industry 4.0 refers to the trend towards digitalisation and automation by utilising various technologies and processes, such as the Internet of Things (IoT), cloud computing and artificial intelligence (AI). The application of technologies related to digital transition with an extended scope beyond organisational boundaries which entails intelligent supply chain and connected customers refers to the supply chain 4.0 (Ghobakhloo, 2019). Some of the applications are as follows (Lu, 2017; Ghobakhloo, 2019):

- E-commerce tools such as Electronic Data Interchange (EDI)
- Advanced Manufacturing Technologies (AMT) such as Computer Aided Design (CAD), Computer Aided Manufacturing (CAM), Computer Aided Engineering (CAE), Industrial Simulation, and Enterprise Resource Planning (ERP); and
- Advanced IDT such as the Internet of Things (IOT), Augmented Reality (AR), Additive Manufacturing, Blockchain, Cloud Computing, and Big Data analytics

In this book chapter, it is discussed that despite the nature of CSC, involvement of various stakeholders and the fragmented nature of the construction industry, collaboration and integration efforts have been perceived as difficult tasks to achieve. The project-oriented and decentralized construction sector in comparison with more interconnected sectors such as manufacturing, requires more societal change measures to be established to improve the collaboration and integration levels and maximise the potential benefits of technology-centric innovation (e.g., BIM) implementation. In order to deepen our understanding of collaboration and integration concepts, discussion of collaborative social networks and context, shared information, integrated data, and interoperable software technologies is imperative.

Existing studies on BIM implementation and diffusion have been drawn in order to systematically unfold the processes and components of collaboration and integration. Hence, initially a conceptual understanding of the key barriers, enablers and stages of BIM-enabled projects were provided. Subsequently, the ways these barriers and enablers influence collaborative working practices in BIM-enabled projects were examined. Integration of all the above-mentioned enablers, barriers and stages has led to the development of a BIM implementation and diffusion model. Collaboration has been considered to play a significant role in any digitally-enabled construction projects such as BIM projects and acts as a mediator between the initial low maturity stages towards higher maturity levels, and to achieve integration and automation objectives. The role of collaboration and the constituent factors in a BIM-enabled project have been further investigated. Collaboration is more than the development of close information exchange based on relationships at an operational level; its implementation in tactical and strategic levels is also required across the supply chain actors. Furthermore, AECO firms need to develop a strategic approach in terms of both intra- and inter-organisational collaboration aspects by a particular focus on relationship management aspects such as interfaces, interactions, interdependencies, and boundaries and move towards alliancing approaches to contracting in BIM-enabled projects. The authors recommend an interface management strategy to be developed for BIM implementation and diffusion process.

Additionally, it is argued that horizontal and vertical collaboration and integration aspects should be addressed and managed by the AECO firms and projects in order to maximise the benefits of adopting any digitally-enabled tool or method such as BIM. Both collaboration and integration may occur at different levels and with multiple components such as BIM tasks, BIM data, culture, resources and infrastructures, business processes, information, vision and strategies, and the like and can be achieved by considering a number of different aspects such as relationship management (management of interfaces, interactions, boundaries and dependencies), standardization of work processes and practices, joint decision makings, joint individual competency assessment and development, relational integration, information and policy (guideline, standards, protocols, manuals, etc.) integration, and technical integration aspects such as compatibility, data integration and interoperability. Among these components perhaps cultural integration which involves integration of norms, values, expectations, willingness, and perspectives is the most difficult one to achieve.

Therefore, what should be significantly highlighted are the differences in attitudes, values, beliefs, norms and expectations towards BIM tools and concepts that generate conflicts and stifle collaborative efforts. Project team members should not only communicate in terms of technology aspects of BIM, but also communication in regard to commitment levels and how a conducive collaboration environment should be established and maintained. BIM project leaders must establish norms and values relating to information sharing and collaboration. Some project members may refuse to share their knowledge and information with others since they may feel through knowledge and information sharing, they may lose their power. Relationships should be established by focusing on relational norms and values to foster a collaborative environment conducive for BIM implementation (Carson et al., 2006). While competitiveness indicates low relational norms and values, coordination and collaboration can only be flourished in an environment where relational norms and values are highly established and maintained (Tangpong et al., 2010). The above-mentioned various perspectives and components related to collaboration applicable in the context of digital-enabled CSC, hopefully, can serve as a basis for developing a systematic "BIM Collaborative Working Model" which contemplates major socio-technical approaches consist of trusted and conducive social networks and contexts, integrated data, effective information sharing, interoperable technical infrastructure, and strong collaboration both within and across organisational boundaries.

ACKNOWLEDGMENT

The authors would like to thank S. Armin Mirhosseini for their work in this publication.

REFERENCES

Abbasnejad, B., Nepal, M., & Drogemuller, R. (2016). Key enablers for effective management of BIM implementation in construction firms. In *Proceedings of the CIB World Building Congress 2016: Volume I-Creating built environments of new opportunities* (Vol. 1, pp. 622-633). TUT–Tampere University of Technology.

Abdel-Basset, M., Manogaran, G., & Mohamed, M. (2018). Internet of Things (IoT) and its impact on supply chain: A framework for building smart, secure and efficient systems. *Future Generation Computer Systems*, *86*, 614–628. doi:10.1016/j.future.2018.04.051

Ahankoob, A., Abbasnejad, B., & Wong, P. S. (2020). The Support of Continuous Information Flow Through Building Information Modeling (BIM). In *The 10th International Conference on Engineering, Project, and Production Management* (pp. 125-137). Springer.

Amin, K. F., & Abanda, H. F. (2017, April). Developing a Business Case for BIM for a Design and Build Project in Egypt. In *International Conference on Engineering, Applied Sciences and System Modeling* (pp. 149-159). Springer.

Ancona, D. G., & Caldwell, D. F. (1992). Bridging the boundary: External activity and performance in organisational teams. *Administrative Science Quarterly*, *37*(4), 634–665. doi:10.2307/2393475

Arayici, Y., Fernando, T., Munoz, V., & Bassanino, M. (2018). Interoperability specification development for integrated BIM use in performance-based design. *Automation in Construction*, *85*, 167–181. doi:10.1016/j.autcon.2017.10.018

Barratt, M. (2004). Understanding the meaning of collaboration in the supply chain. *Supply Chain Management*, *9*(1), 30–42. doi:10.1108/13598540410517566

Bortolini, R., Formoso, C. T., & Viana, D. D. (2019). Site logistics planning and control for engineer-to-order prefabricated building systems using BIM 4D modeling. *Automation in Construction*, *98*, 248–264. doi:10.1016/j.autcon.2018.11.031

Bresnen, M., & Marshall, N. (2000). Partnering in construction: A critical review of issues, problems and dilemmas. *Construction Management and Economics*, *18*(2), 229–237. doi:10.1080/014461900370852

Bryde, D., Broquetas, M., & Volm, J. M. (2013). The project benefits of building information modelling (BIM). *International Journal of Project Management*, *31*(7), 971–980. doi:10.1016/j.ijproman.2012.12.001

Buijs, A., & Silvester, S. (1996). Demonstration projects and sustainable housing: The significance of demonstration projects as a so called second generation steering instrument is examined by the use of case studies. *Building Research and Information*, *24*(4), 195–202. doi:10.1080/09613219608727529

Burritt, R. L., Herzig, C., Schaltegger, S., & Viere, T. (2019). Diffusion of environmental management accounting for cleaner production: Evidence from some case studies. *Journal of Cleaner Production*, *224*, 479–491. doi:10.1016/j.jclepro.2019.03.227

Büyüközkan, G., & Göçer, F. (2018). Digital Supply Chain: Literature review and a proposed framework for future research. *Computers in Industry*, *97*, 157–177. doi:10.1016/j.compind.2018.02.010

Cannella, S., Framinan, J. M., Bruccoleri, M., Barbosa-Póvoa, A. P., & Relvas, S. (2015). The effect of inventory record inaccuracy in information exchange supply chains. *European Journal of Operational Research*, *243*(1), 120–129. doi:10.1016/j.ejor.2014.11.021

Carson, S. J., Madhok, A., & Wu, T. (2006). Uncertainty, opportunism, and governance: The effects of volatility and ambiguity on formal and relational contracting. *Academy of Management Journal*, *49*(5), 1058–1077. doi:10.5465/amj.2006.22798187

Ceschin, F. (2013). Critical factors for implementing and diffusing sustainable product-Service systems: Insights from innovation studies and companies' experiences. *Journal of Cleaner Production*, *45*, 74–88. doi:10.1016/j.jclepro.2012.05.034

Chakkol, M., Selviaridis, K., & Finne, M. (2018). The governance of collaboration in complex projects. *International Journal of Operations & Production Management*, *38*(4), 997–1019. doi:10.1108/IJOPM-11-2017-0717

Chegu Badrinath, A., & Hsieh, S.-H. (2019). Empirical Approach to Identify Operational Critical Success Factors for BIM Projects. *Journal of Construction Engineering and Management*, *145*(3), 04018140. doi:10.1061/(ASCE)CO.1943-7862.0001607

Chen, I. J., & Paulraj, A. (2004). Towards a theory of supply chain management: The constructs and measurements. *Journal of Operations Management*, *22*(2), 119–150. doi:10.1016/j.jom.2003.12.007

Chen, W. T., Merrett, H. C., Lu, S. T., & Mortis, L. (2019). Analysis of Key Failure Factors in Construction Partnering—A Case Study of Taiwan. *Sustainability*, *11*(14), 3994. doi:10.3390u11143994

Cheung, F. K., Rihan, J., Tah, J., Duce, D., & Kurul, E. (2012). Early stage multi-level cost estimation for schematic BIM models. *Automation in Construction*, *27*, 67–77. doi:10.1016/j.autcon.2012.05.008

Cooney, R., & Sohal, A. (2004). Teamwork and total quality management: A durable partnership. *Total Quality Management & Business Excellence*, *15*(8), 1131–1142. doi:10.1080/1478336042000255442

Costa, A. A., & Tavares, L. V. (2012). Social e-business and the Satellite Network model: Innovative concepts to improve collaboration in construction. *Automation in Construction*, *22*, 387–397. doi:10.1016/j.autcon.2011.09.017

Dallasega, P., Rauch, E., & Linder, C. (2018). Industry 4.0 as an enabler of proximity for construction supply chains: A systematic literature review. *Computers in Industry*, *99*, 205–225. doi:10.1016/j.compind.2018.03.039

Davies, A., & Brady, T. (2000). Organisational capabilities and learning in complex product systems: Towards repeatable solutions. *Research Policy*, *29*(7-8), 931–953. doi:10.1016/S0048-7333(00)00113-X

Davies, A., Dodgson, M., & Gann, D. (2016). Dynamic capabilities in complex projects: The case of London Heathrow Terminal 5. *Project Management Journal*, *47*(2), 26–46. doi:10.1002/pmj.21574

Davies, A., & Hobday, M. (2005). *The business of projects: managing innovation in complex products and systems*. Cambridge University Press. doi:10.1017/CBO9780511493294

DeFillippi, R., & Sydow, J. (2016). Project networks: Governance choices and paradoxical tensions. *Project Management Journal*, *47*(5), 6–17. doi:10.1177/875697281604700502

Deng, Y., Gan, V. J., Das, M., Cheng, J. C., & Anumba, C. (2019). Integrating 4D BIM and GIS for Construction Supply Chain Management. *Journal of Construction Engineering and Management*, *145*(4), 04019016. doi:10.1061/(ASCE)CO.1943-7862.0001633

Di Marco, M. K., Taylor, J. E., & Alin, P. (2010). Emergence and role of cultural boundary spanners in global engineering project networks. *Journal of Management Engineering*, *26*(3), 123–132. doi:10.1061/(ASCE)ME.1943-5479.0000019

Dowsett, R. M., & Harty, C. F. (2019). Assessing the implementation of BIM–an information systems approach. *Construction Management and Economics*, *37*(10), 551–566. doi:10.1080/01446193.2018.1476728

Drach-Zahavy, A., & Somech, A. (2010). From an intrateam to an interteam perspective of effectiveness: The role of interdependence and boundary activities. *Small Group Research, 41*(2), 143–174. doi:10.1177/1046496409356479

Du, J., Zou, Z., Shi, Y., & Zhao, D. (2018). Zero latency: Real-time synchronization of BIM data in virtual reality for collaborative decision-making. *Automation in Construction, 85*, 51–64. doi:10.1016/j.autcon.2017.10.009

Eastman, C. M., Eastman, C., Teicholz, P., Sacks, R., & Liston, K. (2011). *BIM handbook: A guide to building information modeling for owners, managers, designers, engineers and contractors*. John Wiley & Sons.

Fang, E., Palmatier, R. W., Scheer, L. K., & Li, N. (2008). Trust at different organisational levels. *Journal of Marketing, 72*(2), 80–98. doi:10.1509/jmkg.72.2.80

Fellows, R., & Liu, A. M. (2012). Managing organisational interfaces in engineering construction projects: Addressing fragmentation and boundary issues across multiple interfaces. *Construction Management and Economics, 30*(8), 653–671. doi:10.1080/01446193.2012.668199

Fox, S., & Hietanen, J. (2007). Interorganisational use of building information models: Potential for automational, informational and transformational effects. *Construction Management and Economics, 25*(3), 289–296. doi:10.1080/01446190600892995

Fredrickson, J. W. (1986). The strategic decision process and organisational structure. *Academy of Management Review, 11*(2), 280–297. doi:10.5465/amr.1986.4283101

Gambetti, R. C., Biraghi, S., Martinelli, E. M., & Tunisini, A. (2018). Supply Chain Communication. The International Encyclopedia of Strategic Communication, 1-6.

Ghobakhloo, M. (2019). Industry 4.0, Digitization, and Opportunities for Sustainability. *Journal of Cleaner Production*, 119869.

Gladstein, D. L. (1984). Groups in context: A model of task group effectiveness. *Administrative Science Quarterly, 29*(4), 499–517. doi:10.2307/2392936

Grilo, A., & Jardim-Goncalves, R. (2010). Value proposition on interoperability of BIM and collaborative working environments. *Automation in Construction, 19*(5), 522–530. doi:10.1016/j.autcon.2009.11.003

Gruber, T. R. (1995). Toward principles for the design of ontologies used for knowledge sharing? *International Journal of Human-Computer Studies, 43*(5-6), 907–928. doi:10.1006/ijhc.1995.1081

Gu, N., Singh, V., Taylor, C., London, K., & Brankovic, L. (2010). BIM adoption: expectations across disciplines. Handbook of research on building information modeling and construction informatics: concepts and technologies, 501-520.

Hargadon, A. B. (1998). Firms as knowledge brokers: Lessons in pursuing continuous innovation. *California Management Review, 40*(3), 209–227. doi:10.2307/41165951

Harty, C. (2008). Implementing innovation in construction: Contexts, relative boundedness, and actor-network theory. *Construction Management and Economics*, *26*(10), 1029–1041. doi:10.1080/01446190802298413

Holweg, M., Disney, S., Holmström, J., & Småros, J. (2005). Supply chain collaboration: Making sense of the strategy continuum. *European Management Journal*, *23*(2), 170–181. doi:10.1016/j.emj.2005.02.008

Hosseini, M. R., Pärn, E. A., Edwards, D. J., Papadonikolaki, E., & Oraee, M. (2018). Roadmap to mature BIM use in Australian SMEs: Competitive dynamics perspective. *Journal of Management Engineering*, *34*(5), 05018008. doi:10.1061/(ASCE)ME.1943-5479.0000636

Hung, R. Y. Y., Yang, B., Lien, B. Y. H., McLean, G. N., & Kuo, Y. M. (2010). Dynamic capability: Impact of process alignment and organisational learning culture on performance. *Journal of World Business*, *45*(3), 285–294. doi:10.1016/j.jwb.2009.09.003

Hutcheson, P., Pearson, A. W., & Ball, D. F. (1996). Sources of technical innovation in the network of companies providing chemical process plant and equipment. *Research Policy*, *25*(1), 25–41. doi:10.1016/0048-7333(94)00805-1

Ireland, R., & Bruce, R. (2000). CPFR. *Supply Chain Management Review*, *1*, 80-88.

Ivanov, D., Sethi, S., Dolgui, A., & Sokolov, B. (2018). A survey on control theory applications to operational systems, supply chain management, and Industry 4.0. *Annual Reviews in Control*, *46*, 134–147. doi:10.1016/j.arcontrol.2018.10.014

Janis, I. L. (1972). *Victims of groupthink: A psychological study of foreign-policy decisions and fiascoes*. Academic Press.

Kahn, K. B., & Mentzer, J. T. (1996). Logistics and interdepartmental integration. *International Journal of Physical Distribution & Logistics Management*, *26*(8), 6–14. doi:10.1108/09600039610182753

Kim, M., & Chai, S. (2017). The impact of supplier innovativeness, information sharing and strategic sourcing on improving supply chain agility: Global supply chain perspective. *International Journal of Production Economics*, *187*, 42–52. doi:10.1016/j.ijpe.2017.02.007

Kotzab, H., Darkow, I. L., Bäumler, I., & Georgi, C. (2019). Coordination, cooperation and collaboration in logistics and supply chains: A bibliometric analysis. *Production*, *29*, 29. doi:10.1590/0103-6513.20180088

Lahti, M., Shamsuzzoha, A. H. M., & Helo, P. (2009). Developing a maturity model for Supply Chain Management. *International Journal of Logistics Systems and Management*, *5*(6), 654–678. doi:10.1504/IJLSM.2009.024796

Lambe, C. J., Wittmann, C. M., & Spekman, R. E. (2001). Social exchange theory and research on business-to-business relational exchange. *Journal of Business-To-Business Marketing*, *8*(3), 1–36. doi:10.1300/J033v08n03_01

Lauer, T., & Franke, K. (2020, February). Behavioral Analysis of Information Exchange Digitalization in the Context of Demand Planning. In *International Conference on Intelligent Human Systems Integration* (pp. 973-979). Springer. 10.1007/978-3-030-39512-4_148

Le, N. Q., Er, M., Sankaran, S., & Ta, N. B. (2020). Perspectives on BIM Profession of BIM Specialists and non-BIM Specialists: Case Study in Vietnam. In *CIGOS 2019, Innovation for Sustainable Infrastructure* (pp. 1223–1228). Springer. doi:10.1007/978-981-15-0802-8_196

Li, C. Z., Xue, F., Li, X., Hong, J., & Shen, G. Q. (2018). An Internet of Things-enabled BIM platform for on-site assembly services in prefabricated construction. *Automation in Construction*, *89*, 146–161. doi:10.1016/j.autcon.2018.01.001

Liao, L., & Teo, E. A. L. (2019). Managing critical drivers for building information modelling implementation in the Singapore construction industry: An organisational change perspective. *International Journal of Construction Management*, *19*(3), 240–256. doi:10.1080/15623599.2017.1423165

Liao, Y., Deschamps, F., Loures, E. D. F. R., & Ramos, L. F. P. (2017). Past, present and future of Industry 4.0-a systematic literature review and research agenda proposal. *International Journal of Production Research*, *55*(12), 3609–3629. doi:10.1080/00207543.2017.1308576

London, K., & Pablo, Z. (2017). An actor–network theory approach to developing an expanded conceptualization of collaboration in industrialized building housing construction. *Construction Management and Economics*, *35*(8-9), 553–577. doi:10.1080/01446193.2017.1339361

Love, P. E., Matthews, J., Lockley, S., Kassem, M., Kelly, G., Dawood, N., & Serginson, M. (2015). BIM in facilities management applications: A case study of a large university complex. *Built Environment Project and Asset Management*.

Lu, P., Guo, S., Qian, L., He, P., & Xu, X. (2015). The effectiveness of contractual and relational governances in construction projects in China. *International Journal of Project Management*, *33*(1), 212–222. doi:10.1016/j.ijproman.2014.03.004

Lu, Y. (2017). Industry 4.0: A survey on technologies, applications and open research issues. *Journal of Industrial Information Integration*, *6*, 1–10. doi:10.1016/j.jii.2017.04.005

Luo, Y. (2002). Contract, cooperation, and performance in international joint ventures. *Strategic Management Journal*, *23*(10), 903–919. doi:10.1002mj.261

Lyytinen, K., & Newman, M. (2008). Explaining information systems change: A punctuated socio-technical change model. *European Journal of Information Systems*, *17*(6), 589–613. doi:10.1057/ejis.2008.50

Ma, K., Pal, R., & Gustafsson, E. (2019). What modelling research on supply chain collaboration informs us? Identifying key themes and future directions through a literature review. *International Journal of Production Research*, *57*(7), 2203–2225. doi:10.1080/00207543.2018.1535204

Manu, E., Ankrah, N., Chinyio, E., & Proverbs, D. (2013). A cognitive approach to trust management in construction. *Proceedings of the Institution of Civil Engineers-Management, Procurement and Law*, *166*(5), 232-239. 10.1680/mpal.12.00006

Manu, E., Ankrah, N., Chinyio, E., & Proverbs, D. (2015). Trust influencing factors in main contractor and subcontractor relationships during projects. *International Journal of Project Management*, *33*(7), 1495–1508. doi:10.1016/j.ijproman.2015.06.006

Matthews, J., Love, P. E., Mewburn, J., Stobaus, C., & Ramanayaka, C. (2018). Building information modelling in construction: Insights from collaboration and change management perspectives. *Production Planning and Control*, *29*(3), 202–216. doi:10.1080/09537287.2017.1407005

Maylor, H., Turner, N., & Murray-Webster, R. (2015). "It worked for manufacturing…!": Operations strategy in project-based operations. *International Journal of Project Management*, *33*(1), 103–115. doi:10.1016/j.ijproman.2014.03.009

Meng, X. (2020). Proactive management in the context of construction supply chains. *Production Planning and Control*, *31*(7), 527–539. doi:10.1080/09537287.2019.1657977

Müller, R., Pemsel, S., & Shao, J. (2015). organisational enablers for project governance and governmentality in project-based organisations. *International Journal of Project Management*, *33*(4), 839–851. doi:10.1016/j.ijproman.2014.07.008

Murphy, M. E. (2014). Implementing innovation: A stakeholder competency-based approach for BIM. *Construction Innovation*, *14*(4), 433–452. doi:10.1108/CI-01-2014-0011

Nam, C. H., & Tatum, C. B. (1997). Leaders and champions for construction innovation. *Construction Management and Economics*, *15*(3), 259–270. doi:10.1080/014461997372999

Nisbet, R. I., & Collins, J. M. (1978). Barriers and resistance to innovation. *The Australian Journal of Teacher Education*, *3*(1), 1. doi:10.14221/ajte.1978v3n1.1

Novais, L., Maqueira, J. M., Ortiz, Á., & Bruque, S. (2019). Strategic simulation models as a new methodological approach: An application to information technologies integration, lean/just-in-time and lead-time. *Central European Journal of Operations Research*, 1–34. doi:10.100710100-019-00654-9

O'Reilly, C. A. III, Caldwell, D. F., & Barnett, W. P. (1989). Work group demography, social integration, and turnover. *Administrative Science Quarterly*, 21–37.

Oesterreich, T. D., & Teuteberg, F. (2016). Understanding the implications of digitisation and automation in the context of Industry 4.0: A triangulation approach and elements of a research agenda for the construction industry. *Computers in Industry*, *83*, 121–139. doi:10.1016/j.compind.2016.09.006

Olorunniwo, F. O., & Li, X. (2010). Information sharing and collaboration practices in reverse logistics. *Supply Chain Management*, *15*(6), 454–462. doi:10.1108/13598541011080437

Oraee, M., Hosseini, M. R., Edwards, D. J., Li, H., Papadonikolaki, E., & Cao, D. (2019). Collaboration barriers in BIM-based construction networks: A conceptual model. *International Journal of Project Management*, *37*(6), 839–854. doi:10.1016/j.ijproman.2019.05.004

Orlikowski, W. J., Yates, J., Okamura, K., & Fujimoto, M. (1995). Shaping electronic communication: The metastructuring of technology in the context of use. *Organisation Science, 6*(4), 423-444.

Osipova, E. (2015). Establishing cooperative relationships and joint risk management in construction projects: Agency theory perspective. *Journal of Management Engineering*, *31*(6), 05014026. doi:10.1061/(ASCE)ME.1943-5479.0000346

Papadonikolaki, E., van Oel, C., & Kagioglou, M. (2019). Organising and Managing boundaries: A structurational view of collaboration with Building Information Modelling (BIM). *International Journal of Project Management*, *37*(3), 378–394. doi:10.1016/j.ijproman.2019.01.010

Papadonikolaki, E., & Wamelink, H. (2017). Inter-and intra-organisational conditions for supply chain integration with BIM. *Building Research and Information*, *45*(6), 649–664. doi:10.1080/09613218.2017.1301718

Prajogo, D., & Olhager, J. (2012). Supply chain integration and performance: The effects of long-term relationships, information technology and sharing, and logistics integration. *International Journal of Production Economics*, *135*(1), 514–522. doi:10.1016/j.ijpe.2011.09.001

Prasanna, S. R., & Haavisto, I. (2018). Collaboration in humanitarian supply chains: An organisational culture framework. *International Journal of Production Research*, *56*(17), 5611–5625. doi:10.1080/00207543.2018.1475762

Raj, S., & Sharma, A. (2014). Supply chain management in the cloud. *Accenture Global Management Consulting*, 1-12.

RICS. (2014). *What is BIM?* Adapted from https://www.rics.org/uk/knowledge/glossary/bim-intro/

Roehrich, J. K., & Lewis, M. A. (2010). Towards a model of governance in complex (product–service) inter-organisational systems. *Construction Management and Economics*, *28*(11), 1155–1164. doi:10.1080/01446191003762249

Rogers, E. M. (2010). *Diffusion of innovations*. Simon and Schuster.

Schweiger, D. M., Sandberg, W. R., & Rechner, P. L. (1989). Experiential effects of dialectical inquiry, devil's advocacy and consensus approaches to strategic decision making. *Academy of Management Journal*, *32*(4), 745–772.

Sherman, J. D., & Keller, R. T. (2011). Suboptimal assessment of interunit task interdependence: Modes of integration and information processing for coordination performance. organisation. *Science*, *22*(1), 245–261.

Singh, V., Gu, N., & Wang, X. (2011). A theoretical framework of a BIM-based multi-disciplinary collaboration platform. *Automation in Construction*, *20*(2), 134–144. doi:10.1016/j.autcon.2010.09.011

Slaughter, S. (1998). Models of Construction Innovation. *Journal of Construction Engineering and Management*, *124*(3), 226–231. doi:10.1061/(ASCE)0733-9364(1998)124:3(226)

Slaughter, S. (2000). Implementation of construction innovations. *Building Research and Information*, *28*(1), 2–17. doi:10.1080/096132100369055

Soosay, C. A., & Hyland, P. (2015). A decade of supply chain collaboration and directions for future research. *Supply Chain Management*, *20*(6), 613–630. doi:10.1108/SCM-06-2015-0217

Strozzi, F., Colicchia, C., Creazza, A., & Noè, C. (2017). Literature review on the 'Smart Factory' concept using bibliometric tools. *International Journal of Production Research*, *55*(22), 6572–6591. doi:10.1080/00207543.2017.1326643

Succar, B. (2009). Building information modelling framework: A research and delivery foundation for industry stakeholders. *Automation in Construction*, *18*(3), 357–375. doi:10.1016/j.autcon.2008.10.003

Succar, B., & Kassem, M. (2015). Macro-BIM adoption: Conceptual structures. *Automation in Construction*, *57*, 64–79. doi:10.1016/j.autcon.2015.04.018

Succar, B., & Poirier, E. (2020). Lifecycle information transformation and exchange for delivering and managing digital and physical assets. *Automation in Construction*, *112*, 103090. doi:10.1016/j.autcon.2020.103090

Tah, J. H. (2012). Virtual planning and knowledge-based decision support. *Construction Innovation and Process Improvement*, 347-362.

Tangpong, C., Hung, K. T., & Ro, Y. K. (2010). The interaction effect of relational norms and agent cooperativeness on opportunism in buyer–supplier relationships. *Journal of Operations Management*, *28*(5), 398–414. doi:10.1016/j.jom.2009.12.001

Taylor, J., & Levitt, R. (2004). Understanding and managing systemic innovation in project-based industries. *Innovations: Project management research*, 83-99.

Thompson, J. D. (2003). Organisations in action: Social science bases of administrative theory. Transaction Publishers.

Tortorella, G. L., & Fettermann, D. (2018). Implementation of Industry 4.0 and lean production in Brazilian manufacturing companies. *International Journal of Production Research*, *56*(8), 2975–2987. doi:10.1080/00207543.2017.1391420

Tsanos, C. S., & Zografos, K. G. (2016). The effects of behavioural supply chain relationship antecedents on integration and performance. *Supply Chain Management*, *21*(6), 678–693. doi:10.1108/SCM-06-2016-0211

Vangen, S., Hayes, J. P., & Cornforth, C. (2015). Governing cross-sector, inter-organisational collaborations. *Public Management Review*, *17*(9), 1237–1260. doi:10.1080/14719037.2014.903658

Vrijhoef, R., & Koskela, L. (2000). The four roles of supply chain management in construction. *European Journal of Purchasing & Supply Management, 6*(3-4), 169-178.

Walker, D. H. T. (2002). Enthusiasm, commitment, and project alliancing: An Australian experience. *Construction Innovation*, *2*(1), 15–31. doi:10.1108/14714170210814667

Wankmüller, C., & Reiner, G. (2020). Coordination, cooperation, and collaboration in relief supply chain management. *Journal of Business Economics*, *90*(2), 239–276. doi:10.100711573-019-00945-2

Whitlock, K., & Abanda, H. F. (2017, April). Making a Business Case for BIM Adoption. In *International Conference on Engineering, Applied Sciences and System Modeling* (pp. 231-246). Springer.

Winch, G. (1998). Zephyrs of creative destruction: Understanding the management of innovation in construction. *Building Research and Information*, *26*(5), 268–279. doi:10.1080/096132198369751

Won, J., Lee, G., Dossick, C., & Messner, J. (2013). Where to Focus for Successful Adoption of Building Information Modeling within organisation. *Journal of Construction Engineering and Management*, *139*(11), 04013014. doi:10.1061/(ASCE)CO.1943-7862.0000731

Wu, G., Zhao, X., & Zuo, J. (2017). Relationship between project's added value and the trust–conflict interaction among project teams. *Journal of Management Engineering*, *33*(4), 04017011. doi:10.1061/(ASCE)ME.1943-5479.0000525

Xu, J. (2019). In the shadow of a negative past: repairing and developing trust in construction projects. *Proceedings of the Institution of Civil Engineers-Management, Procurement and Law*, 1-9.

Xu, L. D., Xu, E. L., & Li, L. (2018). Industry 4.0: State of the art and future trends. *International Journal of Production Research*, *56*(8), 2941–2962. doi:10.1080/00207543.2018.1444806

Yik, F. W., Lai, J. H., Chan, K. T., & Yiu, E. C. (2006). Problems with specialist subcontracting in the construction industry. *Building Services Engineering Research and Technology*, *27*(3), 183–193. doi:10.1191/0143624406bse160oa

Zailani, S., Ariffin, H. A. M., Iranmanesh, M. M., Moeinzadeh, S., & Iranmanesh, M. M. (2016). The Moderating Effect of Project Risk Mitigation Strategies on the Relationship between Delay Factors and Construction Project Performance. *Journal of Science and Technology Policy Management*, *7*(3), 346–368. doi:10.1108/JSTPM-12-2015-0041

Chapter 3
Digitally-Enabled Design Management

Emma Tallet
Northumbria University, UK

Barry Gledson
Northumbria University, UK

Kay Rogage
Northumbria University, UK

Anna Thompson
Turner & Townsend, UK

Drew Wiggett
Turner & Townsend, UK

ABSTRACT

Calls for the digital transformation of the construction sector in part revolve around a need for productivity improvements, with a focus upon project time and cost enhancements. The purpose of this work is to provide a state-of-the-art analysis of design management (DM) usually employed to oversee design quality by coordinating design information, typically on behalf of a construction contractor. DM methods, activities, and processes with respect to the potential and underutilisation of building information modelling (BIM) are discussed. A synthesis of recent research efforts is provided identifying further emerging, disruptive, but underutilised digital tools and technologies, which when integrated with BIM, are capable of supporting DM processes. This chapter will aid practitioners and researchers in the design, implementation, and management of digital tools, and provide greater support to the DM function on modern construction projects. It will also be of use to students for a grounding in BIM and BIM-related technologies.

DOI: 10.4018/978-1-7998-6600-8.ch003

INTRODUCTION

In this chapter, a review of the Design Management (DM) literature is provided, alongside a review of recent research efforts into technology to support DM processes and collaborative working. The chapter provides insight into technologies including Virtual Reality (VR), Point Cloud (PC), blockchain and Artificial Intelligence (AI), which form part of the construction digital transformation landscape. The implementation of these technologies and their application for improving productivity, efficiency and quality outcomes are analysed. The chapter also highlights how technological developments can be integrated with BIM.

The authors discuss BIM, its central role in project delivery and how it is being used to build capacity to deliver outcomes in the construction sector. The authors consider the DM literature, covering the activities and processes of the discipline and further the role of BIM in supporting DM. In the Technology section, the authors discuss recent research and development in BIM and technology to support DM processes. This is followed by a review of technology products available commercially. The authors conclude with some recommendations for the adoption and use of technology for BIM and DM.

BACKGROUND

This section discusses the integration of BIM and DM for Digitally-Enabled Design Management. The authors begin discussing the DM discipline before introducing the reader to BIM and providing a discussion on BIM implementation. This is followed by an introduction to IPD and a discussion on the relationship between BIM and DM.

Design Management

DM is an existing and evolving discipline which emerged due to higher use of design and build procurement methods, increasing contractor involvement in the design coordination process. A demand for DM has also emerged as construction processes, buildability and buildings have become more complex (Eynon, 2013).

In a construction project team, DM supports the related Project Management (PM) and Construction Management (CM) discipline. The PM covers budget and programme while the CM predominantly covers health and safety and site operations. DM overlaps with these roles but has its own principal foci of managing design information, providing buildability analysis, and looking at construction strategy, process and cost (Emmitt & Ruikar, 2013). DM consists of the management of *"activities, people, processes and resources"* (Eynon, 2013, p.2), requiring effective problem solving and communication skills; understood as 'soft skills' (Emmitt & Ruikar, 2013; Tvedt et al., 2019) to ensure information flow and efficiency/value in project delivery in line with specified requirements of cost, time and quality (Eynon, 2013).

DM activities (see Emmitt & Ruikar, 2013 and Emmitt, 2016) operate at strategic organisational and project levels and include management, processes, collaboration, coordination and the integration of design and construction disciplines (Emmitt & Ruikar, 2013; Eynon, 2013). DM activities typically consist of (from Emmitt & Ruikar, 2013, p.65):

- **Design Review:** Pre-construction and construction stage review of design information.
- **Coordination:** Coordination of consultants' design information ensuring flow of information.
- **Tendering:** Managing pre-construction tenders and bids.
- **Design Changes:** Managing design changes, including approval.
- **Requests for Information (RFIs):** Managing and submitting RFIs.
- **Buildability:** Analysing design information to determine buildability and construction process. Considerations include coordination of disciplines and trades, site organisation, health and safety, and efficiencies such as cost.
- **Value Management:** Value-adding value engineering and innovation.
- **Cost Management:** Managing and monitoring the cost of design and construction.

Effective DM relies on lean principles (see Emmitt, 2010, p.163) and management methods to identify buildability problems in design information prior to the commencement of construction. Design changes during the construction process can lead to costs and delays. Ensuring the design is correct before construction progresses reduces the risks associated with uncertainty. The design phase is an opportunity to eliminate waste in the construction phase and create value (Emmitt & Ruikar, 2013).

The communication and understanding fostered by construction site meetings can be effective in achieving lean principles for information flow. The project team is temporary, multi-disciplinary, and in constant change. Integrating the project participants requires good management from the outset (Emmitt, 2010).

Building Information Modelling (BIM)

The UK Government has established BIM as best practice (HM Government, March 2016) and BIM has been mandatory on government projects as of 2016 (Hackett & Statham, 2016).

A BIM is defined as a computerised entity which structures and holds information as a representation of a designed or physical asset, forming a basis for decision making (Antwi-Afari et al., 2018; Emmitt & Ruikar, 2013; Hackett & Stratham, 2016; ISO 19650-1:2018). A central BIM is a more effective way of managing data, supporting drawing creation from a single information source; this eliminates a key source of errors (Sacks, et al., 2018) and supports greater building complexity. A BIM is parametric (Azhar et al., 2012) as modelled components are enriched with intelligent data (Emmitt & Ruikar, 2013; Hackett & Stratham, 2016). Moreover, it is possible to extend BIM beyond 3D to include time and cost dimensions and more (Emmitt & Ruikar, 2013).

BIM is central to a collaborative way of working (Antwi-Afari et al., 2018; Emmitt & Ruikar, 2013; ISO 19650-1:2018). ISO 19650-1:2018 stipulates the use of a central Common Data Environment (CDE) to support collaboration, where federated BIMs are shared throughout production phases (Fig. 1) (ISO 19650-1:2018; Merschbrok & Munkvold, 2015). ISO 19650-1:2018 further sets out a structure and framework for the workflow of data transfer and recommends that each project participant or organisation is responsible for their own information container or BIM (ISO 19650-1:2018). It follows that BIM is defined as a process for the integration of the project team and for the collaborative processing and management of design and construction information (Fig. 1) (Azhar et al., 2012; Piroozfar et al., 2019). When implemented using recommended processes listed in best practice guidance and/or official standards, BIM has capacity to ensure accurate, complete, consistent, well managed and high-quality information is efficiently put together and accessible to the right people at the right time (Hackett & Stratham, 2016; ISO 19650-1:2018).

Figure 1. Traditional information exchange vs information exchange with a central repository/database/CDE
Source: adapted from Chen (2005, Fig. 1) and Azhar et al. (2012, Fig. 2)

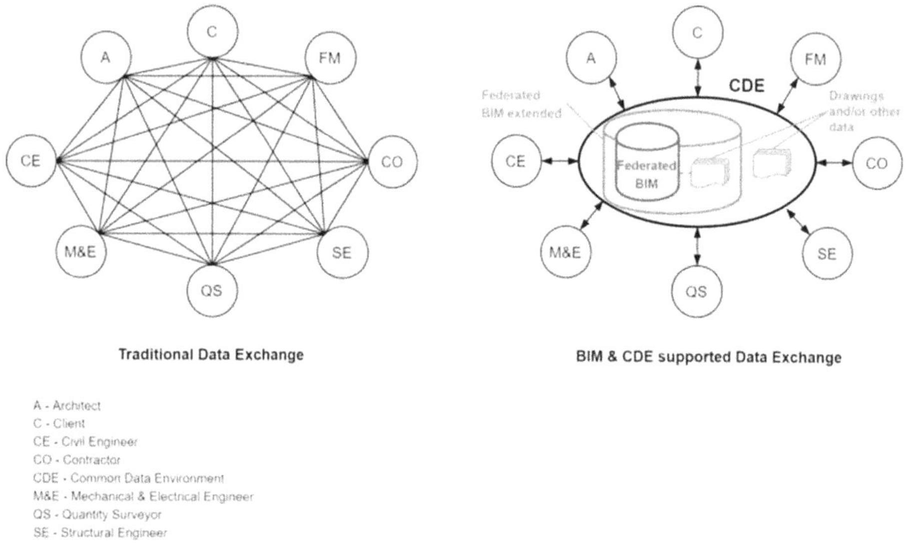

BIM as a process has the capability to reduce costs and add value to project outcomes (ISO 19650-1:2018; Arayici et al., 2010 cited in Koseoglu & Nurtan-Gunes, 2018), with studies suggesting the implementation of BIM has outcomes in productivity, collaboration and relationships (Antwi-Afari et al., 2018; Azhar et al., 2008; Chong et al., 2014; Koseoglu & Nurtan-Gunes, 2018). BIM implementation can ensure information is consistent, coordinated and standardised across the project team (Hackett & Stratham, 2016).

BIM Implementation and Systems Design

BIM is a disruptive innovation, requiring transformative change to working practices and processes (Gledson, 2016). People, management and processes are key in technology implementation, with high-level management and strategy required (Emmitt & Ruikar, 2013; Gledson, 2016; Hackett & Stratham, 2016). BIM can be affected by DM outcomes collaboration, coordination, planning and 3D visualisation among others and vice versa, creating a dual direction relationship between BIM and DM (Antwi-Afari et al., 2018). ISO 19650-1:2018 requires that BIM implementation is managed from the project outset (ISO 19650-1:2018; Piroozfar et al., 2019) in line with the IPD approach (Azhar et al., 2012), and that BIM is integrated with information management processes as one and the same (ISO 19650-1:2018).

Construction is slow to adopt disruptive approaches (Farmer, 2016; Koseoglu & Nurtan-Gunes, 2018) with companies underutilising BIM (Koseoglu & Nurtan-Gunes, 2018; Merschbrok & Munkvold, 2015). Where confidence in BIM is lacking, traditional working processes are maintained alongside new BIM workflows. This duplication can lead to inefficiencies and potential for error. Benefits of BIM can include improved visualisation, design coordination, clash detection and collaboration. However, conflict can exist between users' technological capability and issues around the standard of digital tools, particularly in older and experienced workers who may expect structure and completeness (Gledson, 2016). Innovations such as BIM are continuously developing, particularly as industry moves into its 4th revolution (Alaloul et al., 2020). For example, instead of relying on drawings as the verified source of design information, the objective is to have all design information contained in a BIM. Currently a BIM may only consist of main components, and may not include detailing (Sacks et al., 2018). Although research into measuring BIM benefits and value is ongoing (Abdirad, 2017), studies show BIM achieves outcomes in productivity (Azhar et al., April 2008; Azhar et al., 2011; Poirier et al., 2015). Despite evidence of BIM's value, construction shows a lag between the implementation of digital tools and the availability of those tools. In order to bring its performance in line with other sectors, construction should look to exploit digitalisation (Alaloul et al., 2020).

Construction is not exploiting the digital tools available (Alaloul et al., 2020). BIM implementation requires process, strategy and management (Emmitt & Ruikar, 2013; Gledson, 2016) with collaboration (Antwi-Afari et al., 2018; Emmitt & Ruikar, 2013; ISO 19650-1:2018) and information management at its centre (ISO 19650-1:2018).

BIM & Project Delivery

Traditional modes of procurement and competitive tendering can create waste and contribute toward project-related failure (Boukendour and Hughes, 2014 cited in Uusitalo et al., 2019). Farmer recommends fundamental change to the way project are managed and procured. Fixed and defensive contracts make it difficult to organise and integrate project participants leading to broken up processes and low levels of trust (Oct, 2016), which damage the team's ability to solve problems (Boukendour and Hughes, 2014 cited in Uusitalo et al., 2019). Collaboration requires contributions are treated equally; and as participants are not managed under the same organisation, DM focuses on the quality of participants' relationships or interfaces as sources of value (Emmitt, 2010; Emmitt & Ruikar, 2013), with trust identified as important for problem solving (Uusitalo et al., 2019).

The UK Government has identified collaborative procurement and early involvement of the contractor and supply chain in design as important in attaining efficiencies (HM Government, March 2016). Latham's reports *'Trust and Money'* (1993) and *'Constructing the Team'* (1994) addressed the culture of fragmentation and conflict in construction for greater collaboration and integration in project delivery to deliver value. This literature was later built on by Egan with *'Rethinking Construction'* (1998) and *'Accelerating Change'* (2002) (cited in Emmitt & Ruikar, 2013) Evidence for the success of collaborative procurement can be measured on a project-by-project basis (Bowles and Morgan, 2016; HM Government, July 2014), with the government establishing evidence for best practice for their new forms of procurement (HM Government, March 2016).

The Integrated Project Delivery (IPD) approach consists of early stakeholder involvement for a well-planned design and construction which meets identified requirements. Early stakeholder involvement and organisation is also important in BIM management (Azhar et al., 2012); with IPD addressing and removing barriers to BIM adoption by ensuring collaboration and trust is in place from the start (Piroozfar et al., 2019). As BIM provides the means for collaboration and integration, BIM is a tool/

requirement for IPD. It follows that the two concepts and processes of IPD and BIM are synonymous (Fig. 2) (Piroozfar et al., 2019).

BIM & Design Management Methods

BIM can be integrated with DM activities, processes and methods (Emmitt & Ruikar, 2013). It is important DM follows a clear process or significant waste can occur through constant rework (Eynon, 2013). The high-level DM processes are set-out according to BS 7000-4:2013; and include the alignment of DM processes with BIM standards and processes. ISO 19650-1:2018 stipulates the management of design information through the use of BIM and a CDE. As federated information containers are shared and accessed via the CDE according to process (see ISO 19650-1:2018), integrating the project team (Fig. 1) (Emmitt & Ruikar, 2013; Piroozfar et al., 2019), outcomes can be achieved in DM principles of communication, collaboration, coordination and trust (Fig. 2) (Azhar et al., 2012; Piroozfar et al., 2019).

The application of technology has the capacity to handle processing requirements and to organise and structure design information (Hackett & Stratham, 2016; Koseoglu & Nurtan-Gunes, 2018; ISO 19650-1:2018) supporting complexity (Azhar et al., 2012). In managing building data in 3D, with the use of a centralised source (Sacks, et al., 2018), a BIM-based CDE workflow can have tools to enable design visualisation before construction. Moreover, BIM-enabled tools support coordination and clash detection at work package interfaces (Emmitt & Ruikar, 2013; Koseoglu & Nurtan-Gunes, 2018). Design visualisation is an important aid for communication, collaboration, coordination and thus design development (Emmitt & Ruikar, 2013; Hackett & Stratham, 2016) and can be a focal point for communication and collaborative problem solving among project participants in a synchronous setting (Nørkjaer Gade et al., 2019).

Communication is a relational skill (Hargie et al., 2006 cited in Tvedt et al., 2019) and is developed over time as more information and understanding is transferred (Emmitt, 2010; Uusitalo et al., 2019). In team building, it is important members have developed relationships based on mutual understanding (Tvedt et al., 2019), with research suggesting a link between communication and the success of construction projects (Knotten et al., 2017). As collaborative group decisions can be better than decisions made by individuals, the processes of decision making can be critical in the project team's ability to make decisions which add value to the project (Emmitt, 2010); with research further suggesting the quality of decision making is important in project success (Knotten et al., 2017).

Important in coordination, research suggests information flow is support by trust (Tvedt et al., 2019; Uusitalo et al., 2019) as trust and lean DM reciprocate one another (Uusitalo et al., 2019). Research suggests that trust can provide a foundation for team relationships by creating an environment for mutual understanding (Tvedt et al., 2019). Trust is based on good communication (Emmitt, 2010; Uusitalo et al., 2019) and is important in reducing conflict, increasing collaboration and improving team problem solving (Uusitalo et al., 2019). Generating trust is important in creating working conditions which foster creativity (Hoezen et al., 2006 cited in Ajayi et al., 2019; Edmondson 1999 cited in Uusitalo et al., 2019) and generate social capital (Coleman 1988 cited in Uusitalo et al., 2019) (Fig. 2).

The management framework provided by IPD has the potential to provide opportunities to the DM to reduce stress, resolve low-trust relationships, facilitate team integration and facilitate strong team relationships (Piroozfar et al., 2019), enabling a dual direction relationship between IPD and DM. Piroozfar et al., further finds that in establishing strong communication from a DM perspective, BIM adoption and a high trust working environments reinforce and reciprocate one another, further enabling a dual direction relationship between IPD and BIM (2019) and as such, a dual direction relationship between DM and BIM. According to Tvedt et al., it follows that the DM design process is a learning process and decision making process, digitalised or partially digitalised by BIM (2019) (Fig. 2).

Figure 2. Dual relationships between DM, IPD and BIM based on key concepts of communication and collaboration

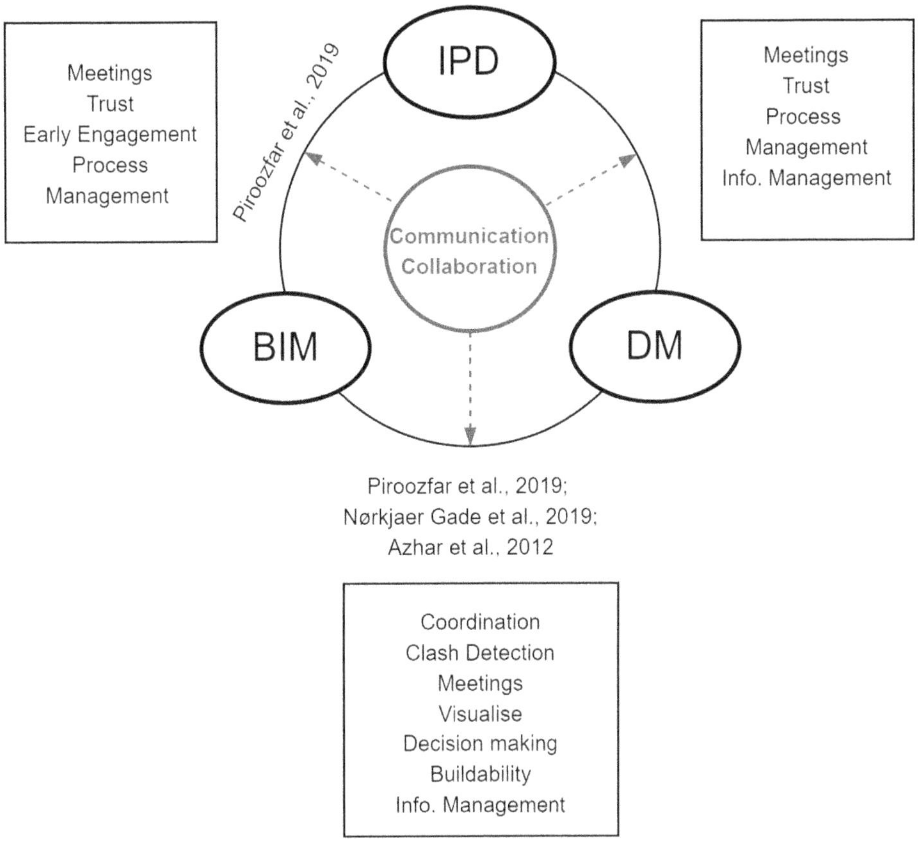

However, there are limitations to BIM-based and other technology-based communication, particularly for remote teams. Despite optimism towards technology take-up, communication is only partly digitalised, and BIM is considered a tool for communication, rather than providing full support (Tvedt et al., 2019). Information sent over digital means is not always clear and understood and does not always reach its destination (Tvedt et al., 2019; Tauriainen et al., 2016 cited in Uusitalo et al., 2019) as understanding is often expressed in facial expression, body language and/or tone of voice (Mehrabian, 2007 cited in Emmitt, 2010; Friedman, 2019).

Communication is subjective, based on human experience and individual personality, with no clear process or standard (Tvedt et al., 2019). Even where business standard communication tools are used, such as video conferencing, these tools still fall short of face-to-face communication in the social presence required for processing social cues (Fox & McEwan, 2017). An overreliance on inappropriate or deficient digitalised communication is cited as a cause of DM failure (Tvedt et al., 2019; Tauriainen et al., 2016 cited in Uusitalo et al., 2019) and might negatively impact the design process. It is important digitalised communication is considered against requirements for relational human interaction (Tvedt et al., 2019).

Organised and structured meetings are an important team building opportunity. Team building can be facilitated by the DM (Tvedt et al., 2019) with research suggesting that a good use of meetings is related to project success (Gorse, 2002 cited in Emmitt & Ruikar, 2013). Meetings are critical in managing information flow and in organising the coordination of information to establish a 'value stream' according to lean principles. Information needs to be accessible and of high quality for efficiency on the construction site. Where information is poor or absent, site will have to request further information, production in construction may be poor, delayed, incorrect or defunct; or design changes may be required (Emmitt, 2010).

Despite best intentions towards collaborative processes, research suggests fragmented working practices can persist. Poor communication, leading to poor quality information and poor information flow, does not only impact on the success or failure of projects, but also causes significant workplace stress (Ajayi et al., 2019). Significant costs and losses in productivity and talent are associated with conflict (Malik & Lenka, 2019) and stress in the workplace (Stevenson & Farmer, 2017 cited in Ajayi et al., 2019), with construction having particularly poor reputation (Farmer, 2016; Monitor Deliotte, 2017 cited in Ajayi et al., 2019). Piroozfar et al. identified poor workplace culture, poor planning and moods as barriers to trust (2019); and as such, barriers to a productive design process (Tvedt et al., 2019) (Fig. 2), with conflict within project teams also identified as a barrier to successful and productive BIM adoption (Piroozfar et al., 2019).

Organisational factors interact with the characteristics of participants to predict destructive behaviour (Hackney & Perrewé, 2018, Malik & Lenka, 2019, see Fig 1; Samnani & Singh, 2016, see Fig. 1) as a phenomenon or process (Douglas et al., 2008 cited in Hackney & Perrewé, 2018). Interventions to destructive behaviours include individual resources, community and relationships, and establishment strategy, planning and communication (Malik & Lenka, 2019, see Fig. 1).

For example, disruption and obstacles to completing tasks can trigger stress and predict destructive behaviours. In line with the frustration-aggression model, work related stress can be displaced into other areas of functioning (Hackney & Perrewé, 2018). In the case of poor information flow, problems can be mitigated with realistic project planning, resource allocation and risk management, and improved team communication (Ajayi et al., 2019). Moreover, as moods such as anger can predict destructive behaviour (Hackney & Perrewé, 2018), the effective DM can influence these 'moods' to create an environment more inducive to learning and creativity (Long and Arroyo, 2018 cited in Tvedt et al., 2019).

IPD supports the DM in facilitating communication and collaboration early in the project team. In establishing strong communication from a DM perspective, BIM adoption and a high trust working environments reciprocate one another (Piroozfar et al., 2019), creating a dual direction between DM and BIM (Fig. 2). Ajayi et al. describe how BIM and IPD have the potential to break down communication barriers by providing increasingly sophisticated digital tools for communication, such as cloud-based federated models (Fig 1), providing information management capabilities which are quicker or potentially increasingly synchronous, reducing or eliminating stress related to information flow (2019).

It is important to facilitate and foster a high trust and well-resourced working environment (Ajayi et al., 2019; Malik & Lenka, 2019) for high-trust and productive workplaces (Hackney & Perrewé, 2018). Future research directions lie in investigating the causes of conflict in project team relationships, and how DMs can intervene and facilitate productive working relationships, creating value in the learning and design process (Tvedt et al., 2019). Further research is also required to investigate the role of BIM and BIM-based digital tools in supporting subjective human experience (Tvedt et al., 2019) when digitalising or partially digitalising DM processes of communication, collaboration and coordination.

Digital Design Management

There are a range of BIM-based technologies available on the market which offer the functionality with which to carry out DM processes (Emmitt & Ruikar, 2013). This section provides a state-of-the-art discussion on recent research and development in BIM and further technologies to support DM processes.

The discussion is broken down into 6 technology areas: collaboration and coordination tools, 4D/5D BIM, Virtual Reality (VR), Point Cloud (PC), Artificial Intelligence (AI) and blockchain. These technologies can be integrated with BIM, with Table 1 showing how each technology area can support DM task areas of: collaboration and coordination, optimisation, as-built validation, and security and trust.

Table 1. Matrix of technology to support DM task areas

	Collaboration Coordination Tools	**4D/5D BIM**	**Virtual Reality**	**Point Cloud**	**Artificial Intelligence**	**Blockchain**
Collaboration Coordination	✓	✓	✓	✓		✓
Optimisation	✓	✓	✓		✓	
Validation	✓			✓	✓	
Security & Trust	✓					✓

BIM-Based Communication & Collaboration

The CDE provides tools for the visualisation of the design (Koseoglu & Nurtan-Gunes, 2018) and has the potential to integrate participants at construction site meetings, supporting communication (Nørkjaer Gade, et al., 2019). The coordination process can be complex, and BIM offers a number of tools to reduce errors and improve efficiency. Digital tools such as Navisworks can partly automate the clash detection activity so that clashes are picked up and resolved before construction, while visual 3D BIMs can be utilised in construction site meetings to discuss coordination issues (Gledson, 2016; Mehrbod et al., 2019).

BIM and Coordination

Annotation, recording and visualisation tools support construction DM processes to resolve problems on the construction site (Mehrbod et al., 2019). The use of BIM for collaboration, coordination and clash detection is well established but adoption remains an issue, with participants often falling back to 2D paper-based methods in meetings (Chu et al., 2018; Mehrbod et al., 2019), leading to inefficiencies and errors (Mehrbod et al., 2019; Wang et al., 2014 cited in Chu et al., 2018). The usability of digital tools can be a barrier to effective use (Mehrbod et al., 2019), and the range of BIM tools and information formats can add to confusion when conducting DM tasks, especially when dealing with large volumes of information (Chu et al., 2018).

The CDE described by ISO 19650-1:2018 has been expanded and can include tools to aid more seamless information access for more efficient DM problem solving, aiding closer collaboration between design and construction professionals. The key tools include:

- **Common Data Environment (CDE):** A CDE is defined as an *"agreed source of information"* (ISO 19650-1:2018, p.5) and consists of a web-based cloud platform where BIMs and drawings can be uploaded.
- **Revision:** ISO 19650-1:2018 recommends the use of a status code and a revision code. The status code reflects the permitted use of the information: 'work-in-progress', 'shared' or 'published'. The design information submitted requires approval to pass through each stage (ISO 19650-1:2018). The revision code will maintain a record or thread of changes to the design information, as recommended by the Hackitt Review (May, 2018). As in ISO 19650-1:2018, this archive is retained should there be legal proceedings or lessons to learn at a later date.
- **Coordination:** Tools to support coordination tasks are as follows:
    - Web-based tools for viewing federated BIMs on a range of devices (Koseoglu & Nurtan-Gunes, 2018) can be essential for navigating the design information and understanding the information from different angles (Mehrbod et al., 2019).
    - Parametrically linking drawings to the 3D model for review if required can be important for, as in Mehrbod et al. (2019), the accessibility of design information. Transitioning between different items in work packages can be cumbersome and inefficient if not properly facilitated.
    - Tools for marking up coordination issues on the 3D model with functionality to attach further information such as a photo can be important (Koseoglu & Nurtan-Gunes, 2018). While sketching on 3D artifacts can be underutilised, annotation is a frequently used interaction when solving coordination issues or construction DM problems (Mehrbod et al., 2019) but can be utilised in communications and documentation (Han & Golparvar-Fard, 2017).
    - In-built and automated revisioning of design information complies with information management procedures, maintaining an archive as in ISO 19650-1:2018 and a 'thread' of information as recommended by the Hackitt Report (May 2018).
- **BIM Collaboration Format (BCF):** Many platforms have developed extended collaboration capabilities using the BIM Collaboration Format (BCF) (see Boton, 2018). These capabilities include:
    - The ability to export 3D mark-ups to BCF format (see Sacks et al., 2018), this format can be imported into Revit, Navisworks or Solibri among other applications for further investigation.
    - The ability to parametrically view a list of mark-ups on a model and assign tasks to individuals in the drawing office(s).
    - The ability to monitor the completion of tasks assigned, see Revizto at Vizerra SA, (2012-2020).
- **Authorisation/Permissions:** It can be important to provide authorisations/permissions so that each participant in the project team only access the information they require, especially where there is a requirement for compliance with ISO 19650-5:2020.

The process of solving a DM problem includes identification, solution building, and solution documentation (Mehrbod et al., 2015 cited in Mehrbod et al., 2019). Solving DM problems requires accessing design information to explain, present, analyse, query and visualise the problem; and document any decisions. However, further research is required around the coordination and the DM problem solving process (Mehrbod et al., 2019).

The tools within an extended CDE support DM activities of design review and coordination. However, communication tends to be supported during construction rather than at preconstruction design stage, meaning the coordination activities supported may be more reactive. The extended CDE tools do not support the design change process or the process of submitting RFIs. These processes are sped up, made informal, and may lack appropriate controls such as approval.

BIM and the Construction Site

Access to design information via a CDE ensures the most up-to-date information is being used and is accessible when required. The extended CDE enables a more seamless connection between the construction site and the design office (Koseoglu & Nurtan-Gunes, 2018). VR can be further applied to resolve interactivity issues (Mehrbod et al., 2019). Mobile devices can be utilised on the construction site as part of lean project delivery to view models and drawings, resolve coordination and DM problems (Koseoglu & Nurtan-Gunes, 2018), monitor progress, validate the as-built construction and provide as-built documentation (Han & Golparvar-Fard, 2017).

Images captured on the construction site are a useful way of monitoring construction (Han & Golparvar-Fard, 2017; Koseoglu & Nurtan-Gunes, 2018), with value gained where visual data is organised with BIM. For example, time lapse data can offer images at any specified stage of the construction for comparison against the schedule. However, obstacles and site conditions affect the quality of the imagery and cameras are prone to movements, which require correction. As it stands, the technology available is still unable to correct the impact of inconsistent light (Han & Golparvar-Fard, 2017). More broadly, aligning visual data with BIM can be challenging and further research is required (Han & Golparvar-Fard, 2017).

New collaboration and coordination technologies better connect the construction site and the design office, streamlining DM tasks around coordination and buildability. CDE and BIM-based digital tools allow for a fluid and visual exploration of the design while on the construction site, aiding collaboration and discussion and DM decision making (Koseoglu & Nurtan-Gunes, 2018; see Revizto at Vizerra SA, 2012-2020). Successes in usability could be extended further with Virtual Reality (VR) capabilities (Mehrbod et al., 2019). Digital tools can be used to validate the construction (Koseoglu & Nurtan-Gunes, 2018), but tools are currently limited in their ability to organise visual as-built data (Han & Golparvar-Fard, 2017). Lastly, the tools are limited in their support of DM activities as they focus on the construction phase rather than the design phase. Emmitt describe that the design changes should adhere to process, and require formal documentation (2010), but the tools described in the extended CDE could act to circumvent process and make design changes informal.

Programme and Cost Management (4D/5D BIM)

Construction projects are well known for unpredictability and overrunning schedule, with this worsened with building size and complexity (Gledson & Greenwood, 2017). Research suggests that poor planning can be a greater cause of delays than unforeseeable events and obstacles (Gonzalez et al., 2013 cited in Gledson & Greenwood, 2017).

The 3D BIM geometry can be assigned to scheduling information, allowing construction tasks to be modelled considering space, resources and time; with 4D BIM further offering automation tools to streamline construction processes. Where scheduling is separate to the BIM, it can be difficult to understand construction processes and the schedule accuracy (Sacks et al., 2018).

4D BIM can simulate and visualise construction sequences virtually, facilitating DM processes of communication, collaboration and decision making (Boton, 2018; Elghaish & Abrishami, 2020; Gledson & Greenwood, 2017; Sacks et al., 2018), while also improving project predictability (Boton, 2018). It follows that 4D visualisation facilitates DM buildability assessments for optimisation and efficiency (Boton, 2018). To establish a 'value-stream', the design information is utilised to plan and schedule construction and is interrogated for buildability and efficiency (Sacks et al. 2018).

5D BIM further incorporates cost, allowing both the 3D geometry and the 4D scheduling information to be utilised for costing activities. 5D BIM automates many costing activities for efficiency and improved accuracy (Vigneault et al., 2019). However, the information contained in BIMs is often inadequate in terms of geometric accuracy and completeness. This can mean that the capabilities of 5D BIM are not fully realised (Smith, 2016). 5D quantity take-off is limited in that quantities need to be exported to a spreadsheet. Full estimates are not fully supported by 5D BIM as they require wider considerations about the construction process. As such, cost estimates are not parametrically linked to the BIM and quantities need to be renewed if there are any design changes (Sacks et al., 2018).

In line with lean principles (Emmitt, 2010), high standards of cost management achieved early allow DM-based discussion and review for efficiencies. BIM is central in supporting interim estimates which highlight budget issues and inefficiencies quickly. The earlier the contractor can participate in the 5D cost management, the greater the confidence and accuracy in cost management outputs (Sacks et al., 2018).

An IPD approach implements early collaboration and planning processes for value generation and waste elimination (Elghaish et al., 2020). Elghaish et al. (2020) integrated 4D and 5D BIM into IPD and developed a Genetic Algorithm (GA) to optimise scheduling and costing. The research utilises Navisworks 4D and 5D BIM tools as a single source for all required digital tools, eliminating the process of importing and exporting in and out of different systems. This required the development of a C# plug-in utilising the Autodesk C# API. The proposed new system integrates Activity-Based Costing (ABC) with the use of an activity library. Multi-criteria for GA based schedule optimisation were available for selection.

BIM is an important tool in ensuring the efficiency and accuracy of scheduling (Sacks et al., 2018) and costing activities, with the capacity to link 3D geometry, scheduling and costing information together (Vigneault et al., 2019). Capabilities in 4D BIM allow construction sequences to be modelled, contrasted and visualised for DM collaborative analysis and discussion (Boton, 2018; Gledson & Greenwood, 2017; Sacks et al., 2018). There have been focused research efforts into schedule optimisation based on multi-criteria (Elghaish et al., 2020).

Virtual Reality (VR), Augmented Reality (AR) and Mixed Reality (MR)

Virtual Reality (VR) can extend the visualisation capabilities of BIM and can be used to explore construction or to promote ideas to clients (Emmitt & Ruikar, 2013). Models of n-dimensions can be visualised with VR (Bolton, 2018), Augmented Reality (AR) (Chu et al., 2018) and Mixed Reality (MR) (Alizadehsalehi et al., June 2019; Wang & Kim, 2019).

AR facilitates the visualisation of a virtual model in the real-world (Alizadehsalehi et al., June 2019) improving navigation and interactivity (Fiorentino et al., 2012 cited in Mehrbod et al., 2019) and can be used to visualise construction sequences, providing an immersive experience in the physical world (Alizadehsalehi et al., June 2019; Chu et al., 2018), improving the quality of DM decisions and the DM decision making process (Chu et al., 2018), particularly as it supports design stage review.

Further to supporting design review through enhanced visualisation, Chu et al., describes how VR capacities in interactivity can be applied to improving information flow. The increasing complexity of construction projects leads to inefficiencies due to increasing amounts of data, information and systems to conduct construction DM activities (2018). The accessibility of design information can be a source of inefficiency (Chu et al., 2018; Mehrbod et al., 2019) because it increases the need to find and coordinate disparate and disjointed pieces of design information (Mehrbod et al., 2019).

VR technology can improve interactivity by improving the efficiency of 2D to 3D information transitions. (Broll et al., 2014 cited in Mehrbod et al., 2019). AR has the potential to provide enhanced functionality, including filters, to drill down into design information, improving the accessibility of information and reducing the effort required to complete DM tasks. Research suggests that AR can improve efficiency in information retrieval in terms of time and accuracy and further can reduce cognitive effort and loading on project participants (Chu et al., 2018). MR builds on AR, allowing virtual content to be manipulated as it is viewed (Alizadehsalehi et al., June 2019). For implementation, MR requires technologies such as Microsoft HoloLens (Alizadehsalehi et al., June 2019; Wang & Kim, 2019) and the Unity Games engine, and can add value to projects when integrated with a lean workflow (Alizadehsalehi et al., June 2019).

VR technology could be a key feature improving usability in applications where BIM is not performing as expected, as is often the case in DM collaborative buildability assessments (Chu et al., 2018; Mehrbod et al., 2019). In a collaborative context, interactivity is shared by participants to achieve common DM tasks and goals (Isenberg, 2011 cited in Boton, 2018), supporting DM decision making processes (Berg & Vance, 2017 cited in Boton, 2018). It follows that VR is able to integrate the project team (Issa, 2005 cited in Boton, 2018). Research has suggested the benefits of utilising VR in the construction sector. However, in a 4D context, VR processes are still limited and are not yet achieving outcomes (Boton, 2018).

Point Cloud

Point cloud has been applied to robotics navigation, automation in construction and construction health and safety. In the context of DM, point cloud has strengths in construction progress monitoring and construction quality assessments (CQA) (Wang & Kim, 2019).

The as-built construction is often different to its design information due to design changes over the course of the construction process (Wang & Kim, 2019). The Hackitt Review has recommended that as-built construction is recorded and validated against the original design information for safety reasons (May 2018). Point cloud technology can be used to create as-built BIMs to perform CQAs and validation throughout the construction process (Han & Golparvar-Fard, 2017; Wang & Kim, 2019) and can also be used to monitor the progress of construction (Wang & Kim, 2019). 4D BIM buildability analysis and validation can be incorporated into construction activities by using point cloud for the timely capture and control of productivity lapses (Omar et al., 2018 cited in Boton, 2018), with 4D point clouds captured and validated against 4D as-designed BIMs to support scheduling and design (Lin & Golparvar-Fard, 2016 cited in Han & Golparvar-Fard, 2017).

There are a number of different methods available in point cloud technology. These include:

- **Laser Scanning:** Laser scanning is strong for accuracy (Rebolj et al., 2017) but is costly and time consuming (Pučko et al., 2018; Rebolj et al., 2017).
- **Photogrammetry:** Photogrammetry is not as expensive as laser scanning and has lower but arguably acceptable accuracy and quality, depending on the use (Rebolj et al., 2017).
- **Videogrammetry:** Similar to photogrammetry, videogrammetry is a lower to mid-range solution and produces acceptable results at a lower cost (Rebolj et al., 2017).
- **Range (Depth):** Depth scanning is mid-range in terms of cost, being expensive compared to photogrammetry but cheaper compared to laser scanning. Results are reasonable, with particular strengths on small projects where distances are small (Omar & Nehdi., 2016 cited in Pučko et al., 2018).

There have been various research efforts into the automation of point cloud scan plans (Rebolj et al., 2017). Researchers have developed an algorithm to identify efficient scan plans considering data properties concerning quality (Tang & Alaswad 2012 cited in Rebolj et al., 2017). Other research has looked at achieving efficiency in the amount of data collected taking into account optimum scanning points (Biswas et al., 2015 cited in Rebolj et al., 2017); with further efforts directed into improving the automation of scan planning (Zhang et al., 2016 cited in Rebolj et al., 2017).

The long-term objective is for as-built scanning and validation to be ongoing throughout the construction process, while attaining good point cloud performance in terms of quality properties: level of detail (LOD), local density, local accuracy and level of scatter (LOS) (Rebolj et al., 2017; Wang & Kim, 2019). Rebolj et al. (2017) developed a method for assessing different point cloud scanning technologies, with the intention of using the workers as mobile scan points for instantaneous, ongoing and real-time CQAs.

Real-time construction monitoring is important so that lapses in construction progress can be picked up quickly and errors in the construction can be addressed and resolved in a timely manner (Pučko et al., 2018; Wang & Kim, 2019). Pučko et al. implemented a system which utilised time intervals to scan, with depth scanning, the relevant section of as-built construction at short-range. The resulting partial point cloud could then be registered, with a time, to a 4D BIM for validation (2018). In realising real-time CQA monitoring, research centres on accurate and suitable automated registration and alignment of as-built point clouds (Lei et al., 2019; Pučko et al., 2018).

Point cloud is valuable in construction progress monitoring (Lei et al., 2019; Pučko et al., 2018; Wang & Kim, 2019). Depth imaging can be used on moving entities to provide short-range incomplete as-built point clouds, which when federated to an as-designed BIM, provide the potential for 4D validation of the construction. However, research challenges remain in the registration and alignment of point clouds (Pučko et al., 2018) for greater levels of accuracy and reliability (Lei et al., 2019). Moreover, although as far as possible, construction is planned and the design information is finalised before construction according to lean principles (Emmitt & Ruikar, 2013), continuously monitoring the construction will allow the DM to identify any process, buildability or coordination issues during construction and provide timely rectification (Pučko et al., 2018; Wang & Kim, 2019)

Artificial Intelligence (AI)

AI has the ability to solve complex and uncalculatable problems from existing datasets to provide predictions and insight (Darko et al., 2020). Machine Learning (ML) falls within the AI subject area and includes Neural Networks (NN), support vector machines (SVM), Gaussian Mixture (GM), Reinforcement Learning (RL), Deep Learning (DL) and probability methods such as naive Bayes. Broadly, ML is the capability a computer has to learn from existing information. NNs achieve this capability by drawing on the biological structure of the brain to conduct processes iteratively or to perform classification tasks by looking at previous successes (Darko et al., 2020).

The construction sector produces high volumes of data in the form of BIMs, site records, post-occupancy evaluation (Chen et al., 2016), visual capture (Rahimian et al., 2020) and IoT (Rogage et al., 2019); with the further the potential to apply analytics to communications (Wang, Thangasamy, Hou, Tiong and Zhang, 2020). The data collected over the course of construction projects presents an opportunity to drive efficiencies and productivity (Bilal et al., 2016 cited in Darko et al., 2020).

Darko et al. (2020) used centrality metrics to identify important and influential areas of research interest, with results showing optimisation receiving a lot of research interest, possibly due to the need to automate and streamline complex processes. Genetic Algorithm (GA) is shown to be the go-to method in optimisation tasks and is a key theme in the literature. NN, ML, Fuzzy Sets (FS) and Fuzzy Logic (FL) are also key themes, with research gaps in the use of methods such as adaptive neuro-fuzzy inference system (ANFIS) and differential evolution (DE) among others.

Research efforts around Key Performance Indicator (KPI) assessments are immature and underdeveloped in their application of AI, with research into cost and risk management making little or poor use of BIM, despite the capability of BIM to provide tools in these areas. As such, there are research gaps in utilising AI with BIM for KPI assessments (Darko et al., 2020). For example, 4D as-built validation can be used to monitor progress and infer productivity in construction (Han & Golparvar-Fard, 2017).

Juszczyk provides an example of how BIM data can be used in implementing a NN for data analytics related to cost (2017). However, much research shows a reliance on data extraneous to the BIM(s). Chen et al. (2016) devised a system to include this data as 'dynamic' BIM data for better integration of BIM and data technology. The system, referred to as CloudBIM, was available to project participants as a CDE. Within CloudBIM, WebGL was used for visualisation, Hadoop Bigtable was used for Big Data storage and MapReduce was used for data processing. The system parses the 3D BIMs and stores the data in a NoSQL database format; further data from construction projects can be stored alongside in the same database for data processing. Zhou et al. (2020) further retained graph-based data relationships in the storage of BIM Big Data.

AI is often used to process Visual Big Data for use in supporting construction tasks (Darko et al., 2020; Han & Golparvar-Fard, 2017). Analytics can be used to automate the alignment of visual data to the BIM (Han & Golparvar-Fard, 2017), perform object recognition (Darko et al., 2020; Han & Golparvar-Fard, 2017; Wang & Kim, 2019) and object classification. CNNs are a form of deep NN often applied to achieve these functionalities in visual data analytics, while also being well known in areas such as Natural Language Processing (NLP). Popular applications for CNNs in DM include construction site safety and structural damage, such as identifying concrete cracks. CNNs are able to process imagery by breaking the pixels down into arrays based on colour, and are able to perform processes in parallel for the identification of arrangements and objects (Darko et al., 2020).

AI and ML are disruptive technologies with strategic management and process design required for implementation and maximum benefit (Darko et al., 2020). Increasing complexity, leading to high volumes of data, presents an opportunity to utilise computing for learning (Bilal et al., 2016 cited in Darko et al., 2020). However, data extraneous to the BIM might be required, and systems have been developed in the literature for the combined storage and processing of BIM and wider data (Chen et al., 2016; see also Zhou et al., 2020). A wide range of methods are available within the AI subject area which are applicable to DM and construction sector problems, with the potential to provide efficiencies. Optimisation has been a key research interest, while research is required in integrating BIM with AI-based KPI assessments. The use of CNNs to organise and understand visual information has also been a key research area, with applications including safety management and the detection of structural defects (Darko et al., 2020).

Blockchain

Blockchain technology is developing as part of the digital transformation of the construction sector, with changes to processes required for its application and implementation (Li et al., 2019; Mathews et al., 2017; Nawari and Ravindran, 2019). As such, Blockchain is a disruptive technology (Li et al., 2019; Mathews et al., 2017; Nawari and Ravindran, 2019; Turk & Kinc, 2017), able to achieve efficiencies and increased productivity, with use cases in collaborative DM, IoT for CQAs, post-evaluation, smart contracts (Li et al., 2019; Nawari and Ravindran, 2019) and smart cities (Li et al., 2019).

Blockchain, also referred to as Distributed Ledger Technology (DLT), is a peer-to-peer network, offering a decentralised approach to the storing and updating of information (Kinnaird et al., 2017; Li et al., 2019; Nawari and Ravindran, 2019). Storing information in a single space, such as a cloud service, can be subject to security and information management concerns (Kinnaird et al., 2017; Nawari and Ravindran, 2019). As soon as participants upload information to a central server, they lose control over that information, whereas a distributed network ensures that contributed information stays on the systems of those who made the contribution (Petri et al., 2017).

The implementation of blockchain in collaborative working practices looks to have the potential to resolve the shortfallings of CDEs by enhancing security and embedding legal rights (Kinnaird et al., 2017; Li et al., 2019), improving DM methods of collaboration and trust (Li et al., 2019; Mathews et al., 2017; Nawari and Ravindran, 2019). In a change to process, blockchain holds the potential to remove lines of management and leadership, as participants can upload their contributions directly. This enhances trust further, making participation more equal and direct (Li et al., 2019; Matthews et al., 2017). In the case of new recommendations set-out by the Hackitt Review (May 2018), blockchain can be applied to the archiving of a 'thread' of information (Li et al., 2019).

Furthermore, blockchain makes it possible to embed contracts in code for the automatic payments of services. Smart contracts functionality can be further embedded into project bank accounts (PBAs) and cryptocurrency functionality, with specific currencies such as EtchCoin (Li et al., 2019) and AECoin (Nawari and Ravindran, 2019) created for the construction sector (Li et al., 2019; Nawari and Ravindran, 2019). The automation of payments could be transformational for the construction sector (Kinnaird et al., 2017; Li et al., 2019), as the government is taking action to improve identified problems with late payments to ensure stability in the supply chain (HM Government, March 2016). Nawari and Ravindran (2019) use a Hyperledger Fabric for greater flexibility with smart contracts functionality to describe and implement a model checking and model validation BIM process. Moreover, Wang, Wang, Hu, Gong, Ren, and Xiao (2020) suggest a smart contract process for the management of precast concrete construction, organising site deliveries and monitoring any conflict or problems.

A distributed peer-to-peer network utilises participants' hardware as servers to hold an authoritative copy of BIMs federated to a single model, which is updated when required. Servers can be different cloud providers and the integration of different cloud services can achieved with existing applications/providers such as CometCloud, which is open-source, or IBM Cast Iron Cloud Integration (Petri et al., 2017). A peer-to-peer network can be implemented without necessarily implementing blockchain. Petri et al. (2017) looked at collaborative data sharing by federating distributed cloud datasources in their 'clouds for coordination' (C4C) framework, supporting 2 APIs and a Revit BIM plug-in. Interactions facilitated by an API allow participants to communicate as 'publishers' or 'subscribers' and identifies participant roles and responsibilities. Another API for rendering the BIM model uses this information, alongside an issue status, to inform access and authorisation to different parts of the federated BIM.

Barriers to the implementation of blockchain include skill shortages and cost. Implementing blockchain for information management and information sharing from one datasource, in place of or enhancing a CDE, requires more fundamental changes to BIM processes and workflows (Li et al., 2019). Factors for deciding on a Blockchain implementation include whether the blockchain needs to be permissioned or not, how scalable the blockchain needs to be, security and privacy requirements, data requirements, and hardware and software requirements for data collection and functionality (Li et al., 2019; Nawari and Ravindran, 2019). Digital tools relying on blockchain might look to utilise open-source or commercial blockchain implementations, rather than provide an embedded implementation (Turk & Kinc, 2017). Li et al. (2019) have proposed a decision tree to help system architects make decisions over the suitability of blockchain for the purposes of an application. This suggests that blockchain is suitable where BIMs are complex, where there is data collection from site, where a digital record of design decisions and changes is required, where automation is required, and where the speed of the network is not important.

Blockchain is a disruptive technology with the capacity to enhance DM methods of trust and collaboration in the construction sector (Li et al., 2019; Matthews et al., 2017; Nawari and Ravindran, 2019). The existing CDEs require participants to upload information to a single shared online space, potentially accessible to everyone on the project team; this is unacceptable from a security perspective (Kinnaird et al., 2017; Nawari and Ravindran, 2019). Blockchain is a distributed, decentralised peer-to-peer system, whereby a consistent copy of the centralised information is kept by each node. This centralised information, or the blockchain, cannot be changed or altered without passing stringent validation and authorisation requirements (Kinnaird et al., 2017; Li et al., 2019; Nawari and Ravindran, 2019). As the blockchain is responsible for validating each transaction/exchange, the automation of payments with smart contracts is a big opportunity to reform the sector's endemic trust issues (Li et al., 2019; Nawari and Ravindran, 2019).

SOLUTIONS AND RECOMMENDATIONS

As BIM is a process (Azhar et al., 2012), BIM should be implemented with IPD and alongside DM methods of communication, collaboration and trust as one and the same (Fig 2). The DM should use early engagement to develop collaborative and trusting relationships among participants in the project team (Piroozfar et al., 2019). BIM can be used as a communication tool (Ajayi et al., 2019) to integrate information and manage information flow (Tvedt et al., 2019) according to the ISO 19650 standard.

Conflict surrounding users' digital literacy and the standard of BIM tools (Gledson, 2016) can be understood in light of the 4th industrial revolution (Alaloul et al., 2020) and the status of BIM as incomplete and a developing innovation (Alaloul et al., 2020; Sacks et al., 2018). This should not deter construction companies from exploiting BIM and digitalisation, and the construction sector needs to do so to bring its performance in line with other sectors (Alaloul et al., 2020).

Alongside BIM implementation (Piroozfar et al., 2019), DMs should use strong social skills, pay attention to sources of conflict (Tvedt et al., 2019) and use good prevention and intervention measures according to Malik & Lenka (2019) to ensure that stress does not become embedded in the project team (see Ajayi et al., 2019 and Samnani & Singh, 2016). Successful collaborative working will establish a good learning, decision and design process (Tvedt et al., 2019) to ensure information flow (Uusitalo et al., 2019) and the 'pull' of value. Further technologies can be integrated with BIM including communication and collaboration tools, programme and cost management tools, VR, point cloud, AI and blockchain. These enhancements can be applied to support digitally-enabled DM tasks and methods.

The following table provides details on DM relevant digital tools available commercially or on an open-source basis. It is further possible to implement bespoke solutions or carry out research and development where commercial and open-source solutions fall short.

CONCLUSION

The construction sector's digital transformation includes BIM integrated digital advances in collaboration and communication, programme and cost management, VR, AI and blockchain. These technologies can be applied to DM activities and methods: collaboration and coordination, optimisation, as-designed and as-built validation, and security and trust (Table 1).

Summary

BIM is a process for the integration of the project team (Azhar et al., 2012; Piroozfar et al., 2019) and is central to a collaborative way of working (Antwi-Afari et al., 2018). BIM adoption and implementation is managed at project outset as part of IPD (Piroozfar et al., 2019). As such, there are dual relationships between DM, IPD and BIM based around communication and collaboration (Fig 2). It follows that the DM design process is a learning and decision making process digitalised or partially digitalised by BIM (Tvedt et al., 2019). Other advanced disruptive technologies can be integrated with BIM to enhance BIM and provide further support to DM methods, activities and processes.

The extended CDE, as described by Koseoglu and Nurtan-Gunes, provides collaboration and coordination tools supporting DM coordination, buildability and design review activities, by creating a more fluid communication channel between the construction site and the design office, also providing validation functionality (2018). Tools which support communication will also foster trust within design team relationships (Uusitalo et al., 2019). However, extended CDE tools focus is on the construction phase. Therefore, extended CDE tools may provide limited support for lean, and may circumvent and make informal some DM processes.

Vigneault et al. describe how 4D and 5D BIM link programming and costing information to the 3D geometry (2019) achieving efficiency and accuracy (Sacks et al., 2018) in DM activities of programme and cost management. According to Boton, 4D BIM enables enhanced visualisation of the construction by including the time dimension in the DM design review activity. VR can be used to enhance and extend visualisation further (2018) and can also be used to improve the usability of BIM in DM collaborative buildability assessments (Mehrbod et al., 2019).

Table 2. This table presents a list of the DM relevant technology products which are available commercially or which are available on an open-source basis.

		Technology Products
BIM Modelling (see Abanda et al., 2015)	Architecture	• Autodesk Revit Architecture: https://www.autodesk.co.uk/products/revit/architecture • ArchiCAD: https://www.graphisoft.com/archicad/ • Bentley MicroStation: https://www.bentley.com/en/products/brands/microstation
	Rendering	• 3ds Max: https://www.autodesk.co.uk/products/3ds-max/overview • Vectorworks: https://www.vectorworks.net/uk/make-the-move • Maya: https://www.autodesk.co.uk/products/maya/ • Autodesk Revit: https://www.autodesk.co.uk/products/revit/overview
	Structures	• Autodesk Revit Structures: https://www.autodesk.co.uk/products/revit/structure • Tekla Structural Designer: https://www.tekla.com/uk/products/tekla-structural-designer • Tekla Structures: https://www.tekla.com/uk/products/tekla-structures
	Civil Engineering	• AutoCAD Civil 3D: https://www.autodesk.co.uk/products/civil-3d/overview • Tekla Civil: https://www.tekla.com/products/tekla-civil • Autodesk Infrastructure Design Suite: https://www.autodesk.co.uk/suites/infrastructure-design-suite/overview • Bentley OpenRoads: https://www.bentley.com/en/products/brands/openroads
	Mechanical & Electrical	• Autodesk Revit MEP: https://www.autodesk.co.uk/products/revit/mep • Bentley OpenBuildings: https://www.bentley.com/en/products/brands/openbuildings • MagiCAD [Autodesk plug-in]: https://www.magicad.com/uk/what-is-magicad-for-mep/ • DDS CAD: https://www.dds-cad.net/
Collaboration & Coordination	Clash Detection	• Navisworks: https://www.autodesk.co.uk/products/navisworks/overview • Solibri: https://www.solibri.com/
	Collaboration & Coordination	• BIM 360: https://www.autodesk.com/bim-360/ • Viewpoint: https://vfpdocs.viewpoint.com/ • Asite: https://www.asite.com/ • GroupBC: https://www.groupbc.com/our-products • Revizto: https://revizto.com/en/ • BIMtrack: https://bimtrack.co/ • BIM Anywhere: https://bimanywhere.com/ • Bluebeam Revu: https://www.bluebeam.com/uk/ • Tekla BIMsight: https://www.tekla.com/tekla-bimsight/ • Dalux: https://www.dalux.com/gb/ • Assemble: https://assemblesystems.com/ • Synchro Field: https://www.bentley.com/en/products/product-line/construction-software/synchro-field
	Facilities Management	• GliderBIM: https://gliderbim.com/ • Twinview: https://www.twinview.com/
Programme & Cost Management	Project Management	• Microsoft Project: https://products.office.com/en-gb/project/project-management-software • Primavera P6: https://www.oracle.com/applications/primavera/products/project-portfolio-management/
	4D BIM (Programme Management)	• Navisworks Manage: https://www.autodesk.co.uk/products/navisworks/overview • Visual 4D Simulation (Innovaya): http://www.innovaya.com/prod_vs.htm • Synchro Pro: https://www.bentley.com/en/Products/Product-Line/Construction-Software/SYNCHRO-PRO • Bentley ProjectWise Manager & ConstructSim Pro: https://www.bentley.com/en/products/product-line/construction-software/synchro-awp • Tekla Structures: https://www.tekla.com/uk • Vico Office Schedule Planner: https://gc.trimble.com/product/schedule-planner
	5D BIM (Cost Management)	• CostX: https://www.exactal.com/en/ • Nomitech: https://www.nomitech.com/ • Innovaya: http://www.innovaya.com/prod_ve.htm • Assemble: https://assemblesystems.com/assemble-sage-integration-request-more-info/ • RIBiTwo: https://www.rib-software.com/en/home • Vico Office: https://gc.trimble.com/product-categories/vico-office-time
Virtual Reality (VR), Augmented Reality (AR) & Mixed Reality (MR)	-	• Navisworks Manage: https://www.autodesk.co.uk/products/navisworks/overview • 3ds Max: https://www.autodesk.co.uk/products/3ds-max/overview • Unity: https://unity.com/ • Wikitude: https://www.wikitude.com/ • Bentley Synchro [Autodesk Revit Plug-In] https://www.bentley.com/en/products/brands/synchro

continued on following page

Table 2. Continued

		Technology Products
Point Cloud	-	• Matterport Cloud: https://go.matterport.com/ • HoloBuilder: https://www.holobuilder.com/ • Trimble Realworks: https://geospatial.trimble.com/products-and-solutions/trimble-realworks • Edgewise 3D: https://www.clearedge3d.com/products/edgewise/ • MeshLab: https://www.meshlab.net/ • LiDAR 360: https://greenvalleyintl.com/software/lidar360/ • MeshLab: https://www.meshlab.net/ • FAROscene: https://www.faro.com/en-gb/products/construction-bim-cim/faro-scene/ • FAROscene Web Share: https://websharecloud.com/ • Leica Cyclone: https://leica-geosystems.com/en-gb/products/laser-scanners/software/leica-cyclone • Leica Pagasus: https://leica-geosystems.com/en-gb/products/mobile-sensor-platforms/capture-platforms/leica-pegasus_two • Flyvast: https://flyvast.com/ • Voxxlr: https://www.voxxlr.com/ • Xtion Pro: https://www.asus.com/3D-Sensor/Xtion_PRO/ • Structure Sensor: https://structure.io/structure-sensor/mark-ii2 • CloudCompare: https://www.danielgm.net/cc/ • Bentley Context Capture: https://www.bentley.com/en/products/brands/contextcapture
Artificial Intelligence (AI) & Machine Learning (ML)	Data Analytics Graphics	• PowerBI: https://powerbi.microsoft.com/en-us/
	Data Processing	• Hadoop: https://hadoop.apache.org/ • MapReduce: https://hadoop.apache.org/docs/r1.2.1/mapred_tutorial.html • Apache Spark: https://spark.apache.org/ • Google Cloud: https://cloud.google.com/products/ai • Tensorflow: https://cloud.google.com/tensorflow-enterprise
Blockchain	Distributed Computing	• ProjectWise: https://www.bentley.com/en/products/product-line/project-delivery-software/projectwise-connection-services • Revit Server: https://knowledge.autodesk.com/support/revit-products/downloads/caas/downloads/content/autodesk-revit-server-2020.html • CometCloud: https://cometcloud.rdi2.rutgers.edu/ • IBM Cast Iron Cloud Integration (part of WebSphere): https://www.ibm.com/uk-en/cloud/integration • BIM Server: https://bimserver.center/en
	Existing Blockchain Implementations	• Ethereum: https://ethereum.org/java/#intermediate-articles • NEO: https://neo.org/ • Hyperledger: https://www.hyperledger.org/ • Storj: https://storj.io/

Real-time CQAs and progress monitoring with point cloud allows the DM to identify process, buildability or coordination issues during construction and provide timely rectification (Lei et al., 2019; Pučko et al., 2018; Wang & Kim, 2019), with real-time depth scanning on moving entities an objective and a solution (Pučko et al., 2018). However, in terms lean principles, CQAs and progress monitoring is reactive and construction phase based. Moreover, as-built point clouds have the potential to provide a 'thread' of as-built information as required by the Hackitt Review (May 2018).

It follows that construction produces high volumes of data (Chen et al., 2016), presenting an opportunity to use data to drive efficiencies and productivity (Bilal et al., 2016 cited in Darko et al., 2020). For AI implementation, there reliance on data extraneous to BIMs, with systems developed to deal with this (Chen et al., 2016). AI also has the capacity to support the processing of visual Big Data (Darko et al., 2020; Han & Golparvar-Fard, 2017) for monitoring and validation purposes (Han & Golparvar-Fard, 2017).

Blockchain can be implemented to resolve CDE security concerns (Kinnaird et al., 2017; Li et al., 2019) with the capacity to enhance DM methods of trust and collaboration (Li et al., 2019; Matthews et al., 2017; Nawari and Ravindran, 2019). Research by Petri et al. (2017) on collaborative data sharing shows how a peer-to-peer network can be used to control BIM visibility and contributions.

Limitations & Future Research

BIM is a disruptive innovation (Antwi-Afari et al., 2018) requiring process for implementation (Azhar et al., 2012) with many companies struggling to implement a full transformation (Gledson, 2016). BIM is still developing, with future research efforts directed toward developing BIM tools which support greater levels of detail and which eliminate the need for drawings (Sacks et al., 2018). BIM is a collaborative process, but the ability BIM has to support communication is limited because BIM and other digital communication tools do not yet reflect the varied, subjective and iterative nature of real world communication. Further research is required to understand group working practices in order to improve communication and collaboration, and reduce conflict (Tvedt et al., 2019). Further research is also required around the coordination and DM problem solving process, as BIM still has limitations in terms of interactivity (Mehrbod et al., 2019). Chu et al. researched AR as a possible solution to data extraction inefficiencies, but further research is required to establish the option as worth the implementation cost (2018). Research further suggests VR processes are problematic in the context of 4D BIM (Woksepp & Olofsson cited in Boton, 2018). Moreover, real-time progress monitoring and CQAs utilising point cloud rely on the accurate registration of as-built point clouds to as-designed BIMs (Lei et al., 2019; Pučko et al., 2018).

AI is a disruptive innovation, requiring management and process for implementation (Darko et al., 2020). The same is true of blockchain where implementation requires new workflows, and where further barriers to implementation exist including cost and skill (Li et al., 2019). Future research in AI is required to better integrate BIM when using AI for KPI assessments (Darko et al., 2020), as data for AI applications often require data extraneous to BIMs (Chen et al., 2016). Future research is also required in optimisation to look at a wider range of AI methods to GA, such as ANFIS, DE (Darko et al., 2020). Furthermore, Han and Golparvar-Fard identify limitations in the ability of BIM tools to organise visual data (2017) with the registration and alignment of as-built point clouds to as-designed BIMs also attracting research interest (Lei et al., 2019; Pučko et al., 2018).

ACKNOWLEDGMENT

This publication forms part of part of a Knowledge Transfer Partnership (Partnership: 1025785) facilitated by Northumbria University and funded by Innovate UK and Turner & Townsend.

REFERENCES

Abanda, F. H., Vidalakis, C., Oti, A. H., & Tah, J. H. (2015). A critical analysis of Building Information Modelling systems used in construction projects. *Advances in Engineering Software*, *90*, 183–201. doi:10.1016/j.advengsoft.2015.08.009

Abdirad, H. (2017). Metric-based BIM implementation assessment: A review of research and practice. *Architectural Engineering and Design Management*, *13*(1), 52–78. doi:10.1080/17452007.2016.1183474

Ajayi, S. O., Jones, W., & Unuigbe, M. (2019). Occupational stress management for UK construction professionals. *Journal of Engineering, Design and Technology*, 819-832.

Alaloul, W. S., Liew, M. S., Zawawi, N. A. W. A., & Kennedy, I. B. (2020). Industrial Revolution 4.0 in the construction industry: Challenges and opportunities for stakeholders. *Ain Shams Engineering Journal*, *11*(1), 225–230. doi:10.1016/j.asej.2019.08.010

Alizadehsalehi, S., Hadavi, A., & Huang, J.C. (2019, June). BIM/MR-Lean Construction Project Delivery Management System. In *2019 IEEE Technology & Engineering Management Conference (TEMSCON)*. IEEE.

Antwi-Afari, M. F., Li, H., Pärn, E. A., & Edwards, D. J. (2018). Critical success factors for implementing building information modelling (BIM): A longitudinal review. *Automation in Construction*, *91*, 100–110. doi:10.1016/j.autcon.2018.03.010

Azhar, S. (2011). Building information modeling (BIM): Trends, benefits, risks, and challenges for the AEC industry. *Leadership and Management in Engineering*, *11*(3), 241–252. doi:10.1061/(ASCE)LM.1943-5630.0000127

Azhar, S., Hein, M., & Sketo, B. (2008, April) Building information modeling (BIM): benefits, risks and challenges. In *Proceedings of the 44th ASC Annual Conference* (pp. 2-5). Academic Press.

Azhar, S., Khalfan, M., & Maqsood, T. (2012). Building information modelling (BIM): Now and beyond. *Construction Economics and Building*, *12*(4), 15–28. doi:10.5130/AJCEB.v12i4.3032

Boton, C. (2018). Supporting constructability analysis meetings with Immersive Virtual Reality-based collaborative BIM 4D simulation. *Automation in Construction*, *96*, 1–15. doi:10.1016/j.autcon.2018.08.020

Bowles, G., & Morgan, J. (2016). An evaluation of the performance of a large scale collaborative procurement initiative in the social housing sector. *Engineering, Construction, and Architectural Management*, *23*(1), 60–74. doi:10.1108/ECAM-03-2014-0036

BSI. (2013). *Design management systems. Part 4: Guide to managing design in construction*. BS 7000-4:2013.

Chen, H. M., Chang, K. C., & Lin, T. H. (2016). A cloud-based system framework for performing online viewing, storage, and analysis on big data of massive BIMs. *Automation in Construction*, *71*, 34–48. doi:10.1016/j.autcon.2016.03.002

Chen, P. H., Cui, L., Wan, C., Yang, Q., Ting, S. K., & Tiong, R. L. (2005). Implementation of IFC-based web server for collaborative building design between architects and structural engineers. *Automation in Construction*, *14*(1), 115–128. doi:10.1016/j.autcon.2004.08.013

Chong, H. Y., Wong, J. S., & Wang, X. (2014). An explanatory case study on cloud computing applications in the built environment. *Automation in Construction*, *44*, 152–162. doi:10.1016/j.autcon.2014.04.010

Chu, M., Matthews, J., & Love, P. E. (2018). Integrating mobile building information modelling and augmented reality systems: An experimental study. *Automation in Construction*, *85*, 305–316. doi:10.1016/j.autcon.2017.10.032

Darko, A., Chan, A. P., Adabre, M. A., Edwards, D. J., Hosseini, M. R., & Ameyaw, E. E. (2020). Artificial intelligence in the AEC industry: Scientometric analysis and visualization of research activities. *Automation in Construction*, *112*, 103081. doi:10.1016/j.autcon.2020.103081

Elghaish, F., & Abrishami, S. (2020). Developing a framework to revolutionise the 4D BIM process: IPD-based solution. *Construction Innovation.*

Emmitt, S. (2010). Managing interdisciplinary projects: a primer for architecture, engineering and construction (1st ed.). Padstow: Spon Press, Taylor & Francis Group. doi:10.4324/9780203885338

Emmitt, S. (2016). The construction design manager–a rapidly evolving innovation. *Architectural Engineering and Design Management, 12*(2), 138-148.

Emmitt, S., & Ruikar, K. (2013). *Collaborative Design Management.* Routledge. doi:10.4324/9780203819128

Eynon, J. (2013). *The Design Manager's Handbook.* Wiley-Blackwell. doi:10.1002/9781118486184

Farmer, M. (2016). *The Farmer Review of of the UK Construction Labour Model, Modernise or Die: Time to decide the industry's future.* Construction Leadership Council (CLC). Retrieved Feb 28, 2020, from https://www.cast-consultancy.com/wp-content/uploads/2016/10/Farmer-Review-1.pdf

Fox, J., & McEwan, B. (2017). Distinguishing technologies for social interaction: The perceived social affordances of communication channels scale. *Communication Monographs, 84*(3), 298–318. doi:10.1080/03637751.2017.1332418

Friedman, H. S. (2019). Introduction to the Special Issue on Theory in Nonverbal Communication. *Journal of Nonverbal Behavior, 43*(2), 107–109. doi:10.100710919-019-00299-x

Gledson, B. J. (2016). Hybrid project delivery processes observed in constructor BIM innovation adoption. *Construction Innovation, 16*(2), 229–246. doi:10.1108/CI-04-2015-0020

Gledson, B. J., & Greenwood, D. (2017). The adoption of 4D BIM in the UK construction industry: An innovation diffusion approach. *Engineering, Construction, and Architectural Management, 24*(6), 950–967. doi:10.1108/ECAM-03-2016-0066

Government, H. M. (2014, July). *New Models of Construction Procurement.* Retrieved May 18, 2020, from https://assets.publishing.service.gov.uk/government/uploads/system/uploads/attachment_data/file/325011/New_Models_of_Construction_Procurement_-_Introduction_to_the_Guidance_-_2_July_2014.pdf

Government, H. M. (2016). *Government Construction Strategy: 2016-2020.* Retrieved March 03, 2020, from https://assets.publishing.service.gov.uk/government/uploads/system/uploads/attachment_data/file/510354/Government_Construction_Strategy_2016-20.pdf

Hackett, M., & Statham, G. (Eds.). (2016). *The aqua group guide to procurement, tendering and contract administration* (2nd ed.). Wiley Blackwell, John Wiley & Sons.

Hackitt, J. (2018). *Building a Safer Future Independent Review of Building Regulations and Fire Safety: Final Report.* Retrieved Feb 28, 2020, from https://assets.publishing.service.gov.uk/government/uploads/system/uploads/attachment_data/file/707785/Building_a_Safer_Future_-_web.pdf

Hackney, K. J., & Perrewé, P. L. (2018). A review of abusive behaviors at work: The development of a process model for studying abuse. *Organizational Psychology Review, 8*(1), 70–92. doi:10.1177/2041386618755724

Han, K. K., & Golparvar-Fard, M. (2017). Potential of big visual data and building information modeling for construction performance analytics: An exploratory study. *Automation in Construction*, *73*, 184–198. doi:10.1016/j.autcon.2016.11.004

ISO, B. (2018). ISO 19650-1:2018. BSI Standards Publication Organization and digitization of information about buildings and civil engineering works, including building information modelling (BIM)-Information management using building information modelling.

ISO, B. (2018). ISO 19650-2: 2018. Organization and Digitization of Information about Buildings and Civil Engineering Works, Including Building Information Modeling (BIM)-Information Management using Building Information Modeling. Part, 2.

ISO, B. (2020). ISO 19650-5: 2020. Organization and Digitization of Information about Buildings and Civil Engineering Works, Including Building Information Modeling (BIM)-Security-minded approach to information management.

Juszczyk, M. (2017). On the Neural Networks Applications in BIM-Based Macro Level Cost Analyzes in Construction Projects. *DEStech Transactions on Economics, Business and Management.*

Kinnaird, C., Geipel, M., & Bew, M. A. (2017). Blockchain technology: how the inventions behind bitcoin are enabling a network of trust for the built environment. London: Arup.

Knotten, V., Lædre, O., & Hansen, G. K. (2017). Building design management–key success factors. *Architectural Engineering and Design Management*, *13*(6), 479–493. doi:10.1080/17452007.2017.1345718

Koseoglu, O., & Nurtan-Gunes, E. T. (2018). Mobile BIM implementation and lean interaction on construction site. *Engineering, Construction, and Architectural Management*, *25*(10), 1298–1321. doi:10.1108/ECAM-08-2017-0188

Lei, L., Zhou, Y., Luo, H., & Love, P. E. (2019). A CNN-based 3D patch registration approach for integrating sequential models in support of progress monitoring. *Advanced Engineering Informatics*, *41*, 100923. doi:10.1016/j.aei.2019.100923

Li, J., Greenwood, D., & Kassem, M. (2019). Blockchain in the built environment and construction industry: A systematic review, conceptual models and practical use cases. *Automation in Construction*, *102*, 288–307. doi:10.1016/j.autcon.2019.02.005

Malik, P., & Lenka, U. (2019). Overcoming destructive deviance: Propositioning an integrated conceptual framework. *The International Journal of Organizational Analysis*, *27*(4), 1148–1167. doi:10.1108/IJOA-06-2018-1441

Mathews, M., Robles, D., & Bowe, B. (2017). *BIM+ blockchain: A solution to the trust problem in collaboration?* Academic Press.

McKnight, D. H., Choudhury, V., & Kacmar, C. (2002). Developing and validating trust measures for e-commerce: An integrative typology. *Information Systems Research*, *13*(3), 334–359.

Mehrbod, S., Staub-French, S., Mahyar, N., & Tory, M. (2019). Characterizing interactions with BIM tools and artifacts in building design coordination meetings. *Automation in Construction*, *98*, 195–213. doi:10.1016/j.autcon.2018.10.025

Merschbrock, C., & Munkvold, B. E. (2015). Effective digital collaboration in the construction industry–A case study of BIM deployment in a hospital construction project. *Computers in Industry*, *73*, 1–7. doi:10.1016/j.compind.2015.07.003

Nawari, N. O., & Ravindran, S. (2019). Blockchain and the built environment: Potentials and limitations. *Journal of Building Engineering*, *25*, 100832. doi:10.1016/j.jobe.2019.100832

Nørkjaer Gade, P., Nørkjaer Gade, A., Otrel-Cass, K., & Svidt, K. (2019). A holistic analysis of a BIM-mediated building design process using activity theory. *Construction Management and Economics*, *37*(6), 336–350. doi:10.1080/01446193.2018.1533644

Petri, I., Beach, T., Rana, O. F., & Rezgui, Y. (2017). Coordinating multi-site construction projects using federated clouds. *Automation in Construction*, *83*, 273–284. doi:10.1016/j.autcon.2017.08.011

Piroozfar, P., Farr, E. R., Zadeh, A. H., Inacio, S. T., Kilgallon, S., & Jin, R. (2019). Facilitating Building Information Modelling (BIM) using Integrated Project Delivery (IPD): A UK perspective. *Journal of Building Engineering*, *26*, 100907. doi:10.1016/j.jobe.2019.100907

Poirier, E. A., Staub-French, S., & Forgues, D. (2015). Measuring the impact of BIM on labor productivity in a small specialty contracting enterprise through action-research. *Automation in Construction*, *58*, 74–84. doi:10.1016/j.autcon.2015.07.002

Pučko, Z., Šuman, N., & Rebolj, D. (2018). Automated continuous construction progress monitoring using multiple workplace real time 3D scans. *Advanced Engineering Informatics*, *38*, 27–40. doi:10.1016/j.aei.2018.06.001

Rebolj, D., Pučko, Z., Babič, N. Č., Bizjak, M., & Mongus, D. (2017). Point cloud quality requirements for Scan-vs-BIM based automated construction progress monitoring. *Automation in Construction*, *84*, 323–334. doi:10.1016/j.autcon.2017.09.021

Rogage, K., Clear, A., Alwan, Z., Lawrence, T., & Kelly, G. (2019). Assessing building performance in residential buildings using BIM and sensor data. *International Journal of Building Pathology and Adaptation*.

Sacks, R., Eastman, C., Lee, G., & Teicholz, P. (2018). *BIM handbook: a guide to building information modeling for owners, designers, engineers, contractors, and facility managers* (3rd ed.). John Wiley & Sons, Inc. doi:10.1002/9781119287568

Samnani, A. K., & Singh, P. (2016). Workplace bullying: Considering the interaction between individual and work environment. *Journal of Business Ethics*, *139*(3), 537–549. doi:10.100710551-015-2653-x

Smith, P. (2016). Project cost management with 5D BIM. *Procedia: Social and Behavioral Sciences*, *226*, 193–200. doi:10.1016/j.sbspro.2016.06.179

Turk, Ž., & Klinc, R. (2017). Potentials of blockchain technology for construction management. *Procedia Engineering*, *196*, 638–645. doi:10.1016/j.proeng.2017.08.052

Tvedt, I. M., & Dyb, K. A. (2019). The Soft Factors in Design Management: A Hidden Success Factor? *10th Nordic Conference on Construction Economics and Organization*, 2, 111-117. 10.1108/S2516-285320190000002032

Uusitalo, P., Seppänen, O., Peltokorpi, A., & Olivieri, H. (2019). Solving design management problems using lean design management: The role of trust. *Engineering, Construction, and Architectural Management, 26*(7), 1387–1405. doi:10.1108/ECAM-03-2018-0135

Vigneault, M. A., Boton, C., Chong, H. Y., & Cooper-Cooke, B. (2019). An innovative framework of 5D BIM solutions for construction cost management: A systematic review. *Archives of Computational Methods in Engineering*, 1–18.

Vizerra, S. A. (2012-2020). *Revizto*. Retrieved May 29, 2020, from Revizto: https://revizto.com/en

Wang, Q., & Kim, M. K. (2019). Applications of 3D point cloud data in the construction industry: A fifteen-year review from 2004 to 2018. *Advanced Engineering Informatics, 39*, 306–319. doi:10.1016/j.aei.2019.02.007

Wang, Y., Thangasamy, V. K., Hou, Z., Tiong, R. L., & Zhang, L. (2020). Collaborative relationship discovery in BIM project delivery: A social network analysis approach. *Automation in Construction, 114*, 103147. doi:10.1016/j.autcon.2020.103147

Wang, Z., Wang, T., Hu, H., Gong, J., Ren, X., & Xiao, Q. (2020). Blockchain-based framework for improving supply chain traceability and information sharing in precast construction. *Automation in Construction, 111*, 103063. doi:10.1016/j.autcon.2019.103063

Zhou, X., Zhao, J., Wang, J., Guo, M., Liu, J., & Shi, H. (2020). Towards product-level parallel computing of large-scale building information modeling data using graph theory. *Building and Environment, 169*, 106558. doi:10.1016/j.buildenv.2019.106558

ADDITIONAL READING

Emmitt, S., & Ruikar, K. (2013). *Collaborative Design Management*. Routledge. doi:10.4324/9780203819128

Eynon, J. (2013). *The Design Manager's Handbook*. Wiley-Blackwell. doi:10.1002/9781118486184

Sacks, R., Eastman, C., Lee, G., & Teicholz, P. (2018). *BIM handbook: a guide to building information modeling for owners, designers, engineers, contractors, and facility managers* (3rd ed.). John Wiley & Sons, Inc. doi:10.1002/9781119287568

KEY TERMS AND DEFINITIONS

Building Information Model (BIM): A computerised entity which structures and holds information as a representation of a designed or physical asset (ISO 19650-1:2018).

Building Information Modelling (BIM): The collaborative process (Azhar, et al., 2012).

Collaboration: Participants from independent organisations coming together to make equal contributions to project activities and deliver outcomes under the management established for a temporary [construction] project (Emmitt & Ruikar, 2013).

Common Data Environment (CDE): A single shared space where project participants can share and access information according to a process (ISO 19650-1:2018). A CDE is usually web or cloud based.

Coordination: The alignment and integration of design information from different project participants.

Data: Structured figures, symbols or numbers stored on a computer (Bennett et al., 2010).

Design Management (DM): The management of people, processes, and resources on a construction project to deliver value in terms of cost, time, and quality (Eynon, 2013).

Industry Foundation Classes (IFC): IFC was developed by buildingSMART (Sacks et al., 2018) to resolve interoperability problems between different BIM providers (Azhar et al., 2012; see Sacks et al., 2018).

Information: Data presented with meaning to aid decision making (Bennett et al., 2010).

Interoperability: Automated data exchange and transfer capability without the need for the manual processing of that data (Azhar et al., 2012).

Key Performance Indicator (KPI): A variable against which to measure a project's performance.

Parametric: Refers to a BIM whose components/entities are enriched with intelligent properties (Azhar et al., 2012).

Process: A structured step-by-step sequence of actions/tasks starting with inputs to achieve given outputs under different defined circumstances (ISO/IEC 19510, 2013).

Trust: The acceptance of vulnerability based on a belief that the other concerning actor will uphold expected standards of conduct or uphold an agreed contract.

Virtual Reality: VR is best defined as "experience of presence in an environment by means of a communications medium" (Steuer, 1995 cited in Boton, 2018, p. 2). VR is further understood in terms of usability and presence-based variables immersion and interactivity (Boton, 2018).

Section 3

Process: Digitally Enabled Health and Safety

Chapter 4
Construction Work Health and Safety Management Using Building Information Modelling

Kerry London
Torrens University, Australia

Gurvin Kaur
Western Sydney University, Australia

Yingbin Feng
Western Sydney University, Australia

Angelica Vårhammar
Centre for Work Health and Safety, Australia

Gabrielle Wallace
Western Sydney University, Australia

Peng Zhang
https://orcid.org/0000-0002-6855-3395
Western Sydney University, Australia

Swapan Saha
Western Sydney University, Australia

Payam Rahnamayiezekavat
Western Sydney University, Australia

ABSTRACT

The high rates of fatalities and injuries in the construction industry highlight the importance of work health and safety (WHS). Building information modelling (BIM) could significantly improve WHS management in construction, but integration has been limited. The chapter explores the application of BIM for WHS management and the rationale for market leaders' role to catalyse a BIM-enabled WHS management ecosystem through procurement strategies and tendering proficiency. An international comparative analysis of BIM WHS is presented along with a BIM drivers and barriers critique, including a clustering into seven areas of market, perceptions and attitudes, leadership, supply chain integration and collaborative practice, procurement strategies and tendering proficiency, decision tools, and technical functionality. A focussed decision framework is outlined. The findings provide insights for researchers and practitioners to understand how a BIM-enabled WHS management system can be facilitated and supported through the project lifecycle.

DOI: 10.4018/978-1-7998-6600-8.ch004

INTRODUCTION

Construction is known as one of the most dangerous industries in which to work and many safety incidents, injuries and fatalities could be prevented through improved design, planning and communication. BIM is an enabler technology that involves the generation and management of digital WHS information in construction. It facilitates the separation of people and hazards by the use of technology and data. The purpose of this chapter is to better understand the application of BIM to support WHS management and, specifically, the role the client can play in enhancing its application. Under this overarching aim, this chapter will explore the use and adoption of BIM for WHS management; briefly examine the barriers and enablers for BIM adoption; identify solutions for integrating WHS management in BIM-enabled project planning, design and delivery; and explore procurement frameworks and the client's role in the use of BIM for WHS management. The chapter presents key results of Phase 1 of a four phase study being conducted in New South Wales a state in Australia:

1. Phase 1: Identify solutions for integrating the WHS aspect in BIM-enabled project planning, design and delivery.
   Understand how and to what extent BIM is used and can be used for WHS management, including identifying the different options of implementation and associated barriers, enablers, limitations, and consequences for WHS risk reduction for the industry, the government and the regulator.
2. Phase 2: Evaluation of WHS management in BIM-enabled project proposals.
   As part of a procurement process and identify preferred procurement models and best practices to evaluate WHS management in BIM-enabled project proposals and recommend best way for government agencies to evaluate the quality of WHS management in BIM-enabled project proposals.
3. Phase 3: Evaluation of the proposed approach for its adoption
   Understand the implications of the recommended solutions for integrating the WHS aspect in BIM-enabled project planning, design and delivery (objective 1) and for enabling the evaluation of WHS management in BIM-enabled project proposals at the procurement stage (objective 2).
4. Phase 4: Transfer of Knowledge and Dissemination.
   Disseminate findings to the industry, to NSW citizens, to the students and to the research community.

However even given the local context the international setting is well considered and presented. It is noted that the funder of the study is the multinational company Lendlease a globally recognised 'digital delivery' innovator and so the study has significant international relevance.

BACKGROUND

The construction industry is well known for its hazardous working conditions and association with a concerning number of injuries and fatalities. The issue of WHS is international, with the construction industry responsible for the highest proportion of the 2.3 million work-related fatalities that occur around the world each year (International Labour Organization, 2020). According to the Bureau of Labor Statistics (BLS, 2019) of the United States (US), construction worker fatalities are increasing (from 965 in 2017 to 1003 in 2018). Similarly, construction workers in the United Kingdom (UK) constituted the highest proportion of occupational fatalities in 2017 and 2018 (Garner-Purkis, 2018).

In Australia, the construction industry employs 9% of the workforce 1.16 million people; (Safe Work Australia, 2017) and has the second highest work-related injury and disease incidence rate among all industries. Numerous statistics support the seriousness of this issue. For example, between 2003 and 2013, 401 construction workers were fatally injured (Safe Work Australia, 2017). This is an average of 40.1 workers per year. Over the five years from 2013 to 2017, a total of 153 work-related fatalities were reported in the construction industry, equalling an average of 31 fatalities per year (Safe Work Australia, 2017). While it is promising that the rates are decreasing, the construction industry remains one of the most dangerous and has been listed as a priority industry for improvement in the Australian Work Health and Safety Strategy 2012–2022 (Safe Work Australia, 2020).

Work-related injuries and fatalities have wide-ranging implications. In addition to direct impacts to victims, work-related injury and disease impose significant losses to victims' families, employers and society (Feng et al., 2015). For example, damage caused to productivity, property, equipment and morale can have a detrimental effect on a construction company's profit and loss statement. Effective investment to improve WHS could therefore bring significant returns, both human and economic (Feng, 2013).

BUILDING INFORMATION MODELLING TO IMPROVE WORK HEALTH AND SAFETY

As it becomes crucial to ensure safer working conditions, the introduction and development of new technologies can result in improved building design, project planning and communication. Over the last decade, the use of BIM has risen across the construction industry. The following section explores the increasing role of BIM in construction, its adoption across countries, the barriers and drivers to its implementation and, most importantly, the application of BIM for WHS management.

Building Information Modelling

BIM is an integrated, model-based process that is used for creating and managing building projects. It has been variously defined over the last decade. According to ISO 19650:2019, BIM is the:

Use of a shared digital representation of a built asset to facilitate design, construction and operation processes to form a reliable basis for decisions. (ISO 19650, 2019).

For the purposes of this book chapter, the following definition from Gu and London (2010) is adopted as it is more expansive:

BIM is an information technology enabled approach that involves applying and maintaining an integral digital representation of all building information for different phases of the project lifecycle in the form of a data repository. The building information involved in the BIM approach can include both geometric data as well as non-geometric data." (Gu & London, 2010).

BIM facilitates design, visualisation, simulation, collaboration and communication of all project information in an integrated model for better decision-making. It allows all stakeholders to access the same information, at the same time via a common data environment. As BIM allows visualisation of the project in advance of its physical realisation, it provides greater cost certainty, facilitates error elimination and reduces risk.

Existing research indicates that BIM is being increasingly adopted by the architecture, engineering and construction industries (Martínez-Aires et al., 2018), although adoption levels have varied across countries, sectors and supply chains (London & Singh, 2013). The adoption or uptake of BIM by industry, known as BIM maturity, is categorised in four levels according to increasing levels of industry collaboration. BIM Level 0 is the lowest level with no cooperation between participants and mainly involves the use of computer-aided design to make 2-dimensional (2D) drawings. BIM level 1 progresses to the generation of a suite of 2D information with sharing of data using a common data environment. BIM level 2 involves collaborative working and the sharing of information between project participants, whereas BIM level 3 involves total and full integration of BIM in a cloud-based environment (NBS, 2020a).

BIM adoption can also differ in the Level of Development (LOD). BIM LOD is an industry standard that details the development stage of different systems in a BIM environment, specifying the design requirements at each stage. It allows professionals to clearly and effectively specify the content of BIM by articulating the expected LOD of an element's geometry and associated information. At LOD 100 (pre-design), the model mainly constitutes of 2D symbols to define an element's existence. By LOD 200 (schematic design), the elements are partially defined by outlining approximate size, shape, quantity and location. At LOD 300 (design development), the elements have accurate dimensions and location. The LOD 400 level (construction stage) provides basic information about the construction of various elements, including fabrication, assembly and installation. By LOD 500 (as built), the model is field-verified and includes real-life functions of elements in the real building or structure.

Cross Country Analysis of Building Information Modelling Adoption

Although the concept of BIM has been around for some time, the levels of adoption vary between countries, between disciplines and even between clients. Coupled with the varying levels of adoption, the way in which organisations approach adoption across both technical and non-technical aspects plays a significant role in increasing the application of BIM (Gu & London, 2010). Technical aspects refer to the digital functionality, such as software, hardware and IT systems. Non-technical aspects refer to the human aspects of adoption, for example, education and training; cultural, organisational and business support factors.

As the use of BIM has been increasing across countries, governments have developed various BIM regulatory and policy initiatives to guide both private and public sector adoption. Particularly important standards include:

1. BS/PAS 1192 series (UK)
2. The National BIM standards (US) (NBIMS, 2015)
3. Common BIM Requirement (COBIM) (Finland)
4. Singapore BIM Guide (Singapore)
5. National Building Specification (NATSPEC) National BIM Guide (Australia) (NATSPEC, 2011)
6. Statsbygg BIM Manual (EU) (Statsbygg,2013)

At the international level, the International Organization for Standardization (ISO) has developed standards based on the UK BS/PAS 1192 series.

United Kingdom

Up until 2016, the UK Government mandated Level 2 maturity BIM adoption for all government funded projects up to a certain project value in line with the British Standards Institution's (BSI) Publicly Available Specifications (PAS). (Burgess et al., 2018). The BSI PAS documents define good practice for products, services and processes, and help establish consistency and integrity through universal approaches across the country. Of particular note for this book chapter is *PAS 1192-6: Specification for integrated and collaborative sharing and use of health and safety information using BIM (BSI, 2019)*, which supports the development of structured WHS information for construction projects using BIM. However, this mandate has since ceased as the UK has moved to promote BIM Level 2 as part of normal operations and is developing strategies for a fully digitised construction industry (NBS, 2020b).

Since 2018, the BSI PAS 1192 series is progressively being replaced by the ISO 19650 series. Three international standards in this series have been published and another two are under development at time of writing:

- **BS EN ISO 19650–1:** Organization of information about Construction works – Information management using building information modelling – Part 1: Concepts and principles
- **BS EN ISO 19650-2:** Organization of information about Construction works – Information management using building information modelling – Part 2: Delivery phase of assets.
- **BS EN ISO 19650–3:** Organization and digitization of information about buildings and civil engineering works, including building information modelling (BIM) – information management using building information modelling – PART 3: operational phase of the assets [UNDER DEVELOPMENT]
- **BS EN ISO 19650–4:** Organization and digitization of information about buildings and civil engineering works, including building information modelling (BIM) – information management using building information modelling – PART 4: information exchange [UNDER DEVELOPMENT]
- **BS EN ISO 19650–5:2020:** Organization and digitization of information about buildings and civil engineering works, including building information modelling (BIM) – information management using building information modelling – PART 5: security-minded approach to information management

The three new standards are founded on the UK's standards for information management using BIM: BS/PAS 1192-1 (principles), BS/PAS 1192-2 (delivery phase) and BS/PAS 1192-5 (security).

To support digitisation of the construction industry, two new programs of work were established in 2016-2017. Firstly, the UK BIM Alliance promotes Level 2 BIM as part of normal operations and, secondly, Digital Built Britain (DBB) promotes new initiatives. In 2019, the UK BIM Alliance, BSI and the Centre for DBB announced the launch of the UK BIM Framework. This framework sets out a guided approach to implementing BIM in the UK and managing information provided by the BS EN ISO 19650 series.

European

According to Ullah et al (2019) in 2014, the European Commission announced directive 2014/24/EU, which recommended member states' use of specific electronic tools such as BIM for public works contracts and design contests (European Parliament, 2014). Research conducted in Germany, France, Brazil and Austria showed that BIM is gaining wide adoption in these countries (Matarneh and Hamed, 2017).

The Scandinavian countries have led BIM adoption (Smith, 2014). BIM was first used in Finland in 2002 and, as early as 2007, it was made mandatory for design software to pass Industry Foundation Class (IFC) certification. IFC is a software file format that can be used across different software applications, allowing teams to share models and collaborate while still working independently. In 2012, Senate Properties migrated the *BIM Requirements for Architectural Design into the Finnish National BIM Guidelines (COBIM)*. Similarly, Norway has been using IFC file formats and BIM in their projects since 2010. In the Netherlands, the Government Buildings Agency mandated the use of BIM for public projects in 2011 (Cheng and Lu, 2015).

UnitedStates

The US was one of the first countries to implement BIM (Wonget al., 2010). However, the lack of national standardisation slowed down its progress. It was in 2003, through the Public Building Services, that the General Services Administration established a 3D/4D/BIM program that released limited guidelines for the construction industry. In recent years, the National Institute of Building Sciences has made considerable progress in ensuring the adoption of BIM in the US.

China

The Chinese construction industry became aware of the concept of Building Information Modeling (BIM) around 2002; although only a few projects embraced BIM until 2011 (Herr and Fischer, 2018) when the Chinese government began to strongly support the adoption of BIM. The adoption of BIM in the Chinese construction industry has been expedited since 2011, when the Chinese government started to promote BIM and plan a series of national-level BIM policies and standards. The 2011-2015 Outline of Development for Informationalisation in the Construction Industry issued by the Ministry of Housing and Urban Rural Development (MHURD) is the first national policy to include BIM as an important aspect of informationalisation development (MHURD, 2011). The 2016-2020 Outline of Development for Informationalisation in the Construction Industry specified BIM as the top of critical information technologies (e.g., BIM, Big Data, Smart Technologies, Mobile Communication, Cloud Computing, Internet of Things) promoted by the construction industry (MHRUD, 2016). A series of national and regional level BIM application guidelines were developed by the central government and local government to comprehensively promote and guide the adoption of BIM in the Chinese construction industry in the last decade (Lin and Zhang, 2018).

Singapore

Singapore's nationwide BIM implementation Roadmap was first laid out in 2011. BIM has since been increasingly used in public sector building-projects, particularly since 2015 when it became mandatory for all building projects larger than 5,000 m2 to use BIM. Another important initiative driving the use of BIM was the Building Information Model Fund administered by the Building and Construction Authority. (BCA, 2013) This initiative aimed to boost the uptake of BIM by construction firms across Singapore by supporting training, consultancy, software and hardware cost for projects.

Hong Kong

The Hong Kong Housing Authority piloted BIM around 2006 and the Construction Industry Council (CIC), a government agency, has since been promoting BIM adoption across the sector. It developed the CIC BIM standards based on international standards like the PAS 1192 and the Singapore BIM guide. Since 2018, all government projects valued over $30 million are required by legislation to incorporate BIM in their design and construction.

Australia

Australia has gradually adopted BIM with various initiatives occurring between 2010 and 2020 that has led to the awareness and adoption during this time period. During this time, a number of key organisations, including the Australasian Procurement and Construction Council (APCC), Australian Construction Industry Forum (ACIF) and the Australasian BIM Advisory Board (ABAB), and the Collaborative Research Centre (CRC) for Construction Innovation promoted the use of BIM in the construction industry. As early as 2003, the Collaborative Research Centre for Construction Innovation led many early BIM initiatives in Australia particularly related to government/academic/industry research studies.

In 2010 the Built Environment Industry Innovation Council, one of many industry focused councils created by the Federal government of the day, which brought together industry, academia and government to respond to key challenges, created the Built Environment Digital Modelling Working Group. The Working Group worked in collaboration with the Information Technology Industry Innovation Council to respond to the emerging need to develop an Australian built environment industry capable of adopting integrated digital modelling technologies to improve efficiency, address environmental and life-cycle issues and increase international competitiveness. The working group brought together industry, researchers, policy-makers and industry associations in the built environment and was chaired by the Department of Industry, Innovation, Science, Research and Tertiary Education, (DIISRTE, 2012).

In 2015, the APCC, in partnership with the ACIF, launched the *BIM Knowledge and Skills Framework* and, in 2017, together with the key standard-setting bodies NATSPEC, buildingSMART and Standards Australia established the ABAB.

The ABAB consists of stakeholders from government, industry and academia who partner to provide expertise and leadership on the adoption of BIM and Project Team Integration (PTI). PTI is an approach that facilitates collaboration among participants at the earliest stages of the project for better efficiency. It supports best practice through advocating a consistent approach to BIM practices, standards and requirements, as well as a holistic approach to the national adoption of BIM in construction. It is through the *Australian Strategic BIM Framework* that all state and territory governments commit to improving procurement methods to allow for the adoption of BIM. As exemplars, they commit to adopt a nationally consistent approach to BIM in the procurement of government-initiated building construction, infrastructure and management projects.

The *Australian Strategic BIM Framework* was launched in March 2017 to incorporate the various guidelines developed since 2010. Included were *the NATSPEC National BIM Guide and BIM Management Plan Template*, which clarify BIM requirements to all project stakeholders, inclusive of clients and consultants, in a nationally consistent manner. Also, in 2017, the buildingSMART National BIM Initiative launched BIMcreds, which provides BIM practitioners with a framework for demonstrating their competence (buildingSMART, 2017). BIMcreds is designed to be used in conjunction with the APCC/ACIF *BIM Knowledge and Skills Framework*.

In 2014, Sustainable Built Environment National Research Council developed a *Towards a National Strategy* that continues to guide the development of an Australian national strategy for BIM adoption. It also developed the *BIM Value Tool* with NATSPEC (2011) to support practitioners who want to understand the benefit and application of BIM and its implementation.

As a result of joint efforts by the ACIF and the APCC, the ABAB has more recently developed the *Framework for the Adoption of Project Team Integration and Building Information Modelling* and *Building and Construction Procurement Guide* with a predominant focus on BIM and PTI. The APCC and ACIF see their role as mediators between industry interests and government requirements, with a focus on legal, procurement, collaboration and process-related topics (ABAB, 2020).

In summary, there is a plethora of BIM documents and guidelines that provide guidance for Australian public and private sector clients and construction industry professionals on BIM implementation. However, regardless of the depth and quality of these guidelines, it is suspected that there is a general lack of adoption of BIM applied specifically to managing WHS. One reason may be that the material is too generic and lacks specificity in relation to WHS management. Given the critical importance of improving WHS in construction, both nationally and internationally, and the increasing global uptake of BIM, it appears the integration of WHS in BIM is an area that should be further explored to facilitate better WHS outcomes. The general approach to BIM adoption has typically pursued the generic pathway. It is perhaps timely to consider the challenge of integration of WHS management systems into the BIM enabled environment.

Barriers and Drivers to Adoption

Despite the numerous benefits, BIM adoption has its own set of challenges. It is useful to understand the significant body of research that has been conducted on the adoption of BIM technologies in construction and the barriers to industry and government uptake when seeking to introduce a BIM-enabled environment. Research studies have categorised barriers according to common characteristics; for example, people, process and product barriers (Lindblad, 2013) or technical, social context, process-related, and work practice issues (Gu & London, 2010). More recent research has added categories such as contractual and legal barriers.

Since 2006 there have been numerous studies that have sought to understand the barriers to adoption of BIM. As early as 2006, London et al in a national broad based e-business adoption study, identified the barriers and drivers as well as other important outcomes including a) barriers and drivers were clustered and many barriers were inter related to other barriers and then barriers were inter related to drivers b) barriers often turned into drivers c) adoption rates varied which was in itself a barrier d) there were three common identifiable hierarchical pathways to adoption (related to perceptions, compatability and communication). There were four main inte related clusters of barriers identified including perceptions and attitudes, communication, compatability and market incentives and within each of these clusters there were various underlying inter related barriers. The overarching barrier of different rates of adoption across the industry largely due to the fragmented industry structure coupled with the project based nature of the industry played out in various ways. For example the differing rate of adoption and the inherent inertia of the industry impacted attitudes and perceptions through a resistance to change, doubts about benefits, an overwhelming feeling of complexity influenced by a lack of technical compatability, a lack of desire to change due to supply chain network; and much of this was influenced by top level management and market leaders in particular client and contractor demand.

Over the last fifteen years various studies have identified similar outcomes. It is useful to acknowledge that there is a depth of understanding in relation to barriers to BIM adoption. The following table is a detailed and comprehensive compilation of barriers and drivers which has been derived from London et al, 2006 and Gu and London 2010; merged with Ullah and Witt (2019) and then updated and complimented with additional literature from more recent studies.

Barriers to the adoption of BIM for WHS management are interlinked with the drivers for its adoption. These, along with developing appropriate responses to encourage uptake, may facilitate future higher levels of adoption of BIM to manage WHS. Developing this deeper level of understanding will contribute to unified action by policy makers, industry and stakeholders alike to achieve the common goal of a safe construction environment.

Understanding the drivers of BIM adoption is important to influence the intention and decision by stakeholders to adopt BIM (Ahmed & Kassem, 2018). Research has identified several examples of drivers, such as the increasing client interest in BIM (Eadie et al., 2013; Hamma-adama & Kouider, 2019), cost savings (Eadie et al., 2013; Hamma-adama & Kouider, 2019), government mandates (Eadie et al., 2013; Hamma-adama & Kouider, 2019), guidance by professional bodies (Oladapo, 2007) and the availability of trained professionals (Badrinath et al. 2016; Macdonald, 2012).

Research indicates that pressure from a variety of sources, for example, government, market competition and clients, are important factors in driving BIM adoption. This demonstrates the importance of government strategies in promoting BIM adoption and implementation, as evidenced in the *Australian Strategic BIM Framework*. BIM adoption is also influenced by many operational drivers; clients and major contractors tend to adopt BIM due to the time and cost savings that can be achieved, particularly through clash detection and improved design quality (Eadie et al., 2013). Over time, each driver has contributed to the uptake and adoption of BIM in the construction industry. These drivers provide important insights for policy makers and construction stakeholders to make informed decisions and policy recommendations that are capable of positively influencing increased BIM adoption.

The barriers and drivers have been clustered into seven thematic areas in Table 1 including; market incentives, leadership, supply chain integration and collaborative practice, procurement and tendering proficiency, decision tools and technical functionality. It is assumed that with leadership from government, public and private sector clients and innovative contractors and subcontractors a conducive environment for adoption will be created and then market incentives, supply chain monitoring and collaborative practice, although longer term initiatives, will follow. Both public and private sector clients can create and implement a BIM enabled environment through various strategies underpinned by their purchasing power on projects and thus detailed attention paid to their expectations and requirements on projects is critical. Procurement strategies and tendering proficiency are critical complimentary implementation tactics towards client's achieving long term outcomes that they desire. There is a delicate balance though between driving market suppliers towards end goals and being prescriptive, regulatory and compliance driven versus descriptive and outcome based.

Table 1. Compilation of Barriers and Drivers BIM adoption 2006-2019

Barriers	Drivers (Research study)	Barriers Research study (Authors/Years)
	MARKET INCENTIVES	
High costs	Cost savings and perceptions of advantage (London et al, 2006; Eadie et al., 2013; Hamma-adama & Kouider, 2019),	London et al, 2006; Yan & Damien 2008; Coates et al., 2010; Azhar, 2011; Ismail et al., 2017
lack of awareness about BIM benefits		Gu and London, 2010; Latiffi et al., 2016; Gerges et al., 2017
doubts about return on investment		London et al, 2006; Gu and London, 2010; Eadie et al., 2013b
PERCEPTIONS AND ATTITUDES		
Resistance to change due to industry culture		Yan & Demian, 2008; Watson, 2010; Abubakar et al., 2014, Ganah and John, 2015; Sahil, 2016
Resistance to change due to industry structure (ie fragmented nature of the construction industry varying adoption patterns)		London et al, 2006
LEADERSHIP		
Insufficient government support	government mandates (Eadie et al., 2013; Hamma-adama & Kouider, 2019)	London et al, 2006; Enshassi et al., 2016
Lack of client demand	Increasing client interest in BIM (Eadie et al., 2013; Hamma-adama & Kouider, 2019),	London et al, 2006; Wang et al., 2015, Sahil, 2016
Lack of support from top management	Management attitude towards innovation London et al 2006	Ganah and John, 2015
SUPPLY CHAIN INTEGRATION AND COLLABORATIVE PRACTICE		
Lack of demand from contractors		Gerges et al., 2017
the lack of collaboration among stakeholders		Wang et al., 2015; Tan et al., 2019;
Subcontractors not interested in using BIM		Hosseini et al., 2016
PROCUREMENT STRATEGIES AND TENDERING PROFICIENCY		
Absence of contractual requirement to use BIM		Ahmed et al., 2008
Quality of model at handover		Gu and London, 2010
legal and contractual issues around ownership and intellectual property (data ownership issues)		London et al,2006; Christensen et al., 2007; Furneaux & Kivvits, 2008; Gu and London, 2010; Oluwole, 2011; Bosch-Sijtsema *et al.*, 2017; Park and Kim, 2017
DECISION TOOLS		
Inadequate training (lack of traning and learning) Lack of expertise in BIM within organisations	availability of trained professionals (Badrinath et al. 2016; Macdonald, 2012). guidance by professional bodies (Oladapo, 2007) Management attitude towards training (London et al 2006)	Gu and London 2010; Azhar & Bahringer, 2013; Zahrizan et al., 2013, Abubakar et al., 2014; Chan, 2014; Eadie et al., 2014; Wang et al., 2015; McAuley et al., 2017; Park and Kim, 2017; Tan et al., 2019;
Lack of standardised tools and protocols		Gu and London, 2010; Zahrizan et al., 2013; Chan, 2014; McAuley et al., 2017; Tan et al., 2019; Wang et al., 2015
TECHNICAL FUNCTIONALITY		
Complexity of the model Longer process		London et al,2006; Ahmed et al., 2018 Ismail et al., 2017
Interoperability between software programs		London et al, 2006; Gu and London, 2010; Park and Kim, 2017
Data management issues including data integrity, storage, version management, security		Gu and London, 2010

Past studies using diffusion theory and BIM adoption indicate that there is clear differentiation in adoption patterns and pathways by innovators and early adopters to that of adoption by late adopters and laggards (London et al, 2006; Gu and London, 2010). In many respects this is why the focus in this study is on exemplars; ie market leaders. In London's (2006) earlier work she identified three common pathways for adoption and three distinct phases ; Perceptions, Compatibility and Communication. Laggards and late adopters moved through the perception and compatibility pathways and late adopters and innovators moved through the compatibility and communication pathways. Key drivers for adoption tend to focus on leadership, collaboration, supply chain integration, managerial tools/competencies (such as procurement and tendering) and less on organisational and economic benefits (as these have been assumed) as well as perceptions and attitudes. Technical functionality challenges that advanced BIM adoption organisations come up against tend to be considered as challenges to solve to develop market competitive advantage. The decision tool framework themes focus on leadership, supply chain integration, collaborative practice, procurement strategies and tendering proficiency.

With WHS management being such a critical compliance component of the design, construction and operations phases it is not an option for clients to not take advantage of the opportunities that a BIM enabled environment can provide. The regulatory obligation is significant but more critically is the social responsibility and ethical obligation of clients to provide every opportunity to create and maintain a safe work environment and reduce WHS risks.

Figure 1. Relationship between barriers and drivers and key conceptual elements of a BIM WHS management Decision Framework

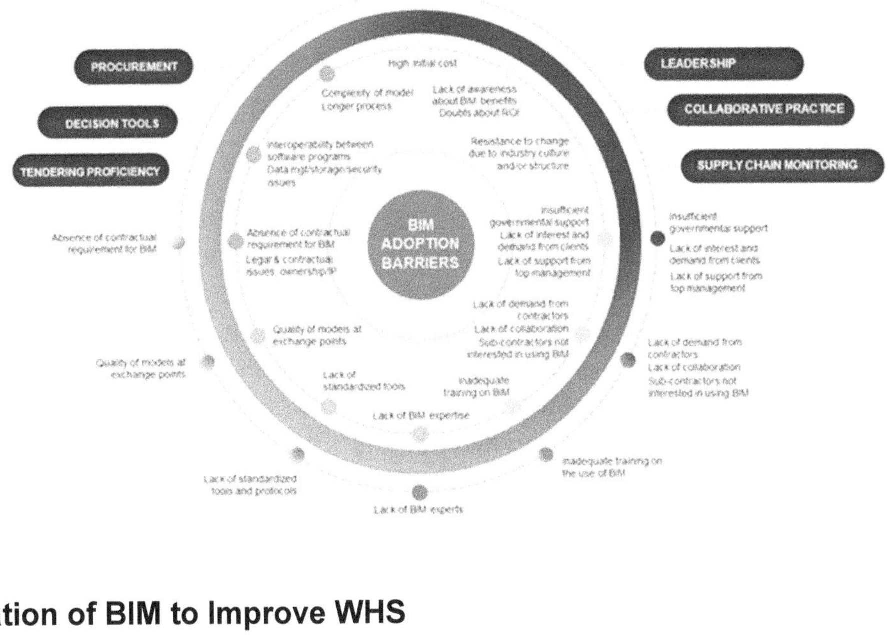

Application of BIM to Improve WHS

WHS is a concern applicable to the entire lifecycle of a project, from planning to design, construction, post-construction management and decommissioning. BIM is an enabler technology that allows WHS data and information to be captured and communicated across all phases and applied in different ways to manage WHS risks (Zhao & Lucas, 2015; Zhang et al., 2013), ultimately preventing injuries and fatalities (Lopez del Puerto & Clevenger, 2010; Riaz et al., 2014). The benefits of using BIM to manage WHS can be summarised in six thematic areas of design, construction and operational phases: (1) Scenario planning; (2) Requirements briefing; (3) Risk assessments; (4) Education and training; (5) Monitoring, surveillance and reporting; and (6) Analysis.

A significant contribution of BIM to WHS management is by virtual application and data management. BIM enables efficient and effective safety planning and analysis by enabling virtual scenario testing. BIM also facilitates education and training by simulating construction workflows; visualisation of hazards; management of safety knowledge (including rule-checking approaches to detect fall hazards and scaffolding requirements); virtual assessment of jobsite conditions to recognise hazards; simulation of the construction process and its relationships to safety professionals when planning safety measures; improvement of site layout planning and the identification of site facilities, along with the inter-relationships between them. These benefits also help optimise the layout and size of facilities along with identifying other constraints on site (Astour & Franz, 2014; Kim & Teizer, 2014; Wang et al., 2015; Zhang et al., 2013; Zhang et al., 2015).

Table 2 outlines exemplars in countries that are at the forefront of research in construction safety using BIM. The US has been a significant contributor in this field and its collaborative links with the UK, Australia, South Korea, Germany and Spain have influenced research efforts in this area (Akram et al., 2019). A review of the literature indicates the research focus combining BIM and WHS has been predominantly within the planning and design phases of the project lifecycle and have primarily focused on scenario planning and risk assessment (Akram et al., 2019). Table 2 provides a summary of relevant studies of BIM and construction safety from these countries. This is not an exhaustive list of research studies in this area, rather it depicts how different researchers and industry practitioners have utilised BIM to ensure better WHS on construction sites.

Hazard recognition and prevention followed by worksite safety and safety planning are the most frequently studied areas in the application of BIM. Also, visualisation followed by safety rule checking are the most common attributes of BIM and construction safety research (Akram et al., 2019). These allow construction personnel to visually observe worksite conditions and check against rules to identify contradictions (Azhar, 2017; Hadikusumo & Rowlinson, 2002; Kim & Teizer, 2014).

These exemplars demonstrate that construction safety practices, along with risk registers and WHS regulations, can be integrated into the BIM environment in the form of repositories of WHS rules. Algorithms and simulations can then be developed to evaluate compliance against these rules on construction projects. Collectively, these studies indicate considerable potential to integrate and assess WHS in the BIM environment. Of the six identified benefits clearly the integration of BIM to WHS thematic areas of scenario planning and risk assessment has dominated. There is still considerable potential to explore the other four thematic areas of: requirements briefing; education and training; monitoring, surveillance and reporting and analysis. This merges well with the previously identified themes of supply chain monitoring, client leadership, collaborative practice, tendering proficiency, procurement methods and decision tools.

Table 2. Summary of the most relevant research on the application of Building Information Modelling (BIM) to enhance Work Health and Safety performance.

Countries	Phase	Benefit/Focus Area	Approach	Authors (year)	Summary
Finland	Construction	Scenario planning/ Fall prevention	Visualisation	Sulankivi et al. (2010)	4D-BIM for construction safety planning.
	Design	Fall prevention	Rule Algorithm/ Case study	Zhang et al. (2013)	Automatic safety checking through construction models and schedules.
Hong Kong	Design	Risk Assessment/Risk detection	Risk recognition mechanism/ Case study	Li et al. (2018)	An automated safety risk recognition mechanism for underground construction at the pre-construction stage based on BIM.
US	Design	Scenario Planning/Fall prevention	Rule checking/ Case study	Melzner et al. (2013)	Automated safety compliance checking to assist fall protection design and planning in BIM.
	Design	Safety planning/Safety hazards	Algorithm/ Case study	Kim et al. (2016)	Integrating work sequences and temporary structures into safety planning: Automated scaffolding-related safety hazard identification and prevention in BIM.
	Design and Construction	Risk Assessment/Risk Detection	Visualisation/ algorithm /Case study	Shen & Marks (2016)	Near-miss information visualisation tool for BIM.
Australia	Design and Construction	Scenario Planning/Fall prevention	Algorithm/ Case study	Shou et al. (2015)	BIM-based dynamic scaffolding design and safety prevention.
	Design	Risk Assessment/Risk Detection	Prevention through Design/ Prototype	Kamardeen (2010)	The system architecture of an 8D-modelling tool for BIM-based prevention through design.
UK	Design and Construction	Risk Assessment/Risk detection	Case study	Health and Safety Executive, (2018)	Improving health and safety outcomes in construction decision making and value propositions of BIM.
Spain	Design	Requirements briefing/ Safety hazards	Visualisation/ Methodology	Cortes-Perez et al. (2020)	BIM-integrated management of occupational hazards in building construction and maintenance.
China	Construction	Risk identification	Hybrid inference/ Case study	Zhang et al. (2016)	A BIM-based Risk Identification Expert System (B-RIES) composed of three main built-in subsystems: BIM extraction, knowledge base management, and risk identification subsystems
	Design and Construction	Automated construction processes	Additive manufacturing / algorithm	Ding et al. (2014)	BIM-based automated construction system (BIMAC) including composition of BIMAC, detail of execution setup, data issues, filling layer algorithm.

STRATEGIES TO INTEGRATE A BIM-ENABLED ENVIRONMENT FOR WHS MANAGEMENT

Clients need to develop an understanding of their own procurement competency in describing how they see BIM being utilized during planning, design, construction and operations phases of facilities; otherwise they will receive what the market wants to push onto them. Thus improved decision tools including protocols and standardized approaches are necessary.

The use of procurement as an instrument to further the adoption of WHS practices using BIM has received scarce attention to date. Although several standards have been developed for WHS, only limited guidance is available on how to incorporate them into procurement. This section seeks to address how WHS and BIM can be integrated into procurement and tendering so that compliance with WHS regulations can be effectively addressed. It focusses on describing procurement and tendering as key catalysts and drivers. It also explores the important role the client can play at the procurement stage to ensure better WHS on construction projects. The following sections describe a proposition for components to a Decision Framework and the potential outcomes that could be produced.

Figure 2 BIM WHS management Decision Framework towards outcomes and deliverables

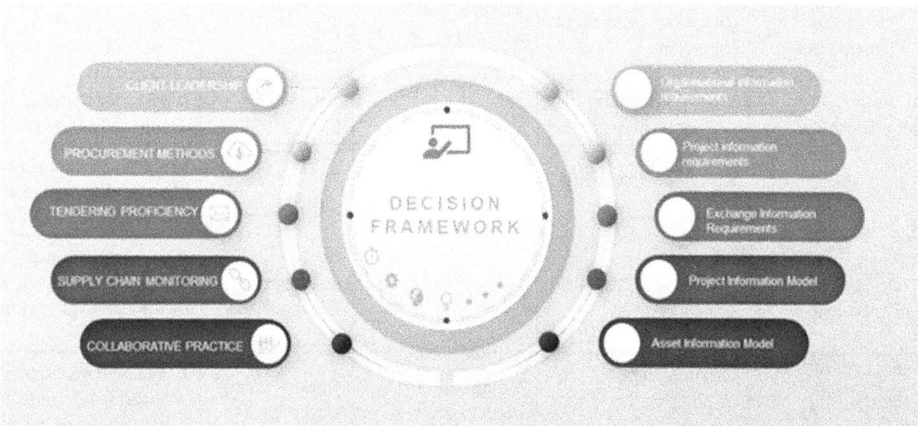

Clients' Leadership role in WHS Management Through Procurement

Construction clients are well positioned to lead WHS management through procurement. In recent years, there has been increasing emphasis on clients to ensure WHS on their construction projects, highlighting the need to consider WHS through procurement as a management strategy. This responsibility is to a large extent dependant on all stakeholders fulfilling their obligations and extends to all participants in the construction supply chain, so that the interests of the client are safeguarded and the legislative requirements met. Clients with a significant portfolio of capital works projects can thus influence WHS behaviours and practices across the entire construction supply chain by taking a leadership role in creating a safe environment and supportive safety culture, along with good processes and procedures to meet their statutory obligations. Similarly, clients can influence BIM adoption (London, 2010; Lingard et al, 2019) as an incentive to effective WHS management. We must not forget that BIM is fundamentally a tool to achieve better communication, collaboration, simulation, monitoring, analysis and education and the key driver to change is that the WHS goals be clearly articulated first and foremost. Following this all the opportunities afforded by BIM can be offered to clients by contractors and subcontractors towards achieving the clients requirements.

Client influence can manifest through communicating safety expectations, selecting consultants and contractors who are competent in designing the appropriate systems and participating in the safety management process (Huang & Hinze, 2006). The International Labour Office Code of Practice (1992) also provides that clients should nominate appropriate persons to coordinate activities relating to construction WHS and provide up-to-date information about any risks to WHS. Bidders should be asked to include the cost of WHS measures in their construction bids.

At the tendering stage, tenders are often awarded to the lowest bidders and there are various justifications for doing so. In some cases, the rationale is that the invitation to bid has been provided to those who have already been assessed and considered to be of a certain quality to provide the services and products or systems at the appropriate standard. It is questionable that this approach is the practice maintained consistently for all projects and all markets and submarkets. Bidding and procurement are complex areas of research and a large body of theory has developed over many years. However, it is useful to discuss briefly some critical concepts to aid understanding.

Despite national and international standards and guidelines of best practice in WHS, in many cases, attention to WHS requirements during procurement occurs in a 'fragmented' manner. A small and focussed study by Deacon and Smallwood (2016) in the South African construction industry explored this problem and established that this fragmented approach resulted in a lack of integration during design, tendering and construction phases, and thus varying degrees of compliance. The lack of WHS integration and legislative compliance across various phases of a project places greater onus of risk on the client, should there be any injuries or fatalities linked to the lack of inclusion of WHS during design and tendering (Deacon & Smallwood, 2016). WHS should be valued by stakeholders on all projects and should be given equal status with cost, quality and time. Equal status can be achieved by the client by ensuring WHS requirements are considered at all levels of decision making (Smallwood, 2019).

To win bids, contractors try to reduce costs for their overall bid and WHS can be penalised in the process (Wells & Hawkins, 2010). This practice may lead to a vicious cycle of cost-cutting during the construction process and eventually lead to injuries and fatalities (Rowlinson, 2004). To ensure adequate WHS, it is critical that the client develops a clear policy for safeguarding WHS on their construction project. WHS requirements should be clearly explained in the invitation to tender documents, with specific requirements detailed. The policy should be widely distributed so that the procuring officers and potential tenderers fully understand the client's priorities. The tender may need to include a site-specific WHS plan, procedures to be adopted to ensure that WHS requirements are met by all, and a system and format for recording incidents and injuries. Smallwood and Emuze (2013) indicated that clients can play an important role in setting the level of standard of WHS on projects, as they are key stakeholders and can be influential through the tone and the form of contract they enter into, from the very beginning. Lingard et al. (2019) concluded that institutional dynamics play a pivotal role in determining how clients ensure WHS safety is considered in the projects they procure. Clients who promote collaborative arrangements can enhance WHS practices on their projects and support innovative strategies that not only promote WHS but helps improve overall project efficiency.

Governments play a major role in preventing work-related fatalities and injuries by promoting, legislating and enforcing WHS regulations in workplaces. Government agencies in Australia provide leadership in WHS through their commitment to the *Australian Strategic BIM Framework* and play a significant role in fostering safe design. As clients on major public sector projects, government procurement processes should be such that they help identify and prevent risks in the workplace. The safe design process aids this by involving decision makers and considering risks throughout the lifecycle of the structure.

Aligned to this, the Office of the Australian Safety and Compensation Council has developed a model of better practice WHS in procurement. The outlined model (including various tools, contract clauses, checklists, and information) acts as a guideline to encourage governments and contractors to incorporate WHS and safe design considerations into their procurement processes (Australian Safety and Compensation Council, 2006). However, the guideline does not specifically discuss project procurement strategies that provide for different contractual arrangements, roles and responsibilities. This might be a useful extension that would assist with the next phase of diffusion of BIM-enabled WHS management implementation. In summary, it is apparent that clients are important change agents and their adoption of best practice for ensuring WHS during procurement can strongly influence the way industry adopts new practices to ensure better WHS management. Clients are therefore expected to use their power and influence to drive change among their project stakeholders.

Procurement Methods and BIM Integration

This section aims to identify the optimal procurement methods for incorporating BIM for effective project delivery. Research in this domain is gaining momentum as integration of BIM is considerably influenced by the method of procurement selected for project delivery. Kuiper and Holzer's (2013) study about the relationship between uses of BIM and procurement strategies revealed that integration across the supply chains and collaborative risk sharing are the ways forward. Their study presented the various procurement strategies on a continuum, from traditional forms to relationship contracting, and indicated that the alliancing strategy yields the greatest opportunity for BIM integration. Alliancing is a collaborative type of agreement between two or more partners, which facilitates aligning the common interests for collective sharing of rewards and risks. In traditional forms, there is individual responsibility of specific risks to the participants who individually manage them. The alliancing approach of collective risk sharing provides a good foundation for collaboration, which is essential in a BIM environment.

In a later study regarding appropriateness of different procurement types that can be applied to BIM projects (for example, Design–Bid–Build, Design and Build [DB], and Integrated Project Delivery [IPD]), Holzer (2015) extended the previous research and revealed that IPD and DB are the optimal procurement routes for integrating BIM as a tool. Design-Bid-Build is also known as Design-Bid-Construct (commonly referred to as 'traditional') and DB is often referred to elsewhere as Design and Construct. IPD is a procurement term that was developed and mainly used in the US. It is considered well-aligned to successful BIM adoption (Holzer, 2015), yet DB is the more commonly used procurement strategy (Bynum, Issa & Olbina, 2013; Holzer, 2015; Sebastian, 2011). IPD is a delivery approach that:

Integrates people, systems, business structures and practices into a process that collaboratively harnesses the talents and insights of all participants to reduce waste and optimize efficiency through all phases of design, fabrication and construction.(American Institutes of Architects, 2007)

The key characteristic for an IPD procurement strategy is that the contractual arrangement calls for an integrated contract between the owner/client, architect and contractor. The arrangement relies upon a risk/reward and value/cost approach, which incentivises innovation and creativity in the early phases and then efficiency of execution in the construction phase.

In Australia, project organisational and procurement strategies such as Public Private Partnerships and alliances have been adapted to accommodate integration of BIM in contracts. Public Private Partnerships are collaborative arrangements between government and the private sector for delivering public projects, such as infrastructure and related services. This type of procurement strategy falls within the umbrella of 'relational contracting' and appears to be more conducive to creating a BIM-enabled environment. To date, the combination of procurement strategy, tendering procedures and outcome has not been evaluated in Australia or for that matter anywhere else in the world so it is difficult to conclusively determine which procurement strategy is most appropriate. One could postulate that any procurement strategy could be conducive to integrating WHS management into a BIM enabled environment so long as client requirements were clearly identified and expectations were well articulated. Perhaps it is the self awareness by clients of the important role they can play that is just as important as actual procurement strategy deployed. Significant effort has been applied to the development of guidelines and it may be worthwhile exploring how to supplement existing guidelines in enabling WHS management using BIM.

Tendering, Specifying and Monitoring for BIM Supply Chains

Regardless of the preferred procurement method, it is important that the client makes clear its BIM environment intentions at the tendering and specification stage. This can be achieved by outlining WHS requirements clearly in the employer information requirements (EIR) documents. The EIR is a document that must be followed by all stakeholders on the project and on which the contractors should develop and submit their initial BIM execution plans (BEPs). In the BEP, the contractor must specify how it proposes to achieve the client's requirements. Together, the EIR and BEP are crucial documents for assessing the quality of tender responses (Mirarchi et al., 2020). The EIR should define what models are to be produced at each project stage and to what level of detail. These models will then provide information to the client at all defined stages of the project lifecycle Including BIM requirements in the tendering documents clearly requires owners/clients to provide more precise information about the required deliverables (Mayo & Issa, 2014).

In more recent times many countries have moved away from utilizing EIR terminology and have shifted towards Organisational Information Requirements (OIRs) and Project Information Requirements (PIRs) one which focusses on the client's needs and one focusing on the project. The creation of the OIR starts with the client organisation firstly identifying what information is needed in order to answer operational decisions; that is, information required for asset management systems and other organisational functions. An OIR would include organisational-level business information requirements, such as whole of government reporting requirements and information required to assess progress against agency strategic plans, rather than asset-level or project-level information requirements. The requirements at the OIR level typically support decisions that need to be made about the asset, including: asset portfolio planning and management (e.g. understand the assets functionality, performance and cost); strategic asset management and regulatory compliance. The Project Information Requirements (PIR) describes the information needed to answer or inform the strategic objectives of infrastructure projects. It conveys the OIR and project requirements to in-house teams and project-specific external suppliers setting out the information to be delivered for planning, design and construction of infrastructure projects. (NSW, 2020). Developing the OIR and PIR are conceptually important phases for a client, whether private or public sector, as it clearly sets out the purpose and use of data and how the client will manage data as an asset. Without this strategic thinking and requirements based approach by clients it is unrealistic and infeasible for both clients and contractors and consultants to have clarity in the tendering phases.

There are a few considerations in relation to tendering that should be considered. The client needs to develop a position on their overall data management and how the particular project, and other projects if they have a program of capital works projects, will be used. This of course relates to not only project portfolio management but asset management. Then once the client's vision on BIM data management within the context of a Common Data Environment has been crafted the vision on the specificity of WHS goals and requirements should be developed. This is when the client articulates what the health and safety issues are during the operations phase which then has implications for the design and construction phases. The Health and Safety Risks and Hazards analysis is typically undertaken. The vision of how BIM will enable WHS management during design, construction and operations phases would then guide a suite of General Requirements for BIM WHS that can be used during tendering and each time of exchange of models from one phase to the next depending upon the procurement strategy. Clearly there are inter relationships between client leadership and how they will set up their unique environment to achieve their own important WHS requirements related to their organizational goals; which then relates to decisions regarding procurement strategies for projects as well as how tenders will be designed, criteria developed and evaluated and then monitored during design and construction phases. Importantly Exchange Information Requirements are critical as this will identify how infrastructure data and information will be transferred between various project team members for various phases. This is one tactic to enable monitoring of the supply chain. The transfer of information from construction to operations is perhaps a critical handover phase but depending upon the procurement strategy undertaken there are other critical potential handover phases as well. The EIR also enables a commitment by both the client and then suppliers (consultants, contractors and subcontractors) to a certain competency and capability level.

It is important to remember the benefits of using BIM to manage WHS which can be summarised across six thematic areas of design, construction and operational phases: (1) Scenario planning; (2) Requirements briefing; (3) Risk assessments; (4) Education and training; (5) Monitoring, surveillance and reporting; and (6) Analysis; when developing the vision for the purpose of WHS data and information management and how BIM will enable that vision.

There are other possibilities of how a BIM enabled environment can assist WHS management which is much about efficient processes. Rule-based software are indeed powerful tools, as they enable clients to customise rules and verify both geometrical and conceptual requirements. This also benefits bidders, who can utilise the tools for self-assessment prior to submission. As an example, Bolpagi (2013) analysed the possible implementation of BIM within public procurement in Finland with particular focus on how model checking can be applied within tendering to verify compliance between the client's requirements and the bid's content. The Solibri Model Checker (SMC), a software that provides the opportunity to customise built-in rules, was used for assessing bidders' proposals and to find possible errors and contradictory information.

More recently, Costa and Grilo (2015) developed an e-procurement system that aids clients to initiate an e-tendering process using a BIM model. In this system, a detailed BIM model is required where every BIM element must include related information. It is easy thereafter to initiate the e-procurement process by selecting BIM elements and asking user questions for the tender. Bidders can also submit tenders using the BIM model. In this way, a bidder may attach all information about products to be used and their costs. This results in improved information management flow and enhances collaboration. However, the level of transparency required by the e-procurement system may be challenging and perhaps naïve in response to how tendering down the construction supply chain actually takes place in extremely competitive tendering environments.

Collaborative Practice: Standards, Guidelines and Decision Tools

To date, it appears that specific advice to clients on how to integrate WHS management in BIM has not been extensively considered in the procurement guidelines, particularly in Australia (refer to Table 3).

Australian government agencies understand the importance of BIM. In partnership with industry organisations, government agencies have therefore developed extensive guidelines and procedures for its application and, through the government procurement process, are encouraging BIM development and adoption in order to harness its potential. Having recognised the increasing benefits and potential for the wider application of BIM, Australian governments have in recent years actively supported its use and integration through the procurement of government-funded construction and infrastructure projects. For approximately 15 years, various government agencies and professional industry organisations have actively promoted BIM awareness and skills development for construction professionals to influence the uptake of BIM and its application across all stages of the project lifecycle.

Although there are numerous guidelines and tendering and procurement documents, there is little evidence of implementation and guidance on WHS management using BIM. A more complete analysis, evaluation and assessment of exemplars and best practice are needed to develop the appropriate enabling environment. Although there is minimal guidance on procurement strategies and tendering criteria and evaluation in relation to BIM these may be internal documents not available to the public. There is also no specific guidance on integration of BIM and WHS management although the thinking may be that this may be too specific and prescriptive an approach to implement. Collaboration and sharing between project stakeholders of structured WHS information across the project lifecycle has been slow to evolve. As Australia moves towards adopting BIM for WHS management, it may gain insights from its international equivalents and adapt and modify to suit the local context.

It is important to understand that while the most recent ISO 45001 sets out the WHS framework at international level, it is the standards, such as the UK PAS 1192-6, that outline the mechanisms by which the ISO 45001 health and safety management system may be applied using BIM. The UK PAS 1192-6 is an exemplar that provides guidance on integrating WHS information through BIM processes and applications, ensuring safer work sites and reducd injuries and fatalities.

PAS 1192-6 models a collaborative approach progressing through the project lifecycle between the client, designer, contractor, commissioner and end user. It places the onus on the client to clearly identify upfront the objectives of WHS requirements in a project. Where clients consider using BIM to assist in the fulfilment of their WHS obligations, they should ensure scope and risk allocation are clearly set out in the EIR.

The UK EIR template has a section for Health and Safety and Construction (Design and Management) Regulations, which assists the client in defining how BIM will support WHS monitoring aligned with the different phases of the project. Details of how all related data can be captured and recorded is documented. Furthermore, the *BIM 4 health & safety* working group has outlined ten questions that a client should include in its EIRs to help prioritise the key WHS issues in line with the PAS 1192-6. The checklist is aimed at developing client understanding of how to integrate BIM and WHS management with the assumption that it is easier to assess WHS management in a BIM-enabled environment when it is integrated in the tender documents in a holistic manner rather than when contained within standalone documents. All in all, there is potential for incorporating BIM enabled WHS management at the procurement stage as clients can more holistically and effectively identify, articulate and ensure WHS integration in the construction lifecycle and workplace.

Table 3. Commentary of in- country Procurement guidelines

Government Jurisdiction	Commentary on Procurement Guidelines
New South Wales	In NSW, Transport for New South Wales (TfNSW) implemented BIM through its digital engineering (DE) framework, having rolled out various transport cluster projects since 2018. Its adoption of DE resulted in best-practice data and information management and its strategy has led stakeholders to make more informed and consistent management decisions. TfNSW has been instrumental in driving adoption of the principles of the ISO 19650 series by applying the standard across its projects. TfNSW's framework specifically outlines procurement as a key area for development in order to ensure a consistent approach to new projects. For example, TfNSW has standard contract templates for use on DE-enabled projects and provides a suite of documents and standards to the contractor. By adopting integrated procurement processes, complexity is reduced, and consistency developed, thus improving tender responses and clarity during project start-up (DE, 2018).
Victoria	The Victoria State Government released the *Victorian Digital Asset Strategy* (VDAS) in 2019 to aid the adoption of BIM. The strategy is tailored to match that of the UK's BIM Level 2 mandate and aims to achieve standardised systems and templates. The VDAS seeks to implement BIM in line with the ISO 19650 series and its procurement and workflows document provide detailed flowcharts, aligning the flow of information between the appointing and appointed parties. This mainly outlines the steps that involve the appointing party defining their organisational information requirements and drafting an EIR, which is then combined with the responsible, accountable, consulted and informed (RACI) scope checklist in the tender invitation document. The tender participants are required to prepare a response by completing a Digital Engineering Execution Plan and RACI scope checklist. The appointing party then responds with a DE execution response and awards the contract accordingly (VDAS, 2019).
Queensland	The Queensland State Government approached BIM adoption with a whole-of-government perspective, enabled by the document: *Digital Enablement for Queensland Infrastructure - Principles for BIM implementation*, which was released in November 2018. The principles apply to all stakeholders involved in the lifecycle of any new major construction project, including the planning, procurement, design, contract management, construction, and maintenance of the assets. The Queensland State Government is committed to a staged adoption of BIM through these principles over five years from 2018-2023 (DSDMIP, 2018).
South Australia	The Government of South Australia introduced BIM through the procurement process for the construction of the Royal Adelaide Hospital, upon which construction commenced in 2011 under the South Australia Public Private Partnership framework. BIM was requested in the tender and tenderers were made aware of the prerequisite to provide all information in an Autodesk Revit format. They were also asked to provide relevant examples of their Revit Library Content. The model provided a centralised database that provided a single source of all information throughout the project lifecycle. Successful completion of this project showed that the construction of large-scale buildings requires a considered collaboration strategy and clearly coordinated plans between design, engineering, construction and information technology professionals. An integrated management system was the key element that helped successfully deliver this large and complex project (Redwood et al., 2017).

With the potential held by BIM, incorporating WHS management at the procurement stage for the lifecycle of a building or infrastructure project is achievable if clients recognise, adopt and promote innovative implementation strategies that fulfil the WHS obligations. All government and industry clients on construction projects could take an active leadership role and promote collaborative arrangements among stakeholders. They could demand the use of BIM from the early design stage, through to the end of the structure's lifecycle however as noted earlier perhaps a balance between a prescriptive vs descriptive approach needs to be developed. An outcomes and performance based strategy is quite achievable. WHS could be included in the OIR and the PIR tender documents to ensure bidders address WHS risks in their BEPs. BIM-enabled WHS management model checking tools could then be used to verify compliance between the client's requirements and the bidder's proposal and automatically find possible errors and contradictory information. Specific, innovative BIM-enabled WHS management approaches, such as simulations and rule algorithms, could also be documented and included as part of the tendering process. Moreover, procurement systems could be evaluated in terms of their impacts on WHS prior to their selection for projects; this would align well with a risk based approach to WHS management. As leaders and exemplars in the procurement of major construction and infrastructure projects, governments should continue to collaborate with industry and professional bodies, working together towards the provision of more comprehensive and tailored approaches for BIM-enabled WHS management.

CONCLUSION

The challenge to all participants in the construction industry is to find ways to reduce WHS incidents and injuries to support a safe workplace. BIM, a process and communication software tool, facilitates the early identification of potential WHS issues and the application of preventative strategies using automated approaches. The strengths of using BIM for WHS management manifest in the embedded processes of visualisation, simulation, analytics, evaluation, and monitoring. It also supports communication, education and training during the design, construction, operational phases and post construction phases. Documenting WHS requirements in plans and monitoring compliance with WHS regulations within a BIM-environment can result in safer working conditions as WHS is identified as an integral part of the construction process, is made more visible to stakeholders, and can be appropriately managed at the various stages of construction.

Australia lacks research that evaluates WHS management in a BIM-environment. BIM-enabled WHS management is rarely considered in construction tender requirements and the evaluation of bidders' proposals, despite the scope for BIM to provide a powerful approach to the management of WHS. The management of WHS is likely to be most effective when considered holistically as part of the project lifecycle and included in a BIM-enabled environment by the construction procurer, with commensurate client response at the construction tender and evaluation stage. This allows WHS requirements to be embedded early in the design and construction process and facilitates increased clarity for the management of WHS across the project lifecycle. The UK, Singapore, Hong Kong and Finland public and private sectors are further advanced in the adoption of BIM and its use for WHS management. They provide valuable research sites that could enrich information on the approaches, tools and outcomes highlighted by the theoretical, prototype and evaluative studies in this area. Sufficient examples of Australian private and public sector BIM-enabled construction projects exist to allow evaluation of BIM, from the tendering stage through to post construction, and the potential creation of management and information systems for monitoring WHS. There also appears to be sufficient experience among industry and government stakeholders to develop models for the adoption of BIM-enabled WHS management systems. Clearly, the brave new world of BIM holds potential for supporting the fulfilment of WHS obligations and it is timely to harness the potential of BIM to ensure safer construction sites.

ACKNOWLEDGMENT

This work was funded by Lendlease Building Pty Ltd as part of an Enforceable Undertaking with SafeWork NSW and the NSW Government's Centre for Work Health and Safety. The Centre for Work Health and Safety also oversaw the work, reviewed the manuscript and approved its publication.

REFERENCES

Abubakar, M., Ibrahim, Y. M., Kado, D., & Bala, K. (2014). Contractors perception of the factors affecting Building Information Modeling (BIM) adoption in the Nigerian construction industry. In R. I. Issa & I. Flood (Eds.), *Computing in Civil and Building Engineering* (pp. 167–178). American Society of Civil Engineers. (Original work published 2014) doi:10.1061/9780784413616.022

Ahmed, A. L., & Kassem, M. J. (2018). A unified BIM adoption taxonomy: Conceptual development, empirical validation and application. *Automation in Construction*, *96*, 103–127. doi:10.1016/j.autcon.2018.08.017

Akram, R., Thaheem, M. J., Nasir, A. R., Ali, T. H., & Khan, S. (2019). Exploring the role of building information modeling in Construction safety through science mapping. *Safety Science*, *120*, 456–470. doi:10.1016/j.ssci.2019.07.036

American Institutes of Architects. (2007). *Integrated Project Delivery*. A Working Definition, Version 2. https://help.aiacontracts.org/public/wp-content/uploads/2020/03/IPD_Guide.pdf

Astour, H., & Franz, V. (2014). BIM-and simulation-based site layout planning. In R. I. Issa & I. Flood (Eds.), *Computing in Civil and Building Engineering* (pp. 291–298). American Society of Civil Engineers. (Original work published 2014) doi:10.1061/9780784413616.037

Australasian BIM Advisory Board (ABAB). (2020). *Australian BIM Strategic Framework*. https://www.abab.net.au/wp-content/uploads/2020/01/ABAB_Australian_BIM_Strategic_Framework_Feb_2019.pdf

Australian Safety and Compensation Council. (2006). *Guidance on Occupational health and Safety in Government procurement*. https://www.safeworkaustralia.gov.au/system/files/documents/1702/guidanceonohsingovernementprocurement_2006_pdf.pdf

Azhar, S. (2011). Building Information Modeling (BIM): Trends, benefits, risks, and challenges for the AEC industry. *Leadership and Management in Engineering*, *11*(3), 241–252. doi:10.1061/(ASCE)LM.1943-5630.0000127

Azhar, S. (2017). Role of visualization technologies in safety planning and management at Construction jobsites. *Procedia Engineering*, *171*, 215–226. doi:10.1016/j.proeng.2017.01.329

Azhar, S., & Bahringer, A. (2013). A BIM-based approach for communicating and implementing a construction site safety plan. In *Proceedings of the 49th ASC Annual International Conference*. Associated Schools of Construction. http://ascpro0.ascweb.org/archives/cd/2013/paper/CPRT43002013.pdf

Badrinath, A. C., Chang, Y. T., & Hsieh, S. H. (2016). A review of tertiary BIM education for advanced engineering communication with visualization. *Visualization in Engineering*, *4*(1), 9. doi:10.118640327-016-0038-6

Bolpagi, M. (2013). *The Implementation of BIM within the Public Procurement: A Model-Based Approach for the Construction Industry*. VTT Technology.

Bosch-Sijtsema, P., Isaksson, A., Lennartsson, M., & Linderoth, H. C. J. (2017). Barriers and facilitators for BIM use among Swedish medium-sized contractors - We wait until someone tells us to use it. *Visualization in Engineering*, *5*(3), 1–12. doi:10.118640327-017-0040-7

British Standards Institution. (2019). *BS EN ISO 19650: Organisation and digitisation of information about buildings and civil engineering works, including building information modelling - Information management using building information modelling*. BSI.

Building and Construction Authority. (2013). *Singapore BIM Guide*. https://www.orenet.gov.sg/media/586132/Singapore-BIM-Guide_V2.pdf

buildingSMART Australasia. (2017). *National Building Information Modelling Initiative*. http://buildingsmart.org.au/wp-content/uploads/2014/03/ NationalBIMIniativeReport_6June2012.pdf

Bureau of Labor Statistics (BLS). (2019). *National Census of Fatal Occupational Injuries in 2017*. https://www.bls.gov/news.release/pdf/cfoi.pdf

Burgess, G., Jones, M., & Muir, K. (2018). *BIM in the UK house building industry: Opportunities and barriers to adoption*. University of Cambridge.

Bynum, P., Issa, R. R. A., & Olbina, S. (2013). Building information modeling in support of sustainable design and construction. *Journal of Construction Engineering and Management*, *139*(1), 24–34.

Chan, C. (2014). Barriers of Implementing BIM in Construction Industry from the Designers' Perspective: A Hong Kong Experience. *Journal of System and Management Sciences*, *4*, 24–40.

Cheng, J. C. P., & Lu, Q. (2015). A review of the efforts and roles of the public sector for BIM adoption worldwide. *Journal of Information Technology in Construction*, *20*, 442–478.

Christensen, S., McNamara, J., & O'Shea, K. (2007). Legal and contracting issues in electronic project administration in the construction industry. *Structural Survey*, *25*(3/4), 191–203.

Coates, P., Arayici, Y., Koskela, L., & Usher, C. (2010). The changing perception in the artefacts used in the design practice through BIM adoption. In P. S. Barrett, R. D. G. Amaratunga, R. P. Haigh, K. P. Keraminiyage, & C. Pathirage (Eds.), *2010 CIB World Congress Proceedings*. International Council for Research and Innovation in Building and Construction.

Cortés-Pérez, J. P., Cortés-Pérez, A., & Prieto-Muriel, P. (2020). BIM-integrated management of occupational hazards in building Construction and maintenance. *Automation in Construction*, *113*, 103–115.

Costa, A. A., & Grilo, A. (2015). BIM-Based E-Procurement: An Innovative Approach to Construction E-Procurement. *The Scientific World Journal*. doi:10.1155/2015/905390

Deacon, C., & Smallwood, J. (2016). The Effect of the Integration of Design, Procurement, and Construction Relative to Health and Safety (H&S) [Paper presentation]. The 2016 International Sustainable Ecological Engineering Design Society (SEEDS) Conference, Leeds Beckett University.

Department of State Development, Manufacturing, Infrastructure and Planning (DSDMIP), State of Queensland. (2018). *Digital Enablement for Queensland Infrastructure - Principles for BIM implementation*. http://www.dsdmip.qld.gov.au/resources/guideline/infrastructure/bim-principles.pdf

Digital Engineering (DE). (2018). *Digital Engineering Framework*. https://www.transport.nsw.gov.au/digital-engineering/digital-engineering-framework

Ding, L., Wei, R., & Che, H. (2014). Development of a BIM-based Automated Construction System. *Procedia Engineering*, *85*, 123–131.

Eadie, R., Browne, M., Odeyinka, H., McKeown, C., & McNiff, S. (2013). BIM implementation throughout the UK Construction project lifecycle: An analysis. *Automation in Construction*, *36*, 145–151.

Eadie, R., Odeyinka, H., Browne, M., McKeown, C., & Yohanis, M. (2013b). An analysis of the drivers for adopting Building Information Modelling. *Journal of Information Technology in Construction*, *18*, 338–352.

Enshassi, A., AbuHamra, L., & Mohamed, S. (2016). Barriers to implementation of building information modelling (BIM) in the Palestinian construction industry. *International Journal of Construction Project Management*, *8*(2), 103–123.

European Parliament. (2014). Directive 2014/24/EU of the European parliament and of the council of 26 February 2014 on public procurement and repealing directive 2004/18/EC. *Official Journal of the European Union*. https://eur-lex.europa.eu/legal-content/EN/TXT/?uri=celex%3A32014L0024

Feng, Y. (2013). Effect of safety investments on safety performance of building projects. *Safety Science*, *59*, 28–45.

Feng, Y., Zhang, S., & Wu, P. (2015). Factors influencing workplace accident costs of building projects. *Safety Science*, *72*, 97–104.

Furneaux, C., & Kivvits, R. (2008). *BIM – implications for government*. CRC for Construction Innovation.

Ganah, A., & John, G. A. (2015). An overview of the feasibility of achieving level 2 building information modeling by 2016 in the UK. *Journal of Civil Engineering and Architecture*, *9*, 885–894.

Garner-Purkis, Z. (2018). Deaths in Construction rise 27% year on year. *Construction News*. https://www.Constructionnews.co.uk/best-practice/health-and-safety/deathsin-Construction-rise-27-year-on-year/10032802.article

Gerges, M., Austin, S., Mayouf, M., Ahiakwo, O., Jaeger, M., Saad, A., & Gohary, T. (2017). An investigation into the implementation of building information modelling in the Middle East. *Journal of Information Technology in Construction*, *22*, 1–15.

Gu, N., & London, K. (2010). Understanding and facilitating BIM adoption in the AEC industry. *Automation in Construction, 19*(8), 988–999.

Hadikusumo, B., & Rowlinson, S. (2002). Integration of virtually real Construction model and design-for-safety-process database. *Automation in Construction, 11*(5), 501–509.

Hamma-adama, M., & Kouider, T. (2019). What are the barriers and drivers towards BIM adoption in Nigeria? In M. J. Skibniewski & M. Hajdu, (Eds.), *Proceedings of the 2019 Creative Construction Conference Budapest Hungary*. Budapest: Diamond Congress Ltd.

Health and Safety Executive (HSE). (2018). *Improving Health and Safety Outcomes in Construction Making the Case for Building Information Modelling (BIM).* https://www.hse.gov.uk/construction/lwit/assets/downloads/improving-health-and-safety-outcomes-in-construction.pdf

Herr, C. M., & Fischer, T. (2019). BIM adoption across the Chinese AEC industries: An extended BIM adoption model. *Journal of Computational Design and Engineering, 6*(2), 173–178. https://doi.org/10.1016/j.jcde.2018.06.001

Holzer, R. H. (2015). BIM for procurement - procuring for BIM. In R. H. Crawford and A. Stephan (Eds.), *Living and Learning: Research for a Better Built Environment: 49th International Conference of the Architectural Science Association* (pp. 237-246). The Architectural Science Association and The University of Melbourne.

Hosseini, M. R., Banihashemi, S., Chileshe, N., Namzadi, M. O., Udaeja, C., Rameezdeen, R., & McCuen, T. (2016). BIM adoption within Australian small and medium-sized enterprises (SMEs): An innovation diffusion model. *Construction Economics and Building, 16*(3), 71–86.

Huang, X., & Hinze, J. (2006). Owner's Role in Construction Safety. *Journal of Construction Engineering and Management, 132*(2), 164–173.

International Labour Organisation. (1992). *Safety and health in Construction. An ILO code of practice.* https://www.ilo.org/wcmsp5/groups/public/---ed_protect/---protrav/---safework/documents/normativeinstrument/wcms_107826.pdf

International Labour Organisation. (2020). *World Statistics*. https://www.ilo.org/moscow/areas-of-work/occupational-safety-and-health/WCMS_249278/lang--en/index.htm

Ismail, N. A. A., Chiozzi, M., & Drogemuller, R. (2017). An overview of BIM uptake in Asian developing countries. *AIP Conference Proceedings, 1903*, 080008. https://doi.org/10.1063/1.5011596

ISO 19650, (2019), ISO BIM - Building Information Modelling - ISO 19650

Kamardeen, I. (2011). E-OHS planning system for builders. *Architectural Science Review, 54*(1), 50–64.

Kim, K., Cho, Y., & Zhang, S. (2016). Integrating work sequences and temporary structures into safety planning: Automated scaffolding-related safety hazard identification and prevention in BIM. *Automation in Construction, 70*, 128–142.

Kim, K., & Teizer, J. (2014). Automatic design and planning of scaffolding systems using building information modeling. *Advanced Engineering Informatics, 28*(1), 66–80.

Kuiper, I., & Holzer, D. (2013). Rethinking the contractual context for Building Information Modelling (BIM) in the Australian built environment industry. *The Australasian Journal of Construction Economics and Building, 13*(4), 1–17.

Latiffi, A. A., Mohd, S., & Rakiman, U. S. (2016). *Potential improvement of building information modeling (BIM) implementation in Malaysian construction projects*. In The 12th IFIP International Conference on Product Lifecycle Management, Doha.

Li, M., Yu, H., & Liu, P. (2018). An automated safety risk recognition mechanism for underground Construction at the pre-Construction stage based on BIM. *Automation in Construction, 91*, 284–292.

Lin, J., & Zhang, J. (2018). Review and Exploratory Text Mining of Building Information Modelling Policies in China [in Chinese]. *Construction Technology, 47*(6), 73–78.

Lindblad, H. (2013). *Study of the implementation process of BIM in construction projects*. http://kth.diva-portal.org/smash/get/diva2:633132/FULLTEXT01

Lingard, H., Oswald, D., & Le, T. (2019). Embedding occupational health and safety in the procurement and management of infrastructure projects: Institutional logics at play in the context of new public management. *Construction Management and Economics, 37*(10), 567–583.

London, K., Mentink, J., Bavinton, N., Egan, G., & Aranda-Mena, G. (2006). *E-Business Adoption in the Construction Industry*. Icon.Net Pty Ltd.

London, K., & Singh, V. (2013). Integrated Construction supply chain design and delivery solutions. *Architectural Engineering and Design Management, 9*(3), 135–157.

Lopez del Puerto, C., & Clevenger, C. (2010). *Enhancing safety throughout Construction using BIM/VDC* [Paper presentation]. The ECOBuild 2010 BIM Academic Forum, Washington, DC.

Macdonald, J. A. (2012). A Framework for Collaborative BIM Education across the AEC Disciplines. In *Proceedings of the 37th Annual Conference of Australasian University Building Educators Association (AUBEA) 2012* (pp. 223-230). https://www.academia.edu/download/22716367/Macdonald_AUBEA2012.pdf

Martínez-Aires, M. D., López-Alonso, M., & Martínez-Rojas, M. (2018). Building information modeling and safety management: A systematic review. *Safety Science, 101*, 11–18.

Matarneh, R., & Hamed, S. (2017). Barriers to the adoption of building information modeling in the Jordanian building industry. *Open Journal of Civil Engineering, 7*(3), 325–335.

Mayo, G., & Issa, R. R. (2014). Processes and standards for BIM closeout information deliverables for owners. In *Proceedings of International Conference on Computing in Civil and Building Engineering* (pp. 673-680). Orlando, FL: Academic Press.

McAuley, B., Hore, A., & West, R. (2017). *Building information modelling in Ireland 2017*. BIM Innovation Capability Programme, CitA Ltd.

Melzner, J., Zhang, S., Teizer, J., & Bargstädt, H. J. (2013). A case study on automated safety compliance checking to assist fall protection design and planning in building information models. *Construction Management and Economics*, *31*(6), 661–674.

MHURD. (2011). *2011-2015 Outline of Development for Informationalisation in the Construction Industry*. Ministry of Housing and Urban Rural Development of China.

MHURD. (2016). *2016-2020 Outline of Development for Informationalisation in the Construction Industry*. Ministry of Housing and Urban Rural Development of China.

Mirarchi, C., Lupica Spagnolo, S., Daniotti, B., & Pavan, A. (2020). Structuring General Information Specifications for Contracts in Accordance with the UNI 11337:2017 Standard. In B. Daniotti, M. Gianinetto, & S. Della Torre (Eds.), Digital Transformation of the Design, Construction and Management Processes of the Built Environment. Research for Development (pp. 103–112). Springer. https://doi.org/10.1007/978-3-030-33570-0_10.

National BIM Specification (NBS) Australia. (2019). *Australia and New Zealand BIM Report 2019*. https://www.thenbs.com.au/bim-report

National Building Specification (NBS). (2020a). *BIM Levels Explained*. https://www.thenbs.com/knowledge/bim-levels-explained

National Building Specification (NBS). (2020b). *Digital Built Britain- Building the Future*. https://www.thenbs.com/knowledge/digital-built-britain-building-the-future

NATSPEC. (2011). *NATSPEC National BIM Guide*. https://bim.natspec.org/documents/natspec-national-bim-guide

NBIMS. (2015). *National BIM Standard - United States® Version 3 - Scope*. https://www.nationalbimstandard.org/

Oladapo, A. A. (2007). An investigation into the use of ICT in the Nigerian construction industry. *Electronic Journal of Information Technology in Construction*, *12*, 261–277.

Oluwole, A. (2011). A preliminary review on the legal implications of BIM and model ownership. *Journal of Information Technology in Construction*, *16*, 687–696.

Park, K. S., & Kim, K. P. (2017). BIM application and adoption in the UK housing sector. In P. Wu, H. Li, & X. Wang (Eds.), *Integrated Building Information Modelling, UAE*. Bentham Science Publishers.

Redwood, J., Thelning, S., Elmualim, A., & Pullen, S. (2017). The proliferation of ICT and digital technology systems and their influence on the dynamic capabilities of construction firms. *Procedia Engineering*, *180*, 804–811.

Riaz, Z., Arslan, M., & Azhar, S. (2014). CoSMoS: A BIM and wireless sensor based integrated solution for worker safety in confined spaces. *Automation in Construction*, *45*(8), 96–106.

Rowlinson, S. (2004). *Construction safety management systems*. Routledge.

Safe Work Australia. (2017). *Construction*. https://www.safeworkaustralia.gov.au/Construction

Safe Work Australia. (2020). *Australian Work Health and Safety Strategy 2012-2022*. https://www.safeworkaustralia.gov.au/about-us/australian-work-health-and-safety-strategy-2012-2022#

Sahil, A. Q. (2016). *Adoption of building information modeling in developing countries: a phenomenological perspective* (Master's thesis). Colorado State University.

Sebastian, R. (2011). Changing roles of the Clients, architects and contractors through BIM. *Engineering, Construction, and Architectural Management*, *18*(2), 176–187.

Shen, X., & Marks, E. (2016). Near-Miss Information Visualization Tool for BIM. *Journal of Construction Engineering and Management*, *142*(4), 04015100.

Shou, W., Hou, L., Wang, J., & Wang, X. (2015). Case studies of BIM-based dynamic scaffolding design and safety prevention. In *Proceedings of 2nd International Conference on Civil and Building Engineering Informatics (ICCBEI 2015)* (pp. 54-56). https://sbenrc.com.au/app/uploads/2013/10/Publication-3-27-2_CASE-STUDIES-OF-BIM-BASED-DYNAMIC-SCAFFOLDING-DESIGN-AND-SAFETY-PREVENTION.pdf

Smallwood, J. (2019). Clients and Construction health and safety (H&S). In *Proceedings of the Creative Construction Conference 2019* (pp. 656-664). doi:10.3311/CCC2019-090

Smallwood, J. J., & Emuze, F. (2013). Assessing the Contributions of a Public Sector Client to Contractor Health and Safety (H&S) Improvement in South Africa. In *Proceedings of the 2nd International Conference on Infrastructure Development in Africa-ICIDA* (pp. 346-354). https://www.researchgate.net/profile/Joshua_Ayarkwa/publication/312175534_Comparative_residential_benefits/links/5874cd6c08ae329d62201a5a/Comparative-residential-benefits.pdf

Smith, P. (2014). BIM implementation - Global strategies. *Procedia Engineering*, *85*, 482–492.

Statsbygg. (2013). *Statsbygg BIM Manual Version 1.2.1*. http://www.statsbygg.no/Files/ publikasjoner/manualer/StatsbyggBIM-manual-ver1-2-1eng-2013-12-17.pdf

Sulankivi, K., Kähkönen, K., Mäkelä, T., & Kiviniemi, M. (2010). 4D-BIM for Construction safety planning. In *Proceedings of W099-Special Track the 18th CIB World Building Congress* (pp. 117-128). https://www.irbnet.de/daten/iconda/CIB_DC24339.pdf

Tan, T., Chen, K., Xue, F., & Lu, W. (2019). Barriers to Building Information Modeling (BIM) implementation in China's prefabricated Construction: An interpretive structural modeling (ISM) approach. *Journal of Cleaner Production*, *219*, 949–959.

Ullah, K., Lill, I., & Witt, E. (2019). An Overview of BIM Adoption in the Construction Industry: Benefits and Barriers. In *10th Nordic Conference on Construction Economics and Organization* (Vol. 2), Emerald Publishing Limited.

Victorian Digital Asset Strategy (VDAS). (2019). *Victorian Digital Asset Strategy*. http://www.opv.vic.gov.au/Victorian-Chief-Engineer/Victorian-Digital-Asset-Strategy

Wang, S., Wang, W., Wang, K., & Shih, S. (2015). Applying building information modeling to support fire safety management. *Automation in Construction*, *59*, 158–167.

Watson, A. (2010). BIM-a driver for change. *Proceedings of the 13th International Conference on Computing in Civil and Building Engineering*. http://www.engineering.nottingham.ac.uk/icccbe/proceedings/pdf/pf69.pdf

Wells, J., & Hawkins, J. (2010). Increasing 'local content' in infrastructure procurement. Part 2. *Proceedings of the Institution of Civil Engineers – Management, Procurement and Law, 163*(2), 71–75. doi:10.1680/mpal.2010.163.2.71

Wong, A. K. D., Wong, F. K. W., & Nadeem, A. (2010). Attributes of building information modelling implementations in various countries. *Architectural Engineering and Design Management, 6*(4), 288–302.

Yan, H., & Damian, P. (2008). Benefits and barriers of Building Information Modelling. *Proceedings of the 12th International Conference on Computing in Civil and Building Engineering (ICCCBE-2008).* https://pdfs.semanticscholar.org/0594/9fac948a129d821f6c9922d92c8dcbb82408.pdf

Zahrizan, Z., Ali, N. M., Haron, A. T., Marshall-Ponting, A., & Abd Hamid, Z. (2013). Exploring the adoption of Building Information Modelling (BIM) in the Malaysian Construction industry: A qualitative approach. *International Journal of Research in Engineering and Technology, 2*(8), 384–395.

Zhang, L., Wu, X., Ding, L., Skibniewski, M., & Lu, Y. (2016). BIM-Based Risk Identification System in tunnel construction. *Journal of Civil Engineering and Management, 22*(4), 529–539.

Zhang, S., Sulankivi, K., Kiviniemi, M., Romo, I., Eastman, C. M., & Teizer, J. (2015). BIM-based fall hazard identification and prevention in Construction safety planning. *Safety Science, 72*, 31–45.

Zhang, S., Teizer, J., Lee, J. K., Eastman, C. M., & Venugopal, M. (2013). Building information modeling (BIM) and safety: Automatic safety checking of Construction models and schedules. *Automation in Construction, 29*, 183–195.

Zhao, D., & Lucas, J. (2015). Virtual reality simulation for Construction safety promotion. *International Journal of Injury Control and Safety Promotion, 22*(1), 57–67.

Chapter 5
Industry Views on BIM for Site Safety in Hong Kong

Allen Chi Kong Wan
Farspeed Group of Companies, Hong Kong

Sambo Lyson Zulu
https://orcid.org/0000-0002-2724-1192
Leeds Beckett University, UK

Farzad Khosrow-Shahi
Victoria University, Australia

ABSTRACT

The construction industry continues to be a high-risk industry in comparison to others. It is imperative therefore that the industry should continue to seek ways to improve its safety performance. Technological developments and in particular building information modelling is seen to be a promising catalyst for improving safety outcomes in the construction industry. The discussion in this chapter focuses on the perception of professionals in the Hong Kong construction industry of the application of BIM for site safety. Given that sizable capital projects in Hong Kong are now using BIM and site safety continues to be a concern for the local construction stakeholders, it is timely to explore their perceptions of the use of BIM for construction safety. The findings show that there is strong support for using BIM to improve safety outcomes. Participants saw potential use in planning, visualization, and communication as the main reasons for supporting BIM for safety. Issues such as data accuracy, BIM competence, and resource implications were raised as some of the concerns with BIM for safety.

INTRODUCTION

The construction industry remains a high-risk industry (Mohammadi et al., 2018; Shi et al., 2018) and has one of the highest fatality statistics of any other sector (Hafsia et al. 2018). Therefore, it should be the case that health and safety performance should be considered one of the key performance indica-

DOI: 10.4018/978-1-7998-6600-8.ch005

tors. While the construction industry has traditionally measured performance based on time, cost, and quality, others have proposed expanding these KPIs over time. Albert and Ada (2004) considered that KPIs to reflect successful construction projects should cover eight different factors: quality specification, commercial value, environmental impact, user expecation, participant's satisfaction, health, and safety performance time, and cost. Gould and Joyce (2009) pointed out that workers expect site safety management practoces in workplaces in the US, including construction sites.

The construction industry is a key sector of Hong Kong, representing between 4.2% to 5.2% of gross domestic product between 2015 and 2019 (Census and Statistics Department, 2020). Unfortunately, and despite all effects, Hong Kong still has a relatively high number of accidents. During 2010-2014, the construction industry accidents accounted for 25% of the total number of industrial accidents in Hong Kong. In fact, the construction industry accounted for 74% of all industry fatalities in Hong Kong over the same period (Labour Department Hong Kong Government, 2015). The high incident rates are also reflected in many other countries. For example, in the UK, the construction industry contributes about 25 per cent of all fatal injuries in the United Kingdom (HSE, 2018). While many initiatives have been proposed to improve health and safety performance, technological innovations offer great potential as a catalyst for health and safety performance. In particular, building information modelling is seen to offer this potential. Therefore, it is unsurprising to see the push by governments and industry leaders to expand the use of building information modelling, including for health and safety purposes. The UK's HSE recognises BIM's potential to improve health and safety outcomes in the construction industry (HSE, 2018). The development of BIM standards for health and safety [PAS1192-6, BSI, 2018] is another demonstration of the potential for improving health and safety outcomes through BIM. BIM for safety has also been a subject of research. See for example Ganah, A., & John, G. A. (2015); Martinez-Aires et al. (2018); Malekitabar et al. (2016); Alizadehsalehi, et al. (2018); Zhang et al. (2018); Kim et al. (2016); Park et al. (2016) among others.

Hong Kong is not an exception in driving the BIM agenda. To "build a better 2030", the Construction Industry Council in Hong Kong issued the challenge of the construction industry manpower, productivity, safer, greener and collaboration including consideration of adopting Building Information Modelling (BIM) for safety McKinsey & Company (2015). The Hong Kong's Construction Industry Council (2020) draft BIM standard also includes mechanisms for the application of BIM to health and safety management on construction projects. The use of BIM for health and safety is an on-going development in the construction industry. This chapter focuses on the perception of industry practitioners in Hong Kong on the potential for BIM to be a catalyst for improved health and safety performance.

IMPROVING CONSTRUCTION SAFETY

The construction sector has a poor health and safety performance record compared to other major industries (Li et al. 2018). The construction industry in Hong Kong is not an exception with an accident rate of about 20% in 2017, a rate similar to other major economies such as the USA, Japan and the United Kingdom (Shafique, & Rafiq, 2019). Chiang et al. (2017) noted that over 21 years from 1995 to 2015, there was an increase in the ratio of fatal accidents per thousand workers with the gross value of work performed per worker. Therefore, it is essential that efforts need to be hard to improve the construction industry's health and safety performance. There are many different tools to improve OSH, including a combination of mandatory and voluntary measures. Ju (2014) listed the tools to include a mix of prescriptive, performance-based and general duties legislation or administrative measures. Other approaches, such as the use of behavioural-based systems, (Li, 2015) have been considered approaches to improving health and safety.

The OSH issues in Hong Kong are multi-faceted, and there are different approaches to manage site safety. Traditionally, the approach was to impose legal controls following serious accidents, which is often referred to as "command and control" (Zhou, Irizarry, & Li, 2013). More recently, the use of risk assessment (RA) is becoming more widespread and evident in human safety, security, environmental protection, product quality, planning and change management processes. This led to the publication of an international standard, the ISO31000 series, with a cycle loop similar to BS8800 involving "mandate and commitment", "design of a framework for managing risks", "implementation of risk management", "monitoring and review" and "continual improvement" (International Organization for Standardization, 2009). The Hong Kong construction industry's introduction of safety management systems was developed to reduce workplace hazards and injuries to persons and property (Yiu, Sze, and Chan, 2018). Evidence shows that the application of safety management systems positively affects safety outcomes (Kim et al., 2019). Many of Hong Kong's legislation or management practices refer to the process of risk management and RA, suggesting that combining RA and other safety initiatives will help improve OSH performances and lower accident injuries (European Union, 2010; Labour Department, 2014).

In parallel, different countries are adopting "safety management" as a strategy to reduce accidents. Safety management systems (SMS) have been adopted for the UK and Hong Kong to manage and reduce site safety risks. Houssin and Coulibaly (2011) pointed out that the root cause of up to 60% of accidents was "design-related". Before ISO45001 – occupational health and safety management system, Hong Kong enacted Safety Management Regulation in 1997, which is now regarded as the critical approach for managing construction safety in Hong Kong. Tam and Fung (1998) found that the best safety management practices to reduce safety risks for Hong Kong's construction industries include safety training, direct workforce, accident investigation, and safety incentive scheme. Cheng, Ryan, and Kelly (2012) revealed that three useful SMS elements are safe inspection, safety training and safe work practices for Hong Kong. Li and Guldenmund (2018), in their review paper of safety management systems, grouped the purposes of safety management systems into two, including control and compliance. They argue that control (of loss, accidents, hazards, and risk) is the primary purpose of safety management systems. Compliance with standards, laws and regulations is another motivation for implementing a safety management system. While there is evidence that some companies may want to use a 'tick-box' approach, company processes and outputs must be compliant, especially when what is at stake is workers lives.

However, Hong Kong's construction industry's safety performance remains a concern and to "build a better 2030", the Construction Industry Council in Hong Kong issued the challenge of the construction industry for manpower, productivity, safer, greener and collaboration. In terms of the need for safer construction environment, McKinsey & Company (2015) recommended the following initiatives:

- Promote "design for safety" via consideration of the Construction Design Management regulation in UK and safety award for stakeholders;
- Promote safety ownership by extending the Pay for Safety scheme to include subcontractors and even frontline workers, and to highlight the different safety responsibilities among multi- tier contractors; and
- Adopting new safety-improving new tools such as Building Information Modelling (BIM) safety from Finland or new construction methods from the US into a safety performance scheme.

Therefore, it is evident that the approach to managing OSH in Hong Kong is mostly associated with RA and SMS. The need to improve health and safety management in Hong Kong is acknowledged. It is, therefore, unsurprising to see proposals of how health and safety management can be improved. See, for example, Yiu et al. (20018); Niu et al. (2019); Getuli et al. (2020); Edirisinghe & Woo (2020) and; Natephra & Motamedi (2019). The integration of BIM processes into health and safety management is also a subject of many investigations. See, for example, Dong et al. (2018) and Niu et al. (2019). If BIM is to be promoted as a tool to enhance health and safety performance, it should be seen to (at least) demonstrate application these two areas in managing constriction site safety.

REVIEW OF BIM APPLICATIONS IN HEALTH AND SAFETY

Like many other industries, such as manufacturing and aerospace, the AEC is looking for ways to improve productivity while keeping cost, time, constructability, environmental concern, and site safety in check. There is evidence to indicate BIM can be a trend for the AEC sector as there are already BIM mandates from the USA, the UK, Singapore and Hong Kong.

British Standards Institution (2013) identified BIM as a digital approach to manage the design, the construction or the operation of a building or infrastructure. Stanford University's Center for Integrated Facility Engineering recognised the BIM advantage to involve seven categories: communication, facility performance, cost, schedule, project delivery, knowledge management, and safety (Kam, Fischer, Rinella, Mak, & Oldfield, 2014).

Royal Institute of British Architects (2012) considered BIM can be overlaid to its "Plan of Work including 4D for time, 5D for cost and 6D for facility management. This study considered the application of BIM to other dimensions such as safety, referred to as 7D. Selected examples are compliance checking (New York City, 2013), simulation of the construction work process and site safety planning (Kiviniemi et al., 2011), worker safety training (Balfour Beatty Construction, 2015) and post-construction and facility management (Mordue & Finch, 2014). Other BIM application areas which would also impact on health and safety performance include logistic planning (Sulankivi, 2010); visualisation and sequencing (Mordue & Finch, 2014); waste reduction (Geldard, 2017); improved communication (Sulankivi, 2010); and schedule progress and monitoring (Mordue & Finch, 2014). BIM for safety research has focused on issues such as planning, identification and virtual simulation, sensor and warning device, and safety management (Wan, Platten, & Briggs, 2013). The integration of BIM with other digital tools also provides a potential area where BIM can impact health and safety performance. Tools such as virtual reality (Swallow & Zulu, 2020), 4D simulation (Jin et al., 2019), drones (Tatum & Liu, 2017) can help improve BIM base safety management. The use of BIM for early warning through hazard identification is also demonstrated as an essential application for using BIM for health and safety management (Abed et al., 2019). BIM's integration with other technical tools such as radio frequency identification technology (RFID), global positioning system, wireless sensor networks, accident investigation from 3D models, animations and scenarios replay, simulation of the construction work process and site safety planning (Kiviniemi et al., 2011) and worker safety training (Balfour Beatty Construction, 2015) can also be used to help health and safety planning and management on construction projects. The tools and process above provide an opportunity to identify potential risks on projects. Thus BIM can be a useful risk identification platform in health and safety (Malekitabar et al., 2016).

INDUSTRY PERCEPTIONS OF BIM FOR SITE SAFETY

While acknowledging the potential for using BIM to improve site safety performance, the authors conducted a focus-group based study to test industry stakeholders' perception of BIM to improve site safety in Hong Kong. Below is a summary of the approach taken, and a discussion of the results.

Methodology: Use of Focus Groups

Focus groups are ideal for collecting in-depth qualitative data to inform an inquiry. Morgan (2012) introduced focus groups to follow a pattern; commonly the focus group meetings start with the facilitator outlining the topic and the purpose of the discussion to the respondents; the facilitator or the moderator will ensure the means of data collection are explained and carried out; then the facilitator will begin to lead participants through the prepared questions and discussions. Freeman (2006) pointed out that the objective of focus groups is to encourage self-discussion among members to capture the group dynamics, even allowing the members to question one another to seek clarification from group/individual statements. Morgan (1993) and Sim (1998) identified three advantages: the focus group not only allows each member to express his/her own view, he/she can also interact with others; in a non-threatening setting, there is little distinction between the management and frontline staff as everyone is allowed to express their ideas freely; and the views from different professions and trades can also be given equal consideration in the focus group.

The focus group meeting allows the generation of data and learning as well as the degree of consensus on a topic from the targeted groups. This is important when different opinions exist between participants and stakeholders. The decision to implement BIM for safety is complicated as it involves various human and contract motivations as well as organisation priorities. Although the focus group study is more an exploratory approach than a number-based one, with adequate statistical sampling, the data can be generalised to a specific population. Since it may not be easy to find ways to talk about BIM implementation, the focus group is a useful tool for different stakeholders to discuss their opinions.

To maintain validity, six different focus groups, representing different spectrums of construction stakeholders, were conducted. More importantly, each member of focus groups or expert had relevant trade qualification. The working experience of respondents ensured each having adequate knowledge in construction, BIM and site safety. Given that over 90% of the working population are Cantonese-speaking in Hong Kong, Cantonese was used in focus-groups meetings, and voice recordings were taken. Besides, notes were taken on the spot. The researchers analysed all voice recordings to translate them consistently into English and validated by third parties. The specific design of the focus groups including the trial, the objective, the sampling and the participants were explained by (S. C. Beyea & Nicoll, 2000; Sim, 1998). All the transcripts were subject to content analysis providing another level of data for discussion.

Construction Stakeholders in Hong Kong

The AEC sector covers various stakeholders, including the client, the designer, the engineer, the contractor and the manufacturer (Saunders et al., 2013). To investigate the potential of using BIM for safety, different parties such as the client (FMI, 2007) and the contractor (Ku & Taiebat, 2011) should be included. This is why six different construction focus groups in Hong Kong were to reflect an industry-wide representation and the potential usage of BIM for site safety is viewed through these construction stakeholders' perceptions, attitudes and views. In theory, stakeholders within the AEC include any combination of:

- Government and professional groups
- Clients and consultants
- Main Contractors, subcontractors and specialists
- Unions, workers and industrial relations

Saunders et al. (2013) believe that the verification of a new concept or tool for the construction industry should involve different players such as the client, the designer, the engineer, the contractor, the manufacturer and other relevant parties. Participants in the focus group represented critical parties to the entire construction supply chain and composed of representation from major construction stakeholders, including clients (most with engineering or construction professional background), contractors (both civil and housing-related), a specialist contractor (for lifting works), construction professionals (inspector grade) and safety professionals; arguably and as a pack, this represents the supply chain of the entire construction sector. Given that this research focused on the construction industry in Hong Kong; all these organisations are based in Hong Kong. Each association from the particular focus group is a good representation of that trade, including the largest safety association, public works representation, contractors from the Housing Department where they construct buildings for about a third of Hong Kong's population, a major lifting specialist, a major civil contractor, and a well-known construction professional association.

Details of the organisations and the focus groups meetings are documented by (Wan, 2017) and included representatives from Hong Kong Federation of Occupational Safety and Health Association [HKFOSHA], representation from the Building New Works Group, a major lift specialist, safety experts from the Department of Works Bureau, a major civil engineering contractor and members of a construction professional association. HKFOSHA is the largest representation of registered safety officers in Hong Kong, representing over 50% of the registered safety personnel in Hong Kong. The Housing Department is the Hong Kong Housing Authority's executive arm, a statutory body, mandated to establish almost all public housing programmes in Hong Kong and the list-building-new works-group two contractors are approved contractors to provide building works for the Housing Department. The major lifting specialist was Farspeed Contractors Limited, which provides most transportation and lifting works for the elevator/escalator trade in Hong Kong. It is involved in almost all urban planning and infrastructure development, Hong Kong's Works Departments includes Buildings Department, Lands Department, Lands Registry, Planning Department, Architectural Services Department, Civil Engineering and Development Department, Drainage Services Department, Electrical and Mechanical.

Services Department, Water Supplies Department. The major civil contractor was Hsin Chong, a listed company in Hong Kong engaged in building, civil engineering, electrical and mechanical engineering, and railway projects. The construction professional association was Hong Kong Institution of Clerks of Works where members involve in superintending of construction works in Hong Kong.

Table 1. Summarised reasons for supporting BIM for safety by focus groups in Hong Kong

Focus groups	Percentage of supporting BIM for safety	Most frequent reason	2nd most frequent reason	3rd most frequent reason
HKFOSHA	100%	Process	Communication	Accident
New Works contractors of the Housing Department	80%	Simulate, visualise Or rehearsal	Training	Planning
Lifting specialist	50%	Assist lifting plan	-	-
Works departments at the Development Bureau	93%	Data provision or coordination	Visualisation or communication	Planning
Major civil contractor	100%	Planning	Visualisation or communication	Assist elements of the safety management system
Construction professional group	75%	Display or explain information	Visualisation	-
Overall/Average	83%	Visualisation or communication	Planning	Data related

Perceptions of Using BIM For Safety

In this chapter, only the overall result and analysis is provided. Discussion of each focus group result is reported in Wan (2017). All six construction relevant focus groups supported BIM for safety. Since this result is based on the conclusions from focus groups in Hong Kong, the BIM for safety direction is most suitable for the construction industry in Hong Kong. In summary, two central research questions were addressed: one concerned about the potential of using BIM for safety with reasons of why and why not as well as which safety management system element(s) has the most BIM application potential; the second was about the potential of using BIM for RA on lifting.

A total of 62 participants in the six different focus groups attended the data collection exercise, covering all the major AEC stakeholders, and all had a background closely related to Hong Kong's construction industry. Other than a few retirees, all were still active in the AEC sector; 95% of the respondents were male. The overall average (from the six different focus groups) supporting BIM for safety was 83%, indicating a significant support level. The most substantial support (100%) came from the safety professional organisation and the major civil contractor. Understandably, safety personnel continue to seek ways to improve safety performance, and they find this potential in BIM. The civil contractor worked for a multi-national railway client with extensive safety management requirements, partly explaining why they supported tools to improve safety further.

The least BIM-for-safety supportive focus group was the lifting specialists where members were mostly operatives with little experience in SMSs, advanced technologies, and constructing lifting plans. Overall support for using BIM for safety is substantial as all groups had at least 75% or more support, except for the lifting specialist group (50%).

It should be noted that within each focus group, there were different representations; for example, different safety associations from various construction organisations came to the HKFOSHA focus group meeting, while different building contractors from different construction projects came to the Housing Department focus group meeting. This is significant because such combined batch of different construction stakeholders represents a heterogeneous mix of Hong Kong's construction industry and trade, which can form a credible basis to generalise the significant findings.

This translates to two key points. First, one can argue that the construction industry in Hong Kong (covering public clients and building and civil contractors) is ready for some form of BIM implementation. This is not a surprise as there are BIM mandates globally already including in the US, the UK, Denmark, Singapore, South Korea, and Hong Kong, mostly for new sizable and public projects (Construction Industry Council Hong Kong, 2014). The focus groups' findings are also consistent with the report that 92% believed the UK organisations would adopt BIM in three years (Royal Institute of British Architects Enterprises Limited, 2015).

The second key remark is that this "readiness" can be further applied to managing safety. This can be attributed to the fact that increasingly, safety is part of the construction key performance indicators (Bealtham, Anumba, & Thorpe, 2004). From the literature, in addition to clash detection, cost estimation, construction planning, and schedule control (Eastman, 2008), there are BIM adoptions for safety including the use of safety compliance checking (New York City, 2013), the simulation of construction work process, and site safety planning (Kiviniemi et al., 2011), for worker safety training (Balfour Beatty Construction), and for post-construction and facility management (Mordue & Finch, 2014). The results here indicate that the construction industry in Hong Kong is ready for using BIM to aid the safety management process. This is because all six different construction stakeholders supported BIM for safety in no small extent. How can BIM contribute to safety? Table 1 lists the summarised reasons for using BIM for safety, namely visualisation or communication, planning and data accuracy. There may be overlaps among these reasons, but they align with the BIM literature (Kiviniemi et al., 2011; Miettinen & Paavola, 2014; Ning & London, 2010). In addition to clients and designers, contractors will also consider BIM for safety as safety rests mostly with contractors; more importantly, safety risk, in theory, is the highest during the construction phase where contractors have the most input. McGraw Hill Construction (2014) revealed that contractors believe top BIM benefits are errors and omissions reduction and collaboration with the client/designer. Looking at the major reasons for supporting BIM for safety, the researchers propose the key is for BIM to hold accurate and updated "safety data" as information or data is critical for managing safety.

Today, we live in an information-packed world. Construction is also about using information and data for budgets, materials, labour, processes, programs and communication, involving clients, designers, contractors, site personnel and more. Similarly, safety is information-dependent covering accident trends, training schedules, incident causes, control measures, etc. The ability of BIM to integrate and display information is so significant that it will self-create more interests, if not already within the AEC sector (Anumba & Wang, 2012). Jung and Joo (2011) argued that the recent BIM upgrades could push towards higher dimensions, such as 5D-cost, 6D–procurement, 7D- safety, and even beyond to nth dimensions. Assuming BIM is a better methodology for information, it is also suitable for safety too.

How information is recognised, inputted, stored, accessed, and retrieved can be a common concern. From Table 2, the consolidated concerns of BIM implementation of safety involved BIM or computer skills and knowledge, data update, quality or operation, and resources implications. The researchers argue that these can be grouped into competency and cost implication as BIM knowledge and data operation are closely related. In the UK, the barriers for more BIM adoption are the lack of in-house expertise, the lack of training, no client demand, and cost (Royal Institute of British Architects Enterprises Limited, 2015). In Hong Kong, contractors' concerns for BIM operation included "Will the contractor be able to obtain a model for further development?" and "what are the risks of owning the data?" The findings of the focus group meetings are consistent with the literature.

Table 2. Summarised concerns about using BIM for safety by focus groups in Hong Kong

Focus groups	Most frequent reason	2nd most frequent reason	3rd most frequent reason
HKFOSHA	BIM/Computer skills	Updating/operating BIM	-
New Works contractors of the Housing Department	Resources	Data updating or quality	Knowledge
Lifting specialist	BIM skills	-	-
Works departments at Development Bureau	Data accuracy	Resource implication	For new projects or not for minor task
Major civil contractor	-	-	-
Construction professional group	-	-	-
Overall/Average	BIM or computer skills and knowledge	Data update, quality or operation	Resource implication

Attempting to overcome the "competency" barrier, Hong Kong is calling to develop BIM standards involving a project execution plan, BIM promotion to industry-wide stakeholders, and BIM training for both construction- and computer-based courses (Construction Industry Council Hong Kong, 2014). In the UK, there was a roadmap to deliver BIM skills to education and training courses (Farzad & Yusuf, 2012). R. Sacks and Pikas (2013) went further in highlighting a BIM framework for construction-related university programmes. Therefore, in time and with better BIM products, standards, specifications, and experience, the BIM competency and the confidence of handling BIM data input, verification, modification, and security can be overcome. The cost implication is a problematic concern to mitigate. On the one hand, there is no standard measurement for "return on investment" yet. More importantly, the return can be positive rather than negative, as McGraw Hill Construction (2014) found the positive gain from BIM can range from 10 to 25%. If BIM implementation is mandated, then the BIM cost will be a part of compliance and merely a cost of doing a specific kind of "construction".

In addition to OSH legislation and economic incentive, a more recent approach to using occupational safety and health management system to improve safety performance has been mandated in Singapore and Hong Kong (Tam & Fung, 1998; Teo & Ling, 2006; Von Ahsen, 2014). There are different SMS standards (British Standards Institution 2000; Standards Australia / Standards New Zealand, 2001). SMS in Hong Kong has fourteen elements (Labour Department, 2002; Labour Department Hong Kong, 2000). If BIM for safety is to be test-run before full implementation, then what process or safety aspects should be conducted first? For countries, companies and sites adopting SMS, the question becomes 'Which SMS element(s) have the most BIM potential'. Table 3 indicates that the summarised BIM potential for SMS elements is safety training, the job hazard analysis (often referred to as risk assessment), and the hazard control programme. The logical findings of this study can serve as the basis for a more industry representative future study also stands here; in order words, the concluding remarks of safety training, job hazard analysis, and hazard control programme can be the foundation of a more quantitative study for which SMS element(s) are most suitable for BIM-for- safety implementation in Hong Kong's construction industry.

Table 3. Summarised SMS elements with the most BIM potential for the construction industry in Hong Kong

Focus groups	Most frequent reason	2nd most frequent reason	3rd most frequent reason
HKFOSHA	Job hazard analysis	Safety training	Hazard control programme
New Works contractors of the Housing Department	-	-	-
Lifting specialist	Safety training	-	-
Works departments at the Development Bureau	Hazard control programme	Job hazard analysis	Accident/incident investigation
Major civil contractor	Safety training	Safety training	Job hazard analysis
Construction professional group	Safety training	Half of all elements	-
Overall/Average	BIM or computer skills and knowledge	Job hazard analysis	Hazard control programme

For construction sites running pay-for-safety schemes in Hong Kong, safety training can be profitable (Choi, Chan, & Chan, 2012). The effectiveness of safety training can be hard to gauge. Sacks, Perlman, and Barak (2013) found that virtual reality (which is BIM relevant) in construction safety training is more effective. It is reasonable to assume that once OSH risks are identified, it is vital to implement control measures to mitigate these risks; this explains why participants selected "hazard control program" as one of the most suitable SMS elements for BIM adoption.

Perceptions of Using BIM For Risk Assessment

Those involved in managing OSH often discuss hazard identification and RA before deciding what control measures should be carried out (International Labour Office - Occupational Safety and Health Branch, 2014). Construction or installation of a project involves many different processes and trades. The researchers selected a familiar, yet hazardous operation - lifting as the example in focus group meetings. On a global scale, there was a significant number of severe lifting accidents in the US (Shepherd, Kahler, & Cross, 2000), in the Netherlands (Aneziris et al., 2008), in Korea (Shin, 2015), in Japan, and Hong Kong (Tam & Fung, 2011).

According to Table 4, there was strong overall support (85%) for using BIM to do the RA with the least support from the Works Departments focus group (50%). Not counting the Works Departments group, the support was at least 78% with three different groups scoring 100%. One explanation is that those working at Works Departments are civil servants, likely to be involved in design or coordination rather than construction activities. Traditionally contractors are perceived to have total safety responsibilities as well as the duty to conduct RA. The cited reasons for supporting BIM for RA were communication, data accuracy, and the ability to expand. The "ability to expand" can be the important reason why many are looking to become involved in BIM as, in theory, there is no boundary for BIM or information. Smith (2009) insisted information and data are important throughout building design and construction and are used by different parties, including the client, the architect, the consultant, the engineer, and the contractor. Jung and Joo (2011) argued this BIM advantage could provide a multi-dimensional platform for 5D, 6D to nth dimensions construction process.

In practice, there is still concern about BIM implementation. In Table 5, the focus groups identified significant concerns to be data accuracy, resources implication, and site conditions. Again, the findings of "what are the reasons and what are the concerns of using BIM to do risk assessment" can be the basis for another significant "quantitative" study on whether or not BIM can be useful for RA for lifting and other relevant construction processes.

Lifting operations are common construction tasks at the site, which may require many lifting visualisations, leading to possible cost implications and implementation bottlenecks. Another valid concern was related to data accuracy as site conditions change quickly and existing BIM may not contain either weather or human factors. The concerns about the legal status of BIM, data security, and provision of hardcopy of RA were also mentioned. Firstly, these are operation issues involving objects, processes and the link to visualisation using BIM software. Currently, about 60% of BIM objects come from in-house creation, project addition, manufacturer or Computer Aid Design (CAD) package (Royal Institute of British Architects Enterprises Limited, 2015). Once the object databases are readily available within the industry, trade or association, the visualisations of these processes are an easy addition to standard computer usage.

There are different BIM training initiatives at various levels in Hong Kong and the UK. Whether or not BIM is implemented at sites, there will always be changes at sites. When the BIM competency framework becomes a norm, it will be a standard practice for the construction industry to adopt BIM in the same way as the CAD evolved. Different organisations set up standardised BIMs; examples are the specification of Construction Operation Building information exchange (Nisbet, 2012), the common data environment (British Standards Institution, 2013), the security arrangement (British Standards Institution, 2015), and regular BIM practices (Construction Industry Council Hong Kong, 2014). More importantly, mobile technologies' verification of changing site conditions can be enhanced (Anumba & Wang, 2012) with laser scanning and live camera/sensor setup (McGraw Hill Construction, 2014). With time, these advance and other cloud systems will assist in resolving the data concern for BIM. The researchers regard the cost implication argument of BIM is similar to the argument of spending more to clean up our environment. The RA process is a part of the SMS, impacting safety performance.

Assuming BIM offers an overall safety improvement, then is it justified for the AEC sector to spend that extra amount to save a life. Interestingly, one of the latest BIM surveys revealed that BIM makes a positive return on investment (including safety consideration) (McGraw Hill Construction, 2014). Guo, Li, Chan, and Skitmore (2012) confirmed that BIM is useful for lifting. There are other examples including the use of BIM for scaffolding (Kim and Teizer, 2014), use of using BIM-Based sensor technologies to improve construction safety (Sidani et al., 2018) as well as the integration with safety management (Benjaoran & Bhokha, 2010). In general, the lifting visualisation can add benefits to RA's process in lifting operations; this is because the visualisation enhances communication among project personnel and identifies relevant hazards. Moreover, the lifting visualisation data can inform a lifting plan, checklist or even development of a lifting game for counterchecking or practice. In theory, the BIM-data setup can be a part of the project submission or process permit application.

Table 4. Summarised reasons for using BIM for risk assessment by focus groups in Hong Kong

Focus groups	Percentage of supporting BIM for safety	Most frequent reason	2nd most frequent reason	3rd most frequent reason
HKFOSHA	100%	Complicated lifting task	Checking or updating lifting	-
New Works contractors of the Housing Department	100%	Facilitate risk management	planning	Training or communication
Lifting specialist	100%	Communication	Data accuracy	-
Works departments at Development Bureau	50%	Ability to expand	Presentation or visualisation	-
Major civil contractor	80%	Visualisation or communication	Apply to large project or the ability to expand	-
Construction professional group	78%	Provide data	communication	-
Overall/Average	85%	Communication related	Data-dependent	Ability to expand

Table 5. Summarised concerns about using BIM for risk assessment by focus groups in Hong Kong

Focus groups	Most frequent reason	2nd most frequent reason	3rd most frequent reason
HKFOSHA	Data details	Time or cost implication	updating
New Works contractors of Housing Department	Scope of quality of data	Cost implication	Site change
Lifting specialist	Data accuracy	-	-
Works departments at Development Bureau	Availability of basic BIM or data	BIM competence	-
Major civil contractor	Reflection of actual conditions	Resource implications	-
Construction professional group	Data-dependent	More research and development	-
Overall/Average	Data dependant	Resource implication	Site conditions

In summary, there is evident support for using BIM for safety in the construction industry in Hong Kong. The support was 83% to adopt BIM for safety and 85% to use BIM for RA in lifting from the six different construction focus groups. Although the investigation was limited in size, the coverage was extensive for the construction stakeholders, totalling six different groups and 62 active construction personnel. This level of support is consistent with other BIM surveys and reports. While the main support of using BIM for safety included visualisation or communication, planning and data quality, the key concerns were BIM or computer skill and knowledge, data updating, and resources implication. Both support and limitation were relevant and in line with the current literature. The reported BIM advantages are relevant for communication and visualisation of the process sequence, which can be useful for coordination, virtual rehearsal, and training. These outweigh the BIM limitations. With time, higher BIM competency and a wider availability of BIM databases will help resolve the concerns.

If BIM was to be added to the SMS set up in Hong Kong, the focus groups considered the most suitable elements would be safety training, job hazard analysis, and hazard control program. Similar to the finding of BIM for safety, there is clear support (85%) for using BIM to do lifting RA. The reasoning and limitation for using BIM for lifting RA followed the similar pattern in "BIM for safety". The primary reasons for using BIM to do RA were communication, data accuracy, and the ability to expand. On the other hand, the concerns with BIM implementation involved data, resources implication, and site conditions. These answers to the question of BIM implementation are consistent with the existing BIM understanding.

The key potential of BIM for safety was investigated and analysed in terms of six different construction stakeholders in Hong Kong, and findings reported above. It is significant to point out that while the qualitative nature of this study may limit the findings to be generalised, the evidence-based conclusion is vital for the starting point of a broader industry study (maybe a quantitative nature) to investigate the potential for BIM implementation in Hong Kong. The argument is powerful for a BIM for safety follow-up study, for which SMS element(s) is the most suitable and for whether or not BIM is useful for conducting RA of lifting.

CONCLUSION

The paper considered the industry views of the application of BIM methodologies to improve construction site safety. Evidence and practice support the application of BIM methodologies for construction safety. It is therefore crucial that the industry should be seen to adopt BIM for safety if we are to see a continued decline in health and safety incidences in the construction industry. This is particularly the case considering that the construction industry is one of the high-risk industries when the number of fatal injuries is examined. The industry should, therefore, welcome initiatives that will help it improve health and safety performance.

We welcome the many studies that demonstrate whether theoretically or with empirical evidence, the application of BIM methodologies to the management of health and safety on construction projects. Like many other countries, Hong Kong is pushing the BIM agenda, including its application to health and safety management. Therefore, industry professionals' perception on the application of BIM to health and safety management was the focus of the discussion and the research carried out on the Hong Kong construction industry.

Studies indicate that the most common strategy of managing OSH often involves a combination of SMS and/or RA. Therefore, the potential of using BIM for safety was reflected under these two elements. Participants for the study were drawn from different industry sectors, including HKFOSHA, New Works contractors of Housing Department, Lifting specialist, Works departments at Development Bureau, Major civil contractor and Construction professional group. The findings demonstrated strong support for using BIM for site safety with an average overall support of 83%, indicating a significant level of support for using BIM for safety in Hong Kong's construction industry. All key stakeholder groups except for the lifting specialist group had at least 75% support. The findings show that there is strong support for using BIM to improve safety outcomes. Participants saw potential use in planning, visualisation and communication as the main reasons for supporting BIM for safety. Issues such as data accuracy, BIM competence and resource implications were raised as some of the concerns with BIM for safety. There was also support for using BIM for risk analysis (RA). In Hong Kong, the 14-elements SMS is a legal mandate for sizable construction companies and sites. The most suitable SMS elements for the BIM implementation were safety training, the job hazard analysis (RA related), and the hazard control programme. The second factor to gauge the BIM for safety potential was related to BIM's ability to assist RAs. The focus was on the RA for lifting. The overall support for using BIM to do RA was at 85%, suggesting a substantial level for using BIM to conduct RA of lifting.

The findings, therefore, demonstrate support for BIM for safety in Hong Kong. The industry views show that this support is high with generic support amongst industry participants for using BIM for safety at 83% and the support of using BIM for Risk Assessment of lifting at 85%. With this level of support, we hope to see an increase in investment in using BIM methodologies for a wide range of construction processes, including health and safety management.

REFERENCES

Abed, H. R., Hatem, W., & Jasim, N. (2019). Adopting BIM Technology in Fall Prevention Plans. *Civil Engineering Journal*, *5*(10), 227–228.

AlBahnassi & Hammad. (2012). Near Real-Time Motion Planning and Simulation of Cranes in Construction: Framework and System Architecture. *Journal of Computing in Civil Engineering*, (Jan/Feb), 54–63.

Albert, P. C. C., & Ada, P. L. C. (2004). Key performance indicators for measuring construction success. *Benchmarking*, *11*(2), 203–221. doi:10.1108/14635770410532624

Alizadehsalehi, S., Yitmen, I., Celik, T., & Arditi, D. (2018). The effectiveness of an integrated BIM/UAV model in managing safety on construction sites. *International Journal of Occupational Safety and Ergonomics*, 1–16. doi:10.1080/10803548.2018.1504487 PMID:30043680

Aneziris, O., Papazoglou, I., Mud, M., Damen, M., Kuiper, J., Baksteen, H., & Oh, J. (2008). Towards risk assessment for crane activities. *Safety Science*, *46*(6), 872–884. doi:10.1016/j.ssci.2007.11.012

Anumba, C. J., & Wang, X. (2012). *Mobile and Pervasive Computing in Construction*. John Wiley & Sons. doi:10.1002/9781118422281

Balfour Beatty Construction. (2015). *BIM for Zero Harm*. Author.

Bealtham, S., Anumba, C. J., & Thorpe, T. (2004). KPIs: A critical appraisal of their use in construction. *Benchmarking*, *11*(1), 93–117. doi:10.1108/14635770410520320

Benjaoran, V., & Bhokha, S. (2010). An integrated safety management with construction management using 4D CAD model. *Safety Science*, *48*(3), 395–403. doi:10.1016/j.ssci.2009.09.009

Beyea, S. C., & Nicoll, L. H. (2000). Learn more using focus groups. *AORN Journal, 71*(4), 897-900. doi:10.1016/S0001-2092(06)62276-X

British Standards Institution. (2000). *Occupational health and safety management systems - guidelines for the implementation of OHSAS 18001*.

British Standards Institution. (2013). PAS 1192-2: *Specification for information management for the capital/delivery phase of construction projects using building information modelling.*

British Standards Institution. (2015). PAS 1192-5: *Specification for security-minded building information modelling, digital built environments and smart asset management.*

Census and Statistics Department, The Government of the Hong Kong Special Administrative Region. (2020). *Table 036: Gross Domestic Product (GDP) by major economic activity - percentage contribution to GDP at basic prices.* https://www.censtatd.gov.hk/hkstat/sub/sp250.jsp?tableID=036&ID=0&productType=8

Chan, L. M., & Yu, S. S. W. (2012). Managing the stress of Hong Kong expatriate construction professionals in mainland China: Focus group study exploring individual coping strategies and organisational support (Author abstract). *Journal of Construction Engineering and Management, 138*(10), 1150–1160. doi:CO.1943-7862.0000533 doi:10.1061/(ASCE)

Cheng, R., Ryan, N., & Kelly, S. (2012). Exploring the perceived influence of safety management practices on project performance in the construction industry. *Safety Science*, *50*(2), 363–369. doi:10.1016/j.ssci.2011.09.016

Chiang, Y. H., Wong, F. K. W., & Liang, S. (2017). Fatal construction accidents in Hong Kong. *Journal of Construction Engineering and Management*, *144*(3), 04017121. doi:10.1061/(ASCE)CO.1943-7862.0001433

Choi, T. N. Y., Chan, D. W. M., & Chan, A. P. C. (2012). Potential difficulties in applying the Pay for Safety Scheme (PFSS) in construction projects. *Accident; Analysis and Prevention*, *48*, 145–155. doi:10.1016/j.aap.2011.04.015 PMID:22664678

Chua, D. K. H., Kog, Y. C., & Loh, P. K. (1999). Critical success factors for different project objectives. *Journal of Construction Engineering and Management*, *125*(3), 142–150. doi:10.1061/(ASCE)0733-9364(1999)125:3(142)

Construction Industry Council Hong Kong. (2014). *Roadmap for BIM strategic implementation in Hong Kong's Construction Industry.* Author.

Construction Industry Council UK. (2013). *Growth through BIM.* Author.

Construction Industry Review Committee Hong Kong Government. (2001). Construction for Excellence Report of Construction Industry Review Committee. *Construction Management and Economics*, *19*(7), 741–751. doi:10.1080/01446190110067037

Dong, S., Li, H., & Yin, Q. (2018). Building information modeling in combination with real-time location systems and sensors for safety performance enhancement. *Safety Science*, *102*, 226–237. doi:10.1016/j.ssci.2017.10.011

Drexler, J. A. Jr, & Larson, E. W. (2000). Partnering: Why project owner-contractor relationships change (Statistical Data Included). *Journal of Construction Engineering and Management*, *126*(4), 293–297. doi:10.1061/(ASCE)0733-9364(2000)126:4(293)

Eastman, C., Teicholz, P., Sacks, R., & Liston, K. (2008). *BIM handbook: a guide to building information modeling for owners, managers, designers, engineers, and contractors*. John Wiley & Sons. doi:10.1002/9780470261309

Edirisinghe, R., & Woo, J. (2020). *BIM-based performance monitoring for smart building management*. Facilities.

Edmunds, H. (1999). *The Focus group research handbook*. NTC Business Books.

European Union. (2010). *Health and safety at work in Europe (1999–2007) A statistical portrait*. Retrieved from http://ec.europa.eu/eurostat/documents/3217494/5718905/KS-31-09-290- EN.PDF/88eef9f7-c229-40deb1cd-43126bc4a946

Farzad, K., & Yusuf, A. (2012). Roadmap for implementation of BIM in the UK construction industry. *Engineering, Construction, and Architectural Management*, *19*(6), 610–635. doi:10.1108/09699981211277531

Fellows & Liu. (2008). *Research methods for construction* (3rd ed.). Wiley Blackwell.

FMI. (2007). *Eighth Annual Survey of Owners*. FMI.

Freeman. (2006). 'Best practice' in focus group research: making sense of different views. *Journal of Advanced Nursing, 56*(5), 491-497. doi:10.1111/j.1365-2648.2006.04043.x

Ganah, A., & John, G. A. (2015). Integrating building information modeling and health and safety for onsite construction. *Safety and Health at Work*, *6*(1), 39–45. doi:10.1016/j.shaw.2014.10.002 PMID:25830069

Getuli, V., Capone, P., Bruttini, A., & Isaac, S. (2020). BIM-based immersive Virtual Reality for construction workspace planning: A safety-oriented approach. *Automation in Construction*, *114*, 103160. doi:10.1016/j.autcon.2020.103160

Gould, F. E., & Joyce, N. (2009). *Construction project management* (3rd ed.). Pearson Prentice Hall.

Grbich, C. (2013). *Qualitative Data Analysis - An Introduction*. SAGE Publications Limited.

Gu, N., & London, K. (2010). Understanding and facilitating BIM adoption in the AEC industry. *Automation in Construction, 19*(8), 988–999. doi:10.1016/j.autcon.2010.09.002

Guo, L., Li, H., Chan, G., & Skitmore, M. (2012). Using game technologies to improve the safety of construction plant operations. *Accident; Analysis and Prevention*, *48*, 204–213. doi:10.1016/j.aap.2011.06.002 PMID:22664683

Hafsia, M., Monacelli, E., & Martin, H. (2018). Virtual Reality Simulator for Construction Workers. *Proceedings of the virtual Reality Conference – Labal Virtual*. Available from: https://dl.acm.org/citation.cfm?id=3234298

Hardin, B., & McCool, D. (2015). *BIM and construction management: proven tools, methods, and workflows*. John Wiley & Sons.

H.K.G. Labour Department. (2002). *Code of Practice on Safety Management.* Author.

H.K.G. Labour Department. (2014). *Occupational Safety and Health Statistics, 2013.* Author.

Houssin, R., & Coulibaly, A. (2011). An approach to solve contradiction problems for the safety integration in innovative design process. *Computers in Industry*, *62*(4), 398–406. doi:10.1016/j.compind.2010.12.009

HSE. (2018b). *Improving Health and Safety Outcomes in Construction. Making the Case of Building Information Modelling (BIM).* Health and Safety Executive, RR 235. Available from: <https://www.hse.gov.uk/research/rrpdf/rr235.pdf>

International Labour Office - Occupational Safety and Health Branch. (2014). *A 5 step guide for employers, workers and their representatives on conducting workplace risk assessments.* Author.

International Organization for Standardization. (2009). *ISO 31000 Risk Management - Principles and Guidelines.* Author.

Ju, C. R. S., & Rowlinson, S. (2014). Institutional determinants of construction safety management strategies of contractors in Hong Kong. *Construction Management and Economics*, *32*(7-8), 725–736. doi:10.1080/01446193.2014.909048

Jung, Y., & Joo, M. (2011). Building information modelling (BIM) framework for practical implementation. *Automation in Construction*, *20*(2), 126–133. doi:10.1016/j.autcon.2010.09.010

Kam, C., Fischer, M., Rinella, T., Mak, D., & Oldfield, J. (2014). Realising the Promise of BIM in Hong Kong's Construction Industry. *Journal of Hong Kong's Construction Industry*, (May), 29–33.

Kidd, P. S., & Parshall, M. B. (2000). Getting the focus and the group: Enhancing analytical rigour in focus group research. *Qualitative Health Research*, *10*(3), 293–308. doi:10.1177/104973200129118453 PMID:10947477

Kim, K., Cho, Y., & Zhang, S. (2016). Integrating work sequences and temporary structures into safety planning: Automated scaffolding-related safety hazard identification and prevention in BIM. *Automation in Construction*, *70*, 128–142. doi:10.1016/j.autcon.2016.06.012

Kim, K., & Teizer, J. (2014). Automatic design and planning of scaffolding systems using building information modeling. *Advanced Engineering Informatics*, *28*(1), 66–80. doi:10.1016/j.aei.2013.12.002

Kim, N. K., Rahim, N. F. A., Iranmanesh, M., & Foroughi, B. (2019). The role of the safety climate in the successful implementation of safety management systems. *Safety Science*, *118*, 48–56. doi:10.1016/j.ssci.2019.05.008

Kiviniemi, M., Sulankivi, K., Kähkönen, K., Mäkelä, T., & Merivirta, M. L. (2011). BIM-based safety management and communication for building construction. *VTT Research Notes, 2597.*

Ku, K., & Taiebat, M. (2011). BIM Experiences and Expectations: The Constructors' Perspective. *International Journal of Construction Education and Research*, *7*(3), 175–197. doi:10.1080/15578771.2010.544155

Labour Department Hong Kong. (2000). *Factories and Industrial Undertakings (Safety Management) Regulation.* Retrieved from http://www.legislation.gov.hk/blis_ind.nsf/WebView?OpenAgent&vwpg=CurAllEngDoc*59.31*0*59.31#59.31

Labour Department Hong Kong Government. (2015). *Occupational Safety and Health Statistics, 2014.* Author.

Leedy, P. D., & Ormrod, J. E. (2010). Practical research: planning and design (9th ed., international ed.). Upper Saddle River, NJ: Pearson Education International.

Li, H., Lu, M., Hsu, S. C., Gray, M., & Huang, T. (2015). Proactive behavior-based safety management for construction safety improvement. *Safety Science*, *75*, 107–117. doi:10.1016/j.ssci.2015.01.013

Li, X., Yi, W., Chi, H., Wang, X., & Chan, A. (2018). A Critical Review of Virtual and Augmented Reality (VR/AR) Applications in Construction Safety. *Automation in Construction*, *86*, 150–162. doi:10.1016/j.autcon.2017.11.003

Li, Y., & Guldenmund, F. W. (2018). Safety management systems: A broad overview of the literature. *Safety Science*, *103*, 94–123. doi:10.1016/j.ssci.2017.11.016

Lingard, G., Graham, P., & Smithers, G. (2000). Employee perceptions of the solid waste management system operating in a large Australian contracting organisation: Implications for company policy implementation. *Construction Management and Economics*, *18*(4), 383–393. doi:10.1080/014461900050024806

Malekitabar, H., Ardeshir, A., Sebt, M. H., & Stouffs, R. (2016). Construction safety risk drivers: A BIM approach. *Safety Science*, *82*, 445–455. doi:10.1016/j.ssci.2015.11.002

Martinez-Aires, M. D., Lopez-Alonso, M., & Martinez-Rojas, M. (2018). Building information modeling and safety management: A systematic review. *Safety Science*, *101*, 11–18. doi:10.1016/j.ssci.2017.08.015

McGraw Hill Construction. (2014). *The Business Value of BIM for Construction in Major Global Markets: How Contractors around the World are driving Innovation with Building Information Modeling*. Author.

McKinsey & Company. (2015). *Building for a better future Vision 2030 for the Hong Kong Construction Industry*. Author.

Miettinen, R., & Paavola, S. (2014). Beyond the BIM utopia: Approaches to the development and implementation of building information modelling. *Automation in Construction*, *43*, 84–91. doi:10.1016/j.autcon.2014.03.009

Mohammadi, A., Tavakolan, M. & Khosravi, Y. (2018) Factors in Fluencing Safety Performance on Construction Projects: A Review. *Safety Science, 109*(October), 382–397.

Mordue, S., & Finch, R. (2014). *BIM for construction health and safety*. RIBA Publishing.

Morgan, D. (1993). *Successful focus groups: advancing the state of the art*. Sage Publications., doi:10.4135/9781483349008

Morgan, D. (2012). Focus groups and social interaction. In J. Gubrium (Ed.), *The Sage handbook of interview research: The complexity of the craft* (pp. 161–176). Sage.

Muzafar, M. (2019). Building information modelling to mitigate the health and safety risks associated with the construction industry: A review. *International Journal of Occupational Safety and Ergonomics*, 1-26.

Natephra, W., & Motamedi, A. (2019) BIM-based live sensor data visualization using virtual reality for monitoring indoor conditions. *Proceedings of the 24th CAADRIA Conference*, 2, 191-200.

Neuendorf, K. A. (2002). *The content analysis guidebook*. Sage Publications.

New York City, U. (2013). *Building Information Modeling Site Safety Submission Guidelines and Standards*. BIM MANUAL.

Nicholas, J. (1989). Successful Project Management. *J. of Systems Management*, *40*, 24–30.

Niu, S., Yang, Y., & Pan, W. (2019). Logistics Planning and Visualization of Modular Integrated Construction Projects Based on BIM-GIS Integration and Vehicle Routing Algorithm. *Modular and Offsite Construction (MOC) Summit Proceedings*, 579-586.

Niu, Y., Lu, W., Xue, F., Liu, D., Chen, K., Fang, D., & Anumba, C. (2019). Towards the "third wave": An SCO-enabled occupational health and safety management system for construction. *Safety Science*, *111*, 213–223. doi:10.1016/j.ssci.2018.07.013

Park, J., Kim, K., & Cho, Y. K. (2016). Framework of automated construction-safety monitoring using cloud-enabled BIM and BLE mobile tracking sensors. *Journal of Construction Engineering and Management*, *143*(2), 05016019. doi:10.1061/(ASCE)CO.1943-7862.0001223

Rajendran & Clarke. (2011). Building Information Modeling: Safety Benefits & Opportunities. *Professional Safety, 56*(10), 44-51.

Reutter & Ford. (1996). Perceptions of public health nursing: views from the field. *Journal of Advanced Nursing, 24*(1), 7-15. doi:10.1046/j.1365-2648.1996.01681.x

Royal Institute of British Architects. (2012). *BIM Overlay to the RIBA Outline Plan of Work*. Author.

Royal Institute of British Architects Enterprises Limited. (2015). *NBS National BIM Report 2015*. Author.

Sacks, P., Perlman, A., & Barak, R. (2013). Construction safety training using immersive virtual reality. *Construction Management and Economics*, *31*(9), 1005–1018. doi:10.1080/01446193.2013.828844

Sacks, R., & Pikas, E. (2013). Building information modeling education for construction engineering and management. I: Industry requirements, state of the art, and gap analysis (Author abstract). *Journal of Construction Engineering and Management*, *139*(11), 04013016. doi:10.1061/(ASCE)CO.1943-7862.0000759

Saunders, L., & Mecca, J., & Schaefer, B. (2013). Analysing the Practice of Life Cycle Assessment Focus on the Building Sector. *Journal of Industrial Ecology*, *17*(5), 777–788. doi:10.1111/jiec.12028

Scott, H., Hofmeister, N., Rogness, N., & Rogers, A. E. (2010). Implementing a fatigue countermeasures program for nurses: A focus group analysis. *The Journal of Nursing Administration*, *40*(5), 233–240. doi:10.1097/NNA.0b013e3181da4078 PMID:20431458

Shafique, M., & Rafiq, M. (2019). An Overview of Construction Occupational Accidents in Hong Kong: A Recent Trend and Future Perspectives. *Applied Sciences (Basel, Switzerland)*, *9*(10), 2069. doi:10.3390/app9102069

Shepherd, G. W., Kahler, R. J., & Cross, J. (2000). Crane fatalities— A taxonomic analysis. *Safety Science, 36*(2), 83–93. doi:10.1016/S0925-7535(00)00017-5

Shi, Y., Jing, D., Ragan, E., Cho, K., & Ma, S. (2018). Social Influence on Construction Safety Behaviours: A Multi-User Virtual Reality Experiment. Construction Research Congress, 174–183.

Shin, I. (2015). Factors that affect safety of tower crane installation/dismantling in construction industry. *Safety Science, 72*, 379–390. doi:10.1016/j.ssci.2014.10.010

Sidani, A., Duarte, J., Baptista, J. S., Martins, J. P., & Soeiro, A. (2018). Improving construction safety using BIM-Based sensor technologies. *2.0 Congresso Português de Building Information Modelling,* (1), 161–169.

Sim, J. (1998). Collecting and analysing qualitative data: Issues raised by the focus group. *Journal of Advanced Nursing, 28*(2), 345–352. doi:10.1046/j.1365-2648.1998.00692.x PMID:9725732

Smith, D. K. T. M. (2009). *Building information modeling: a strategic implementation guide for architects, engineers, constructors, and real estate asset managers.* Wiley. doi:10.1002/9780470432846

Standards Australia / Standards New Zealand. (2001). *Occupational health and safety management systems: general guidelines on principles, systems and supporting techniques* (2nd ed.). Standards Australia Int'l Ltd.

Sulankivi, K., Kähkönen, K., Mäkelä, T., & Kiviniemi, M. (2010, May) 4D-BIM for construction safety planning. In *Proceedings of W099-Special Track 18th CIB World Building Congress* (Vol. 2010, pp. 117-128). Academic Press.

Tam, C., & Fung, I. W. IV. (1998). Effectiveness of safety management strategies on safety performance in Hong Kong. *Construction Management and Economics, 16*(1), 49–55. doi:10.1080/014461998372583

Tam, C., & Fung, I. W. (2011). *Tower crane safety in the construction industry: A Hong Kong study.* Academic Press.

Teo, E. A. L., & Ling, F. Y. Y. (2006). Developing a model to measure the effectiveness of safety management systems of construction sites. *Building and Environment, 41*(11), 1584–1592. doi:10.1016/j.buildenv.2005.06.005

Teo, M. M. M., & Loosemore, M. (2001). *A theory of waste behaviour in the construction industry.* Academic Press.

The Construction Industry Council. (2020). *CIC BIM Standards – General.* Construction Industry Council.

Von Ahsen, A. (2014). The integration of quality, environmental and health and safety management by car manufacturers - A long-term empirical study. *Business Strategy and the Environment, 23*(6), 395–416. doi:10.1002/bse.1791

Wan, A. (2017). *The Application of Building Information Modelling for Safety in the Construction Industry* [PhD Dissertation]. Leeds Beckett University.

Wan, A., Platten, A., & Briggs, T. (2013). Study of Safety Auditors' Views on the Use of BIM for Safety in Hong Kong. *International Journal of 3-D Information Modeling, 2*(1).

Yiu, N. S., Sze, N. N., & Chan, D. W. (2018). Implementation of safety management systems in Hong Kong construction industry–A safety practitioner's perspective. *Journal of Safety Research*, *64*, 1–9. doi:10.1016/j.jsr.2017.12.011 PMID:29636156

Yiu, N. S., Sze, N. N., & Chan, D. W. (2018). Implementation of safety management systems in Hong Kong construction industry–A safety practitioner's perspective. *Journal of Safety Research*, *64*, 1–9. doi:10.1016/j.jsr.2017.12.011 PMID:29636156

Yu, S., Shen, Q., Kelly, J., & Hunter, K. (2006). Investigation of critical success factors in construction project briefing by way of content analysis (Author abstract). *Journal of Construction Engineering and Management*, *132*(11), 1178–1186. doi:10.1061/(ASCE)0733-9364(2006)132:11(1178)

Zhang, S., Sulankivi, K., Kiviniemi, M., Romo, I., Eastman, C. M., & Teizer, J. (2015). BIM-based fall hazard identification and prevention in construction safety planning. *Safety Science*, *72*, 31–45. doi:10.1016/j.ssci.2014.08.001

Zhou, I., Irizarry, J., & Li, Q. (2013). Applying advanced technology to improve safety management in the construction industry: A literature review. *Construction Management and Economics*, *31*(6), 606–622. doi:10.1080/01446193.2013.798423

Section 4
Technology: Enabling Digital Innovation

Chapter 6
Innovative Digital Technologies

Don Amila Sajeevan Samarasinghe
https://orcid.org/0000-0003-0401-2482
Otago Polytechnic, New Zealand

Emma Wood
Otago Polytechnic, New Zealand

ABSTRACT

The construction industry is one of the oldest industries in the world and one that continues to change with client demand. In recent decades, innovation in the construction industry has greatly improved, increasing productivity. Innovation in construction refers to the generation and implementation of new ideas to enhance the performance of construction processes and to gain economic, environmental, and social benefits. Modern innovative digital technologies in construction include application of virtual reality (VR)/augmented reality (AR), blockchain, 3D printing, building information modeling (BIM), and off-site manufacturing. This chapter will explore the application of these innovative digital technologies in construction. It will particularly include recent case studies and examples from the New Zealand construction industry.

INTRODUCTION

In business, innovation is a new idea or a creative thought regarding the development of a product or a service. It is used to make business operations easier, and more effective and competitive (Nambisan et al., 2017). Innovation is not dependent on technology alone; rather, it is defined by the skills and creative thinking of individuals in an organization. Nonetheless, technology is a significant factor in relation to companies' innovative capabilities. Advancing technology has enabled companies to become more productive through the adoption of new and effective methods across all forms of business. As a process, innovation is defined as creating new or changing existing ideas, knowledge, concepts or technologies to renew or significantly improve processes, products or systems. The European Commission (1995) defined innovation as "the transformation of an idea into a marketable product or service, a new or improved manufacturing or distribution process, or a new method of social service". Innovation is also

DOI: 10.4018/978-1-7998-6600-8.ch006

defined as the creation of value by developing new ideas; a combination of creativity and exploitation. It is usually connected to research and development – indeed, innovation is a product of research and development (Manseau, 2019). The construction industry has developed over recent decades as companies have embraced new and innovative methods to make their processes more effective. Such innovation keeps evolving – from 3D printing to hybrid building technologies, the advancements are seemingly endless. Taking advantage of such innovative methodologies is essential if companies are to succeed in the market. The adoption of innovative methodologies is likely to make construction companies more efficient and flexible, while it can also expand possibilities for projects that seemed mundane before. The true definition of innovative construction involves products or techniques that drive a company away from traditional construction practices (Brockmann et al., 2016).

Innovative and advantageous construction methodologies include 3D modeling, the application of virtual reality to monitor health and safety of construction sites, and the use of BIM. For example, hybrid building technologies utilize multiple building materials throughout a structure in order to create a financially responsible or more sustainable project. Through the use of such technology, a company can save money and make effective plans for projects (Ozorhon et al., 2016). Furthermore, the adoption of innovation in construction allows firms to have greener construction processes. Such processes lead to reduced waste and air pollution, and provide increased opportunities to reuse materials. In addition, innovation in construction can result in a decrease in the use of raw materials. Such innovations improve efficiency and increase performance. As new technologies are adopted, a project can proceed at greater speed, lower cost, and with a better overall performance structure. Lastly, being innovative increases a company's competitiveness. Firms can complement any new requirements that may surface by providing innovative solutions and using innovation in their processes (Jackson, 2017).

The construction industry plays a major role in the domestic economy in New Zealand in terms of its economic contribution. The construction industry contributes about 6% – or NZD14 billion – of the country's gross domestic product (GDP). Approximately 15,000 jobs, equating to 9% of the New Zealand workforce, are generated by New Zealand construction firms, making construction the fifth largest sector by employment (New Zealand Construction Industry Council (NZCIC), 2017). The Ministry of Business, Innovation and Employment (MBIE) (2019) projected the demand for construction-related jobs would increase by 56,000 employees between 2016 and 2020. However, the New Zealand construction industry is still characterized by low productivity, a lack of innovation, high unemployment, skill shortages, poor early contractor involvement and poor quality output, even though it is the largest infrastructure sector in New Zealand and currently in the midst of a construction boom (Samarasinghe, 2020; Samarasinghe & Li, 2020; MBIE, 2019; Mirus, Patel & McPherson, 2018; Wilkinson et al., 2017; Noktehdan et al., 2015). NZCIC (2017) reported that the country's construction industry is limited by the development of the national economy, which has not taken full advantage of advances in technology – it has low production efficiency, slow acceptance of new technologies and concepts by industry practitioners, slow technological development, extensive application of traditional construction methods, and long waits for technological improvements. Traditionally, and certainly since the first industrial revolution, the construction industry has been slow to incorporate new technologies into its practice, and it seems this is still the case now in an increasingly digitized world. Due to low productivity, skill shortages, and a lack of innovation in New Zealand construction processes, the country still relies on traditional methods in construction. These challenges have drawn attention to the idea of implementing innovative digital technologies as a way to boost performance. The Stronger Christchurch Infrastructure Rebuild Team (SCIRT), established after the city experienced destructive earthquakes in 2010 and 2011, has recorded more than 600 innovations in its projects. Wilkinson et al. (2015) reported that innovation can bring a number of benefits to New Zealand construction projects: the enhancement of construction safety; a reduction in construction costs; quality improvement; time saving; an improvement in workmanship; and higher productivity and project performance. Badcock (2019) urged the need for digitization in construction (including the use of cloud software, mobile products, and BIM, for example) as a solution to high insolvency, small profit margins, and low levels of productivity in New Zealand construction projects.

This chapter aims to describe how innovative digital technologies such as VR/AR, blockchain technology, 3D printing, BIM and off-site manufacturing influence the construction project performance factors of accuracy, speed, sustainability, health and safety, productivity and profitability. Case studies and examples from New Zealand and other countries such as the UAE, UK, Spain, Denmark and China demonstrate how innovation has significantly improved construction projects.

VIRTUAL REALITY AND AUGMENTED REALITY DELIVERING CONCRETE BENEFITS

VR technology offers immersive simulated experiences that have revolutionized training in a variety of sectors with inherent and elevated health and safety risks, including medicine and the military. It has the potential to do the same for construction, with many opportunities to apply this technology to great effect within the industry. With the widening availability of affordable head-mounted VR devices, the technology has never been more accessible.

The advancement of technology and the continuous enhancement of VR is of great benefit to learning within the fields of design and construction. According to Wang et al. (2018), the use of rapid, as-built scene generation is improving virtual training by enabling the fast conversion of real images into a virtual environment. The level of detail and reality within virtual training systems is increasing, making it possible for learners to gain significant experience of designing and decision-making before entering the workforce. As outlined by Getuli et al. (2020), the rapid, as-built generation technology is a highly-advanced scanning system that very quickly converts real conditions into a virtual environment. Thus, its use has made the development of virtual software for learning easier. In addition to this, the hybrid visualization approach for ubiquitous learning is a major advancement in the education system for design and construction. Bashabsheh et al. (2019) state that student training often requires mobility and concrete interactions while learning about site inspection, heavy equipment, and reviewing design. The use of a hybrid visualization approach has made it possible for learners to acquire actual sensations like hearing and touching with the help of simulation and VR (Otto et al., 2016). Therefore, the need for physical movement of the students and their actual presence on the site is declining due to the implementation of hybrid visualization approaches. Moreover, the ability of the VR-based systems to provide online training and education is a major shift in construction and architecture engineering. As noted by Maghool et al. (2018), the use of online, VR-based education allows students to gain comprehensive learning remotely. Also, it is becoming easier for educators to provide learning and education to off-site students, who do not have to be physically present in order to learn. According to Wang et al. (2018), modern design methods like BIM are highly cooperative as they can be easily made available for online learning. The use of online-based VR learning is an important development for teaching design buildability to students in the most efficient manner. NZCN (2019) reports that VR is an educational reality, boosting construction sector training and employment opportunities in New Zealand.

Samarasinghe, Baghaei and Stemmet (2020) argued that VR is a persuasive technology that can play a role in showcasing emerging or underutilized construction techniques such as earth building techniques. Earth building is a 11,000-year-old practice of architecture using unfired earth material, and some of the oldest of wonders of the world have been constructed this way, including the Great Wall of China. While some of New Zealand's historic buildings feature this method of construction, fewer than 1% of the country's buildings are real earth buildings. Such rare building techniques can be promoted and disseminated through VR programs.

Another advantage is that VR provides practitioners and students the opportunity to virtually visit buildings or sites that would otherwise be inaccessible due to privacy reasons, health and safety restrictions, or a lack of local examples of a particular type of building, structure or practice (Samarasinghe, Abd-Latif & Baghaei, 2019). Even when teachers are able to undertake site visits with learners, these require regulation compliance, correct personal protective equipment (PPE), and close monitoring. However, VR can show them, without any risk, how heavy construction activities occur – such as using tower cranes to install pre-cast concrete slabs. When Construction Management graduates arrive to manage a site, it can be difficult for them to have a full understanding of these practices. Virtual reality is an effective way to gain a closer understanding of what is involved at site-level and how tradespeople put things together. For example, building a reinforced brick wall – in 3D scale, learners can see the proportion of materials required, the correct positions, spacing, dimensions, specifications, and so on.

VR also works well with Building Information Management (BIM) systems. The virtual models within BIM can be translated to VR, allowing clients to see the finished product, or specific aspects of a design or build. They can feel and experience the virtual environment and gain a more realized sense of how a project will look when complete. This technology is put to effective use in the BBC television show, Your Home Made Perfect, which uses VR headsets and visual effects to simulate the outcomes of two competing architects' designs for a home renovation. These inform the clients as they choose their preferred design option. It illustrates that VR is a helpful communication aid that can bring a project to life in clients' minds. New Zealand architects and Engineers are moving to embrace VR in structural design and information sharing (Richardson, 2018) to increase the accuracy and safety aspects of construction projects.

AR is a technology that combines real world objects with computer-generated information. Like VR, AR is finding use in the construction industry. After Ahmed et al (2017) considered the available AR applications and reviewed existing literature, they asserted that AR is useful in project scheduling and tracking progress. AR and VR also help in holding conferences for construction workers. In project planning, AR helps firms to better understand a building, and assists project managers to show both the social and environmental impacts of a project. The use of AR begins in a project's early stages. Throughout a project, and even through to its end, the construction company can showcase its work with accompanying detailed information. By using this technology, the company is in a better position to align everything to the customers' desires. The clients, in turn, can make their decisions based on what they see and whether they like it; this promotes increased customer satisfaction. For building contractors, AR can provide a feedback loop long before problems are encountered on-site. The planned building can appear in front of the construction team (Richardon, 2018). Architects and structural engineers can better understand each other's work and therefore ensure ease in design and implementation. AR also allows for accurate measurement in quantity surveying through the use of smart glasses (Behzadi, 2016). Such a level of accuracy helps to reduce the time spent on making revisions and adjustments when a project is in progress. With correct measurements, a company will acquire the correct quantity of materials, ensuring limited wastage. Managers can also be sure there is no undersupply of materials, which leads to delays in completing work. Another benefit of accurate measurements is the employment of sufficient machines; the managers will have machines that optimize their processes.

Figure 1. Persuasive VR Model for Promoting Earth Buildings in New Zealand
Source: Samarasinghe, Baghaei and Stemmet, 2020 and Samarasinghe, Abd-Latif, and Baghaei, 2019

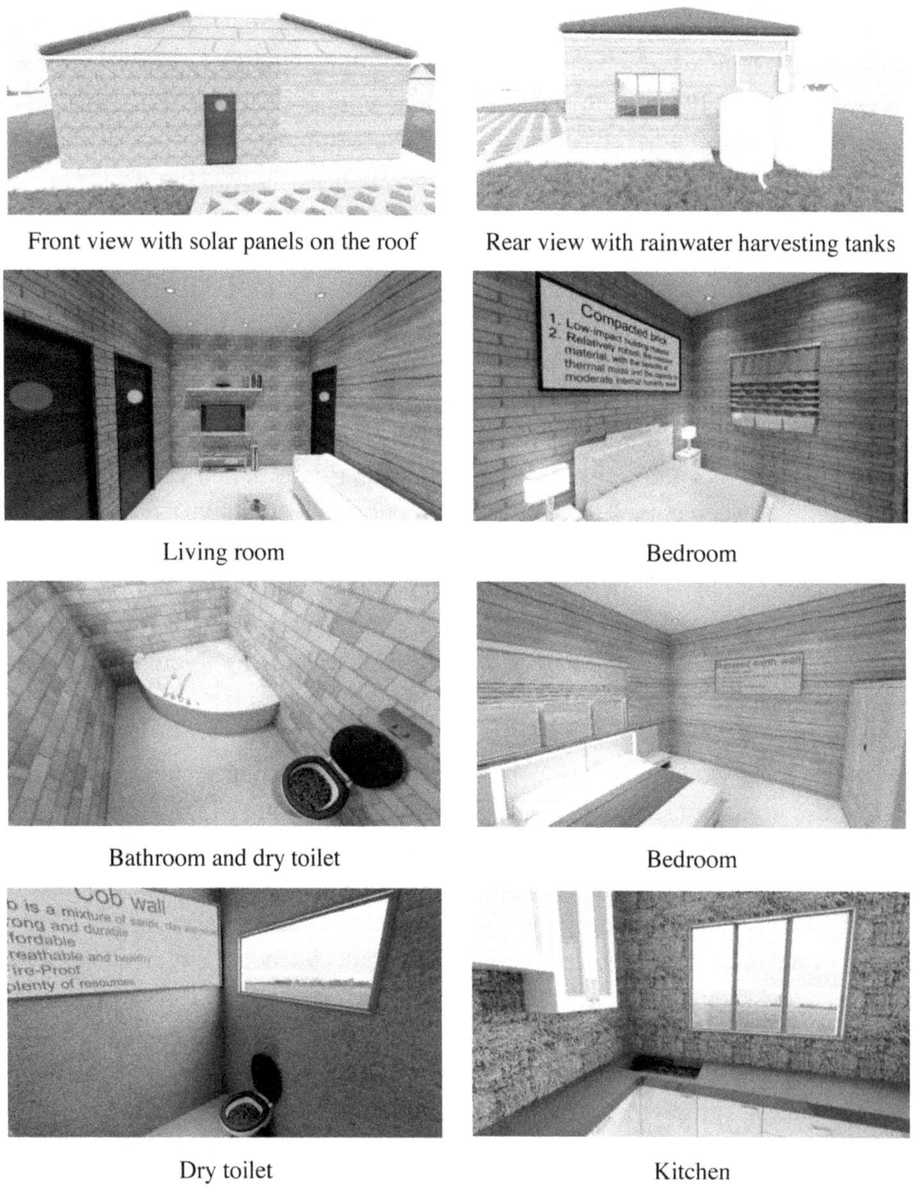

The construction industry is known for being particularly hazardous. In New Zealand in 2018, there were a total of 51,000 claims to the Accident Compensation Corporation (ACC) for construction injuries, at an estimated cost of $133 million. And in 2019, the local industry experienced its highest fatality rate in a decade, with 13 deaths (New Zealand Government, 2019). Because construction practitioners are subject to considerable health and safety risks, efforts to mitigate these are a priority sector-wide. In 2019, the engineering consultancy, BECA, received an ACC workplace injury prevention grant to develop a VR solution to combat the industry-wide issues of health and safety by developing an immersive VR module to improve construction site inductions.

Figure 2. BECA's VR solution for improving health and safety in New Zealand
Source: New Zealand Government, 2019

ACC noted that the use of VR in the construction industry will help prepare workers before they step foot on-site and could lead to a much-needed reduction in injuries in the sector (New Zealand Government, 2019). While VR is still an emerging technology for New Zealand construction firms, there is one notable exception. Fulton Hogan was a finalist in both the 2017 New Zealand Innovation Awards and the 2018 NZCIO (New Zealand Commerce and Industry Office) Awards for using VR to train employees in its boil out procedure for decontaminating bitumen sprayers – the first time VR had been used in the New Zealand construction industry. The New Zealand government is getting on board as well, recently conducting a pilot program using VR to prepare people for jobs in the construction industry. The technology was developed in partnership with business groups and industry to ensure it met the needs of employers. A global formwork and shoring supplier has recently introduced a 3D visualization tool integrated with VR into New Zealand, using the cloud-based advanced VR technology to efficiently share information among the project team (NZCN, 2019).

VR can help teach people how to avoid and respond to hazardous situations, minimizing the potential for harm (University of Nottingham, 2019). In the event of a fire, for example, most people have never used a fire extinguisher before. VR allows them to walk through that process step-by-step, in effect practicing and experiencing it. We can use virtual construction sites to deliver preliminary health and safety training to people before they operate potentially dangerous machinery and tools, and this has the potential to prevent many incidents (Hafsia, Monacelli, & Martin, 2018). Virtual reality is already proving to be a highly effective method for safety training in construction globally, according to BRANZ Build Magazine (Climo, 2018). It reports that Hong Kong-based Gammon Construction employees rate VR training more highly than a traditional lecture-based approach. The company's safety manager is quoted as saying VR has boosted on-site safety and is moving the business closer to its zero-harm goal. Another benefit of VR is it has a far lower ecological footprint than traditional forms of training. It is a great choice in the interests of environmental sustainability because all materials are simulated, and there is no requirement to consume resources such as building materials, electricity, transportation, and water in training.

BLOCKCHAIN, BITCOIN AND BUILDING

In concept, blockchain is a data structure-like chain, composed of data blocks connected according to time logic sequence, that cannot be tampered with by using an asymmetric encryption algorithm (Turk & Klinc, 2017). Blockchain – also known as Distributed Ledger Technology (DLT) – is a digital system that uses a virtual currency, like Bitcoin, to establish private and secure groups within which money can be exchanged and tracked. These networks are wholly transparent and operate as circles of trust – everyone within the group can see every transaction and the party that authorized it. Each private network has its own passcode, and each individual within the network has their own passcode as well. All transactions require the entry of these passcodes for authorization, and all are logged and visible for all group members to see. Cryptography ensures transactions are protected and payment is secure. As Orcutt (2018) asserted, "The whole point of using a blockchain is to let people…share valuable data in a secure, tamperproof way". Blockchain is popular in a range of industries with complex supply chains and those involving a range of international parties, such as transport and logistics, finance, and manufacturing.

Blockchain presents the construction industry with a way to maintain cash flow, save time and cut costs – it bypasses banks to speed up international and domestic payment processes, limiting transaction delays and errors, and reducing the risk of hacking and fraud (San, Choy, & Fung, 2019). Many construction projects have complex supply chains, too, requiring a large number of parties to work together, including the client, advisors, local government and permitting authorities, a head contractor and a range of subcontractors. Among complex supply chains in construction, lack of integration is a prominent issue (Thunberg, Rudberg, & Gustavsson, 2017). Typically, transactions between these parties can be inefficient and fraught with difficulties, including invoice errors, payment errors, bank holidays, and currency conversion fees and variations. When such payment issues arise, the head contractor will find it difficult to pay the subcontractors, which can result in insolvencies, delays and increased costs. A study conducted by Ramachandra and Rotimi (2016) outlined why cash flow is so essential for construction businesses: construction activities take a long time, the products are expensive, and the fact that payments are usually made for work already done, makes cash flow and financial status highly dependent on timely payments from the principals. Cash flow constraints for contractors not only result in business failures but also create flow-on effects in the supply chain. Inserting virtual currency into a project network via blockchain can solve – or at least mitigate – all of these challenges. For this reason, block chain has the potential to speed up construction projects, minimize project delays and save the industry significant amounts of money. While the New Zealand construction industry has not yet embraced blockchain, there is evidence it is becoming a desirable way to pay overseas.

The United Arab Emirates, known for its innovative construction sector, is keen to implement blockchain more widely. The industry loses the equivalent of many millions of dollars each year through payment errors and processing delays, which causes negative flow-on effects, including project delays and cost overruns. The country's rulers believe investing in block chain will help prevent these financial losses (Anthony, 2019). In addition, a considerable amount information can be stored and tracked in blockchain, so project progress can be accurately checked by everyone involved. Walmart uses this technology to manage its supply chain including manufacturing, logistics, and transportation (Mearian, 2019). For example, if a product is damaged, the history of information stored within blockchain can isolate where the issue occurred. Walmart uses blockchain as a tool to track the accuracy of supply chain information. Eventually, blockchain technology could help improve the profitability of mega-companies that use long and complex supply chains.

Figure 3. DL Freight management and payment automation in Walmart
Source: Mearian, 2019

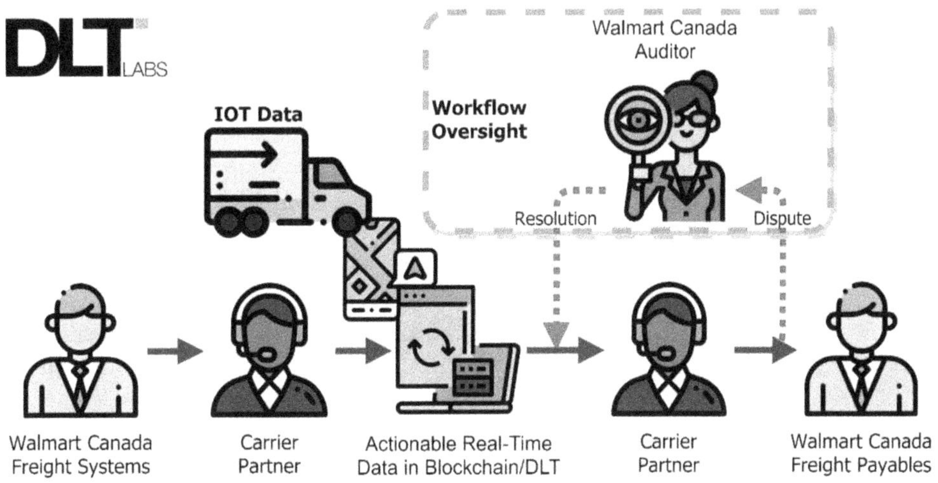

As noted by Penzes (2018, p.19), centralized record-keeping is a major benefit of blockchain: "One of the key advantages of such a platform, apart from being very effective, is the high level of transparency for all parties involved. For capital construction projects, complex contracts, terms and conditions are often applied. Through the whole lifecycle of a project it can be a challenge to ensure collaboration on every level between all the parties according to the contract. With such a system and applied smart contracts it can be ensured, that every action always happens according to the agreed terms". The report also emphasizes the far-reaching consequences that stem from payment issues in construction sector. Late payments and the resulting cash flow issues have been enduring problems in the construction industry. However, the blockchain system provides protection for contractors by promptly reminding them to fulfil their contractual obligations, such as automatic payments, according to the progress of the project. Once information is added to the blockchain, it cannot be amended; a completed contract cannot be altered, preventing unnecessary disputes. Also, the blockchain uses split-led account management, meaning the contract execution process is presented to all protect participants with a high degree of visibility (Penzes, 2018). This also indirectly protects the execution of the contract. In addition, the high-efficiency payment mechanism and irrefutable distributed ledger record information established under the automatic alarm system make any party of the contract have to fulfill its obligations (Penzes, 2018).

Figure 4. The world's first 3D printed bridge in Spain
Source: Valencia, 2017

THE 3D PRINTING CONSTRUCTION REVOLUTION

3D printing or three-dimensional printing is an advanced technology that can produce a 3D model without molds, machine tools, or fixtures. The first 3D printer was created in 1984; it was a complex and costly technology at that time (Hager, Golonka, & Putanowicz, 2016). Since then, 3D printing has been applied to various sectors including the medical, information technology, and space and aeronautics industries (Tay et al., 2017). 3D printers are well known for printing models, toys, various specialist components – and, more recently, face masks in response to the global pandemic caused by the novel coronavirus, COVID-19.

Within the construction industry, 3D printing is still an emerging technology, however an increasing number of structures around the world have been created using it. While larger scale constructions pose unique challenges, 3D printing now represents the future of the construction industry. In about 10 years' time, the global construction industry will change. There will be increased opportunity for utilizing some of today's more innovative construction technologies, including BIM, off-site fabrication and 3D printing. Already, the world is home to a growing number of large-scale 3D printed constructions, including a Spanish bridge in Spain, office buildings in Dubai, and infrastructure in China (Valencia, 2017; Meisenzahl, 2019; Sher, 2019).

Figure 5. The world's first, 500m long, 3D printed river revetment wall in China
Source: Sher, 2019

Figure 6. The world's largest 3D-printed structure in Dubai
Source: Meisenzahl, 2019

3D printers can produce a range of components for use on-site, and even larger-scale projects can be built almost entirely on-site using huge 3D printers. These machines can be 6m tall, 32m long and 10m wide, and feature a large robotic arm. They can be connected to a BIM model to ensure appropriate and accurate use of materials and output. Raw materials – often concrete, polymer or sand – are fed into the machine, which is programmed in accordance with the BIM specifications. The robotic machine can then complete the main frame and structure of a building, laying foundations and building walls by printing them layer by layer (Xu et al., 2018). Other prefabricated elements such as windows, doors, wall coverings, and interior fitting are still required, however. There are three main types of 3D printing technology: material extrusion, powder bonding, and wire arc additive manufacturing (WAAM) (Laubier et al., 2018). There are three large-scale 3D printing processes used in the construction industry: Contour Crafting, D-Shape and concrete printing. Each is developed with different applications in mind, but all three methods use information from digital models to create solid, three-dimensional objects layer by layer. Together, these techniques are also known as additive manufacturing (El-Sayegh, Romdhane, & Manjikian, 2020). A UK study conducted by Lim et al. (2012) found all three methods of additive manufacturing have a proven ability to produce components of significant size suitable for construction and/or architectural projects, and determined that while full-scale construction using these processes is an emerging art, it will soon be a reality. It is a fascinating and exciting area of innovation in the construction industry, and there is great potential for it to play a significant role in the future.

These kinds of machines are not yet used in New Zealand, but countries such as China and Denmark and the United Arab Emirates have already embraced the technology. In fact, in 2016, Dubai's government developed a 3D printing strategy, with the aim that 25% off all new buildings in the city will be constructed using 3D printing technology by 2030 (Bogue, 2018). The New Zealand Government has projected that the construction industry will be valued at $43 billion in 2021 (MBIE, 2019). Yet, it is beset by a number of substantial challenges. There are severe worker shortages and most major projects run over budget. It has high accident rates and is the country's fourth worst industry for workplace fatalities (WorkSafe, 2020). Additionally, construction and demolition account for up to 50% of the country's waste (BRANZ, 2020). Because 3D printing technology is exact, its accuracy reduces costly and time-consuming errors, and ensures inspection processes are much faster. It also results in less wastage, potentially providing some solutions for an industry with significant sustainability challenges (Sakin & Kiroglu, 2017). In addition, 3D printing has been shown to drastically reduce health and safety incidents on site. The research backs this up. A study conducted by Lim et al. (2012) found that the Chinese company Shanghai WinSun, which was the first to build a five-story apartment block and a country villa using 3D printing technology, saved about 60% in building materials and shaved 30% off the time required to build compared with traditional methods. It also noted the construction team was five times smaller, which meant cost savings and a lower risk of work-related injuries. While there is considerable initial investment involved, 3D printing can massively speed up the construction process, improve health and safety, and reduce work and waste costs.

BIM IN CONSTRUCTION

BIM technologies are software or cloud-based programs that use 3D modeling to manage the development of a construction project, from design through to delivery and operation. The technologies allow for detailed information management, from specifications, costings, timelines and beyond. These project management tools connect architects, engineers, suppliers and contractors – even local councils – to provide one source of truth, improving communication and transparency, while reducing duplication, miscommunication and confusion. Australian Institution of Quantity Surveyors (AIQA) (2018) and New Zealand Institution of Quantity Surveyors (ZIQS) (2018) considered BIM as a tool that facilitates effective cooperation and efficient communication between various project stakeholders. It is not only a single process of architectural modeling but also a flexible tool that can be applied throughout the project lifecycle. Nowadays, BIM has been well introduced in many countries and it has been applied in real construction projects for years. Some countries have developed relevant standards for BIM, such as the United Kingdom with the PAS1192 series and Australia with Natspec. BIM is not mandated in many countries, but it is encouraged by a number of governments. BIM is advantageous in that it can:

1. facilitate the coordination between different project elements
2. reduce possible mistakes caused by poor information sharing
3. improve the communication between project stakeholders
4. provide an efficient way for data and facilities management
5. make the process of analysis and simulation easier and more convenient
6. enhance project productivity (BIMinNZ, 2016).

There are clear benefits to streamlining processes throughout a project's entire lifecycle. Among the most lauded of these is that BIM cuts down on the need to rework when factors change – which, in turn, saves both time and money, and minimizes mistakes and errors. BIM allows all modeling and information to be in one central place, so that all parties involved in a construction project are on the same page as the project progresses. For example, when variations are made and approved, these can be easily updated and shared immediately through BIM, meaning real-time access to relevant details and a process that keeps everyone informed. For practical reasons, and to protect matters of commercial sensitivity and security, construction managers can restrict access to the information contained within BIM systems by setting up private channels with distinct permissions for individual players.

One of the potential challenges of BIM technologies is that they require a significant input of time and energy at the start of a project, to ensure all specifications and variables are uploaded and accounted for within the system. Of course, this early time investment pays off later. BIM technologies can be expensive to purchase, therefore, the initial investment costs likely outweigh the potential benefits for many independent contractors or small enterprises (Pham, Skelton & Samarasinghe, 2020). However, when these small businesses or contractors are working as part of a large project, they will directly benefit from BIM technologies where these are utilized. For this reason, many SMEs will find it useful and relevant to learn more about BIM, given that this is a technology that will likely be used more and more widely into the future. Another recent study conducted by Udomdech, Papadonikolaki, & Davies (2018) describes adopting BIM as a radical challenge for small- and medium-sized construction enterprises, and suggests that project-based learning could be the solution.

By centralizing all the information, BIM speeds up the construction process and ensures quality and accuracy is maintained. For large projects, or projects in which there is a large number of clients, contractors and suppliers, BIM ensures the whole process runs much more smoothly. A focus on improved efficiency and productivity is ever-important in an environment where construction is booming and there are many projects to be contested and completed. A recent study conducted by Chen, Liu, & Hua (2017) has asserted that construction industry tends to have a popular image problem, asserting that there are negative perceptions of low innovation, unsafe working practices, slow delivery and high costs in the sector. They attribute this, at least in part, to a fragmented delivery structure, and suggest that BIM presents a possible solution to this issue. By integrating and coordinating the different project phases and stakeholders, they say, enhances project performance and can give construction organizations a competitive advantage. Further research conducted by Papadonikolaki (2017) has noted that the use of BIM has been mandated or strongly recommended by policymakers in countries like the United Kingdom (UK), Germany, France and the Netherlands, among others. As such, the development of BIM technologies is now guided by the International Organization for Standardization (ISO), which published its first international BIM standards early in 2019.

Figure 7. Pukete wastewater treatment plant BIM implementation
Source: BECA, 2020

In New Zealand, the BIM Acceleration Committee is a nationwide alliance of industry and government agencies that work together to encourage the uptake of BIM in the construction sector. It has developed a handbook and a series of resources to assist construction professionals as they adopt – or consider adopting – BIM technologies. It also provides informative case studies from within New Zealand, showcasing construction projects that have successfully used BIM in a local context (BIMinNZ, 2016). New Zealand's first BIMinNZ conference was held in Wellington in 2019, hosted by New Zealand Institute of Buildings (NZIOB). NZCN (2019) reports three exceptional BIM applications: Airways' air traffic control facility in Auckland, Auckland's QT Hotel, and the Christchurch Central Library. These three projects received Master Builders inaugural BIM innovation Awards.

The Pukete wastewater treatment plant in New Zealand was upgraded in 1975 and 2002. However, due to a lack of accurate information within the original documents, asset location could not be accurately confirmed, making the upgrading work very difficult. They now adopt BIM management mode to collect data more intelligently and realize dynamic information modeling. The construction team scanned the entire scene's image surface with unmanned aerial vehicles (UAVs), including the building's exterior and the terrain of the site, and used lidar to scan the interior of the building to construct a complete 3D image (BECA, 2020).

Through the detailed 3D model, designers can accurately understand the properties of the building, such as the space of pipes and the layout of wires. UAV technology can capture model changes during construction, such as foundation excavation. Through the real-time monitoring of UAV, architects can timely understand and adjust the progress of the project. The daily project progress can be seen at a glance under the monitoring of UAV to improve the work productivity and enthusiasm of workers. BIM technicians barcode each part of the model, and field contractors use barcodes to scan applications on tablets to make sure they are using the right device. BIM can predict the impact of time on construction equipment. Through the barcode database, customers can know the operation status and aging status of the equipment in time. These data will be stored in the cloud and can be updated and viewed as required. BIM can also clarify the needs of customers, detailed model will improve customer satisfaction, but also more accurate prediction of costs. If even a tiny screw is used in the construction of the building, BIM will consider it. It makes asset management easier and reduces the probability of rework, increasing the profitability of construction projects.

Ara Institute of Canterbury aimed to build a new building for students majoring in architecture and engineering. This is a three-story building with a total area of about 6,690 square meters and a construction cost of NZD34 million (BIMinNZ, 2019). At the beginning of the project, all advisory groups began to integrate and enter all data into BIM. During the process, the quantity surveyor (QS) brought their model content plan to the table which had been developed by the designers for quantity survey modelling. In the design process, the designer and the consultant team produced 3D models and drawings at each stage and delivered them to the QS to make a cost plan. The cost plan was delivered to the client, with the contractor involved at an early stage to comment on the design and develop a total price tender for the project. In the end, the contractor won this bidding at a more precise price. The challenge is that the QS had to make a cost plan within three days, where traditionally this would take two to three weeks. After using BIM, the QS can produce a cost increase within 20 hours, which greatly shortens project delivery time. The application of 5D BIM ensures that the QS can grasp any changes in quantity and price in the project in real time and keep in touch with the contractor during the design process to ensure the accuracy of the data. As a result, the application of 5D BIM can not only reduce the time required for the project, but also ensure the accuracy of the cost plan, making a collaborative quantity surveying tool.

THE MANY BENEFITS OF MANUFACTURING OFF-SITE

Prefabrication is a technology that contributes to a more adaptive and flexible construction process by manufacturing certain building elements in a controlled environment (mainly off-site) and installing them on the site (Chen & Samarasinghe, 2020). There are five categories of prefabrication, which is classified on the basis of component size: single component, panels, volume, hybrid, and complete building. Prefabrication has been widely applied in construction projects worldwide. It has brought remarkable convenience to the construction industry in various countries such as China, Germany, UK, USA, Australia, and Japan.

Prefabrication is becoming more prevalent in the construction industry, and for good reason. It offers many advantages over traditional on-site construction methods – and it is being adopted by the New Zealand Government as it looks to combat the country's housing shortage (Chen & Samarasinghe, 2020).

Where traditional construction methods involve sourcing building materials and using labor to build on-site, off-site construction involves the installation of prefabricated and precast elements created in a controlled environment in which quality of the manufacturing is monitored and regulated. In the construction industry, off-site manufacturing can vary in scale, ranging from smaller components such as windows and doors, to more complicated elements, such as wall panels, slab panels, roof trusses – even complete buildings. Notably, this approach was adopted recently in China, in the midst of its devastating COVID-19 pandemic, when it made headlines for building brand new hospitals in as few as ten days (Tobias, 2020). According to CNN, this scale and speed of the construction was due to prefabrication, and large teams of workers rostered around-the-clock to install the structures. Even in regular workplace scenarios, the advantages of this approach are widely recognized. Research has shown prefabrication results in cost reductions, improved building quality and consistency, faster timeframes, better worksite safety, and improved environmental friendliness and sustainability. As a result, this technology has the potential to completely transform the global construction industry.

PrefabNZ is a non-profit organization that informs, educates and promotes innovation and excellence in offsite design and construction in New Zealand. It champions the benefits of constructing parts (components, panels, modules, or complete buildings) in a controlled environment, away from inclement weather, where worker and resource use can be closely monitored and enhanced. Its website features a collection of case studies of local projects that have incorporated off-site manufacture. Building contractors are considered among the main beneficiaries of this process, however research indicates there may be a reluctance for these contractors to adopt these technologies. Possible barriers include a high initial cost for factories and machinery, and extra lead-in time needed at the design-stage to ensure a smooth and accurate construction process (Chen & Samarasinghe, 2020). Innovative BIM technologies are a natural fit at this stage of this process. Limited production capacity at New Zealand manufacturing factories is another challenge for an industry looking to increase reliance on prefabricated building components. Consequently, pre-cast components can be expensive. Logistics are also an issue, with some prefabricated components requiring strong cranes to install.

China International Marine Containers Group (CIMC) has worked on several modular construction projects in the United Kingdom and Australia. More than two years after entering New Zealand, CIMC Modular believes that current New Zealand modular buildings are still in the early stages of cultivation from local governments to end users, the extent of local acceptance and recognition of modular buildings is still relatively low. However, increasingly these technologies are being adopted by the industry – even, notably, by the New Zealand Government. It has recognized that off-site manufacturing can play a major role in any solution to the country's housing crisis. In New Zealand, a bank called Westpac has started offering dedicated mortgage products with the aim of helping New Zealanders into prefabricated homes (Dunmore, 2019). In a media statement released late 2019, Housing Minister Megan Woods said the Government needed to find ways to get houses built more quickly. "Offsite manufacturing has the potential to transform the building sector and drive innovative ways to build homes at lower cost, and in faster timeframes," she said. "It can significantly change the way we build homes because builders won't have to contend with external elements like the weather." She also stated that Housing New Zealand believed off-site manufactured new builds could be delivered in half the time as conventional builds. The scale of the KiwiBuild program, especially when aligned with the other arms of the Government's build program, creates a fantastic opportunity to supercharge the use of offsite manufacturing in New Zealand. The demand these program create has the potential for the construction and development industry to make confident investments in their people and new technology. It could play an important role in smoothing out the 'boom-bust' cycle that has plagued the sector in the past and help prevent another housing shortage emerging in the future. NZCN (2019) reports that Fletcher Buildings has opened New Zealand's largest off-site home manufacturing facility that will streamline the design, consenting and manufacturing process.

111 Dixon Apartments is a 20-level residential project located on Dixon Street in Wellington. The project is composed of 228 apartments and each apartment is equipped with a bathroom. In order to ensure cost effectiveness while ensuring quality and reducing construction time, its designer devised 228 prefabricated bathroom sets for the project team to assemble. This greatly shortened the delivery date of the project by about six months compared to traditional methods. The contractor sought prefabricated bathroom contractors in the local and international markets and singled out the two most competitive (one in New Zealand and one in China) for comparison. Finally, after comparing quality and cost performance, the contractor chose the products from the Chinese factory. These were eventually shipped from China to Wellington and transported to the construction site for assembly. The installation was completed by tower crane, and the bathrooms were connected to the plumbing and drainage system of the building by a water end worker. The whole installation process took about one hour. The application of prefabricated bathrooms greatly saves the time required for construction and reduces the labor required for on-site construction (PrefabNZ, 2019).

Figure 8. Ready-to-go bathroom pods being lifted on-site
Source: Prefab, 2019

This can also be very beneficial in terms of safety and health. The traditional construction method requires many subcontractors working on different rooms, increasing quality control risks. Prefabricating bathrooms can effectively ensure that the quality of the product meets local requirements. The installation and transportation of prefabricated products are really only limited by local traffic conditions. There is also the challenge of the contractors' ability to coordinate with the government. However, as a new technology with great potential, it seems set to promote the development of New Zealand's construction industry.

CONCLUSION

In conclusion, the New Zealand construction industry has not made rapid progress in adopting innovations. Compared with large companies, the innovations adopted by small and medium-sized companies are even scarcer. In the context of the global economic recession due to the epidemic, the adoption of construction innovation may be a key factor for construction companies to survive in the future market. Under the impetus of new innovative digital technologies, competitiveness is enhanced, as is the quality of completed projects within the industry. The adoption of new technologies may free the construction industry from its existing image of inefficiency and environmental unfriendliness, thus establishing a more environmentally friendly and sustainable image. Some examples of innovative digital technologies in construction include VR, AR, blockchain, 3D printing, BIM, and off-site manufacturing. The key benefits of innovative digital technologies in construction include higher sustainability, monetary savings, and increased efficiency. Further, the integration of a variety of these technologies will result in the best solutions for enhancing efficiency in the construction industry.

VR and AR can improve performance in risk management. An immersive VR/AR experience enables clients to accurately understand the design work, which can greatly increase the efficiency of communication with clients. Blockchain gives construction contractors a new way of building management and project delivery. It can track the performance of the contract by monitoring the characteristics of the information entered. This information cannot be changed, establishing a fair and transparent supply chain relationship, in which multiple parties to the contract are on the same page. The application of 3D printing in the construction industry can greatly reduce the time required for construction and reduce the waste generated during the construction process. The innovative application of BIM has created a new way of interaction for project stakeholders. It uses 3D models to assist the building process and helps formulate future maintenance plans to reduce building maintenance costs and increase building sustainability, so as to maximize benefits. Off-site construction can recycle surplus materials in the interior and reduce wastage of materials, while helping to reduce project costs through access to discounted materials from suppliers working at scale. Adopting a modular structure also reduces the need to engage contractors and production personnel on-site, who may be unreliable. The precast structure is in a controlled environment and does not need to take into account weather factors and changes to the subcontractors' plans – and compared with traditional architecture, it takes less than half of the time.

It is likely that future trends in the New Zealand construction industry will be driven by increasing and sustained government support of the sector. There will be support for training and education in construction, and for small construction firms weathering and recovering from the impacts of Covid-19. More houses will be built with government input, using off-site manufactured products as a solution to the country's significant housing affordability problem. VR and AR will be engaged in training construction workers, and construction educators will add BIM to their curricula as many large construction companies adopt BIM with government assistance. In time, this will also filter down to smaller companies. Blockchain and 3D printing will continue to generate plenty of discussion within the construction industry but will take considerable time to implement in any significant way.

REFERENCES

Ahmed, S. (2019). A review on using opportunities of augmented reality and virtual reality in construction project management. *Organization. Technology and Management in Construction: an International Journal, 11*(1), 1839–1852. doi:10.2478/otmcj-2018-0012

Ahmed, S., Hossain, M. M., & Hoque, M. I. (2017). A brief discussion on augmented reality and virtual reality in construction industry. *Journal of System and Management Sciences, 7*(3), 1–33.

AIQS & NZIQS. (2018). *Australia and New Zealand BIM best practice guidelines*. Retrieved from https://www.nziqs.co.nz/Portals/9/Documents/Miscellaneous/ANZIQS_BIM_Best_Practice_Guidelines_Nov2018.pdf?ver=2018-11-26-140519-660

Anthony, M. (2019). Blockchain in UAE construction. In Construction week. Auckland, NZ: Balasoglou Books.

Badcock, K. (2019, February/March). Digitisation of the construction industry. *New Zealand Construction News (NZCN)*, 21.

Bashabsheh, A. K., Alzoubi, H. H., & Ali, M. Z. (2019). The application of virtual reality technology in architectural pedagogy for building constructions. *Alexandria Engineering Journal, 58*(2), 713–723. doi:10.1016/j.aej.2019.06.002

BECA. (2020). *Supporting water treatment for a growing city*. Retrieved from https://www.beca.com/what-we-do/projects/water/pukete-wastewater-treatment-plant-bim-implementatic

Behzadi, A. (2016). Using augmented and virtual reality technology in the construction industry. *American Journal of Engineering Research, 5*(12), 350–353.

BIMinNZ. (2016). *Benefits of adopting BIM*. Retrieved from https://bim-in-nz.squarespace.com/benefits-of-adopting-bim

BIMinNZ. (2019). *Ara Institute of Canterbury Kahukura block*. Retrieved from https://bim-in-nz.squarespace.com/casestudies/2017/ara-institute-of-canterbury-kahukura-block

Bogue, R. (2018). What are the prospects for robots in the construction industry? *Industrial Robot. International Journal (Toronto, Ont.)*.

BRANZ. (2020). Material use. *Level*. Retrieved from http://www.level.org.nz/material-use/minimising-waste/

Brockmann, C., Brezinski, H., & Erbe, A. (2016). Innovation in construction megaprojects. *Journal of Construction Engineering and Management, 142*(11), 04016059. doi:10.1061/(ASCE)CO.1943-7862.0001168

Chen, B., Liu, A. M., & Hua, Y. (2017). An exploration of the interaction between BIM technology and the business process of a construction organization in BIM implementation. *WIT Transactions on the Built Environment, 169*, 177–189. doi:10.2495/BIM170171

Chen, H., & Samarasinghe, D. A. S. (2020). *The factors constraining the adoption of prefabrication in New Zealand residential construction sector.* Paper presented at the New Zealand Built Environment Research Symposium, Auckland, New Zealand.

Climo, D. (2018). VR and safety training. *BRANZ Build*, 167.

CSG & NZCIC. (2017). *The NZ construction industry – Briefing to Incoming Government, 2017: How a ministerial champion will benefit all New Zealanders.* Retrieved from https://nzcic.co.nz/wp-content/uploads/2018/10/Construction-Strategy-Brochure-V10.pdf

Dunmore, M. (2019, April/May). Westpac offers prefab choices in New Zealand-first scheme. *New Zealand Construction News (NZCN)*, 17.

El-Sayegh, S., Romdhane, L., & Manjikian, S. (2020). A critical review of 3D printing in construction: Benefits, challenges, and risks. *Archives of Civil and Mechanical Engineering, 20*(2), 1–25. doi:10.100743452-020-00038-w

European Commission. (1995). *Green Paper on Innovation.* Author.

Getuli, V., Capone, P., Bruttini, A., & Isaac, S. (2020). BIM-based immersive Virtual Reality for construction workspace planning: A safety-oriented approach. *Automation in Construction, 114*, 103160. doi:10.1016/j.autcon.2020.103160

Hafsia, M., Monacelli, E., & Martin, H. (2018). Virtual Reality Simulator for Construction workers. In *Proceedings of the Virtual Reality International Conference-Laval Virtual* (pp. 1-7). Academic Press.

Hager, I., Golonka, A., & Putanowicz, R. (2016). 3D printing of buildings and building components as the future of sustainable construction. *Procedia Engineering, 151*, 292–299. doi:10.1016/j.proeng.2016.07.357

Hamilton City Council. (2019, Nov 14). *Pukete wastewater treatment plant.* Retrieved from BIM: https://www.biminnz.co.nz/casestudies/2019/hcc-pukete-wwtp

Jackson, S. (2017). Innovation incentives. *Construction Journal*, 25. Large Cities. *International Journal for Housing Science, 36*(2), 99–107.

Li, H., & Samarasinghe, D. A. S. (2020). *Analysis of factors affecting design Buildability in New Zealand construction Projects.* Paper presented at the New Zealand Built Environment Research Symposium, Auckland, New Zealand.

Lim, S., Buswell, R. A., Le, T. T., Austin, S. A., Gibb, A. G., & Thorpe, T. (2012). Developments in construction-scale additive manufacturing processes. *Automation in Construction, 21*, 262–268. doi:10.1016/j.autcon.2011.06.010

Maghool, S. A. H., Moeini, S. H. I., & Arefazar, Y. (2018). An educational application based on virtual reality technology for learning architectural details: Challenges and benefits. *Archnet-IJAR: International Journal of Architectural Research, 12*(3), 246. doi:10.26687/archnet-ijar.v12i3.1719

Manseau, A. (2019). *Building tomorrow: innovation in construction and engineering.* Routledge. doi:10.4324/9781315261133

MBIE. (2019). *National Construction Pipeline Report 2019*. Retrieved from https://www.mbie.govt.nz/assets/national-construction-pipeline-report-2019.pdf

Mearian, L. (2019). Walmart launches 'world's largest' blockchain-based freight-and-payment network. *Computerworld*.

Meisenzahl, M. (2019). This building in Dubai is the largest 3D-printed structure in the world — and it took just 3 workers and a printer to build it. *Business Insider Australia*. Retrieved from https://www.businessinsider.com.au/dubai-largest-3d-printed-building-apis-cor-photos-2019-12?r=US&IR=T

Mirus, A., Patel, Y., & McPherson, P. (2018). *Prefabrication: New Zealand's golden ticket?* Academic Press.

Nambisan, S., Lyytinen, K., Majchrzak, A., & Song, M. (2017). Digital Innovation Management: Reinventing innovation management research in the digital world. *Management Information Systems Quarterly*, *41*(1), 223–238. doi:10.25300/MISQ/2017/41:1.03

New Zealand Government. (2019). *Using virtual reality to construct a safer workplace*. Retrieved from https://www.acc.co.nz/newsroom/stories/using-virtual-reality-to-construct-a-safer-workplace/

Nickerson, R. S. (1998). Confirmation Bias: A Ubiquitous Phenomenon in Many Guises. *Review of General Psychology*, *2*(2), 175–220. doi:10.1037/1089-2680.2.2.175

NZCN. (2019a). RMD changes the way customers look at designes with new visualization tool. *New Zealand Construction News (NZCN)*, 16.

NZCN. (2019b). Virtual reality becomes an educational reality. *New Zealand Construction News (NZCN)*, 7.

NZCN. (2019c, October/November). Fletcher Building opens New Zealand's largest offsite home manufacturing facility. *New Zealand Construction News (NZCN)*, 8.

NZCN. (2019d, October/November). Technology and collaboration star in award-wining buildings. *New Zealand Construction News (NZCN)*, 8.

O'Sullivan, D., & Dooley, L. (2009). *Applying innovation*. SAGE Publications.

Orcutt, M. (2018). *The block chain issue*. MIT Technology Review.

Otto, M., Prieur, M., Agethen, P., & Rukzio, E. (2016). Dual reality for production verification workshops: A comprehensive set of virtual methods. *Procedia CIRP*, *44*, 38–43. doi:10.1016/j.procir.2016.02.140

Ozorhon, B., Oral, K., & Demirkesen, S. (2016). Investigating the components of innovation in construction projects. *Journal of Management Engineering*, *32*(3), 04015052. doi:10.1061/(ASCE)ME.1943-5479.0000419

Papadonikolaki, E. (2017, June). Aligning BIM Adoption with Implementation In Loosely Coupled construction systems. In *Proceedings of the EPOC-MW Conference*. Engineering Project Organization Society.

Penzes, B. (2018). *Blockchain technology in the construction industry*. Institution of Civil Engineers.

Pham, T. N. T., Skelton, L., & Samarasinghe, D. A. S. (2020). *BIM implementation barriers in the architecture, engineering and construction (AEC) industry in New Zealand.* Paper presented at the 54th International Conference of the Architectural Science Association (ANZAScA) 2020, Auckland, New Zealand.

Prefab, N. Z. (2019). *How to Prefab: A series of New Zealand offsite construction case studies.* Retrieved from https://www.prefabnz.com/Downloads/Assets/13422/1/PrefabNZ%20HOW%20TO%20Prefab%20Book%202019.pdf

Ramachandra, T., & Rotimi, J. O. (2016). The construction contracts act and its implications for payment defaults in New Zealand. *Journal of Legal Affairs and Dispute Resolution in Engineering and Construction, 8*(1), C4515006. doi:10.1061/(ASCE)LA.1943-4170.0000179

Richardon, L. (2018, October/November). Designing in a virtual world. *New Zealand Construction News (NZCN)*, 21.

Sakin, M., & Kiroglu, Y. C. (2017). 3D Printing of Buildings: Construction of the Sustainable Houses of the Future by BIM. *Energy Procedia, 134*, 702–711. doi:10.1016/j.egypro.2017.09.562

Samarasinghe, D. A. S. (2020). *Development of framework for building materials-related decisions: A case study.* Paper presented at the Fifth Australasia and South-East Asia Structural Engineering and Construction Conference (ASEA SEC 5), Christchurch, New Zealand.

Samarasinghe, D. A. S., Abd-Latif, S., & Baghaei, N. (2019). *Virtual reality models for promoting learner engagement in construction studies.* Paper presented at the IEEE EDUCON - Global Engineering Education, Dubai, UAE. 10.1109/EDUCON.2019.8725120

Samarasinghe, D. A. S., Baghaei, N., & Stemmet, L. (2020). *Persuasive Virtual Reality: Promoting Earth Buildings in New Zealand.* Paper presented at the 15th International Conference on Persuasive Technologies, Aalborg, Denmark.

San, K. M., Choy, C. F., & Fung, W. P. (2019). The Potentials and Impacts of Block chain Technology in Construction Industry: A Literature Review. In *IOP Conference Series: Materials Science and Engineering,*. IOP Publishing.

Sher, D. (2019). Winsun completes world's first, 500 meters long, 3D printed river revetment wall. *3D Printing Media Network*. Retrieved from https://www.3dprintingmedia.network/winsun-completes-worlds-first-3d-printed-river-revetment-wall-in-largest-construction-3d-printing-project-yet

Tay, Y. W. D., Panda, B., Paul, S. C., Noor Mohamed, N. A., Tan, M. J., & Leong, K. F. (2017). 3D printing trends in building and construction industry: A review. *Virtual and Physical Prototyping, 12*(3), 261–276. doi:10.1080/17452759.2017.1326724

Thunberg, M., Rudberg, M., & Gustavsson, T. K. (2017). Categorising on-site problems. *Construction Innovation, 17*(1), 90-111. doi:http://dx.doi.org.ezproxy.auckland.ac.nz/10.1108/CI-10-2015-0059

Tobias, M. (2020). Prefabrication: How China built two Coronavirus hospitals in two weeks. *Nearby Engineers*. Retrieved from https://www.ny-engineers.com/blog/prefabrication-how-china-built-two-coronavirus-hospitals-in-two-weeks

Turk, Ž., & Klinc, R. (2017). Potentials of blockchain technology for construction management. *Procedia Engineering*, *196*, 638–645. doi:10.1016/j.proeng.2017.08.052

Udomdech, P., Papadonikolaki, E., & Davies, A. (2018, September). An alternative project-based learning model for building information modelling-using teams. Association of Researchers in Construction Management (ARCOM), 34, 57-66.

University of Nottingham. (2019, September 16). Virtual reality training could improve employee safety. *Science Daily*. Retrieved July 8, 2020 from www.sciencedaily.com/releases/2019/09/190916212516.htm

Valencia, N. (2017). *World's First 3D Printed Bridge Opens in Spain*. Archdaily. Retrieved from https://www.archdaily.com/804596/worlds-first-3d-printed-bridge-opens-in-spain

Wang, P., Wu, P., Wang, J., Chi, H. L., & Wang, X. (2018). A critical review of the use of virtual reality in construction engineering education and training. *International Journal of Environmental Research and Public Health*, *15*(6), 1204. doi:10.3390/ijerph15061204 PMID:29890627

Wilkinson, S., Noktehdan, M., Shahbazpour, M., & Finch, R. (2015). Learning from innovation. *Build*, (149), 86-87. Retrieved from https://www.buildmagazine.org.nz/assets/PDF/Build-149-86-Innovation-Learning-From-Innovation.pdf

Wilkinson, S., Tookey, J., Potangaroa, R., MacGregor, C., Milicich, M., Ghodrati, N., ... (2017). Transforming the Building Industry. *Urban Research Network*. Retrieved from https://www.buildingbetter.nz/publications/SRA6/SRA6_NSC11_State_of_Nation_knowledge_report_2017.pdf

WorkSafe. (2020). *Data centre*. Retrieved from https://data.worksafe.govt.nz/graph/summary/fatalities

Zhang, X., Li, M., Lim, J. H., Weng, Y., Tay, Y. W. D., Pham, H., & Pham, Q. C. (2018). Large-scale 3D printing by a team of mobile robots. *Automation in Construction*, *95*, 98–106. doi:10.1016/j.autcon.2018.08.004

Chapter 7

GIS-Based Watershed Analysis for Water Storage Facilities in Underdeveloped Areas:
Case of a Gravity Hill in Saudi Arabia

Rehan Jamil
https://orcid.org/0000-0002-8639-9571
Imam Abdulrahman Bin Faisal University, Saudi Arabia

ABSTRACT

Digital elevation models (DEMs) are created to study the topography of the area by using point krigging method of gridding. The results show a significant level difference between the start and end points of the valley sloping naturally. A longitudinal profile shows an average slope of 2.6% in a stretch of 15 km inside the area under study. Later, the detailed watershed and runoff analysis have been performed by preparing various maps using IDF curves available for the area. The area is found to experience sufficient rainfall for a 50 to 100-year storm return period. The expected location and amount of runoff water accumulation have also been determined which may help for the construction of water storage or rainwater harvesting facilities. The chapter shows the role of GIS-based techniques for the hydrological analysis of difficult terrains which can be applied for planning and management of water storage facilities in underdeveloped areas.

INTRODUCTION

The estimation of runoff is always a mandatory parameter for the design of infrastructure pertaining to hydrology of the area. Such catchments which are selected to be developed from the scratch, sometimes provide unique challenges to the hydrologic engineers during the planning and designing stage of a new occupational facilitates due to its topography, soil characteristics or various other environmental factors. One of such, difficult to analyze catchments, is that of a gravity hill. A gravity hill is defined as such a location where the layout of surroundings produces an optical illusion which makes a downhill road

DOI: 10.4018/978-1-7998-6600-8.ch007

appear uphill. The major factor which contributes to this illusion is the obstructed horizon. Without a horizon, which is a reliable referencc, it becomes difficult to judge the slope of a surface as it is not visible. There are many such places existing on Earth which have already been discovered. Bressan et al. (2003) has listed various locations of such spots which are famous and exist around the world with around 15 of them only in United States and Canada. All such places are famous either for supernatural activities, magnetic powers or even secret military projects (Benecke, 2002; Polidoro, 2007).

BACKGROUND

Very less literature has been found on the topic of gravity hills alone yet to talk about its hydrologic study. Most of the literature which is available discuss about the topographical aspects of such places and no hydrological study has been found specifically for a gravity hill. Regarding the topographical study of a gravity hill, Proffitt et al. (2001) did a lot of experiments on the perception of slants of hills. Their experiments concluded that the slant perceptions are always grossly overestimated. Bressan et al. (2003) have directly worked on the gravity hill phenomena by recreating the models of various gravity hill spots. They concluded that the gradient of the roads is misperceived due to the absence of a reference line or horizon and also that this phenomena is easy to be recreated artificially without putting any external force of magnetism or antigravity. Stefanucci et al. (2005) has discussed about the appearance of distance on sloping areas and more precisely on the hills. He concluded that the distances appear more on hills but his research is based on pedestrians walking uphill or downhill. Tsuinashi (2008) did an experimental work related with optical illusions by using pitched rooms and pitched floors. His experiments concluded that the vertical gradient of the roads or any surface in view could be misperceived and can work as an illusion. Kitaoka (2015) has discussed illusion on a gravity hill by using diagrams and included a case study of such spot in Radan, Serbia. Jaf (2009) has studied one such spot in Koya region, Kurdistan-Iraq. He performed visual inspections and took photographs to prove the phenomena of gravity hill existing within a small 200m section of road. Hengl et al. (2008) used around 2,000 data points to create the DEM of an area to study its hydrological and land cover properties by using maps. Mitra (2018) performed site suitability analysis to study the possible locations for rain water harvesting systems. She studied an area of West Bengal, India for the purpose. Jamil (2018) has done a detailed research on the topographical characteristics of Wadi Al Baida. He used various data resources to create digital elevation models to study the terrain of the area. Also, he provided logical reasoning to understand the phenomena of a gravity hill. However, no hydrological perspective was provided.

Water has already been declared as one of the important building tool to be considered for new constructions in Saudi Arabia due to the foreseen water shortage in the Gulf region (Shaawat and Jamil, 2014). There are many studies available in which researchers are performing expected water consumption in future to efficiently manage the water resources available at present (Jamil, 2020). Moreover, water is being considered for the sustainable construction as well (Shaawat et al., 2018). Hence study of water conservation is as important as any other survey or studies done before planning a community. Zaigham et al. (2015) prepared the watershed analysis of Rabigh region of Saudi Arabia which is in South of Madinah. They used ArcGIS to create the Digital Elevation Models for the area. A similar kind of study is available for a valley near Riyadh, Saudi Arabia which discusses the flood hazard for that region (Sharif et al., 2014). Subyani et al. (2012) have performed a comprehensive study of watershed of ten valleys situated in the western coastline of Saudi Arabia by using Digital Elevation Models.

Related to the hydrological study of Madinah, few references have been found in literature. Almazroui et al. (2012) have performed the rainfall and temperature analysis of Saudi Arabia for the year 1978-2009. They used three different gridded datasets for the analysis. A similar study was also done by Mashat and Abdel Basset (2011). Abdel Lattif and Al Shamrani (2013) prepared the comparison of the methods for the estimation of runoff coefficients for southwestern region of Saudi Arabia. They used multivariate statistical analysis for the procedure and developed an equation for the estimation of runoff coefficient as a function of catchment area. Abd Rahman et al. (2016) studied the hydrological characteristics of Wadi Al Aqiq in Madinah to determine the hydrological losses in the catchments. Subyani and Al Amri (2015) and Al Saleem (2018) have provided Intensity-Duration-Frequency (IDF) curves for Al Madinah region in their respective studies. The literature related with rainfall in Madinah shows that the area receives comparatively less amount of rainfall than other regions of the country. The data collected from various sources has been compiled and shown in Figure 1 which shows a single event of rainfall during a year from 1972 to 2019.

Figure 1. Maximum Single Rainfall Event during a Year at Madinah (1972-2019)

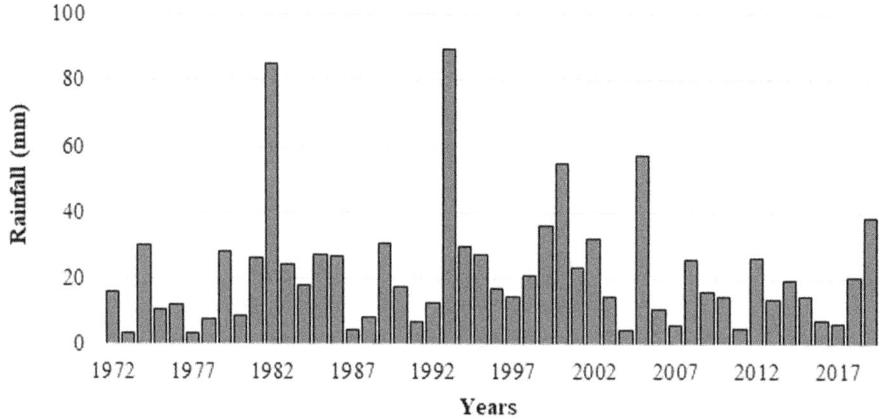

To the best of the author's knowledge, no hydrological study involving geospatial data by using Digital Elevation Modelling was found for any of the gravity hills in the world. The existing studies only consist of visual analysis through photography, existing maps and preparation of physical models, as cited earlier. This chapter focusses on two main objectives; the investigation of the topographical features of the gravity hill to assess the direction of its actual slope and secondly the estimation of runoff in the case of rainfall along with its flow direction for the purpose of suggesting any water storage or rainwater harvesting facilities which could fulfill the water demand of the future occupancy in the area. Both the objectives have been achieved by using the GIS-based techniques of creating DEMs and watershed maps.

Regarding the reliability of geospatial data obtained from Google Earth (GE) Pro 2019 for the study in this research, Mohammed et al. (2013) performed a comparison of the data obtained from GE with the data obtained by using a Global Positioning System (GPS) device. After performing few statistical analysis testing they concluded that the data obtained from GE can be used for investigation and preliminary studies with suitable accuracy. Hapep and Al-Bakri (2020) produced DEM by using Kriging based on GE data. In a similar kind of a study, Rusli et al. (2014) compared the DEM generated by using three different geospatial data sources including GE, Advanced Space-borne Thermal Emission and Reflection Radiometer (ASTER) and Shuttle Radar Topography Mission (SRTM). They created separate DEMs for flat, hilly and mountainous regions and for each section they took around 5,000 data points. After performing the correlation analysis they concluded that the DEMs formed by using the data of GE were closely related to the DEMs generated by the other two sources and the data of GE is completely reliable to be used for any geospatial studies. In another study, Rusli along with her coworker Majid (2012) studied the watershed of Sungai Muar by creating the DEM based on GE data along with ArcGIS. They concluded that GE provides accurate data and the error percentage was found to be only 0.19% which is not significant.

AREA UNDERSTUDY

One of the gravity hill in the world, exist in Saudi Arabia known as Wadi Al Baida (The White Valley). This area is located almost 25km north of the main Madinah city at 24° 40' 56" N and 39° 28' 5" E. As the words 'White Valley' suggest, the name emerged due to its geographical features which consists of light-colored sand in the valley area, which is in contrast to the dark-colored mountains surrounding it and spread over its entire length as shown in Figure 2.

Figure 2. Geological Features of Wadi Al Baida (Google Earth Pro 2019)

Wadi Al Baida is accessible through first ring road around Prophet's Mosque. About 25km along the Uthman bin Affan road, a divergence to North-West marks the start of the study area. About 2km on the North-Western diverging road, Al Khulail dam can be spotted which is built to collect small amount of water during rainy season which helps in irrigating the adjoining date tree fields. The valley is surrounded by dark mountains with few sand patches all along its length and general topography of the area is varying as shown by the contour plan (Figure 3).

The total length of the road of Wadi Al Baida is 15km. The road ends at a roundabout in the midst of the valley. There is no way to move forward as it is blocked by high mountains. The roundabout is constructed for the vehicles to take turn and move out of the valley towards Al Madinah city. The road leading back to the city seems to be flat all along the section which is visible from that point. To observe the gravity hill phenomena, most of the tourists put their cars at neutral gear at this location and the vehicles immediately start to move towards the city gaining speed at the midst due to momentum till it reaches near the Al Khulail Dam. At this point the vehicles start decelerating and engines have to be started to move ahead.

Figure 3. Contour Plan of Wadi Al Baida

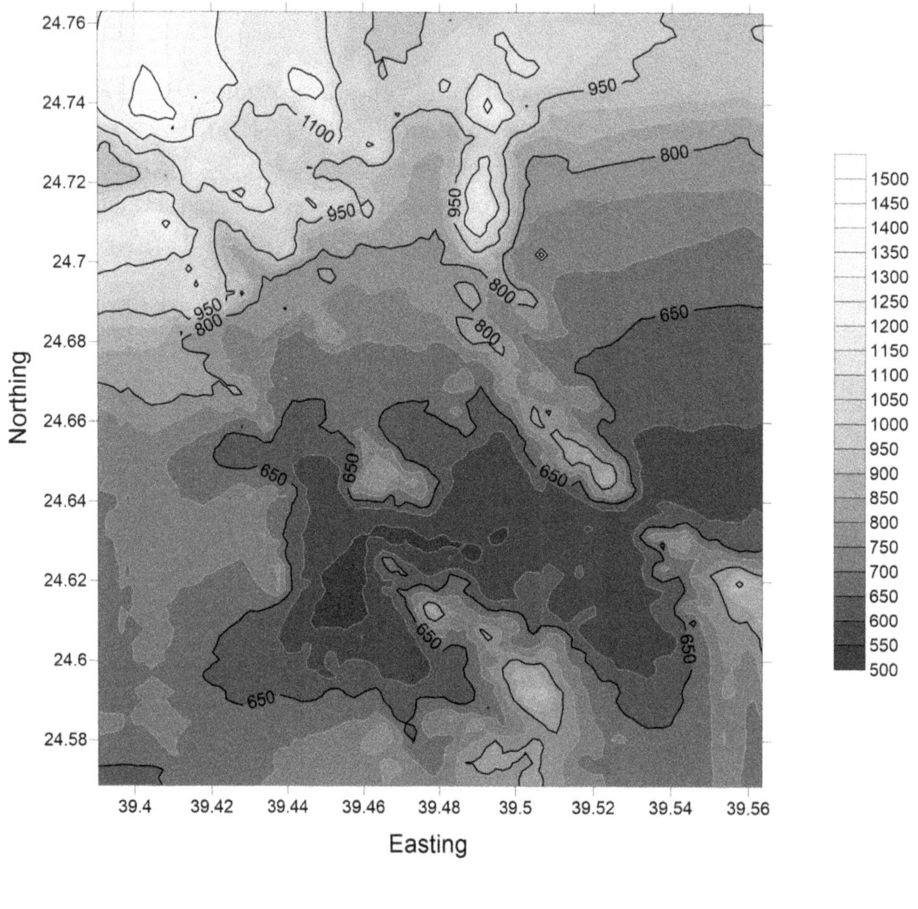

Although this observation was found to be correct, during the site visit, however any type of myths related to this were rejected and found illogical including the one that the valley has some magnetic powers which cause the vehicles to move on their own as nonmetallic round materials, including plastic bottles, also start rolling when put on the road freely.

METHODOLOGY OF INVESTIGATION

The scientific method adopted to study the topography of the gravity hill is the analysis of geospatial data of the area by creating DEMs and 3D surface profiles. A Digital Elevation Model (DEM) is a digital representation of the land surface, its terrain and general topography (Reuter et al., 2009). As mentioned by Nelson et al (2009), there are three methods of obtaining data to create a DEM; ground survey, maps and remote sensing. The source of data for remote sensing technique could be either airborne or situated in space which are satellites. For the purpose of the hydrological studies, catchment basins have been defined and watershed analysis has been performed to estimate the amount and pattern of runoff water.

DATA ACQUISITION

For the purpose of investigating the natural topography, a DEM was prepared for the whole area of Wadi Al Baida. The geospatial data acquisition for this study was done on GE Pro 2019. Approximately 10,500 data points were marked by using the 'Add Path' tool of GE Pro 2019 over the area of 20km by 10km, which makes almost 50 data points per square kilometer as shown in Figure 4. More the data points, more realistic is the elevation model and contour plan (De Smith et al., 2007) also more dense is the elevation data, better the model will depict the real surface (Stefan, 1996).

Each data point consists of three dimensional XYZ characteristic where Latitude is designated by X, Longitude by Y and the Elevation is designated by Z. GE follows World Geodetic System 1984 (WGS84) to maintain coordinates of Earth along latitudes and longitudes whereas for elevation data it follows Earth Gravitational Model 1996 (EGM96) (Mohammed et al., 2013; Ren et al., 2013).

After adding the data points in GE Pro 2019, the data was exported as Keyhole Markup Language (*.kml) file. Using TCX Converter v2.0.32, a conversion tool for various GIS and GPS based data files, the *.kml file was converted to Comma Separated Value (*.csv) file, which is readable in MS Excel format and the values of XYZ for all the data points can be read in separate cells.

GRIDDING AND 3D SURFACE MODELING

3D surface modeling software Surfer v13, developed by Golden Software LLC, Colorado, USA, was used for the preparation of contour maps, development of 3D surfaces for watershed analysis for the area under study. The Surfer 13 converts all the XYZ data in visual form by first creating a grid file and then by converting it into various types of 3D surface models for study and research purpose which gives a good overview of the topography and geospatial characteristics of the area.

For creating grids, Kriging method was selected in Surfer 13. Kriging is an advanced geostatistical procedure which is very useful and flexible method to produce maps when the available data is irregularly spaced, which is an exact case for the current case study. Kriging is of two types namely Point Kriging and Block Kriging. In Block Kriging gridding method the average value of the blocks is estimated. This method is not considered to be a perfect interpolator and develops smoother contours. On the other hand Point Kriging gridding method, estimates the values of the points at the grid nodes and hence produces accurate contours (Isaaks & Srivastava, 1989). The Kriging tool fits a mathematical function to a specified number of points, or all points within a specified radius, to determine the output value for each location (Surfer 13, 2015). Kriging is a multistep process; it includes exploratory statistical analysis of the data, variogram modeling, creating the surface, and (optionally) exploring a variance surface. The general formula for both interpolators is formed as a weighted sum of the data and can be shown as follows (Cressie, 1990).

Figure 4. Location of all data points

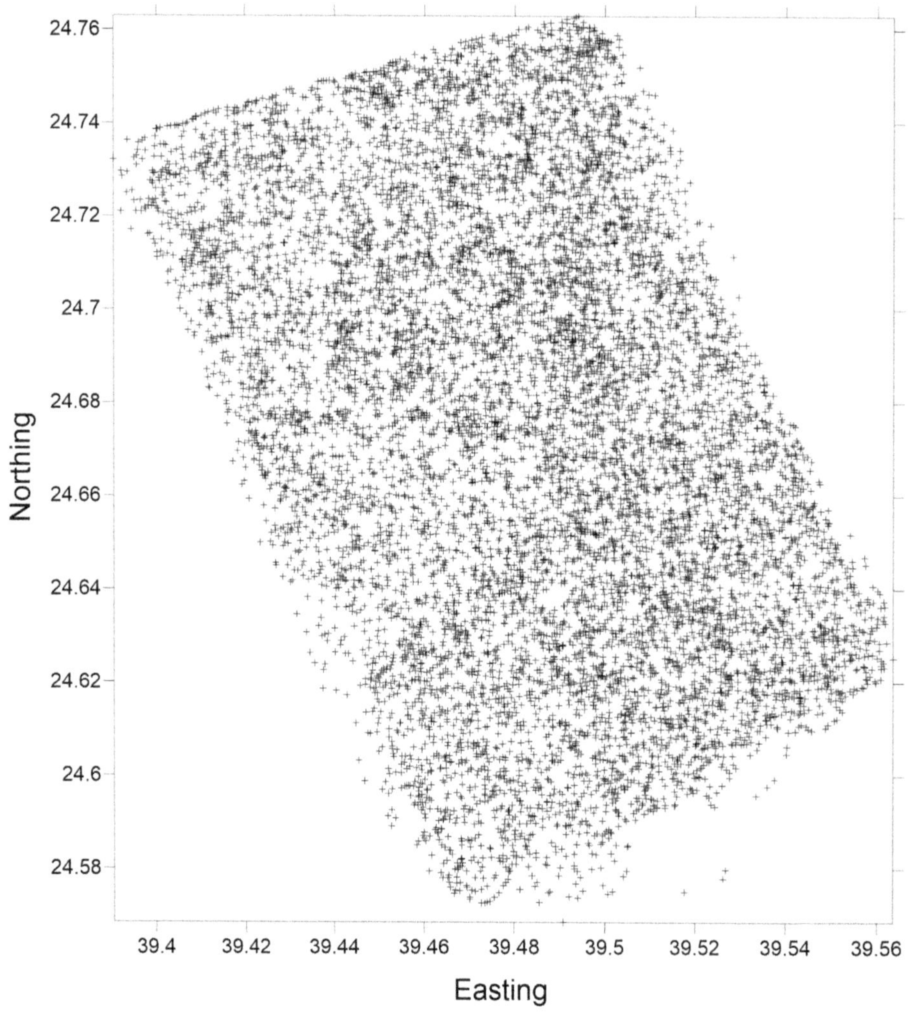

$$\hat{Z}(S_0) = \sum_{i=1}^{N} \lambda_i Z(S_i). \quad (1)$$

Where $Z(S_i)$ is the measured value at the ith location, λ_i is the unknown weight of the measured value at the ith location, S_0 shows the prediction location and N is the number of measured values.

With the Kriging method, the weights are based on the distance between the measured points, the prediction location and on the overall spatial arrangement of the measured points. To use the spatial arrangement in the weights, the spatial autocorrelation must be quantified. Thus, in ordinary Kriging, the weight, λ_i, depends on a fitted model to the measured points, the distance to the prediction location, and the spatial relationships among the measured values around the prediction location. Surfer 13 allows any one of the both options to be selected before creating a grid, Point Kriging being more reliable was selected for gridding of the data in this study. In addition, a longitudinal profile for Wadi Al Baida has also been developed for better understanding of the topography of the area under study.

Watershed Analysis and Runoff Estimation

Watershed maps display the direction that water flows across the grid. The watershed map breaks the grid into drainage basins and streams. Colors can be assigned to the basins and line properties can be associated with the streams in Surfer 13. The process of calculating the runoff quantities in Surfer 13 was accomplished by using the Integrated Volume Command which is a differential and integral operator. It gives the accumulated volume over the grid from the southwest to the northeast corner, or over a grid from southwest to northeast. This method can calculate volumes over grids of any shape. If the rainfall intensity and the runoff coefficient of the area is known for a specific thunderstorm, the integrated volume can be computed by overlaying the watershed boundary map and the amount of water flowing into each stream can then directly be obtained. Mathematically, the integrated volume is defined by the following equation.

$$V_{(x,y)} = \int_{\alpha=x\,min}^{x} \int_{\beta=y\,min}^{y} z(\alpha,\beta)\,d\alpha\, d\beta \,. \qquad (2)$$

For the purpose of calculation of runoff, generally two important parameters are required; rainfall intensity and runoff coefficient specific for the area. Al Saleem (2018) has provided Intensity-Duration-Frequency (IDF) curves for Al Madinah region. According to him the rainfall intensity for a 50-year storm return period is 40 mm/hr for a 30-minute duration of rainfall. Similarly the value has been given as 50 mm/hr for a 100-year storm return period with 30-minute duration of rainfall. Both values were obtained by him using the Gumbel Type-I distribution method for preparing IDF curves and have been taken as reference for the estimation of runoff in this research. A value of 0.0055 has been obtained by Al Hasan and Mattar (2014) as runoff coefficient for the Al Madinah region and has been used in this research.

RESULTS AND DISCUSSION

Digital Elevation Modeling

Point Kriging method was used to create the grid file from the acquired data. The grid having a formation of 100 rows × 89 columns was developed consisting of a total of 8,900 node points with none of them being a blanked node which meant that all nodes were provided with a value and none of the nodes was ignored. Figure 5 shows the DEM of the area when looking towards North (looking from start point towards endpoint of the road). The area where endpoint of the road is located is clearly elevated than the area of the entry of the valley. The height of mountains near the end of the road overtops 1500m of elevation as suggested by the color scale whereas most of the area surrounding the start point of the road is at around 550m of elevation.

Figure 5. Digital elevation model of wadi al baida

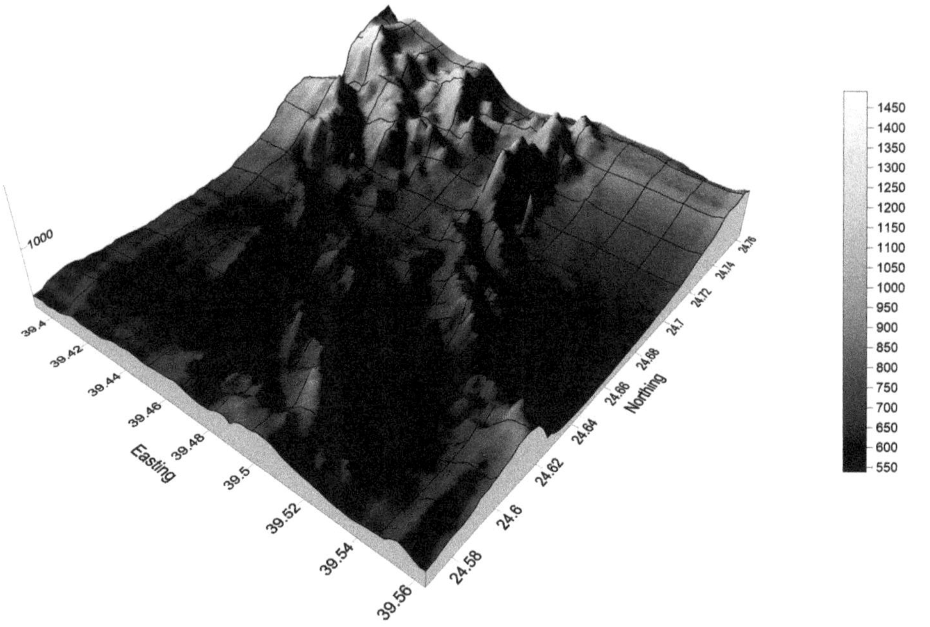

The longitudinal elevation profile of the road inside Wadi Al Baida is shown in Figure 6. It is clear that the endpoint of the road at roundabout is significantly higher than the start point of the road. The elevation of road at the start was found to be 560m whereas the elevation at the end was 949m creating a significant difference of 389m. The slope of road at various sections of road was found to be varying but it was found to be 2.6% on average within the length of 15km.

Figure 6. Longitudinal Profile of Wadi Al Baida

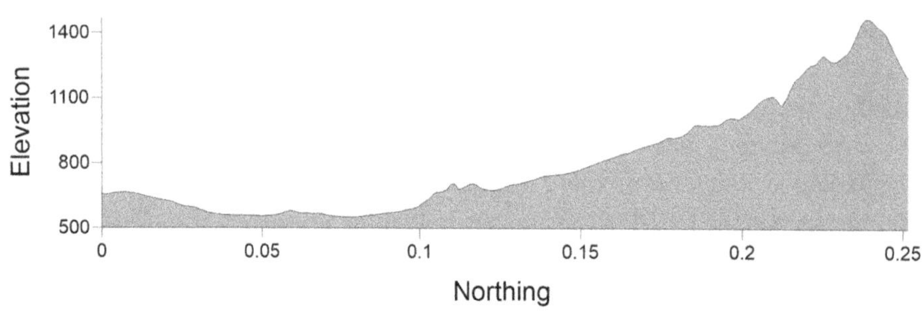

Figure 7. Watershed Map of Wadi Al Baida

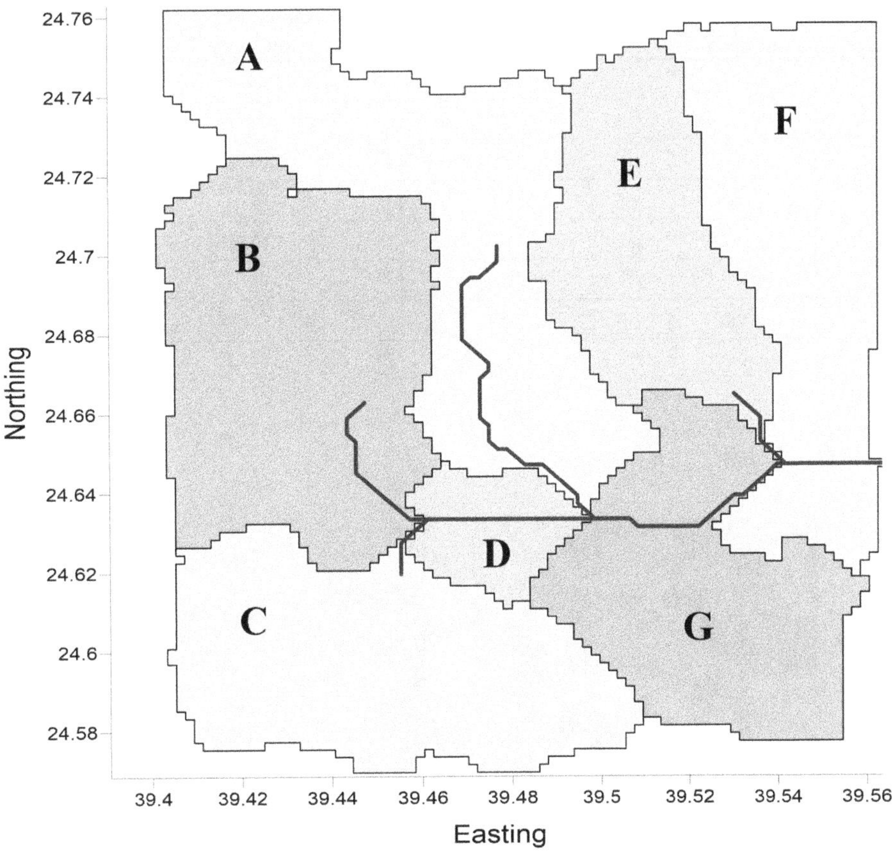

WATERSHED ANALYSIS AND ESTIMATION OF RUNOFF

After all the topographical analysis, the watershed maps can be generated from the software using the DEMs, already created for the purpose. The watershed map for the Wadi Al Baida is shown in Figure 7 along with streamlines and the direction of flow of runoff. The total watershed area has been divided into various small catchment basins based on the topography of each basin.

The whole catchment consisting of a total area of 200 km^2 was divided into seven watershed basins namely A through G. The area of each basin along with its stream length was directly calculated by the software and presented in Table 1. Basin A was the largest area with a value of 38.5 km^2 and basin D came out to be the smallest basin with an area of 6.9 km^2. Basin A contains the longest stream having a length of 6.2 km. The runoff values of Q_{50} and Q_{100} were obtained by using the rainfall intensity for 50 and 100-year storm return periods respectively. The calculations showed that the accumulated runoff values for 50 and 100-year storm return period were found to be 12.2 m^3/s and 15.3 m^3/s respectively and the outfall was found to be happening at eastern side of the valley at 24.65^o N. This location is found to be perfect for creating a water storage reservoir in the form of a lake which can serve the purpose of water supply to the future occupancy within the area. This stored water can be used to irrigate nearby barren lands for growing crops.

Table 1. Catchment Properties and Runoff for 50 and 100 Years Storm Return Periods

Catchment Basin	Area (km²)	Stream Length (km)	Runoff - Q_{50} (m³/s)	Runoff - Q_{100} (m³/s)
A	38.5	6.20	2.35	2.94
B	38.1	2.78	2.33	2.91
C	33.2	1.12	2.03	2.53
D	6.9	2.41	0.42	0.53
E	23.3	1.54	1.42	1.78
F	31.9	1.40	1.95	2.44
G	28.1	3.34	1.72	2.15

Figure 8. Image Map of Wadi Al Baida

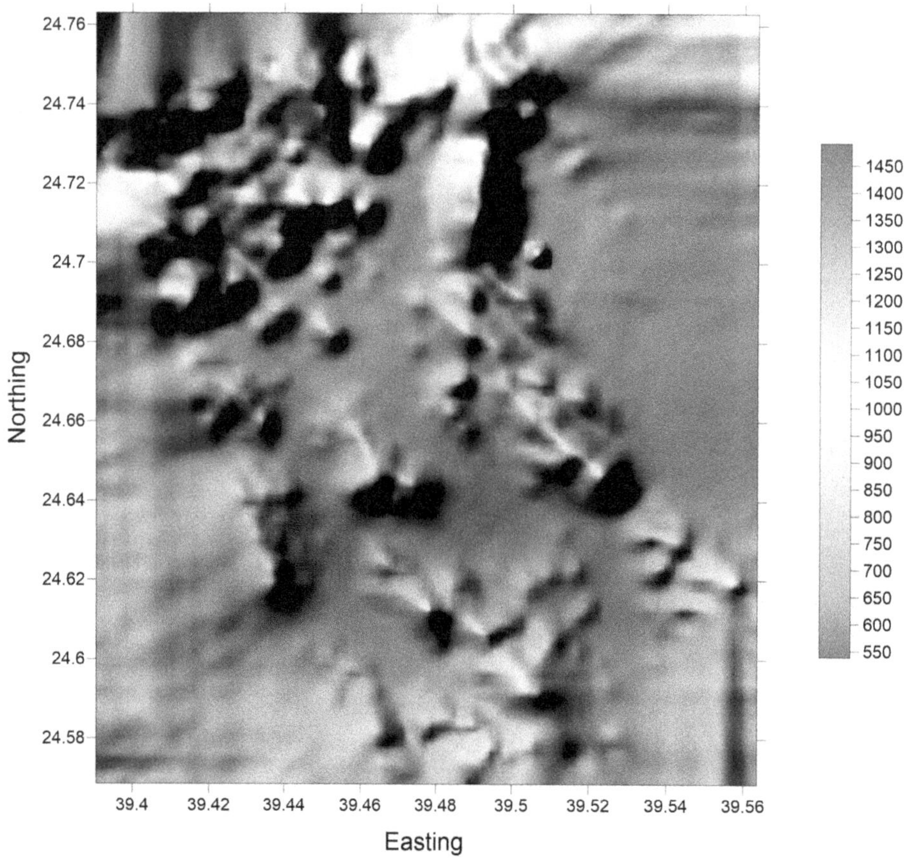

An image map was also generated for the area under study as shown in Figure 8, which is another illustrative way of showing the topography of any surface. This type of map shows all the features on the surface of the earth including mountains, valleys and plateau. Those areas which are flat are visible as smooth areas and an area consisting of steep slopes or mountains are visible as rough areas on the map. The image map shows a lot of rough areas around the valley. The map clearly shows that the whole valley is surrounded by the mountains and lowest-lying area is shown as blue color depicting the accumulation of water suggesting the location of a water storage facility. The watershed map (Figure 7) shows the same region for the accumulation of water which is around basin D and mid of basin G.

CONCLUSION

Built Environment is made useful for the occupants when all of the aspects of living are studied in depth and applied professionally. One of the aspect is the availability of sufficient amount of water. The application of this aspect needs a study of the hydrology of that area before the planning and designing of the infrastructure to assess the suitable location of water reservoirs to store rain water. As water is the basic necessity for a living, the hydrological study combined with topographical study becomes important for the planning of water storage and supply facilitates. There exists few strange and mysterious surfaces in the world having unique topographical characteristics which are known as gravity hills. In a gravity hill, the slopes are misperceived as they actually are. However the flow of water can show the exact situation of the topography of such areas, hence, such an area was put under scientific and hydrological study. 3D surface modeling was performed for Wadi Al Baida by using its geospatial data obtained through GIS-based techniques to assess its topographical parameters along with its watershed analysis for validation of the results. Various DEMs, maps and longitudinal profiles were developed which showed clear results. It was found that the endpoint of the valley was significantly higher in elevation than its start point. By observing the image maps the valley was found to be surrounded by mountains however total flow of water was estimated by dividing the area into various catchment basins and calculating the runoff along with the direction of major stream flow which validated the results obtained by topographical analysis. The study area was found to be experiencing a sufficient amount of runoff ranging from 12.2 m^3/s to 15.3 m^3/s at the lowest point of the basin D and G which also corresponds to the longitudinal profile of the road. The same location has been suggested for the purpose of constructing water storage and supply amenities. This stored rainwater can then be utilized for domestic purposes which shall result in water conservation and sustainability as well as reduction of the cost incurred on the fresh water.

REFERENCES

Abdel Lattif, A., & Al Shamrani, M. (2013). A comparison of methods for estimating the mean annual runoff coefficient at ungauged stream in southwestern region, Kingdom of Saudi Arabia. *International Journal of Environmental Science and Toxicology Research*, *1*(1), 12–20.

Al Hasan, A. A. S., & Mattar, E. (2014). Mean runoff coefficient estimation for ungauged streams in the Kingdom of Saudi Arabia. *Arabian Journal of Geosciences*, *7*(5), 2019–2029. doi:10.100712517-013-0892-7

Al Saleem, S. S. (2018). Rainfall-Runoff Analysis for Sustainable Stormwater Drainage for the City of Madinah, Saudi Arabia. *Arabian Journal for Science and Engineering*, *43*(4), 1955–1967. doi:10.100713369-017-2892-5

Almazroui, M., Islam, M. N., Athar, H., Jones, P. D., & Rahman, M. A. (2012). Recent climate change in the Arabian Peninsula: Annual rainfall and temperature analysis of Saudi Arabia for 1978–2009. *International Journal of Climatology*, *32*(6), 953–966. doi:10.1002/joc.3446

Benecke, M. (2002). Magnetic Mountains: Short Communications. *The Skeptical Inquirer*, *26*(3).

Bressan, P., Garlaschelli, L., & Barracano, M. (2003). Antigravity Hills are Visual Illusions, *American Psychological Society. Psychological Science*, *14*(5), 441–449. doi:10.1111/1467-9280.02451 PMID:12930474

Cressie, N. (1990). The Origins of Kriging. *Mathematical Geology*, *22*(3), 239–252. doi:10.1007/BF00889887

De Smith, M. J., Goodchild, M. F., & Longley, P. (2007). *Geospatial Analysis: A Comprehensive Guide to Principles, Techniques and Software Tools*. Troubador Publishing Ltd.

Hapep, A. M., & Al-Bakri, M. (2020). Comparison of Different DEM Generation Methods based on Open Source Datasets. *Journal of Engineering (Stevenage, England)*, *26*(1), 63–85.

Hengl, T., Bajat, B., Blagojevic´, D., & Reuter, H. I. (2008). Geostatistical Modelling of Topography using Auxiliary Maps. *Computers & Geosciences*, *34*(12), 1886–1899. doi:10.1016/j.cageo.2008.01.005

Isaaks, E. H., & Srivastava, R. M. (1989). *An Introduction to Applied Geostatistics*. Oxford University Press.

Jaf, P. M. (2009). Optical Illusion and the Magnetic Hills in Koya Region (Kurdistan-Iraq), *Journal of Kirkuk University. Scientiae Studia*, *4*(1), 40–47.

Jamil, R. (2018). Digital Elevation Modeling Analysis for Investigation of Gravity Hill Phenomena. *International Journal of 3-D Information Modeling*, *7*(3), 25-38.

Jamil, R. (2020). Hydroelectricity consumption forecast for Pakistan using ARIMA modeling and supply-demand analysis for the year 2030. *Renewable Energy*, *154*, 1–10. doi:10.1016/j.renene.2020.02.117

Kitaoka, A. (2015). Slope Illusion (Magnetic Hill). Art and Its Role in the History: Between Durability and Transient-ISMS, 751-760.

Mitra, S. S. (2018). *Site-Suitability Analysis for the Identification of Potential Sites to Construct Rain Water Harvesting Systems. In Hydrology and Water Resource Management: Breakthroughs in Research and Practice*. IGI Global.

Mohammed, N. Z., Ghazi, A., & Mustafa, H. E. (2013). Positional Accuracy Testing of Google Earth. *International Journal of Multidisciplinary Sciences and Engineering*, *4*(6), 6–9.

Nelson, A., Reuter, H. I., & Gessler, P. (2009). DEM Production Methods and Sources, *Chapter-3. Developments in Soil Science*, *33*, 65–85. doi:10.1016/S0166-2481(08)00003-2

Polidoro, M. (2007). Spook Hills in the Lab: Notes on a Strange World. *The Skeptical Inquirer*, *31*(6).

Pro, G. E. (2019). *Google Earth Pro*. https://www.google.com/earth/

Proffitt, D. R., Creem, S. H., & Zosh, W. D. (2001). Seeing Mountains in Mole Hills: Geographical-Slant Perception. *American Psychological Society. Psychological Science*, *12*(5), 418–423. doi:10.1111/1467-9280.00377 PMID:11554677

Rahman, N. A., Taher, S., Alahmadi, F., & Nasir, K. A. M. (2016). Arid Hydrological Modeling at Wadi Alaqiq, Madinah, Saudi Arabia. *Jurnal Teknologi*, *78*(7), 51–58. doi:10.11113/jt.v78.4516

Ren, S., Li, X., & Liu, X. (2013). The 3D Visual Research of Improved DEM Data Based on Google Earth and ACIS. *Pervasive Computing and The Networked World, Joint International Conference, ICPCA/SWS 2013, LNCS 8351*, 497-507.

Reuter, H. I., Hengl, T., Gessler, P., & Soille, P. (2009). Preparation of DEMs for Geomorphometric Analysis, *Chapter-4. Developments in Soil Science*, *33*, 87–120. doi:10.1016/S0166-2481(08)00004-4

Rusli, M., & Majid, M. R. (2012). Digital Elevation Model (DEM) Extraction from Google Earth: a Study in Sungai Muar Watershed. *Proceedings of the Applied Geoinformatics for Society and Environment 2012 (AGSE 2012),* 24-28.

Rusli, M., Majid, M. R., & Din, A. H. M. (2014). Google Earth's Derived Digital Elevation Model: A Comparative Assessment with ASTER and SRTM Data. *8th International Symposium of the Digital Earth (ISDE8). IOP Conference Series. Earth and Environmental Science*, *18*, 012065. doi:10.1088/1755-1315/18/1/012065

Shaawat, M. E., & Jamil, R. (2014). A Guide to Environmental Building Rating System for Construction of New Buildings in Saudi Arabia. *Emirates Journal for Engineering Research*, *19*(2), 47–56.

Shaawat, M. E., Jamil, R., & Al Enezi, M. M. (2018). Analysis of Challenges in Sustainable Construction Industry by Using Analytic Hierarchy Process: A Case Study of Jubail Industrial City, Saudi Arabia. *International Journal of Sustainable Real Estate and Construction Economics*, *1*(2), 109–122. doi:10.1504/IJSRECE.2018.092254

Sharif, H. O., Al-Juaidi, F. H., Al-Othman, A., Al-Dousary, I., Fadda, E., Jamal-Uddeen, S., & Elhassan, A. (2014). Flood hazards in an urbanizing watershed in Riyadh, Saudi Arabia. *Geomatics, Natural Hazards & Risk*, *7*(2), 702–720. doi:10.1080/19475705.2014.945101

Stefan, W. K. (1996) Using DTMs and GIS to Define Input Variables for Hydrological and Geomorphological Analysis. *Proceeding of the Vienna Conference, April 1996, HydroGIS 96: Applications of GIS in Hydrology and Water Resources Management*, 183-190.

Stefanucci, J. K., Proffitt, D. R., Banton, T., & Epstein, W. (2005). Distances Appear Different on Hills. *Perception & Psychophysics*, *67*(6), 1052–1060. doi:10.3758/BF03193631 PMID:16396013

Subyani, A. M., & Al Amri, N. S. (2015). IDF curves and daily rainfall generation for Al-Madinah city, Western Saudi Arabia. *Arabian Journal of Geosciences*, *8*(12), 11107–11119. doi:10.100712517-015-1999-9

Subyani, A. M., Qari, M. H., & Matsah, M. I. (2012). Digital elevation model and multivariate statistical analysis of morphometric parameters of some wadis, western Saudi Arabia. *Arabian Journal of Geosciences*, *5*(1), 147–157. doi:10.100712517-010-0149-7

Surfer 13. (2015). Full User's Guide. Golden Software LLC.

Tsuinashi, S. (2008). Visual Illusion of a Vertical Gradient: A Pitched Visual Environment and a Pitched Floor. *Japanese Journal of Phycology*, *79*(2), 125–133. PMID:18678062

Zaigham, N. A., Aburizaiza, O. S., Mahar, G. A., Nayyar, Z. A., & Siddique, A. (2015). Watershed Analysis of Rabigh Drainage Basin, Saudi Arabia. *International Journal of Water Resources and Arid Environments*, *4*(2), 138–145.

KEY TERMS AND DEFINITIONS

Catchment Basin: An area of land where precipitation collects and drains off into a common outlet, such as into a river, bay, or other body of water.

DEM: Digital Elevation Model/Modeling, a 3D representation of a terrain's surface created from a terrain's elevation data.

Geospatial Data: The data or information that identifies the geographic location of features and boundaries on Earth, such as natural or constructed features, oceans, mountains, plains and more. Geospatial data is usually stored as coordinates and topology which can be mapped.

IDF Curve: Intensity-Duration-Frequency Curve, a graphical representation of the probability that a given average rainfall intensity will occur. It is given by a mathematical function that relates the rainfall intensity with its duration and frequency of occurrence. These curves are commonly used in hydrology for flood forecasting and civil engineering for urban drainage design.

Runoff: Draining away of water from the surface of an area of land, a building or a structure, etc.

Wadi: Arabic word for valley, a low area of land between hills or mountains, typically with a river or stream flowing through it.

Watershed: An area of land that feeds all the water running under it and draining off of it into a body of water.

Watershed vs. Catchment Basin: A watershed is simply the area of land that catches rain and snow and drains or seeps into a marsh, stream river, lake, or groundwater. Catchment basin is the entire geographical area drained by a river and its tributaries; an area characterized by all runoff being conveyed to the same outlet.

Chapter 8
Blockchains for Use in Construction and Engineering Projects

George Blumberg
Oxford Brookes University, UK

ABSTRACT

This chapter describes blockchains and illustrates this explanation using the results of a prototype project for an industrial application for a construction project. The chapter describes the application and how modular software components can be used to assemble a blockchain solution. The chapter concludes with a design of the system architecture. The background to blockchain technology includes a description of the evolving nature due to communal, open software consortia and an accelerated prototyping of systems. Four recommendations are made in the chapter. These include the need to form consortia for prototyping applications, encouraging government involvement, the need for engagement with the open software development community, and the suggestion that systems should be designed to support Lean production. A final section offers a range of discussion topics on the current state of the technology and where to expect area of increased interest. These are summarized in three areas: Lean management, Industry 4.0 and smart cities, and topics around privacy and security.

INTRODUCTION

Perspectives of The Research and a Summary Of The Contents Of The Chapter

This chapter describes the technology behind blockchains and how, in conjunction with a suite of interrelated modular components, they can be used for a variety of tasks in construction and engineering (C&E) projects (Darabseh & Martins, 2020; Yang, et al., 2020; Hargaden, et al., 2019). There are a number of applications of blockchains that, irregardless of skepticism (Perera, et al., 2020), and the difficulty in scaling-up in size, could provide advantages over current solutions. This is because blockchains provide something novel: the ability to hold critical data that, once written, are immutable, protected,

DOI: 10.4018/978-1-7998-6600-8.ch008

can reside on a distributed network of computers. Immutability provides the data with authenticity, the use of distributed networks allows independence of a central hosting authority (Berg, et al., 2018; Zheng, et al., 2018; Pilkington, 2016), and security of valuable data is made possible by cryptography. Smart contracts, embedded in the blockchains, can help automate processes, most notably, payments. In other words, blockchains enable mutually distrustful parties to transact safely without the intervention of a trusted third party. This, most distinguishing attribute, is a mechanism that institutionalizes trust (Werbach, 2018; Berg, et al., 2017; Weber, et al., 2016; Anon., 2015b).

These features and capabilities are increasingly in demand as mobile arrays of interconnected devices produce volumes of data that can be valorized by artificial intelligence, big-data analytics, and advanced visual displays. Proponents of blockchain technology have been quick to propose applications to support concepts such as the Internet of Things (IoT) (Delgado-Mohatar, et al., 2020; Panarello, et al., 2018), Industry 4.0 (Lee, et al., 2019), smart cities (Swan, 2018; Huckle, et al., 2016), digital twins (Yaqoob, et al., 2020), and edge computing (Xiong, et al., 2018), and BIM. What all these concepts require is reliable machine to machine communications (Afanasev, et al., 2018) on a secure data-layer that is able to overcome the segmentation caused by multiple sources. It was the work of insightful and futuristic thinkers that envisioned the broader application of blockchains across industry (Al-Jaroodi & Mohamed, 2019), government and civil society (Swan, 2015). -

The novelty of the technology has prompted proposals in C&E for: energy management and carbon accounting (Petri, et al., 2020; Wang, et al., 2017), transport monitoring (Knirsch, et al., 2018), contract and document flow automation, and supply chain synchronization. The hope is that the use of blockchains can lead to improved project workflow, time and cost savings, increased transparency and trust amongst stakeholders (Turk and Klinc, 2017).

But there are impediments to the smooth adoption of this new technology, notably the difficulty in scaling up in size, in the management of the networks, and in getting the systems to function correctly in the first place. But there are significant trends that inspire continued interest in this area. Firstly, leading technology companies offer blockchain hosting and partnering solutions. Secondly, start-up companies are developing innovative technology that makes developing and deploying blockchains systems easier and finally, a growing community of enthusiast are collaborating to provide solutions.

For use in C&E projects, systems can be divided into two categories: (1) blockchains for holding critical data such as certificates, authorisations, payments and tokens and (2) blockchains optimised for enacting smart contracts within complex trading environments. Both of these categories would be classified as *industrial blockchains*. They use the same core engineering elements as those in cryptocurrencies such as Bitcoin (Nakamoto, 2008), but have added capabilities that allow for smart contracts and a choice of algorithms for ordering and confirming the authenticity of the data. The following sections will explain these features and describe how they work.

Who is This Chapter for and How Will the Contents Assist in Technology Implementation and Adoption?

Users of the chapter are likely to be those who are faced with a deluge of data emanating from devices in buildings and cities and want to use the properties of the blockchain to obtain improved analysis and problem solving. Amongst this group are likely to be managers who have an interest in automating certain aspects of their administrative, assembly or manufacturing processes and want to be able to integrate these with their suppliers, customers, banks, insurance companies and government regulators.

It is with some optimism that blockchain technology is described in this chapter as there is often hesitance shown by managers working in C&E to adopt new technology (Waterhouse et al., 2019) or for engaging in the level of collaboration required to adopt radical change (Akintoye and Main, 2007). However, there are encouraging signs, particularly if we consider how BIM and project management software, after initial hesitation, are now being used in many projects. Indeed, the adoption of Modern Methods of Construction (MMC) (Raynsford et al., 2016; Pasquire & Connolly, 2002) by contractors is a sign of the changing requirement for more sophisticated approaches to supply chain management, logistics and automation, all of which could be supported by the use of blockchains.

Aims and Objectives of the Chapter and a Summary of the Contents

The long-term aim of the research is to enable authenticity and independence of data in smart cities and to facilitate their use in delivering useful services. The short-term objectives of this chapter are as follows:

Objective 1: To explain private permissioned blockchains and associated modules, collectively referred to as digital ledger technology (DLT).

Objective 2: To illustrate the design principles and implementation process for a DLT using a prototype example.

Objective 3: To show how DLT can be applied to facilitate services in C&E projects.

Objective 4: To propose a system architecture for a DLT solution for C&E projects.

Objective 5: To provide a set of recommendations for those with an interest in seeing blockchains develop further.

Objective 6: To introduce a set of discussion topics on the future technical developments and expanded applications.

BACKGROUND

Basic Description of A Blockchain

Blockchains achieve their unique characteristics by the application of three technologies. These are: cryptography (Kosba, 2016), distributed systems on networks, and consensus algorithms (Wang et al., 2018b).

In its most basic form, a blockchain is a computer file that resides in multiple nodes of a computer network. An example of a fragment of a blockchain file is shown in Figure 1.

In this diagram, the three large vertical rectangles represent the *blocks* from which the blockchain derives its name. At the core of the blockchain is the cryptographic hash function (CHF), a mathematical algorithm that is very useful as it converts data of an arbitrary size (for example, a message or a password) to a string (or *hash*) of a fixed length. There are no passwords involved and, as a one-way (i.e. irreversible) function, it is nearly impossible to reverse. Hashes have useful properties, for example, miniscule difference in the original text makes for significant differences in the hashed versions.

The CHF is used to ensure that the chain of records contained in a blockchain cannot be altered as it would interfere with the propagation of hashes that are used to link one block to the next (Finck, 2018). An example of this is shown in the sixteen-digit hexadecimal hash (representing 64-bit encryption) shown in Figure 1. In addition to the hash of the previous block, each block also contains a schema referred to as the Merkle Tree (not shown in Figure 1), a structure in which historic hashed data are stored to ensure ordering is conserved.

Figure 1. Contains a schematic diagram of a typical blockchain file. In this example, three (3) completed blocks of data, ordered sequentially, are shown. The leftmost (the so-called Genesis block) is the first, with the blocks to the right added on later. The fourth block (shown as the rightmost rectangle that is shaded out) is represented as incomplete, and will only be finalized once the information has been authorised by the consensus algorithm.

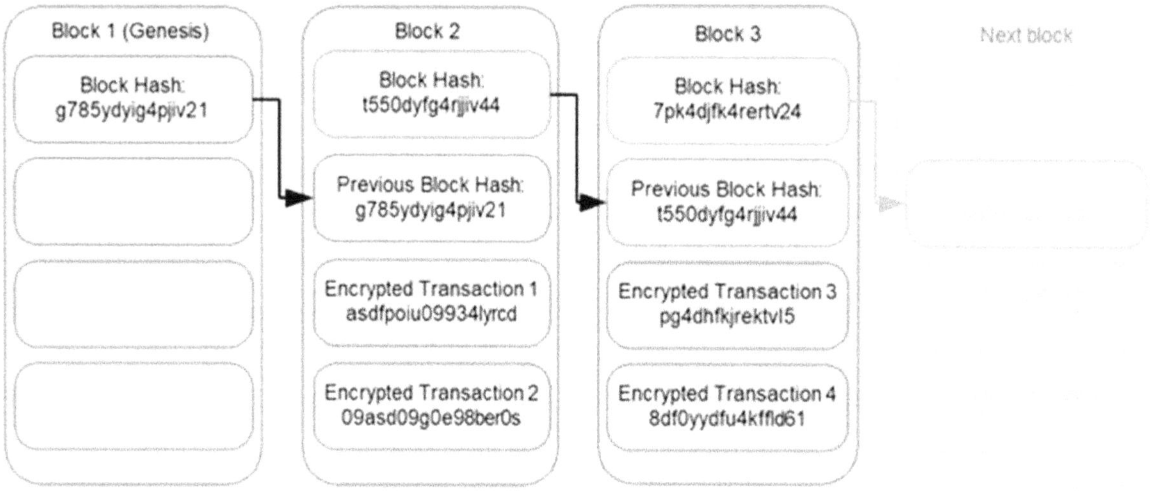

Cryptography, in the form of *asymmetrical key* (see Singh, 2000), is used elsewhere on the blockchain to ensure the security and privacy (Banerjee, et al., 2018; Szabo, 1997) of individual records. This helps protect against attempts to alter the details of the records as is required by current standards for computer security in the built environment (PAS 1192-5., 2015). Users are issued with a unique pair of numbers by a trusted certificate authority (CA). One – the private key – is kept secret, while the other, the so-called public key, is visible to all. The public key of the pair is used to encode a message that can only be decoded with the private key (Rivest, et al., 1978), thus ensuring a level of individually tailored privacy on the data in a shared ledger or blockchain.

Figure 1 illustrates how the blockchain file is constructed, but to determine which records are permitted to be written to the blockchain requires the use of an ordering and consensus algorithm.

Consensus and Ordering Algorithms in Digital Ledger Technology

The second technology needed to make blockchains work is the ordering and consensus algorithm (Cachin and Vukolić, 2017; Swanson, 2015), the main purpose of which is to ensure only legitimate records get added to the blockchain and that they are added in the correct order. This is necessary as data can arrive via any of the nodes in the network.

For cryptocurrencies, like bitcoin, the most common form of consensus algorithm is proof-of-work (Anon., 2015a), a time and energy intensive operation where self-selected members of the network race to perform a routine that confirms the authenticity and order of the transactions. In these blockchains, there can be thousands of nodes, each with their own copy of the blockchain and each recording thousands of transactions per hour. Members who do the computation the fastest are rewarded with cryptocurrency credits.

Fortunately, proof-of-work is not required for industrial blockchains, such as those designed for C&E projects. In these systems, the number of traders is likely to be less than one hundred and a transaction rate less than a hundred per day. In these cases, the ordering and consensus algorithm must ensure that basic trading rules are enforced between members, so that transactions are legal.

More importantly the system must be robust and dependable. This can be achieved by incorporating *Byzantine fault tolerance* (BFT)(Lamport et al., 1982) into the ordering and consensus algorithm. Using BFT, data will be preserved if one or more of the nodes goes out of service or is interrupted.

For example, transaction records can be confirmed as legitimate using a small, embedded programmes called smart contracts. These are configured to verify a if a set of pre-determined requirements, such as entering a digital signature, have been fulfilled. An effective consensus algorithm will reduce the ability to fake, falsify or enter a transaction more than once. In real trading environments, this double entry can be costly and are hallmarks of organized crime (Beare 2007, p43).

Distributed Systems and Digital Ledger Technology

Distributed systems are the final technology that makes blockchains workable. Unfortunately, it is also what makes them so difficult to deploy, maintain and upgrade. Blockchains exist only because they can be shared by nodes across a network using a peer-to-peer application (see Steinmetz & Wehrle, 2005). Key to these working is, of course, the internet, which provides the language and physical systems that permits communication and coordination.

To illustrate these concepts further, consider the diagram in Figure 2. This flowchart maps the transactions of an *asset* between members of a trading network (referred to as *Nodes* in this example). The *asset* traded could be any entity that is able to be represented by digital data, for example, a quantity of material, a confirmation of delivery or a building component.

The initiation of the network starts when Peer node 1 establishes the network and performs a series of tasks, such as setting up a certificate authority for public/private key encryption on the network. This closed, permissioned network is represented by the central octagonal box. In this scenario, the system administrator also sets up a world state database, configures the consensus algorithm and invites members to join the network. This action is represented by the arrow labelled **A** in the figure. The act of establishing a network starts the process where transactions can be recorded. Trading records are sent to the Peer nodes (represented by the arrow labelled **B**) for confirmation via a network consensus and protocol to create the Genesis block. The arrow labelled **C** shows Peer Node 3 writing two separate transactions to the network. Included in this newly written Block 2 includes the hash of the previous block. Block 2 is created only after it has been confirmed and verified by the consensus algorithm. This is shown by the arrow labelled **D**. This process is repeated when Peer node 1 sends 2 transactions to peers to await the creation of Block 3 (**E**). This, in turn, includes the hash of the previous Block 2, again only after consensus confirmation has been confirmed (**F**). At each stage of this process, the nodes within the network contain identical versions of the blockchain file and will continue to do so until the next transaction is proposed.

Figure 2. This diagram shows a simplified schematic of how the core blockchain functions. It is a flowchart describing how transactions are recorded on a blockchain. Arrows show the direction of the records as transactions proceed and are written to the blockchain. The sequence in this flowchart follows alphabetical order starting with A and culminating with the final block being written at F.

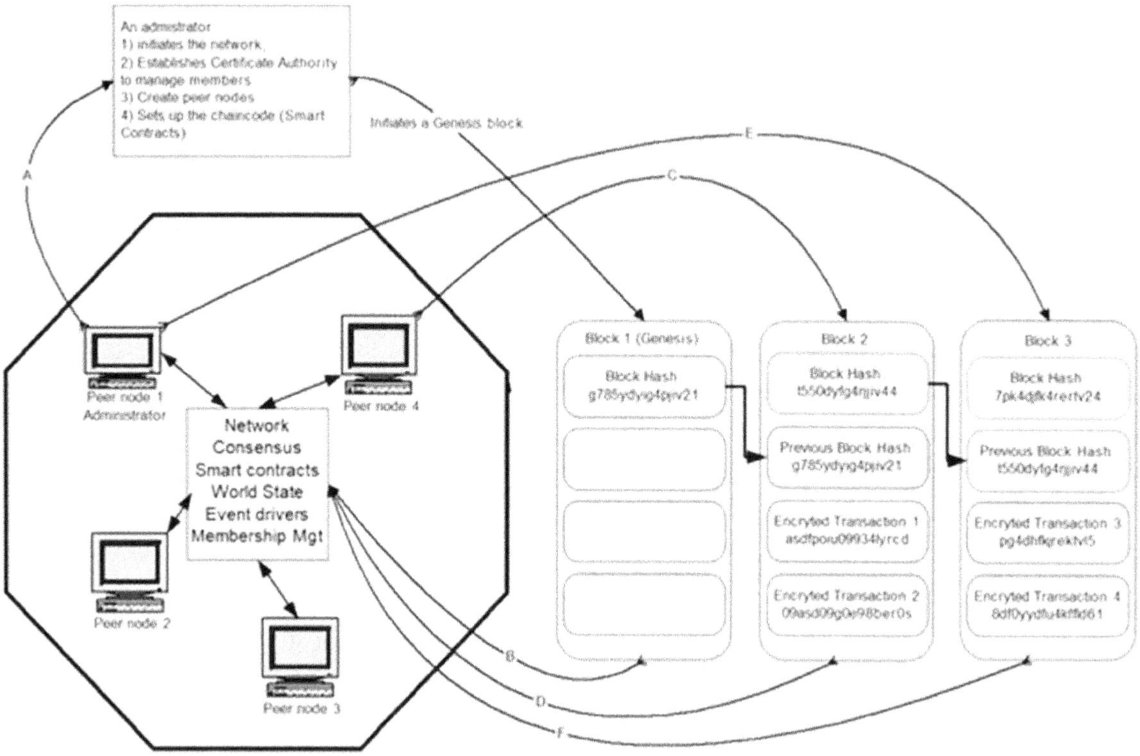

In this example, the basic elements that characterize an industrial blockchain are described. In summary these are:

Item 1: A replicable ledger with the history of all transactions that are added sequentially and have an immutable past held on files that are replicated across a network.
Item 2: Business logic, in the form of embedded smart contracts, are executed along with the transactions.
Item 3: A consensus and ordering service that ensures a decentralized protocol that can be used to control inevitable disruptions and to allow the transactions to be validated.
Item 4: In all these elements, cryptography is used to ensure the integrity of the ledger, the privacy and authenticity of transactions and the identity of participants.

These requirements are best fulfilled by a *private* and *permissioned* blockchain, which requires a central organization to manage the home node of the blockchain, where the administrator can deploy, upgrade, and maintain the network. This doesn't reduce any of the robustness of the systems. Hyperledger Fabric (HLF), developed as an open source project, is ideally suited for industrial applications requiring private, permissioned blockchains (Hyperledger Foundation, 2017; Dhillon, et al., 2017; Vukolić, 2016). HLF is modular and allows tailored and additive solutions. For example, the ordering and consensus algorithm is configurable to allow the business model to be programmed into the smart contracts. This includes a multi-level security setup that ensures transactions are legitimate and ordered correctly.

BLOCKCHAINS USED FOR SECURE RECORDING

The secure handling of data has operational as well as legal requirements (see Yue, et al., 2016, for an example in healthcare). What makes storing information on the blockchain different from hosting it on a secure database is that its storage and retrieval are controllable, but not dependent on a central authority. Case studies of prototype systems have been published, for instance, for digital diplomas and educational qualifications (Jirgensons & Kapenieks, 2018) and other certificate-centered activities, such as commercial debt obligations, could also be accomplished using DLT (Cheng et al., 2018).

Of particularly interest for C&E projects are:

- Building regulations completion certificate,
- Certificate of occupancy,
- Defects certificate,
- Energy performance and carbon reduction certificates,
- Established use certificate,
- Planning permission and,
- Practical completion certificate.

Obtaining approvals represents a burdensome and bureaucratic aspect of commercial life and DLT has the potential to provide considerable cost savings. Indeed, this and other types of non-productive work are noted in Government industrial strategy documents for the sector (see Cable et al., 2013 for an example).

Certification is particularly important in the construction industry due to the inherent risks to the public from dangerous buildings. A good example of this is the heightened concerns about the safety of buildings due to the risk of fire (see Brokenshire, 2018 or Hackitt, 2018). In automated buildings, smart contracts operating on blockchains could act to invalidate a certificate of occupancy should there be a failure to properly monitor or repair safety devices. To provide an idea of what a digital certificate looks like, Figure 3 contains the decrypted version and demonstrates the essential information required for certification.

The certificate represented in Figure 3 is maintained on the blockchain and is readable by anyone with the link and the password. The consensus algorithm validated the certificate using a smart contract to verity the digital signatures of responsible officials.

BLOCKCHAINS USED FOR MANAGING TRADING NETWORKS

Blockchains can also be used to record transaction details (see Chapron, 2017) in complex trading environments that contain multiple layers of suppliers and customers who require a digital paper trail (Hultgren and Pajala, 2018; Penzes 2018). With further enhancement, the system could also be made to allow automation, for example, by making a secure payment (Wang et al., 2018a). Payments could be made independently of any central authority, a feature particularly useful when trading partners are not entirely trusting of each other (Tapscott and Tapscott, 2017; Carroll and Bellotti, 2015) and when delays in payment introduces additional financial strain (Das, et al., 2020). Indeed, further automation of a range of assembly and administration processes could provide broad benefits across the construction industry and to help improve issues of low productivity and profitability (Heiskanen, 2017; Barbosa et al., 2017). These additional elements make up a modular DLT system (Syed, et al., 2019; Xu et al., 2019). Existing technology, such as enterprise resource planning (ERP) software can do the same thing but is expensive and lacks an element of impartiality that the blockchains can provide.

Figure 3. Shows a certificate encrypted and written to a blockchain. This record contains a certification written and authorised by Building Control (a governmental office) to the Clerk of Works that confirms that the component is safe to operate. Further details of this example are provided in this and following sections. The format is consistent with the new IEEE data standard (Li & Sond, 2020).

Unique certification number:	223
Address:	25 Mill Street, Oxford
Time of certification:	14:52:23
Date of certification:	June 6, 2020
State of asset:	Certified
Location:	A46P
Drawing number:	BGH58091
Certification authority:	Building Control
Custodian of certification:	Clerk of Works

Transaction *asdfpou09934yrcd* decoded with key

To clarify the utility of automated data recording systems, consider for a moment the example of an autonomous vehicle moving across a conurbation carrying paying passengers and cargo. As it traverses the city, it uses a combination of public and private toll roads along which it picks up and discharges passengers and cargo. Near the end of the trip, the battery powered vehicle sells its excess power at a favorable rate. What characterizes this imaginary journey is a series of transactions which are shown in Table 1. In this table, the type of asset is listed along with the transactor (or seller) and custodian (the buyer). These transactions are recorded on different blockchains depending on the type of asset.

As a rule, assets can be any tradable entity, such as units of transport, or energy or a measure of work performed. If it can be represented in digital form, then a trade of an entity can be recorded in a blockchain. In the terminology of DLT, transactions would all possess a *state* and the blockchain records *state changes* to the assets. Examples of *state* would be *issued, assigned, sold, rejected* or other qualification that describes the asset. The advantage of this type of recording is that data can be easily streamed for collection and analysis.

Table 1. Contains a list of transactions performed between an autonomous vehicle (numbered 1 in this example) and several clients and customers represented on the blockchain as a transactor and custodian of an asset.

Transaction	Asset:	Transactor of asset:	Custodian of asset:
Vehicle 1 picks up a passenger	Unit human transport	Passengers 1	Vehicle 1
Vehicle picks up cargo	Unit cargo transport	Customer 1	Vehicle 1
Vehicle pays toll on Road 1	Unit use of road	Vehicle 1	Municipality 1
Vehicle 1 picks up a passenger	Unit human transport	Passengers 2	Vehicle 1
Vehicle pays toll on Road 2	Unit use of road	Vehicle 1	Municipality 2
Vehicle discharges excess energy	Unit of electrical energy	Vehicle 1	Local energy collective

Design Principles for Industrial Applications

This section provides a glimpse into the design principles required to integrate data collection via blockchains with databases, messaging systems and websites. The design principle required to implement systems and introduce automation to the process (see Li, et al., 2018 and Li & Sond, 2020 for the data standards) are:

Item 1: The recording of transactions between traders in a way that eliminates data discrepancies and allows simultaneous multi-party collaboration with data accessible to all parties in real-time.
Item 2: The ability to establish multiple trade channels or sub-networks to maintain data privacy.
Item 3: The capability that transactions can trigger events such as sending email messages, automatic invoice creation, payments, and for providing proof of delivery.
Item 4: To support the deployment of smart contracts that can aid in the modelling of business processes.
Item 5: To provide a data streaming service to allow the flow of transaction data with analytical systems, such as an ERP that might use artificial intelligence and business analytics and to maintain control of production and performance.
Item 6: To enable a high degree of security and privacy at a level appropriate for commercial operations. This includes the adherence to General Data Protection Regulation (GDPR) (Trong, et al., 2020).
Item 7: Provide membership management and a certificate authority to ensure that members and peers' identity is authentic and that they are authorized to invoke transactions within a channel in a blockchain network.
Item 8: That it provides a high degree of reliability to guarantee robust operation in industrial settings.

Business Process and Data Modelling

In this example, a scenario is described where building components are designed, ordered, delivered, installed, inspected, certified, and paid for with all transactions recorded by a blockchain. This scenario requires that the business process and the data models are codified. One approach to formulating this mapping process is to use a combination of flow and swimlane charting (Chang, et al., 2019; Auberger and Kloppmann, 2017). In this example, this is done using the artifact-centric business process model

(Damelio, 2016; Nigam and Caswell), where multi-thread and multi-component processes are organized around service provision to online clients (Waller, 2003). Business process mapping (BPM) is seen as one of the fast-growth technology areas as it is an enabling technology underpinning the IoT (Miller, 2019). It is also a process useful in the automation of commercial sites that are accessed by distributed, client-side applications (Viriyasitavat et al., 2018). These use graphical and model-driven tools to design the blockchain business network (Seebacher and Maleshkova, 2018).

Figure 4. Is a flowchart that shows the process of installation of the component that was made off-site. Those involved in this, project manager (PM), Building Control (BC), Contract Administrator (CA), etc are listed in the membership table.

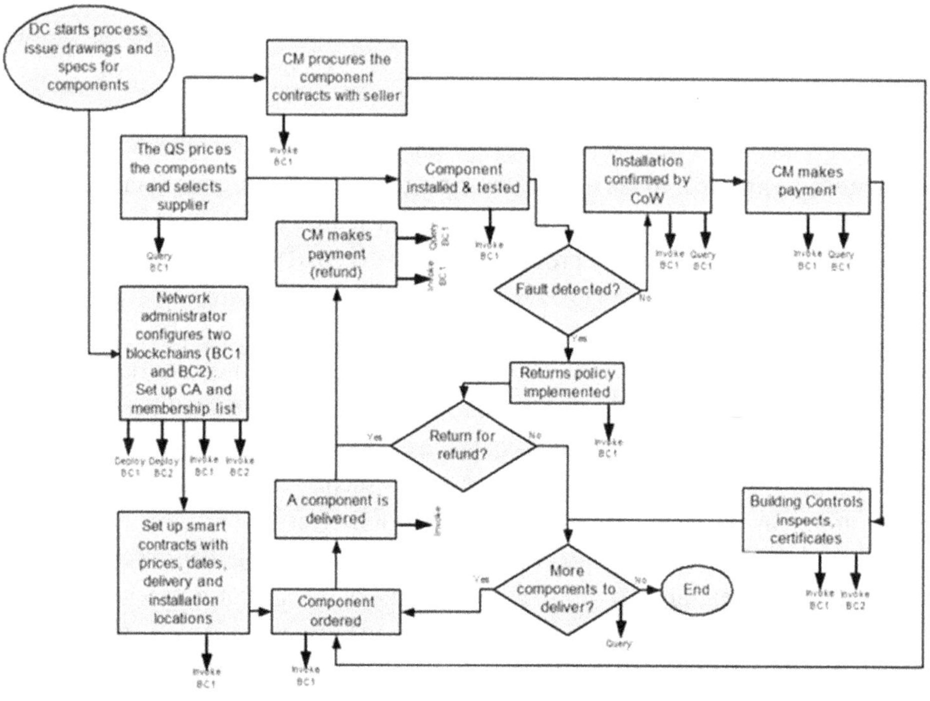

Figure 4 contains a flowchart of a business process describing a building component installation cycle. The flowchart shown here is a simplification of the typical real-life ordering, installation cycles. There are branches not included, for example if the component fails during the warranty period. However, one of the main advantages of DLT is that the transaction variables (or states) can be easily added to cover all eventualities. This amounts to a democratization of the data collection process. In the simplified scenario shown in Figure 4, the *states* of transaction for Blockchain 1 (BC1) are as follows:

State 1: Design approved
State 2: Price agreed
State 3: Component ordered
State 4: Component delivered
State 5: Component installed

State 6: Component inspected
Statc 7: Component Invoiced
State 8: Component paid for
State 9: Component rejected

Figure 5. Contains a swim-lane chart for the example of a DLT in a C&E project that involves the cycle of transactions required to install a series of building components. The rightmost box on the top of the figure represents the blockchain and the arrows indicate that the members of the network, who are also identified in the boxes across the top, write to the blockchain.

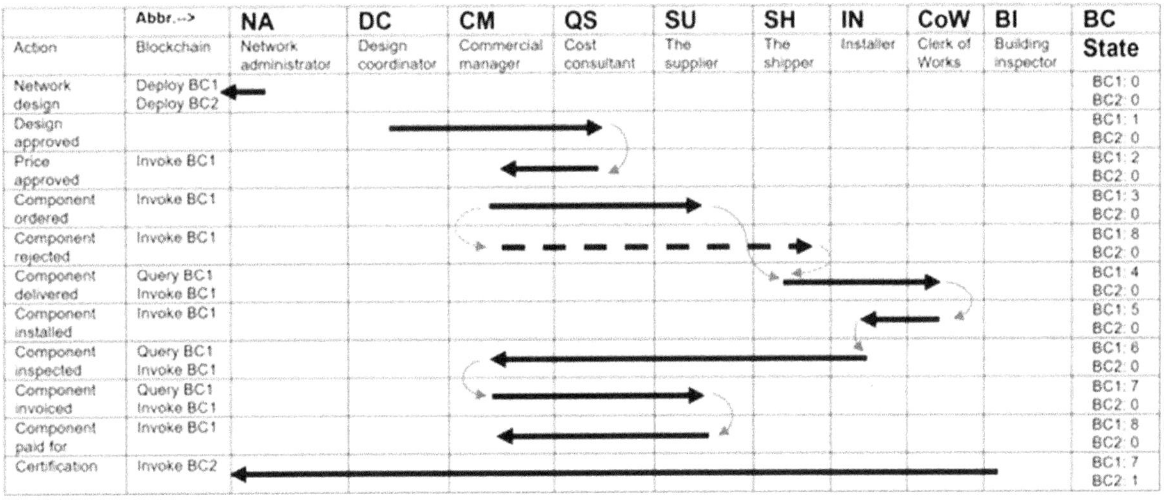

Other states of the BC1 transaction exist in more completely modelled systems, for example, *component failure, warranty period exceeded, payment delayed, delivery delayed, invoice greater than quote*, for example. Any of these states could trigger automatic action in invoking the blockchain. For example, if the *payment delayed* variable exceeds 5 days, then a penalty is automatically added to the invoice.

For Blockchain 2 (BC2), which contains the installation certificate, there are only two states which are:

State 1: Not certified
State 2: Component certified

To help design the business logic required for this example system, a swim-lane chart (after Damelio, 2016), shown in Figure 5, is used to display a chain of transactions that describe the progressively evolving state of the blockchain. Using this in conjunction with the flowchart (Figure 4) allows an iterative approach to modelling to achieve an accurate representation (Garcia-Bañuelos et al., 2017) of the business process leading to the schedule for the DLT coding.

In this swimlane chart, the process proceeds downwards with the state changing at every invocation of the blockchain until the component is invoiced, then paid for. The sequence of transactions starts with the blockchain being *deployed* by the Network Administrator (NA). From this initial stage, the Design Coordinator (DC) submits a set of drawings to a communal repository so that the QS can price the component and set up the terms of the contract. Approval by the Commercial Manager (CM) *Invokes* the transaction to the supplier (SU) to order the component. Once received and confirmed by the Clerk of Works (CoW), the Installer (IN) completes the installation of the component, with the blockchain recording the transaction and informing the Commercial Manager (CM), who in turn informs the SU that the order is ready to be invoiced. Once this is done an automatic payment is made.

Note that two independent blockchains are *Deployed* in this example. The first one mirrors the contracts between the client, shipper, and supplier. The other blockchain (BC2) is used to store official certification documentation. Finally, the Building Inspector (BI) provides certification, a transaction that is stored on another blockchain (BC2, in this example).

To add a record to this the Building Control (BC) regulator *invokes* a transaction to the blockchain with a record that contains certification information that is required for the safe occupancy of the building. This digital certificate is repeated for each component installed done in accordance to the building code and serves as an official mark of compliance. Certificates written to the blockchain can be read by anyone with access privilege.

Using a modular system approach, it should be possible for transactions in the network to trigger signals in the form of a text message or email. This would greatly facilitate automation and communication. The Installer (IN) would receive notification on his mobile device when a component is ready for fixing into place. The exact location (floor and room number) for each component would also be conveyed in the DLT, so that, for example, the crane operator, plumber, and other technicians can ensure the accuracy and timing of their activities.

In the Conclusion of this chapter, the elements that make up an industrial blockchain using modular DLT technology are explained along with the development framework that is used for testing and for prototype systems.

CONCLUSION

This section concludes the chapter with a description of a suitable development framework and architecture for the application of DLT to C&E projects. The first part of this conclusion introduces the HLF framework, describing the open source development environment where professionals in academia, industry and the not-for-profit sector collaborate to advance this complex and novel technology. Essential aspects of the design, such as membership management and the system architecture are detailed and illustrated with the same example as in previous sections. The section finishes with a list of recommendations for readers of this chapter and a brief discussion on future topics for research.

The Hyperledger Fabric Software Framework and Open Source Development Environment

The Hyperledger Fabric (HLF) is a framework implementation based on a series of projects (Androulaki et al., 2018) by the Hyperledger consortium (Hyperledger, 2017) and developed with open software principles. This development was coordinated and partially financed by the Linux Foundation®, which set up the Hyperledger Project in 2015. See Söderberg (2015) for a general overview of the open source movement and Glaser (2017) for a discussion related to open source blockchains. The Consortium is organized around the development of software for use in industrial settings, characterized by the need to reliably record in complex trading environments such as supply chains. The Consortium management has stated that it would not develop applications for a bitcoin-type cryptocurrency.

Consortium members are from a wide range of organizations. These include technology platform companies (such as Cisco, Hitachi, and IBM), banks (ABN AMRO, BNY Mellon, and others), software companies (SAP, IBM, and others) and academic institutions (Columbia, UCLA, and others). In open software development communities' members are often collaborating with each other at one level and then competing on another. The main advantage of open source collaboration is that it speeds up development time, spreads the risk inherent in software projects, encourages a modular approach to problem-solving and spurs innovation.

Ironically, it is the private permissioned DLT development that requires broad collaboration between industry, academic and not-for-profit organizations. Whereas, the widely used and familiar public blockchains (like *Bitcoin, Altcoin,* and *Ethereum*) are developed and maintained by small groups, working mostly with proprietary systems.

Another open source software company, and an original member of the Hyperledger Consortium, is Digital Assets LLC. It contributes to projects using their high-level programming language known by its acronym, DAML for Digital Asset Modeling Language (Kfir, & Fournier, 2019). DAML makes programming of DLT systems easier and cheaper with a language that is specifically designed to model complex and multi-faceted transactional environments. The most significant project to date employs DAML to provide large-scale clearing facilities for the financial services industry (Tsai, et al., 2020).

Membership of Private Permissions Blockchains

Industrial blockchains are permissioned and therefore require that membership is managed. In HLF, membership profiles are controlled by the system administrator (López-Pintado, et al., 2018) who grants access to independent members and to large consortia, who can then control access to their node. Operating through a software development kit (SDK), the system administrator can initiate, build, and maintain the network by adding members and controlling events. The system administrator can create, stop, change the configuration or, if required, delete the peers (Dunphy and Petitcolas, 2018). Members in HLF do not need to be attached to a Peer. For example, the shipper, responsible for delivering components on site would be able to query the DLT (or be prompted by a message) for the expected time of delivery, then once one site, register the delivery with a countersign by the site foreman. This could all be done using RFID tags, digital signatures, drop down menus and tick-boxes on module devices. Indeed, the functionality and intelligence embedded in the system is based on the user interfaces which can limit the choices of how a member interacts with the blockchain.

To understand the basics of membership management, consider again the example project. A description of members, their roles in the project and how they interact with the blockchain is shown in Table 2.

The members shown in Table 2 are linked to the business process as outlined in the flowchart shown in Figure 4 and the swimlane chart in Figure 5. The full logic of this trading network requires hundreds of lines of computer code and extends to define all possible states of the process.

Digital Ledger System Architecture

The example system described in this chapter sets the requirements that dictate a system architecture that is illustrated in Figure 6. This architecture includes the following elements:

Table 2. Showing the members in the network with their roles, activities, and transactions on the blockchain. These are associated with the mobilisation of building projects using a blockchain to record transactions and chaincode in a commercial setting.

Role	Abbr.	Task(s) in the project	Actions on the blockchain
System administrator	SA	Models business and processes, maintains membership, and ensures network operates as designed.	Deploys network, creates channels & smart contracts.
Design coordinator	DC	Delivers as-built design, ensures that drawings are up to date and complete	Evokes BC1: State 1: Design and price approved.
Commercial manager	CM	Procures and orders components, delivery schedule, price, warranty period and other contract details.	Evokes BC1: State 2: Component ordered
Cost consultant	QS	Produces costs and suppliers for procurement	Confirm payment Evokes BC1: State 7: Paid for.
Supplier	SU	Produces the components in accordance with the contract and schedule as provided.	Evokes BC1: State 3: Component delivered
Shipper	SH	Delivers components in accordance to the schedule.	Queries BC1: Times for delivery.
Installer	IN	Receives the building plans from the dc and program from the PC and a signal from the SF when the component is ready for installation.	Queries BC1: to sync with the program, then evokes blockchain to confirm installation.
Clerk of Works	CoW	Confirms and inspects installation of the components.	Evokes BC1: State 5: Component inspected.
Building inspector	BI	Issues certificate of compliance for Building Control.	Deploys BC2. Evokes BC2: certification of compliance.

Item 1: A system administrator that can design and establish the initial configuration of the system, manage deployment of smart contracts, define and set the endorsement, configure the consensus and ordering algorithm, invite members and ensure that the integrity of the data is maintained in a secure private server.

Item 2: An ordering and consensus algorithm to manage the recording of accepted transactional data to the blockchain and to be able to convert sequential data to a more convenient relational format so that is can be used to write reports and other structured documents that are necessary for operational purposes (such as reconciliation, invoicing, and general accounting).

Item 3: Maintain the peer network, through membership recruitment, so that they can hold copies of the distributed ledgers and associated smart contracts.

Item 4: Implement and maintain an events management system as a modular component of the DLT to provide communication and notification.

Item 5: Ensure that a suitable cloud service is maintained so that it can perform the role of a secure data and system repository.

Figure 6. Contains the basic system architecture of the DLT that includes the administrator and developer, smart contracts, database, peers, and events based upon the HLF implementation. Peers are identified by the symbol 'P'. Other components of the system are described in the text.

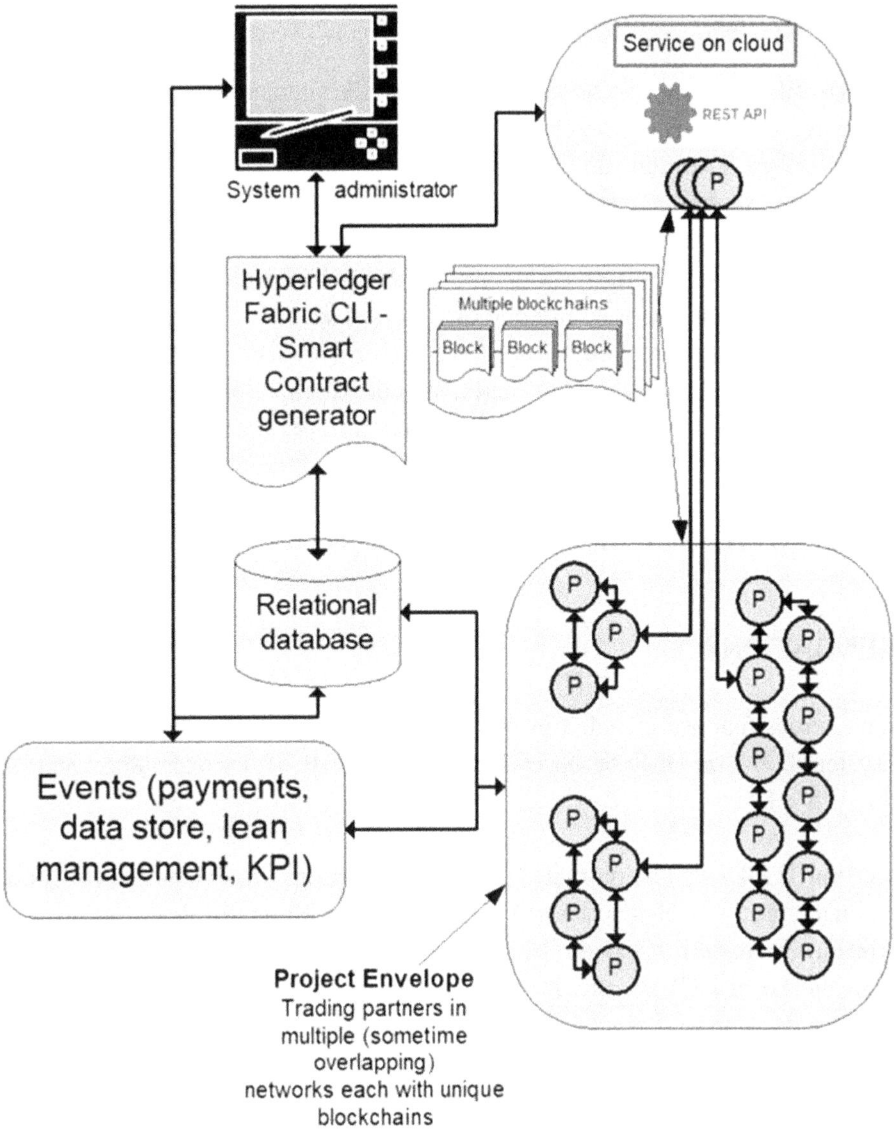

The system administrator, shown near the top of Figure 6, has the responsibility for the deployment and administration of the DLT, which may include multiple channels. These are essential in allowing trading partners to maintain privacy. This protects confidential commercial data but would still allow mission-critical information such as the delivery date, warranty details and maintenance instructions, to be made available for a wider audience. Including in this figure is a representation of a relational database, which is accessed with the standard query language (SQL).

What is not shown in Figure 6 are some of the finer features of the DLT, notably how the consensus algorithm works, and the way that smart contracts are embedded into the core of the blockchain, or how the interface uses pull-down menus, check boxes and other forms of browser-based tools.

Summary of the Objectives of the Chapter

This conclusion fulfils the objectives set out at the start of the chapter. By first describing the three technologies underlying an industrial blockchain and then providing a varied set of applications, a system architecture is described. The rationale for using private permissioned blockchains, with a limited form of centralized control, is demonstrated as necessary in complex, supply-chain based trading networks. This and a set of factors on engineering requirements point to a modular architecture. Hyperledger Fabric provides a good framework for this, but it is the introduction of high-level language implementation, such as DAML, which will make programming of DLTs easier.

Business process and data modelling are needed based on a set of design principles (see Eynon, 2013). These tasks are exemplified by a flowchart and a swimlane graph. To fulfil the additional objectives of the chapter, recommendations are made in the interest in progressing the technology and to ensure engagement with the data-rich and highly interconnected digital future that is predicted to revolutionize manufacturing, commerce and the built environment. Finally, to complete this chapter, a set of discussion topics on areas of active research interest is included.

RECOMMENDATIONS

The recommendations that are included in this section are based on experience gained in designing a prototype system based on HLF, on observations on the state of the technology and its use in other industries. These are:

Recommendation 1: DLT can be promoted by forming consortia to encourage collaboration for data integration along supply chains. These would help to develop the value of digital assets.

Recommendation 2: That efforts should be made to encourage governments and financial institutions to support the use of DLT and that innovative services be approved for use.

Recommendation 3: Support the open source software community in building complex applications through collaborative efforts.

Recommendation 4: Encourage the use of a data rich DLT environment to support Lean management. Lean requires abundant process data of the sort available in a well-monitored trading and supply-chain network.

FUTURE RESEARCH DIRECTIONS

This section provides a brief survey on future research directions that are likely to lead to commercialization for industrial blockchains and is intended to provide a basis for discussion. The following topics are summarized in the following sections:

- Issues around adoption of innovative technology,
- Blockchains for financial services,
- The use of DLT for ERP and Lean management,
- The role of the blockchain in the Industry 4.0, smart cities, and Internet of Things paradigms and
- Issues on privacy, security, and on autonomous data control.

Adoption of Innovative Technology

Ultimately, the adoption of any new system such as DLT by an organization is a commercial decision that is likely to be influenced by the desire to maximize their return on investment on innovation (Christensen & Raynor, 2013). For example, adoption could augment knowledge enhancement and gain a competitive advantage. It could also be used to improve the corporate image as a partner or employer or to provide a basis for carbon accounting. Companies might also be strategizing for brand reinforcement, an approach that could lead to a higher reputational profile. They might also experiment with DLTs to enter new technological ecosystems, a foray that would put them into contact with others who also want to see change. Whatever the motivation is, the introduction of any modern technology, especially one as revolutionary as DLT, is bound to lead to changes inside the organization itself.

Whatever the costs and expected benefits, any adoption must be done with a well-thought-out business justification (Carson et al., 2018) using the standard approach as used in manufacturing industries (Warszawski, 2003). There is also the realization that although adoption might help the industry, the direct benefits to individual companies may be illusive.

For example, the construction industry, prone as it is to sudden downturns and disruptions, renders managers naturally conservative and often reluctant to contemplate introducing technology that might disrupt operations (Waterhouse et al., 2017). Managers in the construction industry get fired for missing deadlines, not for missing out on the promise of a digital revolution.

But the industry could surely use a boost as it is characterized as having low levels of trust (Cerić, 2015) and for being averse to change. It is also known for low profitability (Green, 2016; Davis et al., 2015) and as a locus of crime (Warne, 2016) and questionable business practices (Pontell and Geis, 2007). Optimistically, the industry is so large and strategically important, that it is both driven-to and receptive-for change (Egan, 1998; Latham, 1994).

Blockchains, despite the advantages presented here, may not be the most favored solution and managers need to weigh up the pros and cons of the various alternatives. To begin with, there are alternatives, including doing nothing at all and simply using existing accounting and administrative methods. Other software systems perform similar functions as DLTs, for example, Infrastructure as a service (IaaS), like those offered by Amazon AWS, Google, Oracle, Microsoft Access or open source systems such as MySQL) or ERP systems that can be configured to suit the company. Hand-held devices, of the sort used to track packages (Navon and Berkovich, 2005) could also be integrated into back-office control and management systems that could keep track of supply chain materials and components.

One promising possibility is for governments to offer incentives and other forms of encouragement for firms to accept novel technology that promises to modernize industry. A good example of this is the UK government's strategic plans to implement Building Information Modelling (BIM) (Eadie, et al., 2015; Cable et al., 2013). In the same way that BIM requires a new level of cooperation, installing DLTs could help with collaborative assembly and process improvements (Walasek & Barszcz 2017).

Blockchains and Financial Services in the Built Environment

There are highly publicized projects in the financial services industry that make use of blockchains (McWaters, et al., 2016, McKinsey & Company 2017). The most impressive of these so far is for the clearing (or reconciliation) of international transactions (Meszaros, et al., 2016; CBA-Media, 2016). The advantages of using blockchains in this area are significant (Attaran & Gunasekaran, 2019) as it would help reduce administration costs and speed up the transfer of money. For example, in the age of globalized trade, cross-border payments total around $600 billion annually, with transaction costs somewhere between 2% to 3% percent (and as high as 10% in extraordinary cases). McKinsey and Co. (Higginsonm, et al., 2019), estimate that if blockchains are used for the settlement of cross-border transactions, savings on transaction fees could be on the order of 30% or $4B/year. In 2016, the Canadian ATB Financial Bank successfully used blockchain technology to send 1000 Canadian dollars to Germany in about 20 seconds. Far quicker than the two to three working days that it normally takes to complete using standard methods. Security as well as speed are amongst the benefits of DLT use in financial services (Wüst, et al., 2019; Zhong, et al., 2019).

The provision of novel financial services may attract construction companies and developers into using blockchain technology, for example, by making it easier to raise funding for projects using a cryptocurrency-based approach or by using blockchains to hold commercial debt obligations.

Enterprise Resource and Lean Management

ERP is based around the implementation of enterprise-wide integrated control systems. Lean management (Lean) is a managerial paradigm commonly used in car manufacturing, but increasingly, in other industries (Binder, 2007; Liker and Meier, 2006; Liker, 2004; Womack and Jones, 2003; Womack et al., 1990).

In Lean, data is the driver for continuous improvements, most notably in the coordination of component delivery and assembly. In the C&E industry, the paradigm of Modern Methods of Construction (MMC) calls for off-site production and just-in-time delivery for bathroom and kitchen pods, door sets, and structural insulated panels (SIPS), and precast concrete panels. This gives contractors using MMC all the appearances of a controlled industrial process (Pan and Goodier, 2011; Slaughter, 1998).

Data such as the timing of operations, delivery details, individual costs per item, warranty periods, and aspects of transactional information, are essential for both ERP and Lean systems. These are needed to support a range of organizational–wide business processes, such as payroll, quarterly accounting and reporting, process, and commercial analysis (Morabito, 2017).

Little enthusiasm has been shown by construction managers in implementing large-scale ERP or Lean management of the sort common in automotive, pharmaceutical, or other manufacturing industries. The reason for this is the risk associated with installing large enterprise-wide IT systems. One of the potential advantages of DLT is that it can be installed piecemeal, from the ground up, with minimal risk to the organization. The hidden promise of DLT is that it might provide the capability of a much larger system at a lower cost.

Blockchains in the Industry 4.0, Smart Cities, and the Internet of Things

Automation of construction and other manufacturing forms the core of the Industry 4.0 paradigm (Lee, et al., 2019). This involves the integration of data supplied by IoT devices, such as sensors and RFID tags with operational control. Embedded in this environment, DLT is the keystone technology that provides a secure and immutable data-layer that is accessible from both inside and out of the enterprise (Christidis and Devetsikiotis, 2016).

Automation in this area is needed as IBM, Cisco, and IDC, amongst others, have estimated the number of (partially) connected devices already in use to be in the billions (International Data Corporation. 2020) and that numbers are likely to double within a decade.

As it has been established that construction projects are an industrial process (Koskela, 1992), it is reasonable to think that automation will lead to higher productivity and hence, higher profits (Enshassi et al., 2007). This makes them suitable for computer-based management as called for by Industry 4.0.

Research and development in this area is likely to cover a broad range of topics, for example novel forms of internet-connected digital sensors, analytical engines accessible from the cloud, various forms of autonomous vehicles, better development environments and so forth. There is promise of use. For example, the New York based, DLT Labs implemented an HLF blockchain for Walmart of Canada (Hamilton & Srivastava, 2020) to help manage complex supply chains.

Privacy, Security and Autonomous Data Control

Issues of privacy and security of data abound in data-rich environments such as smart cities, in IoT applications, and in distributed systems. Additionally, it is critical to have control of digital identities and personal data (Zyskind et al., 2015). Not only is the collection and use politicized, it is also controlled by law. Research on the social impacts of privacy and security of data are of great interest to developers of DLTs.

One of the most interesting areas of research is the move away from the centralized control of data towards self-regulating systems. In other words, systems where the data itself can control who has permission to read or write it. This feature could help, for example, to preserve identity in the event of accidental (or intentional) release of confidential information. Blockchains have the potential, with their multi-layer encryption capability, to provide a component in self-regulated data, a feature that would be particularly useful in complex trading environments, such as those encountered in construction projects.

ACKNOWLEDGMENT

This research received no specific grant from any funding agency in the public, commercial, or not-for-profit sectors.

I would like to thank Professor Ramin Keivani and Dr David Valler for supporting this research and the School of the Built Environment for supplying equipment and providing a good base.

REFERENCES

Afanasev, M. Y., Fedosov, Y. V., Krylova, A. A., & Shorokhov, S. A. (2018). *An application of blockchain and smart contracts for machine-to-machine communications in cyber-physical production systems. In 2018 IEEE Industrial Cyber-Physical Systems*. ICPS.

Akintoye, A., & Main, J. (2007). Collaborative relationships in construction: The UK contractors' perception. *Engineering, Construction, and Architectural Management*, *14*(6), 597–617. doi:10.1108/09699980710829049

Al-Jaroodi, J., & Mohamed, N. (2019). Blockchain in Industries: A Survey. *IEEE Access: Practical Innovations, Open Solutions*, *7*, 36500–36515. doi:10.1109/ACCESS.2019.2903554

Androulaki, E., Barger, A., Bortnikov, V., Cachin, C., Christidis, K., De Caro, A., ... Yellick, J. (2018). Hyperledger Fabric: A Distributed Operating System for Permissioned Blockchains. In *Proceedings of the Thirteenth EuroSys Conference* (pp. 30:1–30:15). New York, NY: ACM. 10.1145/3190508.3190538

Anon. (2015a, Jan. 9). How do bitcoin transactions work? How a bitcoin transaction is processed. *The Economist Magazine*.

Anon. (2015b, Oct. 31). The promise of the blockchain. The trust machine. The technology behind bitcoin could transform how the economy works. *The Economist Magazine*.

Attaran, M., & Gunasekaran, A. (2019). Financial Services: The Largest Blockchain Market. In Applications of Blockchain Technology in Business (pp. 21-26). Springer.

Auberger, L., & Kloppmann, M. (2017). *Combine business process management and blockchain*. IBM Tutorial on Blockchain implementation. Available online and accessed on June 15, 2020 at https://www.ibm.com/developerworks/library/mw-1705-auberger-bluemix/1705-auberger.html

Banerjee, M., Lee, J., & Choo, K.-K. R. (2018). A blockchain future for Internet of Things security: A position paper. *Digital Communications and Networks*, *4*(3), 149–160. doi:10.1016/j.dcan.2017.10.006

Barbosa, F., Woetzel, J., Mischke, J., Ribeirinho, M. J., Sridhar, M., Parsons, M., . . . Brown, S. (2017). *Reinventing construction through a productivity revolution*. Tech. rep., McKinsey Global Institute.

Beare, M. E. (2007). The devil made me do it: Business partners in crime. *Journal of Financial Crime*, *14*(1), 34–48. doi:10.1108/13590790710721792

BergC.DavidsonS.PottsJ. (2017). Blockchains industrialise trust. Available at SSRN 3074070.

BergC.DavidsonS.PottsJ. (2018). Ledgers. Available at SSRN 3157421.

Binder, A. K. (2007). *Ward's Automotive Yearbook 2007*. Wards Communications.

Brokenshire, J. (2018). *Building a Safer Future: An Implementation Plan*. Presented to Parliament, Secretary of State for Housing, Communities and Local Government, APS Group on behalf of the Controller of Her Majesty's Stationery Office.

Cable, V., Fallon, M., & Higgins, D. (2013). Construction 2025. Industrial Strategy: Government and Industry in Partnership. In *Industrial Strategy: Government and Industry in Partnership*. Department for Business, Innovation & Skills.

Cachin, C., & Vukolić, M. (2017). *Blockchain consensus protocols in the wild.* arXiv preprint arXiv:1707.01873.

Carroll, J. M., & Bellotti, V. (2015). Creating Value Together: The Emerging Design Space of Peer-to-Peer Currency and Exchange. In *Proceedings of the 18th ACM Conference on Computer Supported Cooperative Work & Social Computing* (pp. 1500–1510). New York, NY: ACM. 10.1145/2675133.2675270

Carson, B., Romanelli, G., Walsh, P., & Zhumaev, A. (2018). *Blockchain beyond the hype: What is the strategic business value?* McKinsey & Company.

CBA-Media. (2016). *Commonwealth Bank, Wells Fargo and Brighann Cotton pioneer landmark blockchain trade transaction.* Retrieved July 25, 2020, from https://www.commbank.com.au/guidance/newsroom/CBA-Wells-Fargo-blockchain-experiment-201610.html

Cerić, A. (2015). *Trust in construction projects.* Routledge. doi:10.4324/9781315747675

Chang, P.-Y., Hwang, M.-S., & Yang, C.-C. (2018). A Blockchain-Based Traceable Certification System. In S.-L. Peng, S.-J. Wang, V. E. Balas, & M. Zhao (Eds.), *Security with Intelligent Computing and Big-data Services* (pp. 363–369). Springer International Publishing. doi:10.1007/978-3-319-76451-1_34

Chang, S. E., Chen, Y. C., & Lu, M. F. (2019). Supply chain re-engineering using blockchain technology: A case of smart contract-based tracking process. *Technological Forecasting and Social Change*, *144*, 1–11. doi:10.1016/j.techfore.2019.03.015

Chapron, G. (2017). The environment needs cryptogovernance. *Nature*, *545*(7655), 403–405. doi:10.1038/545403a PMID:28541332

Cheng, J.-C., Lee, N.-Y., Chi, C., & Chen, Y.-H. (2018). *Blockchain and smart contract for digital certificate. In 2018 IEEE international conference on applied system invention.* ICASI.

Christensen, C., & Raynor, M. (2013). *The innovator's solution: Creating and sustaining successful growth.* Harvard Business Review Press.

Christidis, K., & Devetsikiotis, M. (2016). Blockchains and smart contracts for the Internet of Things. *IEEE Access: Practical Innovations, Open Solutions*, *4*, 2292–2303. doi:10.1109/ACCESS.2016.2566339

Damelio, R. (2011). *The basics of process mapping.* CRC (Taylor & Francis) Press.

Darabseh, M., & Martins, J. P. (2020). Risks and Opportunities for Reforming Construction with Blockchain: Bibliometric Study. *Civil Engineering Journal*, *6*(6), 1204–1217. doi:10.28991/cej-2020-03091541

Das, M., Luo, H., & Cheng, J. C. (2020). Securing interim payments in construction projects through a blockchain-based framework. *Automation in Construction*, *118*, 103211. doi:10.1016/j.autcon.2020.103284

Davis, R., Wilen, A., Crane, T., Bryer, L., Ward, D., Pottier, F., & Blackwell, M. (2015). UK Industry Performance Report-2015. *Construction Industry Key Performance Indicators.*

Delgado-Mohatar, O., Tolosana, R., Fierrez, J., & Morales, A. (2020). *Blockchain in the Internet of Things: Architectures and Implementation.* arXiv preprint arXiv:2004.05817.

Dhillon, V., Metcalf, D., & Hooper, M. (2017). The hyperledger project. In *Blockchain enabled applications* (pp. 139–149). Springer. doi:10.1007/978-1-4842-3081-7_10

Dunphy, P., & Petitcolas, F. A. (2018). A first look at identity management schemes on the blockchain. *IEEE Security and Privacy*, *16*(4), 20–29. doi:10.1109/MSP.2018.3111247

Eadie, R., Browne, M., Odeyinka, H., McKeown, C., & McNiff, S. (2015). A survey of current status of and perceived changes required for BIM adoption in the UK. *Built Environment Project and Asset Management*, *5*(1), 4–21. doi:10.1108/BEPAM-07-2013-0023

Egan, J. (1998). *Rethinking Construction (the Egan Report).* Tech. rep., Department of the Environment, Transport and the Regions, HMSO.

Enshassi, A., Mohamed, S., Mustafa, Z. A., & Mayer, P. E. (2007). Factors affecting labour productivity in building projects in the Gaza Strip. *Journal of Civil Engineering and Management*, *13*(4), 245–254. doi:10.3846/13923730.2007.9636444

Eynon, J. (Ed.). (2013). *The design manager's handbook.* John Wiley & Sons. doi:10.1002/9781118486184

Finck, M. (2018). Blockchains and data protection in the european union. *Eur. Data Prot. L. Rev.*, *4*(1), 17–35. doi:10.21552/edpl/2018/1/6

García-Bañuelos, L., Ponomarev, A., Dumas, M., & Weber, I. (2017). Optimized Execution of Business Processes on Blockchain. In J. Carmona, G. Engels, & A. Kumar (Eds.), *Business Process Management* (pp. 130–146). Springer International Publishing. doi:10.1007/978-3-319-65000-5_8

Glaser, F. (2017). *Pervasive decentralisation of digital infrastructures: a framework for blockchain enabled system and use case analysis.* Academic Press.

Green, B. (2016). *Productivity in construction: creating a framework for the industry to thrive.* Chartered Institute of Building.

Hackitt, D. J. (2018). *Building a Safer Future. Independent Review of Building Regulations and Fire Safety: Final (Hackitt) Report.* Tech. rep.

Hamilton, S., & Srivastava, N. (2020). *Member Webinar: DLT Labs & Walmart Improve Margins, Visibility & Transparency in the Supply Chain.* Video recording accessed on July 17, 2020 on https://www.youtube.com/watch?v=XqXL3kPsBj8

Hargaden, V., Papakostas, N., Newell, A., Khavia, A., & Scanlon, A. (2019). The role of blockchain technologies in construction engineering project management. In *2019 IEEE International Conference on Engineering, Technology and Innovation ICE/ITMC.* IEEE. 10.1109/ICE.2019.8792582

Heiskanen, A. (2017). The technology of trust: How the Internet of Things and blockchain could usher in a new era of construction productivity. *Construction Research and Innovation*, *8*(2), 66–70. doi:10.1080/20450249.2017.1337349

Higginsonm, M., Nadeau, M.-C., & Raigopal, K. (2019). *Blockchain's Occam Problem. McKinsey Quarterly*. Insights on Financial Services.

Huckle, S., Bhattacharya, R., White, M., & Beloff, N. (2016). Internet of things, blockchain and shared economy applications. *Procedia Computer Science*, *98*, 461–466. doi:10.1016/j.procs.2016.09.074

Hultgren, M., & Pajala, F. (2018). *Blockchain technology in construction industry: Transparency and traceability in supply chain*. Academic Press.

Hyperledger. (2017). Hyperledger Announces Production-Ready Hyperledger Fabric 1.0. *Hyperledger Announces Production-Ready Hyperledger Fabric 1.0*. Retrieved from https://www.hyperledger.org/news

International Data Corporation. (2020). IDC FutureScape: Worldwide IT industry 2020 predictions. *IDC FutureScape: Worldwide IT industry 2020 predictions*. Online and accessed on June 20, 2020 from www.idc.com/getdoc.jsp\?containerId=US45599219

Jirgensons, M., & Kapenieks, J. (2018). Blockchain and the future of digital learning credential assessment and management. *Journal of Teacher Education for Sustainability*, *20*(1), 145–156. doi:10.2478/jtes-2018-0009

Kfir, S., & Fournier, C. (2019). DAML: The contract language of distributed ledgers. *Communications of the ACM*, *62*(9), 48–54. doi:10.1145/3343046

Knirsch, F., Unterweger, A., & Engel, D. (2018). Privacy-preserving blockchain-based electric vehicle charging with dynamic tariff decisions. *Computer Science-Research and Development*, *33*(1), 71–79. doi:10.100700450-017-0348-5

Kosba, A., Miller, A., Shi, E., Wen, Z., & Papamanthou, C. (2016). Hawk: The blockchain model of cryptography and privacy-preserving smart contracts. In 2016 IEEE symposium on security and privacy (SP) (pp. 839-858). IEEE.

Koskela, L. (1992). *Application of the new production philosophy to construction* (Vol. 72). Stanford University.

Lamport, L., Shostak, R., & Pease, M. (1982). The Byzantine generals problem. *ACM Transactions on Programming Languages and Systems*, *4*(3), 382–401. doi:10.1145/357172.357176

Latham, S. M. (1994). *Constructing the Team.Joint Review of Procurement and Contractual Arrangements in the United Kingdom Construction Industry. Final Report.* Tech. rep., HMSO.

Lee, J., Azamfar, M., & Singh, J. (2019). A blockchain enabled Cyber-Physical System architecture for Industry 4.0 manufacturing systems. *Manufacturing Letters*, *20*, 34–39. doi:10.1016/j.mfglet.2019.05.003

Li, J., Greenwood, D., & Kassem, M. (2018). *Blockchain in the built environment: analysing current applications and developing an emergent framework*. Academic Press.

Li, M., & Sond, W. (2020). IEEE standard for data format for blockchain systems. IEEE Std 2418.2-2020, Developed by the Standards Activities Board, IEEE Computer Society.

Li, X., Jiang, P., Chen, T., Luo, X., & Wen, Q. (2017). A survey on the security of blockchain systems. *Future Generation Computer Systems*.

Li, Z., Barenji, A. V., & Huang, G. Q. (2018). Toward a blockchain cloud manufacturing system as a peer to peer distributed network platform. *Robotics and Computer-integrated Manufacturing*, *54*, 133–144. doi:10.1016/j.rcim.2018.05.011

Liker, J. K. (2004). *The Toyota Way: 14 Management Principles from the World's Greatest Manufacturer*. McGraw–Hill.

Liker, J. K., & Meier, D. (2006). *The Toyota Way Fieldbook: A practical guide for implementing Toyota's 4Ps*. McGraw–Hill.

López-Pintado, O., Dumas, M., García-Bañuelos, L., & Weber, I. (2018). *Dynamic Role Binding in Blockchain-Based Collaborative Business Processes*. arXiv preprint arXiv:1812.02909.

McKinsey & Company. (2017). Blockchain technology in the insurance sector. Quarterly Meeting of the Federal Advisory Committee on Insurance (FACI). *Blockchain technology in the insurance sector. Quarterly Meeting of the Federal Advisory Committee on Insurance (FACI).*

McWaters, R. J., Galaski, R., & Chatterjee, S. (2016). The future of financial infrastructure: An ambitious look at how blockchain can reshape financial services. In *World Economic Forum* (Vol. 49). Academic Press.

Meszaros, R., Adachi, D., Dharamsi, H., Yetiskin, B., & Thomas, P. (2016). *Blockchain Technology: How banks are building a real-time global payment network*. Accenture Mobility.

Miller, R. (2018). *Camunda hauls in $28M investment as workflow automation remains hot*. Published online in *Tech Crunch* on 5 December 2018. Retrieved June 25, 2020 from https://techcrunch.com/2018/12/05/camunda-hauls-in-28m-investment-as-workflow-automation-remains-hot/

Morabito, V. (2017). Blockchain and Enterprise Systems. In *Business Innovation Through Blockchain* (pp. 125–142). Springer. doi:10.1007/978-3-319-48478-5_7

Nakamoto, S. (2008). *Bitcoin: A peer-to-peer electronic cash system*. Academic Press.

Navon, R., & Berkovich, O. (2005). Development and on-site evaluation of an automated materials management and control model. *Journal of Construction Engineering and Management*, *131*(12), 1328–1336. doi:10.1061/(ASCE)0733-9364(2005)131:12(1328)

Nigam, A., & Caswell, N. S. (2003). Business artifacts: An approach to operational specification. *IBM Systems Journal*, *42*(3), 428–445. doi:10.1147j.423.0428

Pan, W., & Goodier, C. (2011). House-building business models and off-site construction take-up. *Journal of Architectural Engineering*, *18*(2), 84–93. doi:10.1061/(ASCE)AE.1943-5568.0000058

Panarello, A., Tapas, N., Merlino, G., Longo, F., & Puliafito, A. (2018). Blockchain and IoT integration: A systematic survey. *Sensors (Basel)*, *18*(8), 2575. doi:10.339018082575 PMID:30082633

PAS 1192-5. (2015). *Specification for security-minded building information modelling, digital built environments and smart asset management*. Tech. rep., BIM Task Group's security working group of the British Standards Institute.

Pasquire, C. L., & Connolly, G. E. (2002). Leaner construction through off-site manufacturing. *Proceedings IGLC*, 263–266.

Penzes, B. (2018). *Blockchain technology in the construction industry.* Technical report, Institution of Civil Engineers.

Perera, S., Nanayakkara, S., Rodrigo, M., Senaratne, S., & Weinand, R. (2020). Blockchain technology: Is it hype or real in the construction industry? *Journal of Industrial Information Integration*, *17*, 100125. doi:10.1016/j.jii.2020.100125

Petri, I., Barati, M., Rezgui, Y., & Rana, O. F. (2020). Blockchain for energy sharing and trading in distributed prosumer communities. *Computers in Industry*, *123*, 103282. doi:10.1016/j.compind.2020.103282

Pilkington, M. (2016). Blockchain technology: principles and applications. In Research Handbook on Digital Transformations (vol. 225). Edward Elgar Publishing. doi:10.4337/9781784717766.00019

Pontell, H. N., & Geis, G. (2007). *International handbook of white-collar and corporate crime*. Springer. doi:10.1007/978-0-387-34111-8

Raynsford, N., Briginshaw, J., Burke, A., Cook, R., Curtis-Thomas, C., Davies, H., . . . Smith, N. (2016). *Modern methods of construction: Views from the industry.* Technical report based on primary research. Online publication retrieved on 3 Apr 2019 from: www.nhbcfoundation.org/publications/

Rivest, R. L., Shamir, A., & Adleman, L. (1978). A method for obtaining digital signatures and public-key cryptosystems. *Communications of the ACM*, *21*(2), 120–126. doi:10.1145/359340.359342

Seebacher, S., & Maleshkova, M. (2018). A model-driven approach for the description of blockchain business networks. *Proceedings of the 51st Hawaii International Conference on System Sciences*. 10.24251/HICSS.2018.442

Singh, S. (2000). *The Code Book: The Code Book: The Science of Secrecy from Ancient Egypt to Quantum Cryptograph*. Anchor Books.

Slaughter, E. S. (1998). Models of construction innovation. *Journal of Construction Engineering and Management*, *124*(3), 226–231. doi:10.1061/(ASCE)0733-9364(1998)124:3(226)

Söderberg, J. (2015). *Hacking capitalism: The free and open source software movement*. Routledge. doi:10.4324/9780203937853

Steinmetz, R., & Wehrle, K. (Eds.). (2005). *Peer-to-peer systems and applications* (Vol. 3485). Springer. doi:10.1007/11530657_2

Swan, M. (2015). *Blockchain: Blueprint for a new economy*. O'Reilly Media, Inc.

Swan, M. (2018). Blockchain enlightenment and smart city cryptopolis. In *Proceedings of the 1st Workshop on Cryptocurrencies and Blockchains for Distributed Systems* (pp. 48-53). 10.1145/3211933.3211942

Swan, M. (2018). Blockchain Economics: "Ripple for ERP". *Eur. Financ. Rev*, 24–27.

Swanson, T. (2015). *Consensus-as-a-service: a brief report on the emergence of permissioned, distributed ledger systems*. Retrieved from http://www.ofnumbers.com/wp-content/uploads/2015/04/Permissioned-distributed-ledgers.pdf

Syed, T. A., Alzahrani, A., Jan, S., Siddiqui, M. S., Nadeem, A., & Alghamdi, T. (2019). A comparative analysis of blockchain architecture and its applications: Problems and recommendations. *IEEE Access: Practical Innovations, Open Solutions*, *7*, 176838–176869. doi:10.1109/ACCESS.2019.2957660

Szabo, N. (1997). *The God Protocols.* Satoshi Nakamoto Institute. Retrieved from nakamotoinstitute.org/the-god-protocols/

Tapscott, D., & Tapscott, A. (2017). How blockchain will change organizations. *MIT Sloan Management Review*, *58*, 10.

Truong, N. B., Sun, K., Lee, G. M., & Guo, Y. (2020). GDPR-Compliant Personal Data Management: A Blockchain-Based Solution. *IEEE Transactions on Information Forensics and Security*, *15*, 1746–1761. doi:10.1109/TIFS.2019.2948287

Tsai, W. T., Luo, Y., Deng, E., Zhao, J., Ding, X., Li, J., & Yuan, B. (2020). Blockchain systems for trade clearing. *The Journal of Risk Finance*, *21*(5), 469–492. doi:10.1108/JRF-02-2017-0022

Turk, Ž., & Klinc, R. (2017). Potentials of Blockchain Technology for Construction Management. *Procedia Engineering*, *196*, 638–645. doi:10.1016/j.proeng.2017.08.052

Viriyasitavat, W., Da Xu, L., Bi, Z., & Sapsomboon, A. (2018). Blockchain-based business process management (BPM) framework for service composition in industry 4.0. *Journal of Intelligent Manufacturing*, 1–12.

Vukolić, M. (2016). *Hyperledger fabric: towards scalable blockchain for business.* Tech. rep., Tech. rep. Trust in Digital Life 2016. IBM Research. Recovered 1 Apr 2019 from trustindigitallife.eu/wp-content/uploads/2016/07/marko_vukolic.pdf

Vukolić, M. (2017). Rethinking Permissioned Blockchains. *Proceedings of the ACM Workshop on Blockchain, Cryptocurrencies and Contracts*, 3–7. 10.1145/3055518.3055526

Waller, D. L. (2003). *Operations management: a supply chain approach.* International Thomson Business Press.

Wang, J., Wang, Q., Zhou, N., & Chi, Y. (2017). A novel electricity transaction mode of microgrids based on blockchain and continuous double auction. *Energies*, *10*(12), 1971. doi:10.3390/en10121971

Wang, S., Yuan, Y., Wang, X., Li, J., Qin, R., & Wang, F.-Y. (2018). An overview of smart contract: architecture, applications, and future trends. *2018 IEEE Intelligent Vehicles Symposium (IV)*, 108–113. 10.1109/IVS.2018.8500488

Wang, W., Hoang, D. T., Xiong, Z., Niyato, D., Wang, P., Hu, P., & Wen, Y. (2018). *A survey on consensus mechanisms and mining management in blockchain networks.* arXiv preprint arXiv:1805.02707.

Warne, L. (2016). Crime in the Construction Industry. CIOB Policy Document, Chartered Instute of Building.

Warszawski, A. (2003). *Industrialized and automated building systems: A managerial approach*. Routledge. doi:10.4324/9780203223697

Waterhouse, R., Bew, M., Parkinson, K., Malleson, A., May, I., Matthews, A., . . . Hamil, S. (2017). *NBS National BIM Report 2017*. Academic Press.

Waterhouse, R., Highfield, B., Hamil, S., Malleson, A., Kell, A., Tate, G., . . . Ainscough, N. (2019). *NBS Construction Technology Report 2019*. Academic Press.

Weber, I., Xu, X., Riveret, R., Governatori, G., Ponomarev, A., & Mendling, J. (2016). Untrusted business process monitoring and execution using blockchain. *International Conference on Business Process Management*, 329–347. 10.1007/978-3-319-45348-4_19

Werbach, K. (2018). *The blockchain and the new architecture of trust*. MIT Press. doi:10.7551/mitpress/11449.001.0001

Womack, J. P., & Jones, D. T. (2003). *Lean Thinking: Banish Waste and Create Wealth in Your Corporation*. Simon and Schuster.

Womack, J. P., Jones, D. T., & Roos, D. (1990). *Machine that changed the world*. Simon and Schuster.

Wüst, K., Kostiainen, K., Čapkun, V., & Čapkun, S. (2019). PRCash: Fast, private and regulated transactions for digital currencies. In *International Conference on Financial Cryptography and Data Security* (pp. 158-178). Springer. 10.1007/978-3-030-32101-7_11

Xiong, Z., Zhang, Y., Niyato, D., Wang, P., & Han, Z. (2018). When mobile blockchain meets edge computing. *IEEE Communications Magazine*, *56*(8), 33–39. doi:10.1109/MCOM.2018.1701095

Xu, X., Weber, I., & Staples, M. (2019). *Architecture for blockchain applications*. Springer. doi:10.1007/978-3-030-03035-3

Yang, R., Wakeeld, R., Lyu, S., Jayasuriya, S., Han, F., Yi, X., Yang, X., Amarasinghe, G., & Chen, S. (2020). Public and private blockchain in construction business process and information integration. *Automation in Construction*, *118*, 103276. doi:10.1016/j.autcon.2020.103276

Yaqoob, I., Salah, K., Uddin, M., Jayaraman, R., Omar, M., & Imran, M. (2020). Blockchain for digital twins: Recent advances and future research challenges. *IEEE Network*, *34*(5), 290–298. doi:10.1109/MNET.001.1900661

Yue, X., Wang, H., Jin, D., Li, M., & Jiang, W. (2016). Healthcare data gateways: Found healthcare intelligence on blockchain with novel privacy risk control. *Journal of Medical Systems*, *40*(10), 218. doi:10.100710916-016-0574-6 PMID:27565509

Zheng, Z., Xie, S., Dai, H.-N., Chen, X., & Wang, H. (2018). Blockchain challenges and opportunities: A survey. *International Journal of Web and Grid Services*, *14*(4), 352–375. doi:10.1504/IJWGS.2018.095647

Zhong, L., Wu, Q., Xie, J., Li, J., & Qin, B. (2019). A secure versatile light payment system based on blockchain. *Future Generation Computer Systems*, *93*, 327–337. doi:10.1016/j.future.2018.10.012

Zyskind, G., Nathan, O., & Associates. (2015). Decentralizing privacy: Using blockchain to protect personal data. Security and Privacy Workshops (SPW), 2015 IEEE, 180–184.

KEY TERMS AND DEFINITIONS

Block: Is the basic unit that describes how files are organized in a blockchain. Blocks contain an ordered set of transactions that are cryptographically linked to the preceding block, and in turn it is linked to subsequent blocks. Blocks are assembled by the ordering service and then validated and committed to the blockchain by peers that reside on nodes.

Blockchain: A file linked in a block structure that takes the form of a sequential database. It is ubiquitous, immutable, and traceable, and can record transactions with transparent and trusted rules in a peer-to-peer network.

Certification: Is defined as secure data in the form of a signed document that is held on a blockchain. These records can have controlled access through multiple layers of cryptography.

Certificate Authority (CA): Is a modular component of the DLT with the role of issuing encryption keys to network members and other users. The CA issues one root certificate to each member and one enrollment certificate to each authorized user. This should not be confused with using the blockchain to hold official certification by an authority.

Chaincode: Is an alternative name for smart contracts used in the Hyperledger Fabric framework. Using modular features of HFL, smart contracts can be programmed into the system in diverse ways, such as the client interface or through an associated database.

Channel: A name given to an enhanced feature of a DLT that allows a degree of privacy to exist within a subset of a larger trading network.

Consensus Algorithm: Is the process by which the members of a network determine which transaction get recorded onto the blockchain.

Custodian: This is the term used to describe the holder of the asset, receiving it from the Transactor.

Distributed Ledger Technology (DLT): Is the term given to the collection of services, interfaces, software, and associate systems that allow blockchains to be used in industrial settings.

Endorsement: Is defined as the process where specific peer nodes execute a chaincode transaction and return a proposal response to the client application. Endorsement is based on a policy that defines which peer nodes on a channel can execute transactions.

Genesis Block: Is defined as the first block on a chain and represents the configuration that initializes and defines the ordering service.

Hyperledger Fabric: Is primarily aimed at industrial blockchains. It is a quickly evolving framework that contains commands and modules to allow the blockchains to be developed, evaluated, deployed, and initiated.

Invoke: Is when a call is made via chaincode to alter the state (i.e. write) to the blockchain. This requires that the transaction is sent as a proposal to a Peer, which must be endorsed, ordered, and committed to become a permanent record.

Ledger: Is a document that contains records of transactions held in chronological sequence. In modern terms, a digital ledger is defined as containing two distinct parts: the blockchain and the Current State database (or World State). The term Distributed Ledger Technology (DLT) describes copies held by multiple computers (or nodes) across a network.

Membership Service Provider (MSP): Is a set of tasks within the system that provides credentials to clients, and peers that allow them to participate in a HLF network. HLF supports dynamic membership, where members, peers, and ordering service nodes can be added and removed without compromising the integrity of the network.

Orderer Nodes: Are specific nodes on the network that are tasked with ordering the transactions. They ensure the consistency of the blockchain and deliver the endorsed transactions to the peers of the network. The orders provide the Ordering Service that sort the transactions into blocks and then distributes these blocks to peers for validation. The ordering service is independent of the peer processes and orders transactions and in HLF, it supports modular implementations so that the system can be extended and configured.

Organization: Is a collective term used to describe users who can read and write to the blockchain. They are also referred to as members and managed by the MSP, which defines how other members of the network may verify their digital signatures when transacting or reading the ledger access rights of identities within an MSP are governed by policies which are also agreed upon when the organization joins the network. There is no size limit to the organization if they have access to a Peer (the main trading point). If they exist, collections of organizations form a Consortium.

Peer: A network entity that maintains a ledger and runs chaincode containers in order to perform read/write operations to the ledger. Peers are owned and maintained by members and make up the principle nodes in a blockchain network. Peers host ledgers, chaincode and participate in consensus.

Permissioned Blockchain: Infrastructure that is based on a principle of modular architecture. Permissioned describes a DLT that has a controlled and limited membership. This allowed a great deal of flexibility in designing systems as it permits the separation of roles between the nodes in the infrastructure, execution of chaincode and a configurable consensus and membership service.

Permissioned ledger: Is a blockchain network where each entity or node is required to be a member of the network. Anonymous nodes are not allowed to connect.

Policies: Are part of the language used for constructing the layers of encryption in the data blockchain. They are used to control access to data and other resources in a blockchain network, notably who or who cannot read and write to a channer, evoke, query, or deploy chaincode. Policies are defined in the configuration files prior to deploying the network, setting up an ordering service or creating a channel. They can also be specified with instantiating chaincode.

Privacy: Is required by the chain transactors to conceal their identities on the network. While members of the network may examine the transactions, the transactions can't be linked to the transactor without special privilege. Data as well as transaction details can also be held privately.

Private Data: Are confidential information stored by peers on the blockchain but kept separate from other data. Access to this data is restricted to members with permission, while unauthorized organizations will only see a hash of the private data on the channel ledger as evidence of the transaction For an additional level of privacy, these hashes of private data go through the Ordering Service, which keeps it hidden from the Orderer.

Query: Is a call (or invocation) to read from the blockchain ledger. In HLF, chaincode is used to unwrap the blockchain to read certain keys or other data. Queries do not change the ledger state, although the client application can choose to submit a read-only transaction for ordering, validation, and commit, to provide an auditable proof that the blockchain has been read.

Quorum: Is the minimum number of members of the cluster that need to affirm that a transaction is acceptable to write to the ledger. For networks with few members, the central authority may make up the majority vote for acceptance.

Smart Contract: A simple computer program that resides in compiled format within the block data structure across a peer-to-peer network that runs whenever the chain is re-written. A set of instructions in the form of a checklist, executes the instructions held in the blockchain.

Software Development Kit (SDK): Provides a structured environment where the System Administrator can design, deploy, and manage the network. In HLF, the SDK is modular and configurable using standard software tools. Modules, such as the cryptographic algorithms, logging frameworks and others, can be switched in and out by the SA using the SDK. Through the SDK, transaction processing, membership services, node traversal and event handling are deployed. HLF currently uses both Node.js and Java, with two more: Python and Go, in development.

Transactions: Are the official term used to describe the addition of a record to a blockchain. Members cannot write directly but must submit transaction proposals to the consensus and ordering algorithm.

Transactors: Can be either human or a device, for example a shipper delivering a package and registering the transaction with a smartphone, that sends a transaction proposal to the consensus and ordering algorithm.

Validating Peers: All transactions must be validated by Peers. These are networked computer nodes owned either by one of the participant organizations or hosted by a professional service provider. Nodes hold a copy of the blockchains and are responsible for ensuring consensus used to validate transactions. Once validation is complete when all nodes receive an updated version of the blockchain.

World State Database: (also called the *Current State*) is a data store that is permanently attached to the blockchain where the latest and most complete records of transactions are stored. It is more efficient to read and query the blockchain through the World State.

Chapter 9
BIM and IoT for Facilities Management:
Understanding Key Maintenance Issues

Ricardo Codinhoto
University of Bath, UK

Beatriz Campos Fialho
University of Sao Paulo, Brazil

Lidia Pinti
Politecnico di Milano, Italy

Márcio Minto Fabricio
University of São Paulo, Brazil

ABSTRACT

The AEC industry is facing a digital transformation that is improving the efficiency of services involved in designing, building, and operating assets and the users' well-being. Such a transformation towards sustainable smart cities is underpinned by two disruptive technologies: building information modelling (BIM) and internet of things (IoT). In this chapter, the authors present a review of studies that have focused on the identification of BIM and IoT technologies applicable to AEC industry processes, describing suitable strategies for data collection, storage and sharing, and fields of application. They also present data from two case studies that help us to understand how BIM can support better facilities management. From the literature, it was found that process improvements are a predominant research focus, reinforcing that successful BIM and IoT adoption goes beyond the acquisition of technology. The case studies revealed that a framework for prioritising areas is still needed for giving direction to researchers and practitioners concerning where to start the digital transition.

DOI: 10.4018/978-1-7998-6600-8.ch009

INTRODUCTION

Facilities management, much like everything else is undergoing a digital transformation. Technologies that heretofore were restricted to digital products are permeating the post construction building industry. Companies can now create *building data ecosystem grids* to capture and analyse data trails from building users and determine how to enhance their service provision and building performance, as well as their building maintenance practices. Such a transformation has been fostered by several disruptive Information and Communication Technologies (ICT), such as Building Information Modelling (BIM), Artificial intelligence (AI) and Internet of Things (IoT). Within FM, there is a shift towards data ecosystems built around service analytics platforms that ties the ecosystem together. However, the shift is still slow, leading to the creation of various dizygotic digital twins.

The arguments for why adopting disruptive technologies for FM vary. Adoption is often related to perceived economic gains, streamlined well-connected processes that reduce inefficiencies of building operation and maintenance. In this respect, short, medium and long-term gains are achievable due to buildings' long-life cycle. Implementation tends to be focused on technology selection rather than business or societal value proposition.

In terms of benefits to the planet, for instance, the value of digital transformation is immense. For example, the scenarios developed by New et al. (2009) and Buis (2019) predict a global temperature rise of up to 4 Celsius. These conservative models tell us that if we do not change, we will bear the consequences. Concerning global warming, building technology is assisting us to precisely understand building performance in terms of energy consumption and CO_2 emissions. More importantly, it enables to understand and act upon underlying causes of poor performance, being that related to construction materials and process, or user behaviour.

To address the issues related to the use and maintenance phase, previous studies have focused on the identification of BIM and IoT technologies applicable to buildings. These studies tend to describe suitable devices, tools and software for data collection, storage and sharing (Fialho et al., 2019; Ye et al., 2018). As explained later in this chapter, these studies promote digital ways to improve building efficiency and their point of departure tends to be the implementation of technology. This is often achieved through BIM-supported simulation tools that predicts the performance of newly designed buildings.

In this respect, considerably less attention has been given to the currently inefficient existing building stock. It is widely acknowledged in the literature that the volume of resources used during the life cycle of buildings worldwide is the largest of all sectors. Buildings still represent the largest energy-consuming sector in the economy, with over half of global electricity consumption being responsible for approximately one-third of global carbon emissions (Abergel et al., 2017). We also know that the operation phase uses roughly 80% of the cost and resources of a building during its life cycle. However, very little detailed information exists about the sources of inefficiencies within our building stock that can inform value-based digital transformation implementation plans.

The literature is silent with regards to priority areas for the implementation of BIM and IoT solutions. Thus, in this chapter, we advance our work focused on the establishment of contextual knowledge regarding challenges of BIM and IoT systems integration (Fialho et al., 2020) by exploring practical Facilities Management issues that require attention. In other words, this chapter addresses the value proposition of digital transformational within the facilities management field.

BACKGROUND

The general concept of FM is, often misunderstood and its importance underestimated. Traditionally, the facilities management function has been widely described as involving clients, real estate, and AEC teams. It also is planned around a cost factor (to be kept to the minimum) to be spent on activities that are perceived as a non-added value, such as maintenance, janitorial services, and pottering. This narrow view has unwittingly relegated the relevance of FM to the background, away from influencing an organisation's decision-making and strategic thinking. Therefore, it is vital to clarify the concept of FM before proceeding.

In this chapter, facilities management is defined according to the British Standards Institute (2007): "Facilities management is the integration of processes within an organisation to maintain and develop the agreed services which support and improve the effectiveness of its primary activities" (BSI, 2007). Also, according to the Institute of Asset Management ISO 55000 series that defines asset management (AM) as "the systematic and coordinated activities and practices through which an organisation optimally and sustainably manages its assets and asset systems, their associated performance, risks and expenditures over their life cycles to achieve its organisational strategic plan" (BSI, 2008). Both definitions clearly states that both FM and AM contribute to organisations' businesses.

Also, the above concepts reveal an overlap between the concept of asset and facilities management. The differences between these concepts are in that facilities management is restricted to physical assets that are utilised by and support businesses' operations. In contrast, asset management refers to organisations' efforts to measure and realise the value provided by both tangible and intangible assets.

Facilities management is a complex activity that can be organised according to four main activities including i) planned or preventive maintenance, referring to regularly maintaining building fabric and equipment to avoid disruptions to the business; ii) reactive maintenance, referring to fixing problems related to the fabric of the building as they occur; iii) statutory maintenance, concerning mandatory maintenance needed to guarantee the health and safety of building users (e.g. smoke alarms, security cameras and emergency lights); and iv) operations, supporting building use (e.g. security, portering and cleaning).

With regards to standards, the AEC sector sees a move towards better integration of standards related to BIM and facilities management. For example, in the UK, BS 8210: Facilities maintenance management: Code of practice is currently being drafted to support successful maintenance outcomes for a range of facilities. Also, the BS ISO 55000: 2014 series forms the basis for an "asset management system" and addresses the links between facilities and asset management strategies. The BS ISO 55000 provides an overview of asset management and asset management systems and the context for ISO 55001 and ISO 55002 which specifies requirements for an asset management system, and detail sector-specific, asset-specific, or activity-specific technical requirements respectively. Although not explicitly connected, the BS 8219 and ISO 55000 series defines the context for the ISO 19650 series which addresses the organisation and digitisation of information. Also, the PAS 180 series focused on smart cities which also applies to organisations that maintain multi-site building complexes. These various standards are continuously evolving, and although not yet fully integrated, they already provide a push towards digital transformation.

One key aspect of this integration of standards is the recognition of various hierarchical levels of organisational decision-making. As defined within the BS EN ISO 19650-1, the three levels include the establishment of an Organisational Information Requirements (OIR) at the strategic level, linking information management to businesses decision-making; The Asset Information Requirements (AIR) at the tactic level influences both strategic and operation plan; and the Exchange Information Requirements (EIR) at the operational level. Together, these requirements define how the asset information model (AIM) will look like and the level of granularity of information that is required from building components. It refers to the integration of structured information about buildings and its objects in a centralised digital database, thus enabling the collaboration between stakeholders throughout the whole life cycle, reducing rework and increasing the accuracy and availability of information. Part of the AIM is the definition of how information is managed and stored within a Common Data Environment (CDE) when maintenance plans and requests are taking place. This structure is key because it acknowledges the linkages between the three levels and the importance that operational activities have to the success of an organisation, but also it recognises that meaningful integration can only be achieved through a digital transformation. However, it is important to remember that, to date, OIR, AIR or AIM constitute static data sets. In other words, it is a digital descriptive representation of a physical asset(s) gathered at a certain point in time.

It is vital that facilities managers move away from using static models and start using dynamic ones that are linked to live data captured through sensors (dizygotic digital twins). The term dizygotic digital twin is used here to remind readers that a model represents reality for the given purpose, i.e., the model is a simplified abstraction of reality in the sense that it cannot represent all aspects of reality." (Rothenberg et al., 1989). Thus, it is only through the creation of various digital twins (plural) that we can set realistic performance indicators across the three hierarchical levels. The plethora of possible twins is large and ranges from essential aspects of building management such as building fabric integrity, energy consumption and air-quality to optional ones such as patterns of movement. Digital twins enhance managers capacity of predicting problems before they occur and disturb its organisation's business.

However, while knowledge regarding sensor devices is vast, data analytics concerning facilities problems are scarce and troublesome to collate. From the one hand, large complexes with significant amount of data about their maintenance problems and practices tend not to publicise their data. From the other, these organisations are still using methods that does not enable data analytics. Thus, we attempt to generate new insights in this area by presenting a review or research related to the use of sensors for FM focused on reactive maintenance and the results of an analysis of maintenance data from two large organisations. Then we present a discussion regarding gaps in the digital transformation concerning FM.

BIM AND IOT REVIEW

This section presents the results of a systematic literature mapping focused on BIM and IoT applied to FM. The Science Direct (Elsevier) database was considered appropriate for this SLR, thus providing relevant publications on the topic. Keywords, wildcards, and Boolean operators were defined for addressing both individual BIM and IoT concepts and the context for BIM and IoT integration. Filters for selecting innovative works include peer-reviewed journal articles published in Portuguese and English. The review was conducted during spring 2020, thus articles from 2015 to 2019 were included. From 146 identified entries, 75 were deemed relevant and selected for further analysis. These were grouped according to their emphasis on theory building, process change and IT implementation as discussed in the following. Table 1 presents the protocol adopted for structuring the review.

Table 1. Research protocol.

Item	Content
Research Questions	How is FM research addressing the integration between BIM models and IoT devices? What are the problems and challenges of FM-IoT-based systems?
Review Objectives	To develop contextual knowledge concerning the challenges of BIM and IoT systems integration in relation to FM.
Database	Science Direct (Elsevier)
Keywords	("building information modelling" OR "building information modeling" OR "bim") and ("IOT" OR "IoT system" OR "IoT platform") AND ("integration" OR "connection") AND ("Computer Networks and Communications" OR "network") AND "Smart cities"
Filters	Year of publication: 2015 to 2019. Type of publication: peer-reviewed journal articles. Language: English and Portuguese. Inclusion criteria: articles on the integration of BIM and IoT systems.
Selection criteria	Exclusion criteria: Publications neither correlated to BIM and IoT systems integration, nor written in the aforementioned languages.

Concerning theory building, at least 20 literature review articles discuss the role of emerging Information and Communication Technologies (ICT's) in the AEC industry digital transformation. The combination of BIM and IoT is discussed by Tang et al. (2019), who investigated methods for BIM and IoT integration, as well as areas of application. According to the study, in general, BIM tools' APIs and semantic web technologies are used in the Construction and Operation & Maintenance (O&M) stages for activities, such as Operation and Monitoring, Logistics and Management and Health and Safety. Wong, Ge and He (2018) explored digital technological (DT) developments into Facilities Management (FM), including BIM and IoT. Both articles highlight the interoperability of data as a critical issue to be overcome towards improving the integration among ICTs'. Also, they foresee that the way forward is based upon web-based service-oriented architecture that does not yet exist.

The concept of IoT as a disruptive technology has been explored with distinct perspectives. Earlier recent work aimed at promoting smart building industry (e.g. Magruk, 2015). Also, supporting the evolution of AEC industry beyond construction (e.g. Woodhead et al., 2018). later research focuses on the total experience concept to foster the development of business models within the context of smart cities, including circular strategies, smart products, and product-service systems (Alcayaga et al., 2019); specifically over the construction and O&M stages (Jia et al., 2019). IoT ideas from the manufacturing sector are migrating to the construction sector combined with Virtual Reality (VR) and Augmented Reality (IR) (Kerin and Pham, 2019). There is also data security issued been explored through intrusion detection in computer networks (e.g. Kelton et al., 2019). Significant challenges within this area refer to a need for more robust concepts (such as federated machine learning) for integrating performance monitoring and cyber-physical systems (CPS) through IoT.

New emerging concepts in the theoretical field can also be seen through the work of Chaturvedi et al. (2019). They proposed the Smart District Data Infrastructure (SDDI) concept that integrates sensors, simulation tools, and 3D city models for facilitating privacy, security and controlled access to stakeholders. Contributions can also be seen in the work of Yan and Sakairi (2019) who proposed the "Geo CPS" concept by linking the concepts of IoT and CPS applied to geospatial spaces. Both are examples of research that reconceptualise service provision as a result of technological advances. As most, the benefits of such approaches are expected to be positive, however without a defined baseline for measuring such improvements.

The contribution of BIM with regards to facilities management has an operational focus. The new ideas are coming from the integration of existing technologies as well as new models of data connection to support building operations. For example, Gao and Pishdad-Bozorgi (2019) and Araszkiewicz (2017) highlights the integration of building lifecycle data and business's decision-making support. Energy management, as one key area (Araszkiewicz, 2017) is tested through Building Energy Modelling (BEM) methodologies associated with BIM data sets (Pezeshki et al., 2019). Research related to energy addresses mainly two problems related to FM: a) the continuous increase of energy costs affecting maintenance budgets; and b) the reduction of energy consumption in response to future shortages of energy that can impact businesses core activities. Both issues are seen are contributing to a more sustainable environment.

Also with a focus on building operations, Costin et al. (2018) explore BIM to support the development, repair, and expansion of transportation infrastructures. Similarly, Rice and Martin (2018) discuss the advances made in Australia's infrastructure development and operations. Other BIM applications include off-site construction (Yin et al., 2019), network security and data management, combined with Blockchain technologies (Nawari and Ravindran, 2019), and data integration for urban governance (Wang et al., 2019). BIM is also explored in connection with complementary technologies such as GIS (Ma et al., 2017). As previously highlighted, the above research tends to focus on describing areas where improvements can be made without attempting to establish performance baselines for comparison.

With regards to advancing IT to support BIM and IoT applications, the research identified is focused on the development of platforms, systems, interfaces, and new ontologies integrating BIM and IoT. For instance, Dave et al. (2018) proposed the Otaniemi3D web-based platform that integrates built environment data generated by IoT sensors and BIM models. Also, Lokshina et al., (2019) who developed a system design for improvements in data security, data exchange, and information sharing among stakeholders. For as-built information, Zhai et al. (2019) developed an IoT-enabled BIM platform that improves data collection during construction. Zhengdao et al. (2017) develop a radio frequency identification device BIM platform to manage critical schedule factors over prefabrication manufacturing, logistics and on-site assembly construction stages. Subsequently, the same authors developed platforms for the management of prefabricated housing Zhengdao et al. (2016b); and for facilitating the on-site assembly services Zhengdao et al. (2018). Similarly, Zhong et al. (2017) proposed a multi-dimensional IoT-enabled BIM platform for real-time visibility and traceability of building components, remote supervision of construction progress, real-time cost control and information sharing among stakeholders, including building owners. Here, attention is given to improving IT rather than FM practice. There is an implicit assumption that by fixing the current problems of IT will result in better FM. Once again, this approach leaves a gap concerning measurable benefits for FM.

Research has also been developed with a focus on digital transformation potentially impacting building users. One example of BIM and IoT-based solutions in this area include a Real-Time Replay System, which is an interactive environment for building' users to manage electricity consumption data, identify wasteful operations and obtain suggestions for energy conservation (Chiang et al., 2015); Other examples include, a semantic knowledge management service and domain ontology combining BIM and smart water field systems (IoT) for improvements in systems performance has been proposed by Howell et al., (2017); a BIM-based visualisation and fire warning system to support building evacuation (Chen et al., 2018); and a system to identify unsafe worker behaviours (e.g. intrusions) in dynamic environments (Arslan et al., 2019a). Research in this category successfully address issues directly related to FM with a rationale for its development based on operational gains. While this in not incorrect, not invalidate the research, it reinforces the point made that there is a lack of integration across the different hierarchical levels of decision-making.

The development of IoT-based platforms for distinct lifecycle stages is explored. Propositions specifically focused on O&M activities include and a user's communication system that is integrated into facilities management (Ham et al., 2018). For energy management, a sensor management system for smart energy-efficient buildings is proposed by Plageras et al., (2018), as well as an IoT Energy Platform for processing, managing and analysing building energy data is proposed by Terroso-Saenz et al., (2019). Concerning Health and Safety, a holistic and collaborative information integration framework and prototype system for safety monitoring over the AEC industry (Xu et al., 2019) was introduced, as well as a cyber-physical-system-based safety monitoring system for blind hoisting in metro and underground construction and maintenance (Zhou et al., 2019).

BIM-based systems and interfaces were proposed for distinct applications, including an intuitive BIM-based interface to support customers exploring design solutions in mass-customised housing projects (Bianconi et al., 2019), a BIM-based prototype cyber-physical system applied to structural health monitoring (Fitz et al., 2019), a GIS-BIM based urban energy planning system for accessing city infrastructure solutions (Yamamura et al., 2017), and an interactive AR system for Temporal and Spatial analysis of power consumption data integrated with building information models that supports building's occupants managing their energy consumption (Chou et al., 2017).

Finally, with regards to process improvement, contributions have been made through the proposition of approaches, frameworks, methods, and models for the integration of BIM and IoT. For example, Heaton and Parlikad (2019) developed a smart asset alignment framework that integrates information captured at the infrastructure asset level and citizen requirements; Love et al., (2019) proposed a generic business dependency network for the analysis of cause-effect relationships between digital technologies and their implementation processes; and Motamedi et al. (2016) introduced a method for standardising and formally define RFID components for linking BIM and RFID data. Focusing on anticipating building users' needs, Ciribini et al. (2017) developed a workflow to populate BIM models using data generated by remote sensors. Arslan et al., (2019b) designed the 'Occupant Behaviour in Dynamic Environments' framework that combines building occupants' behaviours as contextual information and the dynamic building environment. Zhengdao et al. (2016a) provide a BIM-IoT based method that analyses stakeholder-associated risk factors and effects in prefabrication construction projects.

O&M-oriented solutions are developed by Rashid et al. (2019), who introduced an intuitive framework to control electrical fixtures in a smart built environment. Their framework integrates an ultra-wideband-based indoor positioning system with a BIM-based virtual environment; Also, Edmondson et al. (2018), developed a Smart Sewer Asset Information Model to provide real-time sewer condition data for operation management. For Health and Safety, Liu et al. (2019) developed a safety inspection method that integrates unmanned aerial vehicle camera and dynamic BIM for supplying industrial and residential water; and Peng et al., (2019) proposed a dynamic path planning approach for evacuation planning in large public buildings based on neural networks.

IoT-centred solutions also exist in the literature. For example, a framework for safety management and monitoring in the context of complex dynamic spatial environments was developed by combining IoT and multiagent geo-simulation techniques (Haddad et al., 2019); Also, a big data analytics for IoT-enabled manufacturing shop floor using RFID and machine learning techniques for capturing the real-time production data (Kho et al., 2018); and a semantic knowledge model for smart buildings using the Indoor GML standard (Rubio et al., 2018).

The shift from analogue to digital approaches has also created challenges that fomented research developments. For instance, a Unified BIM Adoption Taxonomy for the classification and integration of BIM adoption drivers and factors was proposed by Louay and Kassem (2018). Also, BIM macro-adoption models, matrices and charts were developed to support BIM diffusion policies (Succar and Kassem, 2015). Similarly, a general recommendation for web-based geometry representations of building data is suggested by Mcglinn et al., (2019), and a conceptual DALTON (Data Linked Through Occurrences Network) model provides checking components for business rules (Fortineau et al., 2019).

BIM-based solutions have also been proposed for demolition stages, which are part of FM provision. For example, a group of BIM expectations for Construction and Demolition Waste (CDW) management is described within Akinade et al., (2018); Chileshe et al. (2019) presents a conceptual model for the reverse logistics supply chain that supports demolition and reprocessing of salvaged materials. Game engines and BIM are proposed for the simulation of demolition processes and the optimisation of waste sortability and reuse (Kunieda et al., 2019). Processual solutions have also been proposed for O&M activities, for instance, a Diagnosis-Aided Historic Building Information Modelling and Management framework is proposed by Bruno et al., (2018); a framework outlining Integrated Knowledge-based Building Management Systems for detecting and diagnosing operational faults (Gha et al., 2019), and an object-oriented city design model for building energy performance assessment (Zachary et al., 2019).

Additional contributions include Good et al., (2017) and their support to the identification of enablers and barriers of using smart grid electric demand-response systems; Novais et al., (2019) investigated the integration between cloud computing and supply chain; Marsal-Llacuna (2018) and Li et al., (2019) investigated Blockchain-based frameworks linked with BIM in Construction; David et al., (2018) and Yang et al. (2018) introduced integrated frameworks for multisector infrastructure asset management; and Hu et al., (2016) developed a Resource Description Framework syntax structure for representing existing database schema information. Last but not least is the contribution made by Simpson et al., (2019) concerning the need for an upskilling framework that addresses the dynamic aspects of learning new disruptive information and communication technologies.

Discussion

This systematic literature mapping analysed relevant articles published in the past five years. Publications on ICT implementation (i.e., BIM and IoT) in the AEC industry has significantly increased over the period, from 4 in 2015 to 35 in 2019. The predominance of journals in construction and engineering domains (Table 3) demonstrates the applicative bias of the studies, concerned with the demands of constructors, managers, and users of the built environment. Regarding the research method, most articles have adopted literature reviews, still required to state a knowledge field, followed by Case Study and experimental methods, demonstrating an orientation to the development of technical and processual solutions for building lifecycle stages.

The bibliometric analysis of the articles included in this review shows a predominance of articles concerning IT propositions (47%) in the first three years (2015-2017) in comparison to theoretical and processual approaches. Research results tend to be endogenous, i.e., an IT solution that is better than the previous IT solution without an assessment of its impact on FM. The focus on IT shifted to process in 2018 and 2019, with a significant increase on processes improvement solutions (42%), evidencing the required transformations imposed by BIM and IoT to organizations for adapting traditional activities to a digital environment. Overall, IT solutions is the second most representative field addressing BIM and IoT integration.

Figure 1. Summary of themes related BIM and IoT research identified through SLR.

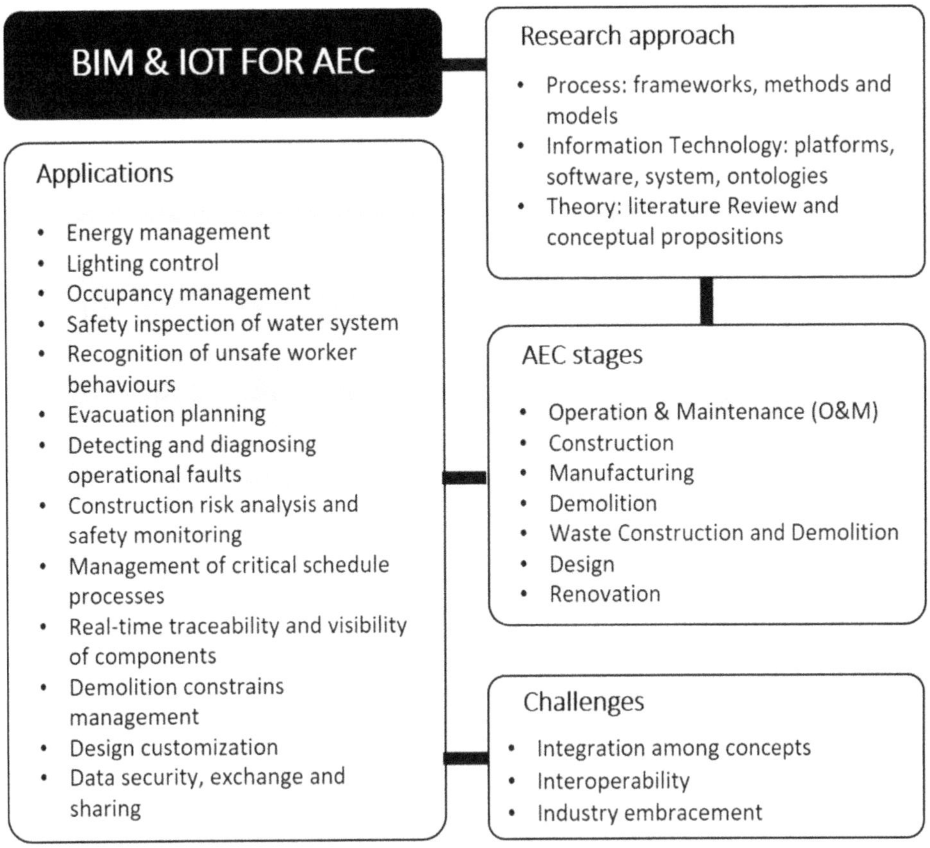

Challenges and problems related to BIM and IoT integration are described in a broad perspective of ICT implementation and can be summarized in one main issue, i.e., interoperability of data over the building lifecycle. Although difficulties have been highlighted, articles usually tend to prioritize positive aspects of the research rather than the negative ones.

Figure 1 summarises the results of the review according to the description above. The results reinforce the argument that there is a disconnect between the hierarchical levels. All research included in this review has demonstrated ideas or results that positively impact FM; however, it is of concern that the extend that these results are significant to FM in specific or to the businesses in general remains uncertain and unmeasured. It is seen as positive that research focus has shifted from IT only to process changes. However, no article focused on people and the adoption of IT-enabled process was gathered through the research protocol. That is also of concern.

UNDERSTANDING FACILITIES MANAGEMENT PROBLEMS

No doubt much research work is being done to advance our knowledge regarding the use of BIM, IoT and various other ICTs. Each piece of research is relevant within the context in which investigations took place. The fact that there are so many problems shows us that there is a need for the development of a mechanism to identify priorities for service provision. Hence, a discussion about data structuring and analysis is proposed with a basis in two case studies.

The scope of research was narrowed to reactive maintenance of educational building complexes (Figure 2). The rationale for this decision was based on the absence of precent studies in any area that could assist with the establishment of a baseline for comparison. Building complexes was chosen for it could reveal issues related to both buildings and infrastructure. Reactive maintenance was chosen for reactive maintenance problems have a time priority over planned maintenance. For both case studies, reactive maintenance service orders were collected for a period of two years. Data included one typical year and one atypical year according to the facilities management team in charge. This approach was chosen for it can reveal unforeseen conditions impacting building maintenance.

Figure 2. Scope of the research

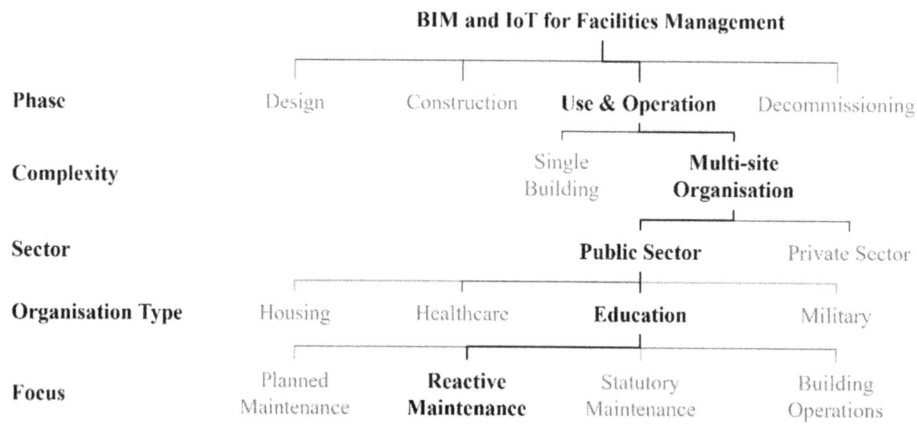

The first case (case study 1 – CS1) is the school portfolio of a large municipality in the North of Italy that counts with over 400 primary and secondary school buildings. The second case (CS2) is a university complex located at the Southwest region of the UK containing 118 buildings with different functionalities including accommodation, education, catering, amongst others that are typical of university campuses. Both organisations have demonstrated interest in the adoption and implementation of BIM for facilities management purposes. For both cases, a descriptive statistical analysis of the reactive maintenance service orders was carried out. The analysis was focused on the characterisation maintenance requests and the identification of priority problems.

Historical data from CS1 was gathered from the years of 2014 to 2016. After data preparation, a total of 17 thousand reactive maintenance service orders described within 889 thousand cells was gathered for analysis. That represents an average of 23 service orders being generated every day. Similarly, for

CS2, historical data was gathered between 2017 and 2018. A total of 66 thousand services orders were compiled for analysis after data preparation, representing 99 service orders per day on average. Data preparation involved the removal of incomplete service orders or their completion whenever data was available. Both cases have a similar framework for reporting and monitoring reactive maintenance services as illustrated in Figure 3. In general, fault detection and report start with a problem as identified by a building user or staff that inform a help-desk system. From this point, the FM team revisit the reported issues and prepare a diagnostic that defines the problem priority and a repair resources plan (including course of action required and the resources necessary for the repair). Once the repair is completed, the case is closed, and feedback is provided to the initiator (user or staff).

Figure 3. Framework of reactive maintenance services

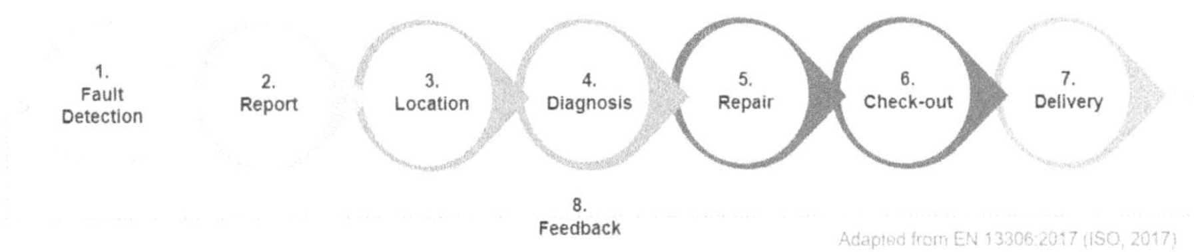

Table 2. Maintenance operational service priorities categories

Category		Response time	Example of problem
	1. Emergency	Immediate response within 2 hours	Fire or risk of explosion Gas leaks Water leaks from plumbing or heating services Loss of electrical supply to rooms Lift breakdown with passengers inside Any incident interrupting teaching or research
	2. Urgent	Less than 24 hours	Blocked drains or internal waste pipes Loss of ventilation and air conditioning Water leaks not causing damage to the building
	3. Important	Less than 3 days	Loss of heating or hot water in residential room Replacement or repairs to light fittings Loss of electrical power Damage to an internal door causing security problems
	4. Standard	Less than 7 days	Loose or missing floor tiles and paving where there is minimal safety risk. Replacement of cracked glass not giving rise to security or safety risk. Replacement or repairs to sanitary ware fittings Adjustment of door closures or floor springs Repairs to joinery items where there is no security risk
	5. Normal	Agreed date with the client	Repairing or purchasing furniture

Figure 4. Maintenance strategic service prioritization variables

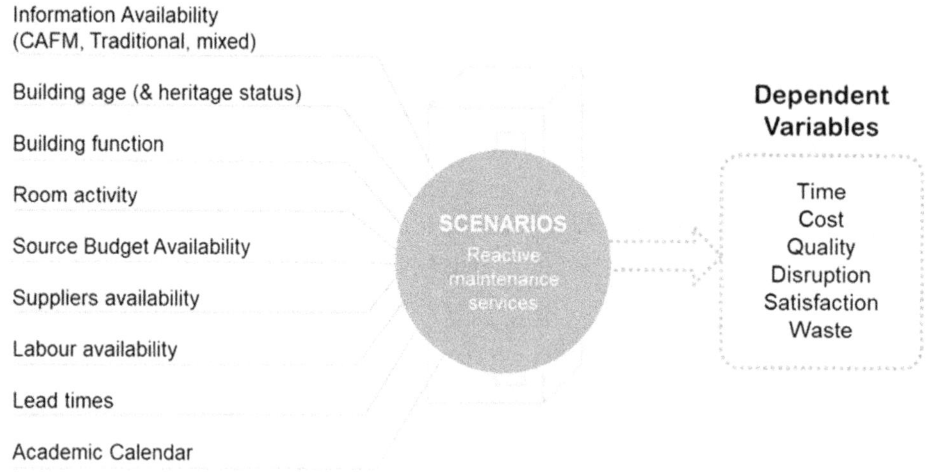

The establishment of service priority relates to two rationales. The first one relates to the risk level of the fault in relation to threatening human life and building structural stability. As shown in Table 1, the levels vary from emergency, prompting a response within 2 hours from the fault being reported, to normal referring to faults that can be fixed in a time frame above 7 days.

Figure 5. Intervention frequency by category

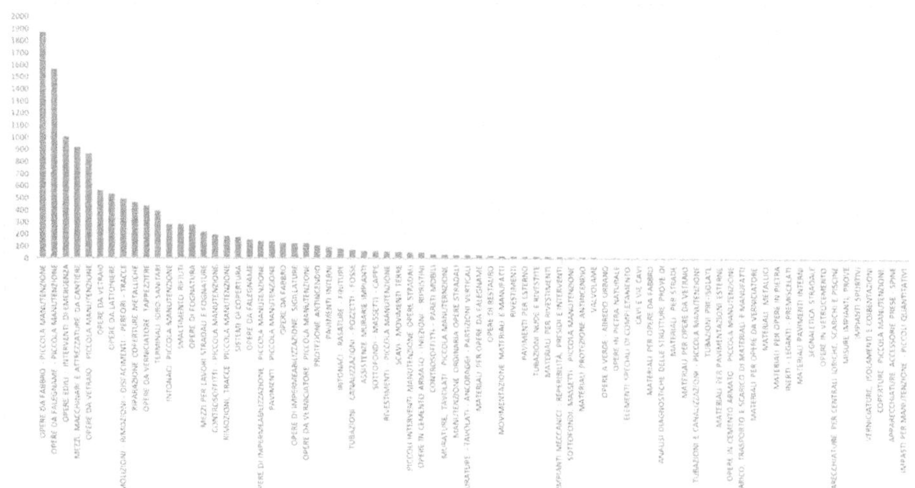

The second one relates to the impact that the fault can have on disrupting the business. In this respect, Figure 4 shows the elicited variables in the context of educational buildings. On the left-hand side, are the variables that are explicitly or implicitly considered in the event of a reactive maintenance occurrence. On the right-hand side are the dependent variables. For example, an electrical fault in a classroom during weekdays will receive a high level of urgency because it can put student's life at risk and also disrupt teaching sessions. Also, a water leakage on a highly graded listed building, will also receive a high level of urgency but below the event in the classroom, since life is not being threatened. One must consider that the decisions concerning the level of urgency are always in a constant state of flux. In other words, if a new fault emerges, the level of all previous identified faults has to be reassessed in relation to its criticality and the level of resources available for fixing the problem.

Figure 6. Pareto analysis of reactive maintenance expenses by category

With regards to the types of fault being reported. The analysed data (*Figure 5*) shows that blacksmith's small maintenance work (oppere da fabro – piccola manutenzione) is the most reoccurring problem. It refers to small repairs to gates, doors and furniture which are considered not urgent if considered on an individual basis.

However, as part of the portfolio of buildings, the frequency of occurrences is so high that makes this a priority type of service to be looked at. That is because such a high frequency imposes pressure on the allocation of shared limited labour resources available. The same applies to carpentry small maintenance work (oppere da falegname – piccola manutenzione). The figures indicates that 3 and 2 faults are reported per day on average within these two categories.

Further analysis of the data is presented in Figure 6, showing the frequency of reactive maintenance services according to the expenses incurred for the conclusion of the service. It can be observed is that the most expensive activity is roof small maintenance (coperture – piccola manutenzione) which in frequency is almost last. That is related to the fact that roof problems, even if small ones, are related to leakages in concrete slabs. Even small faults of such nature means that impermeabilization is compromised. Often, these problems are identified much after they have occurred when there are more serious consequences. As such, a simple roof repair would mean that the layers above impermeabilization have to removed, the impermeabilization redone and the layers above re-installed.

Note that blacksmith's work (opere da fabbro) and carpenter's work (opere da falegname) come second in relation to expenditure. Such configuration is not a surprise, considering that these repairs are small and cheaper, and the increased number of occurrences can be associated to exceptional humid and cold weather patterns that led to an increase in rain and snow during those years. In relation to the literature review, no IoT solution that can be associated with these types of problems was included in the analysis.

Concerning CS2 service orders, the Pareto chart (Figure 7) indicates the frequency of reactive maintenance services according to the category-subcategory and the total cumulative percentage. The most frequent problem is Building-Door with 25,04% of total requests, followed by mechanical-plumbing (22,85%), electrical-lighting (18,58%), mechanical-heat/cool/vent (9,56%) and electrical-power (4,88%). In total, these five categories-subcategories cover 80,91% of the whole requests, thus considered the most relevant services for the organisation. Exploring further the issue through interviews with University's facilities managers, it was found that the increased number of work orders related to doors was an anomaly triggered by an external factor. In June 2017, London, unfortunately, attracted worldwide attention when a fire broke out in the 24-storey Grenfell Tower block of flats in North Kensington, West London, killing 72 of its occupants. This event led to immediate action from the facilities management team to check all doors within the campus with regards to their fire-rate and functionality. Many of the fire containing doors can have their performance reduced after years of use. Although all doors were maintained regularly, the FM team opted for the conduction of an audit to assure students and staff that they are safe within the campus.

Figure 7. Pareto chart – reactive maintenance frequency of services

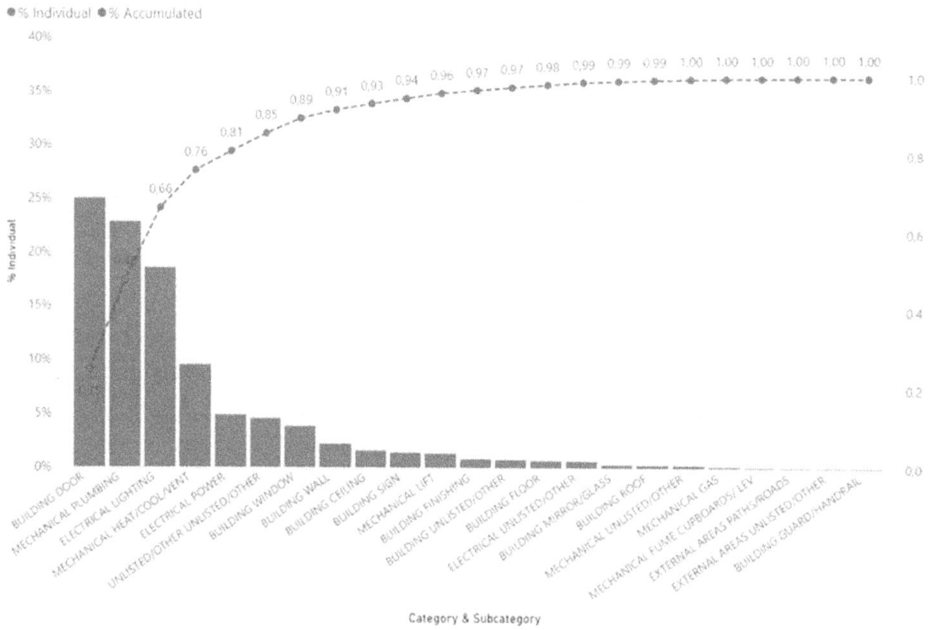

Reactive maintenance cost data is not available for CS2. Upon further investigation it became clear that the cost database is maintained by the finance team. In other words, while the FM is responsible for maintaining buildings, the central finance team is responsible for managing budgets according to building maintenance forecasts created once a year. On the one hand, that is positive because it removes an additional administrative burden from the FM team. On the other, the fragmentation prevents integrated data analytics that can inform strategic decision-making. Because this type of data is absent, data from the two cases cannot be correlated. That is a limitation of the research. In relation to the literature review, IoT solutions that can be associated with the types of problems was included in the analysis. Thus, the next steps of this research are to evaluate the impact of such devices in relation to the current system in place.

However, an additional analysis was also conducted to measure the efficiency of response when a work order is raised. Figure 4 presents the average time spent on each service priority level (as defined within the service orders and classified according to Table 1), and the percentage of requests attended on time (efficiency). The department of estates target is set at 100% response rate for priority 1 services and 85% response rate for priorities 2, 3, 4 e 5. However, the graph shows that most services were not attended on time: 36% for priorities 1 and 2, 32% for priority 3 and 18% for priority 4. 83% of priority 5 services were attended according to the time agreed with the client. The efficiency of Unlisted/Other Services could not be assessed. This result does not show that the FM team is inefficient with regards to the allocation of priority. Rather, the system is based on static data and if a new service order enters the system, the manual reprioritisation of the existing ones is not captured within the reporting system. That is not a surprise considering that circa 90 faults are reported every day. In addition, if a priority 1 service (to be resolved within 2 hours) is resolved in more than two hours, but not much more time was used, still this service order is considered are non-compliant with the set targets. In addition, because service completion data is entered manually, that means that if the person responsible for entering the information did not do so precisely, it is likely that the data will contain errors.

Figure 8. Average time spent on each priority service level.

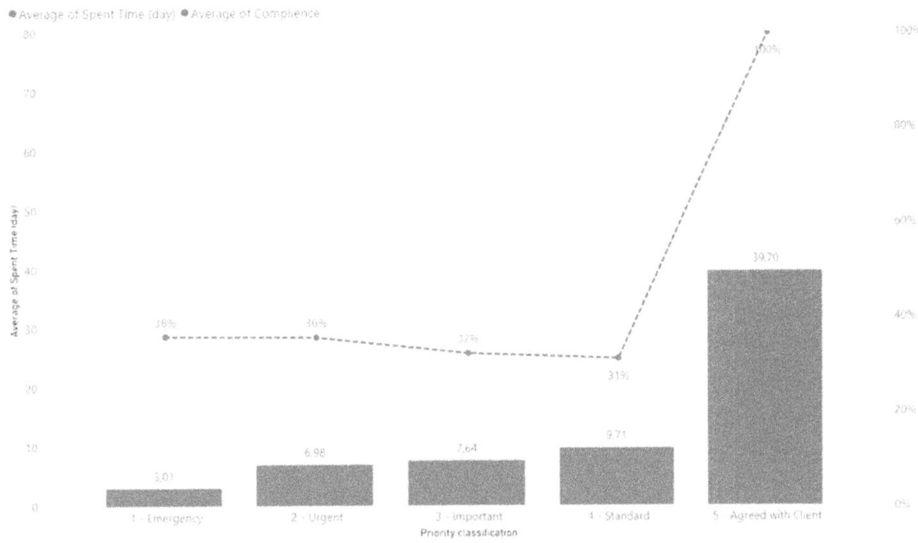

However, this data alone is not sufficient for understanding the reason for efficiency problems. In fact, after interviewing the FM team, it became clear that the system contains priority information as reported to the FM team by building users. Once in the online system, the work requested is assessed by the service engineer that would re-evaluate the problem, the time needed for fixing the problem and the starting date of the repair. In many cases, the priority originally defined by building users is demoted to a less urgent task, and the current system does not capture that (de)classification either.

Such configuration of data leads to another issue, which is the focus of our ongoing research: the definition of a framework for defining reactive maintenance priority. Our initial analysis shows that various criteria could impact on the level of priority of a reactive maintenance service. For instance, a simple jammed door can become a priority if a class is about to start and students and staff cannot enter the classroom (or exit in the occurrence of a fire alarm); Water leaks are always a priority but can become an emergency priority if the leak is near a source of electricity or causing a risk to a heritage-listed asset. Again, once we better understand the problems and the relationship of the problems with the context, we will be able to define better IoT systems that prevent accidents and disruptions more accurately and in the long term prevent or revert (why not) the effect of building users into climate change.

With regards to current IoT devices in use within CS2, none are related to reactive maintenance. Once example is the use of distributed sensors to measure energy consumption within accommodation buildings. The FM team within CS2 is concerned that accommodation buildings may not be compliant with recently released national energy efficiency and CO_2 emissions targets. They are also aware that increases on energy price will significantly impact on the living cost of students living on campus. The data gathered through sensors revealed that most of the energy is used a peak time (when tariffs are higher) for cooking and heating. This information has now been used to define the time of lectures within peak times. This way, it is expected that the consumption of energy will take place at times when the tariffs are cheaper. In addition to that, work is being done to replace devices to more efficient ones and work is being carried out to engage students with more sustainable ways of living.

FINAL REMARKS

We started this chapter proposing a discussion focused on current developments in facilities management research facilitating digital transformation. We argued that digitalisation, in its myriad forms, is beneficial to all stages of the lifecycle of a building, and particularly beneficial during the use, maintenance and operation of buildings. We highlighted that the benefits can be financial, but not if in detriment of sustainable outcomes.

We have also presented the results of a literature review focused on relevant articles published between 2015 and 2019 that relates BIM-FM and IoT. We highlighted the emphasis that is given to the development of technical and processual solutions for buildings lifecycle stages. Process improvement received more predominant attention in most recent years, evidencing that working processes have to change with the adoption of BIM and IoT when organisations decide to shift from traditional activities to a digital environment. IT solutions that address the integration of BIM and IoT devices are abundant for critical areas such as prefabrication, building and infrastructure operation and maintenance, energy management, and health and safety. Challenges and problems faced in the BIM and IoT integration are described in a broad perspective of ICT implementation and can be summarised in three key issues including the interoperability of data over the building lifecycle; the lack of (or perhaps the excess) of standards for information structuring across the various stages of a building lifecycle; and finally, a lack of consensus concerning the establishment of priorities related to facilities management. For that, we presented data from two case studies to illustrate the issues that are emerging with our research.

From what we could see, IoT devices are already available that can support FM through its integration with BIM models. However, as illustrated in both cases, the nature of managing maintenance goes beyond capturing data that represents reality. There is a need for the development of service analytics platforms that ties the data ecosystem together. Such platforms must include mechanism that understand service prioritisation.

REFERENCES

Abergel, T., Dean, B., & Dulac, J. (2017). *Towards a zero-emission, efficient, and resilient buildings and construction sector: Global Status Report 2017*. UN Environment and International Energy Agency.

Akinade, O. O., Oyedele, L. O., Ajayi, S. O., Bilal, M., Alaka, H. A., Owolabi, H. A., & Arawomo, O. O. (2018). Designing out construction waste using BIM technology: Stakeholders' expectations for industry deployment. *Journal of Cleaner Production*, *180*, 375–385. doi:10.1016/j.jclepro.2018.01.022

Alcayaga, A., Wiener, M., & Hansen, E. G. (2019). Towards a framework of smart-circular systems: An integrative literature review. *Journal of Cleaner Production*, *221*, 622–634. doi:10.1016/j.jclepro.2019.02.085

Araszkiewicz, K. (2017). Digital Technologies in Facility Management - The state of Practice and Research Challenges. *Procedia Engineering*. <> doi:10.1016/j.proeng.2017.08.059

Arslan, M., Cruz, C., & Ginhac, D. (2019a). Visualizing intrusions in dynamic building environments for worker safety. *Safety Science*, *120*(July), 428–446. doi:10.1016/j.ssci.2019.07.020

Arslan, M., Cruz, C., & Ginhac, D. (2019b). Understanding Occupant Behaviors in Dynamic Environments using OBiDE framework. *Building and Environment*, *166*(May), 106412. doi:10.1016/j.buildenv.2019.106412

Bianconi, F., Filippucci, M., & Buffi, A. (2019). Automated design and modeling for mass-customised housing. A web-based design space catalog for timber structures. *Automation in Construction*, *103*(March), 13–25. doi:10.1016/j.autcon.2019.03.002

BIM-E Initiative. (n.d.). *Building Information Modelling (BIM)*. Disponível em: https://bimdictionary.com/en/building-information-modelling/1/

Bruno, S., de Fino, M., & Fatiguso, F. (2018). Historic Building Information Modelling: performance assessment for diagnosis-aided information modelling and management. Automation in Construction, 86, 256–276.

Buis, A. (2019). *A Degree of Concern: Why Global Temperatures Matter*. NASA's Global Climate Change Website.

Chaturvedi, K., Matheus, A., Nguyen, S. H., & Kolbe, T. H. (2019). Securing Spatial Data Infrastructures for Distributed Smart City applications and services. *Future Generation Computer Systems*, *101*, 723–736. doi:10.1016/j.future.2019.07.002

Chen, X., Liu, C., & Wu, I. (2018). A BIM-based visualisation and warning system for fi re rescue. *Advanced Engineering Informatics*, *37*(April), 42–53. doi:10.1016/j.aei.2018.04.015

Chiang, C., Chu, C., & Chou, C. (2015). BIM-enabled power consumption data management platform for rendering and analysis of energy usage patterns. *Procedia Engineering*, *118*, 554–562. doi:10.1016/j.proeng.2015.08.480

Chileshe, N., Jayasinghe, R. S., & Rameezdeen, R. (2019). Information flow-centric approach for reverse logistics supply chains. *Automation in Construction*, *106*(June), 102858. doi:10.1016/j.autcon.2019.102858

Chou, C., Chiang, C.-T., Wu, P.-Y., Chu, C.-P., & Lin, C.-Y. (2017). Spatiotemporal analysis and visualisation of power consumption data integrated with building information models for energy savings. *Resources, Conservation and Recycling*, *123*, 219–229. doi:10.1016/j.resconrec.2016.03.008

Ciribini, A. L. C., Pasini, D., Tagliabue, L. C., Manfren, M., Daniotti, B., Rinaldi, S., & De Angelis, E. (2017). Tracking Users' Behaviors through Real-time Information in BIMs: Workflow for Interconnection in the Brescia Smart Campus Demonstrator. *Procedia Engineering*, *180*, 1484–1494. doi:10.1016/j.proeng.2017.04.311

Costin, A., Adibfar, A., Hu, H., & Chen, S. S. (2018). Building Information Modeling (BIM) for transportation infrastructure – Literature review, applications, challenges, and recommendations. *Automation in Construction*, *94*(July), 257–281. doi:10.1016/j.autcon.2018.07.001

Dave, B. (2018). A framework for integrating BIM and IoT through open standards. Automation in Construction, 95, 35–45. doi:10.1016/j.autcon.2018.07.022

David, M., Aubry, A., & Derigent, W. (2018). Towards energy efficient buildings: How ICTs can convert advances? *IFAC-PapersOnLine*, *51*(11), 758–763. doi:10.1016/j.ifacol.2018.08.410

Edmondson, V., Cerny, M., Lim, M., Gledson, B., Lockley, S., & Woodward, J. (2018). A smart sewer asset information model to enable an 'Internet of Things' for operational wastewater management. *Automation in Construction*, *91*, 193–205. doi:10.1016/j.autcon.2018.03.003

Emmitt, S. (2016). BIM and the Future of Design Management. In Construction Manager's BIM Handbook. John Wiley & Sons Ltd. doi:10.1002/9781119163404.ch18

Fialho, B. C., & Codinhoto, R. F. (2019). Trends in BIM and IoT for Reactive Maintenance. *Proceeding of the 36th CIB W78 2019 Conference.*

Fitz, T., Theiler, M., & Smarsly, K. (2019). A metamodel for cyber-physical systems. *Advanced Engineering Informatics*, *41*(May), 100930. doi:10.1016/j.aei.2019.100930

Fortineau, V., Paviot, T., & Lamouri, S. (2019). Automated business rules and requirements to enrich product-centric information. *Computers in Industry*, *104*, 22–33. doi:10.1016/j.compind.2018.10.001

Gao, X., & Pishdad-Bozorgi, P. (2019). BIM-enabled facilities operation and maintenance: A review. Advanced Engineering Informatics, 39, 227–247.

Gha, A. (2019). BIM-integrated knowledge-based building management: Inspecting post- construction energy efficiency. Automation in Construction, 97, 13–28.

Good, N., Ellis, K. A., & Mancarella, P. (2017). Review and classification of barriers and enablers of demand response in the smart grid. *Renewable & Sustainable Energy Reviews*, *72*(January), 57–72. doi:10.1016/j.rser.2017.01.043

Gough, D., Oliver, S., & Thomas, J. (2012). *Systematic reviews* (1st ed.). SAGE.

Gunduz, M., Isikdag, U., & Basaraner, M. (2017). *Integration of BIM, Web maps and IOT for supporting comfort analysis*. ISPRS Annals of the Photogrammetry, Remote Sensing and Spatial Information Sciences. doi:10.5194/isprs-annals-IV-4-W4-221-2017

Haddad, H., Bouyahia, Z., & Chaudhry, S. A. (2019). A Multiagent Geosimulation and IoT-based Framework for Safety Monitoring in Complex Dynamic Spatial Environments. Procedia Computer Science, 151, 527–534. doi:10.1016/j.procs.2019.04.071

Ham, H., Teng, M. A., Wijaya, E., & Wikopratama, R. A. (2018). Integration Citizen' Suggestion System for the Urban Development: Tangerang City Case. *Procedia Computer Science*, *135*, 570–578. doi:10.1016/j.procs.2018.08.210

Heaton, J., & Parlikad, A. K. (2019). A conceptual framework for the alignment of infrastructure assets to citizen requirements within a Smart Cities framework. Cities, 90, 32–41. doi:10.1016/j.cities.2019.01.041

Howell, S., Rezgui, Y., & Beach, T. (2017). Integrating building and urban semantics to empower smart water solutions. *Automation in Construction*, *81*, 434–448. doi:10.1016/j.autcon.2017.02.004

Hu, S., Corry, E., Curry, E., Turner, W. J. N., & O'Donnell, J. (2016). Building performance optimisation: A hybrid architecture for the integration of contextual information and time-series data. *Automation in Construction*, *70*, 51–61. doi:10.1016/j.autcon.2016.05.018

Jia, M. (2019). Adopting Internet of Things for the development of smart buildings: A review of enabling technologies and applications. Automation in Construction, 101, 111–126.

Kelton, A. P. (2019). Internet of Things: A survey on machine learning-based intrusion detection approaches. *Computer Networks*, *151*, 147–157. doi:10.1016/j.comnet.2019.01.023

Kensek, K., & Kahn, W. (2013). *Integration of Environmental Sensors with BIM Seven Case Studies. BESS-SB13 California: Advancing Towards Net Zero*.

Kerin, M., & Pham, D. T. (2019). A review of emerging industry 4. 0 technologies in remanufacturing. *Journal of Cleaner Production*, *237*, 117805. doi:10.1016/j.jclepro.2019.117805

Kho, D. D., Lee, S., & Zhong, R. Y. (2018). Big Data Analytics for Processing Time Analysis in an IoT-enabled manufacturing Shop Floor. *Procedia Manufacturing*, *26*, 1411–1420. doi:10.1016/j.promfg.2018.07.107

Kunieda, Y., Codinhoto, R., & Emmitt, S. (2019). Increasing the efficiency and efficacy of demolition through computerised 4D simulation. *Engineering, Construction, and Architectural Management*, *26*(10), 2186–2205. doi:10.1108/ECAM-11-2018-0492

Li, J., Greenwood, D., & Kassem, M. (2019). Blockchain in the built environment and construction industry: A systematic review, conceptual models and practical use cases. *Automation in Construction*, *102*(March), 288–307. doi:10.1016/j.autcon.2019.02.005

Li, X., Shen, G. Q., Wu, P., Xue, F., Chi, H., & Li, C. Z. (2019). Developing a conceptual framework of smart work packaging for constraints management in prefabrication housing production. *Advanced Engineering Informatics*, *42*(June), 100938. doi:10.1016/j.aei.2019.100938

Li, X., Wu, P., & Yue, T. (2019). Integrating Building Information Modeling and Prefabrication Housing Production. Automation in Construction, 100, 46–60. doi:10.1016/j.autcon.2018.12.024

Liu, D., Chen, J., Hu, D., & Zhang, Z. (2019). Dynamic BIM-augmented UAV safety inspection for water diversion project. *Computers in Industry*, *108*, 163–177. doi:10.1016/j.compind.2019.03.004

Lokshina, V., Greguš, M., & Thomas, W. L. (2019). Application of Integrated Building Information Modeling, IoT and Blockchain of Technologies in System Design of Smart Building. *Procedia Computer Science*, *160*, 497–502. doi:10.1016/j.procs.2019.11.058

Louay, A., & Kassem, M. (2018). A unified BIM adoption taxonomy: Conceptual development, empirical validation and application. *Automation in Construction*, *96*(September), 103–127.

Louis, J., & Dunston, P. S. (2018). Integrating IoT into operational work flows for real-time and automated decision-making in repetitive construction operations. *Automation in Construction*, *94*(April), 317–327. doi:10.1016/j.autcon.2018.07.005

Love, P. E. D., Matthews, J., & Gates, B. (2019). The 'how' of benefits management for digital technology: From engineering to asset management. *Automation in Construction*, *107*(July), 102930. doi:10.1016/j.autcon.2019.102930

Ma, Z., & Ren, Y. (2017). Integrated Application of BIM and GIS: An Overview. *Procedia Engineering*, *196*(June), 1072–1079. doi:10.1016/j.proeng.2017.08.064

Madakan, S., Ramaswamy, S., & Tripathi, S. (2015). Internet of Things (IoT): A Literature Review. *Journal of Computer and Communications*, *3*(05), 164–173. doi:10.4236/jcc.2015.35021

Magruk, A. (2015). The most important aspects of uncertainty on the Internet of Things field – context of smart buildings. Procedia Engineering, 122, 220–227.

Marsal-Llacuna, M.-L. (2018). Future living framework: Is blockchain the next enabling network? Technological Forecasting & Social Change, 128, 226–234.

Mcglinn, K. (2019). Interlinking geospatial and building geometry with existing and developing standards on the web. Automation in Construction, 103, 235–250. doi:10.1016/j.autcon.2018.12.026

Morandi, M. I. W. M., & Camargo, L. F. R. (2015). Revisão sistemática da literatura. In A. Dresch (Ed.), *Design Science Research - método de pesquisa para avanço da ciência e tecnologia* (pp. 141–172). Bookman.

Motamedi, A., Soltani, M. M., Setayeshgar, S., & Hammad, A. (2016). Extending IFC to incorporate information of RFID tags attached to building elements. *Advanced Engineering Informatics*, *30*(1), 39–53. doi:10.1016/j.aei.2015.11.004

Nawari, N. O., & Ravindran, S. (2019). Blockchain and the built environment: Potentials and limitations. *Journal of Building Engineering Journal, 25.*

New, M., Liverman, D., Schroder, H., & Anderson, K. (2011). *Four degrees and beyond: the potential for a global temperature increase of four degrees and its implications.* Academic Press.

Novais, L., Manuel, J., & Ortiz-Bas, Á. (2019). A systematic literature review of cloud computing use in supply chain integration. *Computers & Industrial Engineering, 129*, 296–314.

Pärn, E. A., & Edwards, D. J. (2017). Conceptualising the FinDD API plug-in: A study of BIM-FM integration. *Automation in Construction*, *80*, 11–21. doi:10.1016/j.autcon.2017.03.015

Pauwels, P., Zhang, S., & Lee, Y. (2017). Semantic web technologies in AEC industry: A literature overview. *Automation in Construction*, *73*, 145–165. doi:10.1016/j.autcon.2016.10.003

Peng, Y., Li, S., & Hu, Z. (2019). A self-learning dynamic path planning method for evacuation in large public buildings based on neural networks. *Neurocomputing*, *365*, 71–85. doi:10.1016/j.neucom.2019.06.099

Pezeshki, Z., Soleimani, A., & Darabi, A. (2019). Application of BEM and using BIM database for BEM: A review. *Journal of Building Engineering, 23*, 1–17.

Pinti, L. (2019). BIM-based Facility Management. *Embedded information management for maintenance: Building Information Modelling and database integration for the Italian Public Administration.*

Pishdad-Bozorgi, P. (2017, September). Future Smart Facilities: State-of-the-Art BIM-Enabled Facility Management. *Journal of Construction Engineering and Management*, *143*(9), 02517006. doi:10.1061/(ASCE)CO.1943-7862.0001376

Plageras, A. P., Psannis, K. E., Stergiou, C., Wang, H., & Gupta, B. B. (2018). Efficient IoT-based sensor BIG Data collection – processing and analysis in smart buildings. *Future Generation Computer Systems*, *82*, 349–357. doi:10.1016/j.future.2017.09.082

Ramprasad, B. (2018). Leveraging existing sensor networks as IoT devices for smart buildings. *IEEE World Forum on Internet of Things, WF-IoT 2018 - Proceedings.* 10.1109/WF-IoT.2018.8355121

Rashid, K. M., Louis, J., & Fiawoyife, K. K. (2019). Wireless electric appliance control for smart buildings using indoor location tracking and BIM-based virtual environments. Automation in Construction, 101, 48–58.

Rice, J., & Martin, N. (2018). Smart infrastructure technologies: Crowdsourcing future development and benefits for Australian communities. *Technological Forecasting and Social Change*, (March), 119256.

Rothenberg, J., Widman, L. E., Loparo, K. A., & Nielsen, N. R. (1989). *The nature of modeling* (Vol. 3027). Rand.

Rubio, A. (2018). A common-sense based system for Geo-IoT. *Procedia Computer Science*, *126*, 665–674. doi:10.1016/j.procs.2018.07.301

Simpson, M., Underwood, J., Shelbourn, M., Carlton, D., Aksenova, G., & Mollasalehi, S. (2019). *Pedagogy and Upskilling Network: Evolve or Die-Transforming the productivity of Built Environment Professionals and Organisations of Digital Built Britain through a new, digitally enabled ecosystem underpinned by the mediation between competence supply and demand.* Academic Press.

Succar, B., & Kassem, M. (2015). Macro-BIM adoption: Conceptual structures. *Automation in Construction*, *57*, 64–79. doi:10.1016/j.autcon.2015.04.018

Tang, S., Shelden, D. R., Eastman, C. M., Pishdad-Bozorgi, P., & Gao, X. (2019). A review of building information modeling (BIM) and the internet of things (IoT) devices integration: Present status and future trends. *Automation in Construction*, *101*(February), 127–139. doi:10.1016/j.autcon.2019.01.020

Terroso-Saenz, F., González-Vidal, A., Ramallo-González, A. P., & Skarmeta, A. F. (2019). An open IoT platform for the management and analysis of energy data. *Future Generation Computer Systems*, *92*, 1066–1079. doi:10.1016/j.future.2017.08.046

The British Standard Institute (BSI). (2014). *PAS 180 - Smart cities-Vocabulary*. Disponível em: http://shop.bsigroup.com/upload/PASs/Free-Download/PAS180.pdf

Vandecasteele, F., Merci, B., & Verstockt, S. (2017). Fireground location understanding by semantic linking of visual objects and building information models. *Fire Safety Journal*, *91*, 1026–1034. doi:10.1016/j.firesaf.2017.03.083

Wang, H., Pan, Y., & Luo, X. (2019). Integration of BIM and GIS in sustainable built environment: A review and bibliometric analysis. Automation in Construction, 103, 41–52.

Wong, J. K. W., Ge, J., & He, S. X. (2018). Digitisation in facilities management: A literature review and future research directions. *Automation in Construction.*

Woodhead, R., Stephenson, P., & Morrey, D. (2018). Digital construction: From point solutions to IoT ecosystem. *Automation in Construction*, *93*(March), 35–46. doi:10.1016/j.autcon.2018.05.004

Xu, G. (2018). Cloud asset enabled integrated IoT platform for lean prefabricated construction. Automation in Construction, 93, 123–134. doi:10.1016/j.autcon.2018.05.012

Xu, Q., Chong, H., & Liao, P. (2019). Collaborative information integration for construction safety monitoring. Automation in Construction, 102, 120–134. doi:10.1016/j.autcon.2019.02.004

Yamamura, S., Fan, L., & Suzuki, Y. (2017). Assessment of urban energy performance through integration of BIM and GIS for smart city planning. *Procedia Engineering*, *180*, 1462–1472. doi:10.1016/j.proeng.2017.04.309

Yan, W., & Sakairi, T. (2019). Geo CPS: Spatial challenges and opportunities for CPS in the geographic dimension. *Journal of Urban Management*, *8*(3), 331–341. doi:10.1016/j.jum.2019.09.005

Yang, Y., Ng, S. T., Xu, F. J., & Skitmore, M. (2018). Towards sustainable and resilient high-density cities through better integration of infrastructure networks. *Sustainable Cities and Society*, *42*(April), 407–422. doi:10.1016/j.scs.2018.07.013

Ye, Z. (2018). Cup-of-Water Theory: A Review on the Interaction of BIM, IoT and Blockchain During the Whole Building Lifecycle. *Proceedings of the 35th International Symposium on Automation and Robotics in Construction (ISARC).*

Yin, X. (2019). Building information modelling for o ff -site construction: Review and future directions. Automation in Construction, 101, 72–91.

Zachary, S. (2019). Developing a theory of an object-oriented city: Building energy for urban problems. *Energy Procedia*, *158*, 4210–4217. doi:10.1016/j.egypro.2019.01.807

Zhai, Y., Chen, K., Zhou, J. X., Cao, J., Lyu, Z., Jin, X., Shen, G. Q. P., Lu, W., & Huang, G. Q. (2019). An Internet of Things-enabled BIM platform for modular integrated construction: A case study in Hong Kong. *Advanced Engineering Informatics*, *42*(October), 100997. doi:10.1016/j.aei.2019.100997

Zhengdao, C. (2016a). SWOT analysis and Internet of Things-enabled platform for prefabrication housing production in Hong Kong. *Habitat International*, *57*, 74–87. doi:10.1016/j.habitatint.2016.07.002

Zhengdao, C. (2016b). Schedule risks in prefabrication housing production in Hong Kong: A social network analysis. *Journal of Cleaner Production*, *134*, 482–494. doi:10.1016/j.jclepro.2016.02.123

Zhengdao, C. (2017). Integrating RFID and BIM technologies for mitigating risks and improving schedule performance of prefabricated house construction. *Journal of Cleaner Production*, *165*, 1048–1062. doi:10.1016/j.jclepro.2017.07.156

Zhengdao, C. (2018). An Internet of Things-enabled BIM platform for on-site assembly services in prefabricated construction. Automation in Construction, 89, 146–161.

Zhong, R. Y. (2015). Towards Physical Prefabricated Housing Construction in Towards Physical Prefabricated Housing Construction in Towards Physical Prefabricated Housing Construction in Towards Physical Internet-enabled Prefabricated Housing Construction in. *IFAC-PapersOnLine*, *48*(3), 1079–1086. doi:10.1016/j.ifacol.2015.06.227

Zhong, R. Y., Peng, Y., Xue, F., Fang, J., Zou, W., Luo, H., Thomas Ng, S., Lu, W., Shen, G. Q. P., & Huang, G. Q. (2017). Prefabricated construction enabled by the Internet-of-Things. *Automation in Construction*, *76*, 59–70. doi:10.1016/j.autcon.2017.01.006

Zhong, R. Y., Xu, X., & Wang, L. (2017). IoT-enabled Smart Factory Visibility and Traceability using Laser- scanners. *Procedia Manufacturing*, *10*, 1–14. doi:10.1016/j.promfg.2017.07.103

Zhou, C. (2019). Cyber-physical-system-based safety monitoring for blind hoisting with the internet of things: A case study. Automation in Construction, 97, 138–150. doi:10.1016/j.autcon.2018.10.017

Chapter 10
Transforming Construction Waste Analytics to Support Circular Economy and Carbon Footprint Reduction:
Construction Waste Mapping Protocol CDWiMaP

Vasilios Papastamoulis
University of South Australia, Australia

Robert Crocker
University of South Australia, Australia

Kerry London
Western Sydney University, Australia

Petros Patias
Aristotle University of Thessaloniki, Greece

ABSTRACT

The fourth industrial revolution transforms the built environment through several highly intelligence systems such as drones, 3D printers, robotics, as well as building information modelling (BIM) and geographic information systems (GIS). This transformation has widely been applied in buildings. However, to establish environmental sustainability in the built environment, this transformation needs to be expanded in other sectors that impact climate change such as construction and demolition waste materials. The aim of this chapter is to introduce a new conceptual model that can measure construction and demolition waste in real time and optimise their carbon footprint spatially. A quantitative methodology that embraces a measuring protocol and geospatial research method is proposed for this purpose. The proposed model is capable of measuring the recycling potential result in construction sites to support the circular economy as well as to mitigate the associated carbon dioxide emissions (CO_2) with materials' embodied energy, transportation, and treatment.

DOI: 10.4018/978-1-7998-6600-8.ch010

1. INTRODUCTION: PROBLEM STATEMENT AND ENVIRONMENT

Construction and demolition waste materials (CDW) belong to the solid waste stream (You et al., 2020). The construction and demolition waste stream consists of inert and non-inert materials that result in several residential, renovation, infrastructure as well as demolition projects (Lam et al., 2019). The most commonly used inert CDW materials are asphalt, bricks, concrete, drywall and excavation products such as soil, rocks and gravel (Lam et al., 2019; Wu et al., 2019). Wooden products such as timber and packaging waste materials as well as metals are considered to be the most often used non-inert CDW materials in construction projects (Wu et al., 2019).

The inefficient management of the construction and demolition waste materials is globally responsible for serious harm environmentally, socially and economically (X. P. Wang et al., 2019; Wu et al., 2020; Zhang et al., 2019). Environmental impact is perhaps the most visible and measurable. Critically, it is often proposed that the CDW stream has an impact on climate change. Particularly, CDW materials may contribute to global warming in two key ways; through the released carbon dioxide emissions (CO_2) that are associated with industrial production processes and embodied carbon as well as during their transportation and treatment phase (Giesekam, 2016; Huang et al., 2018; Wu et al., 2020).

Discarding CDW materials unlawfully or in unauthorised sites may induce significant land and water pollution as the non-inert materials contain asbestos, mercury or other toxic substances (Santos et al., 2019; Tam & Lu, 2016; Yu et al., 2018). Neglected construction waste stockpiles consisting of timber, carpets, roofing and other flammable building materials may cause fires harming both the ecosystem and human lives (NSW Government, 2020). Therefore, the approaches regarding discarding waste materials are important and the need to develop better strategies is evident to effectively manage such a global problem.

Despite the magnitude of the problem, CDW materials are still disposed of in landfills with so little thought, limiting their space due to their bulky size (Jain et al., 2019; Srour et al., 2019). This limitation forces authorities to seek for new disposal areas limiting further the open and green spaces, thus, delivering a less sustainable built environment in the future generations. Even worse, the mindless disposal of CDW materials can cost people's lives due to a landslide of a landfill. It is worth noting that 73 and 30 people died in 2015 and 2017 in China and Sri Lanka respectively due to the fatal landslides of the unlawfully discarded CDW materials (Duan et al., 2019, p. 1).

Apart from that, many landfills are usually located in rural areas which are too far away from construction sites in which CDW materials are mainly generated, thus, the latter must be transported a long distance; in many cases, the generated CDW materials must be taken to other cities or states seeking for a suitable or cheaper treatment (H. Wu et al., 2020). As a result, CDW materials contribute to climate change through the released CO_2 from the combustion of the fuels during their transportation and landfilling process (Coelho, as cited in Dahlbo et al., 2015; X. P. Wang et al., 2019; Wu et al., 2020).

Also, CDW materials impact the global warming through their embodied carbon. Wu et al. (2015) estimated the carbon impact of CDW materials in China up to 542.56 $KgCO_2$ per ton of CDW materials generated by using the streamlined Life Cycle Assessment (LCA) method (Wu et al., 2015, p. 637). Considering that CDW materials result in billions of tonnes around the world, it is of great importance to minimise the carbon footprint of the CDW stream to enhance environmental sustainability in the built environment (Akbarnezhad & Xiao, 2017; Gálvez-Martos et al., 2018).

Additionally, the global urbanisation is projected to rise by 68% by 2050 (United Nations, as cited in Aslam et al., 2020). Ahmad et al. (2019) emphasised that the produced carbon dioxide emissions from the construction industry may be dramatically increased due to the rapid urbanization. Taking into account that the generation rate of CDW materials is highly related to urbanisation (Akbarnezhad & Xiao, 2017; Marrero et al., 2017), reducing the carbon footprint of CDW materials has been increasingly brought to the industry's and public's awareness (Akbarnezhad & Xiao, 2017; Xu et al., 2019).

Studies have showed that construction and demolition waste is a global problem that has not been efficiently and effectively managed so far (Ali et al., 2019; Duan et al., 2019; El-Haggar & Samaha, 2019; Huang et al., 2018; London et al., 2014). Waste statistics indicate that landfilling rates regarding CDW materials remain significantly high internationally, with an average rate to be estimated at 35% approximately (Ajayi, as cited in Duan et al., 2019; Kabirifar et al., 2020; X. P. Wang et al., 2019). Due to the ongoing disposal of CDW materials and their bulky size, landfills in major cities in Australia, China and USA are facing space depletion (Crawford et al., 2017; Srour et al., 2019). Jikeka et al. (2020) have also highlighted that CDW materials pressure landfill space in South Africa inducing land pollution due to hazardous substances of materials.

Tables 1 and 2 on page 6 and 7 respectively depict the size of the problem clearly. Specifically, China and India have one of the worst performances regarding construction and demolition waste management; in 2014 the maximum recovery rate was estimated 5% and 10%, respectively (Duan et al., 2019; Huang et al., 2018). As can be seen both from both Tables, the material loss for the global circular economy is dreadful. Especially, in just these two countries almost 1.5 billion of tonnes of CDW materials were disposed of annually (Huang et al., 2018; Jain, Singhal & Jain, 2019). However, it is worth mentioning that the waste generation rate is generally related to the size of the population apart from the growing urbanization (Kabirifar et al., 2020).

The European Union (EU), which currently comprises 27 countries (EU-27) following the recent exit of the United Kingdom (UK), is the second collective waste generator globally following the leader China. In 2018, 835 million tonnes of CDW materials were totally produced in the EU-27 (Eurostat, 2020). Despite the high average recovery rate in the EU-27 and the UK, 89% and 91% respectively, the landfilling rate of CDW materials among its members varies significantly and remains considerably high. For instance, in 2016, the discarded CDW materials in Slovakia, Cyprus, Sweden and France were equal to 47%, 43%, 39% and 29% respectively (Eurostat, 2019; Kabirifar et al., 2020).

The United States of America (USA) is the third major producer of construction and demolition waste as demonstrated by the resulting 600.3 million tonnes of materials in 2018 (US EPA, 2020, p. 20). Construction and demolition waste management performance across the USA seems to be eco-friendly as 76% of CDW materials were diverted from dumpsites in the same year (US EPA, 2020, p. 22). However, this percentage cannot be considered as an average diversion rate nationally, because not all states report their CDW figures consistently (Marcellus-Zamora et al., 2020). Similarly, in Brazil robust data pertaining to CDW materials is absent, which makes their management inefficient as 92% of them were not recycled or recovered in 2015 (Nunes & Mahler, 2020; Paz et al., 2018). Also, Canada has a high landfilling rate, particularly 40% (Environment and Climate Change Canada, 2018).

In a similar fashion to the USA, construction and demolition waste materials in Australia are not regulated at a federal level. Australian states and territories are responsible for recording and reporting the generated or the received quantities within their borders (Pickin et al., 2018; Wu et al., 2020). Although some states have achieved a very high recovery rate of CDW materials such as South Australia (Chileshe et al., 2018), the average landfilling rate of CDW materials was nationally encountered around 27% in 2018 (Ratnasabapathy et al., 2020, p. 1). Crawford et al. (2017) emphasised that if the diversion rate will not be reduced urgently, some of the major landfills in Australia are expected to run out of capacity by 2025.

Table 1 on the next page displays an overview of the generated CDW materials in some of the major waste generators countries and the per capita waste generators around the world. The displayed data is approximate and refers to the latest information that has been recently published in the literature. Although the geographic domain of the present study concerns Australia, China, the EU-27 states, including the former member UK, and the USA, data from Brazil, India are also listed in Table 1. Brazil, China, and India have been chosen to be displayed as representatives amongst the most populated and developing countries globally (Conceição, 2019, p. 348), considering further the total amount of the generated CDW materials as well as the high landfilled rate.

Conversely, recent and accurate data from other populated developing countries such as Indonesia, Mexico and Nigeria are outdated, or the released data is very small compared to their population. For instance, the latest available and accurate data regarding CDW materials in Mexico referred to 2011, particularly 6 million tonnes (Araiza-Aguilar et al., 2019), while the average CDW generation rate in South East Asian countries (i.e., Vietnam, Indonesia, and Malaysia) was 5.37 million of tonnes in 2014 (Hoang et al., 2020, p. 321). In Nigeria, updated CDW data at a national level is also limited; data is indirectly related to the bill of quantities, at particularly 5%-10% (Ogunmakinde, 2019, p. 4).

Table 1. Construction & Demolition Waste overview internationally[1]

Country/ Continent	Year	million tonnes CDW	% of the total solid waste	Inhabitants (in millions)	CDW tonnes/ capita
Globally	2018	10,000	36	7,630	1.31
China	2018	>1,500	30-40	1,430	1.05
EU-27	2018	835	36	446	1.87
USA	2018	600	23-30	327	1.83
India	2016	150	25-35	1,339	0.11
UK	2016	120	64	66	1.82
Brazil	2017	45	63	209	0.22
Australia	2018	20	30	25	0.82

Note. Adopted from Australian Bureau of Statistics (2018); Aslam et al., 2020; Eurostat, 2018; 2020; Jain et al. (2019); Central Pollution Control Board (2017); Nunes & Mahler (2020); Department for Environment, Food and Rural Affairs (2020); Duan et al., 2019; United Nations, 2017; US Bureau of Transportation Service (2017); US EPA (2020); Huang et al. (2018); Shooshtarian et al. (2019)

[1]data concern the major waste generator countries/regions; Although Brazil and India have a small waste generation rate per capita, they produce a significant amount of construction and demolition waste in total, while they present a very high landfilling rate as well.

Table 2 shows clearly that construction and demolition waste materials are not being managed effectively and efficiently internationally. The loss of material potential for the circular economy is tremendously high as billions of tonnes of CDW materials are still disposed of in landfills around the world further damaging the environment. However, discarding the non-hazardous CDW materials can be considered as a potential opportunity. Studies have shown that 50%-80% of them can be reused (i.e., in backfilling, road projects etc) or recycled producing new building materials, depending the type of materials (Chileshe et al., 2018; Hoang et al., 2020; Jikeka et al., 2020; Nakamura, 2020; Yazdani et al., 2021).

Figure 1. ***P****ercentages of Construction & Demolition Waste materials generated globally.*
Note. Generic Figure from Table 1.

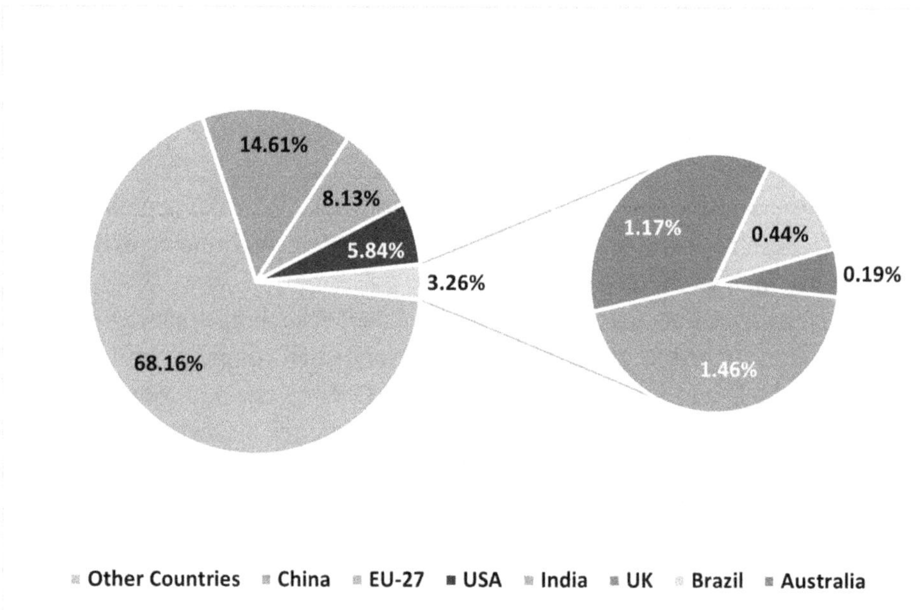

The damage to our economy and environment can be mitigated by diverting the CDW materials from landfills towards supporting waste management strategies, which rely on the '3Re' concept (reuse; recover recycle). The latter constitutes the core of the circular economy (Eberhardt et al., 2019; Tisserant et al., 2017). By doing so, a transformation is further supported from the linear economic model: produce; consume; dispose that the industry promoted over the last 50-60 years (Crocker et al., 2018; Eberhardt et al., 2019) to a more sustainable economic model such as circular economy.

Table 2. Construction & Demolition Waste Landfilled internationally[2]

Country /Continent	million tonnes		% landfilled	million tonnes landfilled
Globally	10,000		35	3,500
China	1,500		90	1,350
EU-27	835		11	92
USA	600		24	144
India	150		70-90	105
UK	120		9	11
Brazil	45		92	41
Australia	20		27	5.5

Note. Adopted from Department for Environment, Food and Rural Affairs (2020); Eurostat (2018); Government of India (2019); Huang et al. (2018); Ratnasabapathy et al. (2020); US EPA (2020).

[2]Data displayed inTable 2 follow the same assumption made for Table 1.

Without having accurate quantitative data, the magnitude of the construction and demolition waste problem is unknown; thus, the risks to human life and the economy can be immediate, visible and equally severe yet somewhat more obscure. The importance of us managing such an important problem to humanity is fast emerging as one of the most critical challenges we face on this earth. Thus, robust and updated CDW data from sources (construction site) is required to map the problem and effectively support decision making in the industry towards supporting the circular economy.

However, recent, detailed and accurate construction and demolition waste analytics are not provided from construction sites to support the circular economy effectively. Such data may simply concern the type and quantity of the resulting recycling potential in construction projects. Instead, aggregated CDW data is released annually or periodically from the authorised waste management facilities around the globe. These data present further inconsistencies as CDW materials are not being measured and reported in a standard and consistent way around the world; nor are they classified under a unified taxonomy system to enable comparisons amongst different jurisdictions (Marcellus-Zamora et al., 2020; Nunes & Mahler, 2020; Paz et al., 2018; Pickin et al., 2018; Sáez et al., 2018; Tam & Lu, 2016).

As a result, there is a high variation amongst the published data in the literature; for instance, the total annually generated amount of CDW materials varies globally from 7 billion to 12.7 billion of tonnes, respectively (Singh, as cited in de Andrade et al., 2018; Duan et al., 2019; Yazdani et al., 2021). This variation impacts the circular economy significantly. It is worth mentioning that data inconsistencies or gaps are evident in the waste stream in the whole towards affecting the efficiency of the circular economy (Tisserant et al., 2017). Also, the European Environment Agency (EEA) highlights the importance for the circular economy to obtain reliable data from the construction and demolition waste sector, while emphasising the data gap on historical buildings (EEA 2020).

Furthermore, the lack of embracing the spatial and temporal dimensions in the quantitative CDW models affects CDW analytics. The construction and demolition waste stream is a dynamic phenomenon which eventuates in a different geographic context throughout its lifespan. That means that CDW materials are on the move from one venue to another, in particular, they are transporting from a construction site to a transfer station, or to a resource centre and a landfill. As a result, CO_2 emissions are produced during the transportation and treatment phases. Most quantitative models embrace the Life Cycle Assessment (LCA) to estimate the environmental impacts which are associated with CDW materials (Simion et al., 2013; Wu et al., 2015; Xu et al., 2019).

Although LCA is an accurate method which has been widely employed in the construction sector (J. Y. Wang et al., 2019), it lacks the spatial and temporal dimensions in the inventory data used for impact assessment; this gap introduces uncertainty in impact results as well as data gaps (Hermann et al., 2007; Martinez-Sanchez et al., 2017). Therefore, monitoring and controlling the resulting material quantities and their environmental impact during the several stages of their life cycle by integrating the spatial and temporal dimensions is essential to support decision-making (Goodchild, 2016; Hao et al., 2010).

In addition, most of the current quantification models in the literature cannot address the construction waste problem beyond the three dimensions (3Ds) of space (X, Y, Z). They are only capable to accurately estimate the weight or volume of the generated CDW materials. Those models which embrace more than 3Ds in their methodology, do not measure and publish the generated CDW materials or their carbon impact in real-time. This data gap affects the performance of construction waste management systems towards impacting the circular economy eventually. Thus, a multi-dimensional approach including i.e., the sustainability dimension is required to achieve effective decision-making in the CDW management field and support the circular economy as well as the environment (Bakchan et al., 2019).

By integrating intelligent models which can encompass more than the 3Ds is fundamental to support the transformation in the CDW management field. Such novel models are Geographic Information Systems (GIS) because they have the ability to predict 'what', 'where', and 'when' is eventuated (Goodchild, 2016). 'What' may refer to the dimensions of the measuring object (i.e., type, quantity of CDW materials, or quantity of the released CO_2 emissions), 'where' corresponds to the spatial dimension and 'when' involves the temporal one, respectively. Apart from that, GIS can significantly contribute to data management due to their high level of interoperability with other types of datasets, i.e., tables, documents etc (Chen et al., 2018; Niska & Serkkola, 2018; Shahrokni et al., 2014).

Consequently, a transformation is needed regarding the construction waste analytics by considering a holistic management approach. This holistic approach should include a transformation in the existing construction and demolition waste definitions by focusing on the word 'material' rather than the term 'waste'. That is essential to complete the shift from the obsolete linear economic model -which was based on the triptych 'produce-consume-discard'- to the modern concept of the circular economy (Crocker et al., 2018). It is worth mentioning that the term 'waste' was also used to describe abandoned materials in remote and isolated venues during the Middle Ages (Barles, 2014). However, the current definitions in the literature pertaining to CDW materials do not embrace the spatial dimension.

The aim of this study is to address the existing poor data management in the literature and industry, towards supporting the transformation in CDW analytics by proposing a new and novel methodological framework, namely, 'Construction and Demolition Waste Mapping Protocol' or 'CDWMaP'. The main objectives of the study are listed below:

1. To underpin the circular economy by measuring, recording, and publishing online the resulting recycling potential from the non-hazardous building materials in each construction site.
2. To support the environmental sustainability in the construction industry by mitigating the carbon footprint that is associated with the transportation and treatment phase, respectively through the employment of the locational intelligence, and
3. To propose a new conceptual model that can be used by state or federal governments as a supervisory tool to monitor the distribution of CDW flows and their impacts among states, and provide updated and detailed CDW data both at a local and national level.

2. METHODOLOGY: AN INTEGRATIVE LITERATURE REVIEW

2.1. Materials, and Methods

An integrative literature review was employed by this study to investigate the rationale of the existing application of GIS in the construction and demolition waste management field over the last decade internationally. Also, quantitative protocols related to construction and demolition waste management were reviewed. The 'Scopus' online database was utilised to obtain the peer-reviewed sources, while Google Scholar, and the online database of the library of the University of South Australia were utilised to locate any non-peer reviewed sources. The 'Scopus' engine was selected as it is the largest scientific online database with 'peer reviewed' journal articles, books (including book chapters) as well as conference proceedings/papers (Jin et al., 2019). Table 3 on page 14 depicts a summary of the designed methodology. Figures 2 and 3 on page 17 illustrate the scientific and geographic domain of the literature, respectively.

The first run was held in 'Scopus' to locate the peer review sources by using the keyword 'construction and demolition waste' in the field 'Article Title', specifically. The search was limited from 2010 to 2021; the first run resulted 721 documents initially, and a second run was implemented using additional keywords to filter the results. In particular, the words 'GIS', 'geospatial', 'quantification', 'impact', 'emissions', and 'protocol' were used in the fields 'Article Title', Abstract' and 'Keyword' in a combination with the Boolean connectors 'AND' and 'OR' with the following order: (TITLE (construction AND demolition AND waste) AND TITLE-ABS-KEY (gis) OR TITLE (geospatial) OR TITLE (quantification) OR TITLE (impact) OR TITLE (emissions) OR TITLE (protocol)) AND PUBYEAR > 2009, and eventually, 39 documents were found in Scopus.

By searching manually, the title and abstract of each source, 16 documents were found to be related to hazardous substances, economic impacts, materials' recycling properties, as well as waste legislation. Also, 2 documents were written in Spanish, so 18 documents were finally removed as irrelevant to the scope of the present study. By reviewing the remaining 21 sources, only 11 documents were finally related to GIS applications concerning the construction and demolition waste field specifically. In addition, by searching the Google Scholar engine and other online databases and websites 38 sources in total were eventually retrieved and reviewed by this study. These documents are listed on Table 4 that is listed on page 15. However, 29 documents were finally found to be related to quantitative protocols and GIS applications as listed in Table 5, particularly on page 16.

The literature showed that GIS have primarily been used in the waste management field to find suitable locations for waste management facilities or to optimise waste transportation routes mainly. However, GIS have been poorly employed to improve CDW analytics. Specifically, real-time data regarding the resulting material potential in multiple construction sites, or the distribution of the CO_2 emissions associated with the transportation and treatment phases of materials is absent from the literature. The literature also disclosed that even though quantitative protocols may ensure data transparency and consistency in the CDW management field, nil protocol is proposed so far to measure both the produced CDW quantities and their associated CO_2 emissions as mentioned above.

2.2. Quantitative Protocols Regarding Construction and Demolition Waste

Quantitative protocols can contribute significantly to data management, particularly in data collection process (Dixit et al., 2012; Formoso et al., 1999; Martin et al., 1995; Shrestha et al., 2014). Although much has been written about the need for quantifying the generated CDW materials accurately, a very few protocols have been developed so far to ensure robust and transparent CDW data. Formoso et al. (1999) were among the pioneers who proposed a quantitative protocol to collect CDW data from three small-size building companies in Brazil, in 1999. However, the current research presents only those protocols that have been developed during the last decade.

The Waste and Resources Action Programme (WRAP) is one of the main bodies for delivering waste management strategies in the UK. In 2010, WRAP introduced a voluntary protocol-guidance to support the diversion of building waste materials from landfill sites towards supporting the sustainable construction management, across the UK. The core of the protocol consisted of three main key performance indicators (KPIs) to standardise the measuring and reporting methods as well as to enable data comparability in the UK construction industry. (WRAP, 2010). However, the protocol measures the resultant amounts of the aggregated CDW materials. As a result, it lacks providing a data breakdown regarding the specific type and amount of each waste material that results in construction sites.

Table 3. Summary of the Design of the Research Methodology

Research Field	Construction and Demolition Waste Management
Research Method	Integrative Literature Review
Period of Research	2010-2021
Scientific Domain	Peer Reviewed Journal Papers, Conference Papers/Proceedings, Books/Book Chapters and Dissertations
Grey Literature	Non-Peer Reviewed Journal Articles, Governmental Websites, Governmental Documents, Governmental Reports, Industry's Reports
Geographic Domain	America, Asia, Australia, and European Union (28 member countries)
Databases-Engines	Scholar Literature: Scopus (Elsevier) Grey Literature: Google Scholar, Online Library of the University of South Australia, Governmental and Organizational Websites
Targeted Australian Organisations (with having published at least one paper regarding CDW management)	Australian Government, Australian Bureau of Statistics, Australian Institute of Architects, Department of the Environment and Energy, Blue Environment Pty Ltd, Edge Consulting, Encycling Consulting, Green Industries SA, Hyder Pty Ltd, The Environment Protection Agency, Local Governments (ACT, NSW, NT, QLD, SA, TAS, VIC, WA), Ramsay and Associates, Waste Management Association of Australia (WMAA), Waste Management and Resource Recovery Association of Australia (WMRR).
Targeted Overseas Organisations	European Commission, European Parliament, Deloitte Limited UK, The UK Department for Environment, Food, and Rural Affairs (DEFRA), Centre for Science and Environment of India, International Council for Research and Innovation in Building and Construction (CIB), International Solid Waste Association (ISWA), The Ministry of Ecology and Environment of China, Government of Hong Kong, The US Environment Protection Agency, The United Nations Environment Programme (UNEP), Global Waste Management Outlook.
Key Words	'construction and demolition waste', 'quantification', 'GIS', 'carbon impact', 'carbon footprint', 'geospatial' 'geographic information', 'protocol'

Note. Generic Table.

The European Network of Construction Companies for Research and Development (ENCORD) developed in 2013 the 'ENCORD Construction Waste Measurement Protocol' -'ENCORD CDWM Protocol' onwards- to measure the generated CDW materials at a project level in accordance with the European Directive 2008/98/EC. The resultant material quantities are classified based on their source activity into three main categories: construction waste, demolition waste and excavation waste, while the data recording is in accordance with the European Waste Catalogue (ENCORD, 2013a, p. 6). To enable comparisons among the released results from construction firms, the 'ENCORD CDWM Protocol' embraces a set of KPIs consisting of two primary and four secondary indices, respectively. The former ones are mandatory for companies, whereas the latter ones are optional (ENCORD, 2013a).

The 'ENCORD CDWM Protocol' is a robust and accurate methodology in agreement with the European waste management law, thus, it can be considered as a best practice regarding CDW management. Despite that, it lacks in embracing the spatial and temporal dimensions. As a result, it cannot optimise the location/allocation problem of the produced quantities of CDW materials. Neither can it monitor the distribution of CDW materials, the recycling potential across multiple construction sites nor the released CO_2 emissions from transportation. Therefore, it fails to address the dynamic nature of the CDW matter and provide real time CDW data from construction sites.

Table 4 .Total Sources reviewed regarding Construction and Demolition Waste Materials

Author	Year	Country	Source	Peer Reviewed
Bengtsson et al	2010	Australia	Report	No
WRAP	2010	UK	Guide	No
Blaschke et al	2011	Austria-Brazil	Journal Article	Yes
Marcellus-Zamora et al	2012	USA	Conference Paper	Yes
Aydi et al	2013	Tunisia	Journal Article	Yes
ENCORD	2013	EU-28	Guide	No
ENCORD	2013	EU-28	Guide	No
Ishikawa	2013	Japan	Journal Article	Yes
Irizarry et al	2013	USA	Journal Article	Yes
Gallardo et al	2014	Spain	Journal Article	Yes
Shahrokni et al	2014	Sweden	Conference Paper	Yes
Miyaoka	2015	Sweden	Dissertation	No
Bilal et al	2016	Globally	Journal Article	Yes
Goodchild	2016	Globally	Web Article	No
European Commission	2016	EU-28	Guide	No
Li et al	2016	Globally	Journal Article	Yes
Wu et al	2016	China	Journal Article	Yes
Sánchez-Lozano et al	2017	Spain	Journal Article	Yes
Chen et al	2018	Globally	Journal Article	Yes
Chileshe et al	2018	Australia	Journal Article	Yes
Niska and Serkkola	2018	Finland	Journal Article	Yes
Paz et al	2018	Brazil	Journal Article	Yes
Seror and Portnov	2018	Israel	Journal Article	Yes
Araiza-Aguilar et al	2019	Mexico	Journal Article	Yes
Ghazifard and Chamchali	2019	Iran	Journal Article	Yes
Madi and Srour	2019	Syria	Journal Article	Yes
Ogunmakinde et al	2019	Nigeria	Journal Article	Yes
Shooshtarian et al	2019	Australia	Conference Paper	No
Srour et al	2019	Lebanon	Journal Article	Yes
Wang et al	2019	China	Journal Article	Yes
Biluca et al	2020	Brazil	Journal Article	Yes
Marcellus-Zamora et al	2020	USA	Journal Article	Yes
Ratnasabapathy et al	2020	Australia	Journal Article	Yes
Shooshtarian et al	2020a	Australia	Journal Article	Yes
Shooshtarian et al	2020b	Australia	Journal Article	No
Wu et al	2020	Australia	Journal Article	Yes
You et al	2020	China	Journal Article	Yes
Yazdani et al	2021	Australia	Journal Article	Yes

Note. Sources retrieved from Scopus, Google Scholar, and the online database of the Library of the University of South Australia. Time period: 2010 - 2021.

Table 5. GIS and Quantitative Protocols regarding Construction and Demolition Waste Materials

Author	Year	Country	Method	Peer Reviewed
Bengtsson et al	2010	Australia	Quantitative Protocol	No
WRAP	2010	UK	Quantitative Protocol	No
Blaschke et al	2011	Austria-Brazil	GIS - Remote Sensing	Yes
Marcellus-Zamora et al	2012	USA	GIS – LEED Buildings	Yes
Aydi et al	2013	Tunisia	GIS-Multi Criteria Analysis	Yes
ENCORD	2013	EU-28	Quantitative Protocol	No
ENCORD	2013	EU-28	CO_{2e} Quantitative Protocol	No
Ishikawa	2013	Japan	Empirical - Questionnaire	Yes
Irizarry et al	2013	USA	GIS - BIM	Yes
Gallardo et al	2014	Spain	GIS - Waste Generation	Yes
Shahrokni et al	2014	Sweden	GIS - Big Data	Yes
Miyaoka	2015	Sweden	GIS – Material Flow	No
Bilal et al	2016	Globally	BIM - Big Data	Yes
Goodchild	2016	Globally	GIS - Big Data	No
European Commission	2016	EU-28	Quantitative Protocol	No
S. Li et al	2016	Globally	GIS - Big Data	Yes
H. Wu et al	2016	China	GIS – Material Flow	Yes
Sánchez-Lozano et al	2017	Spain	GIS-Multi Criteria Analysis	Yes
Chen et al	2018	Globally	Scientometric Review	Yes
Niska and Serkkola	2018	Finland	Data Analytics	Yes
Paz et al	2018	Brazil	GIS - Survey	Yes
Seror and Portnov	2018	Israel	GIS - Geostatistics	Yes
Araiza-Aguilar et al	2019	Mexico	GIS-Multi Criteria Analysis	Yes
Ghazifard and Chamchali	2019	Iran	GIS - Fuzzy Logic	Yes
Madi and Srour	2019	Syria	GIS-Multi Criteria Analysis	Yes
I. Srour et al	2019	Lebanon	GIS-Multi Criteria Analysis	Yes
H. Wang et al	2019	China	GIS - Remote Sensing	Yes
Biluca et al	2020	Brazil	GIS-Multi Criteria Analysis	Yes
Marcellus-Zamora et al	2020	USA	GIS-Material Flow Analysis	Yes

Note. Sources retrieved from Scopus, Google Scholar, and the online database of the Library of the University of South Australia. Time period: 2010 - 2021.

Another notable and relevant protocol to this study is the 'EU Construction and Demolition Waste Management Protocol', 'EU CDWM Protocol' onwards, which was launched in 2016 by the European Commission (EC) and other stakeholders from the construction industry. The aim of this voluntary protocol is to bridge the credibility gap in the European construction market regarding the quality of the recycled building waste towards supporting the circular economy. The structure of the protocol is comprised of five components and eight principles upon which data management issues and waste logistic inconsistencies within the 28 member-countries of the Eurozone can be ameliorated. The 'EU

CDWM Protocol' supports the recycling target set by the Directive 2008/98/EC, which is that 70% of the generated CDW materials be recycled by 2020 (European Commission, 2016).

Figure 2. Scientific Domain of the Integrative Literature Review
Note. Generic Chart based on Table 5.

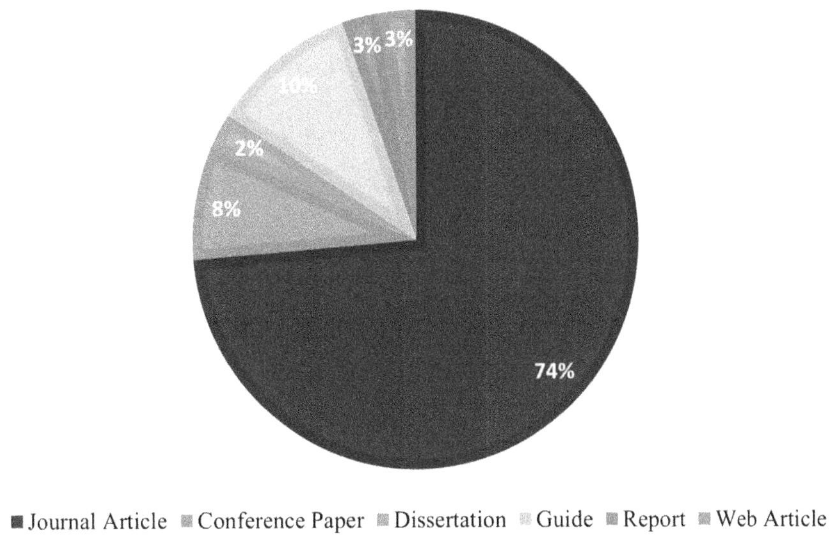

Figure 3. Geographic Domain of the Integrative Literature Review
Note. Generic Chart based on Table 5.

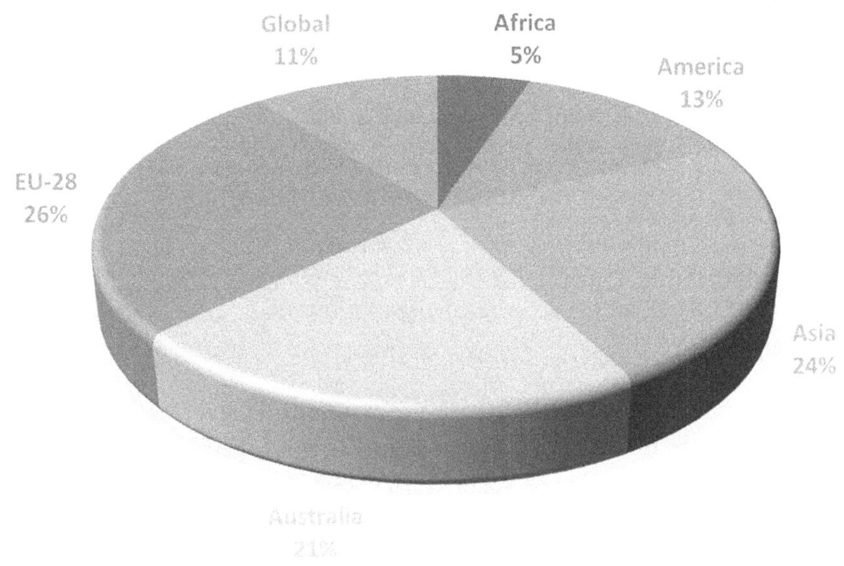

However, the 'EU CDWM Protocol' is primarily considered as best practice in improving logistics and quality management regarding CDW materials, rather than a quantitative methodology. Therefore, it fails to provide quantitative data regarding the amount and type of the recycling materials in construction sites. Nor does it release any information about the environmental impact associated with the CDW stream. Nonetheless, it highlights the importance of proximity and geographical topography in construction waste minimisation. Thus, location is considered amongst the promoted 8 principles of the 'EU CDWM Protocol'. This is very important as considering the proximity factor the produced CO_2 emissions from the transportation of CDW materials can be mitigated.

Table 6 on the next page lists the components and principles of the aforementioned quantitative protocols.

Table 6. Existent Quantitative Protocols regarding Construction and Demolition Waste

Author	Components of Protocol	Country
The UK Waste Recycling Action Programme (WRAP)	Key Performance Indicators (WRAP-KPIs): 1. Total amount of construction, demolition and excavation waste arising in tons generated per £100k of construction output (t/£100k) 2. Total amount of construction, demolition and excavation waste disposed of in landfill sites in tons per £100k of construction output (t/£100k) 3. Percentage of construction, demolition and excavation waste diverted from landfill (%)	United Kingdom
The European Network of Construction Companies for Research and Development (ENCORD)	Primary ENCORD's KPIs (obligatory): 1. Tons of CDW raised (including excavation waste) 2. Percentage (%) of CDW diverted from landfills without energy recovery Secondary ENCORD's KPIs (optional): 1. Construction waste arising in tons/$ (excluding demolition and excavation waste) 2. Tons of waste recovered and reused on the site of production 3. Tons of waste recovered on the site of production, and removed as a non-waste product 4. Waste shipped internationally (in percentage %)	European Union (28 countries)
European Commission (EU CDWM Protocol)	5 Major Components of the EU CDWM Protocol: 1. Waste identification, source separation and collection 2. Waste logistics (promotes 3Re policy: reuse; recycling; recovery) 3. Waste processing 4. Quality management 5. Policy and framework conditions 8 Main Principles of the EU CDWM Protocol: 1. Market-based and promoting competitiveness 2. Ownership by practitioners; support from policy makers 3. Transparency & traceability throughout the CDW management process 4. Promoting certification & audits throughout the entire process (enforceability) 5. No need to reinvent the wheel (is built on existing standards) 6. Location (considers proximity) 7. Respects environmental, health, safety rules & standards 8. Data collection & generation throughout the CDW management process	European Union (28 countries)

Note. Adopted from ENCORD (2013a); European Commission (2016) and WRAP (2010).

Quantitative protocols have also been used to measure and control the released GHG emissions generally. 'The GHG protocol' is one of the most popular protocols developed internationally for accounting and reporting on anthropogenic greenhouse gases (GHG) that are responsible for the climate change issue (Ranganathan et al., 2004). 'The GHG protocol' focuses on measuring and reporting on the six major GHGs covered by the Kyoto Protocol such as carbon dioxide (CO_2), methane (CH_4), nitrous oxide

(N_2O), hydrofluorocarbons (HFCs), perfluorocarbons (PFCs) and sulphur hexafluoride (SF_6) (United Nations, 1998, p. 19). Carbon dioxide, methane and nitrous oxide are the main drivers of the global warming with approximately 75%, 17% and 7% respectively (Ritchie & Roser, 2020).

In terms of the construction industry, the 'ENCORD CO_{2e} measurement protocol' (from hereon in, the 'ENCORD CO_{2e} Protocol') was developed in 2013 based on the methodology of the GHG protocol. Therefore, it measures and reports on the generated GHG emissions from Scopes 1, 2 and 3, as Figure 4 illustrates below. The emissions from activities related to construction and demolition waste stream are included in Scope 3 specifically. However, if a company uses its own vehicles to transport purchased or discarded materials, then the emissions from Scope 1 should also be considered, measured, and reported (ENCORD, 2013b).

Figure 4. Scopes 1, 2 and 3 of the Greenhouse Gas Protocol
Note. From The GHG Protocol (p. 26), by Ranganathan et al., 2004, World Resources Institute (WRI) & World Business Council for Sustainable Development (WBCSD) (www.wbcsd.org). Copyright 2004 by WRI /WBCSD.

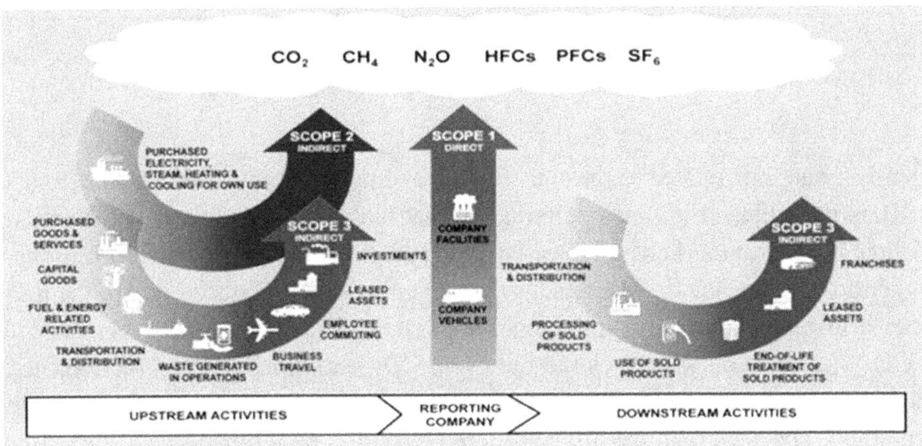

More analytically, the 'ENCORD CO_{2e} Protocol' measures both the direct and indirect CO_{2e} from 12 emissions sources (ENCORD, 2013b, pp. 23-26). These sources are shown on Table 7, which is listed below. The 'ENCORD CO_{2e} Protocol' focuses on the following building materials as these are the most commonly used in miscellaneous construction projects and contain a significant amount of embodied energy and emissions intensity (ENCORD, 2013b, p. 25):

- ferrous metals (e.g., structural and reinforcement steel)
- non-ferrous metals (e.g., aluminium cladding)
- cement (as used in concrete and concrete products)
- brick (ceramics as produced by firing in a kiln)
- glass
- insulation (from non-renewable materials)
- gypsum-based products (as used in plasterboard)
- bituminous products (e.g., asphalt)

Table 7. ENCORD CO_{2e} Protocol: 12 emissions sources

Number of Source	Source
1	Fuel consumption related to construction projects
2	Fuel consumption related to premises
3	Process & Fugitive Emissions
4	Electricity (project)
5	Electricity (premises)
6	Imported Heat & Power
7	Fuel consumption related to vehicles
8	Emissions related to public transport
9	Emissions derive from subcontractors
10	Waste
11	Materials (embodied)
12	Product

Note. Adopted from ENCORD (2013b)

Consequently, emissions related to the construction and demolition waste stream belong to Scope 3 and Source 10, in particular. It is worth pointing out that the embodied carbon of materials purchased belongs to Source 11. These emissions are released during the extraction and manufacturing phase ('Cradle to Gate' boundaries). Also, Source 11 includes the CO_{2e} emissions produced due to the transportation of CDW materials from the factory/supplier to a site ('Gate to Site' boundaries). As can be noticed, the 'ENCORD CO_{2e} Protocol' distinguishes the life cycle CO_{2e} emissions of construction materials with those associated with the CDW materials. This distinction is opposite to a holistic approach, in which the non-hazardous CDW materials are still considered resources rather than waste, and based on which the circular economy is supported.

The Australian Building Products Innovation Council (BPIC) developed in 2010 a protocol to control the environmental impacts associated with the construction materials. In particular, the 'Protocol for the Correct Use of Australian Life Cycle Inventory Data for Construction Materials and Products' or briefly called as 'BPIC Protocol' (Bengtsson et al., 2010). The 'BPIC Protocol' employs the 'Ecopoints' approach that was initially used in the Netherlands and the UK in 1990 (Bengtsson (Bengtsson et al., 2010, p. 52). Australia has developed its own 'Ecopoints' to properly use weighting factors with life cycle data (Bengtsson et al., 2010, p. 54). For instance, 1.5 tonnes of brick includes approximately 0.7 tonnes of CO_{2e} GHGs (Bengtsson et al., 2010, p. 53).

However, the scope of the 'BPIC Protocol' similarly to the 'ENCORD CO_{2e} Protocol' does not focus only on the CDW stream. Thus, detailed CDW data regarding the released CO_{2e} is not provided, especially directly from their sources (construction sites). Also, both methodologies omit embracing the spatial and temporal dimensions to track the distribution of the carbon impact of CDW materials throughout their lifespan, including CO_{2e} emissions associated with the embodied, transportation and treatment phases. Additionally, their scopes are to accurately estimate the released CO_{2e} emissions from several sources rather than to provide real-time CDW data transforming the current poor CDW analytics.

2.3. Applications of GIS in Construction and Demolition Waste Management

Geographic Information Systems (GIS) have become one of the most innovative managerial and powerful decision-making tools (Gallardo et al., 2014; Sánchez-Lozano & Bernal-Conesa, 2017; Wu et al., 2016b). Ishikawa (2013) highlights that 'geospatial science is one of the spatial disciplines that involves a highly sophisticated spatial thinking' (Ishikawa, 2013, p. 637). GIS can efficiently deal with 'big data' and cooperate with other innovative tools to solve complex problems in a multi-dimensional environment due to their spatial prediction abilities; in particular, their ability to predict 'what', 'where', and 'when' (Goodchild, 2016, p. 7). Due to their high data interoperability, and the ability to store, analyse and visualise spatial data in an easy and productive way, GIS can significantly contribute to data management (Chen et al., 2018; Niska & Serkkola, 2018; Shahrokni et al., 2014).

Despite that, few studies have been found in the literature which integrate GIS to improve data analytics in the CDW management field over the last decade. The literature showed that most of studies have employed GIS to find the optimal location of a landfill or resource facility (even a material bank) or to locate illegally disposed CDW materials. Those studies who have engaged GIS in their methodology involve mainly the municipal solid waste stream rather than construction and demolition waste, respectively. Also, these studies utilise other innovative models – apart from GIS – to improve waste analytics. As mentioned above, Table 6 lists the total sources that have been found to be related to the scope of this study.

However, such studies have noted the need for utilising geospatial data, indicating the growing interest in using GIS to reinforce waste analytics. For instance, Irizarry et al. (2013) proposed an integrated model consisting of GIS and Building Information Modelling (BIM) to improve supply chain management through a visualisation process in Georgia, in the U.S.A. The BIM model was utilised to provide managers with accurate measurements of the required material quantities for the procurement phase, while GIS was integrated to support logistics. Specifically, the model was used 'to map the location of suppliers, transportation routes, value adding and no value-adding activities to optimise the logistic costs' (Irizarry et al., 2013, p. 243).

Shahrokni et al. (2014) developed a GIS-based methodology to address inconsistencies in the waste collection routes in the city of Stockholm, in Sweden. Approximately, half a million pieces of data were gathered and analysed for this purpose, consisting mainly of the amount and type of waste; the collection time and location point; and the final disposal site and vehicle licence number (Shahrokni et al., 2014, p. 140). To optimise vehicles' routes, the efficiency index of routes was calculated by dividing the amount of waste generated with the distance travelled, producing also the relevant map. Additionally, waste classification maps were created based on the data variation in terms of densities (Shahrokni et al., 2014, p. 142).

In China, Wu et al. (2016b) designed a quantitative GIS to deal with the generated demolition debris in the city of Shenzhen. Particularly, they projected the resulting amount and composition of bricks, concrete, glass, metals, mortar and others from future demolition projects between 2015 and 2060. The generated mass of demolished materials from buildings was calculated by multiplying the gross floor area of the demolished building with the waste generation index for each type of material; the recycling potential was finally estimated by multiplying that amount with the recycling rate of material type and value of recycling product, while the landfill space by dividing the generated weight of waste with material density (Wu et al., 2016b, p. 498).

Seror and Portnov (2018) utilised GIS and aerial images in addition to identify areas in the Haifa region in Israel that might be exposed to land pollution due to illegal disposal of CDW materials. The research methods they embraced in their methodology were mainly site visit observations and geostatistical analysis. Specifically, they performed a multivariate regression analysis to measure the impacts of variables related to proximity such as the distance from the illegally disposed of areas to the nearby main roads, cities and forests. They finally produced index maps showing the distribution of the illegal dumping as well as the areas at potential risk (Seror & Portnov, 2018, p. 25).

Similarly, Paz et al. (2018) employed geoprocessing tools to map areas in which CDW materials were disposed of illegally in Recife, Brazil. They implemented field surveys to identify these areas and, with the use of Global Positioning Systems (GPS), they recorded their geographical coordinates and entered them into Google Earth. A proximity analysis was performed by drawing buffer zones of 200m diameter from each disposal point to identify suitable areas which can accept up to one cubic metre of waste maximum (Paz et al., 2018, pp. 544-545). Despite that, they did not calculate the environmental impact of the transportation of the illegally disposed of construction and demolition waste stream.

Ghazifard and Chamchali (2019) proposed a GIS-based methodology for the identification of the optimal location of a landfill site in Rudbar, Iran by employing the fuzzy logic method, which is widely used for decision-making. In particular, they assigned values to the area of interest from zero to one, with the former was assigned to the least suitable and the latter to the most appropriate locations respectively. After applying 21 restriction parameters to distinguish improper locations and the deployment of 26 input map layers such, as the digital elevation model, the road and rail network, buildings, etc., and urban and geographical features, they finally produced a 3D map with 56 proper locations (Ghazifard & Chamchali, 2019, pp. 3,14).

Madi and Srour (2019) also employed innovative GIS and remote sensing research methods to manage CDW materials in Syria that result from emergencies, especially in the case of war. Although they considered the gross floor area in their methodology, they estimated the amount of demolition waste in a different manner than Wu et al. (2016b). Specifically, they calculated the total demolished area by multiplying the total built area of the demolished buildings in a specific zone by the number of demolished buildings for the same zone. The total generated amount of waste was determined after multiplying the total demolished area with the concrete production rate for demolition. GIS was recruited to propose a suitable location for the installation of a new recycling facility for CDW materials similar to Ghazifard and Chamchali (2019).

H. Wang et al. (2019) integrated GIS and satellite images to evaluate the vulnerability of the landfill in Shenzhen, in China. Although this method does not measure and track CDW flows, it is worth presenting it briefly as it embraces state of the art technologies, which can contribute to safety and efficiency of CDW management systems. In particularly, satellite images from Google Earth were used to capture the topographic data and extract the elevation of the studied area. The volume of the landfill was calculated by considering the size and variation in height of each pixel of the satellite image (H. Wang et al., 2019, p. 2).

More recently, Biluca et al. (2020) developed a GIS based methodology to identify and rank suitable areas ($5Km^2$) for disposal of inert CDW materials in Brazil. They implemented a multi-criteria analysis considering the hydrography, land use, soil type and slope along with the proximity factor (i.e., the distance from the main roads, residential and education areas etc). The Boolean criterion was employed to accept or discard the potential suitable areas assigning them the value '1' and '0', respectively. They finally classified the identified areas in 'High, Medium and Low' by utilising the 'Jenks Natural Breaks' algorithm, which arranges pixels with a similar value to groups (clusters), and they finally produced the relevant aptitude maps (Biluca et al., 2020, p. 315).

Even though the aforementioned novel studies recruited GIS to monitor and control the material wastage or to solve a decision-making problem, they did not estimate the environmental impact of the transported CDW materials. Tracking the mobility of CDW materials and their associated CO_2 emissions is essential to develop sustainable construction waste management systems (Wu et al., 2020). Actually, Shahrokni et al. (2014) mentioned that the calculation of GHG emissions could be possible if data were available regarding the consumption of the recruited individual trucks for disposal. Also, tracking the diverted CDW flows from landfills and their emissions is very important as landfilling impacts severely the climate chance compared to recycling (Liu et al., 2020).

Conversely, Miyaoka (2015) proposed a methodology to minimise GHG emissions produced due to the transportation of construction waste aggregates from residential projects in Stockholm city, in Sweden. Particularly, the methodology involves GIS and Material Flow Analysis to indicate a strategic location for a new material bank to host for reusing or recycling excavated granular soil, rocks, and sand from future projects (Miyaoka, 2015, p. 5). The ArcMap GIS software was employed to plot and visualise the candidate locations of material banks and to identify clusters of developments with the largest concentration of aggregates (Miyaoka 2015, p. 30). The strategic locations were finally selected after estimating the optimal routes between material banks and development areas by utilising the Google Maps direction tool (Miyaoka 2015, p. 31).

Eventually, 45% of material haulage was avoided, saving 5,248 tons of CO2e emissions (Miyaoka 2015, p. ii). However, the methodology developed by Miyaoka (2015) omits to embrace the lifecycle impact including the embodied carbon and the impact related to their treatment. Miyaoka's study diverges from the study at hand as the former's scope is to optimise the carbon footprint of the CDW materials by proposing a location for the installation of a new processing facility instead of using the existing facilities. In addition, the methodology is limited to construction aggregates, while it does not record the resultant aggregates in accordance with the European Waste Catalogue to enable comparisons among the 27 countries of the European Union.

In conclusion, the literature shows that GIS may significantly contribute to construction and demolition waste management. In particular, it can contribute to solving decision-making problems such as the efficient distribution of materials from manufacturing to sites and from sites to waste management facilities, as they can provide real-time data. In doing so, the burden of landfills can be avoided in favour of the circular economy. Also, studies show that GIS can effectively optimise transportation routes during the distribution of materials, minimising the associated environmental impact. Therefore, in conjunction with the emissions saved due to the diversion of waste materials from landfills, GIS can have an important contribution to the mitigation of global warming.

The literature also disclosed that most of the developed methodologies that have integrated GIS fail to calculate the released CO_2 emissions pertaining to the CDW stream. These studies mainly address the location/allocation problem of CDW materials. Shahrokni et al. (2014) developed the 'efficiency index of routes' to optimise the possible waste disposal paths innovatively. Nonetheless, the index was not successful in estimating the relevant GHG emissions from the transportation phase as data regarding the consumption of the used trucks for disposal was absent.

Although Miyaoka (2015) did estimate and optimise the transportation impact, the methodology lacks the measurement of the embodied carbon of CDW materials and the CO2e emissions that are associated with their treatment. More recently, Paz et al. (2018); Seror and Portnov (2018)[REMOVED

HYPERLINK FIELD] utilised GIS and aerial imagery in their studies to map the areas in which CDW materials are illegally disposed of. Regardless of that, they did not map and calculate the environmental impact which is associated with the transportation of the discarded CDW materials.

Also, very few studies have embraced GIS into their methodology to enhance construction waste analytics. Wu et al. (2016b) designed an innovative quantitative GIS model to estimate the generated demolition debris from future demolished buildings in China. They did provide waste analytics for main demolition waste materials such as concrete, bricks, metals and mortars. Nevertheless, they failed to provide carbon data related to the impact of these materials throughout their life cycle. Also, they did not perform direct measurements to collect real-time data from construction sites as the scope of their research was to evaluate the recycling potential from future demolition projects.

In summary, GIS have primarily been used in the waste management field to find suitable locations for waste management facilities or to optimise waste transportation routes (Gallardo et al., 2014; Tan et al., 2014; Wu et al., 2016b). Little was found in the literature with regards to the contribution of GIS in data management towards improving the existent poor CDW analytics (Bilal et al., 2016; Lu et al., 2016; Wu et al., 2016b). Particularly, Bilal et al. (2016); Lu et al. (2016) utilised big data to improve CDW analytics, while geospatial data was included only in the former research. No methodology has been developed yet to utilise intelligent data management tools such as GIS and drones to transform the CDW analytics by recording, quantifying, monitoring, and publishing real time data regarding the:

- generated CDW flows and recycling potential across multiple construction sites to support the circular economy
- distribution of the carbon impact associated with the lifecycle of CDW materials to underpin environmental management and governance

3. THE CONCEPTUAL MODEL: CDWMAP

3.1. Introduction

The aim of the current study is to introduce a novel conceptual model in the literature, namely, the 'Construction and Demolition Waste Mapping Protocol' (CDWMaP) that can provide real time datasets in order to transform construction waste analytics. The CDWMaP transforms the current conventional data pertaining to the CDW stream from tabular or just numerical to geospatial datasets such as vector and raster data, feature class or shape files and layers. As a result, the CDWMaP transforms the traditional statistical methods to sophisticated ones such as spatial analysis and geostatistics.

The CDWMaP conceptual model integrates the geospatial ontology and epistemology to spatially model the resulting quantities of CDW materials, the construction sites as well as the relevant waste management facilities and suppliers. By doing so, the recycling potential of each types of materials that result in multiple construction sites can be mapped towards supporting the circular economy. By integrating the locational intelligence of GIS, the true distance that CDW materials travel throughout their lifespan can be estimated. Furthermore, the optimal CDW transportation routes can be found to primarily mitigate the carbon impact from the transportation and treatment phases.

Two major components comprise the proposed conceptual model: the first one concerns the development of a quantitative protocol to ensure a robust CDW data flow and transparency from various types of construction projects and different geographically located sites. The second one involves mainly the selection of the data collection and analysis tools as well as the development of a cloud CDW geodatabase to facilitate the publication of real time data across different geographical locations. The rationale of both components are analytically described over the next pages of this section.

3.2. The Quantitative Protocol of the CDWMaP Conceptual Model

The core of the quantitative protocol of the CDWMaP model consists of a specific set of Key Performance Indicators (KPIs), which measure and record in a standard way at least the following five main figures:

1. the resulting quantities for each type of CDW materials
2. the amount of the recycling potential that results in each construction site which is suitable for '3Re' treatment
3. the amount of the CDW materials that ended up in a landfill
4. the contribution of CDW materials to global warming, which is specifically associated with their initial embodied energy, their transportation and treatment, and;
5. the avoided CO_2 emissions due to the diversion of CDW materials from landfills

Introducing such kinds of protocols seems to be essential for underpinning the circular economy and environmental sustainability in the construction industry. Also, data inconsistencies and gaps in the current waste datasets can be bridged by measuring such KPIs. For instance, Pickin et al. (2018) highlighted that the Australian waste sector presents the subsequent issues:

- Many variations exist amongst Australian states and territories regarding the released waste data, and;
- Australia lacks a standard and harmonised waste classification system at a national level.

The generated CDW materials are classified and recorded in a standard way based on a unified and novel taxonomy system. The taxonomy system enables the monitoring of the selected treatment scenarios of CDW materials and their associated CO_2 emissions towards underpinning the circular economy and carbon footprint mitigation. The classification of the resulting types and quantities of CDW materials can be held with two ways (depending on the type of the recorded data). Firstly, manually by the users on sites, who input the weighed CDW quantities in a web mobile interface, which has specifically been designed to record the data according to the developed taxonomy. Alternatively, this can be performed automatically at the office, by performing supervised or unsupervised classification of the raster data (i.e., aerial images) in ArcGIS software (machine learning).

The measured and classified CDW quantities will be then used for the calculation of the KPIs for each construction site in accordance with the protocol's standards. One of these standards concerns the separation of the major resulting building materials into skip bags on sites to ensure accuracy; such materials may be bricks, cement-boards, concrete blocks, iron-bars, steels, plasterboards, and aggregates. The last ones can also be sorted as stockpiles depending on the quantity. The KPIs mainly concern the resultant materials' potential, which is suitable for reuse, recycling, or recovery as well as the released CO_{2e} emissions from the transportation and treatment of CDW materials. The initial embodied carbon of the construction materials is also considered as the CDWMaP emphasises the term 'material' rather than 'waste'.

The CDWMaP (its quantitative protocol part) embraces five main principles, which are built by considering the relevant principles of the GHG Protocol methodology. The five principles of the CDWMaP are listed below:

1. DATA TRANSPARENCY

A standard and cohesive data management 'top-down' approach in accordance with any overarching waste policy or guidance, which is capable of addressing issues regarding the different definitions and classifications pertaining to CDW materials. All stakeholders agree to embrace the scope of the CDW-MaP into the 'Project Charter' of their construction projects to fairly provide the required data regarding the resultant recycling potential and produced carbon footprint. The protocol ensures data transparency through their geovisualisation from the initiation to the completion of projects. The objective nature of the quantitative (numeric) data further ensures transparency in the protocol's results.

2. DATA CONSISTENCY

This principle relies on a robust quantitative methodology which embraces a hierarchically organized CDW taxonomy that classifies and records the generated CDW materials in a standard and unified fashion regardless of their geographical origin (i.e., different region or state or country). The consistency is further supported through the employment of specific KPIs bridging data gaps regarding the generated CDW material flows and their associated carbon dioxide emissions with the initial production of materials (embodied carbon: extraction and manufacturing) as well as the transportation and treatment phases. Data consistency is also achieved through the capability of the CDWiMaP to provide real-time CDW data regardless of the type of project and location of the construction sites and waste management facilities.

3. DATA ACCURACY

The accuracy of the collected CDW data depends on the data collection methods. The CDWMaP conceptual model integrates land and aerial surveying data collection methods (Real Time Kinematic 'RTK method'), and tools such as drones and GPS that can ensure the highest potential precision according to the needs of each project. Regarding the carbon data, the accuracy relates the quality of each LCI database will be used; in this study, the 'Inventory of Carbon & Energy Database' is employed, which has been developed by the University of Bath. However, the carbon data must be in accordance with the LCA to enable comparisons and ensure unbiased results.

4. MODEL'S COMPLETENESS

The proposed conceptual model is able to account for, report on, and track the carbon dioxide emissions associated with the lifecycle of the generated CDW materials across different geographic areas. Also, CDWMaP corporates with several other type of datasets that derive from third databases due to the high interoperability of GIS, thus, making it a flexible and powerful data management tool. Furthermore, by adopting the assumption that the non-hazardous CDW materials remain materials after their first use on sites, CDWMaP promotes a holistic managerial approach towards supporting environmental sustainability in the built environment and circular economy.

5. RELEVANCE

The conceptual model is designed to manage data that is only relevant to the scope of which the CDWMaP is developed; that is to support the circular economy and environmental sustainability in the CDW management field. Thus, the model's geodatabase (geospatial data inventory) stores data from the literature regarding the embodied carbon of the resulting CDW materials, apart from those are collected in the field. Also, the model publishes real-time CDW KPIs to support decision-making policy concerning the impacts of the CDW stream such as environmental (carbon footprint) and economic (transportation costs; material wastage).

3.3. The Geospatial Aspect of the CDWMaP Conceptual Model

The second major component of the CDWMaP conceptual model involves the selection of the geospatial and remote sensing tools for the data acquisition and spatial analysis. In particular, drones, GPS and mobile GIS may be recruited as the main data collection tools, while a cloud geodatabase will host the collected CDW data from different locations. The spatial analysis of the collected data will be held by employing the ArcGIS software in a unified reference system. The type of the geodatabase depends on the number of construction sites. Specifically, it can be a 'file geodatabase' if the data management concerns a project level or a very small sample of construction sites (micro scale); or it can be an 'enterprise geodatabase' if the data management concerns a large sample of construction sites or at a regional and national level.

In the CDWMaP conceptual model, construction sites are represented spatially as 'feature class' files with a specific geometry (polygon) and Cartesian coordinates (X, Y) in the proper projection system, which depends on the state or country in which the model will be used. The CDW materials are spatially modelled, recorded and stored into the cloud geodatabase as 'points' associated with their tabular data, which corresponds to the various classification groups that are related to the materials. Apart from vector data, the conceptual model also employs raster data. This data mainly concerns aerial images captured by drones to measure the volumes of CDW materials as well as to produce the 2D-Orthoimage to be used as an updated basemap for the spatial analysis.

It is noteworthy that in the case of drones not being available to fly over construction sites, the interoperability of the model enables data to be alternatively derived from CAD files or weighbridges and then to be converted into feature class files or thematic layers. Additionally, spatial data sets from external sources may be uploaded for future research (i.e. the digital boundaries of Australia within its regions from the Australian Bureau of Statistics, shape files from other waste or GHG databases etc.).

Figure 5. The Conceptual Model: CDWiMaP
Note. Generic diagram.

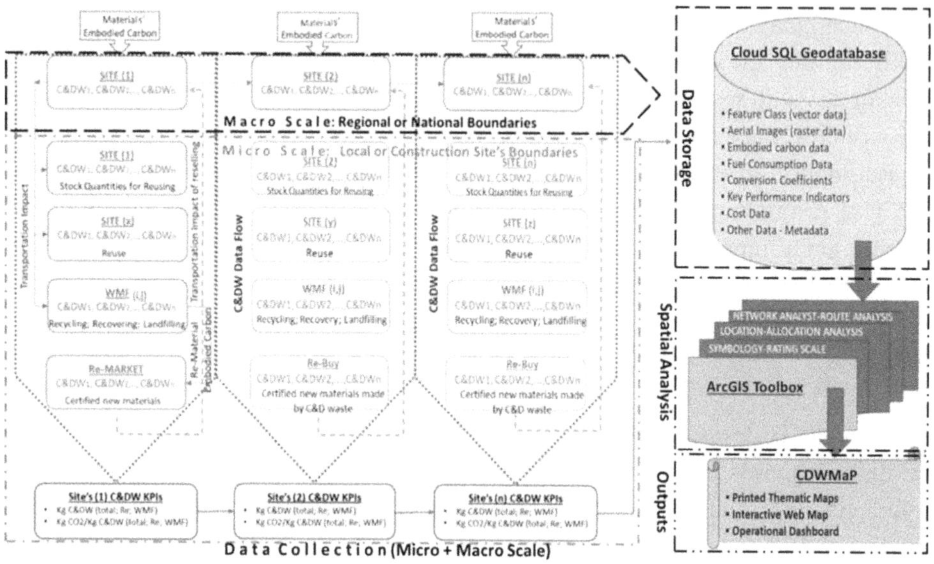

The CDWiMaP allows users to upload both vector and raster data to the cloud geodatabase and share the final outcome on a web map, enabling stakeholders to track the CDW flows and their impacts in real time. Specifically, this can be used to monitor the quantity and type of the resultant materials, as well as their origin and their fate. The origin is related to the construction site that the materials derive from, while the fate refers to the final destination of materials for treatment (reuse; recycling; recovery or landfilling). Also, the CDWiMaP enables the visualisation of the produced carbon footprint that is associated with the embodied carbon of materials as well as with the one from the transportation and treatment of CDW materials.

The scope of the spatial analysis is two-fold: on one hand, it mitigates the carbon footprint that is associated with the CDW stream. This task mainly concerns finding the optimal transportation routes by considering the travelled distance and time. The transportation routes involve the efficient distribution of CDW materials from site to resource facilities and landfill as well as the dispatch of building materials from suppliers to construction site. By finding the shortest routes through which the materials will be efficiently allocated to the suitable locations, the transportation impact is mitigated. On the other hand, it creates an online inventory regarding the recycling potential for each construction site, which is valuable and real time information that can support the circular economy.

However, to be able to analyse and monitor the collected data from different sources, a specific projection map and coordinated system must be specified in advance. All data should be referenced to the same map coordinate system avoiding any errors pertaining to location (georeferencing). After assigning a spatial reference for each feature e.g., sites, roads, waste management facilities (WMFs), suppliers, and eco areas, the generated quantities for each group of CDW materials can be tracked in the same reference system. Then, by utilising ArcGIS tools such as 'Network Analyst', the shortest route or distance between sites and WMFs or suppliers can be calculated, to estimate the impact of the transportation.

Figure 5 on the next page depicts a view of the proposed conceptual model, namely CDWMaP to bridge the research gaps in the literature as mentioned in the previous sections. As can be seen, the proposed conceptual model considers the embodied carbon of materials as an input to support the measurement and monitoring of the lifecycle carbon impact ('Cradle to Cradle'). This assumption also underpins a holistic managerial approach that does not distinguish CDW materials from building materials. Therefore, the 'Cradle to Cradle' carbon footprint for CDW materials is the sum of three major sub-impacts: the materially embodied carbon, the transportation and treatment impact, respectively of CDW materials.

The first boundaries include the released from the extraction and manufacturing phase of building materials as well as from transportation from factories to the relevant suppliers or to construction sites directly, namely 'Cradle to Gate' boundaries (Akbarnezhad & Xiao, 2017; Hammond & Jones, 2008). The second one involve the CO_2 emissions that are produced when materials travel from suppliers to construction sites, and then to waste management facilities (WMFs) for a treatment (i.e., reuse, recycle, recovery or landfilling). It is worth clarifying that this impact may include CO_2 emissions from the handling phase of CDW materials. The third and last one concerns the released CO_2 emissions from the treatment of CDW materials. In doing so, the CDWMaP conceptual model aims to mitigate the carbon footprint of the CDW stream underpinning environmental sustainability in the industry.

Moreover, by monitoring the CDW flows on an interactive map, the circular economy is underpinned. For instance, (n) quantities of CDW materials are generated at site 1. Some of them may remain at the same site for reuse, while some others may be transported to a third site (x) for reuse as well. The resultant quantities are transported from site 1 to the various waste management facilities that exist either for recycling and recovery or for landfill. Those materials that end up at the resource recovery centres become new certified building materials and they are transported again to a new site, and so on. Also, it is worth noting that not all WMFs may accept the same types of materials for a treatment. So, the efficient distribution of CDW materials from sites to the relevant recycling centres is essential to support the circular economy.

To address this problem, CDWMaP may employ the 'Proximity Analysis Toolset' of the ArcGIS software to discover spatial relationships among data, i.e., how close is the resource recovery centre that accepts concrete waste? The ArcGIS Pro software developed by ESRI is proposed to be the main software for performing the spatial analysis as it is considered one of the most sophisticated spatial software. The 'Proximity Analysis Toolset' embraces many tools to analyse spatially both the vector and raster data. The former types of data are feature based, while the second ones are cell based, respectively.

Particularly, the 'Network Analyst' tool will be used to model the main road network (vector data) and find the optimal transportation route to efficiently allocate the CDW materials from sites to WMFs. For this purpose, other vector based distance tools will be used such as 'find the closest facility' and 'Buffer' tools. Regarding the raster data, the 'Spatial Analyst' tool will be used to perform spatial analysis of the images, i.e., 'Clip', 'Extract by Mask' and then to classify them through the 'Image Classification Wizard' tool. Also, the 'Select by Attributes' pane will be used to select features in the relevant layers and create the needing queries on the SQL geodatabase of the CDWMaP.

CDWMaP is a multi-dimensional GIS-oriented model which considers the spatial and temporal dimensions to track the distribution of the CDW flows and their impacts across various geographic areas and different time intervals. As the CDWMaP considers the lifecycle impact of CDW materials it also embraces the sustainability dimension. Thus, its cloud geodatabase embraces cross-related data both from the waste stream and lifecycle inventory data to calculate the KPIs which relate the carbon impact of the CDW stream. In doing so, a carbon inventory regarding the CDW stream can be developed for each project regardless the location of the site. It utilises the ArcGIS online server to monitor in real-time the CDW flows through an interactive web map and the relevant KPIs via an operational dashboard.

Apart from that, CDWMaP may embrace another dimension such as costs. For instance, contractors can be informed about the associated costs of the material wastage (e.g., reason, type, and quantity of CDW materials produced throughout the project). Also, CDWMaP may provide information about cost data that concerns the transportation of the resultant materials based on the several disposal cost scenarios; or costs that relate the subcontractors as well as costs that are associated with the produced CO_2 emissions. The latter also may be very useful to governmental authorities regarding the environmental governance concerning the Climate Change issue particularly, i.e., 'Cap and Trade' (Shooshtarian et al., 2020).

Therefore, the CDWMaP model can be used as a powerful supervisory or enforcement tool from construction companies and local or national governmental authorities regarding CDW management

Generally, CDWMaP can be used either as a 'top down' or a 'bottom up' construction and demolition waste management model. The former concerns the management of the CDW stream at a national level in accordance with an overarching waste legislation or guidance. That is, solid and transparent data regarding the CDW materials and their carbon footprint can be provided to support the existing waste datasets and GHG accounts nationally. The latter refers to the management of CDW at a local level. Gathering real-time construction waste data from construction sites in a unified and standard way may address the inconsistencies in the released construction waste data from different firms and across various regions.

4. CONCLUSION

The current conceptual model transforms construction waste analytics by providing online spatial datasets regarding CDW materials. A local governmental authority can monitor in real-time the recycling potential that results in various construction sites which are located across different geographic areas. This is achievable through the development of an interactive web map which is associated with an operational dashboard displaying real-time construction waste performance indicators (KPIs) to support decision making. Also, through these two tools stakeholders can monitor the distribution of the carbon footprint of CDW materials throughout their lifecycle. Both the web map and the dashboard constitute primary outputs of the CDWMaP.

The proposed conceptual model addresses poor data management in the CDW stream from a geospatial perspective. The existing figures that are released by authorities vary considerably due to the absence of a quantitative methodology to measure CDW materials under a standard construction waste taxonomy among different regions or countries. Apart from that, there is a data gap in national inventories regarding the GHG emissions that are associated with the CDW stream. Therefore, there is a need for developing a methodology, which is capable of collecting, storing, analysing and monitoring data from different sources to create CDW analytics in a spatial format and fill in such data gaps.

The CDW stream has a dynamic nature which takes place in a geographical context with the resulting quantities being moved from one place to another. This change is associated with the release of a significant amount of carbon dioxide emissions due to the ways they are transported and treated. Therefore, the relevant impact must be tracked and controlled throughout the materials' lifetime. By spatially modelling the construction and demolition waste matter, geographic coordinates are assigned to the gathered data (geocoding); as a result, construction sites and waste management facilities are linked to a specific location. In doing so, the true distance between the sites and the resource facilities/suppliers/ landfills is measured, facilitating the calculation of the transportation footprint in an accurate manner.

By embracing the geospatial ontology and epistemology the model is capable of addressing the poor data management regarding the CDW stream both at the local and national levels. The integration of the geospatial and remote sensing technology enables automation and digitation in a multi-dimensional environment, thus supporting the transformation in data management in the CDW stream. Particularly, it transforms the poor CDW analytics by providing real-time and robust CDW data in a map form across multiple locations and different types of materials. This includes the resultant recycling potential that is suitable for the 3Re policy, the produced CO_2 emissions from transportation and treatment phases, or to identify which material contributes the most to the GHG effect, and other useful data in decision-making.

REFERENCES

Ahmad, M., Zhao, Z.-Y., & Li, H. (2019). Revealing stylized empirical interactions among construction sector, urbanization, energy consumption, economic growth and CO2 emissions in China. *The Science of the Total Environment*, *657*, 1085–1098. doi:10.1016/j.scitotenv.2018.12.112 PMID:30677876

Akbarnezhad, A., & Xiao, J. (2017). Estimation and minimization of embodied carbon of buildings: a review. *Buildings, 7*(1), 5.

Ali, T., Mert, G., Ferhat, K., Zhanar, B., Aiganym, S., & Gulzat, S. (2019). A Comprehensive Construction and Demolition Waste Management Model using PESTEL and 3R for Construction Companies Operating in Central Asia. *Sustainability*, *11*(6), 1593. doi:10.3390u11061593

Araiza-Aguilar, J. A., Gutiérrez-Palacios, C., Rojas-Valencia, M. N., Nájera-Aguilar, H. A., Gutiérrez-Hernández, R. F., & Aguilar-Vera, R. A. (2019). Selection of sites for the treatment and the final disposal of construction and demolition waste, using two approaches: An analysis for Mexico City. *Sustainability (Switzerland)*, *11*(15), 4077. Advance online publication. doi:10.3390u11154077

Aslam, M. S., Huang, B. J., & Cui, L. F. (2020). Review of construction and demolition waste management in China and USA. *Journal of Environmental Management*, *264*, 110445. Advance online publication. doi:10.1016/j.jenvman.2020.110445 PMID:32217323

Bakchan, A., Faust, K. M., & Leite, F. (2019). Seven-dimensional automated construction waste quantification and management framework: Integration with project and site planning. *Resources, Conservation and Recycling*, *146*, 462–474. doi:10.1016/j.resconrec.2019.02.020

Barles, S. (2014). History of waste management and the social and cultural representations of waste. In *The Basic Environmental History* (pp. 199–226). Springer. doi:10.1007/978-3-319-09180-8_7

Bengtsson, J., Howard, N., & Kneppers, B. (2010). *Weighting of environmental impacts in Australia*. https://edgeenvironment.com/weighting-of-environmental-impacts-in-australia/

Bilal, M., Oyedele, L. O., Akinade, O. O., Ajayi, S. O., Alaka, H. A., Owolabi, H. A., Qadir, J., Pasha, M., & Bello, S. A. (2016). Big data architecture for construction waste analytics (CWA): A conceptual framework. *Journal of Building Engineering*, *6*(C), 144–156. doi:10.1016/j.jobe.2016.03.002

Biluca, J., de Aguiar, C. R., & Trojan, F. (2020). Sorting of suitable areas for disposal of construction and demolition waste using GIS and ELECTRE TRI. *Waste Management (New York, N.Y.)*, *114*, 307–320. doi:10.1016/j.wasman.2020.07.007 PMID:32688063

Bureau of Transportation Service. (2017). *Municipal Solid Waste and Construction & Demolition Debris*. United States Department of Transportation. https://www.bts.gov/archive/subject_areas/freight_transportation/faf/faf4/debris

Central Pollution Control Board. (2017). *Guidelines on Environmental Management of Construction & Demolition (C & D) Wastes*. Ministry of Environment Forests & Climate Change. http://jkspcb.nic.in/WriteReadData/userfiles/file/cand%20D%20guidelines/guidelines%20for%20cd%20waste.pdf

Chen, J., Su, Y., Si, H., & Chen, J. (2018). Managerial Areas of Construction and Demolition Waste: A Scientometric Review. *International Journal of Environmental Research and Public Health*, *15*(11), 2350. Advance online publication. doi:10.3390/ijerph15112350 PMID:30356018

Chileshe, N., Rameezdeen, R., Hosseini, M. R., Martek, I., Li, H. X., & Panjehbashi-Aghdam, P. (2018). Factors driving the implementation of reverse logistics: A quantified model for the construction industry. *Waste Management (New York, N.Y.)*, *79*, 48–57. doi:10.1016/j.wasman.2018.07.013 PMID:30343778

Conceição, P. (2019). *Human Development Report 2019. Beyond income, beyond averages, beyond today: Inequalities in human development in the 21st century*. The United Nations Development Programme. http://hdr.undp.org/en/content/human-development-report-2019

Crawford, R. H., Mathur, D., & Gerritsen, R. (2017). Barriers to Improving the Environmental Performance of Construction Waste Management in Remote Communities. *Procedia Engineering*, *196*, 830–837. doi:10.1016/j.proeng.2017.08.014

Crocker, R., Saint, C., Chen, G., & Tong, Y. (2018). *Unmaking waste in production and consumption: towards the circular economy*. Emerald Publishing Limited. https://find.library.unisa.edu.au/primo-explore/fulldisplay?vid=UNISA&search_scope=All_Resources&docid=UNISA_ALMA21167209160001831

Dahlbo, H., Bachér, J., Lähtinen, K., Jouttijärvi, T., Suoheimo, P., Mattila, T., Sironen, S., Myllymaa, T., & Saramäki, K. (2015). Construction and demolition waste management – a holistic evaluation of environmental performance. *Journal of Cleaner Production*, *107*, 333–341. doi:10.1016/j.jclepro.2015.02.073

de Andrade, A. R., Gomes, M. C., & Duque, J. (2018). New Approach for Optimization of Construction and Demolition Waste Management: Application to the Lisbon Metropolitan Area. In A. I. F. Vaz, J. P. Almeida, J. F. Oliveira, & A. A. Pinto (Eds.), *Operational Research* (Vol. 223, pp. 1–10). doi:10.1007/978-3-319-71583-4_1

DEFRA. (2020). *Waste Management Plan for England*. Department for Environment Food and Rural Affairs. www.gov.uk/government/publications

Dixit, M. K., Fernández-Solís, J. L., Lavy, S., & Culp, C. H. (2012). Need for an embodied energy measurement protocol for buildings: A review paper. *Renewable & Sustainable Energy Reviews*, *16*(6), 3730–3743. doi:10.1016/j.rser.2012.03.021

Duan, H., Miller, T. R., Liu, G., & Tam, V. W. (2019). Construction debris becomes growing concerns of growing cities. *Waste Management, 83*, 1-5.

Eberhardt, L. C. M., Birgisdottir, H., & Birkved, M. (2019). Potential of Circular Economy in Sustainable Buildings. *IOP Conference Series. Materials Science and Engineering, 471*, 092051. doi:10.1088/1757-899X/471/9/092051

ENCORD. (2013a). *Construction Waste Measurement Protocol. A guide to measuring and reporting waste from construction activities*. European Network of Construction Companies for Research and Development (ENCORD). http://www.encord.org/wp-content/uploads/2013/06/2013-05-ENCORD-Waste-Measurement-Protocol-Issue-V1-Low-Res.pdf

ENCORD. (2013b). *Construction CO2 e Measurement Protocol. A guide to reporting against the Green House Gas Protocol for construction companies*. European Network of Construction Companies for Research and Development (ENCORD). https://ghgprotocol.org/sites/default/files/ENCORD-Construction-CO2-Measurement-Protocol-Lo-Res_FINAL_0.pdf

Environment and Climate Change Canada. (2018). *Canadian Environmental Sustainability Indicators: Solid waste diversion and disposal*. Environment and Climate Change Canada. https://www.canada.ca/content/dam/eccc/documents/pdf/cesindicators/solid-waste/2018/solid-waste-diversion-disposal-en.pdf

Eurostat. (2019). *Generation of waste by waste category,hazardousness and NACE Rev. 2 activity*. European Commission. https://ec.europa.eu/eurostat/web/products-datasets/-/env_wasgen

Eurostat. (2020). *Waste Statistics*. https://ec.europa.eu/eurostat/statistics-explained/index.php/Waste_statistics

Formoso, C., Isatto, E. L., & Hirota, E. H. (1999). Method for waste control in the building industry. *Proceedings IGLC*. https://www.researchgate.net/publication/228646326_Method_for_Waste_Control_in_the_Building_Industry/citation/download

Gallardo, A., Carlos, M., Peris, M., & Colomer, F. J. (2014). Methodology to design a municipal solid waste generation and composition map: A case study. *Waste Management (New York, N.Y.), 34*(11), 1920–1931. doi:10.1016/j.wasman.2014.05.014 PMID:25008298

Gálvez-Martos, J.-L., Styles, D., Schoenberger, H., & Zeschmar-Lahl, B. (2018). Construction and demolition waste best management practice in Europe. *Resources, Conservation and Recycling, 136*, 166–178. doi:10.1016/j.resconrec.2018.04.016

Ghazifard, A., & Chamchali, M. M. (2019). The use of fuzzy logic spatial modeling via GIS for landfill site selection (case study: Rudbar-Iran). *Environmental Earth Sciences, 78*(10), 1–16. doi:10.100712665-019-8296-y

Giesekam, J. J. (2016). *The contribution to UK climate mitigation targets from reducing embodied carbon in the construction sector* [PhD, University of Leeds]. https://core.ac.uk/download/pdf/77023765.pdf

Goodchild, M. (2016). GIS in the Era of Big Data. *Cybergeo: European Journal of Geography*, 1-7. https://journals.openedition.org/cybergeo/27647

Hammond, G. P., & Jones, C. I. (2008). Embodied energy and carbon in construction materials *Proceedings of the Institution of Civil Engineers-Energy.*

Hao, J., Yuan, H., & Shen, L. (2010). *Overview of research in construction and demolition waste management.* doi:10.1680/ener.2008.161.2.87

Hermann, B. G., Kroeze, C., & Jawjit, W. (2007). Assessing environmental performance by combining life cycle assessment, multi-criteria analysis and environmental performance indicators. *Journal of Cleaner Production, 15*(18), 1787–1796. doi:10.1016/j.jclepro.2006.04.004

Hoang, N. H., Ishigaki, T., Kubota, R., Tong, T. K., Nguyen, T. T., Nguyen, H. G., Yamada, M., & Kawamoto, K. (2020). Waste generation, composition, and handling in building-related construction and demolition in Hanoi, Vietnam. *Waste Management (New York, N.Y.), 117*, 32–41. doi:10.1016/j.wasman.2020.08.006 PMID:32805599

Huang, B., Wang, X., Kua, H., Geng, Y., Bleischwitz, R., & Ren, J. (2018). Construction and demolition waste management in China through the 3R principle. *Resources, Conservation and Recycling, 129*, 36–44. doi:10.1016/j.resconrec.2017.09.029

Irizarry, J., Karan, E. P., & Jalaei, F. (2013). Integrating BIM and GIS to improve the visual monitoring of construction supply chain management. *Automation in Construction, 31*, 241–254. doi:10.1016/j.autcon.2012.12.005

Ishikawa, T. (2013). Geospatial thinking and spatial ability: An empirical examination of knowledge and reasoning in geographical science. *The Professional Geographer, 65*(4), 636–646. doi:10.1080/00330124.2012.724350

Jain, S., Singhal, S., & Jain, N. K. (2019). Construction and demolition waste generation in cities in India: An integrated approach. *International Journal of Sustainable Engineering, 12*(5), 333–340. doi:10.1080/19397038.2019.1612967

Jikeka, S., Smallwood, J., Emuze, F., & Simpeh, E. (2020, September 11-12). *Waste Management: The Case of Construction and Demolition Waste in Port Elizabeth.* International Conference of Sustainable Ecological Engineering Design for Society (SEEDS) 2019, Ipswich, UK. https://www.springer.com/gp/book/9783030443801

Jin, F., Liu, X. H., Chen, W. C., Fan, Z. L., & Wang, H. L. (2019). High initial (1, 3) Beta-D-Glucan concentration may be a predictor of satisfactory response of c aspofungin combined with TMP/SMZ for HIV-negative patients with moderate to severe Pneumocystis jirovecii pneumonia. *International Journal of Infectious Diseases, 88*, 141–148. doi:10.1016/j.ijid.2019.08.015 PMID:31442630

Kabirifar, K., Mojtahedi, M., Wang, C. X., & Tam, V. W. Y. (2020). Construction and demolition waste management contributing factors coupled with reduce, reuse, and recycle strategies for effective waste management: A review. *Journal of Cleaner Production, 263*, 121265. Advance online publication. doi:10.1016/j.jclepro.2020.121265

Lam, P. T. I., Yu, A. T. W., Wu, Z., & Poon, C. S. (2019). Methodology for upstream estimation of construction waste for new building projects. *Journal of Cleaner Production, 230*, 1003–1012. doi:10.1016/j.jclepro.2019.04.183

Liu, J., Liu, Y., & Wang, X. (2020). An environmental assessment model of construction and demolition waste based on system dynamics: A case study in Guangzhou. *Environmental Science and Pollution Research International*, *27*(30), 37237–37259. doi:10.100711356-019-07107-5 PMID:31893359

London, K., Zhang, P., Khalfan, M., Maqsood, T., & Siva, J. (2014). *A supply chain management self assessment framework for waste minimisation for the residential sector.* https://find.library.unisa.edu.au/primo-explore/fulldisplay?vid=UNISA&search_scope=All_Resources&docid=UNISA_ALMA51129309780001831

Lu, W., Chen, X., Ho, D. C., & Wang, H. (2016). Analysis of the construction waste management performance in Hong Kong: The public and private sectors compared using big data. *Journal of Cleaner Production*, *112*, 521–531. doi:10.1016/j.jclepro.2015.06.106

Madi, N., & Srour, I. (2019). Managing emergency construction and demolition waste in Syria using GIS. *Resources, Conservation and Recycling*, *141*, 163–175. doi:10.1016/j.resconrec.2018.10.018

Marcellus, K. A., Spatari, S., & Gallagher, P. M. (2012). Utilizing GIS as a geospatial tool to inventory LEED certified buildings and construction and demolition (C&D) waste flows in the United States. *3rd International Symposium on Life-Cycle Civil Engineering, IALCCE 2012*, Vienna, Austria.

Marcellus-Zamora, K. A., Gallagher, P. M., & Spatari, S. (2020). Can Public Construction and Demolition Data Describe Trends in Building Material Recycling? Observations From Philadelphia. *Frontiers in Built Environment*, *6*, 131. Advance online publication. doi:10.3389/fbuil.2020.00131

Marrero, M., Puerto, M., Rivero-Camacho, C., Freire-Guerrero, A., & Solís-Guzmán, J. (2017). Assessing the economic impact and ecological footprint of construction and demolition waste during the urbanization of rural land. *Resources, Conservation & Recycling, 117*(PB), 160-174. doi:10.1016/j.resconrec.2016.10.020

Martin, J. H., Collins, A. R., & Diener, R. G. (1995). A sampling protocol for composting, recycling, and re-use of municipal solid waste. *Journal of the Air & Waste Management Association*, *45*(11), 864–870. doi:10.1080/10473289.1995.10467416

Martinez-Sanchez, V., Levis, J. W., Damgaard, A., Decarolis, J. F., Barlaz, M. A., & Astrup, T. F. (2017). Evaluation of Externality Costs in Life-Cycle Optimization of Municipal Solid Waste Management Systems. *Environmental Science & Technology*, *51*(6), 3119–3127. doi:10.1021/acs.est.6b06125 PMID:28263562

Miyaoka, M. (2015). *Sustainability Assessment for Strategic Material Flows Between Planned Construction Projects in the Stockholm County* [MSc, Halmstad University]. http://hh.diva-portal.org/smash/get/diva2:849623/FULLTEXT01.pdf

Nakamura, S. (2020). Tracking the Product Origins of Waste for Treatment Using the WIO Data Developed by the Japanese Ministry of the Environment. *Environmental Science & Technology*, *54*(23), 14862–14867. doi:10.1021/acs.est.0c06015 PMID:33205952

Niska, H., & Serkkola, A. (2018). Data analytics approach to create waste generation profiles for waste management and collection. *Waste Management (New York, N.Y.)*, *77*, 477–485. doi:10.1016/j.wasman.2018.04.033 PMID:29724480

Nunes, K. R. A., & Mahler, C. F. (2020). Comparison of construction and demolition waste management between Brazil, European Union and USA. *Waste Management & Research, 38*(4), 415–422. doi:10.1177/0734242X20902814 PMID:32043421

Ogunmakinde, O. E. (2019). A Review of Circular Economy Development Models in China, Germany and Japan. *Recycling, 4*(3), 27.

Paz, D. H. F., Lafayette, K. P. V., & Sobral, M. C. (2018). GIS-based planning system for managing the flow of construction and demolition waste in Brazil. *Waste Management & Research, 36*(6), 541–549. doi:10.1177/0734242X18772096 PMID:29776320

Pickin, J., Randell, P., & Latimer, G. (2018). *Improving national waste data and reporting*. D. o. t. E. a. Energy. https://www.environment.gov.au/protection/waste-resource-recovery/publications/improving-national-waste-data-reporting

Ranganathan, J., Corbier, L., Bhatia, P., Schmitz, S., Gage, P., & Oren, K. (2004). *The greenhouse gas protocol: a corporate accounting and reporting standard* (revised edition). https://www.wri.org/sites/default/files/pdf/ghg_protocol_2004.pdf

Ratnasabapathy, S., Alashwal, A., & Perera, S. (2020). Investigation of waste diversion rates in the construction and demolition sector in Australia. *Built Environment Project and Asset Management*. Advance online publication. doi:10.1108/BEPAM-01-2020-0012

Ritchie, H., & Roser, M. (2020). CO_2 and Greenhouse Gas Emissions. *OurWorldInData.org*. https://ourworldindata.org/co2-and-other-greenhouse-gas-emissions

Sáez, P. V., & Santa Cruz Astorqui, J. (2018). Estimation of construction and demolition waste in building energy efficiency retrofitting works of the vertical envelope. *Journal of Cleaner Production, 172*, 2978–2985. doi:10.1016/j.jclepro.2017.11.113

Sánchez-Lozano, J. M., & Bernal-Conesa, J. A. (2017). Environmental management of Natura 2000 network areas through the combination of Geographic Information Systems (GIS) with Multi-Criteria Decision Making (MCDM) methods. Case study in south-eastern Spain. *Land Use Policy, 63*, 86–97. doi:10.1016/j.landusepol.2017.01.021

Santos, A. C., Mendes, P., & Ribau Teixeira, M. (2019). Social life cycle analysis as a tool for sustainable management of illegal waste dumping in municipal services. *Journal of Cleaner Production, 210*, 1141–1149. doi:10.1016/j.jclepro.2018.11.042

Seror, N., & Portnov, B. A. (2018). Identifying areas under potential risk of illegal construction and demolition waste dumping using GIS tools. *Waste Management (New York, N.Y.), 75*, 22–29. doi:10.1016/j.wasman.2018.01.027 PMID:29402616

Shahrokni, H., van der Heijde, B., Lazarevic, D., Brandt, N., Höjer, M., Lago, P., & Wangel, J. (2014). *Big Data GIS Analytics towards Efficient Waste Management in Stockholm*. Academic Press.

Shrestha, S. S., Biswas, K., & Desjarlais, A. O. (2014). A protocol for lifetime energy and environmental impact assessment of building insulation materials. *Environmental Impact Assessment Review, 46*, 25–31. doi:10.1016/j.eiar.2014.01.002

Simion, I. M., Ghinea, C., Maxineasa, S. G., Taranu, N., Bonoli, A., & Gavrilescu, M. (2013). Ecological footprint applied in the assessment of construction and demolition waste integrated management. *Environmental Engineering and Management Journal, 12*(4).

Srour, I., AlZaghrini, N., & Srour, F. J. (2019). Using GIS and optimization to manage construction and demolition waste: The case of abandoned quarries in Lebanon. *Waste Management (New York, N.Y.), 95*, 139–149. doi:10.1016/j.wasman.2019.06.011 PMID:31351599

Tam, V. W. Y., & Lu, W. (2016). Construction Waste Management Profiles, Practices, and Performance: A Cross-Jurisdictional Analysis in Four Countries. *Sustainability, 8*(2), 190. doi:10.3390u8020190

Tan, S., Hashim, H., Lee, C., Lim, J., & Kanniah, K. (2014). *Optimal waste-to-energy strategy assisted by GIS For sustainable solid waste management*. Academic Press.

Tisserant, A., Pauliuk, S., Merciai, S., Schmidt, J., Fry, J., Wood, R., & Tukker, A. (2017). Solid Waste and the Circular Economy: A Global Analysis of Waste Treatment and Waste Footprints. *Journal of Industrial Ecology, 21*(3), 628–640. doi:10.1111/jiec.12562

United Nations. (1998). *Kyoto Protocol to the United Nations framework convention on climate change*. United Nations Framework Convention on Climate Change, Kyoto, Japan. https://unfccc.int/resource/docs/convkp/kpeng.pdf

United Nations. (2017). *World Population Prospects The 2017 Revision. Key Findings and Advance Tables*. Department of Economic and Social Affairs. Population Division. https://www.un.org/development/desa/publications/world-population-prospects-the-2017-revision.html

US EPA. (2020). *Advancing Sustainable Material Management 2018 Fact Sheet*. Office of Land and Emergency Management. The United States Environmental Protection Agency. https://www.epa.gov/sites/production/files/2020-11/documents/2018_ff_fact_sheet.pdf

Wang, H., Zhang, J., & Lin, H. (2019). Satellite-based analysis of landfill landslide: the case of the 2015 Shenzhen landslide. *International Journal of Geotechnical Engineering*, 1-8.

Wang, J. Y., Wu, H. Y., Tam, V. W. Y., & Zuo, J. (2019). Considering life-cycle environmental impacts and society's willingness for optimizing construction and demolition waste management fee: An empirical study of China. *Journal of Cleaner Production, 206*, 1004–1014. doi:10.1016/j.jclepro.2018.09.170

Wang, X. P., Yu, R., Shui, Z. H., Song, Q. L., Liu, Z., Bao, M., Liu, Z. J., & Wu, S. (2019). Optimized treatment of recycled construction and demolition waste in developing sustainable ultra-high performance concrete. *Journal of Cleaner Production, 221*, 805–816. doi:10.1016/j.jclepro.2019.02.201

WRAP. (2010). *Guidelines for measuring and reporting construction, demolition and excavation waste*. The UK Waste Recycling Action Programme. https://www.wrap.org.uk/sites/files/wrap/WRAP%20Waste%20Reporting%20Guidance%20Update%20-%20FINAL1.pdf

Wu, H., Duan, H., Wang, J., Wang, T., & Wang, X. (2015). Quantification of carbon emission of construction waste by using streamlined LCA: a case study of Shenzhen, China. Japan Society of Material Cycles and Waste Management, 17(4), 637-645. doi:10.100710163-015-0404-9

Wu, H., Wang, J., Duan, H., Ouyang, L., Huang, W., & Zuo, J. (2016b). An innovative approach to managing demolition waste via GIS (geographic information system): A case study in Shenzhen city, China. *Journal of Cleaner Production*, *112*(P1), 494–503. doi:10.1016/j.jclepro.2015.08.096

Wu, H., Zuo, J., Yuan, H., Zillante, G., & Wang, J. (2019). A review of performance assessment methods for construction and demolition waste management. *Resources, Conservation and Recycling*, *150*, 104407. Advance online publication. doi:10.1016/j.resconrec.2019.104407

Wu, H., Zuo, J., Yuan, H., Zillante, G., & Wang, J. (2020). Cross-regional mobility of construction and demolition waste in Australia: An exploratory study. *Resources, Conservation and Recycling*, *156*, 104710. doi:10.1016/j.resconrec.2020.104710

Xu, J., Shi, Y., Xie, Y., & Zhao, S. (2019). A BIM-Based construction and demolition waste information management system for greenhouse gas quantification and reduction. *Journal of Cleaner Production*, *229*, 308–324. doi:10.1016/j.jclepro.2019.04.158

Yazdani, M., Kabirifar, K., Frimpong, B. E., Shariati, M., Mirmozaffari, M., & Boskabadi, A. (2021). Improving construction and demolition waste collection service in an urban area using a simheuristic approach: A case study in Sydney, Australia. *Journal of Cleaner Production*, *280*, 124138. Advance online publication. doi:10.1016/j.jclepro.2020.124138

You, Z., Wu, C., Zheng, L., & Feng, L. (2020). An Informatization Scheme for Construction and Demolition Waste Supervision and Management in China. *Sustainability*, *12*(4), 1672. doi:10.3390u12041672

Yu, D., Duan, H., Song, Q., Li, X., Zhang, H., Zhang, H., Liu, Y., Shen, W., & Wang, J. (2018). Characterizing the environmental impact of metals in construction and demolition waste. *Environmental Science and Pollution Research International*, *25*(14), 13823–13832. doi:10.100711356-018-1632-z PMID:29508202

Zhang, N., Zheng, L., Duan, H., Yin, F., Li, J., & Niu, Y. (2019). Differences of methods to quantify construction and demolition waste for less-developed but fast-growing countries: China as a case study. *Environmental Science and Pollution Research International*, *26*(25), 25513–25525. doi:10.100711356-019-05841-4 PMID:31264149

Section 5

People: Education and Training to Facilitate Digital Upskilling/Reskilling

Chapter 11
Digitally Enabled Infrastructure

Pooja Siddhartha Sukhdeve
https://orcid.org/0000-0001-5006-0878
Swami Ramanand Teerth Marathwada University, India

ABSTRACT

Digital infrastructure comes up with an enormous opening and possibility to deepen the knowledge horizon of those who have an instinct for teaching and learning. Likewise, digital infrastructure benefits to reduce information management. Managing the educational system can be categorized into two important parts, namely managing knowledge and managing information. Digital infrastructure facilitates huge opportunities to expand the knowledge and ease of information management, to examine the issues of the infrastructure of ICT and lack of awareness about the new technology about the training and support in education, and to discuss smart education with the implementation of digital infrastructure to the development of education classrooms and institutions. The chapter also presents a framework of ICT infrastructure scenario and smart infrastructure strategies for education. This chapter examines the development of new emerging technologies to build digital infrastructure applications to enhance the education system.

INTRODUCTION

Today, universities are in taking the process of digitalization, movable from application-oriented IT tactics to more digital organizations of holistic visions. In the middle point of globalization in which businesses need to convert and transform, the feature of maintaining management education needs to modify too. In the Overall world, online education is obtaining popularity, so is the plan of IT-enabled learning in schools structure and digital classrooms have obtained importance. These enabled-technology classrooms encourage opportunities for enlightening education teaching and learning by assimilating learning technology, such as audience response technology, computer, listening devices, networking, and video/audio capabilities. And the progression is balanced to obtained additional advantages in time to come. There have been fast changes taking place in multiple sections and sectors, including the education sector. Additionally, increasing competitiveness has made it vital to take in various tactics to impart maintaining education. Advancing nations are finding it hard to construct new "brick-and-mortar"

DOI: 10.4018/978-1-7998-6600-8.ch011

management institutions to keep pace with the knowledge needs. Institutions imparting management education need to the maximum advantage of new technologies and analyze the means for valuable management of knowledge assets and intellectual resources. The development in technology has led to effective innovations in the design and delivery of management courses.

The report (Greg Martin, M. Schutt, Joe McBreen, and Darrin Tams, 2013) Education leaders are seeing for answers. Accelerated technological innovation in the last five years has obscured educational systems that were constructed for an era of teaching and learning that is continually emerging. Outdated systems, networks, and processes are increasingly inadequate to encourage students who own — and want to handle — four to five digital devices, instructors who attempt to grow student learning exceeding the four walls of a traditional classroom, and administrators who are studying to capitalize on chances to decrease expenses and increase positive results for student achievement.

As per the report (Greg Martin, M. Schutt, Joe McBreen, and Darrin Tams, 2013), Education leaders understand that infrastructure is a sticky business. Displace one part and something else crashes. Purchase technology without being planned and lose expensive resources. Execute a system that doesn't work properly with others and everyone has an authentic problem on their hands. Educators are seeing for believed leaders who know their pain points and know their challenges. With fixed budgets and tiny room for error, they require to be assured that the money they spend will place them for at least the next three to five years, as technological innovation and choosing only proceed to rise.

The canter for Digital Education Special the record views at some of the fundamental trends that are reshaping IT infrastructures and provides precise supervision on what intelligent strategies, decisions, and solutions K-20 institutions can execute to assure that they are properly provided to get the benefit of the necessary technological advances coming to their system.

Technology can be a vital enabler for obtaining quality education with the attendant purposes of access, quality, and equity. For this to appear, the strong advancement will need to be executed beyond the spectrum of education, including infrastructure and maintenance, teacher education, and content and curriculum.

Therefore, this study aims to explore the current smart education to implement the digital way of using the infrastructure good educational environment. The outcome of this examination could be set as a baseline for further developments or innovation. The objectives of this study are as follows:

1. To identify that how smart digital infrastructure helps
2. To identify the issues of digital infrastructure in the ruler area infrastructure, lack of awareness about new technology in the classroom
3. To discuss the design and implementation of digital-enabled infrastructure

The outcome from digital-enabled infrastructure study(Knowledge gap, Perception, and practices) could be used as digital infrastructure should be everywhere not only restricted for urban areas but also use in the ruler area, awareness about the new technology and Support for training about new technology implementation So the student can access equal-education, Interactive learning, Ubiquitous learning, Highly efficient management, Safe campus, Teaching Skills Improvement, and Immersive learning anywhere without any interruption and teacher can proactive in knowledge structure, giving not only awareness but help learners identify the individual experience and stimulate personal talents and abilities.

BACKGROUND

Smart Education with Digital Infrastructure

Smart education is a way section in smart city enhancement and development. Support in primary education, more advancement training, and appreciation certification, e-learning infrastructure, universities and community colleges, long-term learning, and modernity and innovation in education technologies are all part of what defines a smart city. For the smart city, the citizens exceed, according to Dr. I-Chang Tsai, Vice President of Digital Education, Taiwan, at (Members of Congress from United, 2018). Smart cities accept the need for "modern education programs creating for graduates with advance knowledge, and experimental skills, at the report (NEF (New Education Forum), 2017) ."

In the previous century, Several technologies were brought up in education, but not all technologies were effective in the classroom. Traditional teaching methods that were not holding to a new generation of universities and colleges students fastly adopt the digital infrastructure. At present, the development of a smart education requires smarter dispersion of digital technologies. To obtain more beneficial use of classroom technologies, educators can turn up to innovations that focus on the students and teacher-student collaboration, at the article (Morning Future, 2019).

Existing classroom technology in schools and other educational benefits is much time obsolete and does not come up with the performance that advances digital and media-rich education should encircle to rise student contribution and learning results (Bee Smart City, 2018). Smart cities need education benefits and school systems which make sure students obtain 21st-century skills, including " digital literacy, innovative thinking, effectual communication, teamwork, and the ability to form high-quality projects. To achieve this elevated goal, the educator needs to center of interest in technology on the way to establish student achievement (Michae W. Firmin and Deanna J.Genesi, 2013). This study aims to evaluate current research studies on how school infrastructure affects Students learning outcomes and to classify important parameters that can inform the design, implementation, and supervision of future educational infrastructure plans.

Smart infrastructure must have drivers to be productive; benefit can be estimated as an accretion of the variety of development in education. Those drivers shall include abroad variety of areas of expertise: applied social sciences, engineering, Earth sciences, and human sciences (Guedes el.al.. and Alvarenga, 2018).

The goal of the smart education structure is to represent necessary elements in a technology-benefited environment that on the one hand supports students to obtain higher study quality and manages innovation and creativity and on the other hand, allows teachers to personalize training. According to, Zhiting Zhu, Yanyan Sun, and Peter Riezebos (2016) explain the three essential elements in a technology-facilitated and a smart educational background are teaching presence, technological presence, and learner presence. The teaching presence is given by both the teacher and students. It represents the teaching part in a smart education system as instructional design, facilitation and primary instruction, and technological assistance. The learner presence is determined by learners' capabilities to being collaborative and autonomous students as well as effective users of technologies. The technological presence interests to what space technologies can build connectivity, provide a universal way to learning sources, and modify to individual needs. Although technological appearance is not immediately connected to learning, it gives support for learning to happen.

As discussed (Zhiting Zhu, Yanyan Sun, and Peter Riezebos, 2016), the Smart Education standard is a combination of Teacher, learner, and technology to raise smart training. When all three core elements associate in a corresponding way a smart education method will develop, as **Figure** 1 is illustrating.

Figure 1. Smart Education

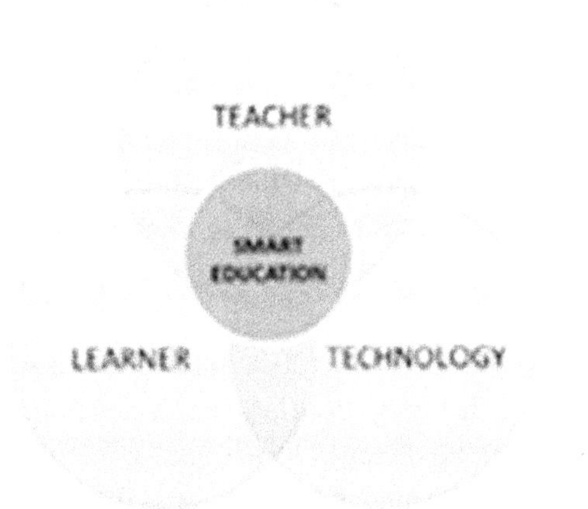

As examined in the study (Zhiting Zhu, Yanyan Sun, and Peter Riezebos, 2016) these core advantages in more excellent detail a smart education structure can be established. As present (Zhiting Zhu, Yanyan Sun, and Peter Riezebos, 2016) **Table** 1 is explaining the three factors of teacher, learner, and technology presence and their distinct functions, which can be used as guidelines for executing smart education within the institutional environment.

Table 1. Smart education Structure

SR. No	*Categories*	*Indicators*
1. Teaching presence	Instructional design	Use student-centered, personalized, and collaborative pedagogical models to design learning
	Facilitation and direct Instructions	Facilitate the learning process by promoting interactions and providing feedback/direct instructions
	Technological supports	Support learners to use technologies
2. Technological presence	Connective	Support social interaction and collaboration
	Ubiquitous access	Ubiquitous access to learning resources
	Personalized	Adapts to personal needs
3. Learner presence	Autonomous learner	Self-directed learners
	Collaborative learner	Abilities to be a collaborative learner
	Efficient technology user	Abilities to learn how to use new Technologies

ICT Infrastructure:

ICT is significant in schools and educational institutions as it supports in bringing out their exercises and purposes such as record keeping, research work, instructional uses, presentations, financial analysis, analysis outcomes supervision, communication, supervision, MIS, teaching-learning exercises, and general school management roles.

According to Peeraer and Petergem (2011), ICT advantages schools in various offers: (i) improving learning in the classroom; (ii) enhancing school administration and related tasks; (iii) enhancing responsibility, performance, and effectiveness in school exercises; (iv) offering the usage of PowerPoint presents and the internet. Keengwe and Onchwari (2011) recommend the view that ICT in schools can commence to high-quality teaching and learning. Others who verify this view are Jhuree (2005), Yusuf (2005), Dzidonu (2010), Higgins and Moseley (2011), and Rebecca and Marshall (2012). Nisar, Munir, and Shafqat (2011) found that the availability and usage of ICT develop the knowledge and learning skills of students.

Additionally, According to Joseph Mukuni (2019) to discuss how ICT Distance education(DE). Increasing access to learning through the appropriation and development of a well-functioning ICT-dependent DE system has numerous advantages to national governments. An advance in technology education in distinct and a growth in the analytical mass of skilled citizenry, in common, can interpret into the presence of a workforce that will provide the human source needs of a nation's economy.

ISSUES AND PROBLEMS

Infrastructure

It is a difficulty that both schools and students encounter. Most schools do not have IT infrastructure that can sustain large-scale eBooks and digital curriculum distribution. Some of the developed countries like the United States has a way to fast internet that can hold large data downloads required for digital books, But most of the developing world still fight with this. As discussed the issues of distance education (Joseph Mukuni, 2019) about Africa facing ICT is still an issue because of the costs. According to the paper Miniwatts Marketing Group (2019), only 37% of the Africa population use the internet. Wealthier regions of the world have high internet usage rates per population. According to Joseph Mukuni (2019), Wealthier areas of the world have high internet usage rates per population. For the situation, the allowances for Europe and North America are 87% and 89% respectively. Interestingly, although Africa has the lowest internet usage rate per population, the country's growth rate in internet usage is very high.

Everyday technology is developing with new inventions and more advance in that field. As per the education is important to everyone in this world as same as the technology is also more important to everyone can get access easily without any interruption and every student can learn things easily. Some of the countries are more developed and enhancing with innovative things and their education system is too strong enough. However, As per discussed (Dr. Shraddha H.Budhedeo, 2016), the challenges and issues are bringing the ICT enabled education in ruler areas of India. Here the remote areas still under developing situations. Although ICT has the power to enhance the education system of a country to a great range, yet it is not the problem of developing countries. Various issues and difficulties are facing the implementation of ICT education in schools and educational institutions in these countries and the difficulties are much more expanded in the case of schools located in remote towns and rural areas. Difficulties arise because of the lack of awareness about the technology and the remote areas still facing the less amount of equipment and not a good infrastructure to access internet facilities because if these major issues students can not reach the technology.

As per discussed in (Joseph Mukuni, 2019) some of the Challenges associating to digital infrastructure in Africa, changing the region's desire of forming societies in which all child despite gender will have admittance to education either through engaging in regular classroom-based learning activities or through use of technology-based learning programs. However, one of the hopeful solutions has been technology-based distance education (DE). although, As Wanzala (2013) examined, developments in information communication technologies (ICT) provided excitement in Africa because it would lessen the costs of education and learning and increase access.

According to Dr. Shradha H. Budhedeo (2016), discussed the issues of the ruler areas about the changes in technology. ICT has an immediate purpose to execute in the education area. It can produce many advantages to schools, educational institutions as well as to the community. ICT in schools continue to knowledge production, information, and communication distribution among the school community. However, According to Arnab Kundu and Dey Kedar Nath (2018), In a developing nation like India, lack of technology, internet access, and unqualified trainers act as the greatest difficulty in implementing ICT-based educational services to the sections, especially to the rural sections.

Technology is only a mechanism and the success of ICTs in improving the distribution of quality education to the needy, without extending the gap, will depend mainly on policy-level interventions that are focused on how ICTs must be expanded in school education. In India, various ICTs have been operated over the years to support primary and secondary education in schools. However, there have been immense geographic and demographic differences in their use. Some states and regions in the country currently have an enabling ecosystem in an area that provides for greater use of ICT for education, whereas others lack such an ecosystem. According to Dr. Shradha H. Budhedeo (2016), As per the 2011 census, nearly three-fourths of the Indian population exists in rural areas covering over 6 lakh villages. The state of rural education in India is though very poor. There are very few government schools in most villages while private schools are largely concentrated in urban areas.

According to Dr. Shradha H. Budhedeo, Students have to walk and travel far away miles to avail of basic education facilities, not to consider the obtaining of ICT skillsets and benefits. Majority schools in rural regions do not produce computer education at all. As per Dr. Shradha H. Budhedeo, The National Policy on Education [NPE] grants for the system of ICT for all rural schools in India.

Measures have also been observed to overcome and eliminate rural-urban differences and encourage diversified and better employment possibilities in rural areas. In the rural context, the central focus of NPE (National Policy on Education) is the implementation of plans and programs, such that predominantly approach the educational needs of rural areas including technical education.

Technology:

The most significant drawback to technological innovation has never been the technology itself, rather it has been the people struck by the change. The teachers, administrators, librarians, and parents are all stuck in traditional ways of education. The difficulty lies in trying to get them to change to digital systems of learning.

As discussed (Amy M. Johnson, Matthew E. Jacovina, Devin G. Russell, and Christian M. Soto, 2016) the issues of the technology change, According to Amy M. Johnson, Matthew E. Jacovina (2016) the challenge introduce of inadequate education related to technology. If teachers are not given an effective professional improvement on new technologies, they will not be proficient in using it to its full potential. As further (Amy M. Johnson, Matthew E. Jacovina, Devin G. Russell, and Christian M. Soto, 2016), discussed the External Challenges to Classroom Technology about training and Support. As per Amy M. Johnson and Matthew E. Jacovina (2016), introduce the challenge of incompetent training associated with technology. If instructors are not implemented powerful expert advancement on new technologies, they will not be able of applying it to its full potential. Also, they discuss parts related to the support restraint. Support obstacles to technology integration introduce inadequate technical support and administrative/peer support.

Also, As discussed in (Mike Tissenbaum and James D. Slotta, 2019) about the k-12 classroom management are not contain new technology such as wide-ranging computing augmented reality, nor also touch surfaces, despite the important influence that such media have done in many other appearances of our lives. Unluckily, classroom environments have not regularly developed to assist students in the new forms of collaboration, view distribution, and analysis that describes the research-based inventions. As responding to the challenge(Mike Tissenbaum and James D. Slotta, 2019), their research was directed by a multi-disciplinary form a team including educational researchers, a high school physics instructor, and technology designers. Mike Tissenbaum and James D. Slotta started on a list of design-based analysis projects to examine a smart classroom infrastructure that scaffolds students and teachers in innovative patterns of collaboration and analysis, including a substantive purpose for large projected presentations and small touch surfaces, as well as a dependency on learners' physical area within the room.

DESIGN AND IMPLEMENTATION:

Smart education identifies itself from traditional education through the system of proactive cooperation. Where traditional education is fundamentally passive guidance – instructor presenting knowledge-based guidance where students 'listen and learn' –, smart education maintains proactive elements such as creative ability and self-directed learning.

In smart education, the teacher is proactive in knowledge structure, giving not only awareness but help learners identify the individual experience and stimulate personal talents and abilities. The students also engage proactively e.g., in arranging learning material, critically reflect, and provide to the overall self and collective learning (group learning). Technology assists help this process. The teacher can apply technology in personalized learning (e.g., monitoring like with e-schoolbag, building material for the flipped classroom method). The students do (adaptive) technology to personalize learning (dynamic).

'Smart Education' for Schools and Institutions

Smart education, a concept that represents learning in the digital age, has grown increased attention. The smart educations in schools and Institutions there are more important components which discuss in a blow.

Figure 2. Smart Education

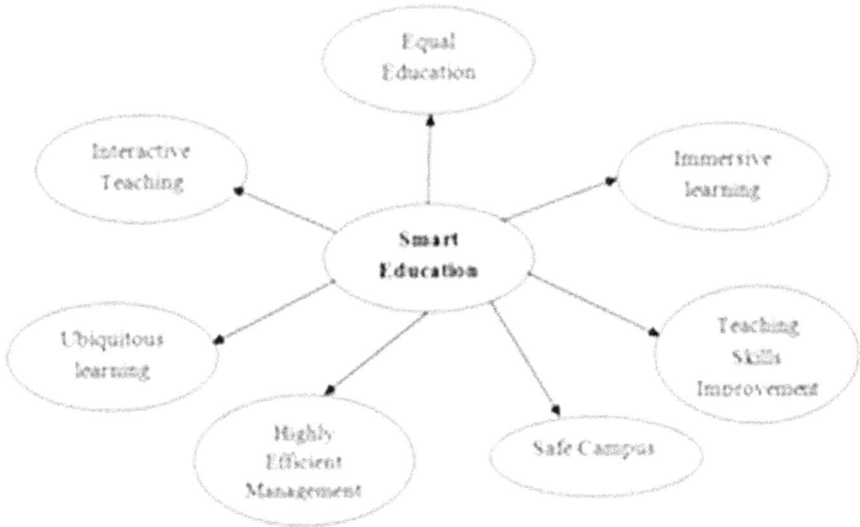

- **Equal Education**: There is an extensive insight that learning today needs to require more than knowledge gain; there needs to be an evenly powerful prominence on skill-advancement, certainly 21st-century skills. technology is a strong source to treat equally in educational learning.
- **Interactive Teaching:** Technology can be a well-built tool for transforming learning. It can assist positive and advanced associations between educators and students, innovate our techniques to learning and collaboration, long-estimation balance shrink and access to the gaps, and change learning experience to meet up with all learners.
- **Ubiquitous learning**: Students can get access to learning everywhere on-campus or off-campus. Everywhere learning resources are available thought it.
- **Highly Efficient Management**: Management is a more vital part of the campus to maintain all the resources. Additionally, on-campus and virtual campuses are highly managed in a particularly efficient way.
- **Safe Campus:** Security is more important when a student studied on campus about the student's data and recourses. And a Strong security campus So no one gets confidential data access from it.
- **Teaching Skills Improvement**: Every time the technology has changed according to the latest modified version of systems so teaching environment will be more enhance and developed.
- **Immersive Learning**: Immersive learning is the procedure of learning with the use of an artificial environment or simulated. It has about come to a greater variety of learners through an environment that includes as many of our five senses as possible, particularly vision, sound, and touch, rather than through the traditional procedure of video, audio, and print. The environment allows the learners to get immersed in the learning and in a way that feels like experiencing an authentic learning environment.

Smart Infrastructure Strategies for Education

- **Managing Devices:** According to, (Greg Martien and Mark Schut, 2013) some skilled scholar support for a way that needs the device registration, that acknowledges IT, administrator, to obtain a big picture sight of all of the devices on the network. A self-service portal can get permit user to register their own devices; electronic device tracing and examination enable IT to link devices to user and IP activity for protection, fulfillment, and reporting. The effortless-to-maintain solution and very less IT interplay to obtain users onboarded as quickly as viable to the network.
- **Virtualizing Desktops:** As mention in the report at (Greg Martien and Mark Schut, 2013) In both K-12 institutions and higher education can use the virtual desktop and students can access the application and data, they need from any device, anywhere, and any time and obtain online learning as well as a cost saver. Even, Virtual desktop makes management much easier; IT can modify applications at the server level, which can be seen persistently by any machine gain access to the virtual desktop.
- **Ensuring Access to the Network**: To make sure students can gain access to the network at the same time, the school will need to raise their current level of network size in both wireline and wireless access. Bandwidth will have to be enlarged to assists a comparatively great number of students taking the tests simultaneously.
- **Embracing Cloud Computing:** Cloud computing is progressively being take in by K-20 education institutions as a key to cost-effectual allow 24/7 way into learning assets both on and off-campus, from any enabled internet device. It permits education leaders to strengthen developing learning domains with fewer assets. Another benefit of cloud computing is its power to have room for various demands that would be complex for a single institution to manage efficaciously.
- **Storing Data:** The report, at (Greg Martien and Mark Schut, 2013) the big data, technology infrastructure question, and deal with more frequently connected to the storage, backup, and recovery, and constant and dependable availability.

ICT Infrastructure

ICT Infrastructure sectors are becoming more applicable to help to teach, learning, assessment, and administration in schools. ICT Infrastructure school planning assets will support schools to arrange for the successful use of digital technologies in their daily experiments to ready students for the requirement of an ever-changing world, to obtain more powerful teaching and learning, and improve learning, teaching, and administration.

A new come from An Roinn Oideachais Agus Scileanna Department of education and skill, Department distributed digital strategy for school 2015-2020, at (An Roinn Oideanchais Agus Scileanna Department of Education and Skills, 2020) to implement ICT infrastructure to obtain the usage in teaching, learning, and assessment by the help of Digital technology.

ICT Infrastructure Includes

- **Cloud-based Tools and Applications:** Cloud-based tools and applications are more progressively to helpful for teaching and learning in schools. The broad scope of cloud-based is available usually through a web browser that is an online and web-based application. Cloud-based tools and applications are using technology usually mention to as cloud computing.

- **Broadband: Broadband is critical but fast and reliable to gain access to online resources for learning. According to, (, 2020) All post-primary Schools have a high-speed Broadband connection about 100mbit/sec.**
- **Mobile Learning Devices/BYOD**: Mobile devices are using increasing numbers of school classrooms in preference to desktop PCs. Even, Tablets are acquiring more increasingly in precedence to the laptop as they allow schools with a different alternate choice to help mobile learning.
- **Wireless Wifi Networks:** Effective learning using to strengthen mobile devices a 'fix for reason' wi-fi network systems connect mobile device online assets through the broadband connection and Such as school wifi, broadband, mobile learning, and cloud computing are all inter-dependent.

ICT Enabled Infrastructure

Figure 3. ICT infrastructure Platform

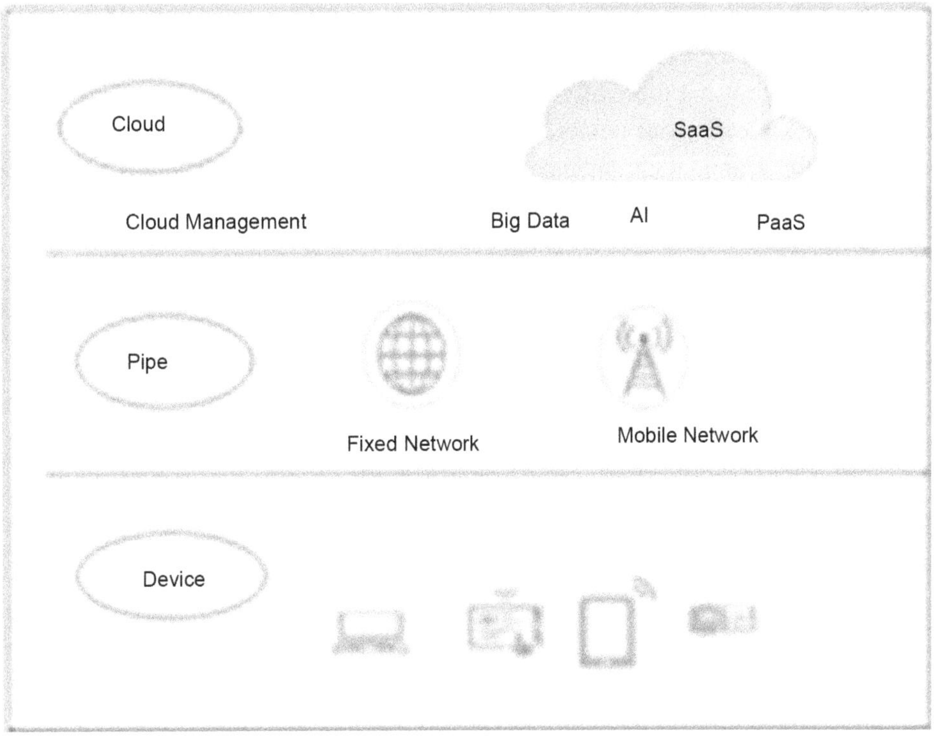

- **Cloud**: Cloud computing gives gain access to computing, storage, and networking assets on demand. As the use of some choosing cloud model services such as Saas, AI, Paas, and Big data.
- **Pipe:** the second phase is Pipeline consist of mobile network and fixed network So any device can access flexibility without ant prevention
- **Devices:** Any device such as mobile, tablets, and laptops can access the network with storing secure data.

Innovation

The progressive innovation has elevated education into a new level. Innovation in education helps teachers and students to examine, research, and use all the new emerging technologies to uncover something new. It involves another approach of studying at problems and solving them. The thinking method that goes into it will assist students to develop their creativity and their problem-solving skills.

- **Intelligence:** Wieldy, Artificial intelligence is enthusiasm, technologists attempt to emulate human knowledge by programming extended rules into the computer, a technique called master system. Today's artificial intelligence depends on machine learning. It is all about to search patterns in profound data and using the pattern to make the decision more intelligently.
- **Simplicity**: A simple architecture can make a more understanding of the structure. The management will not be complex so anyone can handle it very efficiently. In-campus and off-campus management will manage it highly efficient way.
- **Openness**: Openness develop digital innovation in the number of paths that encouraging knowledge and data flows that help innovation. Openness as a suggestion in the form of open content, open license, and open software are enhancing the powerful for addressing obstruction to education. But Still considering a developing digital classify in terms of accessibility and availability of the digital devices for using these resources to developing the educational purpose.
- **Security**: The flexible network is more important for campus management so only campus students, staff members, and faculty members will use it more securely. There is some security solution used in the campus protections.

Figure 4. Innovation

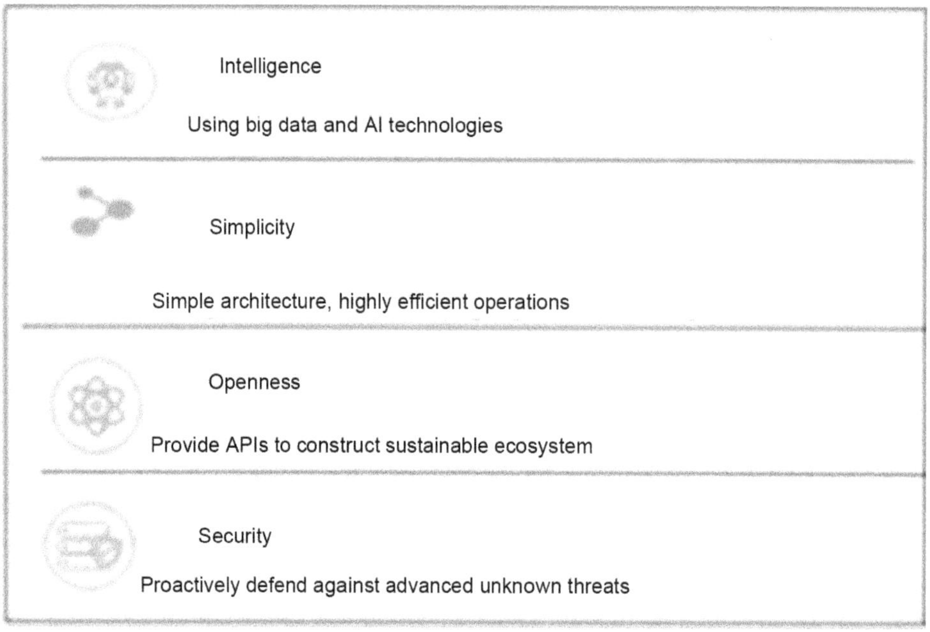

- **IP-based physical security:** This security solution can provide help to send safe surroundings so that students, staff, and faculty can central point on learning.
- **Video Surveillance**: Video surveillancc allow a live recording and recorded visual represents what's incidence occurring around the university and school campus and can willingly access important details during the investigation.
- **Integrated access control**: Integrated access control also make certain school visitor have restricted access and can observe security surveillance. This system can monitor the activity and orientation of students, notify parents/ guardians through emails and SMS, and automatically notifying any potential issues.

Ecosystem

The Sustainable ecosystem that enhancing research and development. A digital ecosystem needs to be strong, self-supporting, and scalable. In the connection of the schools and Institutions, this means a technological platform that helps to learn – course content, homework, grades, shared documents, and research tools – that works concurrently seamlessly.

Figure 5. Ecosystem

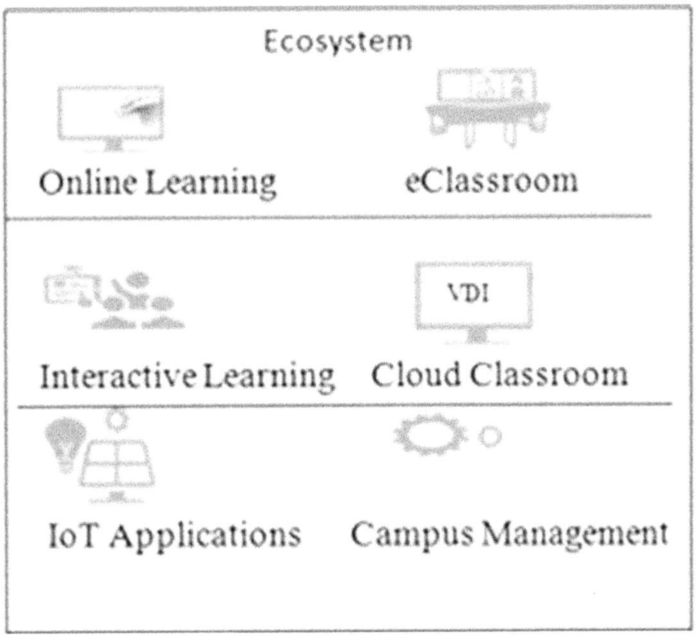

Online Learning

In the last few years, the Online learning platform has impactful for developed and adapted the look toward modern education. online learning is a low cost, more efficacious, and anyone can learn anywhere and anytime with their own time or flexibility.

Due to this pandemic of corona virus is caused to students because the student can not physically present at the class and learn those. However, the student can access some platforms to uplift their knowledge Such as Udacity, Coursera, Eureka, Datacamp. Today Top-ranked institutions offer online learning to attending the dream university. With Courses accessible in every subject, and feasible schedule, Students are growingly turning to online learning as a flexible alternative to on-campus. It is considering to boon due to some reasoning (Indira Dhull and MS Sakshi, 2017):

- **Accessibility:** Online Learning allows accessibility due to which a student can learn from anywhere in the world.
- **Personalized Learning**: Online Learning system allows students to process and determine to learn.
- **Cost-Effectiveness**: Online learning is more cost-effective because of lower money spent on travelling and purchasing books or spending money on college stuff.
- **Promotes Research:** Students are more eager to popularize their work on the online platform when they produce something more high quality.
- **Basic Computer Skills:** Both Off and on-campus students who prefer to study online have an opportunity to obtain technical facilities in using Information Communication Technology.
- **Globalization:** New emerging technologies are shrinking geographical prevention in the process of education. The world has become a little village and the possibilities to know other countries are within our access.

E-Classroom

E-Classroom offers the main structure where the information win be managed to contain each student and faculty accomplishment to help in the enhancement of their learning. Each educational tool or platform has a very unprecedented potential to play a very crucial role in the classroom.

Electronic classroom often includes a teacher's desktop or laptop, computer, wired and wireless speaker, laptop internet connectivity, an LCD Projector, and individual computer for each student to use in the class.

As per the article (Ed Tech Blog), the e-classroom platform in advance and fast-paced world, Multimedia classroom technology is allowing a very unique chance to experiencing the classroom learning atmosphere.

Student performance is more important while they learning or not. If the teacher is teaching to their student but some time student not attentive to attend a lecture so the teacher cannot focus on each of the students. However, if the school use some modern technologies like automatic teaching focus tracing based on image recognition so faculty member to more focus on their students. Also, use remote HD teaching with immersive learning.

Interactive Learning Classroom

Interactive learning classroom: Some emerging technologies are helping a more highly effective way to implement in the classroom interaction.

- **Hololense**

Microsoft Hololense allows viewing and communicating with 3D hologram delivered to live in front. It has different advantages that an educational institution can leverage applying 3D holograms for high-level interactive instruction. Using mixed reality to produce large data sets more accessible to navigate and understand.

According to Edtech Innovation (Mixed Reality For Education) invented the first interactive mixed reality Hololense application for the medical field using Hololense for education provides to shaping education more adaptive and personalized for technical and educational purposes. In the technical sense, Students can observe the simulation in their individually favored approach due to the holographic 3D model projected in the classroom. It would allow the teacher to give lessons to students across the world imagine taking a tour of a historic-3D model of the coliseum right from your desk the potential for new technology in the classroom is pretty amazing but will cause students who are plugged in all the time to forget what it is like to live in the real world.

As discussed in the report (Department of Education, 2016) The University of Wolverhampton is applying a 3D visualization system to enhance more traditional laboratory-based lessons on analysis and examination. Touch screens have improved student contributions by providing them to examine the complicated connections between various parts of the body, which would not unless being possible.

- **Augmented Reality in Education**

Augmented reality, still in a degree of developing process, has been working in the education sector since the beginning. In other areas, such as design, augmented reality is represented as the central tool that will accomplish prototypes created digitally depend on the reality in which the physical elements will be constructed. Within educational processes, augmented reality will allow students to work by enhancing their creativity without fear of building risks and costs (Di Serio, Ibáñez, & Delgado Kloos, 2013). The inexperience risk that importantly increased the problem of carrying out projects with all kinds of consequences could be overcome considerably. Through this technology, a student can show an image of an outcome over a real environment and space, without the need to complete a physical building process.

- **Evaluating Educational benefits** Despite that these positive AR content-creation experiences, an important question remains: Does AR technology enhance learning? constructing and using AR scenes come up with active convoluted problem solving and teamwork to create engaging educational experiences to teach science, math, or language skills and studies have found that this activity improves student motivation, participants, involvement, and engagement (P. O'Shea et al., 2008).
- **Reading comprehension:** The augmented book joins the text-based part and interactive AR sections. (Mark Billinghurst and Andreas Dünser, 2012) stated that they compared how much readers with high and low capability (teacher-provided classifications) recall from the two media formats. From the text-based story segment, those in the high-ability category could retell notably more key story points than the low-ability category (A. Dünser, 2008). However, the two groups showed no important difference in remembering key story points from the interactive AR segment. Arguably then, because they allow a chance for interactive substance engagement augmented books could

benefit students who are less able to apprehend text-based learning content—at present the leading education medium.

- **Teaching topics with a spatial component:** Even, AR can be more beneficial in teaching topics with a robust spatial component. If it is implemented in the classroom so it will most effective learning for students to increase the learnability of any topics. In a study after the reading comprehension evaluation, (Mark Billinghurst and Andreas Dünser, 2012) stated that they differentiate learning results using a print-only text about electromagnetism and an augmented book with in the same content. As Figure 5 shows, the augmented book creates animated 3D models of electromagnetic concepts and permits easy interaction.

Figure 6. Using an interactive augmented book to teach electromagnetism. The user views a 3D model of the earth's magnetic field. Source (Mark Billinghurst and Andreas Duenser, 2012)

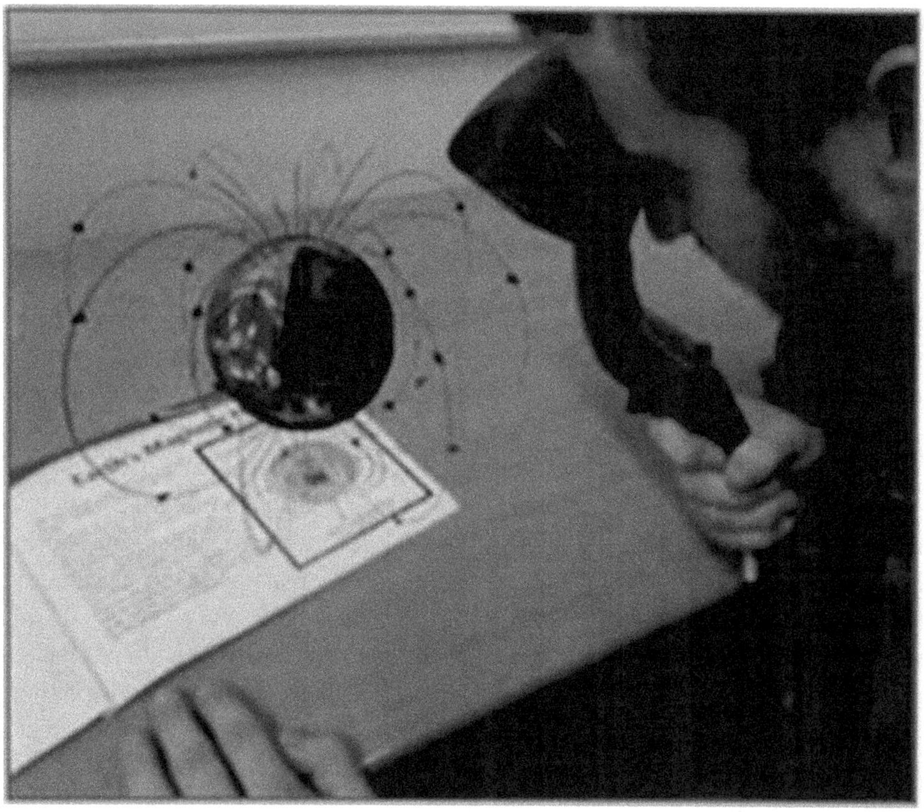

- **Virtual reality**

"Einstein said Imagination is more important than Knowledge." Virtual reality is more visual than ever so it helps the student to expand their imagination virtually (Foton VR). According to the article (Foton VR) discussing here how virtual reality labs will assist the instructor to explain in detail every complicated concept which can not be possible to do in the real-world situation. The teacher interacts with students is significant parts of the teaching method. The usage of fotonVR performance develops the interaction of teachers with students.

Now the current pandemic of the Coronavirus is real cause for more considering amid school closures. But due to this situation, every student and teacher use the E-learning platform and online learning platform. Most of the students learn on the online platform and E-learning platform to enhance their learning ability. And Teachers use the online platform to educate their students. However, some time student cannot access the live response from the teacher and not more collaborative it. According to the article (Eon Reality, March 2020) the universities around the world, Including Australia, searching, have an approach to allow students to learn without being physically present. As per the (Eon Reality, March 2020), The Solution for AR and VR to develop remote, Immersive learning ecosystems presents a possible option for governments and schools to enhance and develop at this critical problem. Presently, AR and VR technology is developed enough to allow remote collaboration classrooms and empower a large number of the student together and learn while lessening the risk of exposure. As discussed in (Eon Reality, March 2020), about the AR and VR is need the high-priced headsets and Visual aids, but EON reality's primary key is to make specifically for classrooms that need to allow AR and VR facilitate without headsets - only requiring a desktop or mobile device.

- **IBM Watson**

The article (Chalapathy Neti, 2018) In classrooms with children and students of all ages and schools around the world, teachers are engaged with the difficulty of addressing every student's different requirements and the rate at which he or she learns. At IBM Watson Education team is concentrated on leveraging AI to enhance learning results and execute solutions that will benefit all students to achieve an outcome. The IBM Watson Education program relies on digital trends and AI to provide teachers the tools they need to be most efficient and support learners present at the top of their intelligence. By using AI as an attached assistant in the classroom, the goal is to support lifetime learning where every person is met with the individualized devices they need to accomplish in school and exceeding.

The IBM Watson Assistant (Luca Benedetto, Paolo Cremonesi, and Manuel Parenti, 2019) is offered by an AI engine that produces different NLP services and Natural language interaction with students and different users, that supports enhanced machine learning capabilities.

Bolton College in the United Kingdom (Aftab Hussain and Dean Baggaley, 2017), The IBM Watson build "Ada" a virtual assistant that developed and enhance learning, teaching, and assessment. According to the [R] Bolton college more captivate with its students and teachers. As the execution of a Chabot, Bolton college is shaping students' learn ability.

- **3D Printing**

Various growing schools, universities, and educational institutions across several regulations have already integrated 3D printing technology into their classrooms and curriculums. 3D printing fuels unlimited creativity and collaboration empowering students to envision, hold and test their designs in real areas.

- **Benefits of 3D printing:**
    - **Practical Hands-On Approach:** 3D printing in academics assists change students from being inactive users to active creators. The current system relies gradually on theoretical learning and very little on the real-world deployment of knowledge. Having access to 3D printing technology at the bottom root level provides more students to utilize a pragmatic hands-on approach.
    - **Engagement and Interactivity:** Educators have already seen the transformative outcome of offering computers that have had in classrooms. 3D printing in education has a similarly transformative prowess and advancing modern technologies in the curriculum stimulate engagement and interactivity.
    - **Promotes Problem Solving Skills:** 3D printing in education shows students a type of learning experience. Observations by leading instructors have recommended that using 3D printers encourages the improvement of problem-solving skills and skills of endurance and prowess to surmount challenges.
    - **Sparks Creativity and Innovation:** Imparting innovative ideas and methodologies is more comfortable doing in the developmental years of children. Empowered students in classrooms and education environments are implemented to resolve tomorrow's difficulties with creativity and innovation. Schools, universities, and other educational institutions ought to start a curriculum based around 3D printing.

Cloud Classroom

Cloud saving students from back problems everywhere but the cloud also gives direct access to date on student study habits by doing all their homework online students provide engagement data on how long they spent on their assignment and whether or not they are taking notes there's a bonus through a major publisher could use that data to judge how effective and engaging certain textbook is so student would be recycling that boring algebra books.

To enlarge the extent of distributing knowledge with students and teachers on the alike online education platform, cloud-based systems allow an excellent environment for a digital/virtual classroom as they allow a seamless way to data, easily shareable data, and encourage a mechanism for following multi-user collaborations. Moreover, there are a host of advantages that a cloud-based system allows. Firstly, since cloud-based applications appear on web browsers and are agreeable with most mobile tools, there is no necessity for costly hardware and schools, and students need not have particular computers or laptops to obtain the material. Also, a low-cost Smartphone can provide students to reach appropriate academic applications. Besides, there is no need to spend in external storage tools since there are diverse platforms open that offer free cloud-based storage services.

As discussed in the article (*Vandis*, 2019) Cloud computing encourages openings for change for all users. Teachers can join their students to various programs and applications, empowering for students to be innovative in their performance of education of standards. For instance, a student could respond to an assignment and exercise by uploading a video recording, taking a picture of the artwork they created or sharing a report that they worked on with companions. The cloud allows students chances to use their voice and opportunity in how they describe their learning and offers them the choice to use many technologies to do so. For the cloud, teachers and students can customize assignments to reach a student's distinct requirements.

- **Accessible and Reliable Cloud Services**

Cloud computing assists schools to reach data security responsibilities to assure student and teacher data is maintained safely. Most cloud providers spend in protection standards that implement a foundation level of safety on their cloud infrastructure. Security devices used in the cloud are accessible for users to drive. The cloud also allows unusual authenticity - when devices crash, information is not always lost as you can save data in the cloud.

- **Cloud-Based Collaboration**

The capability to collaborate efficiently with others is important for students to comprehend. The cloud provides all students to have clear access to sources beyond various platforms, constructing a solid framework for strengthening collaboration skills. For instance, students can cooperate with other students on an assignment while their teacher gives feedback in real-time. Significant feedback is necessary for encouraging students to reach their aims. Cloud computing provides for immediate feedback and evaluation processes, which helps both students and teachers concurrently. The cloud also breaks through difficulties of managing within a group - students no continued need to be in an identical position to work on a group project but can reach an assignment from any area by the cloud. Collaboration is too streamlined for teachers. With the cloud, teachers and administrators can simply distribute lesson programs and can work on them concurrently at any location or time. Through messaging apps and other forms, several schools and areas can break obstacles that stop them from co-operating, which in turn starts to a major global educative association.

IoT Application

The Internet of Things (IoT) has the potential to reconstruct education by extremely modifying how schools, colleges, and universities collect data, interface with users and automate methods. IoT relates to the networking of physical objects into the effectiveness of embedded sensors, actuators, and other devices that can gather and transmit data about real-time campus action. When IoT is linked with technologies such as user mobility and data analytics, it produces a new model in education.

IoT Enables Institutions

- Create new plans for students to read by continuing more personalized and productive learning exercises such as immersive digital textbooks and game-based learning.
- Change how teachers present lessons and test execution with smart audio-visual devices, digital video recorders for lecture taking, and online examination.
- Explain services for school administrators by proactively observing critical infrastructure and designing more efficient, cost-effective rules for HVAC, lighting, and landscape supervision.
- Implement a more reliable environment for students and teachers with digital surveillance cameras and smart door locks.

IoT Scenarios in Education

The report Alcatel-Lucent Enterprise (**ALE**) (2018) IoT solutions ensure to build schools and universities more intelligent as well as more strong at what they do. The IoT can define again how students, teachers, and administrators communicate and connect to technology and devices in classroom environments, supporting improve learning experiences, enhance educational results and lessen costs. Examples of IoT solutions for education include:

- Smart whiteboards and another interactive digital media that can collect and examine data for teachers and students for performance in the classroom—or wherever different, at any time—optimizing guidance and enhancing learning results.
- Solutions such as smart temperature sensors and smart heating, ventilation, and air condition equipment that reduce energy consumption and automate operations management.
- Smart student ID cards, attendance-tracking devices, school bus tracking systems, and parking sensors that monitor the physical locations of students.
- Wireless locks-doors, connected surveillance cameras, and facial recognition systems that produce security for teachers, students, and staff.
- Research programs improved with more advanced and computerized systems in significant areas of study, such as medicine, agriculture, and engineering.

The IoT Compounds Schools' and Universities' Exposure to Cybercrime

The growth of IoT in education also takes an explosion of cybersecurity threats, as the propagation of sensors and connected devices hugely develops the network attack outside. IoT is particularly susceptive because many IoT devices are built without protection in mind, or built by companies that don't know modern security demands. Consequently, IoT systems are frequently the vulnerable link for network security in educational institutions.

Building a Secure IoT Network Infrastructuro

The report (Alcatel Lucent Enterprise (ALE), March 2018)discuss the secure IoT traffic and devices in school and university networks is a challenge that can't be resolved by any single security technology. It needs an imperative way that brings the benefit of various protection safeguards.

To support educational institutions take benefit of the advantages and lessen the risks of IoT deployment, As per (March 2018) Alcatel-Lucent Enterprise (ALE) presents a multi-level protection approach. ALE's approach gives security at each layer of the infrastructure, from the particular user and device out to the network layer itself. It also produces an IoT containment approach to analyze and secure device onboarding and fulfill the right network sources to run the system correctly and efficiently, all in secure surroundings to safeguard institutions from cyberattacks.

CAMPUS MANAGEMENT

Wireless Campus

With the advancements in wireless networking technologies, they are growing a significant part of a campus network. Developments in the bandwidth and protection of Wi-Fi technology execute it feasible for all members of an academic institution to access the Internet and all the assistance of the institution, working their laptops, without obtaining a place to physically join to the campus network, without the menace of an unauthorized user accessing secret data and information systems, and without the danger on a virus-infected computer infiltrating the campus network.

The report (Risto Jurva, Marja Matinmikko-Blue, Ville Niemeta, and Suvi Nenonen, 2020) discuss the 5G technology is required to develop the deployment of smart ecosystems significantly. The features of 5G technology as very high speed, low latency, and reliability possible to connect large numbers of computers, robots, and vehicles. Moreover, data presented by IoT sensor networks concurrently with AI technology and perfect positioning arrangements will more enhance intelligent and autonomous environments to rise.

Smart Campus infrastructure management has newly become extensive research subject to increase in the study. A different intent of the view of the authors in (Walia, J. S., Hämmäinen, H., & Matinmikko, M., 2017)is the operational perspective of the local campus 5G network. Usually, the transformation to a digitalized ecosystem assumes large deployments of ICT and IoT infrastructure. When examining the advantageous execution of the most developed features of 5G technology alike ultra-high data speed and low latency it is considered that the 5G networks are extended indoors or in restricted geographical zones with a solid tiny cell network topology (Matinmikko-Blue, & Latva-aho, 2017). Consequently, the number of terminals, devices, sensors, and robots will grow heavily. Despite these matters, it has not been extensively addressed who could be a relevant actor to serve digital Smart Campus infrastructure and services.

Thorough research has been managed to find alternative operative paradigms and to acknowledge to vertical's developing industry requirements (Walia, J. S, 2017; Jurva, R, 2019; Matinmikko-Blue, 2017; and Ahokangas, P., 2018). The study illustrates that a distinct micro executive is an important character to evolve and to satisfy the operative's expectations of 5G networks and to expedite digitalization (Matinmikko-Blue, M., & Latva-aho, M., 2019). The micro operator idea concentrates on local 5G networks and recommends an ecosystem model and demanded micro operator functions. To extend a 5G network, the micro operator needs permission for the frequency spectrum. Understanding that local licensing has not been approved by settlement with the earlier periods of mobile networks, a distinctive model for local frequency licensing has been improved together with authorities (Matinmikko, M., and Latva-aho, M., 2018). The articles (Walia, J. S, 2017 and Jurva, R, 2019) present cases of campus 5G deployment and an operational paradigm of a micro operator. The micro operator can be also described as a local service provider, which operates the campus digital infrastructure and assistance based on the ecosystem model.

Green Campus

Colleges and universities allow target-rich situations for cybercriminals. From financial, medical, and personal information to government and commercial research, educational networks can hold a stock of data, various access locations, and high amounts of endpoint ways—all of which want unprecedented protection solutions. Institutions can use the data most safely without any prevention. Institutions necessity to specify and execute complete safety planning that implements end-to-end Campus network access control, Automated Internet threat prevention, and BYOD service security.

- **Campus Network Access Control**

The university environment may be protecting the professional's greatest nightmare. At the bottom level, there is a user bottom level that replaces every four months or so, with new students arriving on the network and previous students who shouldn't have access but still make. Staff and faculty may be a little more steady, but they also use various devices and tools from various places. There are allow agencies with access, collaborations with different universities, and outer contractors. And then, of course, there are the alumni and parents who require access to the campus networks. All of these entities require various levels of access permissions.

The Network Access Control (NAC) safety solution will provide campus IT administrators the capacity to arrange a system that manages who has access to what regions of the network. It can also restrain location access – a student device that can be worked on campus may not have access from outside of the location, for instance. NAC control can more ascertain how many devices and tools a user can join to the network or what type of IoT devices would be permitted. A delicate research project may have the approval to set up a safety camera arrangement with network access so it can be controlled remotely, for example, giving recognition that may not be granted to other individuals or departments.

- **Automated Internet Threat Prevention**

Unified Threat Management (UTM) tools produce various security technologies within a particular appliance. Although product capacities vary from merchant to merchant, many UTM solutions introduce firewall capacities, content filtering proxies, and other safety technologies. Intrusion detection and prevention systems (IDS/IPS), meantime, adviser networks and hosts for indications of malicious action. An IDS tool will produce early-warning systems that inform network administrators to mistrusted intrusions, whereas an IPS solution will automatically prevent suspicious traffic before it can reach throughout the campus network.

Every day, the thousands of peculiar devices that are joined to a college or university network produce large volumes of information that can support IT administrators completely know their cyber security aspect. However, this data doesn't do an institution any good unless it's being measured, and the sheer volume of data being gathered makes it unmanageable to prepare manually. Security information and event management (SIEM) systems make together logging data from a type of different devices and security commodities and examine for guides that might show dubious activity.

- **BYOD Service Security**

Today's education ecosystem is well-connected and has large availability and on-demand requirements. Web access is prevalent – everywhere there is a signal, students can be obtained accessing information from their phones, tablets, and smart boards. These smart devices and tools and how students utilize them in their educative recreation include the Internet of Things (IoT) world. Many K-12 school districts are understanding that executing a BYOD system is an efficient way to provide cutting edge technology that allows students to interact, cooperate, and innovate. However, it can serve a big cultural revolution for the learning conditions as well as a difficulty to IT professionals. Safe Connect-enabled BYOD arrangements have produced device perceptibility, strong guest self-enrolment, improved end-user experience, and reliable access for both faculty and staff at many K-12 school districts over the country.

Mobile Campus

The extensive use of mobile phones and other compact and wireless devices has been transforming the landscape of technology-supported learning. Various universities and educational institutions have produced works to improve mobile applications and/or mobile campuses for academic, social, and administrative assistance. Mobile applications enhance accessibility to learning contents and exercises, and questions/answers, particularly for students who are considering e-learning courses. It has trained out that mobile technologies are properly arranged with strategic educational purposes, such as advancing student recognition, comprehension, and achievement, supporting differentiated learning wants, and leading-learners who would otherwise not have the chance to engage in education (Kukulska-Hulme et al., 2009).

Mobile learning has been described as learning helped by mobile devices such as mobile phones, tablet PCs, and personal media players (Herrington & Herrington, 2007; Valk, Rashid, & Elder, 2010) in both formal and informal educational settings (Quinn, 2011; Traxler, 2007, 2010). Advanced awareness of the possibility of mobile learning has developed the body of relevant literature. From the meta-analysis of more than 160 articles advertised between 2003 and 2010, Wu et al. (2012) observed that study has centered (in descending order) on assessing the effects of mobile learning, creating mobile learning systems, examining the affective area during mobile learning, and estimating the importance of learner features on the mobile learning rule. Besides, it has been found that 86% of the subjects that were examined reported positive results (Wu, Wu, Kao, Lin, & Huang, 2012).

RECOMMENDATION

Recommendations to make more development include the following:

1. To implement need-based ICT Education in rural areas special to their skill sets
2. To form strategies to promote widespread access to skills and competencies for learning and adopting ICT
3. To build awareness of ICT Education
4. Government and national education authorities should secure the availability of high-quality internet access to schools and educational institutions

5. The urban-rural divide in terms of access, equity, and devices will proceed to be the main problems that educators will have to address as the requirements of the learning community will improve.
6. The government should assure joint works by software companies and teachers for providing excellent content to help the curriculum and language diversities
7. Give teacher instruction that highlights constructivism and student-centered education;
8. Focus professional improvement efforts toward those which maintain the use of technology in guidance,
9. Involve teachers in the decision-making process when adopting new technologies; Technology combination in the classroom will need the continued collaborative efforts of teachers, educational technology professionals, school administrators, researchers, and educational software personnel.
10. To adopt new technology for classroom development and more awareness about technology

FUTURE RECOMMENDATION

This study has focused only on the strong infrastructure, above discussed the issues about ICT Distance education in Africa and ICT infrastructure in India about ruler areas issues. In this region, there is more research required before implementing a strong smart infrastructure. Hence, Further design and implementation of the ICT Infrastructure, Cloud classroom, Interactive Learning Classroom, and Smart Green Campus framework that can be used by higher educational institutes to advance an integrated sustainability strategy for their campus as such focus group and their government could be used to facilitate in-depth studies on understanding of the area and identify the problem before implement the smart infrastructure.

CONCLUSION

In this research, we have used a literature study, articles, and an expert workshop to gather best practices on sustainability from various educational institutions, schools, and organizations. As per the examine the challenges of infrastructure, implementing new technology in the classroom environment, lack of awareness about the technology, and lack of training and support. After examining these challenges, the study main focus on smart education for the student can access equal-education, Interactive learning, Ubiquitous learning, Highly efficient management, Safe campus, Teaching Skills Improvement, and Immersive learning anywhere without any interference and teacher can proactive in knowledge structure, giving not only awareness but help learners identify the individual experience and stimulate personal talents and abilities. These were consequently examined to provide a Smart Education, ICT Infrastructure, Cloud classroom, Interactive Learning Classroom, and Smart Green Campus framework that can be used by higher educational institutes to advance an integrated sustainability strategy for their campus. To accomplish the necessities of future-proof education, educators and institutions need to integrate IoT platforms, cloud technology, and ICT Infrastructure into science and engineering curricula to support students to develop digital literacy and innovation skills. Furthermore, AI-based data analysis, Interactive learning such as Augmented/Virtual Reality, Hololense, 3D Printing, IBM Watson, and adaptation tactics will be required to give student-centric progress-driven content that enhances the level of perception while increasing the receptivity of the students through multimedia content. Overall, the study touched upon various appearances of education in the digital age: the needed skills now and in the future; development of formal education devices for school education and lifetime learning; and the more spacious role for individuals in society.

REFERENCES

Alcatel Lucent Enterprise (ALE). (2018). *The Internet of Things in Education*. https://www.al-enterprise.com/ko-kr/-/media/assets/internet/documents/iot-for-education-solutionbrief-en.pdf

An Roinn Oideanchais Agus Scileanna Department of Education and Skills. (2020). *Digital Strategy for Schools 2015-2020 ICT Infrastructure Funding*. https://www.education.ie/en/Schools-Colleges/Services/Grants-and-Additional-Support/ict-infrastructure/digital-strategy-for-schools-2015-2020-ict-infrastructure-funding.html

Bee Smart City. (2018). *ViewSonic ViewBoard® Interactive Display Solution for Smart Education*. https://www.beesmart.city/solutions/viewsonic-viewboard-r-interactive-display-solution-for-smart-education

Benedetto, L., Cremonesi, P., & Parenti, M. (2019). *A Virtual Teaching Assistant for Personalized Learning*. https://www.researchgate.net/publication/331343430_A_Virtual_Teaching_Assistant_for_Personalized_Learning

Billinghurst & Duenser. (2012). *Augmented Reality in the Classroom*. https://www.researchgate.net/publication/234793015_Augmented_Reality_in_the_Classroom

Budhedeo. (2016). *Issues And Challenges in Bringing ICT Enabled Education To Rural India*. http://ijsae.in/ijsaeems/index.php/ijsae/article/viewFile/1055/726

Department of Education. (2016). *Realizing the potential of technology in education*. https://assets.publishing.service.gov.uk/government/uploads/system/uploads/attachment_data/file/791931/DfE-Education_Technology_Strategy.pdf

Dhull & Sakshi. (2017). *Online Learning*. https://www.researchgate.net/publication/332833360_Online_Learning

Di Serio, Á., Ibáñez, M. B., & Delgado Kloos, C. (2013). Impact of an augmented reality system on students' motivation for a visual art course. *Computers & Education*, *68*, 586–596. doi:10.1016/j.compedu.2012.03.002

Dünser, A. (2008). Supporting Low-Ability Readers with Interactive Augmented Reality. In *Ann. Rev. CyberTherapy and Telemedicine: Changing the Face of Healthcare* (pp. 41–48). Interactive Media Institute. https://www.researchgate.net/publication/216867629_Supporting_Low_Ability_Readers_With_Interactive_Augmented_Reality

Dzidonu. (2010). *The Role of ICTs in Achieving the MDGs in Education: An Analysis of the Case of African Countries. Accra Ghana*. https://pdfs.semanticscholar.org/957b/2981c7f275f4e8b047f8695173805da0794e.pdf

Ed Tech Blog. (n.d.). *e-Classroom – The Future of Education*. www.edtechnyc.org/e-classroom-the-future-of-education

Firmin & Genesi. (2013). *History and Implementation of Classroom Technology.* https://core.ac.uk/download/pdf/82211949.pdf

Foton, V. R. (n.d.). *Teach abstruse concepts of science by adding Fun with fotonVR.!!* https://fotonvr.com/?gclid=Cj0KCQjwz4z3BRCgARIsAES_OVfkomaWeMp7Pj59byJAkP1vGl5mI687ZsTzPIcwtX-Mg908lZUsPHj8aAvCgEALw_wcB

Future, M. (2019). *Smart Education, Goodbye to Classroom teaching.* https://www.morningfuture.com/en/article/2018/02/09/smart-education-andrea-cioffi/227/

Guedes, A. L. A., Alvarenga, J. C., Goulart, M. S. S., Rodriguez, M. V., & Soares, C. A. P. (2018). Smart Cities: The Main Drivers for Increasing the Intelligence of Cities. *Sustainability*, *2018*(10), 3121. doi:10.3390u10093121

Herrington, A., & Herrington, J. (2007). *Authentic mobile learning in higher education*. In AARE 2007 International Educational Research Conference, Fremantle, Australia. https://pdfs.semanticscholar.org/3c71/e83fb19005cf1c340ceb8af02fe22aa5fae0.pdf

Higgins, S., & Moseley, D. (2011). Teachers' Thinking about ICT and Learning: Believes and Outcomes. *Journal of Teacher Development*, *5*(2), 191–210. doi:10.1080/13664530100200138

Hussain & Baggaley. (2017). *Bolton College used IBM Watson to build a virtual assistant that enhances teaching, learning, and information access.* https://www.ibm.com/blogs/watson/2017/08/bolton-college-uses-ibm-watson-ai-to-build-virtual-assistant-that-enhances-teaching-learning-and-assessment/

Intsiful, J., Okyere, P. F., & Oase, S. (2003). Use of ICT for education, research, and development in Ghana: Challenges, opportunities, and potentials. *The Journal of Technology Transfer*, *25*, 329–342. http://www.ejds.org/meeting2003/ictp/papers/Intsiful.pdf

Jhurree, V. (2005). Technology Integration in Education in Developing Countries: Guidelines to Policy Makers. *International Education Journal*, *6*(4), 467–483. https://ehlt.flinders.edu.au/education/iej/articles/v6n4/jhurree/paper.pdf

Jurva, R., & Matinmikko-Blue, M. (2017). *Stakeholder analysis for digital campus development with 5G micro operators. The ITS Europe Conference.* https://www.econstor.eu/bitstream/10419/205186/1/Jurva-Matinmikko-Blue.pdf

Jurva, R., Matinmikko-Blue, M., Niemeta, V., & Nenonen, S. (2020). *Architecture and Operational Model for Smart Campus Digital Infrastructure*. https://www.researchgate.net/publication/340131385_Architecture_and_Operational_Model_for_Smart_Campus_Digital_Infrastructure

Keengwe, J., Onchwari, G., & Wachira, P. (2011). Computer Technology Integration and Student Learning: Barriers and Promise. *Journal of Science Education and Technology*, *17*(6), 560–570. doi:10.100710956-008-9123-5

Kukulska-Hulme, A., Sharples, M., Milrad, M., Arnedillo-Sánchez, I., & Vavoula, G. (2009). Innovation in mobile learning: A European perspective. *International Journal of Mobile and Blended Learning*, *1*(1), 13–35. doi:10.4018/jmbl.2009010102

Kundu & Nath. (2018). *Barriers to Utilizing ICT in Education in India with a Special Focus on Rural Areas.* https://www.researchgate.net/publication/327671947_Barriers_to_Utilizing_ICT_in_Education_in_India_with_a_Special_Focus_on_Rural_Areas

Martien, G., Schut, M., McBreen, J., & Tams, D. (2013). *Smart Infrastucture.* https://www.commscope.com/globalassets/digizuite/1022-1012-center-for-digital-education-smart-infrastructure-report.pdf?utm_source=ruckus&utm_medium=redirect

Matinmikko-Blue, M., & Latva-aho, M. (2017). Micro operators accelerating 5G deployment. *12th IEEE international conference on industrial and information systems (IEEE ICIIS).* https://www.researchgate.net/publication/323347676_Micro_operators_accelerating_5G_deployment

Members of Congress from United. (2018). *Institute for information industry (III) Attract Multidisciplinary Expert to Make Cities Smarter Through Education.* https://thetechrevolutionist.com/2018/03/institute-for-information-industry-iii.html

Mixed Reality for Education. (n.d.). *Educational Tools and Scenario.* www.mr4education.com

Mukuni, J. (2019). *Challenges of educational digital infrastructure in Africa: A tale of hope and disillusionment.* https://www.researchgate.net/publication/334809878_Challenges_of_educational_digital_infrastructure_in_Africa_A_tale_of_hope_and_disillusionment

Neti, C. (2018). *How Watson Education, Scholastic, and Edmodo are using AI to close the learning gap.* https://www.ibm.com/blogs/watson/2018/06/using-ai-to-close-learning-gap/

New Education Forum (NEF) Brussels. (2017). *Smart Education in Smart Cities and Smart Regions.* http://ciedu.eu/wp-content/uploads/2018/08/NEF_2017_Report.pdf

Nisar, M. W., Munir, E. U., & Shafqat, A. (2011). Usage and Impact of ICT in Education Sector: A Study of Pakistan. *Australian Journal of Basic and Applied Sciences*, *5*(12), 578–583. https://arxiv.org/ftp/arxiv/papers/1206/1206.5132.pdf

O'Shea, P. (2008). Lessons Learned about Designing Augmented Realities. *Int'l J. Gaming and Computer-Mediated Simulations*, *1*(1), 1-15. https://pdfs.semanticscholar.org/6cde/77becd373273bb096e5f7b40543b6b8b8266.pdf

PDST Technology in Education (2020). *Technology Overview in Schools - ICT Infrastructure.* https://www.pdsttechnologyineducation.ie/en/Technology/Technology%20Overview/

Peeraer, J., & Petergem, P. (2011). ICT in Teacher Education in an Emerging Developing Country: Vietnam's Baseline Situation at the Start of the Year of ICT. *Journal of Computers & Education*, *56*(4), 974–982. doi:10.1016/j.compedu.2010.11.015

Quinn, C. N. (2011). *Designing mLearning: Tapping into the mobile revolution for organizational performance.* John Wiley & Sons. https://quinnovation.com/Designing_mLearningSample.pdf

Reality, E. (2020). *How VR and AR can Help in the pandemic.* https://eonreality.com/vr-ar-remote-learning-working-coronavirus/

Rebecca, W., & Marshall, S. (2012). *A New Face of Education: Bring Technology into the Classroom in the Developing World.* Global Economy and Development, Brookings. https://www.brookings.edu/research/a-new-face-of-education-bringing-technology-into-the-classroom-in-the-developing-world/

Tissenbaum & Slotta. (2019). *Developing a smart classroom infrastructure to support real-time student collaboration and inquiry: a 4-year design study.* doi:10.100711251-019-09486-1

Traxler, J. (2007). Defining, discussing, and evaluating mobile learning: The moving finger writes and having writ.... *International Review of Research in Open and Distance Learning*, *8*(2), 1–11. doi:10.19173/irrodl.v8i2.346

Valk, J. H., Rashid, A. T., & Elder, L. (2010). Using mobile phones to improve educational outcomes: An analysis of evidence from Asia. *International Review of Research in Open and Distance Learning*, *11*(1), 117. doi:10.19173/irrodl.v11i1.794

Vandis. (2019). *The Importance of Cloud Computing in Education.* https://www.vandis.com/insights/the-importance-of-cloud-computing-in-education/

Walia, J. S., Hämmäinen, H., & Matinmikko, M. (2017). 5G Micro-operators for the future campus: A techno-economic study. *Conference internet of things business models, users, and networks.* http://jultika.oulu.fi/files/nbnfi-fe201802093294.pdf

Wanzala, W. (2013). Quest for quality and relevant higher education, training, and learning in Kenya: An overview. *Education Journal*, *2*(2), 36–49. doi:10.11648/j.edu.20130202.13

Wu, W. H., Jim Wu, Y. C., Chen, C. Y., Kao, H. Y., Lin, C. H., & Huang, S. H. (2012). Review of trends from mobile learning studies: A meta-analysis. *Computers & Education*, *59*(2), 817–827. doi:10.1016/j.compedu.2012.03.016

Yusuf, M. O. (2005). Information and Communication Education: Analyzing the Nigerian National Policy for Information Technology. *International Education Journal*, *6*(3), 316–321. https://files.eric.ed.gov/fulltext/EJ854985.pdf

Zhu, Z., Sun, Y., & Riezebos, P. (2016). Introducing the smart education framework: core elements for successful learning in a digital world. *Int. J. Smart Technology and Learning, 1*(1). https://www.researchgate.net/publication/305953715_Introducing_the_smart_education_framework_core_elements_for_successful_learning_in_a_digital_world

KEY TERMS AND DEFINITIONS

Ecosystem: The sustainable ecosystem that improving research and advancement. A digital ecosystem requires to be effective, self-supporting, and scalable. In the connection of the schools and Institutions, this means a technological platform that supports learning—course content, homework, grades, shared documents, and research tools—that works concurrently seamlessly.

ICT Infrastructure: ICT Infrastructure sectors are becoming more appropriate to help to teach, learning, assessment, and administration in schools. ICT Infrastructure school planning assets will help schools to arrange for the strong use of digital technologies in their daily experiments. To ready students for the requirement of an ever-changing world, to obtain more powerful teaching and learning, and improve learning, teaching, and administration.

Innovation: Progressive innovation has advanced education to a new level. Innovation in education teachers and students to study, research, and use all the modern emerging technologies to uncover something new.

Smart Digital Infrastructure: Smart education classifies itself from traditional education through the system of proactive assistance. Where traditional education is essentially passive supervision—instructor offering knowledge-based guidance where students 'listen and learn'—, smart education manages proactive parts such as creative ability and self-directed learning.

Chapter 12
The Evolving Integration of BIM Into Built Environment Programmes in a Higher Education Institute

Mark Kelly
Galway-Mayo Institute of Technology, Ireland

Mark Costello
RPS, Ireland

Gerard Nicholson
Galway-Mayo Institute of Technology, Ireland

Jim O'Connor
Galway-Mayo Institute of Technology (Retired), Ireland

ABSTRACT

The recent publication of the Roadmap to Digital Transition for Ireland's Construction Industry 2018-2021 clearly identified training and education as key priorities in the development of core BIM competences to stimulate the transition towards a more collaborative digital working environment. It identified an urgent need for a consistent and coherent digital experience for students in Irish education and industry to help grow capacity and maturity in the use of BIM and other innovative techniques. The higher education sector has a vital role to play. Key to this will be ensuring appropriate graduate knowledge, skills and competences, ongoing professional development and upskilling of higher education staff, and significant collaboration with industry. This chapter will present a best practice example of academic-industry collaboration, which resulted in the delivery of a flexible Level 8 programme for industry, the implementation of a BIM strategy within an academic context, and the establishment of a focused research group, which is currently engaged in ongoing applied research.

DOI: 10.4018/978-1-7998-6600-8.ch012

INTRODUCTION

The World Economic Forum (2016) has identified the need for the construction sector to drive a transformation that will initiate a mindset 'breakthrough' in relation to technology, materials, and tools; processes and operations; strategy and business model innovation; people, organization, and culture; industry collaboration; joint industry marketing; regulation and policies and public procurement. The World Economic Forum (2018) also presents a vision of 'building in a virtual world' where automation, robotics, connected systems and cloud technology permeate daily life and industries of all kinds resulting in digital technologies replacing manual work and Artificial Intelligence (AI) led design and engineering processes creating 7D lifecycle asset models including 3D object data, scheduling (4D), cost (5D), sustainability (6D), and operations and maintenance (O/M) (7D). Building Information Modelling (BIM) is identified not only as a key enabler for collaboration and efficiency but also as a facilitator to utilize other exciting technological applications (3D printing, robotics, augmented and virtual realities and artificial intelligence) in these possible future scenarios. The digital transition presents enormous challenges for the built environment sector, which are further accentuated by the ongoing talent shortage resulting from a failure to innovate, competition from other industries, conservative work cultures and ongoing image difficulties (World Economic Forum, 2018). Historically, the hesitation to fully embrace change and innovation has seen productivity stagnate and even decrease over the past 50 years (World Economic Forum, 2017). In Ireland, the recent publication of *Building Future Skills: The Demand for Skills in Ireland's Built Environment Sector to 2030* (National Skills Council/Expert Group on Future Skills Needs, 2020) has clearly identified the increasing importance of BIM and digital construction (including virtual, augmented or mixed reality; mobile technology; smart sensors; artificial intelligence, blockchain, drones etc.), which is creating an urgent need for additional upskilling and retraining to respond to these trends. The report found that the top three roles that were most difficult to recruit were Quantity Surveyors, BIM Operators/Experts and Mechanical or Electrical Engineers.

However, evidence suggests that the recent evolution of digital technologies and applications has begun to facilitate a move towards a more efficient and collaborative way of working. In Ireland, the government set out their strategy (in November 2017), for the increased use of BIM on a phased basis in the design, construction and operation of public works projects (funded up to 2021) through the public capital programme. The subsequent publication of the *Roadmap to Digital Transition for Ireland's Construction Industry 2018-2021* (National BIM Council, 2017) clearly identified training and education as key priorities in the development of core competences to enable this transition towards a more collaborative working environment. It identified an urgent need for a coherent, consistent, and seamless digital experience for students in Irish education and industry to help grow capacity and maturity in the use of BIM and other innovative techniques. Subsequent government strategy documents including the *Project Ireland 2040 National Development Plan 2018-2027* (Irish Government, 2018) have specifically highlighted the role of BIM and other technologies as key drivers to improve productivity and efficiencies in the construction sector. The higher education sector has a responsibility to respond to these trends, particularly in relation to graduate competency, knowledge, and skills. Higher education can also facilitate fertile learning spaces to experiment with and explore possible digital futures to lead this transformation. Key to this will be ongoing professional development and upskilling of higher education staff across all built environment disciplines and significant ongoing collaboration with industry. This chapter will firstly provide an overview of national BIM drivers and international BIM educational efforts to date. It will then chart the evolution of an academic-industry collaboration over the past seven years, which has utilized a reciprocal learning model to frame the design, piloting, and delivery of a flexible higher education programme to industry, the establishment of a core research team and has positioned Galway-Mayo Institute of Technology (GMIT) as a leader in BIM education.

NATIONAL BIM DRIVERS

The establishment of the Construction Information Technology Alliance (CITA) in May 2001 in the Dublin Institute of Technology[1] (DIT) in association with Waterford Institute of Technology (WIT) demonstrated considerable foresight in identifying the potential role of Information and Communication Technologies (ICT) in the Irish construction industry[2]. One of CITA's key roles has been the dissemination of the latest information on technology trends through workshops, seminars, conferences etc., working in close collaboration with professional bodies and representative organizations. Since 2008, CITA has received annual funding from Enterprise Ireland and Skillnet for its networking and training programmes[3]. High profile events have included the CITA BIM Gathering conferences in 2013, 2015, 2017 and 2019, the CITA Smart Collaboration Challenge 2014, the BIM Innovation Awards 2016 and CITA Tech Live 2018. The CITA BIM Group is a working group of diverse members of the CITA network with a specific interest in how BIM can influence and inform industry best practice.

In 2014, the need to support companies advancing to Level 2 BIM capability was identified in the *Construction 2020 Strategy* document (Irish Government, 2014). Enterprise Ireland responded to this by establishing a funded BIM-Enable and Implement training scheme for industry[4] in late 2014 and invited tenders for the development a BIM Innovation Capability Programme in April 2015. CITA were awarded the tender to undertake the BIM Innovation Capability Programme (BICP), which subsequently evaluated the BIM capability of the Irish construction industry and higher education institutes through a structured programme of consultation with the aim of developing an overall BIM development plan for the Irish construction sector. Key outputs to date have been the *BICP Global BIM Study – Lessons for Ireland's BIM Programme* (Hore et al., 2017a), *BIM in Ireland 2017* report (Hore et al., 2017b) and a suite of design, construction, facilities management, and SME-focused Irish BIM case studies. Through this work, CITA have also been instrumental in influencing the strategic use of BIM across the sector and informing government action in this area.

In mid-2016, the National BIM Council (NBC) of Ireland was convened to set out a roadmap for the successful implementation of the digital design, construction, and operation of built assets in Ireland. In late-2017, the *Roadmap to Digital Transition for Ireland's Construction Industry 2018-2021* (National BIM Council, 2017) was published in line with the Irish Government recommendation that BIM should be used (but not mandated) in the design, construction, and operation of public buildings up to 2021. This demonstrated a recognition by the Irish Government that the public sector should take a leading role in the adoption of digital tools and processes across the whole lifecycle of projects. The *Digital Transition Roadmap* set out a vision up to 2021, where digitally enabled collaborative and integrated ways of working will have become the norm resulting in significant efficiencies around the delivery and operation of the built environment. It also clearly identified training and education as key priorities in the development of core competences to enable this transition towards a more collaborative working environment (Irish BIM Council, 2017) by stating an urgent need for coherent, consistent, and seamless digital learning experiences for both students and industry stakeholders to help grow industry capacity and maturity. As of mid-2020, no government BIM mandate exists in Ireland but interestingly, Covid-19 and its direct consequences, have again highlighted the role of digital tools and processes in facilitating a remote working environment consisting of online meetings, cloud-based collaboration processes, virtual site visits and walkthroughs etc. It is clear, that higher education has a key role to play in this ever-evolving environment to ensure that graduates have the necessary knowledge, skills, and competences to thrive and lead during these uncertain times.

The next section will review international best practice in relation to BIM education before outlining an academic-industry collaboration between the Department of Building and Civil Engineering in the Galway-Mayo Institute of Technology[5] (GMIT) and RPS[6], which has evolved significantly over the past seven years primarily centered on the design, piloting, and development of a flexible industry-focused Level 8 Higher Diploma in Engineering in BIM. This course has recently undergone a significant transformation towards a blended learning approach over the past 18 months to further address industry needs across Ireland.

REVIEW OF INTERNATIONAL BIM EDUCATIONAL DEVELOPMENTS

The role of higher education in supporting the transition towards a BIM-enabled working environment in the construction sector has developed into a significant international research topic. This initially focused on the transition from CAD to BIM (Berwald, 2008; Casey, 2008; Denzer and Hughes, 2008; Kim, 2012; Pikas, Sacks and Hazzan, 2013; Sacks and Barak, 2010; Sacks and Pikas, 2013; Weber and Hedges, 2008), and the resulting impact this had on learning environments (Hedges and Denzer, 2008) to the role of innovative pedagogical approaches to introduce BIM into the curriculum (Burr, 2009; Evans and Counsell, 2009). This work clearly demonstrated that higher education sector was (and still is) cognisant of the key role that BIM can play in a more efficient, collaborative, and low impact construction sector. Early studies (Wong et al., 2011; Jonnides et al., 2012) found that over 50% of American tertiary education institutions already had BIM dedicated courses or had included BIM content into their existing courses[7]. In Ireland, BIM educational initiatives have tended to mirror the US approach, in that they have been specific to individual institutions and disciplines. For example, Kinnane and West (2013) highlighted the difficulty in creating space for BIM in an already crowded engineering curriculum and proposed embedding BIM unobtrusively within civil and structural engineering project-based modules. Thomas et al. (2014) explored a collaborative learning partnership between their own institute and a local main contractor, which engaged architectural technology, construction management and quantity surveying students in the design of a real-world €3m design and build project to inform the design development, which was taking place at the same time. McGovern et al. (2014) outlined the design and development of a new two-year part-time CPD Diploma in Collaborative BIM Technologies and the integration of geomatics within the programme.

These efforts demonstrate the key role education played initially in raising awareness and knowledge with Yalcinkaya and Singh (2015) identifying academic and professional education and training as one of the four main BIM research streams alongside project lifecycle implementation, design-phase analysis, and information exchange/interoperability. However, Underwood and Ayoade (2015) in their review of BIM education in UK Higher Education found that overall levels of BIM maturity awareness were low when examined across the full range of programmes in the built environment discipline. Further analysis also indicated a huge disconnect between built environment disciplines and that, apart from the architecture and construction-related disciplines, low levels of interest were expressed in incorporating BIM into teaching across the built environment disciplines. In addition, current educational developments in the UK appear to be largely focused on the most fundamental aspects of BIM with a surprisingly low level of engagement with industry. This correlates with their finding that 66% of respondents[8] believed that Higher Education Institutions (HEIs) in the UK were not keeping pace with BIM skills requirement and industry knowledge demands. In addition, 40% of respondents still considered themselves not adequately informed on BIM, which demonstrated a considerable level of detachment. Further to this, Puolitaival and Forsythe (2016) suggest that the availability of appropriate teaching and learning resources, the difficulty in balancing theoretical and practical content and the urgent need for ongoing professional development for academic staff were key challenges that need to be addressed.

The importance of multidisciplinary educational responses was highlighted by Bozoglu (2016) who outlined a modular approach that attempted to close the gap between academic disciplinary principles and industry-focused application proficiency through collaborative and coordinated learning modules across different programmes. Comiskey et al. (2016) and Robertson et al. (2017) both expanded on this approach by creating collaborative multidisciplinary learning opportunities for students[9] and facilitating intervarsity, interdisciplinary and collaborative architectural design/management workshops within a community of academics, students and industry practitioners in Denmark, Ireland, and the UK. Shanbari et al. (2016) took a more focused approach by using existing campus buildings as a learning resource to explore the potential application of laser scanning technology in construction management education. The use of BIM as a visualisation tool for use in advanced engineering education was explored by Badrinath et al. (2016), who highlighted the importance of needs, skillsets, educational frameworks and making room for curriculum experimentation while Soemardi and Erwin (2017) explored how BIM could inform teaching for health and safety on site. Wang et al. (2017) tested a Virtual Reality (VR) embedded immersive system for quantity surveying engineering students comparing the use of head-mounted displays/360-degree displays and non-immersive desktop VER to evaluate how this improved their understanding of architectural design, which in turn would improve decision-making and precision in quantity surveying work. This work has led to an awareness and appreciation by students of the value of BIM, who have found the software to be intuitive, user-friendly, and easy to learn (Deniz, 2018; Sánchez et al., 2019). The use of an online project-based course focusing on the BIM process found that students were also receptive to different modes of flexible delivery (Meng-Han et al., 2019) while Deniz (2018) highlighted a student preference for BIM speciality courses in construction management, structure, and mechanical-electrical-plumbing (MEP) disciplines. These findings highlight the potential role of students acting as co-creators in collaboration with academic staff to create BIM learning experiences that will enhance their employability and career development (Aimi, 2019), even if industry is still very much in the early transition phase (Nývlt, 2018).

Several countries have also attempted to take a holistic approach by developing common BIM educational frameworks (BIM Academic Forum UK; Miller et al., 2013; Mills et al., 2013). Mills et al. (2013) outlines a framework for collaborative building design teaching using BIM entitled IMAC (Illustration, Manipulation, Application and Collaboration), which aligns with the different stages of Bloom's Taxonomy[10] (Table 1).

Table 1. A BIM Framework for Collaborative Building Design Teaching (adapted from Mills et al., 2013

Phases	Description	Bloom' Taxonomy Stages
Illustration	Introductory stages where BIM models are used to illustrate key concepts to students.	Knowledge/Comprehension and Receiving/Responding.
Manipulation	Students work in isolation within their discipline to access, view or change aspects of existing BIM models.	Comprehension/Application and Responding/Valuing
Application	Students work within their discipline using BIM processes and tools to solve real world problems relevant to their discipline.	Application/Analysis and Value/ Organising
Collaboration	A fully multidisciplinary working environment to facilitate an integrated and collaborative approach across disciplines.	Synthesis/Evaluation and Characterising.

In New Zealand, an academic (University of Auckland) and industry partnership proposed a set of learning outcomes that encompassed vocational training, university degree programmes and research activities (Miller et al., 2013) (Table 2).

Table 2. Suggested learning outcomes for undergraduate and postgraduate degrees within a BIM educational framework for New Zealand (adapted from Miller et al., 2013)

Framework Element	Proposed Learning Outcomes
Existing Bachelor degree programmes in the AEC sector.	• Understand the key underpinning BIM principles. • Understand 3D modelling, techniques, and an ability to manipulate 3D models. • Appreciate the wide range of benefits from BIM. • Ability to use a variety of BIM-related software. • Ability to use BIM effectively in a wide range of design and construction scenarios.
BIM taught master's degrees	• An advanced understanding of key BIM principles, with an ability to discern what BIM can and cannot do. • Ability to create and manage 3D BIM models and some of the extension applications into 4D, 5D etc. • Knowledge and skills in a range of area including (but not limited to): collaboration; BIM procurement issues; technology management; business improvement; software development; database systems etc. • Ability to undertake research on at least one specialised BIM-related topic. • Ability to use a variety of BIM software. • Ability to introduce BIM into their employer organisations. • Ability to use BIM effectively in a wide range of design and construction scenarios.

In the UK, the BIM Academic Forum (BAF) (2012, 2013) focused on developing learning outcomes to address strategic, management and technical industry needs through the facilitation of knowledge, understanding, practical and transferable skills (Table 3).

Complementary research has identified key competences in relation to BIM basics, legal aspects, project facilitation processes and model management for collaborative working environments (Succar et al., 2013) and highlighted gaps related to technical knowledge, information and communication, cost management, programme management, organizational culture, organizational strategy, personnel management and planning and administration (Murphy, 2014). In addition, Chen et al. (2014) proposed a maturity framework based on four BIM dimensions (technology, process, people, and information).

Table 3. Proposed Learning Outcomes (LOs) to facilitate the embedding of BIM within the taught curriculum in the UK (adapted from BAF, 2013)

LOs	Undergraduate Level	Postgraduate Level
Knowledge and Understanding	• Importance of collaboration. • The business of BIM. • BIM concepts and construction process. • Stakeholder's business drivers. • Supply chain integration. • BIM across the disciplines. • Contractual and legal frameworks/regulation. • People change management.	• Collaborative working, BIM, information management and its application in the built environment. • Commercial implications, contractual, legal etc. • De-risking projects through BIM and risk management. • Understanding the nature of current industry practice. • Client value and Soft Landings. • Business value and RoI/value proposition. • Understanding supply chain management. • Lifecycle management of BIM, asset, performance in use etc.
Practical Skills	• Introduction to technology used across disciplines. • Use of visual representation. • BIM tools and applications. • Attributes of a BIM system. • Technical know-how. • Structures and materials. • Sustainability.	• Demonstrate ability to adopt different platforms. • Critically judge/evaluate various BIM tools and applications. • Protocols/interoperability/standards. • Capability evaluation. • Change in way projects are to be delivered. • Visualization of large data sets. • Lean principles and links to BIM. • Use of BIM enabled technology.
Transferable Skills	• BIM as a process, technology, people, and policy. • Value, lifecycle, and sustainability. • 'Software-as-a-service' platforms for projects. • Collaborative working. • Communication within interdisciplinary teams. • Process management. • How to deliver projects using BIM. • Information and data flows. • BIM protocols/EIR.	• Project level application. • Cross-discipline and team working. • Importance of effective communication and decision-making including human interaction. • Process mapping and business process re-engineering. • Change management and addressing cultural gaps. • Strategic, technical, and managerial thinking. • Ability to assess barriers to BIM at various levels i.e. corporate or project-level.

When attempting to integrate these findings into the higher education sphere, it is useful to situate them within the following levels of engagement (BAF, 2013):

- Absent – BIM should not affect teaching practice and students do not need to know about BIM.
- Aware – BIM should not affect teaching practice, but students should be aware of it.
- Infused – Students should understand how BIM will affect their future and be offered opportunities to engage with BIM in discipline and multi-disciplinary contexts.
- Embedded – BIM is used as a core pedagogical tool in the students' learning experience.

This can be further framed by Rusinko's (2010) integration matrix[11], which outlines opportunities for narrow discipline-specific integration into existing courses and/or programmes or the creation of new discipline-specific courses and/or programmes and broad cross-disciplinary opportunities to integrate into common core requirements or create new cross-disciplinary courses and/or programmes. In 2015, the UKBAF Learning Outcomes Framework was updated to provide consistent information for Level 2 BIM to higher education institutions, training providers and private educators who were developing and delivering training courses to professional in the sector[12]. The main learning outcomes identified were:

- Understand what BIM is, the contextual requirement for BIM Level 2 and its connection to the Government Construction Strategy and Industrial Strategy 2025.
- Understand the implications and value proposition of BIM within your organisation.
- Understand the requirement for the management and exchange of information between supply chain members and clients as described in the 1192 suite of standards and PAS55/ISO55000.

Interestingly, BIM educational initiatives have tended to focus more specifically on competences rather than the broader scope of learning outcomes. Competency as a term includes concepts of knowledge, skills, abilities, behaviours and successful performance (Scottish Futures Trust, 2018). However, there is lack of understanding and consistency in relation to the terms competence, competency and competent within the construction sector. This may be due to the many interpretations of competency, e.g., as an individual, team, project, role, task, process, practice etc. (Zhao, 2017). Despite these challenges, there has been a significant amount of work done to date exploring BIM-related competency development and management. Succar (2007, 2009, 2010a, b, 2012, 2013 and 2014) has carried out extensive work on the development of conceptual constructs (classifications, taxonomies, models, an ontology, and a framework) to build a knowledge structure for the BIM domain, which has informed the introduction of a set of tools and workflows that facilitate BIM assessment, learning and performance improvement. The resulting BIMe initiative aims to develop: a modular language for digital transformation; generate reliable industry-wide competency benchmarks; identify competency gaps; develop competency-based learning resources; facilitate the exchange of knowledge and experience between academia and industry; and developing free-to-use tools and templates to simplify decision-making processes. Other related research has explored the changing roles of clients, architects, and contractors (Sebastian, 2011); the competences of BIM specialists (Barison and Santos, 2011; Uhm, 2017); the comparative skills of project managers and BIM specialists (Rahman, 2016); and the professional development of BIM actors (Bosch-Sijtsema, 2019). Building on this excellent work, the Scottish Futures Trust (2018) are currently working to establish a standardized up-to-date BIM competency framework to identify the knowledge requirements or learning development needs of those who are involved in using BIM. The longer-term aim of this initiative is to develop an easy-to-navigate BIM competency framework to support industry and academia to align curricula, training, and upskilling to a recognized standard of performance, knowledge, and skills. Succar and Sher (2014) have identified the need for the standardization of training and education in BIM to increase consistency, quality and support greater collaboration between education providers. Key to this, will be the ability of individuals to assess their own competence, which can be viewed as currency in the labour market and identify opportunities for personal development. This is supported by more recent findings from Simpson et al. (2019) who identified a lack of 'people-centric' supports and an urgent need to develop an evolving upskilling and reskilling philosophy and ecosystem.

Internationally, there has been some notable efforts at integrating BIM more holistically into the curriculum including in the US where individual institutions, primarily led by architectural schools during the period 2006-2009 started to develop BIM-inclusive curriculum, predominantly at undergraduate level and with a particular focus on software applications (Bercerik-Gerber et al., 2011; Wu and Issa, 2013). Sacks and Pikas (2013) provided an overview of the extensive consultation, which took place in Israel, to identify the BIM educational requirements for the construction and engineering management sector. This resulted in the identification of 39 relevant topics focusing: processes and general knowledge; skills and technologies; functionality and application. Subsequent research in Australia (London, 2015) built on this work to produce a Student Threshold Capability Framework covering five thematic areas: fundamental principles; technical skills; construction project management skills; strategic organizational behaviours; and global market context. London (2015) recommends the use of these threshold capabilities as part of a curriculum redesign process to address the need for BIM to be embedded as a methodological rather than just a technological tool. Challenges remain in the development of holistic and embedded frameworks for BIM educational implementation (Boton et al., 2018, Huang, 2018) with one key barrier being the transition in the interpretation of BIM as a stand-alone topic (usually focused on software applications only) to a more holistic pedagogical approach that feeds into all aspects of the built environment lifecycle. To this end, Hjelseth (2017) explored different pedagogical frameworks (integrated design and delivery solutions; technological pedagogical content knowledge and the building information model/modelling/management trinity) that would enable systematic implementation of BIM across all relevant disciplinary education.

Review of BIM Educational Developments in Ireland and the UK

There is no doubt that BIM awareness has grown exponentially in period between 2011 and 2019 (NBS, 2019) both in Ireland and the UK. Actual adoption rates, however, have stagnated due to issues around different levels of client knowledge; challenges for certain sections of the industry, i.e., one-off housing and refurbishment works; and the current uncertainty around who is responsible for the UK BIM mandate roll-out i.e., within central and local government. NBS (2019) estimate that true BIM adoption rates is closer to 40% in the UK. This trend is also evident in Ireland, where adoption rates are broadly similar, but with variation according to practice size i.e., smaller practices (those employing 15 persons or less) are significantly less likely to have adopted BIM. In addition, BIM is less likely to be used on smaller works i.e., one-off new houses, extensions, conversions, or alteration-type projects (NBS, 2019). In both the UK and Ireland, significant barriers to BIM adoption remain. In Ireland, the main barriers are a lack of in-house expertise, no client demand, a lack of training and no time to get up to speed. These challenges are not unique to Ireland's construction sector but seem to be more prevalent (Underwood and Ayoade, 2015). These findings also reflect the ongoing challenges in higher education both in Ireland and the UK. Irish higher education institutes did respond rapidly to the industry demands in relation to BIM but have traditionally worked in silos (Behan, 2013; Kinnane and West, 2013; Thomas et al., 2013; McGovern and Behan, 2014; Comiskey et al., 2013 and 2015; McDonnell and West, 2015) to provide BIM educational opportunities at different levels of engagement i.e., active, aware, infused and embedded (UK BAF, 2013) and scale i.e., undergraduate, and postgraduate (Table 4).

In January 2017, the BIM Academic Forum Ireland was established with a mission to 'create a dynamic group to develop and promote the training, learning and research aspects of BIM through strong collaboration and cooperation'. An initial assessment of BIM education in Irish HEIs identified a strong demand from industry for BIM-enabled graduates and highlighted the challenges of incorporating BIM into an already concentrated curriculum, the cost of purchasing software and upgrading hardware and the difficulties in changing the culture of academic staff. In addition, recent global reviews have highlighted research gaps in: mobile and cloud computing; laser scanning; augmented reality; ontology; safety rule and code checking; semantic interoperability and automated generation; development of BIM tools; the study of BIM adoption worldwide; energy simulation and BIM-based information as the current hot topics (Zhao, 2017; Santos et al., 2017).Yet, in these challenges lie some clear opportunities to improve knowledge and awareness, address skill and competency gaps and provide evidence-based and research-informed examples of BIM practice across the supply chain. It is clear from the review of international

BIM educational practice that a more coherent and structured approach is required, which requires a collaborative approach between higher education and industry to map out a clear and pragmatic route for BIM implementation across the sector (BIM Academic Forum (BAF) UK, 2013; Mills et al., 2013, Miller et al., 2013; Sacks and Pikas, 2013; London, 2015). This requires a re-imagining of traditional higher education provisions to move towards a more responsive competency-based and demand-led ecosystem of learning opportunities (Simpson et al., 2019).

Table 4. List of Dedicated Undergraduate and Postgraduate BIM Programmes in Irish Higher Education (updated from Hore et al., 2017)

Institute	Course	Award Type
Athlone Institute of Technology	Higher Diploma in BIM for Civil Engineering and Construction (BIM for Construction)	FT Online/Blended Level 8 (60 credits)
Cork Institute of Technology	BSc (Hons) in BIM and Management	PT Level 8 (60 credits)
Galway-Mayo Institute of Technology	Higher Diploma in Engineering in BIM	PT Level 8 (60 credits)
Institute of Technology Sligo	Certificate in BIM and Lean Construction Management	Online Level 9 (35 credits)
Institute of Technology Tralee	Certificate in BIM	FT Level 6 (30 credits)
Technological University Dublin	BSc (Hons) in BIM (Digital Construction) Postgraduate Certificate in BIM Technologies Postgraduate Diploma in Collaborative BIM MSc in Applied BIM and Management	PT Blended Level 8 (60 credits) PT Level 9 (30 credits) PT Level 9 (30 credits) PT Level 9 (90 credits)
Waterford Institute of Technology	BSc (Hons) in Architectural and BIM Higher Diploma in Science in BIM	FT Level 8 (240 credits) FT Level 8 (60 credits)

The next section will chart the evolution of an academic-industry collaborative approach within an Irish context, which resulted in the design and development of an industry-focused flexible Level 8 programme, the application of a strategic industry BIM implementation approach in an academic context and the establishment of a focused research group within the Department.

THE EVOLUTION OF AN ACADEMIC-INDUSTRY BIM EDUCATIONAL COLLABORATION

GMIT first opened its doors to students in September 1972 with the express aim of '*educating for trade and industry over a broad spectrum of occupations ranging from craft to professional notably in engineering and science...*', thus meeting the key recommendation arising out of the government-commissioned technical education report (Mulcahy, 1967). Interestingly, this report also stressed the need for flexibility in educational provision stated that '*we do not foresee any fixed pattern of courses in the colleges. If they are to make the most effective contribution to the needs of society and the economy, they must be capable of continuing adaptation to social, economic and technological changes*[13].' Over the past 40 plus years, the Department of Building and Civil Engineering in GMIT has been guided by these core principles to develop programmes that have met the evolving needs of different stakeholders in the Irish and international construction sector. This has been encapsulated in the gradual progression of construction-related technology provision, which has seen a transition from basic AutoCAD packages (1987 to 2000) to ArchiCAD (2001 to 2005) on to Autodesk Revit Architecture (2005 to present). Fundamental to this was the decision to fully embed 3D building modelling into the Architectural Technology programme in 2005 whilst also retaining a comprehensive delivery of AutoCAD 2D and 3D. In 2012, the Department delivered a year-long certificate in BIM to 19 participants via a Springboard initiative[14], which informed the establishment of a two-year research partnership with a local SME, Carey Building Contractors to explore the application of BIM on small-scale construction projects. This study employed a case study methodology to comparatively analyse the use of traditional construction-phase project management processes versus the use of simulated BIM-enabled project management processes to evaluate their impact on communication, information management, project quality management, material estimating and cost (Nicholson, 2013). This study aligned with the Department's strategic focus on applied research, which aims to inform both curriculum development and industry best practice. Since 2000, this has included national projects on construction and demolition waste management, construction resource efficiency and the circular economy.

In early 2013, an opportunity arose to further develop BIM-related educational provision, when RPS approached the Department with a view to developing appropriate training and continuing professional development (CPD) for their employees. Up until that time (since 2007), RPS had been using 3D visualizations and walk-throughs to assist clients in their decision-making during the early design phases. In 2012, an internal working group was established within RPS to evaluate how BIM could help address the ever-increasing project requirements relating to civil, structural, mechanical services and infrastructural disciplines. An urgent need for upskilling was identified. This need was tempered somewhat by a distinct lack of suitable responsive and flexible industry-focused higher education BIM programmes in Ireland at that time. In mid-June 2013, representatives from RPS and the Department met to formally address this need. A training needs and analysis exercise was undertaken over the subsequent months, which informed a decision to design and pilot a Level 8 accredited Higher Diploma in Engineering in BIM, consisting of the following suite of mandatory and elective modules: BIM Virtual Modelling Fundamentals (Mandatory (M)); BIM Collaboration (M); BIM Project (M); BIM Architecture (Elective (E)); BIM Structure (E); BIM Infrastructure (E); BIM Mechanical, Electrical and Plumbing (E) (Figure 1).

The pilot phase was carried between September 2013 and September 2015, with the outreach delivery of ten mandatory and electives modules to RPS staff in Galway, Dublin, and Cork[15]. This upskilling initiative informed the establishment of core BIM teams in these RPS offices and facilitated the development of evolving in-house BIM protocols and procedures that continue to be used in the design, construction, and operation phases of projects to this day. Key to this was the preparation of an RPS BIM Technical Services Strategy (2014-2018), which provided an organisational roadmap to facilitate a re-evaluation on how RPS projects are procured, delivered and constructed through the incorporation of fully coordinated design models; the development of a new BIM working environment across multiple interoperable design platforms; the production of 4D phasing and 5D costing models; improved planning and coordination; and the minimisation of operational and contractual conflict. This strategy set out a series of milestones related to management, engagement with academic institutions, training and upskilling requirements, industry engagement, core groups and pilot studies that provided a practical template for BIM implementation across a multi-disciplinary organisation. Since September 2015, the Level 8 Higher Diploma in Engineering in BIM programme has been available in flexible[16] part-time mode to all interested industry stakeholders. This offering has directly responded to industry calls for a focus on BIM, lean and green construction (CIF/DKM, 2017), the subsequent creation of new positions such as BIM Manager, Project Information Manager, BIM Modeller etc., and has addressed a significant upskilling gap for mid-career construction workers who have a Level 7 diploma or pass degree[17].

Figure 1. Higher Diploma in Engineering in BIM programme overview

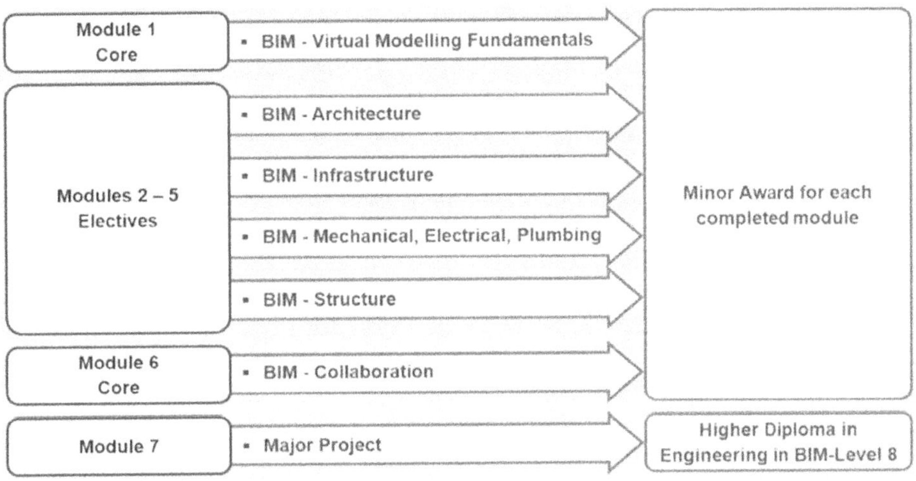

The GMIT/RPS collaboration was also instrumental in the establishment of the BIM Western Hub, which was launched in September 2015 with a remit to raise awareness, facilitate regional engagement and share BIM best practice and experiences. In 2016, RPS were awarded ‘BIM Level 2 Business Systems’ certification by BRE Global, which enabled them to offer professional services on projects in the UK, Europe and globally, resulting in the award of major highways, water and gas projects. In addition, the Director for BIM RPS (co-author Mark Costello) was awarded the BIM accredited professional badge of approval from the BRE Academy. Informed by this ongoing collaboration with RPS and the rapid progression of BIM in industry during this time[18], academic staff within the Department recognised the need to take a more structured approach to build on the good work already done to keep pace and ultimately lead the industry towards a BIM-enabled working environment. The reciprocal nature of the collaboration with RPS and the lessons learned informed the decision to develop a BIM Implementation Strategy for the Department of Building and Civil Engineering, mimicking the approach taken by RPS in developing their BIM Technical Services Strategy 2014-2018. Initially, a draft document was proposed: the establishment of a BIM working group; a full curriculum review; a review of staff BIM literacy and attitudes; a coherent CPD programme for staff; the evaluation of opportunities to embed BIM holistically as a pedagogical methodology across all programmes but especially for multidisciplinary work; and a series of research activities. The draft working document was disseminated to all academic staff within the Department for feedback. Feedback received from staff highlighted the importance of curriculum reform; training; and the need for multi- and trans-disciplinary learning opportunities for staff and students including applied research. This informed a second revision of the BIM strategy, which set out key objectives, proposed actions and milestone dates utilising a roadmap approach (Table 5).

Table 5. GMIT BIM Implementation Strategy 2016-2020

Key Objective	Action	Milestone
Undertake a Curriculum Review	A comprehensive review of curriculum content and learning outcomes within each programme.	2016
Carry out a BIM attitudes and literacy survey of Departmental academic staff	Develop, pilot, and disseminate an online questionnaire survey within the Department to identify BIM attitudes, knowledge, and competences.	2017
Initiate engagement with the Building and Estates Office	Work with the Building and Estates team to integrate BIM working practices into the construction of new buildings the maintenance and refurbishment of existing buildings.	2017-2020
Develop a flexible CPD programme for staff	Present different CPD options i.e. undertaking the Level 8 Higher Diploma in BIM; industry workshops; attendance at conferences; development of reusable learning objects; lunchtime 'bites' series.	2017-2020
Facilitate multi- and trans-disciplinary learning opportunities	A multi-disciplinary project will be piloted over the academic year within the Department focusing on a 'real-world' development.	2017-2020
Benchmark student learning outcomes	Track the impact of BIM-related modules and additional BIM learning interventions on student learning.	2017-2020
Applied research	Explore opportunities to integrate BIM with existing active research and academic practice.	2017-2020

The first step was to undertake a curriculum mapping exercise of the four undergraduate programmes in Architectural Technology, Construction Management, Civil Engineering and Quantity Surveying and Construction Economics within the Department to identify the current level of BIM integration at the time.

Curriculum Review of Undergraduate Programmes in the Department of Building and Civil Engineering at GMIT

The curriculum review used a pragmatic structured approach to identify the level of BIM integration and gaps within the existing suite of undergraduate programmes (Figure 2).

Figure 2. Curriculum review methodology

Table 6. Knowledge, skills, and competence sub-sets for Level 8 programmes

Knowledge *Breadth*	An understanding of the theory, concepts and methods pertaining to a field (or fields) of learning.
Knowledge *Kind*	Detailed knowledge and understanding in one or more specialized areas, some of it at the current boundaries of the field(s).
Know-How and Skill *Range*	Demonstrate mastery of a complex and specialized area of skills and tools; use and modify advanced skills and tools to conduct closely guided research, professional or advanced technical activity.
Know-How and Skill *Selectivity*	Exercise appropriate judgement in several complex planning, design, technical, and/or management functions related to products, services, operations, or processes, including resourcing.
Competence *Context*	Use advanced skills to conduct research, or advanced technical or professional activity, accepting accountability for all related decisions making; transfer and apply diagnostic and creative skills in a range of contexts.
Competence *Role*	Act effectively under guidance in a peer relationship with qualified practitioners; lead multiple, complex, and heterogeneous groups.
Competence *Learning to Know*	Learn to act in variable and unfamiliar learning contexts; learn to manage learning task independently, professionally, and ethically.
Competence *Insight*	Express a comprehensive, internalized, personal world view manifesting solidarity with others.

Since 2016, a detailed curriculum mapping exercise[19] has been undertaken by the (main) author using the following methodology:

- A comparative review of relevant professional body BIM requirements versus individual undergraduate programme learning outcomes.
- Used the National Qualifications Authority Ireland (NQAI) framework to identify if BIM or related terms i.e., ICT were present in the individual programme learning outcomes.
- Identified the presence of BIM as a stand-alone module or as an embedded topic in existing modules available across four undergraduate programmes in the Department.
- Compared the individual module learning outcomes against the UK 2015 BAF Learning Outcomes matrix.

The first step in the curriculum analysis was to identify professional body accreditation requirements for each discipline as this should directly inform the learning outcomes at programme level. Since 1999, one of the key developments in European higher education has been the use of learning outcomes to provide greater clarity in the description of qualifications. Learning Outcomes have become ubiquitous within worldwide curriculum policy in recent years. This move comes with many potential benefits, as it shifts the focus from providers to users of education, and it introduces a common language, addressing issues of progression, transparency, and equity (CEDEFOP, 2009). A learning outcome can be defined as *a statement of what the learner is expected to be able to do on successful completion of a module or programme, to demonstrate their knowledge, understandings, skills and/or competences* (Bowe and Fitzmaurice, ud). They represent a shift towards framing education in terms of learners and their development, rather than in terms of what is to be taught (Biesta & Priestley, 2013). This shift has been associated with tensions between knowledge-rich and learner-centred and competence-based curricula. These tensions have played out in curriculum policy making across multiple international education jurisdictions, in school-based curriculum making, as teachers used to teaching content-driven curricula have been asked to reinvent their practice, and in academic writing on curriculum development (Young and Muller, 2010). The move to a Learning Outcomes based model of education has been reinforced by the publication of competency frameworks by supra-national organizations such as the OECD and the European Union since the turn of the millennium. The NQAI provide a detailed framework for the development, recognition, and award of qualifications from Level 1 Certificates to Level 10 Doctoral Degrees. The framework sets out eight sub-sets of knowledge, skills, and competence, which have been used to generate level indicators that include broad descriptors of learning outcomes at each level (see Table 6 for Level 8 programme example).

Table 7. CIOB and SCSI BIM competency requirements

CIOB Undergraduate Education Framework	Evaluate and apply different project management techniques to complex projects i.e. progress and completion management and decision processes; project evaluation and review technique (PERT); risk analysis; and Building Information Modelling (BIM). Management including: Prevention through design (Building Information Modelling (BIM) and Buildability-Usability-Maintainability (BUM)).
SCSI APC Pathway Guide for Quantity Surveying and Construction	BIM embedded within mandatory competency of communication and negotiation under 'written and graphical communication' and the data management competency.
SCSI APC Pathway Guides for Building Surveying, Building Control and Project Management	Optional competency focusing on: the establishment and management of the information modelling systems on projects e.g. the collaborative process and technological principles involved in implementing Building Information Modelling (BIM) at three levels: Level 1 (knowledge and understanding) - demonstrating knowledge and understanding of the technical, process and collaborative aspects of the use of BIM on projects e.g. understanding of BIM strategies and implementation; understanding of the various technical options and solutions for information modelling; understanding of the collaborative processes necessary for BIM adoption; knowledge of standard classification systems and their use in infrastructure; knowledge of relevant internationally recognized management standards such as Construction Operations Building Information Exchange (COBie). Level 2 (application of knowledge and understanding) - the development and application of management systems to facilitate the use of BIM on projects including unified control and reporting procedures e.g. preparation of a BIM execution plan; design and implementation of a BIM management process; analysis of comparative BIM solutions; maintenance of an information model; agree and implement contractual aspects of BIM such as separate protocols; and facilitate and manage project team members for BIM implementation. Level 3 (reasoned advice and depth of technical knowledge) - show how the knowledge and experience gained in this competency has been applied to advising clients and/or senior management on BIM strategy e.g. analyzing, assessing, evaluating and reporting on options for BIM strategies at a corporate or project level; designing and advising on collaborative strategies for the successful implementation of BIM on projects; advising on the contractual and commercial implications of using BIM on projects; advising on options for software and protocols on BIM projects; advising on technical information systems requirements for BIM at corporate or project level.

A word scan for 'BIM' was carried out in review of the following relevant professional body guidance, which determines accreditation for the four undergraduate programmes in the Department: Quality and Qualifications Ireland (QQI, 2016) Awards Standards for Architectural Technology, Royal Institute of Architects Ireland (RIAI, 2016) Statement on Policy on Education, Chartered Institute of Building (CIOB, 2018) Undergraduate Education Framework, Engineers Ireland (EI, 2014) Accreditation Criteria for Professional Titles, EI 2014 Guidelines for Chartered Engineer, EI Chartered Engineer – 2014 Regulations for the Registered Professional Title, Society of Chartered Surveyors Ireland (SCSI) APC Pathway Guides to Quantity Surveying and Construction, Building Surveying, Building Control and

Project Management. This revealed that only the CIOB and the SCSI have included BIM explicitly in their list of competency requirements (Table 7).

Table 8. BIM presence in the knowledge, skills, and competences sub-sets of the four undergraduate programmes offered in the Department of Building and Civil Engineering at GMIT

Architectural Technology	
Knowledge *Breadth*	None outlined.
Knowledge *Kind*	Detailed knowledge and understanding of the highly technical specialist processes and practices involved in the practice of the Architectural Technologist in the future including Computer Aided Design (CAD), Building Information Modelling (BIM), sustainability, lifecycle analysis, material choice and provenance.
Know-How and Skill *Range*	Comprehensive application of CAD and Building Information Modelling as analytical, research and design tools.
Know-How and Skill *Selectivity*	None outlined.
Competence *Context*	None outlined.
Competence *Role*	None outlined.
Competence *Learning to Know*	None outlined.
Competence *Insight*	None outlined.
Quantity Surveying and Construction Economics	
Knowledge *Breadth*	Information Technology i.e. has knowledge and understanding of the role of ICT and its application to the sub-field of engineering; and has sufficient knowledge and understanding of ICT to adopt it to solve complex engineering problems.
Knowledge *Kind*	Detailed knowledge and understanding of the highly technical specialist processes and practices involved in the practice of the Architectural Technologist in the future including Computer Aided Design (CAD), Building Information Modelling (BIM), sustainability, lifecycle analysis, material choice and provenance.
Know-How and Skill *Range*	Information Technology i.e. can select and use an appropriate technology to meet a stated need; can specify ICT appropriate for complex engineering projects and has an ability to specify the technical performance requirements and develop tools to solve engineering problems.
Know-How and Skill *Selectivity*	Information Technology i.e. can design, select, modify, manage, and use an appropriate technology to solve complex engineering problems; and can develop models of engineering systems using industry standard software tools.
Competence *Context*	None outlined.
Competence *Role*	None outlined.
Competence *Learning to Know*	None outlined.
Competence *Insight*	None outlined.
Civil Engineering	
Knowledge *Breadth*	Has knowledge and understanding of the role of ICT and its application to the sub-field of engineering. Has sufficient knowledge and understanding of ICT to adopt it to solve complex engineering problems.
Knowledge *Kind*	Is sufficiently aware of the concepts behind and the capabilities of ICT as they apply in the sub-field of engineering. Understands the fundamental concepts and the advantages and limitations of the core technologies relevant to the subfield of engineering.
Know-How and Skill *Range*	Can select and use an appropriate technology to meet a stated need and specify ICT appropriate for complex engineering projects.
Know-How and Skill *Selectivity*	Can design, select, modify, manage, and use an appropriate technology to solve complex engineering problems and develop models of engineering systems using industry standard software tools.
Competence *Context*	None outlined.
Competence *Role*	None outlined.
Competence *Learning to Know*	None outlined.
Competence *Insight*	None outlined.
Construction Management	
Knowledge *Breadth*	None outlined.
Knowledge *Kind*	None outlined.
Know-How and Skill *Range*	None outlined.
Know-How and Skill *Selectivity*	None outlined.
Competence *Context*	None outlined.
Competence *Role*	None outlined.
Competence *Learning to Know*	None outlined.
Competence *Insight*	None outlined.

Table 9. Summary of programme learning outcomes related to BIM (as of June 2019)

Programmes	AT	QS	CE	CM
Knowledge *Breadth*		√	√	X
Knowledge *Kind*	√	√	√	X
Know-How and Skill *Range*	√	√	√	X
Know-How and Skill *Selectivity*	X	√	√	X
Competence *Context*	X	X	X	X
Competence *Role*	X	X	X	X
Competence *Learning to Know*	X	X	X	X
Competence *Insight*	X	X	X	X

The next step in the curriculum analysis was to use the NQAI framework to identify if BIM or related terms such as ICT, were present in the programme learning outcomes (Table 8).

It was clear from this analysis, that BIM/ICT did at least feature in three of the programmes, except for Construction Management. However, there was clearly significant gaps in relation to the competency levels (Table 9).

The next step in the curriculum gap analysis was to identify the presence of BIM as a stand-alone module or where BIM was embedded into a module across the four general Level 8 undergraduate programmes in the department (Table 10).

Table 10. Curriculum mapping of GMIT Department of Building and Civil Engineering undergraduate courses as detailed in the programme documentation (as of June 2019)

Programme	Year 1	Year 2	Year 3	Year 4
Architectural Technology	CAD 1 (M) (10 credits)	CAD 2/BIM (M) (10 credits)	CAD 3/BIM (M) (10 credits)	BIM Architecture (M) (10 credits)
Quantity Surveying	BIM for Surveyors 1 (M) (10 credits)	BIM for Surveyors 2 (M) (10 credits)		
Civil Engineering	CAD 1 (M) (5 credits)	CAD 2/BIM (M) (5 credits)		Environmental/Energy Sustainability (M) (5 credits)
Construction Management	CAD/BIM 1 (M) (5 credits)	CAD/BIM 2 (M) (5 credits) BIM for Const. (E) (5 credits)		Project Mng. (M) (10 credits)

The B.Sc. in Architectural Technology[20] demonstrates the most consistent BIM curriculum across the four years with one dedicated technology/software-focused module[21] (each worth 10 credits) in each year. Additionally, BIM topics are included in the following Year 4 modules: use BIM analytical software for daylighting design strategies in the *Innovative in Architectural Technologies* module and BIM legal issues in the *Professional Practice: Contract and Procurement* module. The B.Sc. in Civil Engineering has dedicated technology/software mandatory modules in Years 1, 2 and 4 (each worth 5 credits), while lifecycle analysis and BIM is included as a topic in the *Environmental and Energy Sustainability* module in Year 4. The B.Sc. in Quantity Surveying and Construction Economics has a technology/software focus[22] in Years 1 and 2 with dedicated *BIM for Surveyors 1 and 2* modules that specifically focus on introducing BIM and onscreen measurement applications (Buildsoft, CostX and Autodesk Quantity Take-Off). Additionally, BIM appears as a topic under the communications management element of the Year 4 *Construction Project Management* module. The B.Sc. in Construction Management has a less consistent approach with CAD/BIM modules (each worth 5 credits) in Years 1 and 2. An elective 5-credit module, *BIM for Construction* is also available in Year 2, which expands on the technological/ software applications while also introducing the process. In Year 4, the Project Management module (10 credits) does include collaborative BIM approaches as a topic. This simple curriculum content analysis clearly identified a myopic focus on the software/technology side with gaps evident in BIM processes, people, and policy. This is further demonstrated when examining mandatory credit weighting where dedicated CAD/BIM (software/technology) modules represent only 10 credits (Civil Engineering), 20 credits (Quantity Surveying and Construction Management) and 40 credits (Architectural Technology) of the 240 credits available for each programme.

The next step in the curriculum mapping exercise was to evaluate the current associated module learning outcomes and compare them against the UK 2015 BAF Learning Outcomes Framework (Table 11), which aimed to provide consistent information on Level 2 BIM for educational institutions and training providers who were developing and delivering training courses for professionals in the sector. The framework has 32 learning outcomes, which are grouped under three headings: strategic context; organization and implementation; information management and exchange. Individual module learning outcomes on each of the four undergraduate programmes were mapped against the 32 learning outcomes listed (Table 12).

It was clear from this detailed phased analysis that significant gaps in relation to strategy, organization, implementation, information management and exchange existed across all four programmes[23]. To address this, the authors embedded the BRE BIM Approved Graduate (AG) course[24] into a suite of modules (*Integrated Project*, *Resource Efficiency Strategies for the Construction Sector*, *BIM Architecture* and *BIM Collaboration)* modules in Year 4 of the B.Sc. in Quantity Surveying and Construction Economics, Construction Management and Architectural Technology programmes.

BIM Curriculum Development in the Department of Building and Civil Engineering in GMIT

The BRE BIM AG[25] is an online course that provides a detailed introduction to the BIM process, with the following suite of online learning units: BIM definition and terminology; standards, methods and procedures; advantages of information management; the role of a BIM manager; the information delivery cycle; employer's information requirements (EIRs); pre-contract BIM execution plans; post-contract BIM execution plans; production of information; exchange of information and interoperability; and asset information models. A blended learning approach was undertaken, with students completing weekly online units, supported by facilitated face-to-face workshops during class contact hours. This offered the students an opportunity to gain an extra professional qualification, as well as fast-tracking them towards BRE Global individual certification as a 'BIM Informed Professional' or 'BIM Certified Practitioners'[26], depending on the student's subsequent industry experience after graduation. The integration and facilitation

Table 11. Curriculum mapping exercise using the UK BAF Learning Outcomes Framework

UK BAF Learning Outcomes	AT	QS	CE	CM
Understand what BIM is, the contextual requirement for BIM Level 2 and its connection to the Government Construction Strategy and Industrial Strategy 2025; including an understanding of:				
Background and the need for collaborative working (removing waste, errors and poor quality/in-complete information).	√	√	√	√
The value of whole life and whole estate approach rather than capital-led and single asset.			√	
The concept of Soft Landings/Government Soft Landings (GSL).				
Roles and responsibilities of the supply chain members and clients as part of BIM Level 2 delivery (cultural/behavioural).				
External context for BIM, global, national, standards and support communities.		√	√	
Core and extended suite of standards, documents and deliverables describing BIM Level 2.				
Barriers to successful adoption of BIM Level 2 and how to create the conditions for success.				
The value of high-quality data and the principles of data management.				
The key vulnerability issues and nature of controls required to enable the trustworthiness and security of digitally built assets.				
Understand the implications and value proposition of BIM within your organisation; including an understanding of:				
Implementation implications for the introduction of BIM Level 2 on your organisation and supply chain (e.g. training, management processes and systems).				
Organisational change management considerations in context of the introduction of BIM Level 2.				
Assessment of capability of your organisation and your supply chain (e.g. standard methods of assessment PAS91 Table 8).				
Technical, technology and interoperability requirements of Level 2 BIM (Information Management/ CDE, model-based design and analysis).		√	√	√
The importance of Level 2 BIM as a driver for business process review and improvement.				
Legal and commercial implementation implications for the introduction of BIM Level 2 on your organisation and supply chain (e.g. commercial stakeholders).	√			
How BIM supports the relationship between design, construction, and facilities/asset management.		√		
The potential security threats to built and information assets, and the need for the development of an appropriate and proportionate security risk management approach.				
Understand the requirement for the management and exchange of information between supply chain members and clients as described in the 1192 suite of standards and PAS55/ISO 55000; including an understanding of:				
The purposes for information in the capital and asset phase.				
Requirements for the exchange of information between supply chain members in a collaborative manner as described in PAS1192-2: 2013 and PAS1192.3: 2014 and provided in conjunction with BS1192:2007.			√	√
Roles and responsibilities of the supply chain members and clients of BIM Level 2 and the implications on Scopes of Services.				
BIM Plain Language Questions, Employers Information Requirements (EIR), Organisation Information Requirements, Asset Information Requirements, and the exchange of information between supply chain and client in a collaborative manner in context of PAS1192.2: 2013 and PAS1192.3:2014.				
BIM Execution Plan (BEP) in context of PAS1192.2:2013 - the related concepts, purpose, and implementation principles.				
Digital delivery of information between supply chain members and with clients in context of BS1192-4:2014 (COBie), Digital Plan or Work (DPoW) and classification systems.				
The concept, purpose and implementation principles of Project Information Models (PIM) and Asset Information Models (AIM) and the relationship and interchange between them.				
A Common Data Environment (CDE) as described in the 1192 suite of standards.			√	√
The implications of Level 2 BIM in relation to project team working methods as described in BS1192: 2007.				
The way in which Level 2 BIM can be adopted to benefit decision-making for design management.				
Technologies/methods for creating, using and maintaining structured information.	√	√	√	
Contractual interventions required to support BIM Level 2 and the implications on exiting forms of contract.				
Ownership of information and related issues of IP and copyright, insurances, and potential liabilities.				
Requirements for security-minded policies, processes and procedures which address specific security threats or combinations of threats in a consistent and holistic manner.				

Table 12. Post BRE BIM AG - Curriculum mapping exercise using the UK 2015 BAF Learning Outcomes Framework

UK BAF Learning Outcomes	AT	QS	CE	CM
Understand what BIM is, the contextual requirement for BIM Level 2 and its connection to the Government Construction Strategy and Industrial Strategy 2025; including an understanding of:				
Background and the need for collaborative working (removing waste, errors and poor quality/in-complete information).	√	√		√
The value of whole life and whole estate approach rather than capital-led and single asset.	√	√		√
The concept of Soft Landings/Government Soft Landings (GSL).	√	√		√
Roles and responsibilities of the supply chain members and clients as part of BIM Level 2 delivery (cultural/ behavioural).	√	√		√
External context for BIM, global, national, standards and support communities.	√	√		√
Core and extended suite of standards, documents and deliverables describing BIM Level 2.	√	√		√
Barriers to successful adoption of BIM Level 2 and how to create the conditions for success.	√	√		√
The value of high-quality data and the principles of data management.		√		√
The key vulnerability issues and nature of controls required to enable the trustworthiness and security of digitally built assets.		√		√
Understand the implications and value proposition of BIM within your organisation; including an understanding of:				
Implementation implications for the introduction of BIM Level 2 on your organisation and supply chain (e.g. training, management processes and systems).				
Organisational change management considerations in context of the introduction of BIM Level 2.				
Assessment of capability of your organisation and your supply chain (e.g. standard methods of assessment PAS91 Table 8).				
Technical, technology and interoperability requirements of Level 2 BIM (Information Management/CDE, model-based design and analysis).	√	√		√
The importance of Level 2 BIM as a driver for business process review and improvement.				
Legal and commercial implementation implications for the introduction of BIM Level 2 on your organisation and supply chain (e.g. commercial stakeholders).				
The value, benefits and investment associated with BIM Level 2.		√		√
How BIM supports the relationship between design, construction, and facilities/asset management.	√	√		√
The potential security threats to built and information assets, and the need for the development of an appropriate and proportionate security risk management approach.				
Understand the requirement for the management and exchange of information between supply chain members and clients as described in the 1192 suite of standards and PAS55/ISO 55000; including an understanding of:				
The purposes for information in the capital and asset phase.		√		√
Requirements for the exchange of information between supply chain members in a collaborative manner as described in PAS1192-2: 2013 and PAS1192.3: 2014 and provided in conjunction with BS1192:2007		√		√
Roles and responsibilities of the supply chain members and clients of BIM Level 2 and the implications on Scopes of Services.		√		√
BIM Plain Language Questions, Employers Information Requirements (EIR), Organisation Information Requirements, Asset Information Requirements, and the exchange of information between supply chain and client in a collaborative manner in context of PAS1192.2: 2013 and PAS1192.3:2014.		√		√
BIM Execution Plan (BEP) in context of PAS1192.2:2013 - the related concepts, purpose, and implementation principles.		√		√
Digital delivery of information between supply chain members and with clients in context of BS1192-4:2014 (COBie), Digital Plan or Work (DPoW) and classification systems.		√		√
The concept, purpose, and implementation principles of Project Information Models (PIM) and Asset Information Models (AIM) and the relationship and interchange between them.	√	√		√
A Common Data Environment (CDE) as described in the 1192 suite of standards.	√	√	√	√
The implications of Level 2 BIM in relation to project team working methods as described in BS1192:2007.	√	√		√
The way in which Level 2 BIM can be adopted to benefit decision-making for design management.	√	√		√
Technologies/methods for creating, using, and maintaining structured information.	√	√	√	√
Contractual interventions required to support BIM Level 2 and the implications on exiting forms of contract.				
Ownership of information and related issues of IP and copyright, insurances, and potential liabilities.				
Requirements for security-minded policies, processes and procedures which address specific security threats or combinations of threats in a consistent and holistic manner.		√		√

of the BRE BIM AG course did address some of the gaps in the curriculum but in a stand-alone format as an embedded topic in four existing modules. For example, analyzing the updated curriculum using the UK 2015 BAF learning outcomes framework does illustrate the impact of the BRE course but it can give a false impression of the level of embeddedness across all the undergraduate courses (Table 12).

This detailed curriculum review highlighted the significant challenges of embedding BIM as a pedagogical methodology across the curriculum. It was decided to reframe the implementation strategy using the reciprocal learning model first proposed in 2015 (Kelly et al., 2015). This framework aims to utilize the academic-industry synergies to inform curriculum and pedagogical development, applied research and industry best practice (Figure 3).

Figure 3. GMIT/RPS Reciprocal Learning Framework ©

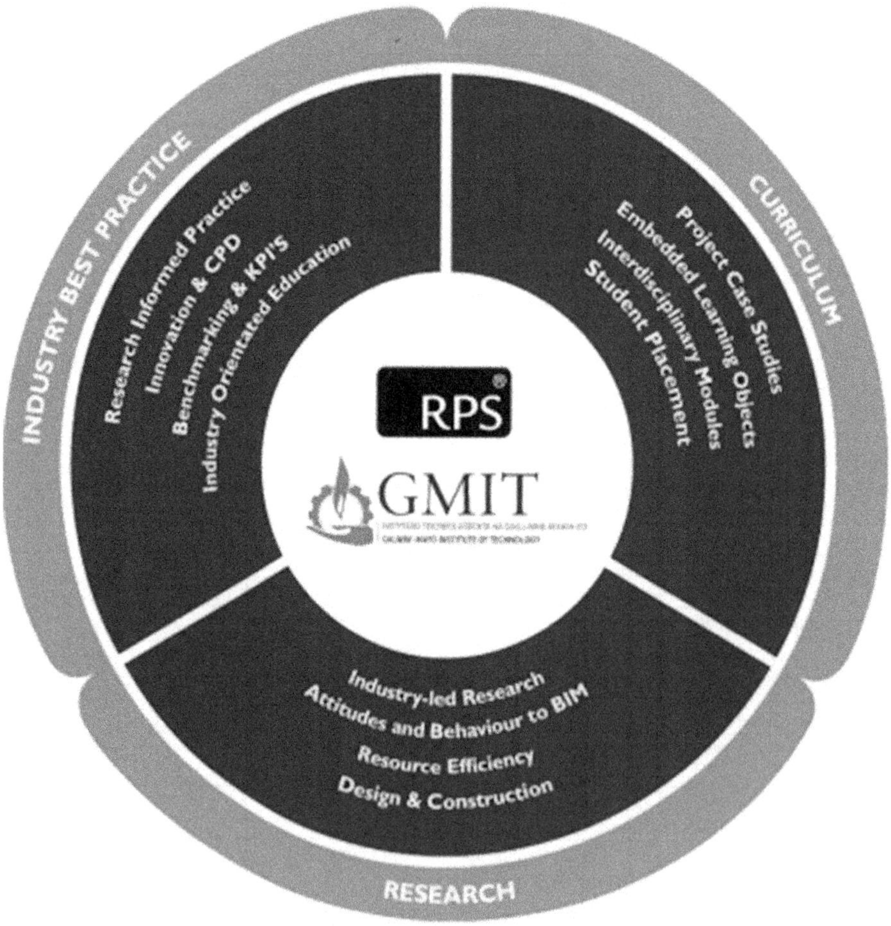

This led to the establishment of a dedicated BIM implementation group within the Department, who were tasked with the development of an evolving BIM strategy.

Establishment of BIM Implementation Group (BIGG) and Programme Transition

A core teaching and research group, the BIM Implementation Group GMIT (BIGG) was established in early 2018 to support the implementation plan, which coincided with a call for submissions under the inaugural Disciplinary Excellence in Learning, Teaching and Assessment (DELTA) Awards by the National Forum for the Enhancement of Teaching and Learning in Higher Education[27]. The award was open to any discipline group who could demonstrate a record of excellence in teaching and learning enhancement and who would commit to an ongoing and clearly articulated process of continued enhancement. The GMIT research group felt that the application process for this award would scaffold the BIM implementation plan and raise awareness of the ongoing academic-industry partnership if successful. A two-phase application process was undertaken, which involved a significant mapping exercise to align ongoing BIM-related teaching, learning and research activities to the DELTA Framework (Figure 4) and the development a capacity-building plan for a subsequent three-year period. This approach demonstrated the linkages between teaching and learning enhancement and other core activities, i.e., international, and national drivers, professional development, student success and teaching and learning in a digital world.

Figure 4. Disciplinary Excellence in Learning, Teaching and Assessment (DELTA) Framework (National Forum, 2020)

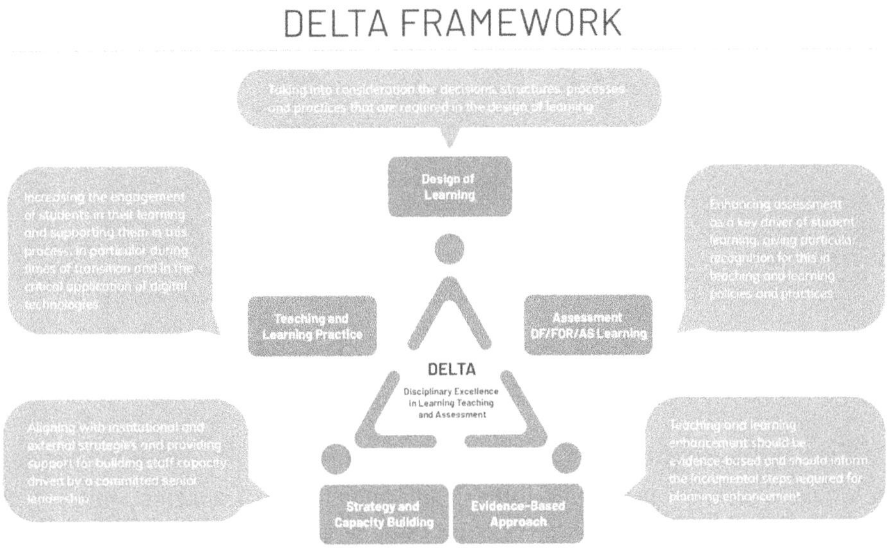

Table 13. Transition of the delivery of the Higher Diploma in Engineering in GMIT pre- and post-2019

Level 8 Higher Diploma in Engineering in BIM 2015-2019				
Module	**Credits**	**Yr**	**Delivery**	**Independent Learning**
BIM Virtual Fundementals	10 (M)	1	4-hour workshops every week for 12 weeks.	4 hours/week
BIM Collaboration	10 (M)	1	4-hour workshops every week for 12 weeks.	4 hours/week
BIM Elective Choice	10 (E)	2	4-hour workshops every week for 12 weeks.	4 hours/week
BIM Research Project	30 (M)	2	4 no. 3-hour workshops every 3-4 weeks over a 16-week period + 16 weeks supervision.	Average of 12 hours/ week over total period
Level 8 Higher Diploma in Engineering in BIM 2019 onwards				
Module	Credits	Yr	Delivery	Independent Learning
Module	10 (M)	1	4-hour workshops in weeks 2,4,6,8,10 and 12 with online facilitation in weeks 1,3,5,7,9 and 11.	4 hours/week
BIM Virtual Fundementals	10 (M)	1	4-hour workshops in weeks 1,3,5,7,9 and 11 with online facilitation in weeks 2,4,6,8,10 and 12.	4 hours/week
BIM Collaboration	10 (E)	1	4-hour workshops in weeks 1,3,5,7,9 and 11 with online facilitation in weeks 2,4,6,8,10 and 12.	4 hours/week
BIM Elective Choice	30 (M)	1	6-week 'Research Design and Methodologies' online unit with an online and face-to-face supervision period of 20 weeks and 3 optional face-to-face recorded workshops	Average of 12 hours/ week over total period

The DELTA Framework provided a structure to prioritise activities in relation to strategy and capacity building including leadership and professional development of staff; teaching and learning practice including blended/online approaches and supporting students in transition; design of learning including curriculum design, collaboration and the learning environment; assessment of/as/for learning; and evidence-based approaches including evaluation, scholarship of teaching and research-informed teaching. In March 2018, the GMIT BIM group (BIGG) were awarded the inaugural engineering DELTA Award, which further reinforced the validity of the academic-industry partnership between GMIT and RPS. During this time, the GMIT/RPS collaboration also won several industry awards[28] as well as numerous other short listings. This raised the profile of the collaboration and encouraged an even deeper engagement, which resulted in the preparation of a research funding application under the National Forum's Teaching and Learning Enhancement Fund 2018. The BIM Futures project was awarded funding and commenced in March 2019. The project aims to design, develop, and pilot a professional development capacity framework for staff, students and industry stakeholders involved in BIM-related practice. The work is ongoing but is currently primarily focusing on providing learning resources for 'non-BIM' academic staff through a BIM Educator digital badge; for students through a BIM Leader digital badge; and for industry 'BIM starters' through a BIM Transitions digital badge. This capacity-building approach aims to embed BIM and digital literacy as a core pedagogical methodology across all programmes in the Department and build on existing regional networks and academic-industry strategic alliances to expand an evolving community of practice and establish GMIT as an international leader in BIM-related teaching and research. It will also inform an evolving national BIM educational framework to provide strategic leadership across the higher education sector by harnessing and building on the synergies between institute policy, professional development provision and curriculum innovation.

This work informed the next phase, which involved the transition of the Level 8 Higher Diploma in Engineering in BIM into a blended format[29] to further expand its reach nationally[30] (Table 13).

The rationale for this move towards a more blended educational approach was threefold:

- To directly respond to a clearly defined industry need for more flexible upskilling and training opportunities in BIM. This also aligned with the Institute's mission as outlined in the Strategic Plan at the time[31], which stated that '*At GMIT, we develop lifelong learning opportunities through our teaching and research; by supporting regional development consistent with national higher education policy.*'
- To directly respond to international and national policy drivers i.e. 2016 UK BIM Mandate and the 2018 'Roadmap to Digital Transition for Ireland's Construction Industry 2018-2021' (National BIM Council, 2018), especially the National BIM Council's call for 'the delivery of a broad awareness and upskilling learning framework for both educators and industry…based on core multi-disciplinary BIM curriculum outlining the skills needed to deliver a 21st century workforce for the built environment sector in Ireland.'
- To further establish GMIT as a national and international leader in the provision of BIM and digital construction education.

The programme team utilized the ADDIE (analysis, design, development, implementation, and evaluation) framework (Dick, Carey, and Carey 2005) and ABC Learning Design (Young and Perović, 2018) model to redesign the programme, which commenced in March 2019. The academic-industry collaboration continues to evolve over the past 6 months with RPS receiving the new international BIM standard ISO 19650 certification in recognition of their best practice in BIM. RPS also launched an internal BIM resource hub for their global offices, which includes a suite of RPS BIM standards, project information standards, best practice case studies, newsletters, and an Autodesk learning platform. To further embed industry expertise into the academic environment, Mark Costello (co-author) was appointed an Adjunct Lecturer in the Department and was also recently appointed as Board Director with CITA.

FUTURE RESEARCH DIRECTIONS

Research in the UK (Underwood and Ayoade, 2015) identified low levels of BIM awareness across built environment-related academic disciplines and more worryingly, a low level of interest in incorporating BIM holistically across teaching practice. This clearly highlights the inertia surrounding the implementation of BIM as a pedagogical methodology in higher education, albeit prior to the enactment of the BIM Level 2 mandate in April 2016. More recently, Simpson et al. (2019) has highlighted the considerable lack of an educational ecosystem that would facilitate curriculum to be more competency-led and responsive to industry needs. The concept of conscious competency evolution in relation to the upskilling of academic staff provides an interesting framework from which to consider future research directions in BIM (Simpson et al., 2019). There is a clear need for higher education to tackle the issue of graduate BIM competency and the need to provide learning opportunities for students to demonstrate the application of acquired knowledge and skills in a real-world environment. Higher education institutions continue to work in disciplinary silos, even within Departments, which do not address the increasing need for interdisciplinary working and the development of leadership capabilities (Cook and Chatterjee, 2015). The lack of a holistic and collaborative approach within departments and across higher education institutes does a disservice to both students and industry as it does not reflect a working environment where students experience what the industry should be moving towards. One significant barrier to this is the lack of research on upskilling, re-skilling or unlearning for academic staff who may be BIM-aware but have no in-depth knowledge of how BIM works in practice. A competency framework needs to be developed for academic staff to enable assess their BIM capability and identify training gap and CPD opportunities. The next phase of the BIM Futures project outlined previously will be to design and pilot a competency framework for academic staff in GMIT and link it to a range of evolving CPD opportunities i.e., BIM Educator Digital Badge, responsive learning units, BRE BIM Approved Graduate Programme etc. Similarly, there is a significant gap in the provision of responsive CPD and training opportunities for industry stakeholders. A clear need exists for step-by-step guidance and training on how to move from BIM Level 0 to BIM Level 1 and 2 for companies in Ireland, especially SMEs. Again, the BIM Futures project will attempt to address some of these gaps by designing and piloting a BIM Transitions Digital Badge and responsive learning units i.e., preparing an EIR, for industry. There is an opportunity to transform and rethink built environment education to move away from stagnant disciplinary silos to holistic multi and transdisciplinary learning environments, where academic staff, students and industry stakeholders can work in synergy within a reciprocal learning framework to embrace authentic collaboration, ever-evolving digital tools and applications and the complexity and challenges of embedding sustainability and lifecycle thinking as integral elements of good design, construction, operation and end-of-life best practice.

CONCLUSION

This chapter has chronologically outlined an academic-industry collaboration between the Department of Building and Civil Engineering GMIT and RPS, which has resulted in the design, piloting and delivery of a highly successful Level 8 Higher Diploma in Engineering in BIM; the establishment of a core research group (BIGG), which is leading a national project on creating professional development capacity opportunities for academic staff, industry stakeholders and students; the development and implementation of strategic plans in both an industry and academic context, which has led to international and national awards and accreditations, and most importantly, has helped create a unique learning environment for students where industry best practice and evidence-based research continues to inform curriculum development and pedagogy. The paradigm shift required to transition towards a fully enabled and collaboration BIM working environment presents many challenges, but also some exciting opportunities to rethink how built environmental educational provision can be more responsive and focused while also encouraging a strategic lifelong learning culture amongst graduates, industry professionals and academic staff. The higher education sector in Ireland has responded well to the BIM transition, albeit still within the limitations of disciplinary boundaries in the main. There remains a clear opportunity for all higher education institutions in Ireland (and the UK) to come together to develop a national BIM educational framework that will inform a holistic embedded approach across all disciplines. The higher education sector needs to be responsive to industry's immediate needs but should also aim to lead industry by ensuring graduates have the necessary competences to work in and facilitate a collaborative digital working environment. but should also aim to lead industry by demonstrating what a collaborative digital working environment would look like. GMIT, in close collaboration with RPS will continue this journey by expanding their network to include other innovative partners who share the same vision of a truly collaborative, efficient, and transparent construction sector that will position Ireland as a global leader.

ACKNOWLEDGMENT

This research was supported by the National Forum for the Enhancement of Teaching and Learning in Higher Education [grant number TL2018GMIT20].

REFERENCES

Aimi, S. (2019). Enhancing the graduates' employability and career development through Building Information Modelling intensive training. *International Journal of Built Environment and Sustainability*, *6*(1-2), 91–99.

Badrinath, A.C., Chang, Y., & Hsieh, S. (2016) A review of tertiary BIM education for advanced engineering communication with visualization. *Visualization in Engineering, 4*(9).

Barison, M.B., & Santos, E.T. (2011). The Competencies of BIM Specialists: A Comparative Analysis of the Literature Review and Job Ad Descriptions. *Computing in Civil Engineering, ASCE 2011.*

Behan, A. (2013). Update on the BIM education of Geomatic Surveyors. *Proceedings of the CITA BIM Gathering Conference*, 163-168.

Bercerik-Gerber, B., Gerber, D. J., & Ku, K. (2011). The pace of technological innovation in architecture, engineering, and construction education: Integrating recent trends into the curricula. *Journal of Information Technology in Construction.*, *V16*, 411–432.

Berwald, S. (2008). From CAD to BIM: The Experience of Architectural Education with Building Information Modelling. *Proceedings of the American Society of Civil Engineers.* 10.1061/41002(328)8

Biesta, M., & Priestley, G. (2013). *Reinventing the Curriculum: New Trends in Curriculum Policy and Practice*. Bloomsbury Academic.

BIM Academic Forum (BAF). (2015). *BIM Learning Outcomes Framework*. Author.

BIM Academic Forum (BAF) UK. (2013). Embedding Building Information Modelling (BIM) within the taught curriculum. York, UK: The Higher Education Academy (HEA).

Bosch-Sijtsema, P. M., Gluch, P., & Sezer, A. A. (2019). Professional Development of the BIM Actor Role. *Automation in Construction*, *V97*, 44–51. doi:10.1016/j.autcon.2018.10.024

Boton, C., Forgues, D., & Halin, G. (2018). *A framework for Building Information Modelling Implementation in Engineering Education. Canadian Journal of Civil Engineering.*

Bozoglu, J. (2016). Collaboration and coordination learning modules for BIM education. *Journal of Information Technology in Construction.*

Burr, K. L. (2009). Creative course design: A study in student-centred course development for a sustainable building/BIM class. In *Proceedings of the 45th Annual Conference by Associated Schools of Construction.* University of Florida.

Casey, M. J. (2008). *Work in Progress: How Building Information Modelling may unify the Civil Engineering Curriculum.* Paper presented to the 38th ASEE/IEEE Frontiers in Education Conference, Saratoga Springs, NY. 10.1109/FIE.2008.4720644

Chartered Institute of Building. (2018). *Undergraduate Education Framework.* Author.

Chen, Y., Dib, H., & Cox, R. F. (2014). A measurement model of building information modelling maturity. *Construction Innovation*, *V14*(2), 186–209. doi:10.1108/CI-11-2012-0060

Comiskey, D., Alexander, G., Eadie, R., Hamill, M., McKane, M., & Weatherup, R. (2013). BIM Education - A Multidisciplinary Analysis. *Proceedings of the CITA BIM Gathering Conference.*

Comiskey, D., McKane, M., Eadie, R., & Goldberg, D. (2015). Providing Collaborative Education with an International Dimension. An Ulster University and Pennsylvania State University Case Study. *Proceedings of the CITA BIM Gathering Conference.*

Comiskey, D., McKane, M., O'Shea, E., Hughes, J., McNiff, S. and Eadie, R. (2016), Collaborative and multidiscipline working - From theory to practice in 48 hours: A case study from BIM Region Northern Ireland. *International Journal of 3-D Information Modelling, 5*(2), 55-71.

Cook, D., & Chatterjee, P. (2015). *RICS Futures: Our Changing World.* RICS.

Deniz, G. O. (2018). Emerging CAD and BIM trends in AEC education: An analysis from students' perspective. *Journal of Information Technology in Construction*, *V23*, 138–156.

Denzer, A., & Hedges, K. (2008). From CAD to BIM: Educational Strategies for the Coming Paradigm Shift. *AEI*, *2008*, 1–11. doi:10.1061/41002(328)6

Dick, W., Carey, L., & Carey, J. O. (2005). *The systematic design of instruction* (6th ed.). Allyn and Bacon.

Engineers Ireland. (2014c). *Chartered Engineer – 2014 Regulations for the Registered Professional Title.* Author.

Evans, N., & Counsell, J. (2009). Web-mediated student peer group assessment of Building Information Modelling performance. Paper presented at VIZ09 – Visualisation in Built and Rural Environments, Barcelona, Spain. doi:10.1109/VIZ.2009.34

Hedges, K. E., & Denzer, A. S. (2008). *How Collaborative Architecture influences Structural Engineering Education*. Paper presented to the Structures Congress 2008, Vancouver, Canada. 10.1061/41016(314)147

Hjelseth, E. (2017). BIM understanding and activities. *WIT Transactions on the Built Environment*, 3–14. doi:10.2495/BIM170011

Hore. (2017a). *BICP Global BIM Study – Lessons for Ireland's BIM Programme*. Academic Press.

Hore. (2017b). *BIM in Ireland 2017 report*. Academic Press.

Huang, Y. (2018). Developing a Modular Advanced BIM Course in Construction Management. *Journal of Building Construction and Planning Research*, *6*(4), 198–214. doi:10.4236/jbcpr.2018.64014

Engineers Ireland. (2014a). *Accreditation Criteria for Professional Titles*. Author.

Engineers Ireland. (2014b). *Guidelines for Chartered Engineer*. Author.

Irish, BIM. (2017). *Council*. Roadmap to Digital Transition for Ireland's Construction Industry.

Irish Government. (2014). *Construction Strategy 2020*. Author.

Irish Government. (2018). *Project Ireland 2040 National Development Plan 2018-2027*. Author.

Joannides, M. M., Olbina, S., & Issa, R. R. A. (2012). Implementation of Building Information Modelling into Accredited Programmes in Architecture and Construction Education. *International Journal of Construction Education and Research*, *8*(2), 83–100. doi:10.1080/15578771.2011.632809

Kim, J. (2012). Use of BIM for effective visualisation teaching approach in construction education. *Journal of Professional Issues in Engineering Education and Practice*, *138*(3), 214–223. doi:10.1061/(ASCE)EI.1943-5541.0000102

Kinnane, O., & West, R. (2013). BIM introduction into the curriculum of Civil and Structural Engineering students: A project-based active learning approach. *Proceedings of the CITA BIM Gathering Conference*, 175-184.

London, K. (2015). *Global Passport through Co-Integration of Construction Immersive Environments*. RMIT Learning and Teaching Investment Fund Final Report.

McDonnell, P., & West, R. P. (2015). The adoption of Building Information Modelling (BIM) to improve existing teaching methods and support services within a Higher Education Institution in Ireland. *Proceedings of the CITA BIM Gathering Conference*, 271-278.

McGovern, E., & Behan, A. (2014). *Geomatics and developments in BIM education in Ireland*. Paper presented at the FIG Congress 2014 'Engaging the Challenges - Enhancing the relevance, Kuala Lumpar, Malaysia.

McGovern, E., Behan, V., & Furlong, K. (2014). *Geomatics and Developments in BIM Education in Ireland*. FIG Congress 2014, Engaging the Challenges - Enhancing the Relevance, Kuala Lumpur, Malaysia.

Meng-Han, T., Chen, K.-L., & Chang, Y.-L. (2019). Development of a Project-Based Online Course for BIM Learning. *Sustainability*, *11*(5772).

Miller, G., Sharma, S., Donald, C., & Amor, R. (2013). Developing a Building Information Modelling Educational Framework for the Tertiary Sector in New Zealand. In A. Bernard, L. Rivest, & D. Dutta (Eds.), *Product Lifecycle Management for Society. Springer*. doi:10.1007/978-3-642-41501-2_60

Mills, J., Tran, A., Parks, A., Kelly, S., Smith, E., & Jupp, J. (2013). *Collaborative building design education using Building Information Modelling (CodeBIM), Final Report 2013*. Australian Government Office for Learning and Teaching.

Murphy, M. E. (2014). Implementing innovation: A stakeholder competency-based approach for BIM. *Construction Innovation*, *14*(4), 433–452. doi:10.1108/CI-01-2014-0011

National, BIM Council. (2017). *Roadmap to Digital Transition for Ireland's Construction Industry*. Author.

National Forum for the Enhancement of Teaching and Learning in Higher Education. (2020) *DELTA Framework*. Author.

NBS, National BIM Report 2019. (2020). *National Skills Council/Expert Group on Future Skills Needs*. NBS.

Nicholson, G. (2013). Investigating the Application of BIM on Small-Scale Construction Projects. *Proceedings of the CITA BIM Gathering Conference*, 49-60.

Nývlt, V. (2018). The role of managing knowledge and information in BIM implementation processes in the Czech Republic. *9th International Scientific Conference Building Defects (Building Defects 2017)*.

Perović, N., & Young, C. (2016). Rapid and Creative Course Design: As Easy as ABC? *Procedia: Social and Behavioral Sciences*, *228*, 390–395. doi:10.1016/j.sbspro.2016.07.058

Pikas, E., Sacks, R., & Hazzan, O. (2013). Building Information Modelling Education for Construction Engineering and Management. *Journal of Construction Engineering and Management*, *139*(11). Advance online publication. doi:10.1061/(ASCE)CO.1943-7862.0000765

Puolitaival, T., & Forsythe, P. (2016). Practical challenges of BIM education. *Structural Survey*, *34*(4/5), 351–366.

Quality and Qualifications Ireland (QQI). (2016). *Awards Standards for Architectural Technology*. Author.

Rahman, R., Suleiman, A., Pingbo, T., & Ayer, S. K. (2016). Comparing Building Information Modelling Skills of Project Managers and BIM Managers based on Social Media Analysis, International Conference on Sustainable Design, Engineering and Construction. *Procedia Engineering*, *V145*, 812–819. doi:10.1016/j.proeng.2016.04.106

Robertson, F.J. (2017). Tri-varsity, interdisciplinary BIM Modelling/Management (BIM) workshop: An action research international example. *International Journal of 3-D Information Modelling*, *6*(4), 37-56.

Royal Institute of Architects Ireland (RIAI). (2016). *Statement on Policy on Education*. Author.

Rusinko, C. (2010). Integrating sustainability in higher education: A generic matrix. *International Journal of Sustainability in Higher Education, 11*(3), 250–259. doi:10.1108/14676371011058541

Sacks, R., & Barak, R. (2010). Teaching Building Information Modelling as an integral part of freshmen year civil engineering education. *Journal of Professional Issues in Engineering Education and Practice, 136*(1), 30–38. doi:10.1061/(ASCE)EI.1943-5541.0000003

Sacks, R., & Pikas, E. (2013). Building Information Modelling Education for Construction Engineering and Management 1: Industry Requirements, State-of-the-Art and Gap Analysis. *Journal of Construction Engineering and Management, 139*(11), 04013016. doi:10.1061/(ASCE)CO.1943-7862.0000759

Sánchez, A., González-Gaya, C., Zulueta, P., & Sampaio, Z. (2019). *A Comparative Study of the Use of Building Information Modelling in Teaching Engineering Projects. IEEE Access Journal.*

Santos, R., Costa, A. A., & Grilo, A. (2017). Bibliometric analysis and review of Building Information Modelling literature published between 2005 and 2015. *Automation in Construction, 80*, 118–136. doi:10.1016/j.autcon.2017.03.005

Scottish Futures Trust. (2018). *Developing a BIM Competency Framework: Research and Key Principles report.* Author.

Sebastian, R. (2011). Changing roles of the clients, architects and contractors through BIM. *Engineering, Construction and Architectural Management, 18*(2), 176-187.

Simpson, M., Underwood, J., Shelbourn, M., & Mollasalehi, S. (2019). *Evolve or Die: Transforming the productivity of Built Environment Professionals and Organisations of Digital Built Britain through a new digitally enabled ecosystem underpinned by the mediation between competence supply and demand.* Pedagogy and Upskilling CDBB Network Report.

Society of Chartered Surveyors Ireland (SCSI). (n.d.). *APC Pathway Guides to Quantity Surveying and Construction, Building Surveying, Building Control and Project Management.* Author.

Soemardi, B., & Erwin, R. G. (2017). Using BIM as a Tool to Teach Construction Safety. *MATEC Web of Conferences, 138*(4), 05007. doi:10.1051/matecconf/201713805007

Succar, B. (2010a). Building Information Modelling Maturity Matrix. In J. Underwood & U. Iskidag (Eds.), *Handbook of Research on Building Information Modelling and Construction Informatics: Concepts and Technologies, 65-103, Information Science Reference.* IGI Publishing. doi:10.4018/978-1-60566-928-1.ch004

Succar, B. (2010b). *The Five Components of BIM Performance Measurement.* Paper presented at the CIB World Congress, Salford, UK.

Succar, B. (2014). *Building information modelling: conceptual constructs and performance improvement tools* (Ph.D. thesis). University of Newcastle.

Succar, B., Agar, C., Beazley, S., Berkemenier, P., Choy, R., Rosetta, D. G., Donaghey, S., Linning, C., Macdonald, J., Perey, R., & Plume, J. (2012). *BIM in Practice – BIM Education.* Paper by the Australian Institute of Architects and Consult Australia.

Succar, B., & Sher, W. (2014). A Competency Knowledge Base for BIM Learning. *Australian Journal of Construction Economics and Building Conference Series*, *V2*(2), 1–10. doi:10.5130/ajceb-cs.v2i2.3883

Succar, B., Sher, W., & Aranda-Mena, G. (2007). *A proposal Framework to Investigate Building Information Modelling through Knowledge Elicitation and Visual Models*. Paper presented at the Australian Universities Building Education (AUBEA 2007), Melbourne, Australia.

Succar, B., Sher, W., & Williams, A. (2013). An integrated approach to BIM competency assessment, acquisition and application. *Automation in Construction*, *35*, 174–189. doi:10.1016/j.autcon.2013.05.016

Thomas. (2014) Collaborative BIM Learning via an Academia-Industry Partnership. *International Journal of 3-D Information Modelling (IJ3DIM), 3*(1), 40-48.

Thomas, K., Chisholm, G., Dempsey, B., Graham, B., & Stubbs, R. (2013). Collaborative BIM learning via an academic-industry partnership. *Proceedings of the CITA BIM Gathering Conference*, 201-206.

Uhm, M., Lee, G., & Jeon, B. (2017). An analysis of BIM jobs and competence based on the use of terms in the industry. *Automation in Construction*, *81*, 67–98. doi:10.1016/j.autcon.2017.06.002

Underwood, J., & Ayoade, O. (2015). *Current Position and Associated Challenges of BIM Education in UK Higher Education*. Prepared for the BIM Academic Forum UK.

Weber, D., & Hedges, K. (2008). *From CAD to BIM: The Engineering Student Perspective. Proceedings from the American Society of Engineers*.

Wong, K., Wong, F. K., & Nadeem, A. (2011). Building information modelling for tertiary construction education in Hong Kong. *Journal of Information Technology in Construction, 16*, 467–476.

World Economic Forum. (2016). *Shaping the Future of Construction: A Breakthrough in Mindset and Technology*. Prepared in collaboration with the Boston Consulting Group.

World Economic Forum. (2017). *Shaping the Future of Construction: Inspiring innovators to redefine the industry*. Prepared in collaboration with the Boston Consulting Group.

World Economic Forum. (2018). *An Action Plan for solve the Industry's Talent Gap*. Prepared in collaboration with the Boston Consulting Group.

Yalcinkaya, M., & Singh, V. (2015). Patterns and trends in Building Information Modelling (BIM) research: A Latent Semantic Analysis. *Automation in Construction*, *V59*, 68–80. doi:10.1016/j.autcon.2015.07.012

Young, M., & Muller, J. (2010). Three Educational Scenarios for the Future: Lessons from the sociology of knowledge. *European Journal of Education*, *45*(1), 11–27. doi:10.1111/j.1465-3435.2009.01413.x

Zhao, X. (2017). A scientometric review of global BIM research: Analysis and visualisation. *Automation in Construction*, *80*, 37–47. doi:10.1016/j.autcon.2017.04.002

KEY TERMS AND DEFINITIONS

BIM Competences: Relates to concepts of knowledge, skills, abilities, behaviours and successful performance (Scottish Futures Trust, 2018), which can be interpretated within different contexts i.e., as an individual, team, project, role, task, process, practice, etc.

BIM Education: In the context of this chapter is defined as any module or programme that has BIM-related learning outcomes, curriculum content and/or associated assessment activities.

BIM Learning Outcomes: Can be defined as a statement of what the learner is expected to be able to do on successful completion of a BIM-related module or programme, to demonstrate their knowledge, understandings, skills and/or competences.

Building Information Management: Is a process for creating and managing information on a construction project across the whole lifecycle (NBS, 2016).

Building Information Model: Is a digital description of every aspect of the built asset drawing on information assembled collaboratively and updated at key stages of a project (NBS, 2016).

ENDNOTES

1 Now amalgamated with IT Tallaght and IT Blanchardstown to form the Technological University of Dublin as of January 2019.

2 Membership is open to all stakeholders in the Irish construction sector, who wish to advance the use of ICT to improve working practices.

3 Since 2008, CITA has facilitated an average of 300 trainees per year through the Skillnet programme.

4 Enterprise Ireland offer's BIM-Enable, which is a 7-day strategic consultancy programme to assist companies to set out a digital roadmap and consider how they can apply BIM within their organization and BIM-Implement, which focuses on training and the learning and implementation of BIM across the organization.

5 Referred to as the 'Department' from now on.

6 RPS is a leading international multidisciplinary consultancy with over 5,000 employees located in the UK, the Netherlands, USA, Canada, Brazil, Africa, the Middle East, Australia, and South East Asia. It is the largest infrastructural, built environment and planning consultancy in Ireland

7 This percentage is likely to be much higher now in the eight years since.

8 The questionnaire was sent electronically to academic and professional networks across the UK. Respondents were divided into three major groups: those with academic, industry and both academic and industry experience in BIM.

9 In Ulster University and Pennsylvania State University.

10 Bloom's Taxonomy in its original form consisted of six main categories: Knowledge, Comprehension, Application, Analysis, Synthesis, and Evaluation. Knowledge was understood to be the pre-requisite for all other categories, which were presented as skills and abilities. In 2001, a revised taxonomy was proposed replacing the original nouns with verbs, which described the cognitive processes by which persons engage and work with knowledge i.e., remember, understand, apply, analyse, evaluate, and create with the knowledge still the basis of these cognitive processes.

11 Rusinko's (2010) matrix was initially proposed for embedding sustainability across the curriculum.

12 The more detailed UKBAF Learning Outcomes Framework is presented in Table 10.

13 This core mission was further reinforced by the subsequent Regional Technical Colleges Act of 1992 and the Institute of Technologies Act of 2006.

14 The Springboard initiative offers free courses at undergraduate and postgraduate levels for persons seeking employment. It is co-funded by the Irish Government and the European Social Fund as part of the ESF Programme for Employability, Inclusion and Learning 2014-2020.

15 The number of RPS staff who undertook the following modules are BIM Architecture (13); BIM Infrastructure (41); BIM Collaboration (22); BIM Research Project (22); BIM MEP (3); BIM Virtual Modelling Fundamentals (74).

16 Modules were delivered in face-to-face evening workshops supported with resources hosted on the Virtual Learning Environment (VLE), Moodle.

17 The development of the Level 8 Higher Diploma in Engineering in BIM identified these key issues as part of the extensive pre-pilot review and subsequent piloted outreach initiative, where several participants who completed the course had a previous Level 7 qualification. By offering the Higher Diploma as a Level 8 qualification, it covers the widest range of construction workers who may have previous Level 7 and 8 qualifications.

18 Primarily led by large main contractors, who began to invest heavily in BIM in 2015/2016.

19 Focusing on the following four undergraduate programmes on offer in the Department: Construction Management, Architectural Technology, Construction Economics and Quantity Surveying and Civil Engineering Level 8 programmes.

20 All undergraduate programmes referred to are Level 8.

21 All modules are mandatory unless otherwise stated.

22 This also introduces the BIM process.

23 It should be noted that the BAF learning outcomes framework does not include the specific technical aspects of BIM i.e., use of software and related tools, which represent the BIM modules present in the department. This may be one of the reasons that concerns have been raised about the currency of this framework due to the lack of its comprehensive adoption (despite widespread awareness) across the higher education sector (Scottish Futures Trust, 2018).

24 This was also embedded into BIM Collaboration module on the Higher Diploma in Engineering in BIM by two co-authors, Mark Costello (RPS) and Gerard Nicholson (GMIT).

25 The GMIT is licensed by the BRE to deliver this programme.

26 Students needed to achieve a total grade of 70% to be put on the individual certification route.

27 Henceforth referred to as the 'National Forum'.

28 2017 Irish Education 'Business Collaboration' award; 2017 Irish Construction Industry 'Excellence in Training and Education' award; 2018 Irish Construction Excellence 'Third-Level Course – Postgraduate, Special Purpose or Distance Learning' award and the 2018 HR Leadership and Management 'Excellence in Education and Training'

29 Consisting of online delivery (75%) and face-to-face workshops (25%).

30 This was supported by Springboard funding, which was awarded in 2018.

31 GMIT Strategic Plan 2013-2016.

Chapter 13
BIM Education in the Dominican Republic:
A Framework to Guide the Implementation of BIM Into University Curricula

Ana Karina Silverio
University of Wolverhampton, UK

Subashini Suresh
University of Wolverhmpton, UK

David Heesom
https://orcid.org/0000-0003-1762-8681
University of Wolverhampton, UK

Renukappa Suresh
University of Wolvehampton, UK

ABSTRACT

Lack of BIM skilled professionals and lack of education and BIM training are key challenges to BIM implementation. The provision of BIM education in higher education institutions is seen as the main solution to these challenges. This chapter is part of an on-going Ph.D. research about the implementation of BIM in the Dominican Republic, a country with interest in implementing BIM but suffering from many challenges including lack of BIM education. This study aims to present the development of a framework to guide the integration of BIM in university curricula in the D.R., which is part of a comprehensive framework, outcome of the Ph.D. research. The framework was developed from a thorough literature on BIM education as well as an analysis of existing BIM Education frameworks. Moreover, findings of the research about BIM education in the D.R. are presented to describe the presence of BIM education in the country.

DOI: 10.4018/978-1-7998-6600-8.ch013

INTRODUCTION

The Dominican Republic (D.R.) is a Caribbean country whose construction industry is one of the sectors that more contributes to the economy's growth in the country (Soler et al., 2013). The Dominican construction sector has grown significantly over the last years thanks to the low rate loans and mortgages, and the investment from the public and private sector (Banco Central de la Republica Dominicana, 2015, 2016, 2017, 2018, 2019). Despite this continuous growth, the Dominican construction industry is lacking, in many cases, qualified workforce and innovative and efficient construction techniques (Soler et al., 2013). Other barriers affecting the sector include issues related to the quality of the materials, documenting agreements and procedures, understanding the necessities of internal and external clients, well-defined team focus, and communication between teams (Senior & Rodriguez, 2012). Recommendations to improve this sector include the use of technologies to enhance communication among parties, facilitate the management of the construction process, and implement sustainable principles. The introduction of these technologies involves a process of change and adaptation of the different methods that take part in the construction of buildings, including the knowledge, abilities, attitudes, and activities of the personnel involved (Soler et al., 2013). In the presence of these needs and due to advent and proven benefits of BIM implementation worldwide, BIM comes into view as an appropriate method to improve current practices and open the way for the digitisation of the Dominican construction industry. Currently, the implementation of BIM in the Dominican construction industry is undocumented, and there are fundamental questions that need to be addressed to understand how the implementation of BIM can be realised, including the delivery of BIM education.

BIM education, training and research are vital to propel and motivate the implementation of BIM (Smith, 2014). Sharag-Eldin & Nawari (2010) further considered that education and training are the basis of BIM evolution. To keep the impetus of BIM, it is necessary to effectively prepare the industry's labour force to respond to current supply-demand (Bozoglu, 2016). Nonetheless, lack of BIM skilled professionals (Eadie, Browne, Odeyinka, McKeown, & McNiff, 2013; Lee & Hollar, 2013; Wu & Issa, 2013; Gardner, Hosseini, Rameezdeen, & Chileshe, 2014) and lack of education and training are one of the most important barriers hindering the implementation of BIM in the industry (Wong & Gray, 2019).

This chapter aims to present the status of BIM education in the Dominican Republic and the development of a BIM education framework to guide the integration of BIM into university curricula in this country. The work presented is a part of a wider researcher entitled "Implementation of Building Information Modelling in the Dominican Republic construction industry".

This chapter first introduces the topic of BIM education and its importance for the implementation of BIM in the construction industry. Then, several topics to understand the field of BIM education and aid in the development of the framework are reviewed from the literature. Challenges hindering the integration of BIM in academia as well as strategies to overcome these challenges and enhance this integration are presented. The collaboration of academia and industry players is highlighted as crucial for the integration of BIM in academia and several strategies are shown in this regard. Moreover, a thorough analysis of BIM education drivers is presented. Furthermore, key findings of the presence of BIM education in the D.R. are presented followed by the study of several BIM education frameworks, which informed the development of the framework. Lastly, the development of the framework is thoroughly explained, and solutions, recommendations and future research directions are posited.

BIM EDUCATION

BIM education can be defined as "*a continuous learning process that covers the knowledge required for individuals to be capable of being part and understand what a BIM process is. Such learning process needs to include the essential aspects of BIM implementation, the performance of the participants and the technical skills individuals need to acquire depending on the discipline they belong to and their role in the construction team*" (Silverio et al., 2016; 2017a, b). BIM education is seen as the solution to speed the BIM learning curve; thus, future graduates would be BIM-ready when they start their professional careers (Wu & Issa, 2013).

There are several stakeholders involved in the delivery of BIM education: Academia, construction firms, software companies, BIM training centres, professional associations, BIM learners, etc. (AIA-CA, 2012; CIC, 2013; Rooney, 2019). Academia is par excellence the main provider of BIM education; whereas the rest provide BIM training, which is a complementary part of BIM education (Silverio et al., 2017a, b).

It is widely believed that the primary education required to overcome the lack of BIM skilled needs to be cultivated in higher education institutions (Miller, Sharma, Donald, & Amor, 2013; Hardin & McCool, 2015). Efforts to integrate BIM in educational institutions are new to a certain extent (Abbas et al., 2016). Many higher education institutions around the world are incorporating BIM education in their AEC programmes in undergraduate and postgraduate level. USA, Chile, Finland, Sweden, Netherlands, UK, Singapore, and Australia are creating many courses and subjects. Also, vocational institutions are delivering BIM education to the professionals of the industry (Rooney, 2019).

Nonetheless, many institutions have been criticised for their poor approaches and because of the quality of many academic programmes has not met the industry and students expectations (Wu & Issa, 2013). The high demand for BIM skilled professionals, and other aspects of BIM such as the increasing implementation of BIM, its constant evolution of BIM and the fast development of BIM processes and technologies and present challenges to develop and plan BIM educational programmes (Suwal & Singh, 2018). AEC educational institutions need to take advantage of the opportunities BIM offer and overcome the challenges related to BIM to continue being relevant and up-to-date (Bozoglu, 2016).

Challenges of integrating BIM in academia

Research has documented numerous challenges that hinder the integration of BIM in academia, which include:

- **Curriculum development.** Lack of space (Sabongi, 2009) and time (Witt & Kähkönen, 2019) in the curriculum to integrate BIM-related education; and uncertainty on how to adapt new subjects into an already full curriculum (Macdonald, 2012). Lack of support from faculty administrators and colleagues to modify the curriculum is also present (Sabongi, 2009; Becerik-Gerber, Gerber, & Ku, 2011). The insertion of BIM is usually carried out in simple, intensive, and isolated courses (Lee & Hollar, 2013; Magiera, 2013). Uncertainty on whether BIM should be included in current curricula of AEC disciplines or be provided as a subject on its own (Miller, Sharma, Donald, & Amor, 2013). Incapacity to connect the traditional silos that are present in the architecture, engineering and construction schools, where there is limited to no collaboration between disciplines (Macdonald, 2012). Balancing theory and practice; technology and processes; and traditional and new methods has represented an important hurdle (Puolitaival & Forsythe, 2016). There is a lack of standardisation. There is a discrepancy in courses at all education levels (BIM Plus, 2018).

- **Accreditation.** There are different standards for graduates from different areas concerning inter-disciplinary collaboration skills (Witt & Kähkönen, 2019). Moreover, the insertion of BIM can affect the accreditation status of the curriculum (Kocaturk & Kiviniemi, 2013; Suwal, Jäväjä, Rahman, & Gonzalez, 2013).
- **Academics' preparation for BIM.** The insertion of BIM in academia has represented a big challenge for academics. Lack of knowledge and experience of academics in BIM has been reported as a significant barrier by many researchers (Becerik-Gerber et al., 2011; Gardner et al., 2014; Kugbeadjor, Suresh, & Renukappa, 2015; Puolitaival & Forsythe, 2016; Rooney, 2019) and it has been difficult for academia to enable professional development for academics in this field (Puolitaival & Forsythe, 2016). There is also concern on how the current personnel will adapt to the new skills and understanding demanded by BIM (Kocaturk & Kiviniemi 2013). Moreover, academics have the pressure of being skilled and stay up to date with current BIM software (Puolitaival & Forsythe, 2016; Witt & Kähkönen, 2019) and processes as BIM is constantly evolving (Puolitaival & Forsythe, 2016). The time responsibilities of faculty staff to get knowledge of BIM software is also significant (Witt & Kähkönen, 2019).
- **Teaching resources.** Shortage of books and other training materials was a challenge identified by several authors (Sabongi, 2009; Becerik-Gerber et al., 2011; Magiera 2013; Gardner, Hosseini, Rameezdeen, & Chileshe, 2014;). Nonetheless, Puolitaival & Forsythe (2016) pointed out that this challenge has turned into a new one: the difficulty of finding suitable teaching and learning resources for BIM among a large amount of material available. The development of BIM models also represents a challenge. There is a lack of proper BIM models to be used in BIM Courses (Puolitaival & Forsythe, 2016). Lastly, classrooms may not be properly designed and equipped to undertake collaborative work (Witt & Kähkönen, 2019).
- **Investment.** BIM demands the use of state-of-the-art IT equipment and high-priced software, and it can be difficult for universities to get the correct IT environment (Hogle, 2013; Magiera, 2013). Like in the industry, there is mistrust among the professionals in academia and questions such as "who is responsible for" and "who will pay for" multidisciplinary courses can arise (Macdonald, 2012).
- **Cultural resistance.** There is cultural resistance from both, academics and the students.
    - **From academics.** Several academics still see BIM as a CAD software that students should learn by themselves in their own time (Becerik-Gerber, Jazizadeh, Li, & Calis, 2011). There is resistance to change traditional teaching methods which have been implemented over many years to incorporate BIM (Sabongi, 2009; Macdonald, 2012). There is also reluctance from experienced academics to adopt a new concept where they are not experienced or be trained again in a field that they are not familiar with (Macdonald, 2012). Academics might object to modify the curriculum to include BIM since there is the possibility that other technologies replace BIM in a few years (Sabongi, 2009). There is also uncertainty on whether a creative artistic expression can be achieved within a collaborative practice and if professionals can preserve and defend their values in this innovative, cooperative and democratic pluralism (Kocaturk & Kiviniemi, 2013).

  - **From the students.** Lack of interest or reluctance from students to scrutinise the new technology (Sabongi, 2009). Students' perception that the workload of BIM-related education is bigger than conventional curricula (Witt & Kähkönen, 2019).
- **Teaching methods.** BIM programmes focus mostly on software skills (Lee & Hollar, 2013; Rooney, 2019). BIM is usually treated as a design tool only (Wong, Wong, & Nadeem, 2011). Moreover, BIM needs a determined knowledge of business practice and workflow which is hard to simulate in a teaching course (Magiera, 2013). The use of real projects can be overwhelming for the students due to their technical complexity. Also, evaluating students in a BIM-enabled education is challenging because it is difficult to fairly motivate all students to do collaborative work (Witt & Kähkönen, 2019).
- **Technology aspects.** BIM technologies tend to evolve quickly, and academics that have been kept out of the industry for some time may feel overwhelmed when attempting to keep updated (Macdonald, 2012). This has also brought uncertainty on which BIM platform (e.g., Revit, Bentley, etc.) will become predominant (Sabongi, 2009) and what academics should teach (Witt & Kähkönen, 2019). There are also interoperability issues which can occur between different software packages and between different versions of the same software (Witt & Kähkönen, 2019).
- **Limited student's knowledge of construction.** As per Puolitaival & Forsythe (2016), the discipline-specific knowledge and understanding of construction systems and processes of undergraduate students are still limited to learn BIM. Furthermore, lack of knowledge on construction sequence leads to improper model development (Puolitaival & Forsythe, 2016).
- **Student's skills.** BIM skills vary significantly between individual students and students from different areas (Witt & Kähkönen, 2019).

Enhancing the integration of BIM in higher education

Faced with these challenges, researchers and BIM educationalist have proposed several strategies to overcome these challenges and improve how BIM education is currently delivered in higher education:

- **Curriculum development.** Full programs should be reevaluated and reformulated concentrating on the changes that BIM implies. Almost all the AEC disciplines need to change their current curricula to integrate BIM as one of the main subjects. A common framework should be developed to specify the requirements of which BIM topics could or should be integrated into university curricula and how (Suwal, Jäväjä, Rahman, & Gonzalez, 2013). This academic framework should be informed by BIM research, BIM professionals and other members of the industry (Bozoglu, 2016). Core curriculum development (Puolitaival & Forsythe, 2016), to properly define the core content of the curriculum and focus on fundamental aspects of a subject (Karjalainen, Alha, & Jutila, 2006). BIM education should be integrated at the program level rather than in an isolated course (Pikas, Sacks, & Hazzan, 2013). BIM education should also pay attention to both, managerial and technical skills (Jin et al., 2019). Active research on BIM curriculum development processes and emerging BIM trends worldwide to respond to industry's needs (Puolitaival & Forsythe, 2016) and prepare BIM-ready graduates worldwide (Badrinath et al., 2016). Moreover, it is necessary to define the skill sets and competencies required for BIM (BIM Plus, 2018) to increase the quality, consistency and relevance of the education/training (Bush & Robinson, 2018)

- **Accreditation.** International accreditation or, alternatively, national accreditation should be sought to incentive academic institutions to introduce BIM in their curricula (Suwal et al., 2013) and assure uniformity in the provision of BIM education (Badrinath, Chang, & Hsieh, 2016; BIM Plus, 2018)
- **Collaborative approaches.** Higher education institutions need to provide collaborative learning spaces to allow multidisciplinary/collaborative team working and equipped with ICT and BIM tools (Shelbourn, Macdonald, McCuen, & Lee, 2017). Collaborative practices with other disciplines should be developed so students understand the different roles in the industry and their implications in a project. This can be achieved through collaborative teaching and activities such as student competitions (Shelbourn et al., 2017). Inclusion of collaboration requirements and BIM technical skills within the new/modified BIM curricula to enhance the abilities of graduates from different disciplines of the industry (Gardner et al., 2014). Establishment of multidisciplinary schools and BIM educational institutions supported by the government and the industry to enable BIM learning across industry and academia (Badrinath et al., 2016).
- **Academics' preparation.** Self-training along with constant peer support and bespoke workshops as a cost-effective option to enable the professional development of academic staff (Puolitaival & Forsythe, 2016). Academics also need to receive training on new software technology (Badrinath et al., 2016).
- **Teaching methods.** Integration of practical implementation of subjects and problems reported in real situations after training is provided (Kumar, 2015). Project-based teamwork approaches can be integrated into BIM courses to reduce the differences in the perception towards BIM between students and professionals in the industry (Zou et al., 2019). Other educational approaches such as experiential learning and problem-based learning are also suggested (Wang et al., 2020). BIM should not be seen as a standalone digital technology but could be taken as part of the Internet of Things and Industry 4.0. to educate the next generation of AEC professionals. The most recent practices (e.g. drones) or research movements related to BIM (e.g. interoperability to improve information exchange) could be implemented in education to trigger practice-based and research-informed teaching (Wang et al, 2020).
- **Teaching resources.** To solve the problem of lack of proper BIM model, Puolitaival & Forsythe (2016) suggested two approaches. The first option is the use of models obtained from the industry following industry standards, which can be: (1) models used in real projects and (2) models created by contracting model authoring professionals or companies based on 2D documentation. The second option is the use of models developed internally by (1) staff (either full time or contract-based) from 2D documentation, and (2) by students, from both, 2D documentation and their design projects.
- **Students involvement.** Students should collaborate in the process of creating a BIM programme and developing a BIM curriculum (Wong et al., 2011). For instance, the perception towards BIM from students is very important in the development of BIM education. Academics can assess their pedagogical work through students' feedback and make appropriate changes if necessary (Jin et al., 2019). Moreover, motivate and advise students to look for resources on their own to solve the issue of selecting appropriate learning materials for BIM due to the number of materials and fast-changing nature of this topic (Puolitaival & Forsythe, 2016).
- **BIM diffusion.** Diffusion of BIM through workshops and short courses directed to the AEC community outside academia (Badrinath et al., 2016).

Collaboration Between Academia and Industry Players

Collaboration between academia and industry players is highly recommended to enhance the benefits of BIM implementation (Lee & Hollar, 2013), overcome the lack of BIM skilled professionals in the industry (AIA-CA, 2012; Hogle, 2013), and keep academics up-to-date (Macdonald & Mills, 2011). Academic staff teaching BIM-related programmes should keep in contact with industry players to integrate and redesign the content of the BIM programmes following industry tendencies and practices (Lee & Hollar, 2013) and bridge the gap between theory and knowledge with the contribution of industry BIM experts (Wu & Issa, 2013).

A closer relationship between academia and the industry would provide multiple benefits to BIM education. This collaboration would improve higher education institutions and assist industry partners to hire BIM talents (Wu & Issa, 2014); assist in developing a greater perception of BIM education areas (e.g. BIM for lifecycle management) (Jin, et al., 2019); and would allow academics to check industry's needs to manage the balance between theory and practice; technology and process; and traditional and emerging methods (Puolitaival & Forsythe, 2016).

In terms of curriculum development, the specifications of BIM-related careers/courses/modules and the expectations of BIM education learning outcomes can be better defined with the participation of academia and the industry (Wu & Issa, 2013). BIM teaching material can also be created through the collaboration of academics, industry experts and software developers. An online collaboration platform would be advantageous for this type of initiative (Suwal et al., 2013). Academics and industry players can join through knowledge transfer partnerships to create training programs that meet industry demands. Another important benefit of this alliance is the creation of frameworks to integrate BIM in existing curricula and create new BIM curricula (Byrne, 2015).

Students can also be given opportunities to research new concepts and technologies within the field of BIM, a task that the industry struggles to do on its own due to lack of time and resources (Hogle, 2013). Moreover, incentives to attract and motivate students in BIM-oriented career paths can be provided through this alliance. For instance, formal BIM programmes such as internships and coops; interdisciplinary BIM education; intercollegiate BIM education collaboration; and lab facilities (Wu & Issa, 2013). Activities to drive BIM education can also be developed. For instance, direct pedagogic contributions by BIM experts from the industry; knowledge sharing between academics and professionals through conferences or workshops (Wu & Issa, 2013); site visits; case studies; visiting lecturers from the industry (Hogle, 2013) and the participation of industrial speakers (Shelbourn et al., 2017).

The presence of industry players adds value to BIM teaching practices. For instance, practical learning methods can be integrated into BIM curricula through modules supported by the industry (Byrne, 2015). Industry partners can be part of the BIM courses to complement the skills and knowledge of academics (Puolitaival & Forsythe, 2016) and assess students work (Jin et al., 2019). Industry players can also indicate the significance of learning material allowing students to become betters hires (Hogle, 2013). If academic institutions want to use BIM models of real projects, a close relationship with external property developers interested in the use of BIM can be beneficial to avoid IP issues (Puolitaival & Forsythe, 2016).

BIM Education Drivers

This section will discuss the main drivers for BIM education namely Central government and public sector, academia, and construction industry players. These were identified by studying different initiatives that have driven BIM education worldwide.

The Central Government and Public Sector

The central government and public sector can support the provision of BIM education and training to develop BIM skills in the industry. As per Sanchez et al. (2014), government mandates would give the necessary context for BIM education/training providers to collaborate in the development of new courses to improve the skills of the professionals of the construction industry. For instance, the UK government created the former BIM Task Group (currently the Centre for Digital Built Britan (CDBB)) to lead the BIM mandate in the country (HM Government, 2013). The BIM task Group set different approaches to providing resources, education, and training to the government and the industry (McGraw-Hill Construction, 2014). In 2011, they created the BIM Academic Forum (BAF), a group of representatives from several UK universities that aimed to support the educational aspects of BIM. This group focused mainly on the creation of a BIM academic framework: Learning Outcomes Framework (LOF) (Underwood et al., 2013). The LOF was developed to guide the creation of BIM training programmes (Cheng & Lu, 2015). The first version was created in 2012 (BIM Task Group, 2012) and the second version in 2015 (Underwood et al., 2015).

There are other examples of BIM education frameworks developed by the central government and public sector. For instance, the Building Construction Authority (BCA) Academy in Singapore created a training framework to ensure that all the stakeholders in the industry receive BIM knowledge (BCA, 2011). In Scotland, the Scottish Futures Trust (SFT) along with CITB (Construction Industry Training Board) appointed the Construction Scotland Innovation Centre in 2017 to research on the development of a BIM Competency Framework. A report was published with a proposed approach for this framework (Bush & Robinson, 2018), but the complete work has not been published yet. In Chile, Plan BIM, which is the 10-year programme that proposes the implementation of BIM by 2020 (Plan BIM, 2018), created in 2018 the *Seminario Internacional de Formacion de Capital Humano en BIM* (International Seminar on BIM Human Capital, in English) in which was introduced the *Matrix of BIM* Roles, a document that guides academic institutions regarding the different skills that professionals need to cultivate (Rooney, 2019). The central government and public sector can also act as educators if they have qualified personnel for this purpose. For instance, the Hong Kong Construction Industry Council (CIC) delivers BIM courses to a wide range of professionals. They also work with training institutions to increase BIM competencies throughout the industry and with industry players to organise BIM promotional activities (Rooney, 2017).

Academia

Many higher education institutions around the world have gotten involved in providing BIM education either driven by a BIM mandate or by own initiative (Silverio, 2020). For instance, in the USA by 2019, there were 68 public and 39 private universities providing formal Architecture and Engineering Programmes in the USA (Rooney, 2019). Academic institutions in late adopters BIM countries are also catching up with BIM leader countries. In Chile, the numbers of higher education institutions delivering

BIM education programmes is growing significantly. In 2018, Fernandez Gonzalez (2018) reported that 41% of universities in Chile where delivering BIM knowledge mainly in Architecture. Later, Rooney (2019) reported that by 2019 there were 42 universities in Chile delivering BIM training, which represents an increase of 81% throughout the country. 84% of the 106 programmes delivered in 2018 have integrated BIM in compulsory courses on their curricula.

Industry

The industry can contribute to the provision of BIM education through the participation of different players: BIM training centres, software developers, professional bodies/industry associations and even construction organisations.

BIM Training centres and registered training organisations are important providers of BIM training within the industry. They offer services to the industry in general as well as customised services to construction organisations (AIA-CA, 2012). Moreover, software developers provide BIM training for their software solutions which can be included in the software package or have an additional fee (CIC, 2013; Rooney, 2017).

Professionals bodies and industry associations can play an important role in promoting BIM in academia (AIA-CA, 2012). As indicated by Bozoglu (2016), professional bodies advocate BIM professionals and create interesting job positions; therefore, another important role for them is to guarantee that BIM is a career option for those who are skilled in the subject. Professional bodies also accredit degree courses delivered by higher education institutions and monitor that they are meeting their mandatory published standards. Important organisations are providing BIM training to the industry. For instance, the Associated General Contractors of America (AGC) in collaboration with leading BIM users, technology companies and academics, created a BIM education Program to prepare professionals at all levels (CM-BIM) (Rooney, 2019). The global industry body, buildingSMART is supporting openBIM training through the 'buildingSMART Professional Certification Program which is believed will have a significant impact on the delivery of BIM education worldwide (buildingSMART, 2020a).

BIM-centred professional organisations are organisations created solely to drive, promote, and support the implementation of BIM in the industry. Their main role is to create networks as they can involve different stakeholders from the government, academia and the industry (Silverio, 2020). In terms of BIM education, these organisations have made significant contributions worldwide. For instance, the Hong Kong Institute of Building Information Modelling (HKIBIM) has guided the creation of BIM careers and BIM training (HKIBIM, 2018). Another example is the UK BIM Framework which is the overall approach to implementing BIM delivered by the UK BIM Alliance (UKBIMA), BSI and the CDBB. In July 2020, they launched 'the UK framework Learning Outcomes Edition 1' to assist in the development of training content according to the new BIM international standards BS EN ISO 19650 series. The framework was created by the UK BIM Framework member organisations, academics, and UK construction industry education professionals and is based on the LOF developed in 2012 (Castello, 2020).

Construction organisations can also get involved in the delivery of BIM training. Driven by market demands, construction organisations are heavily investing to upskill their staff by providing training through either external training providers or in-house training (AIA-CA, 2012; CIC, 2013).

METHODOLOGY

This section will present how the study of BIM education in the Dominican Republic was conducted and how the existing BIM frameworks education frameworks were selected and analysed for the development of the framework of this study.

Investigation of BIM education in the Dominican Republic

There is a paucity of information about the construction industry in general and the presence of BIM in the Dominican Republic. The first indications of BIM education in the D.R. were identified in 2016 in the first phase of the aforementioned research: a training centre (Training Centre A) that provides a type of "BIM seminar" in the last phase of a software training course in Vectorworks and a university interested in integrating BIM into the Civil Engineering curriculum (University B). These findings suggested the need to explore this subject further. Therefore, the presence of BIM education was investigated in the second phase of the research, which took place in December 2016-January 2017. A qualitative approach was adopted since it would provide the in-depth analysis needed to explore this topic which had never been studied before (Creswell, 2014) in the context of the D.R. Semi-structured interviews were the method selected for the data collection, while content analysis was method selected for the data analysis.

To recruit potential candidates, an online search was carried out. Keywords related to BIM education and training were used (in Spanish). Two Diploma courses, a BIM conference, courses on BIM software for architectural design, and an architect committed to the diffusion of BIM knowledge were identified. Moreover, keywords were also used in the websites of universities offering careers that belong to the construction industry and the professional networking platform LinkedIn. Nonetheless, there were no new results (Refer to Table 1).

Table 1. Results of the online search for BIM education

Keyword combination	Results
BIM + Educacion + Republica + Dominicana	BIM Diploma course and BIM informative talk in *University A*
	A conference dedicated to BIM and IPD in *University B*
BIM + Diplomado + Republica + Dominicana	BIM Diploma course in *University B*
	Three Revit courses and one ArchiCAD course
BIM + Cursos + Republica + Dominicana	Webpage of an Architect dedicated to the dissemination of BIM knowledge
BIM + Universidad + Republica + Dominicana; BIM + Formacion + Republica + Dominicana	No results

A lecturer from University A (Interviewee C) and the previously mentioned architect (Interviewee G) could be contacted through the online search. Also, a representative (Interviewee E) of a professional body (professional body A) took part of this research because it was identified that the institution hosted a BIM talk conducted by Interviewee G, whose video is available online. The rest of the participants were identified through the researcher's networking (Interviewee E) and the snowball sampling strategy. In total, 9 people participated in the investigation of the presence of BIM education in the D.R. Refer to Table 2 to see the profile of the participants. Findings on the study of BIM education in the D.R. are presented in the following section.

Development of the BIM Education Framework

A thorough investigation of BIM education was undertaken in the first phase of the research project. Literature findings presented the development of a framework for BIM education as an excellent strategy to assist in the integration of BIM education in academia. An initial BIM education framework was developed based on the literature review on BIM education and the analysis of several BIM education frameworks. This initial version was conceived under the name *"BIM education framework for clients and professionals of the construction industry"* where guidance to elaborate a BIM education plan is given not only to academia but also to the industry. For its development, several BIM education to guide the delivery of BIM education in general, not for specific disciplines/courses, were gathered from the literature (Silverio et al., 2016; Silverio et al., 2017a,b). The following frameworks were selected: the New Zealand's national draft framework, IMAC and Collaborative BIM education framework from Australia, and the Learning Outcomes Framework from the UK, directed to academia; and Kumar's framework and roadmap from the UK, directed to the industry.

Table 2. Profile of the participants of the study of BIM education in the country

Int.	Profile
A	Civil Engineer and MSc in BIM.
B	Non-graduate. He was a student of Architecture.
C	Civil Engineer. Professor and Researcher at *University A*.
D	Architect. Founder member of *Training Centre C*.
E	Architect. Director of the Continuing Education Department of the Headquarters of *Professional Body A*.
F	Architect. Professor at *University C*, CEO of *Training Centre A*, and exclusive Vectorworks provider
G	Freelance Architect.
H	Civil Engineer. Professor at *University A*.
I	Architect.

After the collection of more robust primary data on BIM education in the Dominican Republic, it was determined that the initial framework had to be modified to meet the needs of the country. This revised framework would cover academia only, while the industry section would be part of another framework developed in the research project. Therefore, the study of frameworks directed to the industry were disregarded. To update the framework, more recent frameworks were included in the study, namely, the 201in Competency Table by BIMe, UK BIM Framework Learning Outcomes Edition 1, the Matrix of BIM Roles by Plan BIM in Chile, and the Learning Outcomes Framework from the 'buildingSMART Professional Certification Program'.

Important aspects of the frameworks were analysed such as their objectives, components and practical application. Moreover, their advantages and limitations were identified to support the development of the final framework. Both, the advantages and limitations of the frameworks were taken from the descriptions provided by their developers and the researcher's evaluation of their content. This process will be discussed in the sections 'Study of BIM education frameworks' and 'Development of the BIM education framework for the Dominican Republic'.

BIM EDUCATION IN THE DOMINICAN REPUBLIC

The key findings of the investigation of the presence of BIM education in the D.R. include (1) Shortage of BIM experts, (2) Lack of BIM education: the presence of training, (3) Dissemination of BIM knowledge, and (4) Plans on BIM education and training (Silverio et al., 2017b).

Shortage of BIM Experts

Gardner et al. (2014) indicated that there is a small group of academics committed to disseminating BIM knowledge across the industry because BIM is still novel, and its implementation is still progressing. Hence, it was expected to find the same scenario in the D.R., where BIM is new, and its implementation is still limited. This shortage of BIM experts was first detected in the difficulty of recruiting participants. Also, only six of participants are BIM knowledgeable. Interviewees C, D, and E do not deliver any BIM knowledge themselves; they are just advocating its dissemination (Refer to Table 3).

Lack of BIM Education: The Presence of BIM Training

Thus far, BIM education is starting to be developed in Dominican higher education institutions (Refer to section 'Plans on BIM education and training') and BIM training is currently the main approach to obtain BIM knowledge in the country. BIM training is primarily focused on software skills and delivered in the form of software courses and diploma courses by BIM training centres and professionals of the construction industry, referred also as BIM training providers in this research. As confirmed by Interviewees G, H, I, construction organisations are starting to demand in-house training and consultancies to implement BIM. Also, BIM knowledge is being currently disseminated through educational activities and a BIM informative forum has been created.

The lack of BIM education in the country is also evident through the profile of the participants of this study. As shown in Table 5, many of them are autodidacts in the area of BIM. Only interviewees A, C, 45/H and 46/I took a Diploma course in the country that is not delivered permanently. Interviewee 19/A did a master's degree in BIM in the UK.

Software Courses

Software courses are the most popular mode of BIM training in the country. Most of them do not focus on BIM. Their descriptions/content usually emphasise that they merely teach the software. This issue was identified by several interviewees (19/A, B, G, 45/H and 46/I) who pointed out their limited focus on BIM and their poor teaching approaches: *"That's the other issue, that the courses delivered in this country are normally focused on software and the trainers, I won't say all of them, but some of them are software providers, and they restrict themselves to teach you the software. They do not give you the vision of implementing BIM in a project"* – Interviewee 19/A.

Table 3. Academic background of the participants and the BIM training/knowledge they deliver

Int.	Academic background on BIM	Modes of delivering BIM knowledge and training
19/A*	BIM Diploma course at University A MSc in BIM in the UK	- Coordination of the BIM Diploma course in University A - Talks.
B	No education. An autodidact with books and the internet.	- Talks - Revit courses oriented to BIM - Revit in-house training and BIM champion of the company he works for
C	BIM Diploma course in University A. Does not teach about BIM; he is just a supporter.	- Coordination of the BIM Diploma course and talks in University A - Coordination of the insertion of two BIM modules in the Civil Engineering curriculum of University A
D	No education. Just software courses in Revit. Does not teach about BIM; he is just a supporter.	- Development of conferences where BIM talks and panels have been included - Coordination of Revit courses oriented to BIM.
E	No education. Does not teach about BIM; he is just a supporter.	- Host talks - Interest in delivering courses on BIM.
F	No education. An autodidact with books and the internet.	- Talks - Vectorworks courses oriented to BIM - Software consultancy to companies
G	No education. An autodidact with the internet.	- Revit courses oriented to BIM - BIM Diploma courses - Revit in-house training and BIM implementation strategies consultancy to companies - Talks
45/H and 46/I*	BIM Diploma course at University A.	- Coordination of the BIM Diploma course in University A - BIM Diploma courses - Participation and organisation of talks, conferences and forums in University A and Training Centre D - Training and consultancy to companies - Revit training courses oriented to BIM.

*This naming convention was used for all the interviewees that participated in the research

Diploma Courses

Diploma courses were the most formal mode of BIM training identified. These courses are not delivered regularly and usually focused on software skills. By the time of the investigation, only two Diploma courses had been delivered in 2013: one in University A and one in University B (Refer to Table 3). The diploma course of University A was coordinated by academic staff and delivered by educators from Brigham University (USA), through an agreement between both universities. Interviewee C indicated that the course primarily included an introduction to BIM and modelling lessons in the software Revit and Navisworks. On the other hand, the diploma course of University B was delivered by a BIM training provider, not an academic staff from the university. The course included knowledge on IPD and BIM processes, use of Revit and Navisworks and the development of a BIM project (UNIBE, 2013). Information about this diploma course was obtained through the internet since the facilitator of the course could not be contacted for this study.

During and after the collection of primary data, other diploma courses were identified. Interviewee G indicated to have prepared a diploma course to teach BIM and the software Revit. The course was planned to start being delivered since January 2017 in two universities located in the capital and the north of the country (University C and D). After the first delivery, the course has been delivered continuously on both universities. Moreover, Interviewees 45/H and 46/I got involved in the delivery of one diploma course in Training Centre D. The course encompassed three modules: introduction to BIM, project's coordination with BIM and implementation of BIM. Skills on the software Revit were indicated as an entry requirement. In September 2019, University B took the initiative of developing a diploma course with academic staff directed to architecture students and architects, under the name "Diploma course of Innovation in Architecture". In the description is specified that the students will be introduced to design through programming, IPD and digital fabrication (UNIBE, 2019).

In-House Training and BIM Consultancy to Companies

Several interviewees stated to have provided this type of services to construction organisations. Interviewee B introduced the implementation of BIM in the construction organisation he works for. Interviewee E has provided software consultancy to several organisations through Training centre A. Interviewee G has introduced BIM to several organisations and provided in-house training in the software Revit. Interviewees 45/H and 46/I have provided consultancies in BIM implementation strategies and in-house training in the software Revit and Navisworks.

Dissemination of BIM Knowledge

Educational events such as conferences, seminars and forums about BIM were identified. These activities have been mostly held in universities across the country, followed by professional bodies and training centres. As presented in Table 5, all the interviewees have participated in this type of activities either as organisers or facilitators. The popularity of these activities is increasing progressively.

In December 2016 five professionals of the construction industry along with the researcher created the first informative BIM forum of the country called ProBIM. The forum started to disseminate information about BIM through "quick read" articles in their initial web page. Nowadays, the forum is just present in social media platforms: Instagram (user: @ProBIM_ORG), LinkedIn and YouTube (user: ProBIM), from which Instagram is the most important channel. Short articles are written and published through Instagram about different topics: general BIM knowledge, emerging tendencies, case studies of BIM projects, etc. Also, live videos are broadcasted to cover specific topics and interact with their followers. Both, the publications and videos are shared in the other platforms (ProBIM, 2016; ProBIM, 2017b; ProBIM, 2019a). The forum has contributed significantly to the diffusion of BIM knowledge among professionals of the industry. It has grown in terms of members and scope. In the last two years, it has hosted important educational activities in different universities and in Professional Body A to raise BIM awareness in the industry. They have also delivered one-day workshops in specific areas such as "Modelling from Point cloud data". The forum has stated their commitment to driving the OpenBIM approach in the Dominican construction industry. For instance, their latest event "2da Jornada OPEN BIM" (2nd OPEN BIM Conference, in English) was around this topic (ProBIM, 2019b).

Professional body A has also expressed their interest in delivering BIM knowledge to the industry. Interviewee E stated that the institution is interested in delivering BIM courses. There was already an attempt to planning a BIM Diploma course that failed due to lack of BIM experts (Silverio et al., 2017b). In terms of activities, interviewees confirmed that Professional body A has hosted educational activities related to BIM: an introductory BIM talk delivered by Interviewee G, seminars held by Interviewee

45/H and 46/I. They have also collaborated with ProBIM in educational activities since August 2017 (ProBIM, 2017a).

Plans on BIM Education and Training

Interviewees were asked about their plans concerning the delivery of BIM knowledge and BIM education and training. The most interesting was the plan of one university to integrate BIM within a career's curriculum. As per interviewees 19/A and C, there was a curriculum reform in the career of Civil Engineering in University A which will take effect for students enrolled from August 2016 onwards. The modifications include the improvement of one module and the insertion of a new module, both to be delivered in the last year of this career, which is three-year long. Therefore, the new curriculum will be effective from September 2019. The modified module is called "Technical drawing for Civil Engineering" and students will be taught the software Revit instead of AutoCAD. On the other hand, the new module is called "Computer-Aided Design" which is planned to be delivered by Interviewee 19/A. Interviewee 19/A gave some details about the module, including the intentions of teaching students the software Navisworks. The aim of both modules is that students acquire the necessary skills to use models for different tasks such as Design analysis, Clash detections, Cost Estimations, construction schedules, construction simulations, etc.

The strategy implemented by this university is defined by Suwal & Singh (2018) as Individual BIM, in which a stand-alone course is provided as part of CAD courses or as a substitute of existing CAD courses. The project of University A consists of inserting two modules, both with a strong focus on software skills, in the last year of the Civil Engineering career. The approach of inserting BIM in academia through teaching specific software has been identified even in BIM leader countries (Rooney, 2017). Nonetheless, it has been reported that training in OpenBIM concepts and other topics such as collaborative BIM environments, BIM management, Sustainable BIM, BIM for FM and Life Cycle Costing seems to be growing in countries such Australia, Canada, Finland, Hong Kong, Japan, Norway, Singapore, Sweden, Switzerland and UK (Rooney, 2019). This infers that introducing BIM through software should not be seen completely as a wrong approach in the D.R., but as the first steps to get into more BIM-related processes areas, which are the core knowledge needed for successful BIM implementation.

These key findings suggested the need for guidance for the successful integration of BIM in university curricula. Therefore, the development of a framework to guide the integration of BIM in university curricula in the Dominican Republic was considered. The following section presents the study of several BIM education frameworks that were analysed to support the development of the framework.

STUDY OF EXISTING BIM EDUCATION FRAMEWORKS

The need for frameworks to facilitate the integration of collaborative design and BIM education in higher education institutions has been posited before by academics (Macdonald & Mills, 2011; Macdonald, 2012; Macdonald & Granroth, 2013). This section will present the content, practical application, when available, and a critical analysis of the frameworks studied in this research.

New Zealand's National Draft Framework

The University of Auckland in New Zealand elaborated a draft BIM education framework in which knowledge domains were classified and mapped to the NZQF (New Zealand Qualifications Framework). The framework includes three types of education: (1) vocational training and continuing professional development, (2) degree programmes (undergraduate and masters), and (3) research (including PhDs). It also specifies the learning outcomes demanded in three areas of BIM implementation: Architecture, Engineering, and Construction; Software and Technologies; Business, Enterprise, and Management. The framework proposes the insertion of BIM in an existing bachelor's degree's curricula and the creation of Integrated BIM masters (Miller et al., 2013). The framework remained as a draft and no examples of its application have been found.

IMAC Framework – Code BIM project, Australia

The Australian Office of Learning and Teaching (OLT) funded a project called Code BIM for the elaboration of a framework that could assist AEC academics in the insertion of collaborative programs using BIM tools and procedures (Macdonald & Granroth, 2013). The framework has two components: a benchmarking tool and a separate guide to implementation. The framework is divided into four stages, which describe different levels of accomplishment: (1) Illustration, (2) Manipulation, (3) Application, and (4) Collaboration. The framework does not indicate in which academic year each stage should be inserted and 'practice by doing' and 'teach others' are proposed as the preferred delivery methods (Macdonald & Mills 2011; Macdonald, 2012; Macdonald & Granroth 2013).

Application

Macdonald (2012) explained that the benchmarking tool of this framework has been implemented in the three universities that are part of the Code BIM to benchmark existing courses project and outline goals for future curriculum improvements. Table 4 presents two cases where the framework was implemented in two of these universities:

Table 4. Practical implementation of IMAC framework (Tran et al., 2012; Rooney, 2019)

University of South Australia (UniSA)	University of Technology Sydney: UTS
Purpose: Benchmarking of two courses	**Purpose:** Development of a class
Construction in the Workshop course and Construction on-site course: The benchmarking aimed to identify if the courses were useful in developing collaborative design competencies in the students. Afterwards, it was recommended to make the classes more collaborative and the insertion of BIM tools.	**IPD studio class:** Multi-disciplinary students had to work during one semester on a project implementing BIM tools and processes and learning how to work collaboratively in a team and the information required from their partner disciplines.

Collaborative BIM Education Framework – EWG, Australia

The BIM Education Working Group (EWG) in Australia created the Collaborative BIM education framework which consists of six components: (1) Identifying BIM Competencies, (2) Classifying BIM Competencies, (3) Arranging competencies and designing BIM learning modules, (d) An industry framework for professional development, (e) An academic framework for BIM education, and (f) the BIM Institute. In the first three components, the framework proposes how BIM learning modules can be designed, a process that starts with the identification of BIM competencies. Such competencies need to be classified and organised to create learning modules. For that, and the elaboration of BIM learning materials the creation of an online learning hub where BIM education providers and other stakeholders can work collaboratively is suggested. The following two components recommend the creation of two frameworks: an industry and an academic framework. The last component consists of the creation of a National BIM Institute to promote BIM learning and support the implementation of the Collaborative framework (AIA-CA, 2012; Succar & Sher, 2014). No examples of the application of this framework were found in the literature.

201in Competency Table v2.1, BIMe initiative

The BIMe Initiative is a non-profit project based on the BIM Excellence (BIMe) approach which consists of a set of online and offline templates to evaluate and enhance the competency of organisations, teams, projects and individuals. One of its objectives of the BIMe initiative is to develop competency-based learning methods, tools and materials appropriate for higher education, vocational training and professional development (Change Agents, 2019). This has been achieved with the Competency Table, based on the published research of Dr Succar (including the Collaborative BIM Education Framework) and other international collaborators. The table can be used to organise competencies and offer a structure to develop assessment modules, competency-based certification regimes and accreditation programmes, learning units and competency-based education programmes and identify competency profiles of different roles throughout markets and disciplines.

The Competency Table is organised under three Competency Tiers: (1) Core Tier, (2) Domain Tier, and (3) Execution Tier. For instance, the Domain Tier has 8 *Competency Sets,* which consists of four primary competency sets (Managerial, Functional, Technical and Supportive) and four secondary competency sets (Administration, Operation, Implementation and Research & Development). The Domain Tier also has 55 standards *Competency Topics*. Each standard contains from 10-100 *Competency Items* which are a phrase/sentence to describe an ability, activity or learning outcome that can be assessed, learned or applied. Competency Items are applied at specific organisational scales and granularity levels and are used to establish assessment modules and training lessons (BIMe initiative, 2019).

Application

According to BIMe Initiative, the table has been used to evaluate more than 60 companies in the UK and used UK-specific competency items (Bush & Robinson, 2018). Nonetheless, no examples of its application in the creation of BIM training/education content were found in the literature.

Learning Outcomes Framework (LOF) – BIM Academic Forum, UK

In 2012, the BAF developed a framework to define the learning outcomes (LOF) needed to implement BIM level 2 successfully. The LOs were specified for undergraduate (4, 5 and 6) and post-graduate level (7). The learning outcomes were outlined in the following topics: Strategic, Management, and Technical (BIM Task Group, 2012).

In 2015, an updated version of the LOF was created. This version uses the term BAF ILO (Intended Learning Outcomes). The learning outcomes, which derived from knowledge and understanding, practical skills and transferable skills, are grouped into the following themes:

1. Understand the essence BIM, BIM Level 2 requirements established and its relationship with the Government Construction strategy 2025.
2. Understand the implications and value of BIM in the organisation.
3. Understand the requirements for information management and exchange described in the 1192 suite of standards PAS55/ISO 55000 (Underwood et al., 2015; BIM Level 2, 2016).

Application

An example of the application of the later version of the LOF is a 3-year plan of Loughborough University to BIM-upgrade 26 priority modules in undergraduate and post-graduate degrees. To embed BIM into the priority modules, the BAF ILOs were mapped into the LOs of the existing modules. The BAF ILOs guided the creation of new BIM-specific modules:

- **BSc: CVB042 3D BIM Auditing and Coordination.** A module related to the coordination of 3D BIM, which includes the database drive auditing of BIM data and clash detection/avoidance;
- **BSc and MSc: CVC045 Collaborative. BIM Design Project.** Two modules where digital workflow is connected to collaborative role-playing (Adamu, 2016).

In their study, Bush & Robinson (2018) found that there was broad awareness of the framework and that several organisations have used it to different extents in the UK. Nonetheless, participants of the study also expressed to have concerns regarding this framework such as acceptance and lack of update since 2015, lack of clarity in terms of responsibility and ownership; lack of global adoption; and lack of examining the compliance with the framework.

UK BIM Framework Learning Outcomes Edition 1

Created in 2020 based on the LOF, this updated framework aims to assist in the creation of training content following the new international BIM standards: BS EN ISO 19650 series (Castello, 2020). The framework presents the required learning outcomes for introductory training in information management implementing BIM. It was created to be used by training providers to support their course development and as an evaluation resource for individuals looking for training. The LOs of the framework are organised in five sections which, in turn, present in detail the knowledge to be delivered to achieve each LO:

1. Why information management implementing BIM is needed and its context in the UK.

2. The significance and benefit of information management using BIM for organisations
3. How information requirements are recognised and informed.
4. How information is shared between organisations
5. Understanding the legal and security repercussions of the UK BIM Framework and its interfaces with other significant processes (e.g. health and safety).

The framework refers to the BS EN ISO 19650-1, 19650-2, 19650-3, 19650-5, as well as BS 1192-4 (COBie) and PAS 1192-6 (UK BIM Framework, 2020). It is aligned with the buildingSMART UK Chapter's Professional Certification (PCERT), which is to be launched in Autumn, 2020 (Castello, 2020). This framework was released recently; therefore, there are no examples of its implementation.

Matrix of BIM roles – Plan BIM, Chile

The Matrix of BIM roles defines five BIM roles (i.e. BIM management at an organisational level, BIM revision, BIM modelling, BIM coordination, BIM management at a project level) along with 42 skills that needed to implement BIM throughout the life-cycle of projects. This matrix aims to assist the industry in achieving collaborative work with the provision of a uniform definition of roles; organisations and companies to develop learning strategies and evaluate the abilities of their team members; and educational institutions to update their curricula, ensuring that their programmes are following the Industry 4.0 requirements (Plan BIM, 2019).

The matrix consists of a table in which the five roles are in columns, while the 42 abilities of each are in rows, which are grouped in 15 themes. Each role has specific verbs which describe what the individual needs to do and know on each ability (Plan BIM, 2017).

Application

As per Rooney (2019), 25 new continuing education courses had been developed following the Matrix of roles. Nonetheless, there were not found any examples of how educational institutions have adopted the matrix to create new BIM-related courses.

Learning Outcomes Framework - Individual Qualification, 'buildingSMART Professional Certification Program'

This framework is part of the 'buildingSMART Professional Certification Program' which allows learning organisations to provide education and certifications in compliance with a well-known global learning framework. The goals of this programme are to standardise and encourage openBIM training content; assist and accredit training organisations; and assess and certify individuals (buildingSMART, 2020a). It has been reported that the programme will have a significant impact on BIM education worldwide. It will set a benchmark for professional competency and will help to increase the level of BIM delivery throughout the global industry (Baldwin, 2018).

The programme has two phases. Phase 1 'Individual Qualification' was launched in 2017. For this phase, a 'learning outcome framework' (LOF) that outlines the extent of BIM training for individuals was created. Training providers can follow the framework while still employing their course materials, provided that the course content is in line with the framework's requirements. They can also include country-specific material in an additional module (buildingSMART, 2017). Training providers that want to deliver a buildingSMART approved course need to apply through their local buildingSMART chapter. The review and approval process consists of four stages application, assessment, approval, and monitoring:

LOs are the minimum learning required and it is assumed that approved courses include other content than the one described within the framework. All content within the LOF can be covered in a minimum of 6 hours and is designed to be adopted by training courses with a duration of 1-2 days. The Basic Learning Outcomes are 28, which are classified into the following categories:

1. Understand what BIM is, why it is needed and identify its specific terminology.
2. Acknowledge the advantages of BIM compared to traditional project delivery.
3. Understand the project information development cycle (and key terms); particularly how project information is specified, produced, exchanges, and maintained.
4. Acknowledge the need for open and interoperable solutions.
5. Determine an organisations capability in working with BIM (Baldwin, 2018).

Important international standards such as ISO 19650-1 have been consulted to define the LOF for Individual Qualification. The LOs are available to the public but their learning content, knows as 'Body of knowledge' is given only to registered training providers (buildingSMART, 2020b).

Phase 2, named 'Professional Qualification' includes the certification of companies using BIM and is planned to be launched in 2021 (buildingSMART, 2017; buildingSMART, 2020a). The programme does not cover higher education; however, HE institutions may integrate any of these phases in their curricula (buildingSMART, 2020a). It is currently available in specific countries of North America, Europe, Asia and Oceania (buildingSMART, 2020b).

Based on their aim, the studied frameworks could be classified into two categories: (1) those which set strategies on how to integrate BIM in university curricula (i.e. New Zealand's national draft framework, IMAC framework), and those which establish a set of competencies to guide the BIM training/education content (i.e. 201in Competency Table v2.1, LOF, UK BIM Framework learning outcomes Edition 1, Matrix of BIM roles, Learning outcomes framework – Individual Qualification). Both types offer good guidance to educational institutions, but it could be noticed that there is a tendency of developing these frameworks is focused on competencies. Literature has shown that is beneficial to set BIM competencies in BIM education frameworks. As per Baldwin (2018), there is a huge discrepancy in the skills of industry professionals that are working in BIM projects. General agreement on the use of standardised terms and processes as well as a benchmarking system to evaluate the competence of individuals is necessary for successful BIM. Bush & Robinson (2018) also stated that the inclusion of BIM competencies in a single framework would give a universal approach to organisations that accredit, fund, develop and deliver training and BIM education. It would also assist these organisations in terms of time and resources. Having transparent and available core and role-specific competencies would also help businesses and individuals to evaluate the appropriateness and veracity of their training/education options. A good example of establishing and organising BIM competencies is the Matrix of BIM Roles by Plan BIM. The competencies are listed on each of the five roles selected by Plan BIM and words are used for each competency to understand the extent of knowledge needed from the individuals (e.g. plan, understand, validate, N/A). Moreover, the BIM competencies are within themes for better comprehension.

Some of these frameworks also aim to guide the delivery of BIM education and training at diverse levels. For instance, the New Zealand's national draft includes vocational training to research. Whereas others do not specify their scope in this regard, which infers they can be adopted to design BIM programmes at any level (e.g. Matrix of BIM roles and 201in Competency Table). One important advantage identified in the study is the global applicability of the Learning outcomes framework by bSI, which could benefit any country, especially those that do not have any type of framework to develop BIM training courses.

The analysis of the frameworks has shown that although each of framework has its merits, they are not without limitations. One important limitation identified was the lack of update. Some of these frameworks were developed before 2015 and they appear to be obsolete for the current state of the construction industry. The field of BIM and digital construction is constantly evolving; therefore, frameworks of this type need to be a continuous development rather than just a finished product. The Competency Table by BIMe initiative is a good example in this regard, which is still an on-going project that keeps updating continuously.

Another important limitation is the lack of examples of their application. Some of these frameworks have not been applied or do not have examples of their application available in the literature. For instance, literature shows that the LOF has been implemented considerably in the UK. However, there is a need for more detailed examples of how courses have been developed with these frameworks to demonstrate their application and practicality to organisations interested in adopting them. The UK BIM Framework Learning Outcomes and the Learning outcomes framework by bSI are the most recent frameworks; nonetheless, their applicability scope is BIM training. Refer to Table 5 to see other advantages and limitations of these frameworks.

DEVELOPMENT OF THE BIM EDUCATION FRAMEWORK FOR THE D.R.

The development of the initial framework was informed mainly by the New Zealand's National Draft and the Collaborative BIM education framework by EWG because they provide guidance on the integration of BIM in academia, which is the aim of the framework. This framework had two sections, one part directed to academia and the other directed to the industry (Refer to Appendix 1). For its development, a series of strategies from the three frameworks and solutions from the literature on how to overcome the challenges on BIM education were listed and classified according to the domain they were going to be implemented: academia or the industry. The strategies were divided into 3 sections, which represent the stages for BIM education planning: Strategy, Implementation and Revision. Each section presents the actions that need to take place on each stage. In the implementation stage, the framework presents not only the actions but lists the participants that should be involved.

After a more in-depth collection of primary data on BIM education in the Dominican Republic, it was determined that this initial framework had to be modified to respond better to the status of BIM education in the country. The changes made to the framework include: (1) the framework focuses only academia; (2) the framework comprehends four stages: Strategy, Preparation, Realisation and Revision; (3) more recent BIM education frameworks were studied to update its content; and (4) the framework was simplified graphically to facilitate its understanding (Refer to Figure 1). In terms of content development, the updated framework follows the same approach used in the initial framework described above, but including aspects related to the status of BIM education in the country and recommendations in this regard. The updated framework is explained below.

Table 5. Analysis of the BIM education frameworks

Framework	Release year	Advantage	Limitations
New Zealand's national draft framework	2013	- Encompasses several types and levels of education - Specifies a wide range of topics to be taught in vocational training and CPD - Specifies the BIM knowledge and skills that students would be able to acquire at undergraduate and post-graduate level	- No examples of application - Out of date. Its applicability needs for the current state of the industry needs to be re-assessed (emerging technologies, international standards)
IMAC Framework	2012	- Can be applied to different types of education delivery - Examples of its successful application to benchmark and improve courses in terms of collaborative work and BIM implementation. - Explanation on how students can build their BIM knowledge gradually	- The stages are broad and might not give enough guidance for specific disciplines - Out of date. Its applicability for the current state of the industry needs to be re-assessed (e.g. emerging technologies, international BIM standards)
Collaborative BIM education framework	2012	- Offers good guidance to develop BIM modules through the identification, classification and collection of competencies - Focus on both, the industry (professional development) and academia (BIM education)	- Only guides the planning for BIM education. It does not specify knowledge, skills or learning outcomes - No examples of application - Out of date. Its applicability needs for the current state of the industry needs to be re-assessed (emerging technologies, international standards)
201in Competency Table v 2.1	First publication (v1.4): 2016 Last version (v2.1): 2019	- It is constantly updated - Available in English, French and Spanish - Encompasses abilities, activities and learning outcomes -	- It is not completely available. Only the Domain Tier is published. - Major focus on top management skills
Learning Outcomes Framework	Initial version 2012 Update: 2015	- Aligned with the UK BIM Level 2 requirements - Detailed learning outcomes to be attained in undergraduate levels (level 4, 5, and 6) and post-graduate level (level 7).	- Out of date. It is built on the PAS 1192 series which is being gradually replaced by the BS EN ISO series - Country-specific adoption - Directed to higher education
UK BIM Framework Learning Outcomes	2020 – 1st Edition	- Up to date: Includes the BS EN ISO series. - Detailed learning outcomes on information management using BIM	- Focus on introductory training - Country-specific aspects (e.g. Government Soft Landings).
Matrix of BIM roles	2017	- A detailed explanation of the roles and the extent of the skills needed within each BIM role - Covers skills needed in all the stages of a project.	- It is specific to five roles. - Limited presence of Industry 4.0 concepts, although it is specified to be part of its objectives
Learning Outcomes Framework – Individual qualification (bSI)	2017 (Part of an ongoing programme)	- Allows global applicability - Up to date - OpenBIM approach - Considers the BS EN ISO series - Monitors course's compliance with the framework	- The scope does not include tertiary education - Designed for short courses (1-2 days). - Use of the framework is based on being registered or paying a course approval fee, which could represent a challenge for some training providers.

Figure 1. Framework to guide the integration of BIM into university curricula

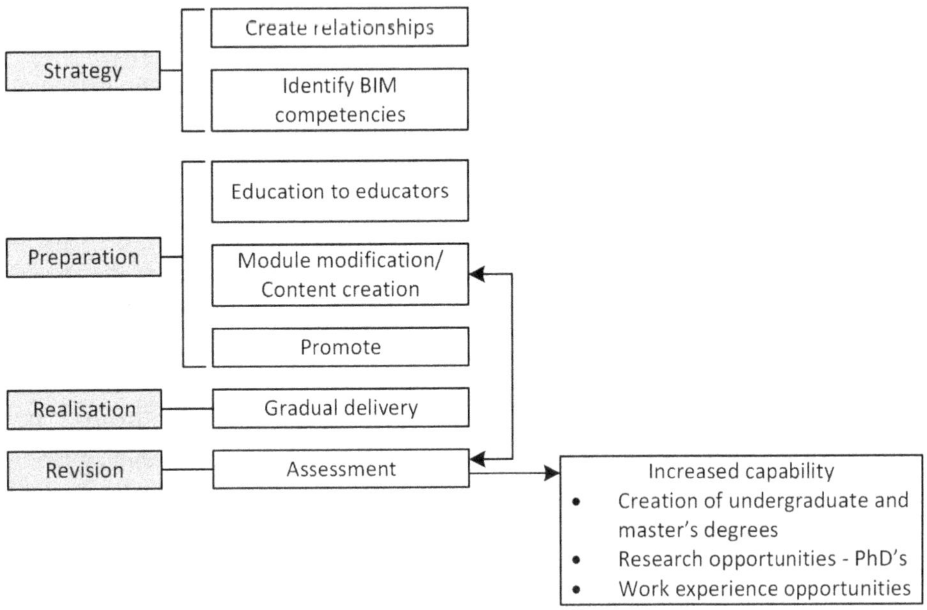

Stage 1: Strategy. The first stage consists of the development of the strategies to be followed. For that, it is necessary to establish a lead. The best option for this is to establish formal relationships between academics and industry players interested in the integration and improvement of current approaches to BIM education. In the D.R., industry players that can be part of this process comprehend professional bodies, BIM training centres, BIM training providers, BIM adopters (professionals and construction organisations), and BIM informative forums. Support from the government is desirable, but in the context of the D.R., this is not attainable yet (Silverio, 2020).

The players taking part in this strategy need to identify the BIM competencies required in disciplines and courses of current curricula. These competencies can be identified, selected, classified and organised in a way that could allow the creation of learning modules as suggested by AIA-CA (2012). Competencies could be presented following the structure of the Matrix of BIM roles because of its comprehensive approach of presenting both, BIM roles and competencies. Current BIM practitioners in the D.R. could be of great help in identifying the BIM competencies needed in the Dominican construction industry. BIM competencies also need to be up to date. Therefore, current topics around BIM must be included such as Industry 4.0 and its related technological advancements, openBIM and the BS EN ISO series. Alike the Learning Outcomes Framework of the bSI Individual Qualification, the resulting BIM competencies document should be used as a benchmarking tool to guide, evaluate, approve and monitor the BIM modules created by Dominican HE institutions, etc. This way, the quality of BIM content delivered can be ensured. Accreditation schemes could be developed by professional bodies if they get involved.

Lastly, the BIM competencies need to be regularly checked and updated per the new advancements taking place in the construction industry. Regular meetings need to be held to evaluate the appropriateness of the competencies over the years.

Stage 2: Preparation. Once the competencies are identified, proper education needs to be provided to prospect educators. In the D.R. was observed that most of the stakeholders involved in BIM education and training are autodidact. Even though these players are already contributing significantly to enhancing BIM skills in the country, they may require proper preparation before getting involved in BIM education. After educators receive the required training/education, they will be able to take part in the development of BIM modules and the selection and preparation of teaching material. The BIM competencies document must be used as guidance in this process. It is the responsibility of educational institutions to make sure that the BIM competencies are correctly mapped into the courses they are developing.

While doing all this process, BIM needs to be promoted through different activities directed not only students but to current practitioners (Wu & Issa, 2013; Badrinath et al., 2016). In the D.R. this is being done mainly to raise BIM awareness but with more participation from the industry than academia itself.

Stage 3: Realisation. This stage consists of the delivery of the modules that have been modified to be aligned with BIM. It is suggested to deliver modules gradually for a smooth transition (Miller et al., 2013). This gradual integration can be organised by disciplines. For instance, start with Architecture courses, which is the discipline most advanced in terms of BIM in the country (Silverio, 2020), and the move to the remainder disciplines. It is recommended to start the integration and/or modification of modules with senior year students who are more experienced and then work with lower levels.

Stage 4: Revision. This stage consists of the assessment of the first module(s) delivered. It is recommended that different players participate in this assessment needs to be performed by different players. Firstly, it is important to get feedback from the BIM learners about the module(s) programme, learning materials provided and learning attained (Jin et al., 2019). Furthermore, educators need to supervise the delivery of the modules thus they will be able to evaluate different aspects such as progress and performance of the students; the need for extra training/education for them; and successfulness of the module programme. In this stage, input from industry players is recommended to guide educators as to which improvements may be necessary after the assessment has been carried out (Puolitaival & Forsythe, 2016).

After the first attempt of integration is revised, the process can start again until attaining the final milestone. Then, the lead group will have the necessary capability to take further steps such as:

- Creation of BIM-based undergraduate and master's degrees (Miller et al., 2013, Underwood 2015)
- Research opportunities – PhDs (Hogle, 2013; Miller et al., 2013; BIMe initiative, 2019)
- Offer work experience opportunities to BIM learners: internships, coops, etc. (Wu & Issa, 2013).

The inclusion of BIM competencies was recognised as important and is stated in the proposed framework; however, the framework does not contain a set of BIM competencies. As Bush & Robinson (2018) explained, developing a BIM competence framework is a serious task that would require a significant investment of time and resources from numerous organisations. That could also be confirmed with the competency-based frameworks studied in this research, which were the result of the effort of several organisations and researchers. That lack of resources to undertake this task is one of the limitations of this research; therefore, it has been put forward as a suggestion for further research.

This framework was validated through member checking (Creswell & Poth, 2017). The validation process took place in March 2019 with participants that took part in the research "Implementation of Building Information Modelling in the Dominican Republic construction industry". Through a questionnaire, the participants gave their opinion about the content, relevance, and suitability of the framework. Overall, the framework received positive feedback. Participants that specifically took part in the investigation of BIM education in the country expressed they would be willing to adopt and participate in the different collaborative strategies suggested in the framework (Silverio, 2020).

SOLUTIONS AND RECOMMENDATIONS

This study has presented the importance of the provision of BIM education in academia and the numerous challenges hindering higher education institutions to teach BIM. Many academics and educationalists around the world have worked to overcome these challenges and offer the required education to implement BIM successfully. Nonetheless, in countries like the Dominican Republic, where BIM is at its infancy, these challenges can be more difficult to overcome.

Thus far, the country is interested in BIM but is suffering from a lack of BIM education. Covering this lack represents a difficult task since the country also has a shortage of BIM experts. There is only BIM training, which is mainly delivered by training centres and professionals of the construction industry. This is a good initiative; however, some of the professionals delivering these courses have not received formal education and are autodidact and these efforts might be counterproductive if the knowledge delivered is not appropriate. BIM educational activities have also been developed with the collaboration of professionals of the industry, professional bodies, BIM informative forums and higher education institutions. The involvement of academia is also evident with the plans of University A to teach BIM in the career of Civil Engineering. Although it is very recent, this action will most likely motivate other universities to teach BIM. Lastly, there is no support from a key BIM education driver: the central government. Hitherto, the Dominican government has not expressed its interest in implementing BIM yet nor are not contributing to the delivery of BIM education.

Given the current situation, BIM education in the Dominican Republic will continue to be driven by academia and the construction industry. The framework to integrate BIM into university curricula has been developed to guide higher education institutions in the D.R. to plan BIM curricula development. Considering the current lack of guidance on BIM in the country, the approach of this framework of explaining the process of BIM curricula development through a set of sequential stages is suitable to start building up the delivery of proper BIM education in academia in the future. The need for guidance of this type is reflected in the efforts of University A, whose current approach is inserting, rather than integrating BIM in their curricula, which is also known as 'Individual BIM approach' (Suwal & Singh, 2018). Guidance like the one provided in the proposed framework would instruct these institutions into properly integrating BIM in their curricula.

Different players of the construction industry are contributing significantly to the diffusion of BIM knowledge in the country. However, a different approach would be needed to guide them in the development of appropriate BIM training courses for the Dominican construction sector. The establishment of BIM competencies would be advantageous for this purpose; nonetheless, this task was not part of the scope of this research and has been suggested for further research in the area. The creation of a group of academics and professionals involved in BIM education and BIM training as well as current professionals from companies in which BIM is being implemented is recommended to undertake this type of project.

FUTURE RESEARCH DIRECTIONS

The status of BIM education in the country has been explored and a framework to guide the integration of BIM in university curricula has been created and validated by the participants of the research. Nonetheless, there is a breadth to carry out further research based on the work presented.

First, it would be valuable to test the practical implementation of this framework with the Dominican universities interested in integrating BIM in their courses. It is also important to continue monitoring the current development of BIM education in the country. It would be valuable to assess the successfulness of the approach of teaching BIM in university A and investigate if other universities are planning to integrate BIM in their curricula.

Moreover, as previously stated, it would be beneficial to undertake an analysis of the BIM competencies needed in the Dominican construction industry. The establishment of BIM competencies would guide HE institutions and training providers, which are currently contributing significantly to the diffusion of BIM in the country, in the development of BIM courses and content. An evaluation of the demanded BIM roles in the Dominican construction industry would also be beneficial in this regard.

CONCLUSION

This study has broadly discussed the subject of BIM education. Key findings from the investigation of BIM education in the Dominican Republic were subsequently presented, namely, shortage of BIM experts, lack of BIM education: the presence of BIM training, dissemination of BIM knowledge in the country and plans on BIM education and training. BIM training is delivered in the country with a strong emphasis on software skills, which is mainly driven by professionals of the industry. Many educational events have been organised in the last years, which is contributing to raising awareness in the industry. These events have included the participation of universities, BIM training providers and professional bodies and a BIM informative forum. The first steps of inserting BIM into university curricula have been taken by an important university in the capital of the country. The efforts are in two complementary subjects through an Individual BIM approach that, even though is limited, represent an important initiative in the country in terms of BIM education. These findings, especially the first effort of teaching BIM in a higher education institution, suggests the need for guidance for the successful integration of BIM in academia, which is the main BIM education provider.

Existing BIM education frameworks were studied to assist in the development of the BIM education framework to guide the integration of BIM in academia into university curricula in the Dominican Republic. Subsequently, the development of the framework was illustrated. The framework consists of four stages: Strategy, preparation, realisation and revision. The framework was validated with participants of the research that took place in the Dominican Republic. Recommendations and suggestions for further research have been put forward.

REFERENCES

Abbas, A., Din, Z. U., & Farooqui, R. (2016). Integration of BIM in construction management education: An overview of Pakistani engineering universities. *Procedia Engineering*, *145*, 151–157. doi:10.1016/j.proeng.2016.04.034

Adamu, Z. (2016). How universities are teaching BIM: A review and case study from the UK. *Journal of Information Technology in Construction*, *21*(8), 119–139.

AIA-CA. (2012). *BIM education - BIM in practice. A position paper by the Australian Institute of architects and consult Australia*. Consult Australia and Australian Institute of Architects.

Badrinath, A. C., Chang, Y., & Hsieh, S. (2016). A review of tertiary BIM education for advanced engineering communication with visualization. *Visualization in Engineering*, *4*(1), 9. doi:10.118640327-016-0038-6

Baldwin, M. (2018). *buildingSMART Professional Certification. PHASE 1: individual qualification. Overview and Sponsorship Plan*. Retrieved from education.buildingsmart.org

Banco Central de la Republica Dominicana. (2015). *Informe de la Economia Dominicana*. Enero-Diciembre 2014 Retrieved from https://www.bancentral.gov.do/

Banco Central de la Republica Dominicana. (2016). *Informe de la Economia Dominicana*. Enero-Diciembre 2015 Retrieved from https://www.bancentral.gov.do/

Banco Central de la Republica Dominicana. (2017). *Informe de la Economia Dominicana*. Enero-Diciembre 2016 Retrieved from https://www.bancentral.gov.do/

Banco Central de la Republica Dominicana. (2018). *Informe de la Economia Dominicana*. Enero-Diciembre 2017 Retrieved from https://www.bancentral.gov.do/

Banco Central de la Republica Dominicana. (2019). *Informe de la Economía Dominicana*. Enero-Junio 2019. Retrieved from https://www.bancentral.gov.do/

BCA. (2011). *The BIM issue*. Retrieved from https://www.bca.gov.sg/publications/BuildSmart/others/buildsmart_11issue9.pdf

Becerik-Gerber, B., Gerber, D. J., & Ku, K. (2011). The pace of technological innovation in architecture, engineering, and construction education: Integrating recent trends into the curricula. *Journal of Information Technology in Construction*, *16*(24), 411–432.

Becerik-Gerber, B., Jazizadeh, F., Li, N., & Calis, G. (2011). Application areas and data requirements for BIM-enabled facilities management. *Journal of Construction Engineering and Management*, *138*(3), 431–442. doi:10.1061/(ASCE)CO.1943-7862.0000433

BIM Level 2. (2016). *BIM learning outcomes framework*. Retrieved from http://bim-level2.org/globalassets/pdfs/learning-outcomes-framework.pdf

BIM Task Group. (2012). *The BIM task group publishes the initial BIM learning outcomes framework*. Retrieved from www.bimtaskgroup.org

BIMe Initiative. (2019). *BIMe initiative. 201in Competency Table v2.1*. Retrieved from https://bimexcellence.org/resources/200series/201in/

Bozoglu, J. (2016). Collaboration and coordination learning modules for BIM education. *Journal of Information Technology in Construction*, *21*(10), 152–163.

buildingSMART. (2017). *Individual Qualification Scheme Launched*. Retrieved from www.buildingsmart.org

buildingSMART. (2020a). *buildingSMART Professional Certification Program. The global benchmark for openBIM training*. Retrieved from https://education.buildingsmart.org/

buildingSMART. (2020b). *buildingSMART Professional Certification. Frequently Asked Questions*. Retrieved from education.buildingsmart.org

Bush, R., & Robinson, M. (2018). *Developing a BIM Competency Framework. Research and key principles*. Scottish Future Trust. Retrieved from https://www.scottishfuturestrust.org.uk/

Byrne, C. (2015). *Building information modelling in Australia, lessons from the UK*. Higher Education and Skills Group.

Castello, S. (2020). *UK BIM Framework publishes new resource to support BIM training providers*. Retrieved from www.cdbb.cam.ac.uk

Change Agents. (2019). *What is BIM Excellence?* Retrieved from https://changeagents.com.au/bim-excellence.html

Cheng, J. C., & Lu, Q. (2015). A review of the efforts and roles of the public sector for BIM adoption worldwide. *Journal of Information Technology in Construction*, *20*(27), 442–478.

CIC. (2013). *BIM planning guide for facility owners. version 1.02*. Retrieved from https://bim.psu.edu/resources/owner/bim_planning_guide_for_facility_owners-version_2.0.pdf

Creswell, J. W. (2014). *Research design: qualitative, quantitative, and mixed methods approach* (4th ed.). SAGE Publications.

Creswell, J. W., & Poth, C. N. (2017). *Qualitative inquiry and research design: Choosing among five approaches*. Sage Publications.

Eadie, R., Browne, M., Odeyinka, H., McKeown, C., & McNiff, S. (2013). BIM implementation throughout the UK construction project lifecycle: An analysis. *Automation in Construction*, *36*, 145–151.

Fernandez Gonzalez, A. (2018). *BIM strategy and education in chile*. Universidad de Chile.

Gardner, J. C., Hosseini, M. R., Rameezdeen, R., & Chileshe, N. (2014). Building information modelling (BIM) education in South Australia: Industry needs. *5th International Conference on Engineering, Project, and Production Management (EPPM 2014)*, 293-302.

Government, H. M. (2013). *Industrial strategy: Government and industry in partnership*. Retrieved from https://assets.publishing.service.gov.uk/government/uploads/system/uploads/attachment_data/file/210099/bis-13-955-construction-2025-industrial-strategy.pdf

Government, H. M. (2016). *Press release: Launch of Digital Built Britain*. Retrieve from https://www.gov.uk/government/news/launch-of-digital-built-britain

Hardin, B., & McCool, D. (2015). *BIM and construction management: Proven tools, methods, and workflows*. John Wiley & Sons.

Hogle, L. (2013). *Industry + academia: The perfect partnership*. Academic Press.

Jin, R., Zou, P. X., Li, B., Piroozfar, P., & Painting, N. (2019). Comparisons of students' perceptions on BIM practice among Australia, China and UK. *Engineering, Construction and Architectural Management.*

Karjalainen, A., Alha, K., & Jutila, S. (2006). *Give me time to think: Determining student workload in higher education; has been written as part of the project titled" five years, two degrees", funded by the ministry of education, 2004-2006.* Oulu University Press.

Kocaturk, T., & Kiviniemi, A. (2013). Challenges of integrating BIM in architectural education. *eCAADe 2013: Computation and Performance–Proceedings of the 31st International Conference on Education and Research in Computer Aided Architectural Design in Europe.*

Kugbeadjor, W., Suresh, S., & Renukappa, S. (2015). BIM awareness and readiness of postgraduate built environment students in west midlands universities, UK. *CIB International Conference Proceedings. Going North for Sustainability: Leveraging Knowledge and Innovation for Sustainable Construction and Development*, 23-25.

Kumar, B. (2015). *A practical guide to adopting BIM in construction projects.* Whittles Publishing. Retrieved from http://lib.myilibrary.com?ID=883866

Lee, N., & Hollar, D. (2013). *Probing BIM education in construction engineering and management programs using industry perceptions.* Paper presented at the 49th ASC Annual Int. Conf. Proc., California Polytechnic State Univ., San Luis Obispo, CA.

Macdonald, J., & Granroth, M. Multidisciplinary AEC education utilising BIM/PLIM tools and processes. *10th Product Lifecycle Management for Society (PLM)*, 663-674.

Macdonald, J. A. (2012). A framework for collaborative BIM education across the AEC disciplines. *37th Annual Conference of Australasian University Building Educators Association (AUBEA)*, *4*, 4-6.

Macdonald, J. A., & Mills, J. E. (2011). *The potential of BIM to facilitate collaborative AEC education.* Paper presented at the American Society for Engineering Education.

Magiera, J. (2013). *Integrated teaching for integrated engineering practice-BIM on campus.* Paper presented at the 20th International conference on computer methods in mechanics, Poznan, Poland.

McGraw-Hill Construction. (2014). *The business value of BIM owners.* Retrieved from https://group.skanska.com/48f2cd/globalassets/about-us/building-information-modeling/business-value-of-bim-for-owners-smr-2014.pdf

Miller, G., Sharma, S., Donald, C., & Amor, R. (2013). Developing a building information modelling educational framework for the tertiary sector in New Zealand. *IFIP International Conference on Product Lifecycle Management*, 606-618.

Pikas, E., Sacks, R., & Hazzan, O. (2013). Building information modeling education for construction engineering and management. II: Procedures and implementation case study. *Journal of Construction Engineering and Management*, *139*(11), 05013002.

Plan BIM. (2017). *Roles BIM and Matriz de Roles BIM.* Comite de Transformacion Digital (CORFO). Retrieved from planbim.cl.

PlanBIM. (2018). *Plan BIM*. Retrieved from https://planbim.cl/

Plan BIM. (2019). *Estandar BIM para proyectos publicos. Intercambio de Informacion entre Solicitante y Proveedores*. Comite de Transformacion Digital (CORFO). Retrieved from planbim.cl.

ProBIM. (2016). *Equipo ProBIM.do. 1ra reunión de coordinación*. Retrieved from https://www.instagram.com/p/BOAv_cojY4c/

ProBIM. (2017a). *Conferencia Agosto 10*. Retrieved from https://www.instagram.com/p/BXDl_kgDdmu/

ProBIM. (2017b). *ProBIM. promoviendo una construcción más BIM y LEAN*. Retrieved from https://www.linkedin.com/company/probim-do/

ProBIM. (2019a). *¿Qué es BIM?* Retrieved from https://www.youtube.com/watch?v=SBwgYAihh1o

ProBIM. (2019b). *2da Jornada OPEN BIM*. Retrieved from https://www.instagram.com/p/B45vPgEH0dh/

Puolitaival, T., & Forsythe, P. (2016). Practical challenges of BIM education. *Structural Survey*, *34*(4/5), 351–366.

Rooney, K. (2017). *BIM education - global - 2017 update report*. Retrieved from https://www.canbim.com/single-post/2017/05/29/NATSPEC-Global-BIM-Education-Report-2017

Rooney, K. (2019). *BIM education – global – 2019 updated report*. ICIS Integrating Construction. Retrieved from https://www.icis.org/wp-content/uploads/2019/05/BIM_Education_Global_2019_Update_Report_V6.0.pdf

Sabongi, F. J. (2009). The integration of BIM in the undergraduate curriculum: An analysis of undergraduate courses. *Proceedings of the 45th ASC Annual Conference*, 1-4.

Sanchez, A. X., Kraatz, J. A., & Hampson, K. D. (2014). *Towards a national strategy. research report 1*. Retrieved from http://www.sbenrc.com.au/wp-content/uploads/2013/10/National-Strategy-research-report.pdf

Senior, B. A., & Rodriguez, T. A. (2012). *Analyzing Barriers to Construction Productivity Improvement in the Dominican Republic. In 20th Annual Conference-International Group for Lean Construction (IGLC-20)*. Montezuma Publishing.

Sharag-Eldin, A., & Nawari, N. O. (2010). BIM in AEC education. *Structures Congress 2010*, 1676-1688.

Shelbourn, M., Macdonald, J., McCuen, T., & Lee, S. (2017). Students' perceptions of BIM education in the higher education sector: A UK and US perspective. *Industry and Higher Education*, *31*(5), 293–304.

Silverio, A., Suresh, S., Suresh, R., & Heesom, D. (2016). *BIM Education Framework for Clients and Professionals of the Construction Industry*. Glasgow Caledonian University.

Silverio, A. K. (2020). *Implementation of Building Information Modelling in the Dominican Republic construction industry* (Unpublished doctoral dissertation). University of Wolverhampton, Wolverhampton, UK.

Silverio, A. K., Suresh, S., Heesom, D., & Renukappa, S. (2017a) BIM Education Framework for Clients and Professionals of the Construction Industry. *International Journal of 3-D Information Modeling, 6*(2), 57-79.

Silverio, A. K., Suresh, S., Suresh, S., & Heesom, D. (2017) Building information management (BIM) education in the Dominican Republic: An empirical study. In *International Conference on Sustainable Futures*. Applied Science University.

Soler, J., Salcedo, N., & Núñez, Y. (2013). *Estudio prospectivo de la formación profesional del sector de la construcción civil*. INFOTEP. Retrieved from https://www.oitcinterfor.org/node/5539

Succar, B., & Sher, W. (n.d.). A competency knowledge-base for BIM learning. *Australasian Journal of Construction Economics and Building-Conference Series, 2*(2) 1-10.

Suwal, S., Jäväjä, P., Rahman, M. A., & Gonzalez, V. (2013). Exploring BIM-based education perspectives. *Aubea–2013*.

Suwal, S., & Singh, V. (2018). Assessing students' sentiments towards the use of a building information modelling (BIM) learning platform in a construction project management course. *European Journal of Engineering Education, 43*(4), 492–506.

Tran, A., Mills, J., Morris, D., & Phillips, M. (2012). *"All hands on deck": Collaborative building design education for architects and engineers*. Paper presented at the Conference Innovation, Practice and Research in Engineering Education, Coventry University, UK.

Underwood, J., Ayoade, O., Khosrowshahi, F., Greenwood, D., Pittard, S., & Garvey, R. (2015). *Current position and associated challenges of BIM education in UK higher education*. BIM Academic Forum.

Underwood, J., Khosrowshahi, F., Pittard, S., Greenwood, D., & Platts, T. (2013). *Embedding building information modelling (BIM) within the taught curriculum: Supporting BIM implementation and adoption through the development of learning outcomes within the UK academic context for built environment programmes*. Academic Press.

UNIBE. (2013). *Especialista de BIM y AUTODESK imparte conferencia sobre implementación de procesos BIM e IPD en proyectos de construcción*. Retrieved from https://www.unibe.edu.do/es/noticias/thu-2013-07-18-1525/especialista-bim-y-autodesk-imparte-conferencia-implementacion-procesos-bim-e-ipd-proyectos-construccion

UNIBE. (2019). *Diplomado de innovación en arquitectura*. Retrieved from https://www.unibe.edu.do/programa/diplomado-de-innovacion-en-arquitectura/

Walters, A. (2017). *Press Release: Government Announce New Centre at Cambridge*. Retrieved from https://www.cdbb.cam.ac.uk/news/2017NovPressRelease

Wang, L., Huang, M., Zhang, X., Jin, R., & Ynag, T. (2020). Review of BIM Adoption in the Higher Education of AEC Disciplines. *Journal of Civil Engineering Education, 146*(3).

BIM Wiki. (2015). *BIM task group*. Author.

Witt, E., & Kähkönen, K. (2019). BIM-enabled education: A systematic literature review. *10th Nordic Conference on Construction Economics and Organization*, 261-269.

Wong, K., Wong, F. K., & Nadeem, A. (2011). Building information modelling for tertiary construction education in hong kong. *Journal of Information Technology in Construction*.

Wong, S. Y., & Gray, J. (2019). Barriers to implementing building information modelling (BIM) in the Malaysian construction industry. *Materials Science and Engineering*, *495*(1), 012002.

Wu, W., & Issa, R. R. (n.d.). BIM education for new career options: An initial investigation. Paper presented at the BIM Academic Workshop, Washington, DC.

Wu, W., & Issa, R. R. A. (2014). BIM education and recruiting: Survey-based comparative analysis of issues, perceptions, and collaboration opportunities. *Journal of Professional Issues in Engineering Education and Practice*, *140*(2).

Zou, P. X., Xu, X., Jin, R., Painting, N., & Li, B. (2019). AEC students' perceptions of BIM practice at Swinburne university of technology. *Journal of Professional Issues in Engineering Education and Practice*, *145*(3).

KEY TERMS AND DEFINITIONS

BIM-Centred Professional Organisations: Organisations that have been created exclusively to drive, promote and give support to the industry concerning BIM (Silverio, 2019).

Higher Education (HE) Institutions: Institutions that offer degree-level courses.

APPENDIX 1

Table 6.

Academia			
Strategy stage	Implementation stage		Revision stage
	Actions	Participants involved	
• Creation of groups with members from Academia and the industry to lead the strategy; • Identification of the BIM abilities needed per discipline from the educators; • Identification of the BIM skills per discipline required from the students; • Categorisation of these abilities by themes, according to the knowledge disciplines of the construction industry.	Provision of education and training required for the educators.	• Academics • BIM education providers • Industry	Assessment of the programs created by getting feedback from: • BIM learners: to express their opinion about program and learning outcomes; • Academics: by analysing the progress and performance of the students • Industry: by the evaluation of the BIM learners skills when practical implementation activities are programmed.
	Elaboration of the learning modules: • To be embedded in current programs; and • For the creation of new programs.	• Academics • Industry • BIM education providers • BIM learners • Researchers • Government (as support)	
	Creation of BIM learning materials for new modules inserted and programs modified.	• Academics • Industry • BIM providers • BIM learners • Researchers	
	Practical implementation of the knowledge acquired: • Knowledge transfer partnerships; • Internships; • Co-ops.	• Academics: for planning; • Industry: to provide opportunities to the BIM learners; • Government: to promote this practice in BIM education.	
	Diffusion of other BIM type of education and activities for knowledge boosting and acquisition of transferable skills through workshops, conferences, seminars, etc.	• Academics • BIM education providers • Industry • Government	
	Establishment of minimum knowledge requirements to undertake higher education degrees such as masters and research. That is to avoid teaching new things to students that have some gaps in knowledge.	• Academics • Industry: to identify problems in practice and propose research in the field.	

Industry				
Strategy stage	Implementation stage			Revision stage
	Actions		Participants involved	
• Designation of a BIM Champion to lead the strategy; • Identification of the BIM scope of the company; • Identification of the BIM abilities required per role in the organisation; • Identification of the training to be taken by the staff depending on their role in the team and requirements of the type of projects: - Managerial levels; - Remainder staff; - The client, when needed.	BIM education for the staff	Higher education: Large funding needed. *Suitable for large companies with considerable budget dedicated to training*	Provided by: • Academia	• Evaluation of the performance of the staff and the client; • Comparisons of project outcomes and performances before and after the strategy.
		Acquisition/upskill of software abilities:	Provided by: • Software developers; • Expert(s) hired by the organisation; or • Skilled staff from the company designated by the BIM champion.	
		Acquisition of different transferable skills: • CPD programs • Attend to workshops, conferences, seminars, etc.	Provided by: • Organisation; and • BIM education providers	
	BIM education to the client	To receive consultancy in the following areas: - Introduction to BIM; - Establishment of BIM goals for the project; - Documentation management.	BIM consultant from: • The organisation; or • An independent BIM consultant.	

Chapter 14
Development of a Building Information Modelling Threshold Capability Framework to Enable Global Curriculum Co-Integration

Kerry London
Torrens University, Australia

Nicola Willand
RMIT University, Australia

Peng Zhang
https://orcid.org/0000-0002-6855-3395
Western Sydney University, Australia

ABSTRACT

The construction and project management graduates are entering an ever-changing workforce that will require a smarter way of working. Creation, use, and management of building information modelling (BIM) models is a critical part of this smarter world. The aim of this research is to develop a threshold capability framework within the context of a broader digital construction project management curriculum to enable the global integration of BIM into a construction management curriculum. This chapter reports the evaluation of the framework through an analysis of 21 interviews with key stakeholder groups. Much effort is required to guide Australian construction practitioners to embrace a greater use of BIM in practice. This research identifies that the concept of graduate resilience skills in students is critical for the success of such a transition. Infusing construction management digital literacy is a long term and evolving exercise, and confidence in delivery capability must be simultaneously built.

DOI: 10.4018/978-1-7998-6600-8.ch014

INTRODUCTION

The delivery of construction projects has undergone a major shift internationally with growing utilisation of advanced information technologies. Worldwide project teams are increasingly benefitting from faster communications, more robust digital modelling tools and more powerful computers towards integrated delivery processes, all of which seemed unimaginable just a few years ago (McGraw Hill Construction, 2014). A critical part of this 'smarter' world is the creation, use and management of Building Information Modelling (BIM) models. As a set of tools, processes and practices (Demian & Walters, 2013; Eastman et al., 2018), BIM is being increasingly adopted by the architecture, engineering and construction (AEC) sector (Martínez-Aires et al., 2018; Hong et al. 2020) although adoption levels have been varied across, countries, sectors and supply chains (London & Singh, 2013, London et al, 2021). Importantly, BIM can be integrated with other advanced information technologies to leverage their advantages, including Internet of Things (IoT) (Dave et al., 2018; Teizer et al., 2017; Riaz et al., 2014), Robotics (Health and Safety Executive, 2018; London et al., 2018; Oesterreich & Teuteberg, 2016), and Cloud Computing (Fang et al., 2016; Park et al., 2017; Zou et al., 2017; Mutis & Paramashivam, 2019).

Worldwide the construction and project management graduates are entering an ever-changing workforce that will require a 'smarter' way of working. For an industry wide transformation to eventuate it is imperative that universities take a leading role in delivering graduates who are skilled in BIM (Hosseini et al. 2018). This is a very complex challenge for education providers. In Australia, it particularly represents an industry and nationwide shift for all built environment stakeholders. Higher education, industry training and professional development for BIM has the attention of numerous countries involved in infrastructure design, construction, use and management because the diffusion of this new technology and approach does not limit itself to the boundaries of one country. The holistic and all pervasive change in graduate attributes is being realised by many in industry, academia and government alike as BIM is being increasingly viewed as not simply a technological innovation or tool but a way of thinking. As Gu and London (2010) prophetically suggested, BIM is an IT enabled approach that is systemic and involves people, culture, processes, business decisions, market economics, work practices and regulatory and policy frameworks. Significantly, varying adoption rates by individual organisations influence diffusion in various sectors (London & Bavinton, 2006). It is noted that the varying industry adoption rates can also result in a symbiotic relationship with the curriculum redesign in specific localities. However the uptake particularly in the higher education sector in Australia has been challenging and Universities are not preparing graduates to lead a digital construction industry (Casasayas et al; 2021). Although a very limited study including 18 respondents to a survey and only five interviews across multiple disciplines, Casasayas et al (2021) identified key barriers in adoption of BIM education have been identified and include four key themes; 1) change management challenges, (2) curriculum and content limitation, (3) educators' problems, and (4) disconnect with the industry. Curriculum and content limitation barriers included three major subthemes including; 1) curriculum content was at capacity 2) difficulty of designing BIM related content and 3) difficulty of designin horizontal and vertical curricula. Curriculum design challenges is a worthwhile exploring barrier to explore in detail.

A conclusion was made by these researchers that professional associations, government, industry and academia need to collaborate to change the curriculum and the suggestion was that educators are not completely empowered to change curriculum. This proposition is challenged and is a limited perspective of the dual leadership and learning role that higher education institutions (HIE) should and can play when designing,.cvvvvvvvv][developing and delivering innovative curriculum. That is not to say that we should not collaborate with all stakeholders but HIE should not abrogate their responsibility to educate graduates for a digital future.

There is an emerging globalisation of construction practice and education. Not only are there multinational project teams but academic institutions are also collaborating to provide graduates with more opportunities to study in different countries. Incorporating BIM learning and teaching into curricula to account for the practice of working across borders has not been considered to date. In order to support student career options after graduation, students in construction management need to be prepared for increased internationalisation, be able to lead projects that are delivered through an integrated model and be comfortable with the increased adoption of virtual design and construction through BIM models and decision making through augmented reality (McGraw Hill Construction, 2014). Despite the apparent absence of formal BIM accreditation criteria, academic institutions have realised the need to incorporate BIM to prepare their graduates for managing constructions projects in the digital age (Wong et al., 2011). BIM is increasingly appearing in construction management courses in the US, UK, Scandinavia, Singapore, Hong Kong and Australia (Wong et al., 2011; Joannides et al., 2012; Wu & Issa, 2014). However, conventional approaches to curriculum design may no longer be suitable for nurturing future construction professionals. There is a need to rethink the type of professional the higher education providers are seeking to develop and with a much greater emphasis on analytic and decision-making capabilities. Key to this change is a transformation of teaching and learning of the various uses of BIM to accommodate the emerging international trend of advanced information technology adoption in the industry. This research sought to guide curriculum redesign in BIM education with broad applicability to disciplines with external professional accrediting bodies and recommendations of principles specific to construction management program delivery.

The goal of the higher education providers is to ensure that graduates will have the theoretical knowledge, technical skills and capabilities to enable them to work effectively as construction management professionals. To critically engage with contemporary construction management theory and situate the graduates' learning in the wider industry context, the higher education providers need to establish some form of commitment to BIM. Therefore, the aim of this research is to develop a framework to enable the global integration of BIM into a construction management curriculum that is grounded in theory and practice. A Threshold Capability framework is developed and the authors will report on the evaluation and refinement of this framework through empirical evidence in this chapter. The overarching question that guided the investigation is:

How may a construction project management curriculum be structured to support the global trend towards increasing use of BIM and related technological advances?

The aim of this chapter is to contribute to the facilitation of BIM integration in curricula and also shift the discourse on pedagogical design and developments for the field.

Theoretical Framework

The literature review provides an overview of BIM education in construction with a focus on the approaches to curriculum design. This research identified the theory and practice underpinning curriculum redesign to incorporate BIM into programs, the wider discourse in curriculum design and the nature and response to these changes in the local market in Australia. In recognition of the need to integrate BIM as a construction management process requiring multidimensional learning concepts and outcomes, the authors developed a theoretical model of curriculum design based upon the educational Threshold Concept theory.

Literature Review

Significant steps have been made by various institutions and industry bodies to support the digitisation of the construction industry and BIM adoption. For example, in 2019, the UK BIM Alliance, British Standards Institution and the Centre for Digital Built Britain announced the launch of the UK BIM Framework. This framework sets out a guided approach to implementing BIM in the UK and managing information provided by the BS EN ISO 19650 series. As a global professional association, buildingSMART aims to drive the digital transformation of the built asset industry. In 2017, the buildingSMART National BIM Initiative launched BIMcreds, which provides BIM practitioners with a framework for demonstrating their competence. Also, as a major worldwide development across all 16 international chapters of buildingSMART, 6 modules within a foundation professional certification programme is in development.

However, although very laudable initiatives they have little tangible and direct impact on the curriculum in the Australian Universities, as these organisations are not responsible for accrediting undergraduate Construction Management program in Australia. The main Accreditation bodies are the Australian Institute of Quantity Surveyors (AIQS), Chartered Institute of Building (CIOB), Royal Institution of Chartered Surveyors (RICS), and Australian Institute of Building (AIB).

We now turn to a brief consideration of how BIM is assessed by accrediting organisations. BIM is not required in the latest AIQS Course Accreditation Policy (AIQS, 2017). BIM is also not one of the threshold outcomes that CIOB expects that the core undergraduate programme modules should meet (CIOB, 2018). RICS requires the programme to map to the mandatory and core competencies of a RICS pathway however BIM is not included in the mandatory and core competencies list (RICS, 2019). Communication is one of the required General Graduate Attributes from AIB. One of the characteristics of Communication is 'understanding of the use of information and communication technology in the construction industry (e.g. CAD, BIM)' (AIB, 2016) and this means that BIM is not necessary if CAD has been included in the curriculum.

Universally accepted curriculum guidelines for undergraduate degree programs in BIM for construction management programs are still not prevalent. As noted earlier BIM is being taught in more and more construction management courses. In the US in particular construction schools are increasingly including BIM content into courses whereby a survey as far back as in 2012 found that a large percentage of the respondent schools were planning to fully integrate BIM into their curriculum (Joannides et al., 2012). Nonetheless, a universal approach or national framework for teaching BIM in AEC programs in the US seemed to be missing (Sacks & Pikas, 2013). There is no national framework in use in Australia although the 'Collaborative building design education using Building Information Modelling (CodeBIM)' (Mills et al., 2013) was a national Learning and Teaching research project funded by the Australian Government Office for Learning and Teaching (OLT) that sought to raise awareness within the Australian context.

Previous research has synthesised current approaches of integrating BIM into construction management programs. The attitudes to and extent of introducing BIM teaching into current curricula ranged from the mere acknowledgement of the presence of BIM to the embracement of BIM into all elements of the program and across disciplines where BIM evolves into a 'lingua franca' (Pikas et al., 2013; Sacks & Pikas, 2013; Zhao et al., 2015; Huang, 2018; Olowa et al., 2019; Abdirad & Dossick, 2016). While the aims, conceptual frameworks and approaches to curriculum design of various initiatives differ, there is an agreement that BIM should foster interdisciplinary collaboration. BIM education should also cover all aspects of BIM, namely people/culture/ project team roles (psychological and social skills), processes/ management processes (conceptual skills) and technology/software, in a balanced manner. Construction courses should stress BIM as a methodological rather than a mere technological tool (Barison & Santos, 2010; BIM Academic Forum UK, 2013; Macdonald, 2011; Mills et al., 2013; Sacks & Pikas, 2013). Concerted efforts to introduce BIM into construction management programs have addressed the cross-disciplinary dimension of BIM by introducing inter-disciplinary modes of teaching delivery (Barison & Santos, 2010; Becerik-Gerber et al., 2011; BIM Academic Forum UK, 2013; Mills et al., 2010). In addition, a comprehensive approach to content and curriculum design with BIM as the vehicle is espoused by leading academics and industry (Gier, 2008; Lee et al., 2013; Lee & Hollar, 2013; McCuen, 2014; Mills et al., 2010; Mutai & Guidera, 2010; Sacks & Pikas, 2013), yet evidence for a successful incorporation of BIM capabilities around core aspects of construction management theory in construction management curriculum is missing to date.

There appeared to be consensus that BIM education should be delivered through sequential learning, allowing the continuous and progressive build-up of levels of learning as students advance through their degree (Azhar et al., 2010; BIM Academic Forum UK, 2013; Maghiar, 2014; Pikas, Sacks & Hazzan, 2013). Prerequisites for any BIM course should be a basic understanding of construction drawing principles and building technology, science, professional practice, programming language and data modelling concepts (Barison & Santos, 2010), with the focus in intermediate and finals years shifting to BIM collaboration (Becerik-Gerber et al., 2011). Real life learning through interactions with industry were also recommended (Barison & Santos 2010; BIM Academic Forum UK 2013; Mills et al., 2013; Molavi & Shapoorian, 2013; Sacks & Pikas, 2013). BIM in the context of internationalisation was, however rarely a subject of discussion; from either the student mobility aspect or the implications of international project teams. With regards to the process of developing curricula in universities, researchers agreed that each institution would need to develop its own approach, based on current BIM content in the curriculum, available infrastructure and current staff skills (Williams & Lees, 2009, p8 as cited in BIM Academic Forum UK, 2013). Learning outcomes and targets levels should also be determined based on local industry needs and the students' existing knowledge (Sacks & Pikas, 2013).

BIM Threshold Capability Framework Introduction

Discourse in Curriculum Design

In keeping with the predominance of current outcome-based education in higher education curricula (Allan, 1996), the reviewed curricula were discipline bound, promoted sequential learning with the aim to provide 'cognitive coherence' (Ensor, 2004) and proposed program and course specific learning outcomes based on progressive learning levels. This approach to learning and curriculum development is based upon a concept referred to as Outcome Based Education (OBE). OBE addresses individual achievements at hierarchical levels and focusses on pre-specified learning outcomes and behavioural objectives (Bloom, 1956). However, considering the potential of rapid and uncertain transformation of the building industry in this particular knowledge domain (Azhar, 2011; McGraw Hill Construction, 2014; Philp & Thompson, 2014) which requires cognitive, psycho-social communication and negotiation skills, proactive behaviour and self-directed learning in response to continuous innovation (Newton,

2011), it might be timely to question this approach. OBE is considered to explicitly exclude affective outcomes (Spady, 1994), prevent wider learning (Harðarson, 2012) and inhibit autonomous learning (Maher, 2004). Conventional approaches to structuring learning outcomes may not meet the challenge of different learning styles among the increasingly diverse student cohorts at tertiary education institution which in part is a result of the internationalisation of higher education (Ramburuth & McCormick, 2001). The Threshold Concept theory has offered an alternative approach to learning outcomes when structuring a curriculum in higher education.

Threshold Concept theory approaches learning as an epistemological process and addresses the "way of thinking and practising" (Meyer & Land, 2003) that is particularly relevant to the call for embracing BIM as a methodological tool. First advocated by Meyer and Land (2003), threshold concepts are "critical points in a student's learning" (Barradell, 2012, p. 268), events in understanding that may shift the learner's perception and world-view (Land et al., 2014; Meyer & Land, 2003). The Threshold Concept is particularly relevant to the call for embracing BIM as a methodological tool as it addresses the dynamic environment of this emerging technology and the dual role of academics as adopters, teachers and carriers of new construction management practices (Anderson, 2016). Threshold concepts focus on the attributes and quality rather than on the subject of learning. They help to move the discourse in higher education from emphasising the 'what' to being concerned with 'how' and 'why' in the context of a dynamic and difficult to predict future (Barradell & Kennedy-Jones, 2015). It is believed that using pre-defined and sequential threshold concepts as the framework for curriculum design may facilitate a more efficient and succinct curriculum that is mindful of the non-linear nature of learning (Cousin, 2006). In response to the criticism of arbitrariness and lack of rigour in the identification of thresholds (Barradell, 2012; O'Donnell, 2012; Tight, 2014), a 'transactional curriculum inquiry' that includes all stakeholders, i.e. teaching staff, students and external stakeholders has been proposed (Barradell, 2012). This is not that different to the Accreditation process by the various professional associations and School and/or Program Industry Advisory Boards that support many of the construction management programs in most countries around the world today.

The Threshold Concept Theory has been applied in disciplines including the sciences, health care, literature business, law and social studies (Tight, 2014). For example, an Australian team of educators at the University of Western Australia combined the threshold concept with capability and variation theory into the Threshold Capability Integrated Theoretical Framework (Baillie et al., 2012). This Framework was underpinned by a series of questions to guide a curriculum design process and used to develop a curriculum of a two-year engineering foundation course. The Threshold Concept had the advantage of making apparent implicit learning concepts which facilitated the focus on the most important steps in the students' learning progress, the most appropriate educational experience and the best assessment methods (Baillie et al., 2012). Capability theory is concerned with teaching students the capabilities to deal with unforeseeable future events and problems (Bowden, 2004). Capabilities were merged with the threshold concept as the elements that were deemed troublesome were not always disciplinary concepts but included capabilities to cope with new situations (Baillie et al., 2012). While in the threshold concept theory the disciplinary knowledge content is driving the structure of a curriculum, curriculum design based on capability theory focusses more on situational and progressive learning and the requirement of fundamental cognitive attitudes that are seen to be essential in becoming successful professionals (Åkerlind et al., 2014; Bowden, 2004).

The extension of the threshold concept into the integrative model developed by Barradell and Kennedy-Jones (2015) promised the delivery of a curriculum that was in keeping with Australian Learning and Teaching Academic Standards requirements. In this integrative model the elements of threshold concepts, student learning and curriculum were combined, and the outcomes of ways of thinking and practising, liminality, meaningful learning and meta-learning were seen to happen at the intersections of the elements. The researchers argued that only the integration of the content-focused thresholds into the other essential constructs of education, namely desired learning outcomes and learning processes, would provide students with the necessary educational foundation for their future professional life (Barradell & Kennedy-Jones, 2015).

In 2011, the national project "Learning and Teaching Academic Standards Project (LTAS)", funded by the Office of Learning and Teaching addressed discipline specific competencies. The project developed Threshold Learning Outcomes (TLOs) for the Building discipline (Newton, 2011).

The national Building Discipline Threshold Learning Outcomes was also instrumental to this study. All TLOs had relevance to curriculum redesign, however, importantly the TLO1 organises curriculum into four main streams including; technology, management, economics and law, and TLO4 is specifically focused on interpreting and negotiating construction information.

In keeping with Diamond's (2008) two phase approach, the curriculum redesign should be presented an iterative and inclusive process that accessed perspectives from teaching staff, students and external stakeholders as well as being informed by the literature.

Diamond (2008) splits the process of curriculum design into two phases: the revision of the program curriculum (Phase I) and the development of the curriculum at the course level (Phase II). On the premise that there is agreement on student-centred learning, he proposes to start the process of curriculum revision with a diagram illustrating the structure, flow and content of the curriculum called curriculum audit or curriculum mapping. The next step, data gathering, should address the nature of the students, the requirements and priorities of society and the institution itself, the focus of the discipline related knowledge and be collected from employers, past graduates, student recruiters and the literature. The data analysis is supposed to reveal variations in perceived priorities and in the perceived effectiveness of the current curriculum. Based on the findings, an ideal curriculum should be designed with focus on sequence and cohesion. The final stage of the curriculum design should be the adjustment of the ideal to the possible program based on the necessary extent of the revisions and resource constraints. In Phase II, learning outcomes for individual courses should be developed, educational methodologies, activities and assessments assigned and tested. Continuous evaluation of curriculum outcomes should be used to revise and adjust the planned curriculum (Diamond, 2008).

Threshold Capability Framework

In the curriculum redesign, this research sought to address the strained relationship between the Threshold Concept theory to its application to outcome-based curricula by adopting a hybrid approach. The starting point for developing a BIM Threshold Capability Framework is the idea of the Threshold Capability as developed by Baillie et al. (2012). It merged knowable disciplinary threshold learning concepts with the capability theory and the need for students to be able to deal with unforeseeable future events and problems. This approach complemented the finding of the researchers' review of the BIM educational literature and the research team's research and knowledge about BIM adoption in Australia that the adoption of BIM was not static and that the graduates who would be entering the workforce would encounter quite diverse attitudes to adoption and varied BIM policy, process and practices. The authors' aim is to be mindful that students would need to be able to cope with new situations and be resilient to changes and challenges. A hybrid approach in the curriculum redesign could balance the threshold concept theory, in which disciplinary knowledge content is driving the structure of a curriculum, and curriculum design based on capability theory, which focuses more on situational and progressive learning and the requirement of fundamental cognitive attitudes that are seen to be essential in becoming successful professionals.

This research started with the Threshold Concept, as it promised a way to frame authors' investigation into exploring the challenges of teaching BIM and integrating it into the curriculum. Key to the authors' efforts was to focus on the first two questions in the Baillie et al. (2012) Threshold Capability Integrated Theoretical Framework:

1. What should the learner be capable of doing at the end, given the need to deal with an unknown future?
2. What threshold concepts are important to understand to enable the development of such capability?

A preliminary Student Threshold Capability Framework (Figure 1) is developed based upon the literature review with the view to challenge, test and refine this framework through an empirical study. The threshold concepts in relation to different learning categories are considered (Baillie et al., 2012), for example, learning to think like a BIM construction manager, understanding and learning to become a BIM construction manager and shaping the world as a BIM construction manager. In the first instance, the authors are guided by past work that identified critical points and thresholds in BIM adoption in Australian firms (London et al., 2006).

The preliminary Threshold Capability Framework is arranged according to the four threshold concepts of cognition, compatibility, connectivity and integration. The first threshold in the Framework is Introductory and termed the 'Cognition Level', which is concerned with learning to think about BIM environments and developing basic capabilities to operate within a BIM environment as a construction manager. The next level is considered 'Advanced' and is termed the 'Compatibility Level', which is concerned with learning to think, understand and act like a construction manager who integrates people, systems and processes within BIM environments, i.e. an understanding of the ability of various systems to be able to coexist harmoniously. The third level is 'Application', and is termed the 'Connectivity Level', which is concerned with learning to effectively collaborate with others involved in BIM projects and demonstrate intellectual independence and autonomy to solve problems in BIM environments. The researchers then departed from London's model and introduced the final level which is 'Self-applied' and is termed the 'Integration Level', which is concerned with learning about what it means to lead BIM projects and understand the organisational environments required to shape the world for multiple and diverse connections across organisations. These four levels may correspond broadly to the four years of instruction in Universities' program. Further to this philosophical conception of these four threshold concepts of cognition, compatibility, connectivity and integration, the Framework identifies specific BIM themes and content which the authors have that were synthesised and distilled from the literature and which may assist the learner in achieving the threshold. The authors have synthesised the BIM education literature content, reflected upon University's strategic goals in relation to international education (as this is the University in which the Project Leader was employed at the time of the study) and identified the following key five thematic content areas:

- Fundamental principles.
- Technical skills.

- Construction project management skills.
- Strategic organisational behaviours.
- Global market context.

The following Figure 1 presents the conceptual model of a Threshold Capability Framework for BIM in a construction project management curriculum with an initial consideration of some of the detailed content topics within each of the five thematic areas.

Figure 1. Preliminary Student Threshold Capability Framework (developed for this research)

Threshold Capability for BIM in a Construction Project Management Curriculum

Threshold concepts and capabilities are certain concepts and capabilities within a field that are required to be held to ensure mastery of a particular knowledge domain.
Threshold concepts and capabilities are transformative, troublesome, irreversible and integrative.

THEMES	**Introductory: Cognition Level** Learning to think about BIM environments and develop basic capabilities to operate within a BIM environment as a construction manager. **Year 1**	**Advanced: Compatability Level** learning to think, understand and act like a construction manager who integrates people, systems and processes within BIM environments. **Year 2**	**Application: Connectivity Level** Learning to effectively collaborate with others involved in BIM projects and demonstrate intellectual independence and autonomy to solve problems with in BIM environments. **Year 3**	**Integration - Self Applied: Capstones** Learning to lead BIM projects and organisational environments by shaping the world for multiple and diverse connections of people, organisations and systems . **Year 4**
FUNDAMENTAL PRINCIPLES	Generic: Concept, Definition, Trends , Stakeholders, Implementation, Project Phases, Scope, Purpose, Conflicts	CM /QS Discipline principles application: safety, quantities, visualisation and communication, scheduling, clash detection, site planning logistics, Project Method Statements, tendering and procurement, constructability	Examples of emerging theories	final year capstone experience
TECHNICAL SKILLS	Skills: Opening, Notations, Sharing, Importing, Exporting, Software: AutoCAD, Google Sketch, REVIT, Navisworks, Infraworks	Skills: Clash detection analysis, BoQ generation, Generation PMS, Proof of Concept/Re-engineering/VE, Design>Prefab, Construction sequencing, MEP trade coordination model sharing. Software: Autodesk Quantity Take off, Bentley, Solibri, VICO, Navisworks, Infraworks	Exemplars of major construction projects	
CONSTRUCTION PROJECT MANAGEMENT SKILLS	Cognition Level : Simple team environments: Collaboration within the team and organisation, managing the environment	Compatability Level: Complex collaborative environments: large scale projects or multi project contexts, leading the environment, virtual teams	Exemplars of construction organisations in networks that are highly connected (Connectivity Level) examples of entrepreneurship	
STRATEGIC ORGANISATIONAL BEHAVIOURS	CBA of BIM implementation, simple decision making, identification of purpose and small firm strategy, simple Model Ownership & IP rights, BIM Management Project Plans	Strategic Business Case Planning, Complex decision making, diverse strategies of Model Ownership, staff training - project start up (Level 1) , systems integration (Level 2) and knowledge management (level 3); BIM Operational Plans	exemplars of HR training plans, global systems integration and business models of BIM implementation	
GLOBAL MARKET CONTEXT	Data types and management, data and information regulatory and policy framework, standards and protocols, national organisations (BuildSmart), Australian accrediting bodies standards/expectations, regional norms, values and accepted practices; beginning to understand cultural contexts and how this shapes BIM implementation	Inter-operability protocols (history of STEP/BIM/), international organisations, international accrediting bodies expectations/standards, reflexive approach to balancing the tension between local context, different ways of doing things in different countries but still operating in a multi country project team environment with various standards;	Case studies of companies that have a corporate social responsibility policy that explicitly demonstrates respect and enhancement of diversity	

"A threshold concept can be considered as akin to a portal, opening up a new and previously inaccessible way of thinking about something."

RESEARCH DESIGN

A phenomenographic approach was adopted to validate and refine the Student Threshold Capability Framework. Phenomenography is a research approach specifically "designed to answer certain questions about thinking and learning" (Marton, 1986, p. 28). Phenomenography explores the variations in understanding of a phenomenon among different groups of people (Larsson & Holmström, 2007) and is frequently used in research in higher education (Åkerlind et al., 2014; Entwistle, 1997). The focus of a phenomenographic approach is on the collective rather than on the individual which is the key difference to a phenomenological research approach. This helps to convey the common experiences or perceptions of a phenomenon within a group and to compare these to the common experiences or perceptions of other groups of people to find categories or relationships that are linked logically and often hierarchically (Åkerlind, 2012).

In this study, a draft Preliminary Student Threshold Capability Framework was developed to articulate the phenomenon of the BIM curriculum redesign and conceptual learning to relevant stakeholder groups in construction education. In keeping with the best practice approach in curriculum development of incorporating peer-review (Diamond, 2008), the researchers engaged with the construction education stakeholders including staff, students, employers and representatives of the professional bodies, to validate and improve the draft framework. The objectives were to initiate a dialogue on the integration of BIM, to foster communication and cooperation among staff and with the industry and to collaboratively review and optimise the draft framework with the aim of producing a curriculum that would reflect the goals and visions of the academic staff and the profession.

Data Collection and Analysis

Qualitative data was collected through semi-structured interviews and focus groups. Recruitment used a combination of purposeful sampling framework to cover all accreditation bodies and maximise diversity in terms of stakeholders and companies, and the snow-balling method to reach participants with BIM experience who were deemed hard to find in the Australian construction industry. Hence, the research team invited 26 eligible persons known to them or identified from accreditation body websites, and respondents and interviewees suggested further potential participants. Data was collected from 21 participants who had a basic understanding of higher education construction programs and some experience with BIM. The participants included employers of graduates, representatives of accrediting bodies, recent graduates, an international educator, staff teaching into programs and a Head of School (note: one participant represented an employer as well as two professional associations simultaneously). The interviews were conducted face to face or online via skype with the interviewsfollowing an interview protocol and running for no more than 90 minutes. To provide context, interviewees were firstly asked to introduce themselves and share their current experiences with BIM, expectations and attitudes to BIM education. Subsequently, participants were asked to comment on the Premininary Student Threshold Capability Framework which had been provided to them prior to the interviews. The data analysis was deductive following Halcomb et al.'s (2006) approach of using a coding matrix that was completed by each member of the research team immediately after the interview. This process was followed by the research team reflecting on the interview and its interpretation to reach a consensus. After the team discussion, the individual matrices were combined into one integrated matrix. Further to this, themes were identified, summarized and compared across stakeholder groups. This paper reports specifically on the participants' responses to the draft Threshold Framework.

RESULTS

Overview

The results presented are summaries of the responses to the Threshold Capability Framework from 21 research participants across the various stakeholder groups including past graduates (4), employers (7), national and international accrediting bodies (3), academics (6) and an international educator (1). Table 1 provides a summary of the participants interviewed alongside their roles in their organisations.

Table 1. Summary of participants interviewed

	Participants	Role	Organisation
1	Employer 1	Senior Business Analyst/Estimating Systems	Subsidiary of an international contracting company in the infrastructure, building and mining sectors
2	Employer 2	General Manager, National Building and Business Improvement	Residential building company with about 3000 homes built per year
3	Employer 3	BIM Manager	Subsidiary of an international contracting company in the infrastructure, building and mining sectors
4	Employer 4	Owner	Tender management platform used regularly by 500 contractors/subcontractors and suppliers; handles 75,000 tenders/year
5	Employer 5	Engineering interface manager	Pipeline construction only package
6	Employer 6	Innovation manager	International contracting company in the infrastructure, building and mining sectors
7	Employer 7	BIM practice lead for New Zealand and Australia	International company providing architectural, design, engineering and construction services
8	Accreditation body 1	Chair	Chartered Institute of Building (CIOB) BIM working group
9	Accreditation body 2	AIQS rep on Standards Australia, BIM Working group	Royal Institute of Chartered Surveyors (RICS); Australian Institute of Quantity Surveyors (AIQS)
10	Accreditation body 3	AIB National Vice President	Australian Institute of Building (AIB)
11	Graduate 1	Product manager	Tender management platform used regularly by 500 contractors/subcontractors and suppliers; handles 75,000 tenders/year
12	Graduate 2	Assistant Development Manager	Residential volume builder listed on stock exchange
13	Graduate 3	Quality Engineer in OSM	Tier 2 builder that produces modular construction units
14	Graduate 4	Quantity Surveyor	Quantity surveying company that works for small and medium sized, low rise residential projects
15	International educator	Associate Professor and Graduate Program Chair	An USA University
16	Academic 1	Lecturer	The Australian University
17	Academic 2	Lecturer	The Australian University
18	Academic 3	Lecturer	The Australian University
19	Academic 4	Lecturer	The Australian University
20	Academic 5	Lecturer	The Australian University
21	Academic 6	Head of School	The Australian University

Responses to the Threshold Capability Framework

Matrices were developed for each stakeholder group in terms of their reflections on the Threshold Capability Framework. Participants were asked to provide their thoughts on 1) the integrated model underpinning the framework, 2) the content topics across the five thematic areas and 3) ideal delivery or teaching methods for BIM. A summary discussion of the participants' responses to the framework is now provided.

Responses to Integrated Model

In general, the 21 participants responded positively to the threshold capability framework. The response of the past graduates, in particular, was very positive as they saw the validity of the framework and liked the themes as well as how the curriculum was broken up over the four years within each of those themes. The following is an example quote in relation to the participants' response to the framework:

Five areas make sense, range of technical through to strategic components are perfect, important to have market context. (Employer 2)

The participants were also in general agreement about the philosophy of an integrated approach to teaching BIM rather than a stand-alone one. One participant explained that the threshold diagram was useful as it reflected the teaching of BIM as a complement, rather than a substitute, to project management and construction management skills. This promised to produce graduates who are project and construction managers with BIM capabilities rather than BIM technicians.

Participants raised the significance of increasing student awareness for the fundamental principles of BIM. The "bigger picture of BIM" needed to be explained to students so that they "[knew] the why right from the start" and "how BIM could add value" to their professional practices. At different points of the various interviews nearly all participants stressed that BIM should not be taught like a tool or software. In order to make BIM education successful, a holistic approach should be taken to enable students to learn the consequences of their professional legacy. The employer group suggested that one way of achieving the holistic approach was by demonstrating to students the relationship between BIM and their careers and more specifically how BIM could benefit them in their career development. The employer group also stressed that the curriculum should reflect the broad applications of BIM as the B for 'building' in the acronym may be misconstrued as implying 'building works only'.

Responses to Thematic Content Areas and Topics

The discussion was quite wide-ranging about the skills of the students and topics to be covered as part of the integrated BIM curriculum. Four key themes were identified including technical skills; graduate resilience; construction project management skills and BIM in the global context which are now discussed.

Technical Skills

Some positive observations regarding the framework were made in that teaching technical skills about BIM to students was important. A need was also highlighted to consider specific technical aspects beyond the conventional design-focused applications of BIM which would enable the collection of feedback from site onto the model in areas such as concrete shrinkage and how this may affect on-site dimensioning works. Even though the idea of equipping graduates with strong technical skills related to BIM was largely supported, one participant warned against a preoccupation with specific tools, by highlighting the dynamics of technological developments:

Do not fixate too much on tools, as they will change. (Employer 7)

Nonetheless, students should understanding how to design in BIM and be familiar with the ability to use pdf drawings (2D) and BIM (3D). Students should also be encouraged to complete a partially designed BIM project, e.g. a sub-model of a services lay out. In addition, the curriculum should explose students to a variety of modelling softwares and disciplinary applications, including civil engineering works. In addition, an elective on advanced BIM should be offered.

Graduate Resilience

An important discussion took place in relation to the term "student resilience". It was elaborated that students should be made aware that they might not be able to apply their BIM skills immediately after graduation. A recent graduate stressed that students should be prepared for the reality of limited adoption of BIM in the Australian construction industry:

You don't want [graduates to know] BIM is the only thing they use - they would be shocked [when they land within the industry]. (Graduate 1)

Students should be conscious that the BIM approach was "really powerful" (Employer 2), however that its significance may only become evident in a later stage of their careers. Nonetheless, graduates of the BIM curriculum should not be disillusioned by their employer's potential lack of engagement as pointed out by a participant:

Resilience is really important... For their first…or two...or four or five years, maybe they are just going to accept the fact that they are going to meet the organisational imperatives right now. But I am not going to lose sight of the fact that I've been trained in something that is really powerful. At some stage, I want to see that realised. So that when I start to manage people, or I have my own business, or I am leading a business, or whatever it is, there is a world out there that can be done a whole lot better. The resilience to say 'I am just going to meet my organisational imperatives for now, I am going to do what I am told today, and tomorrow and the day after…Resilience is a good one. (Employer 2)

Certainly the graduates interviewed were in agreement as they too highlighted the importance of providing students with resilience skills in relation to intergenerational communication or "combatting reluctance by older and influential people and how to communicate or show them the benefits of BIM" (Graduate 2).

Construction Project Management Skills

Participants also mentioned that students should learn about project planning and control, integrated project delivery and scope management on the basis that these were important for all stages of the project life cycle. Other topics included change management and impact upon time and cost, BIM modelling work flows, and BIM management project plans. On the operational side, it was highlighted that setting projects up for asset management was becoming increasingly important:

We need to know what facility managers want from the model. (Employer 5)

Suggestions for other areas that were deemed important with regards to construction project management skills included:

- Understanding of all stakeholders' role in creating, using and managing a BIM.
- Interacting with other people including BIM designers during the process.
- Integration and lean construction.
- Procurement and construction using the model and its potential use in housing.
- Site operations modelling and site logistics; i.e. what is around the building; streets, roads, energy supply.
- Materials tracking through radio-frequency identification (RFID) scanning and linking it to BIM and scheduling.
- Understanding BIM information input and how that could become more efficient.
- Use of BIM for the Off-site Manufacturing processes within the industry.

BIM in the Global Context

Since this Australian University is a global university of technology and design, the participants agreed that students should learn the global matkey context of BIM as part of the program. However, there was some confusion of how to integrate this theme within the framework because working in a global market was not covered in the current construction management curriculum. In general, though, participants recommended that the content of the early years in Strategic Organisational Behaviours and Global Market Context should be more exploratory, and that advanced concepts and applications should be introduced in the latter part of the undergraduate degree in year 3 and year 4 or be left for postgraduate studies. Nonetheless, participants suggested that the framework should include an understanding BIM related technical issues with international perspective and not to be too Australia specific.

Suggestions for Ideal Teaching/Delivery Methods

Participants also made suggestions for advantageous teaching methods, including case studies, T-shaped and action learning, and raised expected challenges in implementing BIM into the curriculum.

Case Study Models

Academics, employers and representatives of accreditation bodies agreed that it was critical that academia and industry should collaborate so that best practice examples could be demonstrated. Students should also understand the business and enterprise aspect of the industry, in particular with regards to the internationalisation of construction, as stated by one participant:

It is so important to have business context. I guess that is the global market context and the strategic organisation behaviour part in the diagram…For our industry, in these construction project management skills, that collaboration piece is... try to get some really powerful examples, best practice examples, that would be really useful…This is critical. (Employer 2)

Case studies would help in walking students through different stages of a project including investment planning, business case, concept stage, as well as the design, construction and operational phases. The participants were in general agreement that the program should provide exemplars and real world examples to articulate the concepts and uses of BIM as "they are easier to remember than something abstract". In group work, sending work packages from one group to another would also help students to learn how to communicate clearly.

Participants in all groups offered valuable recommendations on how to provide students with a "hands on approach" by bringing "industry into the classroom":

- "Pull in an alumni and their experience with BIM." (Academic 5)
- "Connect different courses with one real project...get one student from project management, construction management and architectural students in one course that pulls together all you have done, work as a team." (Academic 5)
- "We just have to get a model and share it across courses." (Academic 1)
- "Give students a BIM project and do price take-off." (Graduate 1)

It was also noted that one of the University's latest buildings had 3D design model/BIM which academics could utilise to explain the nature and content of the model and its usefulness for the varied disciplines. As suggested by a participant from academia: "We could use the model to explain the different things that make up a wall for example; and turn things off in a BIM model that are dumb objects that do not have properties assigned, and see how it was created in the first place".

Models/Approaches to Learning

An industry body representative suggested the adoption of T-shaped learning. T-shaped learning combines the acquisition of in-depth disciplinary knowledge, as represented in the vertical bar, with cross-disciplinary competencies, as represented in the horizontal bar. In this mode, the first year of the program should focus on team building amongst students from different pathways to be followed by a a specialisation in later years. A couple of suggestions were to start teaching the basics of BIM within the current 'Communication course' and to introduce BIM gradually into later, discipline-specific courses at an advanced level.

In addition, some participants suggested that BIM should be taught through action learning projects which would help students promote BIM transitions in their prospective workplaces, for example:

Help craft transition, or be involved in the implementation of it, or diagnose it in a business who is thinking about it. (Employer 2)

Challenges to Teaching BIM

It is also important to note that most of the academics interviewed identified the lack of fundamental student IT skills and resources as a major challenge for teaching BIM. There were some concerns raised about the training that the School could offer to the teaching staff. Associated with this were challenges in the provision of infrastructure support namely in two areas including; the lack of IT support and capability at the University level, and the limited number of BIM models that could be used as teaching examples. Perceived challenges included the limited staff capabilities to build the scenarios and BIM related games using the available off the shelf programs and software. In addition, some academic who had explored BIM models in the past raised the issue of incomplete data. Hence, trying to conduct analyses or simulations by exporting incomplete data into other programs was considered to be challenging.

Educators stressed that these real and practical problems needed extensive investigation and resolution if curriculum was to change.

Refinement of the Threshold Capability Framework

In response to the stakeholder feedback, the draft framework was refined. The modified Threshold Capability framework is presented in Figure 2. Table 2 documents a summary of the changes made to the framework.

Table 2. Summary of changes to refine the Threshold Capability Framework

Overview • The title of the framework was revised from Threshold Capability for BIM in a Construction Project Management Curriculum to Student Threshold Capability Framework for Digital Construction Project Management Curriculum; • Enhanced and refined definitions of each of the Threshold conceptualisations at each year (Level); in particular Year 3 introduction of 'Collaboration' as the focus in the anticipation that cross disciplinary practice could be introduced and students may be able to take part in some cross-program activities and be shown industry best practice; Year 4 introduces student resilience to cope with varying adoption patterns in the industry.
Fundamental Principles • Clearer alignment with current construction project management curriculum in Year 2 and Year 3. Discipline foundation principles and application focussed on in Year 2 and then more complex skills and capabilities leading to managing projects and across disciplines in Year 3.
Technical Skills • Clarity on developing fundamental data management skills in Year 1 to underpin digital management skills; • Clarity on Year 2 and Year 3 on focussed alignment of software usage to School's and industry's practices as well as the introduction (Year2) and development/consolidation of (Year3) construction management capabilities and skills being concurrently taught.
Construction project management skills • More detailed description at each Threshold level that aligns with current curriculum and industry practice; • For e.g. instead of 'team management' in Year 1 introduce concepts of supply chain management and design management and reinforce subcontractor management; • In Year 2 align more closely with procurement and contract management and implications of Model ownership; • In Year 3 introduce relationship of innovative construction project management environments for e.g. off-site manufacturing and client management
Strategic organisational behaviours • Introduction to the value proposition for BIM adoption for different industry stakeholders in Year 1; • In Year 2, including the concept of change management during BIM introduction within an organisation; • Introduction of advanced concepts of BIM and construction business.
Global Market Context • More detailed description of use of BIM within international context at each Threshold level that aligns with current curriculum and industry practice; • In year 3, use of BIM while managing project participants across different countries.

Figure 2. Final Student Threshold Capability Framework for Digital Construction Project Management Curriculum

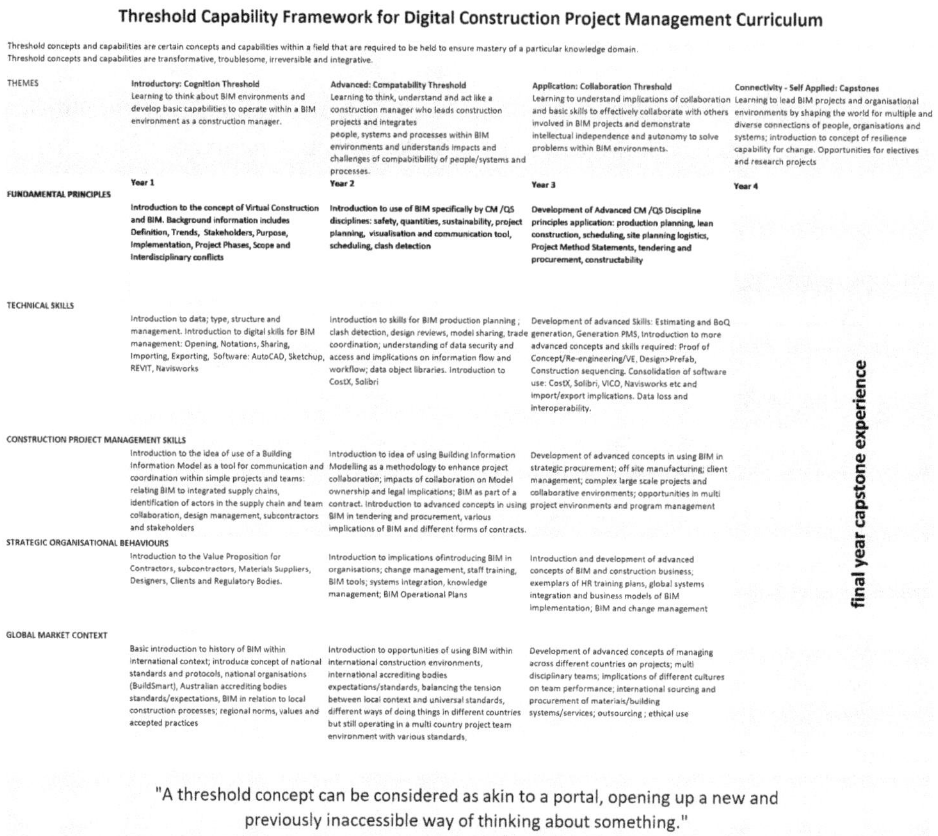

Threshold Capability Framework for Digital Construction Project Management Curriculum

Threshold concepts and capabilities are certain concepts and capabilities within a field that are required to be held to ensure mastery of a particular knowledge domain.
Threshold concepts and capabilities are transformative, troublesome, irreversible and integrative.

THEMES	**Introductory: Cognition Threshold** Learning to think about BIM environments and develop basic capabilities to operate within a BIM environment as a construction manager.	**Advanced: Compatability Threshold** Learning to think, understand and act like a construction manager who leads construction projects and integrates people, systems and processes within BIM environments and understands impacts and challenges of compabitibility of people/systems and processes.	**Application: Collaboration Threshold** Learning to understand implications of collaboration and basic skills to effectively collaborate with others involved in BIM projects and demonstrate intellectual independence and autonomy to solve problems within BIM environments.	**Connectivity - Self Applied: Capstones** Learning to lead BIM projects and organisational environments by shaping the world for multiple and diverse connections of people, organisations and systems; introduction to concept of resilience capability for change. Opportunities for electives and research projects
	Year 1	**Year 2**	**Year 3**	**Year 4**
FUNDAMENTAL PRINCIPLES	**Introduction to the concept of Virtual Construction and BIM. Background information includes Definition, Trends, Stakeholders, Purpose, Implementation, Project Phases, Scope and Interdisciplinary conflicts**	**Introduction to use of BIM specifically by CM /QS disciplines: safety, quantities, sustainability, project planning, visualisation and communication tool, scheduling, clash detection**	**Development of Advanced CM /QS Discipline principles application: production planning, lean construction, scheduling, site planning logistics, Project Method Statements, tendering and procurement, constructability**	**final year capstone experience**
TECHNICAL SKILLS	Introduction to data; type, structure and management. Introduction to digital skills for BIM management: Opening, Notations, Sharing, Importing, Exporting, Software: AutoCAD, Sketchup, REVIT, Navisworks	Introduction to skills for BIM production planning ; clash detection, design reviews, model sharing, trade coordination; understanding of data security and access and implications on information flow and workflow; data object libraries. Introduction to CostX, Solibri	Development of advanced Skills: Estimating and BoQ generation, Generation PMS, Introduction to more advanced concepts and skills required: Proof of Concept/Re-engineering/VE, Design>Prefab, Construction sequencing. Consolidation of software use: CostX, Solibri, VICO, Navisworks etc and import/export implications. Data loss and interoperability.	
CONSTRUCTION PROJECT MANAGEMENT SKILLS	Introduction to the idea of use of a Building Information Model as a tool for communication and coordination within simple projects and teams: relating BIM to integrated supply chains, identification of actors in the supply chain and team collaboration, design management, subcontractors and stakeholders	Introduction to idea of using Building Information Modelling as a methodology to enhance project collaboration; impacts of collaboration on Model ownership and legal implications; BIM as part of a contract. Introduction to advanced concepts in using BIM in tendering and procurement, various implications of BIM and different forms of contracts.	Development of advanced concepts in using BIM in strategic procurement; off site manufacturing; client management; complex large scale projects and collaborative environments; opportunities in multi project environments and program management	
STRATEGIC ORGANISATIONAL BEHAVIOURS	Introduction to the Value Proposition for Contractors, subcontractors, Materials Suppliers, Designers, Clients and Regulatory Bodies.	Introduction to implications ofintroducing BIM in organisations; change management, staff training, BIM tools; systems integration, knowledge management; BIM Operational Plans	Introduction and development of advanced concepts of BIM and construction business; exemplars of HR training plans, global systems integration and business models of BIM implementation; BIM and change management	
GLOBAL MARKET CONTEXT	Basic introduction to history of BIM within international context; introduce concept of national standards and protocols, national organisations (BuildSmart), Australian accrediting bodies standards/expectations, BIM in relation to local construction processes; regional norms, values and accepted practices	Introduction to opportunities of using BIM within international construction environments, international accrediting bodies expectations/standards, balancing the tension between local context and universal standards, different ways of doing things in different countries but still operating in a multi country project team environment with various standards,	Development of advanced concepts of managing across different countries on projects; multi disciplinary teams; implications of different cultures on team performance; international sourcing and procurement of materials/building systems/services; outsourcing ; ethical use	

"A threshold concept can be considered as akin to a portal, opening up a new and previously inaccessible way of thinking about something."

DISCUSSION AND CONCLUDING REMARKS

This research responded to the shift in construction management practices towards more virtual environments and BIM. Increasingly, immersive technology is being utilised to blur the lines between the physical world and the virtual world to design, visualise, simulate, deliver and manage the built environment. It is necessary to equip the construction management graduates with compatible theoretical knowledge and technical skills to enable them to work effectively in a project environment that is increasingly global and that embraces the greater use of the BIM models.

Guided by principles on curriculum redesign, this research has developed a Student Threshold Capability Framework for Digital Construction Project Management Curriculum which builds on disciplinary, foundational skills and knowledge, accounts for global integration and reflects the experiences and expectations of 21 key stakeholders.

The suggestion of using a comprehensive approach to content and curriculum to teach BIM is not new and indeed has been espoused by leading academics and industry professionals for nearly a decade (Gier, 2008; Lee et al., 2013; Lee & Hollar, 2013; McCuen, 2014; Mills et al., 2010; Mutai & Guidera, 2010; Sacks & Pikas, 2013), however, evidence for a successful integration of BIM in construction management curriculum is still lacking to date. Although this study represents only a small step towards achieving successful implementation of an integrated BIM approach in construction curriculum design, this research has started to develop an understanding around this complex issue. Specifically the review of the Threshold framework by stakeholders has shed some light on the key issues to effective implementation and offered some practical guidance on ways in which BIM capabilities can be woven into core aspects of a construction management curriculum.

For example, the holistic approach reflected in the framework and approved by the stakeholders is supported by the literature where past work in this area has suggested that BIM should be seen as a methodological rather than a mere technological tool (Barison & Santos, 2010; BIM Academic Forum UK, 2013; Macdonald, 2011; Mills et al., 2013; Sacks & Pikas, 2013). Similarly, the significance of real life learning through interactions with the industry has been strongly recommended by past researchers (Barison & Santos, 2010; BIM Academic Forum UK, 2013; Mills et al., 2013; Molavi & Shapoorian, 2013; Sacks & Pikas, 2013). In addition, the stakeholder-recommended T-shaped curriculum model agrees with literature that favours a combination of specialised and transferable skills and knowledge over siloed educational frameworks, as the T-shaped model promises to better equip graduates to rapidly changing contexts, technologies and markets in their chosen professions (Bierema, 2019; Saviano et al., 2016). Finally, the recommended action learning approach, which involves an experiential approach to problem solving, confirmed the validity of the BIM curriculum framework that was built on the threshold concept and capability theories as a way of overcoming the limitations of the outcome based education (OBE) approach. Thus, the threshold framework appeared as a highly appropriate way to guide students' transition into BIM education.

The study was partly subject to selection bias, as particpants other than those representing the accreditation bodies were likely to perceive BIM as an important part of a construction management curriculum, and snow-balling may have explained the consensus on certain topics among industry participants. In addition, the study only covered the views of academics of the one University as the framework was aligned with a specific curriculum and set of graduate attributes . Further research could investigate the perceptions of stakeholders of other higher education organisations. While the participants validated and helped enhance the Threshold Capability Framework, further refinement and fine-tuning is an ongoing and iterative process through implementation and accreditation of the program.

Nonetheless, the proposed Threshold Capability Framework has delivered some important messages to construction management educators in general. Firstly, educators need to understand that they need to teach a basic understanding of information management before they embark on any ambitious program of advanced information management technologies and capabilities that are required for BIM. Educators should also focus on a skills approach that could be utilised by students in a globalised environment. Secondly, a redesign of construction management curricula in Australia must acknowledge that the local construction industry may not be ready to accommodate graduate skill sets as proposed in the Threshold Capability Framework. Much effort is required to guide the construction practitioners to embrace a greater use of BIM in practice. In this aspect the concept of Graduate Resilience skills in students may be critical for the success of such transition. Thirdly, educators must not lose sight of the importance of core graduate outcomes and the academic staffs' ability to teach core construction management courses. Infusing Construction Management Digital Literacy is a long term and evolving exercise, and it is essential to build confidence in delivery capability. Finally, the results of this study should be read with due caveat that thematic areas described in the Threshold framework do not necessarily mean that these are 'courses'. They are simply content areas that could be taught in a range of different locations which should be refined in various future curriculum planning stages.

REFERENCES

Abdirad, H., & Dossick, C. S. (2016). BIM curriculum design in architecture, engineering, and construction education: a systematic review. *Journal of Information Technology in Construction*, *21*(17), 250-271.

Åkerlind, G., McKenzie, J., & Lupton, M. (2014). The Potential of Combining Phenomenography. *Variation Theory and Threshold Concepts to Inform Curriculum Design in Higher Education*, *10*, 227–247.

Åkerlind, G. S. (2012). Variation and commonality in phenomenographic research methods. *Higher Education Research & Development*, *31*(1), 115–127.

Allan, J. (1996). Learning outcomes in higher education. *Studies in Higher Education*, *21*(1), 93–108.

Anderson, T. (2016). Theories for Learning with Emerging Technologies. In G. Veletsianos (Ed.), *Emergence and Innovation in Digital Learning: Foundations and Applications* (pp. 35–50). Athabasca University Press.

Australian Institute of Building (AIB). (2016). *Standards for the Accreditation of Building & Construction Management Programs- Academic Standards for Course Accreditation*. https://aib.org.au/wp-content/uploads/2014/10/AIB-Accreditation-Academic-Standards-Document-Version-4-5-April-2017-1.pdf

Australian Institute of Quantity Surveyors (AIQS). (2017). *AIQS Course Accreditation Policy*. https://www.aiqs.com.au/sites/default/files/uploaded-content/field_f_content_file/aiqs_course_accreditation_policy_nb_v2.pdf

Azhar, S. (2011). Building Information Modeling (BIM): Trends, Benefits, Risks, and Challenges for the AEC Industry. *Leadership and Management in Engineering*, *11*(3), 241–252.

Azhar, S., Sattineni, A., & Hein, M. (2010). *BIM Undergraduate Capstone Thesis: Student Perceptions and Lessons Learned* [Paper presentation]. The 46th ASC Annual Conference, Boston, MA.

Baillie, C., MacNish, C., Tavner, A., Trevelyan, J., Royle, G., Hesterman, D., Leggoe, J., Guzzomi, A., Oldham, C., Hardin, M., Henry, J., Scott, N., Doherty, J., & Male, S. (2012). Engineering Thresholds: an Approach to Curriculum Renewal. Final Report 2012, Australian Office for Learning and Teaching.

Barison, M. B., & Santos, E. T. (2010). *BIM teaching strategies: An overview of the current approaches* [Paper presentation]. *International Conference on Computing in Civil and Building Engineering*, Nottingham, UK.

Barradell, S. (2012). The identification of threshold concepts: A review of theoretical complexities and methodological challenges. *Higher Education*, *65*(2), 265–276.

Barradell, S., & Kennedy-Jones, M. (2015). Threshold concepts, student learning and curriculum: Making connections between theory and practice. *Innovations in Education and Teaching International*, *52*(5), 536–545.

Becerik-Gerber, B., Gerber, D. J., & Ku, K. (2011). Pace of Technological Innovation in AEC Education: Integrating Recent Trends into the Curricula. *Journal of Information Technology in Construction, 16*, 411-432.

Bierema, L. L. (2019). Enhancing Employability through Developing T-Shaped Professionals. *New Directions for Adult and Continuing Education, 2019*(163), 67–81.

BIM Academic Forum UK. (2013). *Embedding Building Information Modelling (BIM) within the taught curriculum: Supporting BIM implementation and adoption through the development of learning outcomes within the UK academic context for built environment programmes*. The Higher Education Academy.

Bloom, B. (1956). *Taxonomy of educational objectives: Handbook I: Cognitive domain*. David McKay Company.

Bowden, J. A. (2004). Capabilities-driven curriculum design. In C. Baillie & I. Moore (Eds.), *Effective Learning and Teaching in Engineering, Effective learning and teaching in engineering* (pp. 36–47). RoutledgeFalmer.

British Standards Institution. (2019). *BS EN ISO 19650: Organisation and digitisation of information about buildings and civil engineering works, including building information modelling - Information management using building information modelling*. BSI.

Casasayas, O., Hosseini, M. R., Edwards, D. J., Shuchi, S., & Chowdhury, M. (2021). Integrating BIM in Higher Education Programs: Barriers and Remedial Solutions in Australia. *Journal of Architectural Engineering, 27*(1), 05020010.

Chartered Institute of Building (CIOB). (2018). *CIOB Undergraduate Education Framework*. https://www.ciob.org/ciob-education-framework

Chen, K., & Lu, W. (2021). University–industry collaboration for BIM education: Lessons learned from a case study. *Industry and Higher Education, 34*(6), 401–409.

Cousin, G. (2006). An introduction to threshold concepts. *Planet, 17*, 4–5.

Dave, B., Buda, A., Nurminen, A., & Främling, K. (2018). A framework for integrating BIM and IoT through open standards. *Automation in Construction, 95*, 35–45.

Demian, P., & Walters, D. (2013). The advantages of information management through building information modelling. *Construction Management and Economics, 32*(12), 1153–1165.

Diamond, R. M. (2008). *Designing and Assessing Courses and Curricula. A Practical Guide* (3rd ed.). John Wiley & Sons, Inc.

Eastman, C., Teicholz, P., Sacks, R., & Liston, K. (2018). *BIM handbook: A guide to building information modeling for owners, managers, designers, engineers and contractors* (3rd ed.). John Wiley & Sons.

Ensor, P. (2004). Contesting discourses in higher education curriculum restructuring in South Africa. *Higher Education, 48*, 339–359.

Entwistle, N. (1997). Introduction: Phenomenography in Higher Education. *Higher Education Research & Development, 16*(2), 127–134.

Fang, Y., Cho, Y. K., Zhang, S., & Perez, E. (2016). Case study of BIM and cloud-enabled real-time RFID indoor localization for construction management applications. *Journal of Construction Engineering and Management, 142*(7), 05016003.

Gier, D. M. (2008). *What Impact Does Using Building Information Modeling Have on Teaching Estimating to Construction Management Students?* [Paper presentation]. 44th Annual Conference of the Associated Schools of Construction, Auburn, AL.

Gu, N., & London, K. (2010). Understanding and facilitating BIM adoption in the AEC industry. *Automation in Construction, 19*(8), 988–999.

Halcomb, E. J., Patterson, E., & Davidson, P. M. (2006). Evolution of practice nursing in Australia. *Journal of Advanced Nursing, 55*(3), 388–390.

Harðarson, A. (2012). Why the aims of education cannot be settled. *Journal of Philosophy of Education, 46*(2), 223–235.

Health and Safety Executive. (2018). *Improving Health and Safety Outcomes in Construction Making the Case for Building Information Modelling (BIM).* https://www.hse.gov.uk/construction/lwit/assets/downloads/improving-health-and-safety-outcomes-in-construction.pdf

Hong, Y., Hammad, A., Zhong, X., Wang, B., & Akbarnezhad, A. (2020). Comparative modeling approach to capture the differences in BIM adoption decision-making process in Australia and China. *Journal of Construction Engineering and Management, 146*(2), 04019099. doi:10.1061/(ASCE)CO.1943-7862.0001746

Hosseini, M. R., Martek, I., Papadonikolaki, E., Sheikhkhoshkar, M., Banihashemi, S., & Arashpour, M. (2018). Viability of the BIM manager enduring as a distinct role: Association rule mining of job advertisements. *Journal of Construction Engineering and Management, 144*(9), 04018085. doi:10.1061/(ASCE)CO.1943-7862.0001542

Huang, Y. (2018). A Review of Approaches and Challenges of BIM Education in Construction Management. *Journal of Civil Engineering and Architecture, 12*, 401–407.

Joannides, M. M., Olbina, S., & Issa, R. R. A. (2012). Implementation of Building Information Modeling into Accredited Programs in Architecture and Construction Education. *International Journal of Construction Education and Research, 8*(2), 83–100.

Kline, A. R., Leite, F. L., & Isatto, E. L. (2020). *The perspectives of three university's building information modeling course development* [Paper presentation]. *ASEE Virtual Annual Conference, Virtual, Online.*

Land, R., Rattray, J., & Vivian, P. (2014). Learning in the liminal space: A semiotic approach to threshold concepts. *Higher Education, 67*(2), 199–217.

Larsson, J., & Holmström, I. (2007). Phenomenographic or phenomenological analysis: Does it matter? Examples from a study on anaesthesiologists' work. *International Journal of Qualitative Studies on Health and Well-being, 2*(1), 55–64.

Lee, N., Dossick, C. S., & Foley, S. P. (2013). Guideline for Building Information Modeling in Construction Engineering and Management Education. *Journal of Professional Issues in Engineering Education and Practice, 139*(4), 266–274.

Lee, N., & Hollar, D. A. (2013). *Probing BIM education in construction engineering and management programs using industry perceptions* [Paper presentation]. *49th Annual International Conference Associated Schools of Construction*, Windsor, CA.

London, K., & Bavinton, N. (2006). A Reflexive Capability Model for Sustainable E-Business Environments in Construction Supply Chain. In K. Brown & K. Hampson (Eds.), Clients Driving Construction Innovation (pp. 313-336). Icon.Net Pty Ltd.

London, K., Bavinton, N., Mentink, J., & Egan, B. (2006). *e-Adoption Profile Report*. Cooperative Research Centre for Construction Innovation, Icon.Net.

London, K., Feng, Y., & Ang, K. (2018). Collaborative Robotics in Construction: A Literature Review for Landcom NSW. Academic Press.

London, K., & Singh, V. (2013). Integrated construction supply chain design and delivery solutions. *Architectural Engineering and Design Management, 9*(3), 135–157.

Macdonald, J. A. (2011). *BIM – Adding Value by Assisting Collaboration* [Paper presentation]. The LSAA 2011 Conference, Sydney, Australia.

Maghiar, M. (2014). *Planning and scheduling curriculum integration of BIM with industry input* [Paper presentation]. *BIM Academic Symposium*, Washington, DC.

Maher, A. (2004). Learning Outcomes in Higher Education: Implications for Curriculum Design and Student Learning. *The Journal of Hospitality Leisure Sport and Tourism, 3*(2), 46–54.

Martínez-Aires, M. D., López-Alonso, M., & Martínez-Rojas, M. (2018). Building information modeling and safety management: A systematic review. *Safety Science, 101*, 11–18.

Marton, F. (1986). Phenomenography—A Research Approach to Investigating Different Understandings of Reality. *Journal of Thought, 21*(3), 28–49.

McCuen, T. (2014). *The Challenges of Advancing BIM in the Curriculum while Addressing Current Accreditation Standards for Construction* [Paper presentation]. *BIM Academic Symposium*, Washington, DC.

McGraw Hill Construction. (2014). *The Business Value of BIM in Major Global Markets. How Contractors Around the World Are Driving Innovation With Building Information Modeling*. McGraw Hill Construction.

Meyer, J. H. F., & Land, R. (2003). Threshold Concepts and Troublesome Knowledge: linkages to ways of thinking and practising within the disciplines. In C. Rust (Ed.), *Improving Student Learning – Ten Years On* (pp. 412–424). OCSLD.

Mills, A., Ashford, P., & McLaughlin, P. (2010). *The value of experiential learning for providing a contextual understanding of the construction process* [Paper presentation]. The 31st Australasian University Building Educators Association Conference, Sydney, Australia.

Mills, J., Tran, A., Parks, A., Kelly, S., Smith, E., Jupp, J., Macdonald, J., Williams, A., & Sher, W. (2013). Collaborative building design education using Building Information Modelling (CodeBIM). Final report 2013, Australian Government Office for Learning & Teaching, Canberra, Australia.

Molavi, J., & Shapoorian, B. (2013). Implementing an Interactive Program of BIM Applications for Graduating Students [Paper presentation]. ICSDEC 2012: Developing the Frontier of Sustainable Design, Engineering, and Construction, Fort Worth, TX.

Mutai, A., & Guidera, S. (2010). *Building Information Modeling in Construction: Current Practices and their Implications for Construction Engineering Education* [Paper presentation]. 117th American Societyfor Engineering Education (ASEE) *Annual Conference*, Louisville, KY.

Mutis, I., & Paramashivam, A. (2019). Cybersecurity Management Framework for a Cloud-Based BIM Model. Advances in Informatics and Computing in Civil and Construction Engineering, 325-333. doi:10.1007/978-3-030-00220-6_39

Newton, S. (2011). *Building and Construction. Learning and Teaching Academic Standards Statement.* Australian Learning and Teaching Council.

O'Donnell, R. (2012). *Profession of Engineering Education: Advancing Teaching, Research and Careers* [Paper presentation]. *23rd Annual Conference of the Australasian Association for Engineering Education*, Melbourne, Australia.

Oesterreich, T. D., & Teuteberg, F. (2016). Understanding the implications of digitisation and automation in the context of Industry 4.0: A triangulation approach and elements of a research agenda for the construction industry. *Computers in Industry*, *83*, 121–139.

Olowa, T. O. O., Witt, E., & Lill, I. (2019). *BIM for Construction Education: Initial Findings from a Literature Review* [Paper presentation]. The 10th Nordic Conference on Construction Economics and Organization, Tallinn, Estonia.

Park, J., Kim, K., & Cho, Y. K. (2017). Framework of automated construction-safety monitoring using cloud-enabled BIM and BLE mobile tracking sensors. *Journal of Construction Engineering and Management*, *143*(2), 05016019.

Philp, D., & Thompson, N. (2014). *Built Environment 2050: A Report on Our Digital Future*. Construction Industry Council.

Pikas, E., Sacks, R., & Hazzan, O. (2013). Building Information Modeling Education for Construction Engineering and Management. II: Procedures and Implementation Case Study. *Journal of Construction Engineering and Management*, *139*(11).

Ramburuth, P., & McCormick, J. (2001). Learning diversity in higher education: A comparative study of Asian international and Australian students. *Higher Education*, *42*, 333–350.

Riaz, Z., Arslan, M., Kiani, A. K., & Azhar, S. J. A. (2014). CoSMoS: A BIM and wireless sensor based integrated solution for worker safety in confined spaces. *Automation in Construction*, *45*, 96–106.

Royal Institution of Chartered Surveyors (RICS). (2019). *RICS Global Accreditation –Policy and Process*. https://www.rics.org/globalassets/rics-website/media/upholding-professional-standards/standards-of-qualification/rics-global-accreditation-policy-and-processes.pdf

Sacks, R., & Pikas, E. (2013). Building Information Modeling Education for Construction Engineering and Management. I: Industry Requirements, State of the Art, and Gap Analysis. *Journal of Construction Engineering and Management*, *139*(11), 04013016.

Saviano, M., Polese, F., Caputo, F., & Walletzký, L. (2016). *A T-shaped model for rethinking higher education programs* [Paper presentation]. The 19th Toulon-Verona International Conference Excellence in Services, Huelva, Spain.

Spady, W. G. (1994). *Oucome-Based Education: Critical Issues and Answers*. American Association of School Administrators.

Teizer, J., Wolf, M., Golovina, O., Perschewski, M., Propach, M., Neges, M., & König, M. (2017). *Internet of Things (IoT) for integrating environmental and localization data in Building Information Modeling (BIM)* [Paper presentation]. The 34th International Symposium on Automation and Robotics in Construction and Mining (ISARC 2017), Taipei, Taiwan.

Tight, M. (2014). Theory Development and Application in Higher Education Research: The Case of Threshold Concepts. *Theory and Method in Higher Education Research II (International Perspectives on Higher Education Research*, *10*, 249-267.

Wong, K. A., Wong, F. K. W., & Nadeem, A. (2011). Building information modelling for tertiary construction education in Hong Kong. *Journal of Information Technology in Construction*, *16*, 467-476.

Wu, W., & Issa, R. R. A. (2014). BIM Education and Recruiting: Survey-Based Comparative Analysis of Issues, Perceptions, and Collaboration Opportunities. *Journal of Professional Issues in Engineering Education and Practice*, *140*(2), 04013014.

Zhao, D., McCoy, A. P., Bulbul, T., Fiori, C., & Nikkhoo, P. (2015). Building Collaborative Construction Skills through BIM-integrated Learning Environment. *International Journal of Construction Education and Research*, *11*(2), 97–120.

Zou, P. X. W., Lun, P., Cipolla, D., & Mohamed, S. (2017). Cloud-based safety information and communication system in infrastructure construction. *Safety Science*, *98*, 50–69.

Chapter 15
Digital Transformation and Construction:
Education Case Studies of Construction Programs in the United States, Great Britain, and Ireland

Tamera McCuen
University of Oklahoma, USA

Mark Shelbourn
Birmingham City University, UK

Glenda Mayo
University of North Carolina at Charlotte, USA

Lloyd Martin Scott
https://orcid.org/0000-0001-6305-3749
Technological University Dublin, Ireland

ABSTRACT

In recent years, the construction industry has experienced the introduction of emerging technologies developed to improve project outcomes and labor productivity, including BIM. Construction companies feel the pressure to adopt these technologies and the importance that all project stakeholders have an understanding of the technologies and their impact on industry change. In many ways, BIM is the connection between these technologies and provides a single source of information for sharing within the project team and across the facility life cycle. As the digital transformation proliferates all aspects of the BE, higher education is challenged to prepare design, engineering, and construction students who are ready to enter the workforce. The challenge is complex as educators strive to meet the criteria set forth by their institution, policy decisions/regulation, accreditation standards, and industry expectations. While workforce requirements typically respond to the national economy, digital transformation in the construction industry is global.

DOI: 10.4018/978-1-7998-6600-8.ch015

1. BACKGROUND

1.1 History of Digital Transformation

The key challenges to enabling a Digital Transformation at both the industry and academic levels have been defined in many countries around the world. Some countries are further forward than others, with some saying the United Kingdom (UK) is leading with its Digital Built Britain strategy (UK Government, 2015). The strategy aims to ensure that the UK is able to harness new technologies and digital connectivity to transform the built environment and deliver real social and economic benefits to its citizens. Further initiatives such as the National Digital Twin (NDT) programme (and the NDT Hub where digital twins are hosted), the Gemini Principles, and the supporting Information Management Framework have been established to help drive the digital agenda in the UK. They are hosted and managed by the Centre for Digital Built Britain, information can be found at https://www.cdbb.cam.ac.uk/what-we-do/national-digital-twin-programme.

In Finland, a recent study (Aksenova et al., 2018) assessed the Digital Transformation journey of the Architecture, Engineering, Construction and Operation (AECO) sector in that country and concluded that overemphasis on technological capabilities in practice and the extensive portfolio of national R&D initiatives targeting mainly the productivity and efficiencies with technological developments did not lead to the systemic change in the established regime nor did it lead to the emergence of the new Digital Business Transformation. As a lesson learnt, the government of Finland has established a new programme in 2016, named KiraDigi (Törrönen, 2017). KiraDigi has brought about a new experimental platform to coordinate the discussions between the government departments, the city stakeholders, the AECO industry stakeholders and the complementary industries to support the emergence of the digital business ecosystem, while "The €16M programme's vision is to develop an open, interoperable information management ecosystem for the built environment" (Törrönen, 2017).

The relationships between the professional institutions, universities, government, the industry and the other stakeholders for the competence management ecosystem are, therefore, critical if the competent workforce is to mitigate the risks of continuing with the status quo. For example, the conclusion from a number of workshops in the UK (PUN, 2019) suggested that Funders and Insurers could be critical partners to ensure that real change happens within the UK Digital Built Environment Ecosystem, particularly away from projects covered by the 2016 UK Government BIM mandate. Therefore, it is required to assess the impact and implications for the various stakeholders that are part of this digital transformation, for example:

- clients, and by extension those involved in procuring projects,
- those representing entities (either company or individual) being engaged to deliver projects (who in turn must be able to demonstrate their organisations have the collective competence, capability and capacity to discharge their contractual obligations),
- the myriad of professions who are involved in the delivering the project at any stage of the asset lifecycle; including commissioning, designing, constructing, maintaining, operating, altering and decommissioning of Built Environment assets,
- those responsible for educating and upskilling current (and future) professionals,
- those responsible with assuring the competence of practitioners,
- those responsible for the regulatory framework and standards development,

- the citizens that interact and use the built environment for all aspects of their day-to-day life, not just business

In order to be successful, it is essential to reach out beyond the boundaries of the traditional built environment professionals and including experts from other domains and industries that can complement the Digital Built Environment with new propositions (e.g. manufacturing, finance, IT) agile business practices, and competence management.

1.2 Emerging Technologies

Although visual technologies have been in existence for years, their use continues to expand and evolve. The categories of visual technologies as outlined by Wang et al. (2018) include Virtual Reality (desktop, game-based, and immersive), Mixed Reality, and Augmented Reality. These technologies have been areas of exploration for programs working to add a visual context to enhance the learning experience. Wang et al. (2018) reviewed the use of VR in construction engineering and training and found that almost 50% of the publications about VR applications in construction education are related to architectural visualization. However, instructors have also used the virtual platform to showcase safety (Xiao, et al., 2018) and Finite Element Analysis (Fiorentino et al., 2009, Patil, et al., 2019).

Drones are an emerging technology that has been added to both industry and curriculum use. The use of drones is expected to create 1,000 jobs by 2025 according to the Association for Unmanned Vehicle Systems International (AAUVSI, 2020). Drones provide solutions for construction site monitoring, roof inspections (Bown and Miller, 2018), and various Civil Engineering applications for surveys. One of the primary benefits for curriculum inclusion is the ability to use a safe method to collect and use data.

The use of LiDAR (laser) scanning is a topic that has very broad applications with its use by architects, engineers, and constructors on projects across project phases from pre-design through construction. LiDar scanners are being used to document existing conditions. Such modelling is a process in which a 3D model of the existing conditions for a site, facilities on a site, or a specific area within a facility is created from the point cloud import to BIM software (Computer Integrated Construction Research Program, 2011). The existing conditions BIM model created can then be used for design, site planning, real-time quantity verification, pre-disaster planning, post-disaster record, and visualization (Computer Integrated Construction Research Program, 2011). Contractors can compare scans to verify accuracy of floors, walls and floor flatness. Post-construction, scanning can provide an "as-is" representation in lieu of the contract as-builts. Additionally recent research has included its use for energy modeling (O'Donnell, et al., 2019). For these types of technologies, the contract and project delivery methods taught in curriculum must now be updated to showcase the emergence of technology uses. This can be a difficult task due to funding, training and more importantly, the lack of room in the curriculum.

As BIM use expands in the construction industry, it is expected that each of these emerging technologies will become ubiquitous with preconstruction and construction BIM. Virtual reality provides an interactive environment valuable for design and constructability reviews during preconstruction. Whereas drones and LiDAR scanning support on-site data collection for documentation, inspection, and analysis during preconstruction and construction activities.

1.3 Stakeholder Expectations

In the context of the growing organizational complexity of the Digital Built Environment, driven by adaptation to high uncertainty, and the central role of collaboration, a new transformational approach to the Digital Built Environment is becoming of greater importance. A multidisciplinary move towards competency-based management across sectors, disciplines, professions, etc., matters more than targeting the productivity improvement with rapidly changing technologies. There is a significant risk that if the stakeholders of the Digital Built Environment do not re-design its institutional and industrial landscape in a system-based manner that acknowledges the increasing granularity, complexity and turbulence of the environment as well as placing the individual at the centre of a competency-based management approach, the current initiatives will have limited impact in terms of innovation (Autio and Thomas, 2014; Clarysse et al., 2014; Aksenova et al., 2018; Russell and Smorodinskaya, 2018). An individual centred competency-based management approach to sustain innovation requires a close industry-academic partnership, such as the one created by the BIM Body of Knowledge (BOK) to be utilized by a multidisciplinary audience of BIM users (Wu et al., 2018)

1.3.1 Industry Practitioners

Although BIM has been in practice for more than 20 years, it was only a few years ago that the stakeholders of the AECO sector were attracted by BIM potential which promised to makeover the design, construction and operation procedures of buildings in a more rationalized and efficient way (Coates et al., 2010). Across the sectors, employers have started to insist on architects, other related designers, and contractors adopt BIM to achieve the utmost sustainability and cost-efficiency of unique construction outputs (Mihindu and Arayici, 2008). So, it is reasonable to identify that the AECO sector holds a greater impact on global economic strategies and produces extremely unique outcomes compared with other industries. However, more recent studies have revealed the AECO sectors extreme effort in implementing certain modifications to upsurge the main three criteria of:

- efficiency, value, and sustainability of productions,
- reduce lifespan overheads, and
- overcome interoperability among participants in construction projects (Noor, 2008).

Consequently, it is inevitable to notice that many deficiencies have been caused by the late adoption of new technologies to the construction industry where it is still heavily depending on the labor force. Accordingly, many governments have recognized the critical factors that may affect the efficiency of the construction industry (Guillen et al., 2016). As a result, different directorial standards have been hosted that either validate or mandate the use of BIM as a scheme to address the lack of productivity. The compulsory mandating of BIM Level 02 by the UK Government can be identified as a recent uprising of BIM deployment to protect the significance of the construction industry (Kassem, et al., 2015). Because of mandating BIM Level 02, the UK government expected to minimize the 80% of construction waste due to accidental decisions, while also declare BIM's prominent nature towards the image of the construction industry as a unique and global economic impact (Noor, 2008).

Research by Rathnasinghe and Kulatunga (2019) found that respondents believed that skilled project members are essential for implementation as BIM is a high technical related phenomenon where the proper learning of it is needed. As BIM is a modern evolution to the traditional orthodox of construction, there is an expectation from professionals in the industry to be guided by the standards for effective implementation. BIM-related standards by various leading institutions perform this service, the Rathnasinghe and Kulatunga (2019) research agreed the positive effect of BIM standards for a successful BIM implementation is inevitable. Therefore, when respondents were questioned on the awareness of the BIM Protocol and its content, all respondents answered positively that understanding this issue is key for successful implementation of BIM. Hence, BIM Protocol by the UK Construction Industry Council has mandated the appointment of an 'Information Manager' which was also known by all the respondents. Previous research on the development of expertise and digital transformation in construction has mostly focused on observation 'from the outside'. The Rathnasinghe and Kulatunga (2019) research includes case study research showing the relationship of learning in construction to the use of knowledge management systems. This research project sought to understand the nature of expertise and how that expertise can impact knowledge. Each practitioner in the study offered their own positioning and their own narrative about their knowledge reproduction and expertise development. Individual stories offer their versions of that reproduction and expertise development and about their unique perceptions of the nature of their learning and its translation into their professional practice in construction.

1.3.2 Construction Students

There has been much debate about how BIM is currently employed in undergraduate curricula around the world; is BIM included as a stand-alone subject in a program, or an underlying theme across the program? In the majority of universities in US, Europe and Australia, AECO students continue to be educated in separate schools, departments or divisions, with little or no integration or collaboration between the disciplines (Scott, 2015). Often the first time that students from each AECO discipline are exposed to working with team members from other disciplines is in the workplace after graduation. It is important for graduates to have an understanding of the roles played by other AECO professionals and the impact that their decisions have on projects overall. However, the isolated manner in which they are currently educated does not provide this understanding. It is not only students of the separate AECO disciplines working in isolation from each other, students have also highlighted the lack of integration in several studies.

Research studies by Adamu and Thorpe (2015), Olowa, Witt, & Lill, (2020), Pugacheva et al., (2020) found the feedback between teacher and student can provide dynamic learning and the students reported some interesting observations or 'wish list' of features they thought would improve their learning and teaching. Some notable commentaries include: students wanting software (e.g. 4Projects) to allow the "*linking of Revit files*" (as done on desktop software) "*for Level 3 BIM*" and being able to "*view and mark up of 3D models*". Additionally, there appears to be some impact of social media on how students prefer to use BIM technology, for example, a few students have requested "*instant messaging functionality*" within BIM software.

Adamu and Thorpe (2015) also reported on remote modelling through desktop-sharing and found similar interests in the use of instant messaging for collaboration. Students in this case, argued that "*a discussion forum is not the best form of communication*" because essentially, "*it is the same as sending an email*". It would appear there is a direct relationship observed between the satisfaction with the use of BIM and the use of the tool, but the 3D tool could be a barrier for the BIM implementation. This result has also been found by other authors in academic and professional, non-academic settings. Adamu and Thorpe (2015), Olowa, Witt, & Lill, (2020), Pugacheva et al., (2020) all reported the incorporation of BIM for voluntary use by students and the choice of the degree for which it has been proposed as a mandatory subject appear to be successful, and the results are encouraging.

Before moving into the educational standards it is important to understand the terminology being used in this chapter. The UK National Building Specification organization defines BIM as "…*BIM or Building Information Modelling is a process for creating and managing information on a construction project across the project lifecycle*…" (NBS, 2021). The National BIM Standard – United States® Version 3 (NIBS, 2015) defines BIM as a term which represents three separate but linked functions. First is building information modeling, which is defined as a business process for generating and leveraging building data to design, construct and operate the building during its lifecycle. The second function is building information model, which is the digital representation of physical and functional characteristics of a facility that serves as a shared knowledge resource for information about a facility. Building information management is the third function of BIM, which is defined as the organization and control of the business process by utilizing the information in the digital prototype to effect the sharing of information over the entire lifecycle of an asset. In Ireland, the National Standards Association of Ireland (NSAI) (2021) describe BIM as "…*a process for creating and managing information on a construction project across the lifecycle of the project*…". The definitions here show the complexity of defining what exactly BIM is as it means different things to different people and organisations. This is increasingly difficult when BIM technology (ie modelling software), information management or methodologies are introduced into the discussion.

For the purposes of this chapter the approach has been taken to look at both the 'use' in terms of technologies as well as the 'implement' indicating the methodologies and information management aspects. When looking at the learning and teaching of these aspects this chapter will be considering all aspects.

2. CONSTRUCTION EDUCATION STANDARDS

This section addresses various construction education standards and the requirements from each for student learning of digital technologies from the perspective of four construction programs in the U.S., Great Britain, and Ireland. A brief overview of the standard criteria and requirements is given below.

2.1 United States

For many BIM professionals their BIM education and training begins in academia. Education, particularly in universities, is where new outlooks and exposures to new knowledge are most effective (Hardin 2009:298). BIM education is a viable solution to accelerate the BIM learning curve, allowing companies to recruit skilled BIM talent when students graduate (McGraw-Hill Construction 2008). Accordingly, the effective integration of BIM into educational programs is both a practical and pedagogic goal in preparing future BIM talent for AECO industry employment (Crumpton and Miller 2008; McGraw-Hill Construction 2009).

The classic divide between academia and the AECO industry also occurs in college BIM education. Although BIM is pervasive in college curricula, academia initially struggled to meet the expectations of students and the AECO industry on content coverage and problem-solving skills gained through course work (Clevenger and Rush 2011; Wu and Issa 2014). BIM is most effective when taught in a collaborative and multidisciplinary setting. However, in academia, departments are traditionally segregated, making it near impossible to replicate this work environment, and students are usually educated with an emphasis on technical skills. This absence of educators with professional experience in the areas of critical workflow and management attributes of BIM projects is a severe weakness of prevailing courses in BIM (Wu, Mayo, McCuen, et al., 2021).

2.1.1 American Council for Construction Education

The University of Oklahoma (OU) Construction Science program is accredited by the American Council for Construction Education (ACCE). The *Standards and Criteria for Accreditation of Bachelor's Degree Construction Education Programs* includes 20 student learning outcomes (SLOs) for which programs provide evidence through student work to demonstrate learning. Subject areas included in the SLOs are construction, communications, mathematics, physical science, business and management. Bloom's taxonomy is used to define the SLOs, which for bachelor's degree programs are: understand, apply, analyze, and create. To avoid restating ACCE information, only the SLO explicitly stating the outcome relevant to the digital transformation of construction education is discussed in this chapter. To meet ACCE's SLO #10 requires students: *Apply electronic-based technology to manage the construction process.* The criterion is open ended for the program to select the technologies for student learning and demonstration of their ability to apply the technology to manage the construction process. In doing so ACCE acknowledges the variety of technology available and is flexible in allowing programs the freedom to select and modify selections as technologies continue to emerge.

2.1.2 Accreditation Board of Engineering and Technology

The University of North Carolina Charlotte Engineering Technology and Construction Management (ETCM) program is accredited by the Accreditation Board for Engineering and Technology (ABET). Additionally, UNC Charlotte's program is also accredited through both the Applied and Natural Science Accreditation Commission (ANSAC) and Engineering Technology Accreditation Commission (ETAC) programs. Similar to ACCE, the criterion provided by each Commission includes 6 statements that can be interpreted and adapted broadly by programs. The criterion focuses on math, science and engineering problem solving and the most relevant criterion related to digital transformation states, "*An ability to design systems, components, or processes meeting specified needs for broadly-defined engineering problems appropriate to the discipline.*" Additionally, like ACCE, the criterion simply requires some form of demonstrated evidence that students have mastered an assigned task in this category. An additional issue that has been discussed is that most programs include a Building Information Modelling course in the curriculum, but due to faculty expertise it is not as common to see BIM integrated into other classes, such as estimating and scheduling. BIM has been a catalyst for changing how the industry conducts business since it has changed contracts and delivery processes. However, in terms of the accrediting bodies and their outcomes, they are much too broad to be used as a means to describe goals for preparing the workforce for digital transformation.

2.2 Great Britain

Successful evolution within any industry or sector is often seen as a combination of People, Process and Technology (Arayici, 2011a & 2011b). However, it is difficult to quantify the relevant proportion of investment into People, Process and Technology investments that the Construction Sector and Government strategies and reports suggest. McGraw Hill (2014) report stated that 61% of 'highly engaged Contractors', and 41% of the sector, rate 'Training' in Building Information Modelling (BIM) as a high priority. However, the Institution of Structural Engineers BIM Survey (2017) reported that 38% of respondents said their firm provided poor or no training. Of those receiving training only 10% reported principal staff and/or Directors being trained in BIM; the majority being either Technicians or Graduates. It is a reasonable statement that 'BIM training' can either mean 'learning to use technology' (the type often being undertaken by junior staff) or 'process focused' with the majority based strongly on the PAS 1192-2:2013 standard (STF, 2018), although this has been superseded by the ISO19650 suite of standards in the early 2020's.

Ever since the publication of Peter Senge's The Fifth Discipline (1990), organizations have sought to become 'learning organizations' that continually transform themselves. In our era of digital transformation, this goal is more important than ever; however, making real progress in this area even for the best organizations still presents a significant challenge. A key problem is that organizations have been focused on the wrong thing, whereby the problem is not learning, but in fact, it is unlearning. In every aspect of business, there are mental models that have grown outdated or obsolete, from strategy to marketing to organization to leadership. To embrace the new logic of value creation embracing all that digital allows, we have to unlearn the old one. The misconception is that unlearning is about forgetting, whereas it is about the ability to choose an alternative mental model or paradigm. When we learn, we add new skills or knowledge to what we already know. When we unlearn, we step outside the mental model in order to choose a different one. In this time of (digital) transformative change, we need to be conscious of our mental models and ambidextrous in our thinking (HBR, 2016). The UK Digital Built Britain vision (UK Government, 2015) is far more-wide ranging than that envisaged with the introduction of the 2016 UK Government BIM Mandate, asked for in 2011. It follows that although 'BIM Training' is an important first step towards the Digital Transformation Journey, success will require an evolved Upskilling philosophy of industry as well as a change in educational practices.

The major accrediting bodies in the UK that publish competencies in the Built Environment are the Royal Institution of Chartered Surveyors (RICS), the Chartered Institute of Building (CIOB) and the ICE, the Institute of Civil Engineers. Each of these have taken a slightly different approach to their inclusion of digital. After a brief introduction, this section will discuss the approaches taken by these three professional institutions.

There is considerable history and inertia regarding the professions of the built environment, in particular the entrenchment of 'roles' and 'occupations'. This has led to established professions and particularly Professional Institutions having a tendency towards protectionism, resistance to change, the reinforcement of silos and the preservation of hierarchies. (Morrell, 2015). 'Occupations' and 'professions' have in turn become a proxy for competence demand (CEDEFOP, 2013). For example, in the majority of large Built Environment projects, workforce planning is based on 'we need 'n' roles (e.g. project managers)'.

The latest requirements published by the RICS were in August 2018, their 'RICS Requirements & Competencies Guide' (RICS, 2018). In this document it describes the mandatory and optional competencies required for each pathway of career. There are also three different levels of competency, from

1 to 3. Of all the competencies there is little or no mention of digital. There is a competency on data management. There is no definitive competence to demonstrate any digital transformation awareness or management, there are mentions of BIM management across the different pathways. This could be classed as digital transformation in some quarters. There are also wider competences for leadership and some references to change but nothing specific to digital transformation.

Much like the RICS the CIOB has no specific competencies describing digital or digital transformation. It does have competencies in 'Managing Information', and 'Innovation' that display competencies that one would need to enable the change needed for digital transformation in projects and organizations (CIOB, 2020).

The Institute of Civil Engineers (ICE) have similar competencies but they are broader in their description. They include: 'Technology Management', and 'Information and Knowledge Management' in their 'Practice Competency' section (ICE, 2020). In the 'Elements of Behavioral Competency' they have competencies that include 'Collaborative Working', and 'Teamworking Skills'. Again, all of these are not direct references to digital transformation, but they are certainly contributors to enable digital transformation.

Another initiative developed by the ICE is 'Project 13' (ICE, 2018). This initiative is an attempt "to improve the way high-performing infrastructure is delivered and managed". Project 13 blueprint is summarized in Figure 1. As part of the blueprint published in 2018, (ICE, 2018) section 5c describes 'Skills, Behaviors and Leadership' needed to enable the vision to be realized. In the blueprint there are a number of references where the behaviors and skills needed to enable digital transformation are described. These include:

- "...be open to innovation and alternative solutions through digital transformation"
- "...have the ability to understand and leverage technology and digital opportunities"
- "...have strong digital leadership across owner and partners, empowered to drive change. Coherent digital strategies in place"

Figure 1. Project 13 and digital transformation blueprint summary

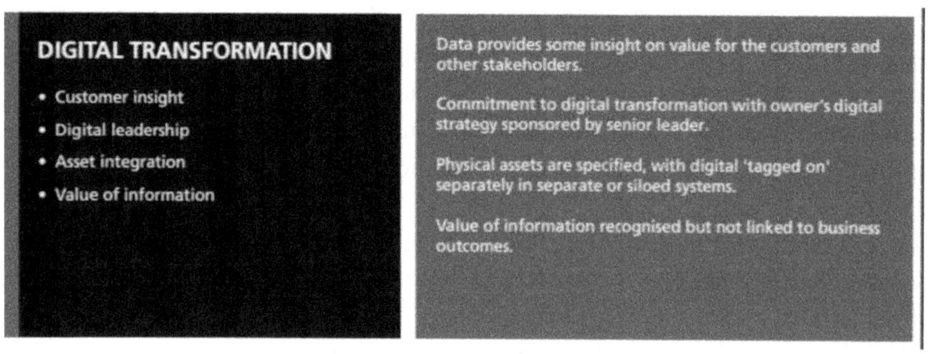

In many of the UK professional accrediting bodies documents there is little to no reference to digital transformation being described. To enable digital transformation there will be a need and reliance on people having the right skills/competencies and behaviors to enact the change needed for digital transformation (PUN, 2019). The work begun by the ICE in Project 13 is a step in the right direction in moving forward in enabling digital transformation.

2.3 Ireland

Across the world the advancement of digital technologies are disrupting markets and challenging businesses to reassess and realign how they operate, compete, and ultimately shape the built environment, and Ireland is no exception to this. Project clients look to the sector to combine deep domain knowledge and new technologies with strategies that sharpen performance, uncover new business models, and ensure that they thrive in this new, much connected world. While many regard the AECO sector in Ireland as fragmented there has begun a greater initiative for more cohesion between the stakeholders and the adoption of BIM processes has influenced this.

A holistic, multi-disciplinary approach to the design, construction, production and operation of the built environment is likely to require changes in the way the process is arranged, resourced and managed as we plan and implement change. Built Environment graduates of the future will need to be highly technical, adaptable, good communicators along with being lifelong learners. Much of the disruption in the area of digital technologies has been influenced by the early adopters but also some coherent academic engagement . While some report that this has not been quick enough others are more realistic and have wanted a coherent adoption to ensure that a balanced and informed journey is travelled.

As part of the balanced approach, The National BIM Council (NBC) was set up and a strategic focus was to prepare the Roadmap to Digital Transition of the Irish Construction Industry 2018-2021, a very important step in growing digital technologies in the AECO sector in Ireland. The four pillars of the NBC Roadmap include:

1. Leadership – The Irish Government must lead the way through the formation of a Centre of Excellence envisaged in the NBC Roadmap which will support the rollout of digital technologies in Ireland.
2. Standards – The role of the National Standards Authority of Ireland (NSAI) is critically important by ensuring appropriate standards are introduced to provide clarity and confidence to the sector.
3. Education – All stakeholders need to be educated as to the benefits of digital technologies and the significant contribution it can make to expatiating infrastructure, commercial and housing delivery in Ireland.
4. Procurement – Introduce contractual frameworks that will support the rollout of digital technologies across the AECO sector in Ireland.

There is a commitment that the above pillars will be supported by a research programme focused on digital technologies rollout. The AECO contractors national representative group, the Construction Industry Federation have called for careful consideration to be given to engaging with internationally positioned initiatives, such as the Construction Innovation. In addition, to the four pillars mapped out, the sector is of the view that there should be a single point of reliable, independent and factual information developed.

In Europe many countries have shown a particular enthusiasm for BIM particularly since 2013 with the introduction of the EU Procurement Directive (2013) which showed its support for BIM through its ratification at the European Parliament. There is at present a rapidly evolving suite of national digital programmes across the EU as a result of this. Ireland has certainly played its part in the development of this EU Directive.

As part of the commitment to a future digital technologies orientated AECO sector, in 2016 Enterprise Ireland in association with the Construction Information Technology Alliance (CitA), reported that of the 100 construction related organisations surveyed in excess of 80% of them have experienced an increased demand for BIM from their clients (Enterprise Ireland and CitA, 2016). More than half (57% of them) reported that awareness among clients had increased and the knock on effects of that was an increased use of BIM in projects. 57% of the sample reported that their awareness of BIM had improved significantly and were now using BIM on a number of projects. A further 33% reported that their awareness had slightly improved and they were using BIM on some of their projects. Only 4% of the sample from last year claimed that their awareness had not changed. It also found that 66% of the sample believe that Ireland should follow in the footsteps of the UK and mandate BIM. This was in comparison to the 50/50 breakdown in 2015. Whilst there is evidence of an increasing level of collaboration amongst industry professionals in Ireland in recent years, BIM use remains very immature generally, in particular, among the SME business community.

The Construction IT Alliance (CitA) has been promoting the benefits of digitisation in the construction and engineering industry in Ireland since its formation back in 2002. Whilst the Alliance has always strived to ensure that its members are exposed to a wide spectrum of contemporary digital tools and processes, the concept of Building Information Modelling (BIM) has dominated the focus of its activities over the past five years. Education also proved a key feature in the countries reviewed with increased evidence of third level colleges integrating BIM into their curriculum offerings but also the setting up of specialist education working groups to help create a more consistent BIM education experience. There was also a theme of developing training awareness programmes for government agencies, departments, academia and industry. The adoption of open BIM standards has been prevalent in those jurisdictions where buildingSMART Alliance chapters were active.

Table 1. Professional body representation in the Irish AECO sector

	RIAI	**CIOB**	**SCSI**	**Engineers Ireland**
Architect	X			
Construction Manager		X	X	X
Building Surveyor			X	
Civil Engineer		X		X
Structural Engineer		X		X
Quantity Surveyor		X	X	
Site Foreman		X		
BIM manager	X	X	X	X

The major Built Environment professional bodies in Ireland that address competencies in the AECO are the Society of Chartered Surveyors Ireland (part of RICS), the Chartered Institute of Building (CIOB), The Royal Institute of Architects of Ireland (RIAI) and the Engineers Ireland as shown in table 1 above. Each of these have taken a slightly different approach to their inclusion of digital knowledge, skill and competences. The Irish AECO sector is not too dissimilar to that in the UK in turning to BIM as a solution to help drive this transformational change to the sector as there is considerable history between Ireland and the UK. This is particularly as regards professional bodies and more particularly with the CIOB and RICS there is an aligned protocol. Engineers Ireland and RIAI, who are more focused across the island of Ireland, have adopted an approach to link with the universities and membership to develop their policies. There has been ongoing development with the Construction Information and Technology Alliance (CITA) leading on the integration of professions. The BICP roadmap report (2017) has identified the key competencies required for those professionals associated with digital transformation in the Irish AECO sector. CITA's activities have featured prominently over the past number of years, including the 2015, 2017 and 2019 CITA BIM Gathering, the CITA Smarter Cooperative Building Series', the BIM Innovation Capability Programme (BICP), and the CITA Regional events.

3. METHODOLGOY

The case studies in this paper offer a snapshot of four construction programs. The case studies focus on the integration of digital technologies within each program and the relationship between the technologies, regulations, standards, and stakeholder expectations. As previously discussed, there is a digital transformation of the construction industry to which construction programs must respond. Construction is a global industry, therefore this chapter includes two U.S. programs under different accreditation standards, one program in Great Britain, and one program in Ireland. The authors sought to answer, *"How do construction programs and educators meet mandated criteria?"* The current-state of technology integration was analyzed in terms of the mandated criteria guiding the program. The data was collected from authors experiences and knowledge and general conversations with faculty staff in the institutions. There was no formal data collection activities undertaken. The authors realise that this may not be a true reflection of current thinking, but do feel their collective knowledge and understanding does provide a snapshot of current thinking. The current-state of technology integration was analyzed in terms of the mandated criteria guiding the courses/programs.

4. CASE STUDIES

4.1 University of Oklahoma

In 2008 the OU Construction Science program began its pursuit to integrate BIM across the curriculum, which has continued with some modifications. In addition to BIM, other technologies in support of the digitization of construction have been integrated in multiple courses to coincide with the level of Bloom's taxonomy – understand, apply, analyze, and create – appropriate for the course. Technologies are introduced in courses for students to utilize in context with the subject area and their construction knowledge. The integration of construction related electronic technologies in the curriculum begins with *Computers in Construction*, a required course for all first year students. In this first course students are introduced to the fundamental 'tools' through hands-on use of various software, such as Excel™, SketchUp Pro, and Revit. Due to their lack of general construction knowledge, students also receive instruction on the basics of architectural graphics and project related uses for each software. To fulfill

ACCE SLO #10, the program requires *BIM for Constructors,* a course required for all seniors. Students are assessed using a final project through which each student must:

- Produce construction building information models validated using the drawings and specifications from project contract documents.
- Use the appropriate BIM technology to create a construction model for: visualization, cost estimation, 4D phase planning/sequencing, scheduling, and 3D coordination.
- Apply BIM for the purpose of communication with project stakeholders, problem solving, and improved project productivity.

The ACCE criteria requirement for the application of electronic-based technology to manage the construction process is broad and open for the program's selection of technologies to meet the criteria. Within the U.S. National BIM Standard [NBIMS] version 3 (NIBS, 2015) are multiple BIM use cases from which construction educators can select as the basis to create instruction and learning activities specific to the application of BIM for construction. The OU construction program selected to assess three BIM Uses: Visualization, Cost Estimation, and 4D Planning.

4.2 University of North Carolina

The UNC Charlotte Engineering Technology and Construction Management (ETCM) bachelor's degree program has tied the Building Information Modeling course (CMET2135) to accreditation through the following Student Learning Outcomes (SLOs). The current ties to the criterion are loosely connected due to the broad list of outcomes.

Course Objectives: To provide the student with a working knowledge of creation, management, and application of building information models to the construction, operation, and maintenance of a facility.

Course Learning Outcomes: Upon completion of the course, the student will be able to:

- Select an appropriate Level of Detail (LOD) to develop BIM contents.
- Determine responsibilities of Model Element Authors (MEAs) for a BIM project.
- Develop a complete building information model, including architectural, structural, and MEP components.
- Assess the technical and economic impacts of building information modeling on design, construction, and operation of facilities, through:
    - 4D construction sequence simulation,
    - BIM-based cost estimate,
    - Clash detection, and
    - Energy simulation

ETAC of ABET Criterion 3, Outcome 5: "an ability to apply written, oral, and graphical communication in both technical and nontechnical environments; and an ability to identify and use appropriate technical literature."

ANSAC of ABET Criterion 3, Outcome 3: "an ability to develop and conduct experiments or test hypotheses, analyze and interpret data and use scientific judgment to draw conclusions."

UNC Charlotte is currently reviewing ABET requirements and potential curriculum modifications to incorporate BIM across a broader scope of courses. A typical issue is the need for numerous courses to cover content but one solution is to incorporate BIM first into an introductory Graphical Communications course that replaces AutoCAD. This is viable because students no longer need CAD Drafting once they use BIM since 2D drawings can be generated from a model (Huang, 2017). As early as 2012, Lee and Dossick stated that just providing students BIM tools and teaching modeling practices is not a true BIM education. The objective is to efficiently manage information throughout the construction process and enable better construction collaboration. Reviewing accreditation requirements in conjunction with the need to work toward digital transformation is in the very early stage of discussion and planning.

4.3 Birmingham City University

As a university in the UK that provides both undergraduate and postgraduate courses nationally and internationally across the built environment, Birmingham City University has a reputation for providing graduates with the skills and competencies needed to make an impact in the organisations they join. All of the courses within the Department for Built Environment (DBE) at the University have been accredited by at least one relevant professional body, some of these bodies have been discussed in this chapter.

As discussed within this chapter the current requirements from these professional institutions is limited in providing graduates with knowledge and understanding of the global digital transformation in the AECO sectors. That is not to say that the Department at BCU does not teach their students about digital transformation, it does it through osmosis by following the requirements of the professional bodies. This means that graduates are aware of the issues and challenges that are affecting the AEC industry, but they are not directly taught:

- This is what digital transformation is…
- This is how to do it…
- This is how much it will cost…
- This is how long it will take…

If built environment provision at universities is to reflect the needs of the AECO industry, then the professional bodies need to take a lead in specifying that need in the documents that have been described in this chapter. It would be a brave university to design a suite of courses that are innovative in their thinking and structure that meet the needs of digital transformation, to then not get them accredited by the professional bodies, ultimately meaning little or no recruitment onto those courses.

4.4 Technological University Dublin

As a university in Ireland that provides both undergraduate and postgraduate programs nationally and internationally across the built environment, TU Dublin has a reputation for providing graduates with the knowledge, skill and competence required to make an immediate impact in the AECO organisation they sign up with once graduated. CM students have access to Revit and BIM modules which provides comprehensive investigations into graphic representation and communication with REVIT for Architecture. Students learn the skills, procedures and commands required to produce 2D/3D virtual building models using building modelling software to professional standards; presentation in 2D formats and fully rendered photorealistic 3D images; use of BIM as a design and production information tool in developing parametric models; and collaborative BIM drawing information coordination and integration within the design team. In the later stages of the program students are exposed to digital practices and structures impact on the management process. The CM curriculum includes aspects of digital transformation and the programme learning outcomes reflect an aspiration to achieving this. There is no individual module or course devoted to this and the main way students are exposed to this is through devoted academics who bring in cutting edge developments. Some development has been achieved through collaboration across the disciplines and this initiative has proven to be positive in leading the way for a more collaborative curriculum.

5. APPLIED DIGITAL TRANSFORMATION

The examples and case studies in this chapter provide an international perspective of different efforts to implement BIM in construction programs. Each case study highlights the current criteria to evaluate students' competency with digital technology is limited to metrics established by the academic accreditors. However, construction education is generally classified as an applied science in which educators emphasize '*learning by doing*' through practical activities designed with consideration for industry's expectations of the competencies graduates will have entering the workforce. Thus it is important that educators be aware of the expected competencies. The Academic Interoperability Coalition (AiC) BIM Body of Knowledge (BOK) includes a list of the BIM job tasks graduates are typically assigned to complete in the role of designer, constructor, or facility manager. The competencies relate to tasks commonly performed in a variety of BIM use cases (Wu, Mayo, McCuen, et al., 2021).

Competencies are not the only potential for addressing digital transformation. Technologies are also a contributing factor. Software packages and technologies such as LiDAR and Drones also assist to expedite project goals and these technologies are quickly being adopted by contractors to supplement BIM efforts and to improve their processes.

In addition to competencies and technologies, the primary difference in the UK and the U.S. is that there are no mandates in the U.S. to implement BIM or advanced technologies. Additionally, accrediting bodies do not specify any specific competencies for BIM technology use. For example, the ABET accreditation for construction programs state that students must show, "An ability to formulate or design a system, process, procedure or program to meet desired needs." But in order for accrediting bodies to contribute to the effort to mandate specific levels of competencies, they must be more explicit. This statement could mean that the use of Excel for estimating is meeting the criteria. However, it does not meet the needs for industry.

6. DISCUSSION AND CONCLUSIONS

This chapter addressed digital transformation in construction education through the lens of four programs located across the globe. Institutional, regulatory, and industry requirements were presented and the respective impact on the curriculum. To best answer "*How do construction programs and educators meet mandated criteria?*", the four case studies highlighted each program's approach to the integration of digital technology to not only meet the requirements, but facilitate student application of technology to perform construction business processes. Evident in the U.S. case studies were defined criteria to meet accreditation, whereas the UK and Ireland case studies revealed a more interpretive approach for programs to define criteria in alignment with professional organizations. While each program has a unique approach, they all have a common goal for the digital transformation of construction education and industry.

Moving forward lessons can be learnt from developments defined by those organisations that are leading the digital transformation. The UK BIM Framework published a document in July 2020 that showcased a set of learning outcomes that could be used as a basis for the learning and teaching the understanding of digital transformation.

Digital transformation is a global challenge. Educational institutions must introduce students to this challenge to enable:

- Industry to have confidence in the graduates they employ understand current challenges in the global industry;
- Students entering the market are employable, and they have the competences needed for the 21st century global industry;
- Students understand that digital transformation is more than implementing technology; people and processes are more important – the why we do it question needs reinforcement in them; and
- Consistency of the message students leave our institutions with is crucial. Educators have a duty to work with accreditors of our courses and programs to ensure the knowledge and understanding students leave with is consistent.

A final thought and summary is:

"...digital technologies are transforming every aspect of the built environment. Helping BE education take advantage of this new and connected world is important to the toolkit for the future BE professionals..."

REFERENCES

ABET Accreditation Criteria and Supporting Documents. (2020). *Applied and Natural Science Accreditation Commission (ANSAC). 2020-2021 Criteria.* https://www.abet.org/accreditation/accreditation-criteria/

Adamu, Z. A., & Thorpe, T. (2016). How universities are teaching BIM: A review and case study from the UK. *Journal of Information Technology in Construction*, *21*, 119–139.

Aksenova, G., Kiviniemi, A., Kocaturk, T., & Lejeune, A. (2018). From Finnish AEC knowledge ecosystem to business ecosystem: Lessons learned from the national deployment of BIM. *Construction Management and Economics*, 1–19.

Arayici, Y., Coates, S. P., Koskela, L. J., Kagioglou, M., Usher, C., & O'Reilly, K. (2011a). BIM adoption and implementation for architectural practices. *Structural Survey*, *29*(1), 7–25. doi:10.1108/02630801111118377

Arayici, Y., Coates, S. P., Koskela, L. J., Kagioglou, M., Usher, C., & O'Reilly, K. (2011b). Technology adoption in the BIM implementation for lean architectural practice. *Automation in Construction, 20*(2), 189–195. doi:10.1016/j.autcon.2010.09.016

Association for Unmanned Vehicle Systems International. (2020). https://www.auvsi.org/our-impact/economic-report

Autio and Thomas, (2014). The Fifth Facet: the Ecosystem as an Organisational Field. *Proceedings of the DRUID Society Conference 2014.*

Bown, M., & Miller, K. (2018). The Use of Unmanned Aerial Vehicles for Sloped Roof Inspections – Considerations and Constraints. *Journal of Facility Management Education and Research, 1*(2), 12–18. doi:10.22361/jfmer/93832

CEDEFOP. (2013). *Quantifying skill needs in Europe: Occupational skills profiles: methodology and application.* Author.

Chartered Institute of Building (CIOB). (2020). *Professional Review Guidance – Assessment Criteria for Candidates.* Available from https://www.ciob.org/sites/default/files/Professional%20Review%20Guidance%20Notes%20for%20Industry%20Candidates_5.pdf

Clarysse, B., Wright, M., Bruneel, J., & Mahajan, A. (2014). Creating value in ecosystems: Crossing the chasm between knowledge and business ecosystems. *Research Policy, 43*(7), 1164–1176. doi:10.1016/j.respol.2014.04.014

Clevenger, C., & Rush, M. (2011). Collaborating with Industry and Facilities Management to teach BIM. In *EcoBuild America 2011: BIM Academic Forum.* National Institute of Building Sciences.

Coates, P., Arayici, Y., Koskela, L., Kagioglou, M., Usher, C., & O'Reilly, K. (2010). The key performance indicators of the BIM implementation process. *Proceedings of the ICCBE Conference.*

Computer Integrated Construction Research Program. (2011). *BIM Project Execution Planning Guide – Version 2.1.* The Pennsylvania State University.

Crumpton, A., & Miller, B. (2008). Building Information Modeling: State of the A&D Industry and BIM integration into Design Education. *Proc., Interior Design Educator's 2008 Annual Conference.*

Enterprise Ireland & CitA. (2016). *Leadership in Transition Survey Report.* Author.

Fiorentino, M., Monno, G., & Uva, A. (2009). Interactive "touch and see" FEM Simulation using Augmented Reality. *International Journal of Engineering Education, 25,* 1124–1128.

Guillen, A., Crespo, A., Gómez, J., González-Prida, V., Kobbacy, K., & Shariff, S. (2016). Building Information Modelling as an Asset Management Tool. *IFAC-PapersOnLine, 49*(28), 191–196. doi:10.1016/j.ifacol.2016.11.033

HBR. (2016). *Why the Problem with Learning Is Unlearning.* Available from https://hbr.org/2016/11/why-the-problem-with-learning-is-unlearning

Huang, Y. (2017). American Society for Engineering Education (ASEE). *ASEE Annual Conference.* https://www.asee.org/public/conferences/78/papers/17619/view

Institute of Civil Engineers (ICE). (2018). *P13 Blueprint*. Available from http://www.p13.org.uk/wp-content/uploads/2018/06/P13-Blueprint-Web.pdf

Kassem, M., Kelly, G., Dawood, N., Serginson, M., & Lockley, S. (2015). BIM in facilities management applications: a case study of a large university complex. *Built Environment Project and Asset Management, 5*(3), 261-277. Accessed 8 Dec. 2019 at: http://www.emeraldinsight.com/doi/pdfplus /10.1108/BEPAM-02-2014-0011

Lee, N., & Dossick, C. S. (2012). *Leveraging Building Information Modeling Technology in Construction Engineering and Management Education. American Society for Engineering Education*. ASEE.

Li, X., Yi, W., Chi, H.-L., Wang, X., & Chan, A. P. (2018). A critical review of virtual and augmented reality (VR/AR) applications in construction safety. *Automation in Construction, 86*, 150–162. doi:10.1016/j.autcon.2017.11.003

Liu, R., & Issa, R. (2012). *3D visualization of sub-surface pipelines in connection with the building utilities: Integrating GIS and BIM for facility management. In Computing in Civil Engineering*. ASCE Publications.

McGraw-Hill Construction. (2008). *Building Information Modeling (BIM): Transforming Design and Construction to Achieve Greater Industry Productivity*. McGraw-Hill Construction.

McGraw-Hill Construction. (2009). *The Business Value of BIM: Getting Building Information Modeling to the Bottom Line*. McGraw-Hill Construction.

Mihindu, S., & Arayici, Y. (2008). Digital construction through BIM systems will drive the re-engineering of construction business practices. In *Proceedings of the International Conference Visualisation*. IEEE Computer Society. 10.1109/VIS.2008.22

Morrell, P. (2015). *Collaboration for Change*. The Edge Commission Report on the Future of Professionalism. Available at http://www.edgedebate.com/?page_id=2829

National Institute of Building Sciences (NIBS). (2015). *buildingSMART alliance*. National BIM Standard – United States® Version 3. Available at https://www.nationalbimstandard.org/

NBS. (2012). *BIM Legal Roundtable Discussion* [Webcast]. NBS Website. Retrieved from https://www.thenbs.com

NBS. (2021). *What is BIM?* NBS Website. Retrieved from https://www.thenbs.com/knowledge/what-is-building-information-modelling-bim

Noor, K. B. M. (2008). Case study: A strategic research methodology. *American Journal of Applied Sciences, 5*(11), 1602–1604. doi:10.3844/ajassp.2008.1602.1604

NSAI. (2021). *National Standards Authority of Ireland - ISO 19650-2 BIM - Building Information Modelling*. Available from: https://www.nsai.ie/certification/management-systems/bim-building-information-modelling/

O'Donnell, J., Truong-Hong, L., Boyl, N., Corry, E., Cao, J., & Laefer, D. (2019). LiDAR Point-Cloud mapping of building facades for building energy performance simulation. *Automation in Construction, 107*, 102905. Advance online publication. doi:10.1016/j.autcon.2019.102905

Olowa, T., Witt, E., & Lill, I. (2020). Conceptualising building information modelling for construction education. *Journal of Civil Engineering and Management, 26*(6), 551–563. doi:10.3846/jcem.2020.12918

Patil, G., Chen, D., Mayo, G., Smithwick, J., & Barclay, N. (2019). Implementing BIM and Finite Element Analysis in a Structural Integrity Investigation of Curtain Walls. *Journal of Facility Management Education and Research, 3*(1), 27–37. doi:10.22361/jfmer/112193

Pedagogy & Upskilling Network (PUN). (2019). *Evolve or Die: Transforming the productivity of Built Environment Professionals and Organisations of Digital Built Britain through a new digitally enabled ecosystem underpinned by the mediation between competence supply and demand*. Available from https://ciat.org.uk/uploads/assets/uploaded/101f628c-061d-445d-817368422197b626.pdf

Pugacheva, N., Kirillova, T., Kirillova, O., Luchinina, A., Korolyuk, I., & Lunev, A. (2020). Digital Paradigm in Educational Management: The Case of Construction Education Based on Emerging Technologies. *International Journal of Emerging Technologies in Learning, 15*(13), 96-115. Retrieved March 10, 2021 from https://www.learntechlib.org/p/217607/

Rathnasinghe, A.P., & Kulatunga, A., (n.d.). *Investigation on comparative roles of major stakeholders for BIM implementation in a traditional construction project environment*. Academic Press.

Royal Institution of Chartered Surveyors (RICS). (2018). *RICS Requirements and Competencies Guide*. Available from https://www.rics.org/globalassets/rics-website/media/qualify/pathway-guides/pathway-guides-requirements-and-competencies-v1.3-oct-2018.pdf

Russell, M. G., & Smorodinskaya, N. V. (2018). Leveraging complexity for ecosystemic innovation. *Technological Forecasting and Social Change, 136*, 114–131. doi:10.1016/j.techfore.2017.11.024

Scott, L. (2015). The Changing Landscape of Construction Higher Education. *International Journal of Construction Education and Research, 11*(2), 78–78. doi:10.1080/15578771.2015.1022454

Senge, P. M. (1990). *The fifth discipline: the art and practice of the learning organization*. Doubleday/Currency.

STF. (2018). *Developing a BIM Competency Framework: Research & Key Principles*. STF.

Törrönen, A. (2017). Automating Construction Processes by Making BIMs Machine-Readable. In A. Heiskanen (Ed.), *On the fast KIRA-digi track. AEC-Business*.

UK Government. (2015). *Digital Built Britain. Level 3 Building Information Modelling – Strategic Plan*. UK Government. Available from https://assets.publishing.service.gov.uk/government/uploads/system/uploads/attachment_data/file/410096/bis-15-155-digital-built-britain-level-3-strategy.pdf

U.S. Institute of Building Documentation (USIBD). (2016). *C120 (v3) Guide for Level of Accuracy (LOA) Specification for Building Documentation*. http://www.usibd.org/products/level-accuracy-loa-specificationversion

Wang, P., Wu, P., Wang, J., Chi, H., & Wang, X. (2018). A Critical Review of the use of Virtual Reality in Construction Engineering Education and Training. *International Journal of Environmental Research and Public Health, 15*(6), 1204. doi:10.3390/ijerph15061204 PMID:29890627

Wu, W., & Issa, R. R. A. (2014). BIM Education and Recruiting: Survey-Based Comparative Analysis of Issues, Perceptions, and Collaboration Opportunities. *J Prof Iss Eng Ed Pr, 140*(2), 04013014. doi:10.1061/(ASCE)EI.1943-5541.0000186

Wu, W., Mayo, G., McCuen, T., Issa, R., & Smith, D. (2018). Building Information Modeling Body of Knowledge I: Background, Framework, and Initial Development. *Journal of Construction Engineering Management*, (144). . doi:10.1061/(ASCE)CO.1943-7862.0001518

Wu, W., Mayo, G., McCuen, T., Issa, R., & Smith, D. (2021). *Developing BIM Talent: A Guide to the BIM Body of Knowledge with Metrics, KSAs, and Learning Outcomes*. Wiley.

Chapter 16
Simulation-Based Collaboration Training:
Strengthening the Industry's Capacity to Collaborate

Kerry A. London
Torrens University, Australia

Zelinna Pablo
Torrens University, Australia

ABSTRACT

Collaboration is a key challenge in the construction sector. While many innovative solutions have been explored, collaboration is persistently hampered by barriers such as the dominance of short-term project-based work and the geographical dispersion of actors. The chapter presents an approach for strengthening collaborative capacity in the construction industry. Building collaborative capacity requires the systematic identification, development, and assessment of specific knowledge, skill, and behavioural competencies. To address the challenge of identification, the authors propose 133 competencies based on a collaborative practice model developed from six qualitative case studies. To address the challenge of development, the authors propose the mobilization of simulation-based training (SBT) grounded in detailed scenarios linked to their model. Finally, for the challenge of assessment, the researchers propose an assessment instrument and a novel collaborative practice index. The authors also provide discussions on digitizing the SBT as part of future work.

INTRODUCTION

Collaboration is a key challenge in the construction sector. Studies on the construction sector show a long history of researchers exploring various configurations for organizing work collaboratively, among them "partnering" in projects (Chan et al., 2004), "integration" in teams (Baiden, Price, & Dainty, 2006) and "collaborative project procurement arrangements" (Walker & Walker, 2015). For decades there has

DOI: 10.4018/978-1-7998-6600-8.ch016

been a trail of innovative project delivery approaches such as Alliance Contracting, Partnering, Early Contractor Involvement and Integrated Project Delivery (IPD). More recently, digital technologies and approaches such as Building Information Modeling (BIM) and Digital Twinning have been gaining traction in the sector. These innovations are also underpinned by the premise that there is a critical need to achieve tighter cohesion, better communication and coordination among supply chain actors (Poirer, Forgues, & Staub-French, 2016; Zhang, Huang, & Peng, 2018).

The interest in collaboration and related strategies such as supply integration and partnering can be seen in part as responses to persistent problems in the construction industry. Construction is dominated by project-based work leading to alliances that are temporary, adversarial and lacking in trust (Bechky, 2006; Chan et al. 2004) and these challenges have triggered the search for collaborative solutions (Blanco, Janauskas, & Ribeirinho, 2016). Researchers from other disciplines have also noted collaboration's other benefits, arguing that leveraging the resources and capabilities of inter-organisational networks is key to responding to increasing complex and turbulent environments (Gulati, 2007; Hardy, Phillips, & Lawrence, 2003; Hoffman & Schlosser, 2001). Construction and project management researchers have thus embarked on theoretical and empirical research to understand collaboration systematically. Some researchers, for example, have sought to systematize understandings by developing definitions specifically suited for construction domains (London & Pablo, 2017; Pablo & London, 2016; Poirer et al., 2016; Schöttle et al., 2014; Walker & Walker, 2015).

In this chapter, the authors move from definitions to focus on the specific competencies linked to collaboration. Similar work has been done by Walker and Walker (2015) in a comprehensive study on collaborative knowledge, skills, attributes and experience for collaborative project procurement arrangements. The study underpinning this book chapter focuses on case studies mainly from the housing construction sector where professional project management and collaboration research has had much less attention. *The aim of this book chapter is to present an approach for identifying, developing and assessing competencies that strengthen the capacity to collaborate within the construction industry.*

The first component of the authors' approach involves *identifying competencies*. Researchers address the question, What are knowledge, skills and behaviours linked to collaborative capacity? They propose 133 competencies, based on a collaborative practice model (CPM) they developed from six qualitative case studies.

The second component of their approach involves *developing competencies.* Researchers address the question, How can increasing levels of competency be achieved? At the core of the authors' approach is the notion of developing collaborative capacity by motivating behavioural change through the use of simulation-based training (SBT). SBT is an approach that involves situating a learner in a scenario that amplifies, mimics or replicates many aspects of a real-life setting (Holland et al., 2014). Researchers have argued that SBT supports active learning in a range of situations, including team-based contexts involving complex problem-solving scenarios (Salas, Rosen, Held, & Weissmuller, 2009), making it ideal for learning the complex knowledge and skill base associated with collaboration. In this study, the authors focus on presenting their experience with piloting a face-to-face SBT, but they also explore the affordances and limitations of enhanced, hybrid and online formats.

The third component of the authors' approach involves *assessing competencies.* Researchers address the question, How can these competencies be assessed through quantitative means and/ or qualitative descriptors? The authors present two tools: a tool for assessing training outcomes *after* the simulation and a needs assessment tool to be used *before* the training. To support both, the researchers propose a customised Collaborative Practice Index.

The remainder of the chapter is structured as follows. In Section 1, the researchers describe the barriers to building collaborative capacity in construction, laying the argument for the need to systematically identify, develop and assess the specific competencies that comprise collaborative capacity. The process of identification is critical, since competency development and assessment are meaningful only if the appropriate competencies are identified. In Section 2, the methodology used to identify key competencies is described. The conceptual device that is central to the work is a Collaborative Practice *Model* comprising nine Collaborative Practice *Elements*. The model was developed from six detailed qualitative case studies and a rich database of 42 semi-structured interview. The model and the elements are underpinned by considerable interpretive flexibility and are mobilised to develop a range of training and development resources, explained in Section 3. Specifically, the authors use the model to identify (1) 133 competencies (knowledge, skills and behaviours) comprising collaborative capacity; (2) two simulation-based training scenarios and (3) two tools for assessing collaborative capacity. In Section 4, the authors discuss future work, which primarily focuses on digitising the SBT.

COLLABORATIVE CAPACITY IN CONSTRUCTION

Collaboration has been defined as "a group of autonomous stakeholders of a problem domain engage in an interactive process, using shared rules, norms, and structures, to act or decide on issues related to that domain" (Wood & Gray, 1991, p. 146) and as the pooling of "resources and appreciations" by multiple stakeholders to address problems (Gray, 1985, p. 912). The need for collaboration is pronounced in domains like construction, which are dominated by project-based work undertaken by multiple organisations, often in geographically dispersed locations (Erdogan, Anumba, Bouchlaghem, & Nielson, 2008). Construction projects involve numerous actors, for example architects, engineers, designers, developers, builders and trade subcontractors who need to communicate and align their activities and objectives frequently. Failure to do so, even with minor matters such as correcting a defect, can result in significant costs in times of time, rework and safety (Blanco et al., 2016).

One trend that is indicative of the increasing primacy of collaboration is the heightened interest of researchers and practitioners in IT-supported collaborative solutions (Erdogan et al., 2008). Digital collaborative technologies are being mobilized to support the management of design, contracts, defects and performance (Blanco et al., 2017). Interestingly, it has been found that many of these IT initiatives fail to meet their performance objectives. A key cause of these failures is the overemphasis on technology at the expense of human and organizational factors (Erdogan et al., 2008).

This study focuses on the human side of collaboration. Specifically, the authors propose an approach to strengthening individuals' knowledge, skill and behavioural competencies linked to collaboration, in order to build what Agranoff and McGuire (2003) refer to in another discipline as "collaborative capacity". The authors thus draw on training literature, as training involves "the acquisition of skills, concepts and attitudes that results in improved performance in an on-the-job environment" (Goldstein, 1980, p. 230). There are three critical phases in training: training needs assessment, which involves an analysis of organisations, people and task design or selection of the mode of instruction and the evaluation methodology (Goldstein, 1980). In this section, the authors focus on one key question under each issue:

Table 1. Training issues and research questions

Key Training Issue	Question Addressed by Author
Needs assessment	Identifying competencies: What are the competencies linked to collaborative capacity?
Design/ selection of mode of instruction	Developing competencies: How can these competencies be developed using a specific mode of instruction?
Evaluation methodology	Assessing competencies: How can these competencies be assessed (measured OR described qualitatively in ways that can capture progress)?

Identifying Competencies

The starting point for strengthening collaborative capacity is defining specific competencies. In training, this process is often understood as needs assessment. Within a given organization, formal needs assessment can be undertaken in a number of ways: testing individuals, observing them, having them self-assess, interviewing key people within an organization, conducting peer review or analysis of skill gaps (Grant, 2002). The interest of the researchers for this study, though, is identifying a set of competencies that is robust and flexible enough to be relevant across the construction industry. Identifying industry-wide competencies requires a needs assessment framework starting at a higher level of abstraction, for example a conceptual definition or a model. Cheng and Sturtevant (2012) have mobilized this approach. In their work on forest management, they built a systematic understanding of "collaborative capacity" by starting with concepts and definitions. From these definitions, they identified four key components, organized these into a framework and identified six arenas of collaborative action, "labels that were parsimonious yet descriptive of processes and attributes that define how a collaborative body functions" (p. 680).

As researchers, the authors believe that this model-based approach to competency identification is critical, since it addresses two goals: ongoing theoretical development in collaboration research and practical implications through competency identification. Traditional needs assessment techniques do not address the former. However, the researchers also acknowledge that starting with conceptual definitions and models of collaboration can be challenging. Moving from models to competencies is not a straightforward process. Once a model is developed, further work must be done to translate these forms of conceptual knowledge into knowledge that is actionable for and relevant to practice.

In the domain of construction, claims have also been made in the past that this is a big point of failure and that "[t]here are almost no studies available about what works best for translating research to practice in construction" (Chapman, 2013, p. 82). The universal theme of the application of research findings to improve both professional practice and integrate to the higher education sector teaching of built environment professionals is highly relevant and yet often overlooked in knowledge transfer and translation activities. However, there is now more evidence showing successful efforts that bridge the evidence-to-practice gap, for example research-industry linkages in the UK (UK Knowledge Transfer Partnerships) and in Australia (the Australian Research Council funding Linkage Projects). In this study, one aim is to translate a conceptual model of collaboration into resources that can achieve day-to-day behavioural change for people "on the ground" (Lavis et al., 2003). This is a bottom-up approach that draws from research on the related phenomenon of partnering, which researchers have found to be very difficult to engineer in a top-down manner (Bygballe & Sward 2019, p. 163).

Developing Competencies

Once new target behaviours associated with collaborative capacity have been identified, another challenge is how to motivate people to increasingly adopt these behaviours. Systematically facilitating behavioural transformation is difficult in a sector like construction, which is known to be resistant to change (Blanco et al., 2016; Harty, 2008; Lines, Sullivan, Smithwick, & Mischung, 2015). However, techniques for modifying behaviour are not new to the construction industry. For example, numerous strategies are deployed on a regular basis to change behaviours in the area of work health and safety. Common strategies for modifying behaviours include goal-setting, feedback mechanisms, communication and reward systems and training (Duff, Robertson, Phillips, & Cooper, 1994) and these have all been mobilized, with varying results.

The focus of this study is on training. Training can take the form of on-the-job training (internships, apprenticeships) or off-the-job training (classroom type sessions, lectures). Empirical work suggests a link between training and construction project quality (Tabassi & Abu Bakar, 2009). However, some training programs have been criticized for their lack of effectiveness, particularly those that employ minimal levels of learner engagement (Burke et al., 2006). To address this problem, researchers have advocated for conducting "less training in an artificial classroom and more in the workplace" (Wilkins, 2011, p. 2013) and making training more experiential (Holland et al., 2014).

One experiential learning technique that allows participants to train and engage within an environment that replicates, mimics, amplifies, or replaces many elements of real-life situations is simulation-based training (SBT) (McAllister et al., 2013). Instead of the uncontrolled real-life scenarios of apprenticeships, trainers using simulations can guide learners through carefully crafted situations that allow for substantial interaction. One of the common assumptions people make about simulation-based training is that it is 'role' play. The critical element of SBT, however, is that people are put into well-crafted scenarios where they act as they would in their current employment role. Many of these interactions can be immersive, set up such that a participant feels like s/he are in the real world (Lateef, 2010). Such simulations, when executed properly, allow for strong engagement and a holistic experience of complex topics. They can also be safe, economical and easy to reproduce (Van der Zee, Holkenberg, & Robinson, 2010). Robust simulation-based training in construction and engineering can capture rich aspects of activities such as building a facility, capturing minute details such as resource availability, weather, environment and manpower. Specialists can thus experiment with wide-ranging, realistic scenarios involving costs, time and risks, with high levels of precision, even without being physically on site (AbouRizk, 2010).

There are two points to note about research on SBT in construction. First, much of the research currently focuses on operational activities like health and safety. That said, the use of SBT in more managerial and strategic areas is gaining interest in construction. In Australia, for example, the Master Builders' Association of Victoria, a leading industry association, has a training arm known as the Business Leadership Simulation Centre. The centre's offerings include traditional training packages on onsite safety and incident investigation as well as simulation-based training on leadership, communication and industrial relations. It is not difficult then, to consider collaboration-focused simulations, since simulations have been known to be effective in developing skills for engaging with complex, multifaceted dilemmas in the context of a team (Salas et al., 2009).

Second, SBTs in specific areas of construction are increasingly supported by digital tools. Virtual reality, gaming, augmented reality and Building Information Modelling are increasingly used to support simulations to improve work, health and safety for construction (Kassem, Benomran, & Teizer,

2017). While the use of such sophisticated tools may not be as urgent in training programs for topics like collaboration, the role of digital tools in SBTs for collaboration must nevertheless be considered. A US-based report indicates that 69.3% of training was being done in digitally-enhanced, hybrid or fully online formats (Training Industry Report, 2018). It is plausible that the ongoing global pandemic might strengthen the need for delivering training online.

Assessing Competencies

Determining the extent to which skills are developing requires some form of assessment. There are two approaches to conducting skills assessment. The first, the "economic" view, assumes that skills can be measured through precise, quantifiable tools. The second, the "sociological" or "situated" view, is based on the premise that "[t]he social setting in which cognitive activity takes place is an integral part of that activity…[and that the] knowledge, attitudes or abilities needed for a certain job can be understood only within that particular working context" (Stasz, 2001, p. 388). When using sociological/ situated methods of assessment, evaluators use ethnographic methods and qualitative case study techniques. The situatedness of this second approach is consistent with the philosophy underpinning SBTs, that of immersing participants in real life situations. However, the researchers discuss in later sections how they mobilised both approaches, providing a strategy for "measuring" collaborative capacity as a tool for facilitating behavioural change in the industry, as well as a tool with qualitative descriptors to capture nuanced behaviours. In the next section, the authors show how they identified critical knowledge, skill and behavioural competencies through an empirically-grounded Collaborative Practice Model.

METHODOLOGY

The authors' approach to strengthening collaborative capacity through the identification, development and assessment of competencies was developed as part of a larger research study. The authors undertook two three-year research studies, both federally funded by the Australian Research Council (ARC) in the national competitive grant scheme. The second research project is now in its final year.

The focus of the first project was the development of a model of collaboration specifically for supply chain clusters undertaking offsite manufacturing (OSM) innovations. The underlying premise of the grant was that the housing construction sector is siloed and fragmented and that OSM is a construction methodology that requires integration of supply chains. Transforming how things are "currently done" requires collaboration. OSM contexts are rich settings for examining collaborative phenomena since they require exceptional levels of collaboration (Pablo & London, 2020). The authors examined housing construction supply chains undertaking innovations such as prefabricated flooring systems, components, walls, or even entire houses manufactured offsite then transported to their final location. Case studies involved "downstream" construction supply chains mainly comprising developers, builders, designers, engineers and regulators and a model of collaboration (the Collaborative Practice Model or CPM) for OSM was developed.

The second project was built on a similar premise: that the broader housing industry, not just housing construction, faced challenges that required transformational change, which in turn could only be fuelled by collaboration. In this project, the team sought to test and enrich the CPM by assessing its explanatory power in cases beyond housing construction and OSM. The two new case studies conducted so far involved innovations beyond prefabrication: one supply chain was a novel housing arrangement called a pop-up shelter, where temporary housing was set up in an empty hotel. Another case study involved a tablet-based application for monitoring construction quality. More case studies are in progress. In this second project, the researchers sought to examine if the model of collaboration would change, given that actors, the nature of innovation, and the project mandate (economic goals vs social goals) were all different. Analysis is ongoing but based on the first two case studies included in this chapter, the key elements of the Collaborative Practice Model remained the same, but with nuances within elements are emerging. Details of these case studies are shown in Table 2.

Table 2. Case study details

	Innovation	Actors	Location	Size	Interviews
Case 1	Cassette floor (a prefabricated flooring system to replace manually-installed flooring systems which required working at heights)	Lead actor: national developer of commercial, residential and investment properties. Other actors: builder, designer, engineer, regulators, project managers (downstream actors)	Victoria, Australia	Multi-national company	8
Case 2	Transportable house (prefabricated then trucked to site)	Lead actor: small builder of traditional as well as transportable housing. Other actors: designer, trades, suppliers (downstream actors)	South Australia, Australia	Small/ medium enterprise	6
Case 3	Prefabricated housing components for houses with complex designs and ambitious sustainability targets	Lead actor: small design/ build organization. Other actors: builders, designer, engineer, regulators, project managers, manufacturing facility (downstream actors)	Tasmania, Australia	Small/ medium enterprise	6
Case 4	Prefabricated wall and flooring systems manufactured using German technology	Lead actor: manufacturer of prefabricated housing systems. Other actors: builder, designer, engineer, regulators, project managers (downstream actors)	Victoria, Australia	Startup	6
Case 5	Pop up shelter, or the use of empty structures awaiting development approval as temporary housing	Lead actor: developer. Other actors: developer, NSW planning, community service, services like free grocery, laundry and clothing (upstream actors)	NSW, Australia	Multi-national company	6
Case 6	Tablet-based quality control application to support on-site inspections	Lead actor: national developer of residential, commercial and investment properties. Other actors: construction managers, supervisors, IT personnel, customer strategy, business development (upstream + downstream actors)	NSW, Australia	Multi-national company	10
TOTAL NUMBER OF INTERVIEWS CONDUCTED					42

The authors grounded their case study method in the interpretive tradition, a methodology that is premised on the core assumption that much of reality emerges from people's shared meanings and the patterns of interactions that arise from these meanings (Lincoln & Guba, 2016). Interpretivists assume that much of social phenomena, including the phenomenon of "collaboration", are socially constructed, complex, nuanced and cannot be detached from the people that create them (Orlikowski & Baroudi,

1991). These assumptions drive core methodological decisions. For example, interpretivist research is often, though not exclusively, linked to qualitative research data collection and analysis methods. Qualitative researchers prioritise prolonged immersion in empirical settings and the use of tools such as semi-structured interviews, observations and in-depth documentary as data collection tools. These tools, it is claimed, can capture rich details about phenomena in ways that structured surveys, strictly defined variables and quantifiable, law-like relationships cannot (Gioia, Corley, & Hamilton, 2012). The use of qualitative tools has been useful for enriching understandings of collaboration, which in the past had been examined primarily using quantitative approaches that obscured nuances in phenomena (Hardy et al., 2003). The ability of qualitative tools to capture textured, complex processes as they unfold in their natural settings is also an advantage when data is being gathered not just for analysis, but for the design of scenarios for SBTs. This point will be explored shortly.

Interpretive research and qualitative studies have significant potential, but just like quantitative/positivist studies they have their own limitations. A persistent point of debate is the issue of interpretivists' use of a few case studies, which raises the question on generalisability. These questions have been addressed cogently by Eisenhardt (1989), who argues that theory can be developed from single case studies; Gioia et al. (2012) who argue that detailed data structures and detailed narratives of a case study can give rise to "portable principles" and by Flyvbjerg (2006), who argues that cases can be strategically selected to make them more generalisable. In this study, Table 2 shows that the authors have chosen case studies involving collaboration around various types of innovation (Column 2), actors (Column 3), sizes (Column 4) and location (Column 5), illustrating what Flyvbjerg (2006) refers to as generalisability through maximum variation.

Data Gathering

For qualitative case studies, data can be gathered through methods such as ethnographic. tools, semi-structured interviews or document analysis, but "the heart of these studies is the semi-structured interview" (Gioia et al., 2012, p. 19). Semi-structured interviews provide genuine opportunities for participant engagement and the semi-structured nature of the questions allow for rich narratives to emerge. The researchers had a list of broadly-framed initial questions anchored on the research aims, which included gaining a rich understanding of participant experiences of collaboration, identifying barriers and drivers to collaboration and exploring what effective collaboration was to interviewees. The main interview questions included "Who did you collaborate with?" "What was that experience like?" "What helped collaboration in this specific situation?" "What made it difficult?" "What specific strategies did you find most helpful?" While this standard list was available as a guide, the process of interviewing was not rigid. Interview questions evolved, "following the twists, turns, and roller-coaster rides involved" (Gioia et al. 2012, p. 19). The authors have so far conducted 42 interviews, each one lasting an average of one hour each. Each interview was transcribed and the full qualitative database so far includes more than 1400 pages of text.

Data Analysis

The authors conducted their analysis by drawing on the method prescribed by Gioia et al. (2012), an approach that focuses on conceptual development from inductive studies with a specific focus on transparency and qualitative rigour. The process is well-regarded and accepted and has been published in a journal article that has been cited more than 4000 times in Google Scholar. Guided by this method, researcher conducted analysis in four main stages: (1) a first level analysis of data resulting in participant-driven *themes*; (2) a second level analysis resulting in researcher-driven *categories;* and (3) the development of relationships between key themes. The authors then take the analysis further into a fourth stage, (4) the development of narratives.

First Level Analysis: Participant-Driven Themes

The authors initially conducted first-level analysis of interview data, a process which yielded 102 participant-driven "themes", or clustered of related texts. Examples of these themes were drivers of collaboration (co-location, detailed standards, good quality personnel, strong leadership) and barriers of collaboration (geographic dispersal, the view that collaboration required too much time, high turnover, too much focus on a single organisation's goals). The labels of these themes were developed to map as closely as possible to terms used by participants. Beginning with participant-driven themes is important to help researchers ensure that the integrity of the full qualitative database is retained. Researchers using this approach are committed to categorizing all units of text meaningfully, without discarding data prematurely on the basis that it does not fit with a predetermined theoretical lens (Gioia et al., 2012).

Second Level Analysis: Researcher-Driven Categories

The authors then conducted a second-level analysis where these 102 themes were grouped through an iterative process into bigger "categories". For example, five first-order themes that emerged included the following enablers and barriers to collaboration:

Theme 12: Lack of strong leadership; absence of a unifying leadership team
Theme 65: Unclear identity, reputation, branding of innovation or innovator
Theme 45: Exercise of situational leadership
Theme 69: Champion with credentials and support; compelling mentor
Theme 99: Strategic position of influence in supply chain; reputation of key firm

These themes were all related to the ability, or the lack of the ability, to exercise influence over actors, and thus these first-order themes have been grouped under a second order category, "Leadership". From the 102 themes, nine second order categories emerged. These categories are presented as the nine critical elements of successful collaboration and are referred to onwards as Collaborative Practice Elements. Gioia et al. (2012) argue that this "tandem reporting" of participant and researcher voices are part of the process for achieving rigor and are one hallmark of high-quality qualitative research.

Table 3. Nine second order categories, referred to henceforth as Collaborative Practice Elements (adopted from previous work, citation masked)

Collaborative Practice Elements
1: Leadership: Creating a leadership environment for change
2: Shared Goals: Envisioning a compelling project that addresses clear problems
3: Expertise: Designing a well-qualified project team
4: Change: Strategising for change and for managing resistance
5: Investment: Investing in people and assets
6: Shared Space: Creating and sustaining productive face-to-face interactions
7: Problem Solving: Pursuing participative, flexible and anticipatory solutions
8: Organising Mechanisms: Formalising team characteristics to sustain productive patterns of work
9: Technical Standards: Documenting and disseminating clear and accurate information and committing to precision

Proposition/ Model Development

Once researcher-driven categories were identified, the researchers moved from a static list to a model proposing meaningful links between categories. These links are propositions inferred by the researchers and are an important part of making the study generalisable. Qualitative researchers are divided as to whether or not such propositions and models are necessary research outcomes when undertaking interpretive work, but Gioia et al. (2012) note that propositions are the means by which transferable principles can be framed in succinct ways then reapplied to other contexts. In this study, the nine Collaborative Practice Elements were linked and assembled into a process model. Put another way, The Collaborative Practice *Elements,* along with the proposed links between them, make up the Collaborative Practice *Model.* Details of this model were discussed in earlier work (Pablo & London, 2020). In this book chapter, elements are often listed as a set of separate constructs but this is for clarity of discussion. The model is organic. Elements are linked to one another, so all elements can be discerned in almost any collaborative situation, even if at some points one element may be foregrounded more than others.

Development of Narratives

The analysis has so far demonstrated a progressive move from detailed data to a few increasingly abstract concepts. In interpretive research, however, it is important that abstract concepts (elements like "Leadership", even the Collaborative Practice Model as a conceptual device) do not lose the richness and texture of the data from which they emerged. The authors did offer basic definitions of the elements in Table 3, but it must be noted that a core assumption of interpretive work is that there is no single, tightly-defined definition of the Collaborative Practice Elements. The definitions offered function as sensitizing devices and are able to catalyse and trigger contextualized meanings. Therefore the definitions are not derived as limiting boundaries. Both the model and the elements are thus endowed with interpretive flexibility.

Capturing this interpretive flexibility can be achieved by writing narratives. Each of the nine Collaborative Practice Elements was supported by "an informative story that is driving toward some new concept development and theoretical discovery with the careful presentation of evidence" (Gioia et al.,

2012, p. 24). The process of generating narratives is different from the earlier process of identifying themes and categories. In the earlier process, the outcomes included themes and categories that were tightly bound to the data. In narrative development, analysis is undertaken in ways that are more "unbounded", "inductive" and "generative" (Wolcott, 1994, p. 23), but not arbitrary. As the authors explained in previous work on narratives, "[t]he outcome is not a strict and confining system of codes, but a rich story, still linked to data, but with coherence, vividness and texture. Narratives are thus endowed with a level of flexibility that lends itself to the process of translation into actionable forms of knowledge for other contexts. Revisiting the data and bearing in mind the broad story line encapsulated in the process-based framework, we thus generated narratives for each of the nine elements on collaboration." (Pablo & London, 2017, pp. 600-601).

The next section discusses how the model, the Collaborative Practice Elements and the underlying narratives provided a robust platform for identifying, developing and assessing competencies linked to collaborative capacity.

STRENGTHENING COLLABORATIVE CAPACITY in construction

Figure 1. Moving from the model to training resources

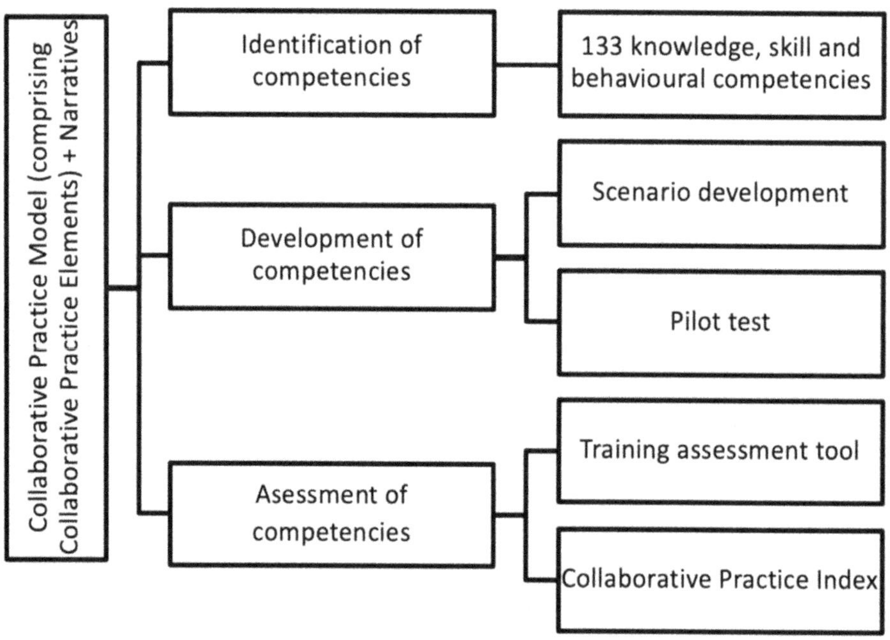

Figure 1 shows how the Collaborative Practice Model and Collaborative Practice Elements became the basis for a suite of resources for developing collaborative capacity, with narratives as the bridge. This approach is drawn from Torrell (2006, p. 248), who claims that "a storied report of past events is the frequent first step in training design." To understand how narratives can be a bridge, an excerpt for the "Leadership" narrative is provided in Appendix 1 for illustrative purposes.

Identifying Competencies Through Models

Using the narrative for "Leadership", the authors identified a total of 15 competencies shown below.

Table 4. Sample collaborative competencies for "Leadership"

Collaborative Practice Element 1: Leadership
Knowledge
1. Expertise in formal strategic planning processes
2. Expertise in informal strategic planning processes
3. Technical knowledge on developing feasibility studies related to the innovation
4. Principles of organisational and project leadership
Skills
1. Scans systematically external environment and identifies opportunities and threats related to the innovation
2. Scans systematically internal environment and identifies relevant organisational strengths and weaknesses
3. Identifies a viable strategic position (cost leader, niche, etc.) for one's firm in the industry and communicates how innovation supports this strategy
4. Mobilises research to support recommendations related to innovation
5. Communicates key elements of a compelling vision related to innovation and translates vision into concrete feasibility studies
6. Establishes a track record for success with new products or markets related to the innovation
Behaviours
1. Presents a compelling vision for innovation to industry and to organisational leaders
2. Demonstrates prospecting behaviours that support the exploration of new OSM products and markets
3. Facilitates empowerment strategies and mentor staff
4. Adopts a flexible approach to leadership
5. Demonstrates openness to the potential of untried solutions related to innovation

The competencies above can be linked, but are not identical, to the narrative. Many of the key competencies can be traced back directly to the data and to specific quotes. For example, the idea of demonstrating "a track record for success" emerged in case studies where leaders took an incremental approach to change, setting up small innovations that empowered the supply chain to eventually take larger, more risky innovations. The use of a viable strategic position was linked to data as well. However, there is no one-to-one correspondence between competencies and quotes, nor does there have to be. As explained earlier, narratives emerge from a process of analysis that is unbounded and generative (Wolcott, 1994). Other items under "Leadership" were identified because they were closely linked to findings about other elements ("openness to the potential of untired solutions" was closely linked to the element of Change). The researchers also included, in a considered manner, their own proposed competencies based on the knowledge of the full qualitative database and of literature in the area. For example, "environmental scanning" was not a term that were used by participants, but the researchers argued that interviewees mentioning "opportunities for OSM" and saying "we saw OSM was being used in Europe" could be classified as "scanning the external environment".

Developing Competencies

Scenario Development

Narratives were also used to develop two training meeting "scenarios", carefully-planned semi-scripted situations that captured the need for collaboration within a hypothetical multi-stakeholder construction team. The role of narratives in scenario-development cannot be discounted: strong narratives support authenticity, provide a rich basis for integration with target skills and can drive higher levels of knowledge transfer (Gegenfurtner, Quesada-Pallarès, & Knog, 2014).

A key point to note about the current scenarios is that they have deliberately been developed around the need to collaborate around dilemmas linked to the introduction of novel offsite manufacturing (OSM) techniques in residential projects. This is because the scenarios were developed incrementally and were written when most of the completed case studies were those that were OSM-centred. Future work will involve the development of scenarios involving more generic innovations. The carefully-formulated dilemmas were crafted to draw out certain types of behaviours linked to collaborative capacity. Scenario 1 was designed to simulate the introduction of OSM in a Sydney project to a multi-stakeholder project team, while Scenario 2 was designed to simulate OSM's expanded use to a Melbourne project after achieving mixed results in the Sydney project. Both scenarios were designed to include four to six participants. The scenarios were a compilation of key activities and events from the case studies that were uncovered during the data collection phase.

The simulations were designed to initially take place at the Business Leadership Simulation Centre (BLSC) in Melbourne, Australia. The BLSC is a custom-designed training facility, one of three in the world, fully equipped to support simulation-based training. The other two are in Leeuwarden (the Netherlands) and in Coventry (UK). The facility has meeting rooms, video panels and audio and video equipment that allow viewers to observe the scenario training from a control room. Two organizational psychologists were hired to assess participants according to assessment criteria linked to the behavioural indicators. These indicators will be discussed shortly. Preparations for the simulation also included the hiring of three professional actors who would play roles meant to elicit responses to collaborative dilemmas.

The two scenarios were designed to be undertaken sequentially. At the beginning of the training day, participants would be given an overview of the collaborative practice model. Participants then would undertake Scenario 1, receive feedback, undertake Scenario 2, then receive feedback again. An extract from Scenario 1 is shown in Figure 2 in the Appendix.

Scenarios were initially developed from conversation structures (see Figure 3 in the Appendix), hypothetical exchanges between actors who were assumed to have opposing views on OSM. The script was developed such that resistance and objections to OSM would be embedded in the flow of the conversation. Ideas for resistance and objections were distilled from actual interviews. Each line in the scenario was deliberately crafted to target key competencies and the overall scenario was shaped to target all nine collaborative practice elements.

Pilot Test

The pilot test of the SBT was scheduled over a two-day period, but the SBT itself, once stabilized, is envisioned to span only one day. The two-day period covered by the pilot provided time for more extensive feedback sessions about participants and about the training program itself.

Of the three professional actors recruited by the BLSC for the pilot, two were trained to play the role of architects with conflicting OSM objectives and a third was trained to play the role of a project manager resistant to the idea of OSM. On Day 1, these actors were given a detailed script designed to "lead" the meeting discussions in certain directions. Actors were briefed collectively by BLSC staff and then were further oriented by the research team. The scenarios were more complex and nuanced than previous roles that they had played and so wanted to understand the parameters of their roles.

Non-actor participants were recruited by the research team and BLSC executives and were asked to come in on Day 2. The final pool of non-actor participants for the pilot session included executives from a developer organization, a manager from a manufacturer of panelized systems, two suppliers and an executive from HR. The non-actor participants were given information packs with descriptions of their roles, details of their organization and their key organizational priorities. Participants were given role descriptions that mirrored their real-life professions. In this sense, they were not "role playing" because their participation was grounded on their real-life expertise. The mix of actors and non-actors was meant to simulate a certain type of real-life small group: residential construction teams spanning different organisations working together to implement an innovation for the first time.

Assessing Competencies

Training Assessment

Two organizational psychologists were hired by BLSC staff to conduct assessment of the two meetings, which ran for 45 minutes each. These psychologists were given an instrument developed by the research team. The detailed assessment was based on the 133 competencies. Selected competencies were developed into more fine-grained descriptions, to reflect "high", "acceptable" and "low" levels of attainment. These variations of "high", "medium" and "low", like the competencies, were developed based on different foundations. First, empirical data from case studies showed concrete examples of competencies at different levels. Case study 1, for example, involved a highly skilled project director who demonstrated "high" levels of Knowledge Competency 1 (see Table 5 below). On the other hand, Case Study 2 showed an organization with a new non-technical general manager who was struggling to implement change, but was met with limited success because of failure to explain the implications of new innovations to stakeholders outside and even sometimes within the organization. Apart from descriptors directly linked to data, the authors also proposed descriptors that were plausible, even if they were not encountered in a given case study, by making inferences from the full data set and from investigating literature. Again, this is consistent with Wolcott's (1994) notion of interpretation that is more generative.

Organisational psychologists were also instructed to provide an overall score for the entire team, which is vital as collaboration is a group-based dynamic. This was a novel approach for the BLSC training and added complexity to the assessment by the psychologists. Examples of detailed assessment criteria for two competencies are shown in Table 5. In the debrief session led by organizational psychologists, participants were also asked to conduct self-assessments.

Table 5. Sample competencies linked to different levels of attainment

Expertise		
KNOWLEDGE COMPETENCY 1: Extensive knowledge of general construction, new processes and discrete work tasks and/ or understanding of the impact of new technology on one's role.	High	(For technical specialists) Explains implications of technology-driven change in relation to construction methodologies. (For non-technical specialists) Clarifies impact of new technology on one's own role as well as those of other team members, including those that may not seem directly involved (e.g. HR).
	Acceptable	(For technical specialists) Explains new work processes in sound but general terms, focusing on more "obvious" changes such as changes in work arising from new technology. (For non- technical specialists) Clarifies impact of new technology on one's own role as well as some team members, but fails to engage others who are not perceived to be directly involved.
	Low	(For technical specialists) Explains new work processes in general terms. (For technical specialists) Clarifies impact of new technology on one's own roles but not on those of others. Translates these into skill sets for future members.
Leadership		
KNOWLEDGE COMPETENCY 3: Identifies a viable strategic position (cost leader, niche, etc.) for one's firm in the industry and communicates how new technology supports this strategy	High	Explains why the change (involving new technology) could be a successful strategy for one's own organisation and the organisation of other members involved in this initiative in a convincing manner.
	Acceptable	Explains why the technology-driven change supports a viable strategy for own organisation and may be able to explain how it could be a viable proposition for the other organisations.
	Low	Explains why new technology could support strategy for their own organisation. Has limited understanding of other organisations' strategic objectives in relation to new technology and is at times sceptical about the value of new technology.

Collaborative Practice Index

Specific competencies, and even specific model elements, will not always have equal importance in every situation. In a supply chain that is launching an innovation, for example, "Leadership" and "Shared Goals" might be important. In a contrast, a supply chain that is stabilizing change may find that "Organizing Mechanisms" and "Technical Standards" are vital. Differences might also emerge relative to individual positions: executives from a developer that is pushing for non-traditional methods may be expected to demonstrate the ability to bring a team to "Shared Goals", but may not be expected to contribute much to "Technical Standards".

The researchers' proposed approach to addressing this challenge of variability is to have a system by which weights can be assigned to specific competencies. This weighting process paves the way for collaboration to eventually be measured through a customisable Collaborative Practice Index (CPI). The CPI is a novel innovation that paves the way for the possibility of a flexible mechanism for measuring collaborative capacity in construction. The authors envision that in future SBTs, the CPI will bookend the entire training initiative, serving as both the tool for needs assessment before the training as well as a basis for evaluation at the end of the training.

An initial training needs assessment can be conducted by asking a given group of people to identify critical items from the list of 133 competencies, then assigning weights to them. A choice can also be made to select only some knowledge or skill competencies under a Collaborative Practice Element. A customized SBT can then be developed using the results of this needs assessment. At the end of the

SBT, evaluation will be aligned to these selections and will focus on the critical competencies that were previously identified.

The CPI, as an "index", involves assigning quantifiable weights to competencies during needs assessment before a training program. However, using weights at pre-training for needs assessment does that mean that post-training assessment must inevitably be quantitative. The CPI can simply be used for competency selection. Once competencies are chosen, these competencies can then be linked either economic and sociological assessment techniques (Stasz, 2001). Use of the CPI can be limited to choosing the most important competencies, which can then be linked qualitative descriptors as assessment tools, as the authors did here. In another situation, they can be be linked to a system where descriptors are linked numerical scores (Low = 1, High = 5), leading to a quantitative measure of performance.

FUTURE RESEARCH DIRECTIONS: DIGITISING THE SBT

The pilot of the SBT was conducted in face-to-face mode in 2018. Following this, the research team initiated the commercialisation of the training framework and this process is now in its final stages. Under the terms of the commercialisation agreement, the training program can be run in Victoria, Australia, exclusively by the research team and the BLSC, for a period of three years.

The Collaborative Practice Model and the training materials were presented by the research team to the Australian Housing Supply Chain Alliance, a group of representatives from state-based, national and multinational companies in housing. Representatives came from different organisations: engineering firms, development companies, suppliers of materials and industry associations. Collectively they represent the majority of the housing supply chain. The Alliance, which has been in existence since 2009, upholds a mandate to conduct research projects to improve the culture and behaviour to support innovative products and processes. The first research project involving the OSM case studies was one of the projects that they directly supported through a co-funding scheme with the federal government. Selected members of the Alliance were appointed as members of the research project's steering committee, thus detailed and regular industry feedback was received throughout the process of developing the model and the training materials.

One of the recommendations proposed by the Alliance was to create an online version of the SBT, mainly for industry partners who wanted to deliver the training in countries where they had factories. The research team has been undertaking preparatory work for this, noting that there are four broader trends that further strengthen the argument for digitisation. First, there is a general trend towards increasing digital support in training, a point that was mentioned early (Training Industry Report, 2018). Second, projects are becoming unfettered from tightly clustered geographic locations. Capital and labour in the industry are becoming internationalised (Utama, Kwan, Sesmiwati, & Khan, 2019). Global firms that have international standards for labour qualifications would favour training programs that cut across locations because they are either portable or accessible. Digitisation of training programs such as these address both. Third, the uptake of digital collaborative tools is also increasing (Blanco et al., 2017). Even if this uptake is particularly slow in the housing construction sector and empirical work still shows a strong preference for actors coming together face to face to utilise these tools (Pablo & London, 2020), face-to-face collaboration may not always be the norm. The current global pandemic has already significantly challenged this. Finally, simulations can be expensive and digitising SBTs may make these more economical. For example, geographical dispersion of participants may make it costly for trainees to travel to a single location (Kapp & Stahl, 2018). Ongoing face-to-face simulations may also require repeatedly hiring actors, which can be expensive. In contrast, online simulations eliminate the need for co-location, support "recorded" scenarios and in addition may provide a safe way of exposing trainees to potentially risky situations (Cobbett & Snelgrove-Clarke, 2016).

All these do not necessarily mean that the SBT must be fully digitised, simply that digital options must be seriously considered as a way forward. The rest of the section explores five issues: levels of digitisation, and the impact of digitisation on design, effectiveness, assessment and on participant interactions and the CPM itself.

Table 6. Possible directions for digitising the training

Possible Format	Key Features of SBT
Traditional face-to-face SBT	Face-to-face, as discussed in the paper. All participants, actors, organisational psychologists and trainers are in a single location. Simulations are done live.
Enhanced SBT	Use of digital tools is limited. All participants, actors, organisational psychologists and trainers are in a single location for most of the training period. Pre-recorded, non-interactive scenarios may be given to participants for analysis prior to the training. Participants will still undergo simulation in a face-to-face meeting in a single location. Another alternative is to have one or two participants who connect to the main meeting using virtual tools. ADMINISTRATIVELY: Pre-recorded lectures may be provided, or psychologists may conduct assessments remotely.
Hybrid SBT	Use of digital tools is significant, but the SBT is not in completely online format. This arrangement is envisioned to require a library of pre-recorded meeting segments to cover multiple "branching" scenarios. For example, training participants may me shown an initial scenario (S-0) requiring a response to collaborative dilemma. Participants frame a response as a team. Their response will lead the trainer to present another scenario (S-1) which can vary depending on the contents of their recommendations (e.g. there can be an S-1a, S-1b, S-1c). The selection of scenarios therefore follows a dynamic branching trajectory that is dependent on participant responses (Kapp & Stahl 2018). ADMINISTRATIVELY: Pre-recorded lectures may be provided, or psychologists may conduct assessments remotely.
Online SBT	Synchronous, digitally mediated training. The SBT is implemented in the same way as it is implemented in traditional mode, except sessions, simulations and assessments are conducted through a tool like Zoom.

Issue 1: Levels of Digitisation

Literature suggests that there are different degrees of digitisation: *traditional* face-to-face formats, requiring no digitisation, as discussed above; an *enhanced* format where the program is delivered primarily in face-to-face format but with limited use of digital tools; a *hybrid* format where the program is primarily, but not purely, in digital format and finally an *online* format where the entire SBT is delivered virtually (Duke University, n.d.). The need for changes in content may emerge, as research indicates that unreflective shoehorning of face-to-face content into online formats can be problematic (Shaw, 2001). When content is transformed in a considered manner, both face to face and digital formats can be equally effective (Johnson et al., in Alonso Diaz & Blázquez Entonado, 2009). Table 6 summarises the possible formats of the SBT ranging from traditional to fully online.

Issue 2: Impact of Digitisation on Design/ Delivery

The current contents of the scenarios and assessment tools were designed for face-to-face delivery and can readily be mobilised to support full online delivery, using synchronous tools such as Zoom. In the fully online context, facilitators will deliver the lectures via webinars, share slides using screen share functions and manage discussions through meeting tools. Actor and non-actor participants will undergo the simulation scenario through digitally-supported interactions. Zoom is one possibility; it has been found to be effective insofar as it possesses a degree of media naturalness and supports spontaneous user interaction (Blau et al., 2016). Organisational psychologists will continue to observe these meetings as they did from the control room. This arrangement addresses the issue of geographic dispersion of people involved in the SBT and does not require the creation of a new pool of materials. The format retains the dynamism of team-based SBTs, with conversations being organic and free-flowing despite key scripted elements. The key disadvantage of the fully online format is that it is still dependent on live contributions, one implication being that actors still have to be hired.

The shift to enhanced and hybrid formats will require the creation of an expanded repository of video-based scenarios and will require more planning and design. The hybrid format in particular will require a full range of possible branching story lines. However, material for the design of these additional scenarios is already available in the research team's large qualitative database. Real-life examples from interview data can readily be obtained, for example, about resistance to innovation on different grounds. Scripts can be developed to create situations where participants will have to address resistance to innovation on different grounds: lack of skill, up-front costs, restrictive financial regulations, concerns about regulators. Scripts can also be created to simulate hurdles to collaboration coming from different sources: poor communication, gaps between specialised disciplines, the lack of a common pool of standards, the inability to come to consensus around goals. In short, a range of real-life examples of these situations can already be pulled from transcripts.

Issue 3: Impact of Digitisation on Training Effectiveness

Researchers are considering that shifting to digitally-supported modes may have impact on training effectiveness, either generally or with respect to the acquisition of specific skills. For example, research has indicated that online SBTs may be better for building abstract knowledge and less effective when the goal is to strengthen concrete forms of knowledge (Alonso Diaz & Blázquez Entonado, 2009; Hui, Hu, Clark, Tam & Milton, 2008). However, this finding must be taken alongside the studies such as those done by Salas et al. (2009), mentioned earlier. Salas et al. (2009) suggest SBTs are well suited for team-based, problem-driven scenarios that mimic real life. In other words, the affordances of limitations of the medium (for example Zoom) must be considered alongside the affordances and limitations of SBTs as a training approach.

Issue 4: Impact of Digitisation on Assessment

The researchers do not expect significant impact on the assessment process of organisational psychologists. Under face-to-face arrangements, psychologists already "observed" participant interactions through screens from a control room. This screen-based observation does not significantly change when they begin observing in videoconferencing settings such as Zoom. That said, organisational psychologists would do well to acknowledge that they may be losing some signals in both situations. Participants may be interacting online or face to face, but in both cases observers are removed from the actual meeting room and are separated from participants. Observers may miss eye contact, some body language and microfacial expressions in digital interactions (Rudkovska & Sui, 2020).

Issue 5: Impact of Digitisation on Interactions and on the Collaborative Practice Model

A final point to note about moving to digital formats is that collaborative interactions between participants themselves may fundamentally change. Earlier it was mentioned that the research team undertook two research projects and that the aim of the second project was to explore if the Collaborative Practice Model would change if the case study attributes changed. A key element of this second research project is the selection of case studies that are different because they use digital technologies to collaborate. Work is ongoing to analyse if the nine elements will change significantly, or will simply be more detailed and nuanced due to digitisation. For example, one thing the research team is exploring is if goal negotiation (part of "Shared Goals") and "Problem-Solving" change if digital three-dimensional models replace 2D drawings in a collaborative design dilemma. There is the possibility that collaboration in digital environments will lead to the recreation of a new model. If this is the case, then this model change could have a cascading effect on the competencies identified, the interactions between elements, the scenario design and the assessment tools. Examining these possible changes are all part of the research team's agenda.

CONCLUSION

This study puts forth an argument for SBT as a strategy for building collaborative capacity in construction. Learning the skill sets linked to collaborative capacity is a complex process. Collaboration involves a broad swathe of knowledge, skill and behavioural competencies that must be mobilised in the context of teams, often with the goal of resolving complex problems and dilemmas. SBT is an approach that allows training participants to be immersed in team-based, problem-driven scenarios that mimic real life (Salas et al., 2009). When properly constructed, SBT enables participants to effectively learn, as well as reflect on, new behaviours in ways that more passive learning approaches cannot.

While SBTs are core to this initiative, it should be noted that "SBT" should not be narrowly interpreted to mean "the training program." The researchers have developed an entire suite of resources: behaviours, competency lists, the idea of an index, qualitative assessment tools, scenarios, scaffolds for scenarios and narratives. The researchers refer to this comprehensive suite of resources as the "Collaborative Practice Training Framework". Identifying, developing and assessing competencies for collaborative capacity requires the mobilisation of the entire framework. This entire framework is now being mobilised as the research team moves in the direction of digitisation.

ACKNOWLEDGMENT

This research was supported by the Australian Research Council [DP 180101178]. The authors would also like to acknowledge the research project's industry partners who are members of the Australian

Housing Supply Chain Alliance: Frasers Property Australia, FMG Engineering, Master Builders Association of Victoria and Metricon.

REFERENCES

AbouRizk, S. (2010). Role of simulation in construction engineering and management. *Journal of Construction Engineering and Management, 136*(10), 1140–1553. doi:10.1061/(ASCE)CO.1943-7862.0000220

Agranoff, R., & McGuire, M. (2003). *Collaborative public management: new strategies for local governments*. Georgetown University Press.

Alonso Díaz, L., & Blázquez Entonado, F. (2009). Are the functions of teachers in e-learning and face-to-face learning environments really different? *Journal of Educational Technology & Society, 12*(4), 331–343.

Baiden, B. K., Price, A. D. F., & Dainty, A. R. J. (2006). The extent of team integration within construction projects. *International Journal of Project Management, 24*(1), 13–23. doi:10.1016/j.ijproman.2005.05.001

Bechky, B. A. (2006). Gaffers, gofers, and grips: Role-based coordination in temporary organizations. *Organization Science, 1*(17), 3–21. doi:10.1287/orsc.1050.0149

Blanco, J. L., Janauskas, M., & Ribeirinho, M. J. (2016). *Beating the low-productivity trap: How to transform construction operations*. https://www.mckinsey.com/industries/capital-projects-and-infrastructure/our-insights/beating-the-low-productivity-trap-how-to-transform-construction-operations#

Blau, I., Weiser, O., & Eshet-Alkalai, Y. (2016). Face-to-face versus one-way and two-way videoconferencing: How medium naturalness and personality traits influence achievement and perceived learning? *11th Iberian Conference on Information Systems and Technologies*. doi:10.1109/CISTI.2016.7521581

Burke, M. J., Sarpy, S. A., Smith-Crowe, K., Chan-Serafin, S., Salvador, R. O., & Islam, G. (2006). Relative effectiveness of worker safety and health training methods. *American Journal of Public Health, 96*(2), 315–324. doi:10.2105/AJPH.2004.059840 PMID:16380566

Bygballe, L. E., & Sward, A. (2019). Collaborative project delivery models and the role of routines in institutionalizing partnering. *Project Management Journal, 50*(2), 161–176. doi:10.1177/8756972818820213

Chan, A. P. C., Chan, D. W. M., Chiang, Y. H., Tang, B. S., Chan, E. H. W., & Ho, K. S. K. (2004). Exploring critical success factors for partnering in construction projects. *Journal of Construction Engineering and Management, 130*(2), 188–198. doi:10.1061/(ASCE)0733-9364(2004)130:2(188)

Chapman, L. J. (2013). *Literature review and environmental scan for better translation of research to practice in residential construction.* Retrieved from https://www.cpwr.com/sites/default/files/publications/residential_construction_r2p_literature_review_chapman.pdf

Cheng, A. S., & Sturtevant, V. E. (2012). A framework for assessing collaborative capacity in community-based public forest management. *Environmental Management, 49*(3), 675–689. doi:10.100700267-011-9801-6 PMID:22209779

Cobbett, S., & Snelgrove-Clarke, E. (2016). Virtual versus face-to-face clinical simulation in relation to student knowledge, anxiety, and self-confidence in maternal-newborn nursing: A randomized controlled trial. *Nurse Education Today*, *45*, 179–184. doi:10.1016/j.nedt.2016.08.004 PMID:27537670

Duff, A. R., Robertson, I. T., Phillips, R. A., & Cooper, M. D. (1994). Improving safety by the modification of behaviour. *Construction Management and Economics*, *12*(1), 67–78. doi:10.1080/01446199400000008

Duke University. (n.d.). *Comparison: online learning environment vs. classroom-based learning environment.* Online teaching guide. https://sites.duke.edu/onlineguide/why-online-learning/comparison/

Eisenhardt, K. M. (1989). Building theories from case study research. *Academy of Management Review*, *14*(4), 532–550. doi:10.5465/amr.1989.4308385

Erdogan, B., Anumba, C., Bouchlaghem, D., & Nielsen, Y. (2008). Collaboration environments for construction: Implementation case studies. *Journal of Management Engineering*, *24*(4), 234–244. doi:10.1061/(ASCE)0742-597X(2008)24:4(234)

Flyvbjerg, B. (2006). Five misunderstandings about case-study research. *Qualitative Inquiry*, *12*(2), 219–245. doi:10.1177/1077800405284363

Gegenfurtner, A., Quesada-Pallarès, C., & Knog, M. (2014). Digital simulation-based training: A meta-analysis. *British Journal of Educational Technology*, *45*(6), 1097–1114. doi:10.1111/bjet.12188

Gioia, D. A., Corley, K. G., & Hamilton, A. L. (2012). Seeking qualitative rigor in inductive research: Notes on the Gioia methodology. *Organizational Research Methods*, *16*(1), 15–31. doi:10.1177/1094428112452151

Goldstein, I. L. (1980). Training in work organizations. *Annual Review of Psychology*, *31*(1), 229–272. doi:10.1146/annurev.ps.31.020180.001305

Grant, J. (2002). Learning needs assessment: Assessing the need. *British Medical Journal*, *324*(7330), 156–159. doi:10.1136/bmj.324.7330.156 PMID:11799035

Gray, B. (1985). Conditions facilitating inter-organizational collaboration. *Human Relations*, *38*(10), 911–936. doi:10.1177/001872678503801001

Gulati, R. (2007). Silo busting: How to execute on the promise of customer focus. *Harvard Business Review*, *85*(5), 98–108. PMID:17494254

Hardy, C., Phillips, N., & Lawrence, T. B. (2003). Resources, knowledge and influence: The organizational effects of interorganizational collaboration. *Journal of Management Studies*, *40*(2), 321–347. doi:10.1111/1467-6486.00342

Harty, C. (2008). Implementing innovation in construction: Contexts, relative boundedness and actor- network theory. *Construction Management and Economics*, *26*(10), 1029–1041. doi:10.1080/01446190802298413

Hoffman, W. H., & Schlosser, R. (2001). Success factors of strategic alliances in small and medium sized enterprises: An empirical survey. *Long Range Planning*, *34*(3), 357–381. doi:10.1016/S0024-6301(01)00041-3

Holland, P. J., Allen, B., & Douglas-Lenders, R. C. (2014). *Building a better workforce: assessing the impact of the experiential and simulation-based learning in the Australian building and construction industry*. Monash University.

Hui, W., Hu, P. J. H., Clark, T. H. K., Tam, K. Y., & Milton, J. (2008). Technology-assisted learning: A longitudinal field study of knowledge category, learning effectiveness and satisfaction in language learning. *Journal of Computer Assisted Learning*, *3*(24), 245–259. doi:10.1111/j.1365-2729.2007.00257.x

Kapp, K., & Stahl, B. (2018, January 10). *3 lessons from converting a face-to-face role-play into an online simulation.* Association for Talent Development. https://www.td.org/blog-post/3-lessons-from-converting-a-face-to-face-role-play-into-an-online-simulation

Kassem, M., Benomran, L., & Teizer, J. (2017). Virtual environments for safety learning in construction and engineering: Seeking evidence and identifying gaps for future research. *Visualization in Engineering*, *5*(16), 2–15. doi:10.118640327-017-0054-1

Lateef, F. (2010). Simulation-based learning: Just like the real thing. *Journal of Emergencies, Trauma and Shock*, *3*(4), 348–352. doi:10.4103/0974-2700.70743 PMID:21063557

Lavis, J. N., Robertson, D., Woodside, J. M., McLeod, C. B., & Abelson, J.Knowledge Transfer Group. (2003). How can research organizations more effectively transfer research knowledge to decision makers? *The Milbank Quarterly*, *81*(2), 221–248. doi:10.1111/1468-0009.t01-1-00052 PMID:12841049

Lincoln, Y. S., & Guba, E. G. (2016). *The constructivist credo*. Routledge. doi:10.4324/9781315418810

Lines, B. C., Sullivan, K., Smithwick, J., & Mischung, J. (2015). Overcoming resistance to change in engineering and construction: Change management factors for owner organizations. *International Journal of Project Management*, *33*(5), 1170–1179. doi:10.1016/j.ijproman.2015.01.008

London, K., & Pablo, Z. (2017). An actor–network theory approach to developing an expanded conceptualization of collaboration in industrialized building housing construction. *Construction Management and Economics*, *35*(8-9), 553–577. doi:10.1080/01446193.2017.1339361

McAllister, M., Levett-Jones, T., Downer, T., Harrison, P., Harvey, T., Reid-Searl, K., Lynch, K., Arthur, C., Layh, J., & Calleja, P. (2013). Snapshots of simulation: Creative strategies used by Australian educators to enhance simulation learning experiences for nursing students. *Nurse Education in Practice*, *13*(6), 567–572. doi:10.1016/j.nepr.2013.04.010 PMID:23707534

Orlikowski, W. J., & Baroudi, J. J. (1991). Studying information technology in organizations: Research approaches and assumptions. *Information Systems Research*, *2*(1), 1–28. doi:10.1287/isre.2.1.1

Pablo, Z., & London, K. (2016, November). A collaborative practices typology for Australian prefabricated housing networks: Convergence, alignment and coordination. In R. Amoêda, & C. Pinheiro (Eds.), *Proceedings of the 2016 International Conference on Sustainable Housing Planning, Management, and Usability* (pp. 725-734). Green Lines Institute for Sustainable Development.

Pablo, Z., & London, K. (2017, June). Developing collaborative capacities in industrialized building: roadmap for knowledge transfer'. In E. Fidelis and M. Behm (Eds.), *Proceedings of the Joint CIB W099 and TG59 International Safety, Health, and People in Construction Conference: Towards better Safety, Health, Wellbeing, and Life in Construction* (pp. 593-602). Department of Built Environment, Central University of Technology, South Africa.

Pablo, Z., & London, K. (2020). Stable relationality and dynamic innovation: Two models of collaboration in SME-driven offsite manufacturing supply chains in housing construction. *Engineering, Construction, and Architectural Management, 27*(7), 1553–1577. doi:10.1108/ECAM-07-2019-0346

Poirer, E., Forgues, D., & Staub-French, S. (2016). Collaboration through innovation: Implications for expertise in the AEC sector. *Construction Management and Economics, 34*(11), 769–789. doi:10.1080/01446193.2016.1206660

Rudkovska, A., & Sui, W. (2020, April 29). *Why FaceTime can't replace face-to-face time during social distancing*. The Conversation. https://theconversation.com/why-facetime-cant-replace-face-to-face-time-during-social-distancing-136206

Salas, E., Rosen, M. A., Held, J. D., & Weissmuller, J. J. (2009). Performance measurement in simulation-based training: A review and best practices. *Simulation & Gaming, 40*(3), 328–376. doi:10.1177/1046878108326734

Schöttle, A., Haghsheno, S., & Gehbauer, F. (2014, June 25-27). Defining cooperation and collaboration in the context of Lean Construction. In *22nd Annual Conference of the International Group for Lean Construction* (pp. 1269-1280). Oslo, Norway: IGLC.

Shaw, K. (2001). Designing online learning opportunities, orchestrating experiences and managing learning. In J. Stephenson (Ed.), *Teaching and learning online: Pedagogies for new technologies* (pp. 53–66). Stylus.

Stasz, C. (2001). Assessing skills for work: Two perspectives. *Oxford Economic Papers, 53*(3), 385–405. doi:10.1093/oep/53.3.385

Tabassi, A. A., & Abu Bakar, A. H. (2009). Training, motivation, and performance: The case of human resource management in construction projects in Mashhad, Iran. *International Journal of Project Management, 27*(5), 471–480. doi:10.1016/j.ijproman.2008.08.002

Torrell, B. (2006). The role of narration in high stakes decision-making: translating knowing into telling. In G. Schreyogg & J. Koch (Eds.), *Knowledge management and narratives: organizational effectiveness through storytelling* (pp. 247–272). Erich Schmidt Verlag GmbH & Cp.

Training Industry Report. (2018). *Training Magazine*. Retrieved from https://trainingmag.com/sites/default/files/trn-2018-industry-report.pdf

Utama, W. P., Chan, A. P. C., Sesmiwati, S., & Khan, H. Z. A. (2019). Internationalization of construction enterprises: An overview of motivation. *International Journal of Technology, 10*(1), 36–46. doi:10.14716/ijtech.v10i1.1558

Van der Zee, D. J., Holkenberg, B., & Robinson, S. (2010). Conceptual modelling for simulation-based serious gaming. *Decision Support Systems*, *54*(1), 33–45. doi:10.1016/j.dss.2012.03.006

Walker, D., & Walker, B. (2015). *Collaborative project procurement arrangements*. PMI Inc.

Wilkins, J. R. (2011). Construction workers' perceptions of health and safety training programmes. *Construction Management and Economics*, *29*(10), 1017–1026. doi:10.1080/01446193.2011.633538

Wolcott, H. F. (1994). Transforming qualitative data. *Sage (Atlanta, Ga.)*.

Wood, D. J., & Gray, B. (1991). Towards a comprehensive theory of collaboration. *The Journal of Applied Behavioral Science*, *27*(2), 139–162. doi:10.1177/0021886391272001

Zhang, L., Huang, S., & Peng, Y. (2018). Collaboration in Integrated Project Delivery: The effects of trust and formal contracts. *Engineering Management Journal*, *4*(30), 262–273. doi:10.1080/10429247.2018.1498259

ADDITIONAL READING

Gioia, D. A., Corley, K. G., & Hamilton, A. L. (2012). Seeking qualitative rigor in inductive research: Notes on the Gioia methodology. *Organizational Research Methods*, *16*(1), 15–31. doi:10.1177/1094428112452151

Hensen, J. L. M., & Lamberts, R. (2019). *Building Performance Simulation for Design and Operation* (2nd ed.). Routledge. doi:10.1201/9780429402296

Holland, P. J., Allen, B., & Douglas-Lenders, R. C. (2014). *Building a better workforce: assessing the impact of the experiential and simulation-based learning in the Australian building and construction industry*. Monash University.

Lincoln, Y. S., & Guba, E. G. (2016). *The constructivist credo*. Routledge. doi:10.4324/9781315418810

London, K., & Pablo, Z. (2017). An actor–network theory approach to developing an expanded conceptualization of collaboration in industrialized building housing construction. *Construction Management and Economics*, *35*(8-9), 553–577. doi:10.1080/01446193.2017.1339361

Smyth, H., & Pryke, S. (2008). *Collaborative relationships in construction: developing frameworks and networks*. Wiley-Blackwell. doi:10.1002/9781444301069

Torrell, B. (2006). The role of narration in high stakes decision-making: translating knowing into telling. In G. Schreyogg & J. Koch (Eds.), *Knowledge management and narratives: organizational effectiveness through storytelling* (pp. 247–272). Erich Schmidt Verlag GmbH & Cp.

Wolcott, H. F. (1994). Transforming qualitative data. *Sage (Atlanta, Ga.)*.

KEY TERMS AND DEFINITIONS

Collaborative Competencies: Knowledge, skills, and behaviours linked to the nine elements of the Collaborative Practice Model, that are linked to collaborative capacity

Collaborative Practice Element: One of the nine components identified by an extended empirical study as being critical to collaboration in construction. Nine elements are organically linked in a collaborative practice model.

Collaborative Practice Index: A proposed measure of collaboration that is obtained by assigning weights to specific collaborative competencies or to specific elements of the collaborative practice model. The weights can vary depending on type of organization, the stage of a given project or the position of an individual in a project.

Collaborative Practice Model: An explanatory framework comprising nine elements including: leadership, shared goals, expertise, change, investment, shard space, problem solving, organizing mechanisms and technical standards. The model is the basis of all elements in the collaborative practice training framework.

Collaborative Practice Training Framework: A system of interventions including a model, behavioural competencies, Collaborative Practice Indicators, scenarios, and assessment criteria which collectively support simulation-based training for building collaborative capacity in the industry.

Experiential Learning: An approach to developing knowledge and skills by engaging the learner in direct experiences.

Hybrid Training: Delivery of training with significant use of digital tools, without moving into fully online format.

Online Training: Delivery of training using purely synchronous and/ or asynchronous digital tools.

Scenario: An outline of a script or scene embedded in simulation-based training, designed in this case to motivate simulation-based training participants to exhibit specific collaborative competencies.

APPENDIX

Figure 2. Sample narrative.

> *Leadership involves developing a compelling vision for the innovation at the organisational and industry level. The leader communicates this throughout the supply chain through formal and informal strategic planning mechanisms. All relevant staff are engaged with the vision and involved as appropriate in strategic and operational planning processes. The leader capitalises on their position, or on the position of the organization, to influence the supply chain to embrace and test the innovation. Resistance and risk are managed through compelling feasibility studies that present a viable project and through evidence of successful cases in other contexts. Demonstrating project feasibility rests on strong technical expertise which the leader either possesses, or has access to through close partnerships.*

Figure 3. Conversation structure to scaffold scenarios

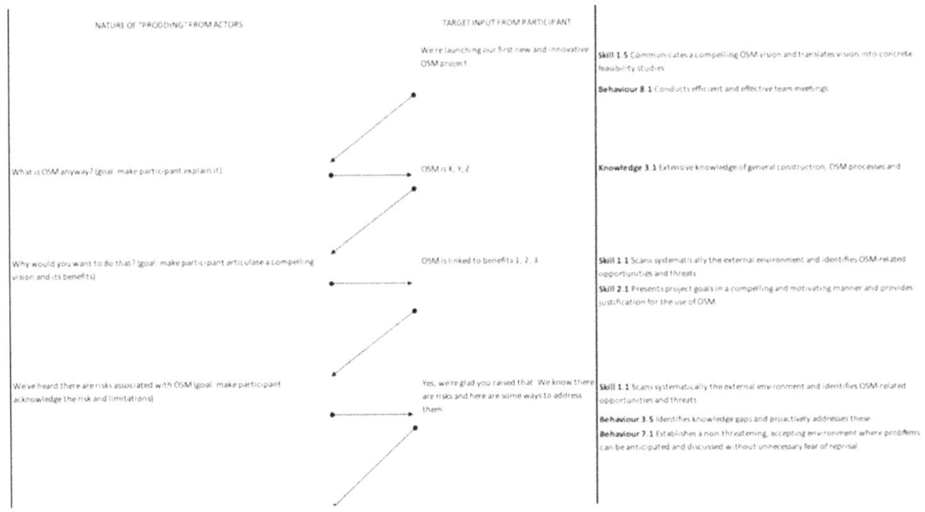

Chapter 17 Transforming the Productivity of People in the Built Environment: Emergence of a Digital Competency Management Ecosystem

Jason Underwood
https://orcid.org/0000-0002-3193-1620
University of Salford, UK

Mark Shelbourn
Birmingham City University, UK

Debbie Carlton
Dynamic Knowledge, UK

Gang Zhao
Intelartes, UK

Martin Simpson
Digital BE Ltd., UK

Gulnaz Aksenova
https://orcid.org/0000-0002-3261-0917
University College London, UK

Sajedeh Mollasalehi
University of Salford, UK

ABSTRACT

This chapter explores how we create and support a digitally enabled, agile, competent, and ultimately, productive workforce and determines the key research questions that need to be addressed if Digital Built Britain (DBB) is to provide return on investment and succeed as the catalyst for evolving the manner in which we conceive, plan, design, construct, operate, and interact with the built environment. The proposed vision is a digital competency management ecosystem where interdependent stakeholders are incentivised to work together in coopetition to create, capture, infer, interpret, specify, integrate, accredit, apply, use, monitor, and evolve competence as a working (data) asset. This needs to be in a consistent, objective, explicit, and scalable manner, with end2end transparency and traceability for all stakeholders that overcome the challenges of competency management. Moreover, a core element must be an ecosystem organised around digital infrastructure of competency frameworks and other knowledge sources of competence, so that competency frameworks are in digital operation and dynamic context.

DOI: 10.4018/978-1-7998-6600-8.ch017

INTRODUCTION

A successful evolution within any industry or sector is often seen as a combination of People, Process, and Technology (Arayici, 2011). However, it is difficult to quantify the relevant proportion of investment into People, Process, and Technology suggested by the Construction Sector and Government strategies and reports. The Digital Groundwork report (PlanGrid, Inc., 2019) stated that 50% of industry professionals say that improving digital skills[1] will be a key focus for their business over the next three years. However, in the latest NBS Annual BIM Report (NBS, 2020), 63% attribute the barrier to BIM use to a lack of in-house skills/expertise, and 61% to the lack of training; both continuing to remain two of the top three reported key barriers to the successful adoption of BIM and digital transformation over the last decade. The Construction Manager Annual BIM survey 2020, similarly, reports that poor digital skills are holding back the adoption of BIM with 64% attributing the main organisational barrier to the adoption of BIM, or further adoption, to the lack of digital skills (Construction Manager, 2020). Furthermore, when considering the main barrier to productivity in their business, industry professionals most commonly point to a lack of skills amongst employees (40%) and talent shortages (39%) (PlanGrid, Inc., 2019).

When we consider the construction sector is comprised of a significant proportion of micro and small and medium sized enterprises (SME's) that are essential to the whole built asset life cycle (over 99% (BIS, 2013; Designing Buildings Wiki, 2019)), the development of their digital skills is critical to facilitate the ongoing digital transformation (CLC, 2019). Improving digital skills and the ability to use digital tools of SMEs could add £9.9 billion, i.e. 0.5% of total UK GDP (Oxford Economics, 2020). However, according to the 'Bricks, Mortar and Digital Transformation' report, SMEs within the UK construction sector risk being left behind if they do not embrace the digital transformation journey (Zen Internet, 2019).

The implementation of BIM not only demands the use of computer-aided design, modelling, and collaborative tools, but also requires a shift in/or having a flexible mindset (sometimes referred to as metaskills) to embrace these techniques in all construction processes. Therefore, BIM training can either mean 'learning to use technology' or 'process focused' with the majority based strongly on the PAS/ISO standards (SFT, 2018). Thus, Barley (2019) asks, do we do more than, "*give people the process and the technology and call those who reject it luddites or laggards*"?

The digital transformation picture is complex and fragmented; however, there is little evidence of industry-wide initiatives that support the 'People-centric' focus required to successfully deliver 'transformational change' on a scale that is unprecedented. The COVID-19 pandemic offers an unprecedented opportunity for not only new digital ways of working but also the digitalisation related to work, people, and competences. There is an urgent need for digital talent in the built environment sector to drive BIM and digital transformation not only in the UK but across the world.

The Digital Built Britain (DBB) vision is far more-wide ranging than that envisaged for the 2016 UK Government BIM mandate. It follows that, although BIM training and certification is an important first step towards the digital transformation journey within the built environment, success will require an evolved upskilling[2]/reskilling philosophy to develop a digitally competent and agile workforce capable of supporting the emerging digitally-enabled processes, purpose driven information flow, and the collaborative behaviours that underpin them. The post pandemic crisis further supports the need for such an approach to facilitate the sector's future recovery in the emergence of a 'new normal'. The Client Leadership Council (CLC) launched a Recovery Plan in June 2020 to drive the recovery of the construction and built environment sectors following the COVID-19 pandemic and the resulting economic downturn

(CLC, 2020), through a three phase strategy to restart, reset, and reinvent the sector. The modernising of upskilling, training, qualifications, accreditation, and certification systems for construction to ensure that this is fit for purpose; supporting the delivery of a competent workforce required by the future industry, and the multi-skilling of the construction workforce to increase flexibility and adaptability, are stipulated as part of the reinvent/transform phase. The training, certification, and qualification systems need to be driven by dynamic data so competency demand is explicit. The problem has been the unsuccessful focus on supply-side interventions when policy needs to be focused on demand-led interventions. The use of qualifications or 'cards' as a proxy for supply, from a competency management perspective, is neither granular, multi-dimensional, nor dynamic enough.

It is vital to ensure that individuals, teams, and organisations have the right knowledge, skills, abilities, experience, behaviours, and attitude at work, i.e. 'competencies', to carry out tasks/activities they are appointed to perform. It is also vital for Clients, when appointing individuals, teams and organisations, to undertake construction operations, to ensure they are 'competent' to be able to conduct the work appropriately (ICM, 2020). The Dame Judith Hackitt (DJH) (Hackitt, 2018) report entitled, "*Building a Safer Future*" that followed the Grenfell Fire tragedy, concluded that the current system of building regulations and fire safety is not fit for purpose and that competence development and culture change are required to support the delivery of buildings that are safe now and in the future. These systemic failures occur in the delivery of all projects within the construction sector and the wider built environment. Therefore, a competence problem exists resulting from the lack of effective strategies for competence definition, development, management and, importantly, assurance. The sector, unlike Rail, has not invested in competency management systems. Although, there are positive indications, such as in Social Housing Maintenance Sector, that is developing a UKAS TPS 69 Standard for a Competency Management System (NHA, 2016).

The subsequent DJH 'Raising the Bar' interim report (CIC, 2019) and the recommended 'overarching system for overseeing competence' validates the need for a systemic approach and many (but not all) of the stakeholders. However, it does not address the extent of the 144 competency supply-demand interrelationships and activities (Figure 2), or how the system will be fully implemented or digitally-enabled, including a new 'framework of standards', such that competence truly acts as the currency of the labour marketplace. In countries such as Canada, where the industry is highly regulated, competency acts as a basis for transaction and comparison.

Furthermore, the DCOM 'A Call to Digitise' report (DCOM, 2019), states that, so far, there has been no meaningful suggestion of either the digitisation of the regulations or compliance systems despite the concerns raised by the Hackitt Review around responsibility and departure from regulations being a systemic problem. However, this call to action does not explicitly recognise the need to digitise knowledge sources of competences (e.g. competence frameworks) and the competency assurance system, despite a core recommendation of a new 'framework of standards'.

The final report 'Setting the Bar' (CIC, 2020), which should be read in association with the draft Building Safety Bill (2020), sets out a new system that includes a comprehensive set of competence standards underpinned by rigorous third party assessment for individuals and companies and third party accreditation of those who carry out the assessment. The report neither acknowledges, nor promotes a digital vision for a functioning competency system, or where a framework of competency frameworks sits within this. This despite Hackitt's call to action for the urgent need for radical changes in culture, competencies, and processes, and that digital will be the norm (NBS Summit, 2020).

The Cambridge Centre for Smart Infrastructure & Construction 'Flourishing Systems' report (CCSIC, 2020) propose a people-focused 'infrastructure industry' that recognises infrastructure as a complex, interconnected system of systems, and which considers value in terms of outcomes and human flourishing. In addition, it states that "*the long-term viability of infrastructure requires it to be sustainable, which includes the people who sustain the system of systems themselves being supported in their wellbeing and safety, their employment security, and in the development of their capability*". However, an ecosystem approach to competency (and its management) is not explicitly considered as a key system within the system of systems or the need for digital infrastructure to support this. The CDBB recognise that a people-centric approach to competency evolution and supporting the upskilling/reskilling to ensure a competent and digital workforce requires digitalisation.

Similarly, a digital ecosystem approach to competency and its management and assurance (to ensure there is a competent supply and 'good work' now and in the future (RSA, 2019)) is not considered within the CLC Roadmap to Recovery plan to 'reinvent' the construction and built environment sectors, and drive a more capable, professional, productive, and profitable sector post-COVID-19. In addition, 'skills' rather than 'competence' is the choice of language despite the increased forensic lens of competence assurance post Hackitt. While there is continued investment targeted at more general digital literacy and skills frameworks, through the DfE, CITB, CDBB, etc., such initiatives still take a classical information model (pdf, static, narrative format, schematic in hierarchy, and time bound) approach to competency frameworks when a systems model is required.

The lack of clear and consistent information management strategies and the operationalisation of competency management mean that finding the right information to make timely decisions by the right person is often impossible. Information management is part of digital competence requirements (CDBB, 2021). In the 'Building a Safer Future' report, the term the 'Golden Thread' of information is used. Of these two, the 'Golden Thread' is likely to be achieved through 'Technology' and 'Process'. Consequently, this may be easier to achieve than the 'Competence' issues, which, enabled through technology, will present a much greater challenge. The Steering Group on Competences for Building a Safer Future that was set up by the Construction Industry Council in response to the DJH report (Hackitt, 2018), intends to ensure a joined-up approach to achieve a comprehensive, coherent, and robust framework for the competence of all those creating, maintaining, and managing higher risk residential buildings.

There is little argument that successful projects are characterised by the right people making the right decisions at the right time with the right information. In the context of "Building a Safer Future", the 'Golden Thread' means having appropriate information at hand to make informed decisions within the expected timeframes. However, 'competence' ensures there are the right people available to specify the information required to make those decisions, and correctly interpret the received information. Furthermore, it can be argued that productive work only occurs when competent people specify the work that must be done, which is in turn, carried out by other competent people. The productivity of people comes from effective mediation between competency supply and demand, and enabled by the flow of appropriate and correct information. The productivity problems of the construction sector and wider built environment is therefore likely to be due to as much a manifestation of the failures relating to the mediation of competency supply and demand, as the failures relating to the flow of purpose driven information.

This chapter presents the findings, and further research work, from the Pedagogy and Upskilling Network (PUN, 2019) report. PUN is a network of over 50 collaborators drawn from research, practice, and other relevant areas, commissioned by the Centre for Digital Built Britain (CDBB) to explore how we create and support a digitally enabled, agile, competent and ultimately, productive workforce. Results from a series of network workshops proffers key research questions that need to be addressed if DBB is to both provide return on investment, and succeed as the catalyst for evolving the manner in which we conceive, plan, design, construct, operate, and interact with the built environment (CDBB PUN, 2018). Over the course of the network, two primary questions emerged:

- *What is required to facilitate the emergence of a digital ecosystem for the effective mediation of competency supply and demand within the construction sector and wider built environment that supports digitalisation and productivity?*
- *How does the sector support the competence development, management and assurance of individuals, teams, and organisations throughout their career in the built environment?*

While competency frameworks are obviously an important knowledge element of a digital competency management ecosystem, they were not the prime focus of the research. Furthermore, no conceptual distinction of potential or actual is made between competence(s) and competency(ies). The latter is considered as a collective noun of the former.

These are broad topics and the findings suggests a number of questions and recommendations relevant to PUN that need further attention at the next stage of the CDBB initiative. In addition, a successful response will require different combinations of leadership from industry and government, fundamental research, and pilot projects to rigorously measure impact and effectiveness. The proposed vision is to imagine a ***digital ecosystem of competency management where interdependent stakeholders are incentivised to work together in coopetition to create, capture, infer, interpret, specify, integrate, accredit, apply, use, monitor, and evolve competence as a working asset just like any other data asset***. Furthermore, this needs to be in a consistent, objective, explicit, and scalable manner; with end2end transparency and traceability for all stakeholders that face the challenges of competency management. With an increasing overlap of competences across sectors, professions, trades, roles, and disciplines, a core element must be an ecosystem of competency frameworks, and competency frameworks in digital operation and dynamic context; as opposed to the current traditional informational model of competence. In addition, a digital competency management ecosystem relies on robust lifelong learning opportunities that empower individuals to take responsibility for managing their own career, i.e. ***Conscious Competence Evolution***, as well as, allowing employers to both recognise and reward innovation.

ECOSYSTEM STAKEHOLDERS

The key challenge of DBB is to ensure that the UK is able to harness new technologies and digital connectivity to transform the built environment and to deliver real social and economic benefits to its citizens. Therefore, it is necessary to assess the impact of, and implications for, the various (Digital Built Environment) stakeholders that comprise the ecosystem, including (but not exhaustive):

- Clients, and by extension, those involved in procuring projects;
- Those representing entities (either a company or individual) and engaged to deliver projects (who in turn, must be able to demonstrate their organisations have the collective competence, capability, and capacity to discharge their contractual obligations);

- The myriad of professions who are involved in delivering projects at any stage of the asset life-cycle, including the commissioning, designing, constructing, maintaining, operating, altering, and decommissioning of Built Environment assets;
- Those responsible for educating and upskilling current (and future) professionals;
- Those responsible with certifying and assuring the competence of practitioners;
- Those responsible for the regulatory framework and the development of standards such as the Engineering Council that is aiming to get more protected professional status without operating in a transparent and digitised manner; and
- The citizens who interact and use the built environment for all aspects of their day-to-day lives, not just for business.

In order to be successful, we should ensure that we reach out beyond traditional built environment professionals and include experts from other domains and industries who can complement the ecosystem with new propositions, e.g. manufacturing, finance, IT, agile business practice and competence management. On full consideration of the wide range of stakeholders and their interactions within the ecosystem, the classification of initiatives by People, Process and Technology is particularly nebulous when considering 'People'. This classification specifically misses the critical distinction between the 'individual as a stakeholder' and the 'organisation (that comprises People) as a stakeholder'. Interventions can be aimed at either the individual, the organisation, or through intermediaries for work and workers (e.g. many individuals are not employees, but are other types of workers). Furthermore, there are increasingly different types of work and workers, which makes mediation more complex (RSA, 2019). To make this relationship more explicit, the classification of initiatives in Figure 1 was proposed.

Furthermore, the proposed initiatives are classified as having 'Research' focus (TRL 1→5) teams or an 'Implementation' focus (TRL 6→9). While the next stage of CDBB will have a research focus, the implementation-focused initiatives also need attention, as success relies on the interdependence of all initiatives across the ecosystem. Furthermore, research focused initiatives should consult with industry to assess whether any complementary implementation focused initiatives are required to ensure that long-term impact is possible. This includes partnering with industry domains that are the thought leaders in digital Human Resource Management.

DIGITAL COMPETENCY MANAGEMENT ECOSYSTEM

The ecosystem has a critical function as it sets out the norms of interactions between individuals and organisations, and organisations with other organisations (many-to-many relationship of competency demand and supply); process and technology are the means by which these interactions take place. A recent study assessed the digital transformation journey of the Architecture, Engineering, Construction and Operation (AECO) sector in Finland and concluded that an overemphasis on technological capabilities in practice and the extensive portfolio of national R&D initiatives, which mainly target productivity and efficiencies with technological developments, did not lead to the systemic change in the established ecosystem, nor led to the emergence of the new Digital Business Ecosystems (Aksenova *et al.*, 2018). As a lesson learnt, the government of Finland established a new programme in 2016-2018, named Kira-Digi. Kira-Digi has brought about a new experimental neutral platform to coordinate the discussions between the government departments, city stakeholders, AECO industry stakeholders, and complementary industries to support the emergence of the digital business ecosystems, while "The €16M programme's vision is to develop an open, interoperable information management ecosystem for the built environment" (Heiskanen, 2017).

Figure 1. Proposed alternative to People: Process: Technology (Figure on ecosystem is adapted from Moore (1997) and the concepts on an ecosystem of competency mediation between demand and supply are adapted from Zhao (2017))

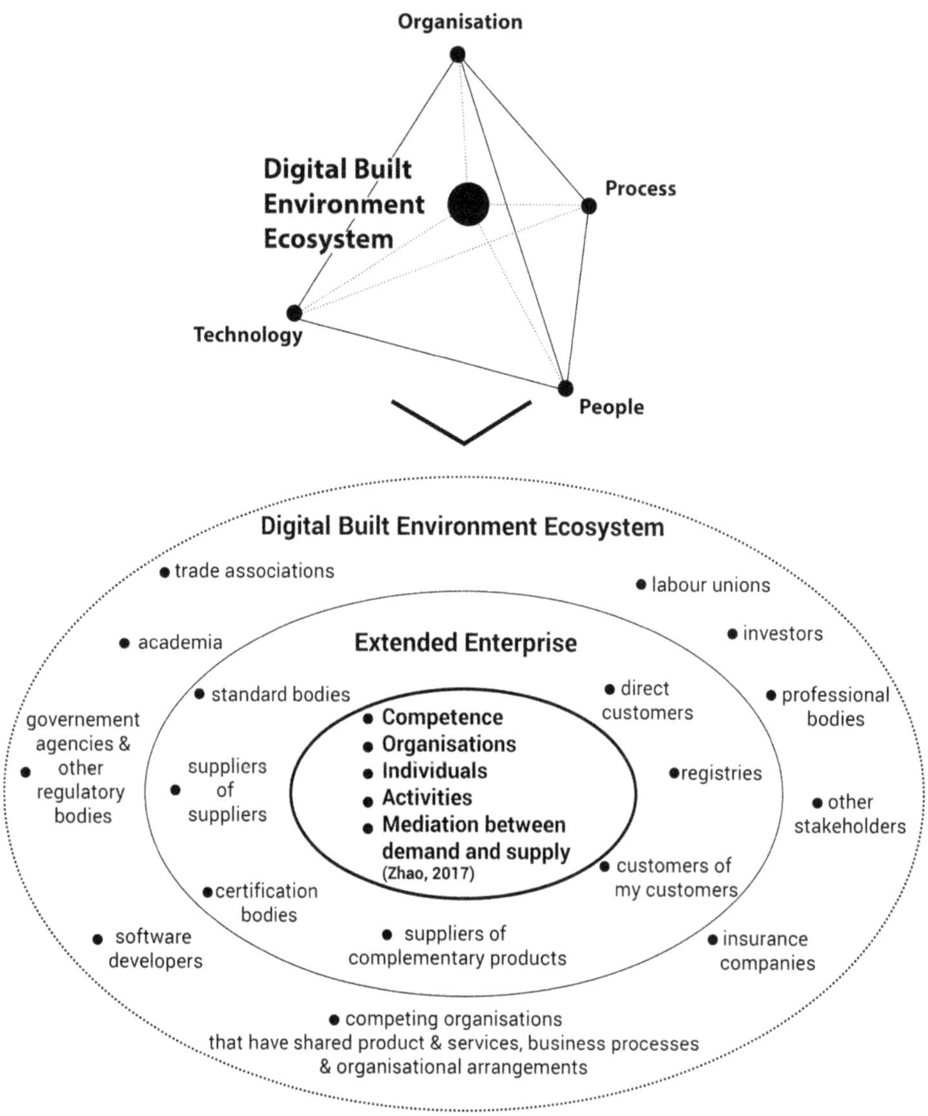

Clarysse *et al.* (2014) analysed a knowledge ecosystem around the region of Flanders that aimed to stimulate knowledge creation in technology hotspots on the assumption that these knowledge networks will lead to the region's competitiveness. Despite the financial support network, and a 100% publicly-funded and well-structured knowledge ecosystem, the emergence of new businesses has been almost non-existent. This posed serious implications for policy makers in the need to learn that investment in the creation of knowledge ecosystems does not necessarily lead to the development of business ecosystems, because the value creation processes for knowledge and businesses are fundamentally different, while local actors might struggle to attract global players to support the emergence of new business ecosystems. Different types of ecosystem require specifically tailored policies and power recognition. Furthermore, Rinkinen and Harmaakorpi (2017) have recognised that policy development for ecosystem evolution is an important field for future empirical studies; hence, the policy positioning on ecosystem creation is in its infancy. There have been successful skills projects to stimulate ecosystem growth in the automotive sector in Limburg Province that address the cyclical nature of work and keep the skills in the province.

In the context of the growing organisational complexity of the Digital Built Environment, which is driven by an adaptation to high uncertainty, and the central role of collaboration, the ecosystem approach is becoming more important. A multidisciplinary approach to competency management across sectors, disciplines, professions, etc., matters more than targeting productivity improvement through rapidly changing technologies. There is a significant risk that, if the Digital Built Environment does not re-design its policies, institutional and industrial landscape in an ecosystem-based manner that acknowledges the increasing granularity, complexity, and turbulence of the environment, as well as placing the individual at the centre of the competency management, current initiatives will have limited impact in terms of innovation (Autio & Thomas, 2014; Clarysse *et al.*, 2014; Aksenova *et al.*, 2018, Russell and Smorodinskaya, 2018).

Therefore, the relationships between the professional institutions, universities, government, the industry, and other accreditation and certification stakeholders of the digital competency management ecosystem are critical if a competent workforce is to mitigate risks. For example, funders and insurers could be critical partners to ensure that real change happens within the Digital Built Environment, particularly away from projects covered by the 2016 UK Government BIM mandate. In addition, many of the traditional paradigms and structures are not fit-for-purpose and should be re-examined within the context of the emerging digital interconnected world. There is a considerable risk that, if we attempt to embark on a digital transformation journey with the handbrake on, we will not get far!

COMPETENCY SUPPLY AND DEMAND

The mediation of competency supply and demand directly impacts productivity (Zhao, 2017) as much as the flow of timely and appropriate information, i.e. an Information Transaction. Information is used in the broadest sense of data, which is formatted to allow the inference of meaning. However, does the mediation of competency supply and demand mirror the information transactions that occur between parties or individuals?

OECD (2017b) identified within the UK, a high level of mismatch between the skill supply and skill demand, and that this is due to poor mediation in conjunction with ineffective supply focused interventions facilitated by the UK government (IPPR, 2017). Therefore, a transformation in mediation between demand and supply is a critical issue for competency management. The two key stakeholders in competency mediation are those who demand competency (i.e. asset owners and employers) and those who supply competencies (workers); they are the primary stakeholders. The secondary stakeholders (government, professional bodies, trade associations, etc.) are mediators between the supply demand relationships. Research on an ecosystem approach to mediation of competency demand and supply has established that there are 144 competency supply-demand inter-relationships between 9 stakeholders types (Figure 2) (Zhao, 2017a, b; 2018). This involves many-to-many relationships, activities, and transactions depending on who is demanding competency (employers and importantly, owners), who is supplying the competencies (workers), those who are overseeing competence (regulators, etc.) and those facilitating competency by providing training, assessment, validation, verification, accreditation, certification, or other services (Zhao, 2017a, b; 2018). A critical component of this ecosystem is that workers are a key stakeholder, i.e. it is people-centric.

Figure 2. Ecosystem of Competency Demand & Supply (Zhao, 2018)

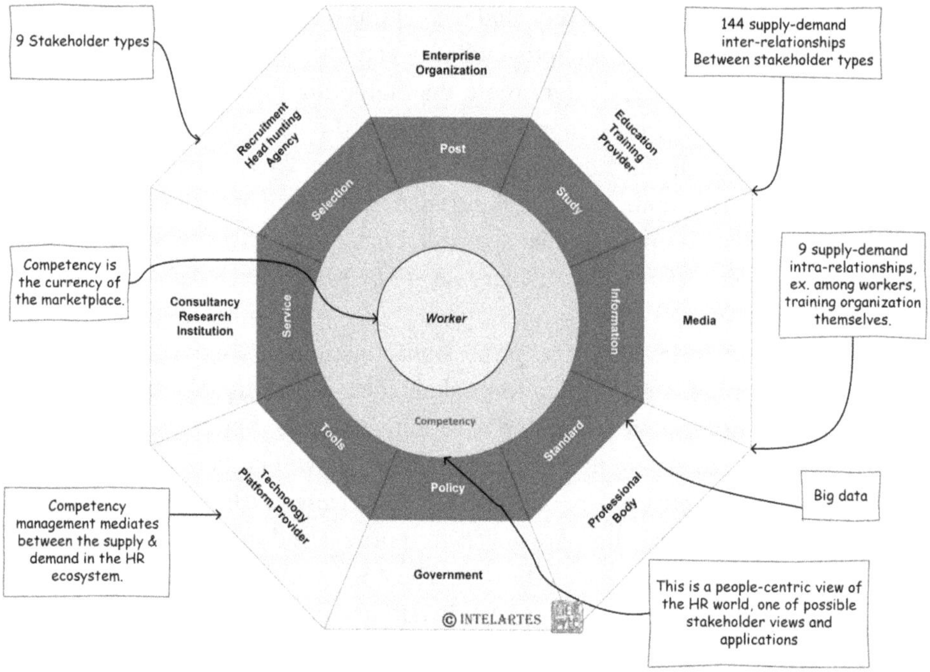

There is a need to build competency profiles of individuals to understand competency supply, and to conduct a more intelligent analysis of work and the value of work to understand demand and produce target competency profiles. There is an increased UK Government focus on meaningful (good) work following the Taylor Review (2017). The innovation and challenge is how to make the value and meaning of work and competency explicit to enable this. Currently, the proxy for competency demand is occupation and role. The proxy for competency supply is qualifications or 'cards'. The better articulated demand side of competency could potentially explain the systemic cause behind the low productivity of the UK workforce. Ultimately, this is a cascading relationship throughout the entire project/asset supply chain where the primary stakeholders are the clients and asset owners. Competence is not just about knowledge and skills, but is the application of a combination of such dimensions as knowledge, skills, abilities, experience, behaviours, attitude, etc. at work (Zhao, 2015a, b). In addition, competence is only truly gained through its repeated application and verification at work. Moreover, Susskind & Susskind (2015), in their definition of competence/practical expertise, also recognise machines and systems – *"practical expertise which should not only include formal knowledge, know-how, expertise,*

experience, and skills of traditional professionals, but also the relevant output of various machines and systems". This is critical as it underlies why the solution to the productivity problems of the industry does not lie in 'training' alone, in that:

- Education and training can never plug the competence gap as they only ever provide the fundamental knowledge, foundations skills, or theoretical behaviours that individuals should exhibit in their work. Education cannot substitute the experience that comes from having applied any of this in 'the real world';
- Current Continual Professional Development (CPD), including re-validation (BSI, 2020), is not fit-for-purpose (ICE Skills Review, 2018). While most Professional Bodies' codes of conduct for CPD focus on remaining competent to undertake work activities, current CPD processes do not actually provide any assurance of this;
- Current certification schemes, in the UK and across the world appear to be limited towards BIM and skills, focusing on particular technologies, rather than adopting broader schemes for digital competence that include consideration of the ISO/IEC 17024 scheme development for the certification of competent persons;
- Qualifications gained through education and training (including CPD) are a poor proxy for competence as they only provide, at best, the confidence that an individual has a foundation knowledge or skills, and not the experience of having applied it in 'the real world';
- The likelihood of increased certification of products, process, systems, organisations, or persons and increased regulation (such as a Product Regulator in the UK) will impact the contextualising factors that will affect competency requirements; and
- Knowledge and data about competences is for both humans and machines.

Moreover, could emerging Immersive Technologies challenge the pre-eminence of the 'real world' as the only place to get experience? The professions associated with construction and the wider built environment has evolved to their current state over many years. Traditional project delivery methods have failed the construction industry and its clients due to their lack of confidence in shared data and their incorrect assessment of how to minimise risk and liability. These systemic problems are manifestations of one or more of the following:

- A failure to engage early with the parties who have appropriate competence to both supply and interpret information on which decisions are to be based. This particularly involves the failure to engage construction and operation specialists early enough in design;
- **Inconsistent definition of activities** leading to inconsistent information requirements and programming;
- Inconsistent exchanges of information at the project inception, e.g. requirement capture, briefing, proposals, and pricing;
- Inconsistent, incomplete, ambiguous, and ultimately unverifiable information deliverables leading to subjective decisions across the project stakeholders and lifecycle;
- Lack of standards or codes relevant to offsite construction methods;
- Inconsistent and non-transparent cost data leading to inconsistent pricing;

- Deadlines associated with information deliverables are allowed, and almost expected, to slip inconsistently in the notification of the project, both in terms of the information and the timeliness of the notification.
- Inconsistent communication across the project; both digitally in terms of the interoperability and overreliance on written/verbal instructions and feedback, which cannot be audited.
- **The lack of any consistent framework or digital infrastructure to facilitate the mediation of competency supply and demand.**
- The lack of both a system/infrastructure for the capture and utilisation of 'lessons learned' from project to project and the lack of a 'lessons learned/shared' culture within the industry.

Any one or combination of the above can undermine the efficiency and efficacy of the asset during the delivery, in-use, and end-of-use phases (Bernstein, 2018). If existing traditional (document-centric) processes are just simply digitised, does this risk constraining the potential to find solutions to the underlying issues facing the construction sector and wider built environment? This is reflected in the distinction between Digitisation (digitising current processes, systems, and products) and Digitalisation (creating new processes, systems, and products unfettered by the constraints inherited from the document-centric world). Those highlighted were within the remit of PUN, while it was recommended that the other points are acknowledged and covered by other CDBB or wider industry initiatives.

Inertia in the Paradigms of the Construction Sector and the Wider Built Environment

There is considerable history and inertia regarding the professions and trades of the built environment, and in particular, the entrenchment of 'roles' and 'occupations'. This has led to the development of established professions, and particularly, professional and trade institutions, with detractors in their conduct, and practice tending towards protectionism, alongside resistance to change, the reinforcement of silos, and the preservation of hierarchies (Morrell, 2015). Furthermore, the Hackitt report (2018) stressed the requirement for each body (professional, trade, or skills body) to define the competences that are unique to the activities over which they have primacy, thereby confirming that activities are central. 'Occupations' and 'professions' have, in turn, become a proxy for competence demand (Cedefop, 2013), e.g. in the majority of large built environment projects, workforce planning is based on 'we need 'n' roles (e.g. project managers)'. In reality, the boundaries between professions are becoming blurred with the emergence of digital innovation. Theoretically, any organisation can participate in a design or construction project in any location (Langford & Male, 2008) and digitalisation blurs the professional roles and responsibilities for information creation. If we are willing to accept that the current information flow processes are neither dynamic nor granular enough to capitalise on the potential of the emerging digitally connected world, and we can show that information transactions mirror the mediation between competency supply and demand, then it follows that the proxies we use for competency supply and demand are also neither granular nor dynamic enough for the changing world of work. The OECD (2017b) identified in the UK:

- High level of mismatch between skill-supply with skill-demand due to poor mediation in conjunction with ineffective supply focused interventions facilitated by the UK government;
- Low demand for higher order competences;

- Increasing polarisation between high-end skills and low-end skills, which increases mobility challenges.

Ultimately, competence assurance, as recommended by Hackitt (2018), requires the capability (including technologies and processes) to intelligently manage competences, people, and work. *However, do employers actually know enough about the competency of their workers at any point in time?*

Over the past twenty-five years, the UK skills policy has primarily focused on boosting the supply of skilled or qualified labour. Despite significant progress on this front, as policy in this area is relatively underdeveloped in the UK, its productivity continues to lag behind those other OECD countries (Brinkley 2017; Payne, 2008). UK employers spend less on training than other major EU economies and less than the EU average. Participation in job-related adult learning has fallen significantly in recent years to the lowest ranks (Brinkley & Crowley, 2017). However, the UK has record-high employment levels and very low jobless rates compared to most OECD countries (OECD, 2017b). The UK sector's focus on the 'shortage of skills' implies both the recognition that there is a problem and an appetite for policy to tackle the challenge. However, despite initiatives in this area, there has been little progress on productivity in the last 30 years (Schouten, 2016; IPPR, 2017; ICE Skills Review Group, 2018). It seems reasonable to conclude that initiatives focused on training alone will not solve productivity challenges. Therefore, future initiatives should consider how to address BOTH competency supply and, in particular, competency demand, and how to improve mediation to reduce mismatch. The continuous emphasis in the UK on 'skills gaps or shortages' belies a more complex and multi-dimensional challenge around competency mismatch (OECD 2017a). As skills are only one dimension of competence, then a best-fit decision to employ an individual based on a skills profile alone may be different to a decision taken on an entire competence profile. Thus, *does a 'skill-centric' view of a person's suitability for employment have implications for diversity and social mobility? Skills also have an increasing short shelf-life and automation will only accelerate this.*

Furthermore, digital change shifts the composition of demand and supply with concern raised over the increasing gap between digital and non-digital, and between high and low skilled workers. By 2022, no less than 54% of all employees will require significant reskilling and upskilling (World Economic Forum, 2018a). As workforce transformations accelerate, the window of opportunity for the proactive management of this change is closing fast. If the transition towards digitalisation is managed poorly, it poses the risk of a widening competency gap, greater inequality, and broader polarisation (World Economic Forum, 2018a). Furthermore, the industry lacks the dynamic capabilities to build the necessary competencies for a digital transformation (World Economic Forum, 2018b), while missing the human dimension and the need for individuals to flourish in the system. Therefore, a greater shift from the inward-looking goals of individual organisations to an ecosystem with a coopetition approach is required. The traditional structures of 'roles' and 'occupations' could therefore act to constrain the skill/competence composition of the employer, its contribution to job growth, and ultimately to productivity.

To address these challenges, two further key questions need to be addressed: 1. *what does upskilling look like through the lens of dynamic competency management?* and 2. *what is the digital infrastructure required to support the mediation of competency supply and demand?*

The link Between Competence and Activity at Work

All individuals have competency, which can be expressed in terms of an individual's knowledge, skills, experience and behaviours related to the work activities they have done, can do, are qualified to do, are certified to do, want or willing to do. This profile, in part, signifies an individual's effectiveness in the day-to-day activities required as part of their role, job, or task. This is not how effective they are at their role, job, or task, as other evidence sources are required to provide assurance of effective performance. Other sectors, such as Oil & Gas, are more mature in the implementation of organisational competency management to drive productivity and ensure compliance.

A role can be an aggregation of all the activities an individual carries out, but often two individuals with the same role may carry out different activities. Likewise, different roles in different organisations require individuals to carry out different activities or associated duties. Profiles are the pivot of business logic for the nine key stakeholders of a competency management ecosystem. There is currently no construct of competency profiling to implement a framework or frameworks post Hackitt.

Activities are the fundamental building blocks of work, whereby, for the purpose of this chapter, an 'activity' has a definable and intentional result. When we look at built environment projects, which involve multiple organisations collaborating and contracting with each other, fundamentally, projects are also a collection of activities. ***The activity, the information required to carry out that activity, and the competency of the person to carry out that activity, are all intrinsically linked***. The potential that consistent structured information could have for productivity is well recognised. For example, the Industry Foundation Classes (IFC) is a non-proprietary schema/standard for defining how information can be structured to describe the built environment. However, the built environment has no such schema/standard to define how activities can be consistently and semantically linked. Of the 'BIM' frameworks that currently exists (SFT, 2018), many are now out of date and have not been maintained, some are for specific user groups and are not suitable for wider application, are internationally based and have not been tested in the UK context, or have been adopted by specific users, but do not have a broader uptake.

Furthermore, although the report by Bush and Robinson (2018) for the Scottish Futures Trust investigated the challenges of "Developing a BIM Competency Framework" based on traditional occupational roles, the scope for DBB is significantly more comprehensive than simply BIM moving towards a digital transformation within the Digital Built Environment. The challenge lies within the demand side, in that this needs to become a more responsible and intelligent actor in this mediation. Current policy does not recognise the (digital competency management) ecosystem explicitly as one of the systems, while IPPR (2017, pp. 3) noted, "*Demand for skills (particularly high-level competences) among employers is low. Employer investment has fallen in recent years and there is a large investment gap with the EU average*". Clear client requirements and value definition in projects have been missing, whilst ineffective old procurement models have been maintained (Mosey et al., 2016). Consequently, clients do not know what competences to demand, nor do they have the capabilities to lead and define the value they need. Therefore, clients themselves require certain competences to fulfil the activities associated with their role.

While post Grenfell has sadly put competence in the spotlight, the forensic lens of competence assurance has not been entirely understood by the sector. The Hackitt Report (2018, pp. 74) states, "*an existing approach to competence which is fragmented, encompassing a range of disciplines and different competency frameworks even within one discipline and without reference to other interacting disciplines. This results in people working within the system, focusing on their individual specialisms without giving due consideration to how their work may interact with the work of others*". Furthermore, "*the JCA [Joint*

Competent Authority] will become more astute at interrogating the work undertaken by these actors, completing the competence loop and ensuring that the skills, knowledge and experience of each of the actors is mutually reinforcing" (Hackitt Report, 2018, pp. 74). This highlights the need for collaborative competency management such that the interactions of work activities, actors (worker), actions, and outcomes can be understood more explicitly in terms of competency and information management.

It also implies that, "*competence assurance is set within the context of competency demand and supply mediation; primarily concerned with understanding people@work risks more dynamically within an organisation or ecosystem, where the need, existence, extent, currency, validity, and meaning of competency can be understood more dynamically and securely shared at a granular level and relevant to the next in line process or work activity*" (Carlton, 2019). This is compounded by the lack of knowledge on the impact of a competence gap, shortages, and mismatches on productivity, labour mobility, and diversity. Simply focusing on skills versus all the dimensions of competence, limits diversity. However, the OECD (2017) has started to address this lack of knowledge through its Skills for Jobs Indicators (Figure 3) by understanding competency demand and supply in more granular and multi-dimensional terms. In addition, the UK ONS is also looking at more a granular 'activity' classification of work.

Figure 3. The structure and components of the OECD Skills for Jobs Indicators

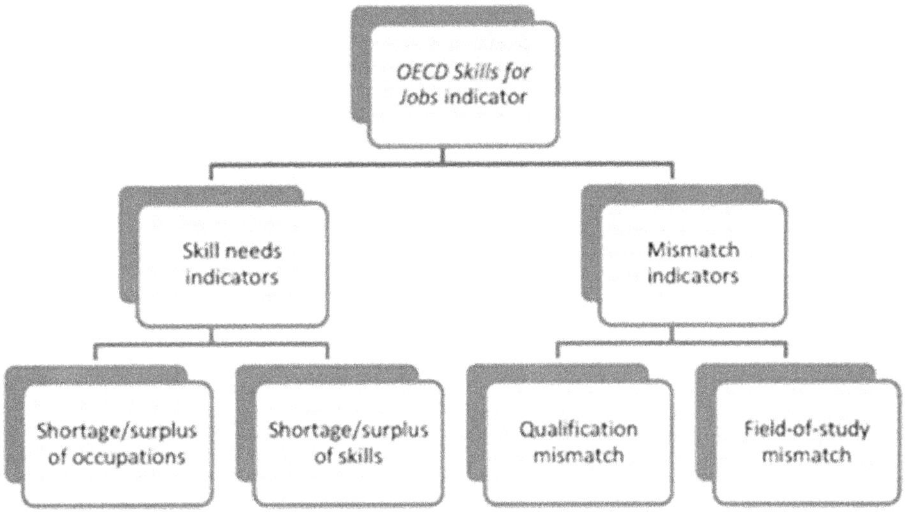

Activity Semantics (Zhao, 2017a) is a proven technique that could solve this challenge, but that requires testing, and perhaps adapting, for Digital Built Environment applications (Figure 4). These semantic links would underpin Competency Frameworks and other competence knowledge sources, which in turn, would underpin any Competency Definition. There is increasing overlap of competencies across sectors, professions, trades, roles, and disciplines. A core element must be a digital ecosystem organised around competency frameworks in digital operation, and dynamic context; as opposed to the current traditional static informational model of competence (Zhao, 2019). A fundamental aspect of Competency Demand is Competency Definition.

Figure 4. Adapted from Core of Competency [© Interlates, see Zhao, 2017a]

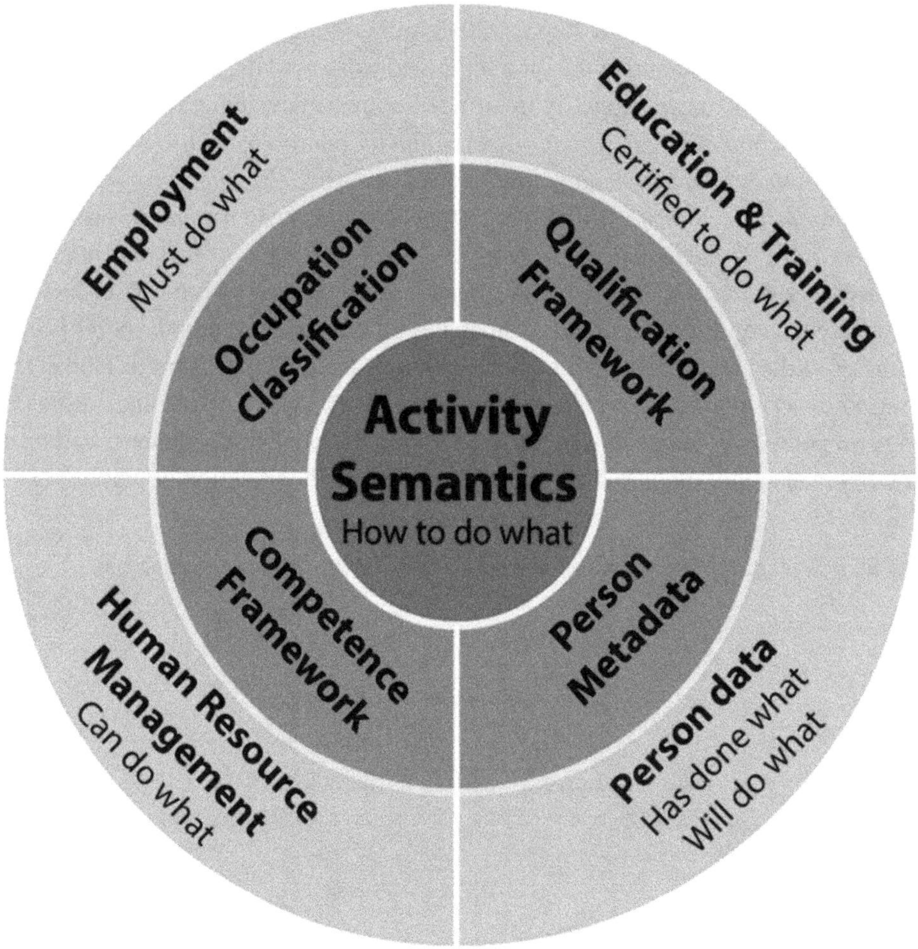

To define the competency profile required for an activity, we must be able to articulate: *Why that activity is needed (its purpose); What is required (the activity and the deliverables); How the activity should be undertaken, and finally by whom and with whom* (Zhao, 2012). The result is that the value, the meaning of work, and the competences required are all explicitly linked. Value (benefit divided by resources used) is directly linked to productivity. Furthermore, we have the potential to definitively and consistently answer: '*What are we being asked to do?*'; '*Why are we being asked to do it?*'; '*When do we need to deliver?; How do we deliver?*' Then assemble a competent team based on these requirements to assure the competence of the team to the client (or the stakeholder to whom competence is being supplied). In comparison, the workforce planning on the majority of large built environment projects can be summarised as, 'we need 'n' roles (e.g. project managers)'. The key disruptive factor required is the change in granularity by using the meaning of competency as the computational unit (Zhao, 2012).

If the competency profile required for an activity can be consistently defined then the activity can effectively be matched against the competency profile of an individual or team. This can be scaled up the aggregation of activities required for a role, team, work package, project, organisation, and even across the whole industry/sector, which could provide far better data to support policy to address the shortage, supply, and mismatch of skills (or competence). 'Skills mismatches' are where individuals are mismatched to their jobs in terms of their competencies, qualifications, or field of study (discipline). The mismatch can be classed as over-skilled/qualified, or under-skilled/qualified or due to skills or competency decay. The difference between the competencies required (for an activity) and the competencies offered (by an individual) is often context dependent. Activities are often cross-disciplinary and cross-role. This will become more prevalent as technology blurs the boundaries between which role (or profession) carries out which activity. However, most training (both initial professional and CPD) is generic, meaning there is a considerable waste (time and money) in the trainee attending courses when only a small subset of the content is actually required. Moreover, unless knowledge/skills is quickly applied and reinforced by experience, the knowledge, skills, and competencies are lost; with skills having a shelf-life of 4.5 years and shortening.

The Competency Profile of the individual is not static; certain competencies can grow and others diminish through a lack of use. This is not immediately obvious to either the employer or the individual. A digital infrastructure that allows the individual to transact on the basis of their competencies would, by necessity, also allow the individual to manage their career. This is important, from a historical perspective, to manage their experience and portfolio, and from a future perspective, to manage their career. Developing a digital infrastructure that places the needs of the individual at the centre of the initiative would be a new paradigm. Upskilling (and reskilling) is the conscious migration of one competency profile to another though education and/or training and/or experience (Zhao, 2018). The competency profile is the pivot of the business logic for competency demand and supply ecosystem stakeholders (Zhao, 2017b).

Competency management is therefore required to identify, assess, match, foresee, control, and assure competency at work. Competency Management is equally applicable to upskilling and reskilling, and both of these are required to address potential future imbalances. Upskilling occurs when an individual is migrating from one competency profile to another within the same profession, e.g. a graduate structural engineer upskills to enable them to become a Chartered Engineer. On the other hand, reskilling tends to occur when someone enters the construction industry from a completely different profession, or moves within the industry from one profession or discipline to another. There are efforts to define competencies across sectors and countries, which overlap such as European Skills/Competences, Qualifications and Occupations (ESCO, 2020). There are also classification efforts of competence topics and subjects such as BIMe (BIMe Initiative, 2020) that is being operationalised through the Quebec Construction 4.0 Initiative (Quebec Construction 4.0 Initiative, 2020) among others. Competency management can potentially target this issue such that these efforts are coordinated across the ecosystem whereby a universal matrix of competencies across ecosystem stakeholders would promote the knowledge of what competencies are required in the labour market.

Competency as Currency in the Labour Market

Productivity concerns the competency of the current and future workforce and effective mediation of competency demand and supply. If individuals and employers have the digital infrastructure within which to transact on the basis of competencies, then competency would become a currency within the labour market. Competencies across all dimensions could become liquidities across the ecosystem. The changing focus from the skilled individual to the composition of competencies that an individual possesses as a computation unit can become the new currency within the market. This could potentially become a disruptive factor across sectors targeting the pressing issues of a changing labour landscape and the increasing granularity through digitalisation and capabilities, such as activity semantics (Zhao, 2012).

There must be a shift from traditional professional disciplines, roles, and occupations that evolve through silo-based career paths to the support for a career portfolio based on competency management. However, as the current infrastructure is based on the occupational roles, employers only rely on the reputations of educational institutions and professional bodies to assess the competencies of its employees. The current pressing issue in the changing labour landscape is that competent individuals are increasingly responsible for managing their own career in moving from one profile to another. A further key issue is that individuals/employee have no infrastructure with which to effectively transact in the marketplace, or to manage their career portfolio data (Zhao, 2014), and neither is the individual a stakeholder in the changing labour landscape. Competent people are a source of innovation and productivity, and therefore, they are a foundational block in the domain of DBB. However, competent people are usually a lost or unrecognised asset for many organisations. Competency management must therefore be delivered with a people-centric approach (Zhao, 2018).

The key is to make the value and the meaning of work explicit; where attributes of competence in relation to why, how, and what become critical to meaningful work, engagement, and self-actualisation, while improving productivity and reducing risks, as noted by the Skills Review Group (2018). The CPD, professional, trade, and education bodies are currently not fit for this purpose, as the evidence shows (Ball, 2008; Hughes & Hughes, 2013; Morrell, 2015; Uff, 2016; Skills Review Group, 2018). Furthermore, it would empower individuals to take responsibility for their Conscious Competence Evolution and allow employers to both recognise and reward innovation. This needs to be underpinned by a shift from the predominant supply-side policies and interventions to a people-centric approach to improve the mediation and matching of people to meaningful work to improve productivity.

A consensus on foundational definitions and guiding values must be established to begin enabling an alignment on the ecosystem approach to competency management across the built environment. A new digital competency management ecosystem underpinned by the mediation between competency supply and demand would enable a paradigm shift, which would transform the productivity of built environment professionals and organisations of DBB.

Infrastructure Required to Facilitate a New Digitally Enabled Ecosystem

As discussed, the activities, the information required/delivered by the activity, and the competencies to carry out the activity are all intrinsically linked (Simpson & Carlton, 2019). A new digital competency management ecosystem underpinned by the mediation between competency supply and demand, would require a new infrastructure (and underpinning research) incorporating:

- Competence Frameworks to enable consistency and a definition of Competency for a particular activity, function or role;
- The ability for individuals and teams to assess and evidence their personal Competency Profile and to use this to plan their development and careers;
- The ability for organisations and projects to describe their workforce or requirements in terms of Competency Profiles;

- The ability for individuals, teams, and organisations to identify gaps, deficiencies, redundancies (and duplications), and adequacy within their Competency Profiles and, thus, identify how they can best migrate/grow their profiles;
- The ability to provide dynamic career pathways (which should be simple and intuitive to use);
- The ability for education providers to offer courses to enable effective and efficient strategies for the Competency Profile Evolution. This requires competency-based curriculum management;
- The ability for professional institutions and trade organisation to take a more informed view as to the upskilling requirements of their members and to better plan policy, support, and interventions;
- The ability for those responsible for assuring a competent current and future workforce to fully understand and mitigate the risks of gaps and deficiencies in the profile of the workforce. The management and mitigation of such risks would be more dynamic within an organisation and ecosystem where the need, existence, extent, currency, validity, and meaning of competency can be understood and securely shared in a transparent and auditable manner; and
- The ability for government to understand the true imbalances in competence and to plan effective policy to address and imbalances (shortage, oversupply. or mismatch) within the sector and analyse the impact of interventions and policy.

When referring to the digital infrastructure for an ecosystem of competency management we are referring to leveraging an **ecosystem technology architecture** that is dedicated to developing a competency management **platform** that enables a coopetitive e-marketplace of multiple stakeholders, transaction and interaction models, open data and information, and services concerning competency insight, assessment, analysis, planning, development, certification, and assurance.

At times of profound change in the economy, business, professions, and corporations, the disruptive evolution from IT to **ecosystem technology** (ET)[3] has begun. Though success stories are found in multiple sectors from telecom to sales, customer journey to health, ET is still in its infancy. Its application in the new drive for people-centric human resource management (HRM) has bright prospects, particularly as types of work and workers (including dynamic career pathways and upskilling), how work is defined, coordinated, conducted, and assured becomes more complex and dynamic.

Currently, the digital ecosystem competency management infrastructure to enable this does not exist nor the underpinning knowledge base required to underpin this. However, once the infrastructure exists then Competency Analytics (Zhao, 2018) is possible. Competency analytics is a set of technologies and methodologies for competency management, and a decision support system to bi-directionally mediate competency supply and demand. It enables the analysis of the competency state and evolution in order to facilitate a match in competency supply and demand, the fulfilment of competency needs by competency development, competency planning, migration, and assurance. Furthermore, this approach is ideally suited to Artificial Intelligence (AI) and Machine Learning (ML), which could enable unbiased approaches to enable the definition, collection, comparison, extension, and evolution of competences, and is scalable to other contexts (Zhao & Carlton, 2015a, b). Traceability, accountability, and verification/assurance are required to provide confidence in the infrastructure. The AI analytics of competencies present opportunities for DBB organisations to enable an environment in which competent workers are not left without meaningful work, and can transact more effectively in the labour market. In the current infrastructure landscape, the individual is not yet truly a stakeholder.

Moreover, the infrastructure can provide the opportunities to increase the transparency of competency data on individuals and organisations; currently available data is mostly non-transparent, unstructured, and distributed. In fact, organisations know very little about their workers and the competences that need to be developed to provide future capability. The link between the needs and interests of workers and what they know, can do, and will do, in the context of work opportunities and what value (and to whom) that delivers, is unclear. Demand forecasting in the UK is currently only based on occupational classifications and predominantly survey or statistical datasets. However, an effective ecosystem infrastructure can enable workers to transact effectively in the labour market based on their competencies, and competency can act as a true currency for work, professional development, training, education, credentials, careers, labour planning, certification, and assurance.

PROFESSIONAL DEVELOPMENT LIFELONG LEARNING FOR THE CONSCIOUS COMPETENCY EVOLUTION

Evidence shows that the environment of work is becoming more turbulent, dynamic, and complex; the boundaries of professional occupations and roles are becoming more blurred as architects have started to compete with surveyors and general contractors (Langford & Male, 2008; World Economic Forum, 2018a; World Economic Forum, 2018b). However, competent people are a source of innovation and productivity and are therefore fundamental to the success of DBB. Whether competent people are a lost or unrecognised asset for organisations was debated during the PUN workshops. Workers that better use their skills are more likely to have greater job satisfaction, earn higher salaries, and are more prepared to adapt to changes in the nature of work. Employers benefit from a more productive and innovative workforce, enabling them to maximise business performance and profitability (OECD, 2017a).

The current pressing issue in the changing labour landscape within the built environment is best supported through a digital competency management ecosystem that empowers competent individuals, through the Conscious Competency Evolution, to be responsible for managing their own career. In turn, the Conscious Competency Evolution relies on robust lifelong learning opportunities as well as the enabling digital infrastructure. This aligns with OECD's (2017b) recommendations that stronger incentives must be put in place to encourage lifelong learning among adults and that these initiatives and incentives should be tied to individuals.

In relation to the UK built environment sector, lifelong learning opportunities are typically split into three phases (Figure 5):

Phase 1:

- Pre-18 Formal Education in schools and Further Education Institutes (FEI)
- Level 1: GCSE grades 3→1 or D→G and NVQ Level 1
- Level 2: GCSE grades 9→4 or A*→C, NVQ Level 2, Nat. Dip. L2 etc.
- Level 3: A Level, AS Level, Tech Level, NVQ Level 3, Nat. Dip. L3 etc.

Phase 2:

- Post-18 Formal Education in FEI and Higher Education Institutes (HEI)
- Level 4: HNC, NVQ Level 4
- Level 5: HND, NVQ Level 5
- Level 6: Bachelor's Degree, NVQ Level 7
- Level 7: Master's Degree, NVQ Level 8
- Level 8: Doctorate

Phase 3:

- Continued Professional Development (CPD) and re-qualification

Figure 5. Phases of lifelong learning

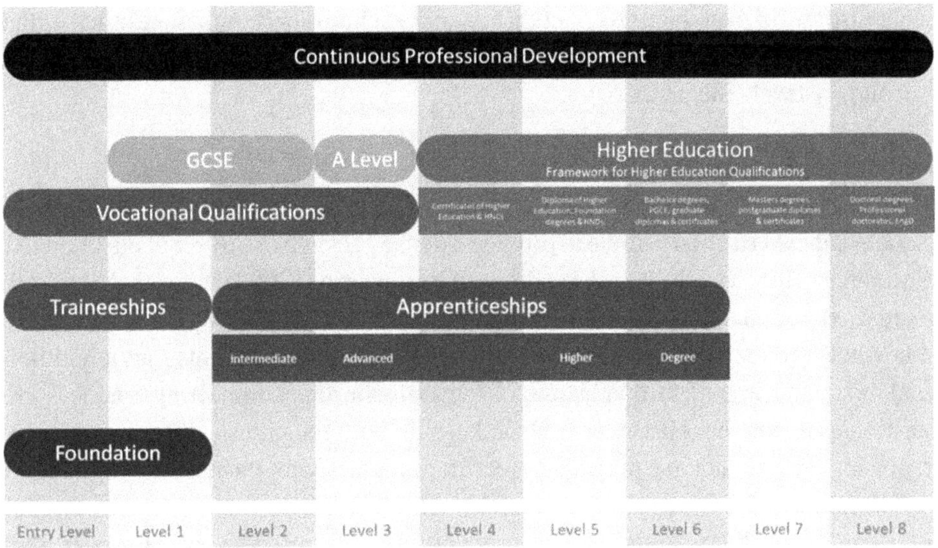

When examined through the lens of competency, both the Post-18 formal education and CPD phases of lifelong learning are found to be out-of-date, siloed, and not fit-for-purpose. The education system is still focused on producing a workforce for previous industrial revolutions, whilst industry attempts to embrace and embark on Industry 4.0; the forth industrial revolution.

Pre-18 Formal Education

In terms of Pre-18 formal education and the development of digital competencies to facilitate the sector's digital transformation, much more needs to be done to address the poor image of the industry and prevent the talents of the next generation from being siphoned from the built environment professions to other more perceived attractive sectors, e.g. big tech, medicine, law, etc. Key areas for action include:

1. improved career advice to address the awareness and recognition of the construction/built environment sector as a viable proposition for a professional career;
2. initiatives that deliver construction/built environment specific curriculum and qualifications, e.g. Design Engineer Construct!® (DEC, 2020);
3. greater involvement and engagement of construction professions with schools to support the previous areas for action and support improving the image of the construction/built environment sector.

In this context, a much deeper rethink is required on whether Pre-18 formal education is actually setting students up to get the most from their continued learning opportunities at degree level and beyond. National standards and national curriculum, enforced by high-stakes testing, can at best teach learners what is prescribed. As a result, learners talented in other areas never have the opportunity to discover those talents, while those learners with broader interests are discouraged and not rewarded. The system results in a population with similar skills in a narrow spectrum of talents. However, particularly in today's society, innovation and creativity are needed in many areas; some as yet undiscovered. Primary education in particular, is based on a Victorian model that is cheap and efficient, rather than one based on educational reasons. An example is the fact that primary schools have remained stuck with the system of generalist teachers, the class teacher, rather than the more expensive model of subject specialists, as happens at secondary level (Baker, 2020).

Post-18 Formal Education

An undergraduate degree is a good vehicle to provide a dense package of pre-requisite knowledge prior to embarking on a career. However, the Royal Academy of Engineering (2007) stated that, whilst the industry is generally satisfied with the engineers it recruits, there are concerns about the ability of graduates to apply their knowledge to real industrial problems. The PUN report highlights that it had become more acute in recent years and was identified as one of the skill shortages impacting business growth. This raises the question as to whether HEIs were able to keep up with the industry's fast pace of change, with employers reporting they are not in a position to recruit graduates with the relevant digital competences. Furthermore, HEIs are perceived to reinforce the silo mentality of industry, the nature of which is widely seen as a significant contributor to the lack of productivity within the sector. However, the system of education in the health sector could provide useful lessons to learn from, e.g. trainee doctors undertake a degree in general medicine prior to specialising in a particular branch. In addition, built environment specific courses should be split into a 'generalist' undergraduate degree followed by a 'specialist' post-graduate level. While there is much merit in this as an approach, it is also recognised that the specialist colleges in medicine are just as impacted by a silo mentality as the construction sector. This has led to some calls for a new medical 'specialism' that takes a systems integration approach to coordinate the different specialists. Similarly, Cook and Chatterjee (2015) suggest that there is an increasing need for interdisciplinary working and leadership capabilities in the built environment.

It should be noted that the silo mentality exhibited by HEI's is a direct response to the accreditation of their courses by specific professional intuitions. Furthermore, it is accepted that HEIs have limited scope for innovation if their courses are to be accredited, while the accreditation criteria defined by the professional bodies has, in the main, lacked incorporating requirements for digital competencies. Whilst there is desire to innovate, the procedures for major change to programmes and modules results in a risk-averse strategy with considerable pressure to meet accreditation requirements of the various professional bodies within the constraints of the current system. In addition, this is further exacerbated by the lack of technologies in curriculum management, where competence demand can be understood more explicitly. Many HEIs pursue both teaching and research agendas and traditionally recruit research professionals to ensure they meet the requirements for the Research Excellence Framework (REF). *However, will the introduction of the Teaching Excellence Framework (TEF) mean the same focus on teaching and learning in the future?*

This is further reinforced through promotion routes that favour those with research success in winning grants and the amount of published works. In many HEIs, research specialists are also expected to deliver teaching modules; whilst this may be acceptable for theoretical modules, built environment courses often have significant elements of vocational training that require knowledge and experience of current and evolving industry practice. As we move towards a more digitally-enabled future, the 'human-centric' competencies involving multi-disciplinary teamworking, creativity, innovation, and design will become increasingly more important as the traditional technical aspects will be taken over by AI and ML. The pressure from industry to increase these aspects on courses are met by the constraints of suitable spaces, staff-student ratios, and the competence and experience of staff to lead such activities, in addition to meeting the accreditation requirements of the relevant professional bodies. Therefore, it is important to consider that there needs to be alignment between the transformation of built environment education and the accreditation of built environment programmes in order to prepare students with the necessary entry-level competence for their future careers, and to build capacity for a transforming industry (Farmer, 2016), i.e. future proofing.

Learning outcome frameworks such as the UK BIM Framework's (UKBF, 2020) and potential work by CDBB to establish digital competencies required by new graduates to the built environment professions aim to facilitate consistency in the development and delivery of digitally-enabled built environment training and education. Similarly, the UK BIM Academic Forum (BAF) proposed an academic roadmap to a longer-term vision that embeds digital construction learning at the appropriate levels within discipline-specific HE undergraduate and postgraduate education (BAF, 2013) and further work continuing to break down and establish the potential learning outcome requirements at each level of HE (i.e. 4-7). However, adopting such learning outcomes/frameworks within education curricula also requires a change to the culture and mindset of academics alongside the development of their competence to drive change in the current curricula and align with the needs of future generations of learners (BIM2050, 2014). As is similarly evident in industry, the understanding, acceptance, and importance of digital transformation amongst HE academics within the built environment, engineering, architecture, etc. is still considerably low (NATSPEC, 2014, 2015, 2016, 2017; 2018; Underwood, 2014; BAF, 2015). Therefore, changing the culture and mindset that exists among many academics poses a significant challenge to the transformation of built environment education from one that currently reinforces a silo mentality and continues the development of disciplinary-specific (silo-minded) professionals; driven by the accreditation requirements (BIM2050, 2014). Moreover, this is still considerably short of a digital ecosystem that would facilitate demand-led competency-based curriculum management. In addition, there is the question of how we ensure the academics who develop and deliver courses also assure their own competence; academics similarly require, and have the need for a conscious competence evolution as much as industrialists.

Amongst the case studies that demonstrate an exemplary engagement between industry and universities (RAEng, 2010), there is a lack of data to indicate whether this is a consistent story or whether, for some, interactions only occur through occasional meetings with their Industrial Liaison/Advisory Committees. The question of whether universities should play a more significant role in the lifelong learning journey beyond the provision of degree programmes was explored. This is not to say that HEIs have little to contribute to lifelong learning, but rather the way the learning provision is consumed is at odds with the evolving requirements of a highly agile and digitally connected built environment workforce. While digital construction is becoming more widespread across various levels of education, in the main, the approach is still tending to be ad hoc and without consistency. Furthermore, rather than being set as a strategic objective at a school/department or institutional level, this is being driven by individual academics or schools/departments that have a particular interest in the area and recognise its importance in the education of current and future professionals, as opposed to being demand-led, adaptive to the demand, and driven in a work context, i.e. bottom-up (Underwood, 2014; BAF, 2015). However, the introduction of Degree Apprenticeships is a welcome concept to bridge the gap between industry and academia (Universities UK, 2019). It remains to be seen as to whether Degree Apprenticeships deliver the potential for a true collaboration between industry and academia in the upskilling/reskilling of the workforce, or simply provides a day-release course under another name.

CPD and Professional Institutes/Institutions

Professional Institutes/Institutions have evolved in a manner that reinforces the silo thinking within the Built Environment. This is a direct response that both protects their members and acts as a body of knowledge to serve their members (Egan, 1998; Wolstenholme *et al.*, 2009; Morrell, 2015; Mosey *et al.*, 2016; All Party Parliamentary Group for Excellence in the Built Environment, 2017). The exigencies of DBB transcend the objectives of professions and demand a broader view on the evolution of professionalism as professions create cognitive frameworks within their jurisdiction to only seek control (Hughes & Hughes, 2013).

Furthermore, the Professional Institutes/Institutions set the requirements for the degrees they accredit both directly and, for some, indirectly through their input to the UK Standard for Professional Engineering Competence (UKSPEC). As such, they influence the FEI and HEI provision of lifelong learning, as well as the provision of Continued Professional Development (CPD). CPD is the means by which the Professional Institutes/Institutions assure that their members remain competent, as beyond the Initial Professional Review, there are no further checks. However, the ICE Skills Report (ICE, 2018) states that the current CPD model is not-fit-for assurance, and that "*civil engineers who fail to keep abreast of changes affecting their areas of activity are simply unfit to practice*". The challenges that the ICE and other Built Environment Professional Institutes/Institutions face are not unique, as CPD is problematic in multiple ways:

- CPD is not typically competence-based, either in assessing the requirement for CPD, or in demonstrating how CPD has made a discernible difference to the competence of the receiving individual;
- CPD requirements are specified as a number of days, meaning that quantity is valued, rather than quality (or even applicability);
- There are no means of comparing one CPD provision over another; and
- There is a lack of consistency and transparency that would limit the effectiveness of mandatory auditing.

Therefore, CPD requirements should be urgently reviewed to establish a more robust system that ensures a member's qualification remains relevant to their work and aspirations and up-to-date throughout their professional career. Professional Institutes/Institutions and trade organisations are the key players that specify both the technical requirements for lifelong learning and the criteria by which this can be assessed for suitability. However, there are too many voices and the silo nature appears too entrenched to provide any industry-wide leadership in this area. The multiplicity of institutes/institutions and their nature means they cannot agree on a common issue even when the public good represents a common struggle within the industry. This high fragmentation creates challenges for the Construction Leadership Council in addressing multiple voices; consequently, it takes a neutral position that leads to no change. The professions have been poor at collaboration at an institute/institutional level (Morrell, 2015), and the characteristics of their business models are contradictory to what the Construction Leadership Council hopes to achieve. Thus, Hughes and Hughes (2013) and Morrell (2015) have called for a significant reassessment of the importance of professionalism in the society. However, there are positive signs of change, as seen in the recent Hackitt Implementation Plan, that engineering institutes/institutions are beginning to collaborate in order to address the issue of competency assurance. Sectorial collaborative competency management is the obvious way forward, but there is no underlying (digital) infrastructure to enable this.

CPD Provision

The provision of CPD is fragmented and disjointed. Furthermore, there is no consistent framework, which in turn, means that it is impossible for individuals to compare one provider's course against another when they have identified specific competence development requirements, and for providers to develop courses tailored to specific and identifiable gaps in both current and emerging competence. This is not to disparage CPD provision, but rather, to illustrate why it is currently not-fit-for purpose. If the underlying infrastructure to facilitate comparison does not exist, then providers cannot be blamed for their attempts to respond to market forces. The lack of a digital infrastructure and ecosystem to support this infrastructure is problematic in that many course participants are only attending for a fraction of the content that is relevant to their specific requirements. As different attendees have different requirements, CPD provision often takes a 'shotgun' approach in trying to cover a wide range of topics with the hope that all attendees will take home something they find useful. There are currently no real examples of dynamic, competency-based CPD planning, although the ICE has undertaken some foundational work in this area (Zhao and Carlton, 2015a, b, 2019).

It is well documented that, unless an individual applies the skills and knowledge they gain from a training course, both immediately and consistently in their day-to-day activities, their knowledge and skill retention decays. Therefore, it is questionable as to whether any CPD that is not directly relevant to their day-to-day activities is of benefit. CPD provision is usually provided in multiples of half days, with many courses taking a full day. This time actively discourages individuals who are sole traders and/or from micro SMEs, as the time to attend courses is taken directly from their fee-earning capacity. However, a significant proportion of the workforce of the built environment (both in terms of construction and professional services) is comprised of micro SME/sole traders (BIS, 2013; Designing Buildings Wiki, 2019). Moreover, it is unclear as to what real effect or change will be brought about by the mandatory CPD auditing and reporting required from January 2019.

Unlearning: Selecting an Alternative Mental Model/ Paradigm for Conscious Competency Evolution

Ever since the publication of Peter Senge's "*The Fifth Discipline*" 25 years ago, organisations have sought to become 'learning organisations' that continually transform themselves. In our era of digital disruption, this goal is more important than ever. However, making real progress in this area, even for the best organisations, still presents a significant challenge. A key problem is that organisations have focused on the wrong thing. The problem is not learning, but in fact, unlearning. In every aspect of business, we are operating with mental models that have grown outdated or obsolete, from strategy to marketing to organisation and leadership.

To embrace a new logic of value creation by embracing all that digital allows we have to unlearn the old logic to facilitate digital transformation. The misconception is that unlearning is about forgetting, but instead, it is about the ability to choose an alternative mental model or paradigm. When we learn, we add new skills or knowledge to what we already know, whereas when we unlearn, we step outside the mental model in order to choose a different model. Therefore, in a time of digital transformative change and empowering competent individuals to be responsible for managing their own career through the Conscious Competency Evolution, individuals need to be conscious of their mental models and ambidextrous in their thinking (HBR, 2016).

Organisational Performance and Learning

New approaches or methods that emphasise a 'data-to-learning-to-action' process (Flinn, 2018) that focuses on improved evidence-based decision making and continuous performance improvement need to be considered as part of the upskilling agenda. This approach recognises the key to improving business performance is improving decision-making. Furthermore, it is the decisions that are aligned to the most important drivers of the value of an organisation that make the difference. There is great claim of the power of big data but the key is how it helps improves decision-making. There is no intrinsic value in data unless it serves as the basis of enhanced learning or has the potential to (including unlearning), and then to action. Learning is not just a human endeavour it is learning by humans and machines. It is NOT all about data but it is ALL about decision-making and learning. Businesses that are not knowledge, learning, and decision-intensive are much more likely to be automated and routine.

WHAT NEEDS TO HAPPEN NEXT?

The focus of the work presented in this chapter explores how we create and support a digitally-enabled, agile, competent, and ultimately, productive workforce and determine the key questions that need to be addressed if DBB is to both provide return on investment, and succeed as the catalyst for evolving the manner in which we conceive, plan, design, construct, operate and interact with the built environment. The vision is that we need to imagine a digital competency management ecosystem where all stakeholders are incentivised to work together in coopetition to create, capture, infer, interpret, specify, integrate, accredit, apply, use, monitor, and evolve competence as a working asset just like any other data asset. Moreover, this needs to be in a consistent, objective, explicit and scalable manner; with end2end transparency and traceability for all stakeholders that overcome the challenges of competency management. *With an increasing overlap of activities and competences across sectors, professions, trades, roles, and disciplines, a core element must be an ecosystem organised around digital infrastructure of competency frameworks and other knowledge sources of competence, so that competency frameworks are in digital operation and dynamic context; as opposed to the current traditional, static informational model of competence.*

Table 1 presents the fifteen key research (and subset) questions derived to guide further research. By addressing these questions in future research activities, the pedagogical and upskilling/reskilling aspects, through the vision outlined above, DBB could be effectively realised. The questions are structured to demonstrate:

- Urgency of the research needed (from the completion of the CDBB PUN) in relation to the immediate and short-term, medium-term of three to five years, and long-term of five to ten years, based on the previous research being achieved;
- Current level of maturity of research (i.e. the current research knowledge-base in terms of being novel or some level of research existing);
- Ascertaining whether the proposed further research activities require industry, academia, and other complementary, but important, institutional stakeholders, who relate to the value proposition to work together. From previous experience, it can be clearly demonstrated that, when the cooperation extends beyond a single industry-breaking institutional silo, this leads to the achievement of a more successful outcome.

Based on the key research questions from the PUN network, the following provides a summary of the key findings and associated recommendations:

- Central to productivity are people; the competency of the current and future workforce, which need to be treated as a working asset just like any other data asset in the built environment.
- The purposeful language of Dame Judith Hackitt concerning competence, competent, competency, and competency assurance has not been fully adopted by the sector.
- Competence is not just about competence frameworks; there are other sources of knowledge of competence as well as competency data (unstructured and distributed), which we also need to digitise. Competency does not currently act as the unifying currency of the labour market. Occupations are used as a proxy for demand and qualifications as a proxy for supply, which is no longer granular or multi-dimensional for the dynamic world of work.
- There is increasing overlap of competencies across sectors, activities, professions and trades, roles, disciplines, etc. and this will only increase with automation, digitalisation and increased regulation. A new model for the post professional society is required.
- There needs to be recognition that knowledge of, and data about, competence is for machines as well as humans
- Boosting productivity also comes from the effective mediation of competency demand and supply; skills imbalances are linked to low productivity.
- A digital competency management ecosystem can facilitate the mediation of competence demand and supply in DBB; people-centred, community-based that connects all the inter-dependent stakeholders in coopetition within and across sectors, occupations, and disciplines. A digital marketplace facilitating and mediating supply and demand of data, information, and knowledge about competences, occupations and roles, disciplines, upskilling/reskilling and learning, qualifications, certification, standards, regulations and policies.
- The upskilling/reskilling of people and their competencies is simply the evolution (acquisition, application, appreciation, and depreciation) of one competency profile to another. One off static qualification or certifications are no longer fit for purpose.
- Upskilling/reskilling must be competency based and granular enough to address change to work duties and activities, and how work is conducted; across all dimensions of competence.
- Lifelong learning journeys of individual learners must be relevant to competency demand and the evolving competency needs of individuals as well as industry.

- In relation to the pedagogy and upskilling/reskilling aspects in the context of lifelong learning and achieving the long-term vision outlined, the logical order for the sector would be to define what upskilling/reskilling needs to mean in terms of competence, and in particular, the phrase, conscious competence evolution in the first instance, followed by then defining, creating, and growing the digital competency management ecosystem to support upskilling/reskilling, and enable the conscious competence evolution, and finally, define the pedagogy that best suits the required upskilling/reskilling.

CONCLUSION

This chapter has presented a proposition for the digital transformational change of people and their competencies to facilitate the change required to enable the built environment to evolve into the 21st Century and the future. Change has begun with Government mandates generating a number of BIM training activities, as an important first step. However, continued success will require an evolved upskilling/reskilling philosophy to develop a digitally competent and agile workforce capable of supporting the emerging digitally-enabled processes, purpose driven information flow, and the collaborative behaviours that underpin them.

It is vital to ensure that individuals, teams, and organisations have the necessary appropriate knowledge, skills, abilities, experience, behaviours, and attitude at work, i.e. 'competence', to carry out tasks/activities they are appointed to perform. It is also vital for clients, when appointing individuals, teams, and organisations, to undertake construction operations, to ensure they are 'competent' to be able to conduct the work appropriately. The chapter has highlighted a need for training and qualification systems that need to be driven by dynamic competency demand data.

Whilst there is continued investment being targeted at more general digital literacy and skills frameworks, through the DfE, CITB, CDBB, etc. such initiatives still take a classical information model approach to competency frameworks when a systems model is required, i.e. a digital competency management ecosystem. It is imperative that this systems approach is adopted to enable the sector to demonstrate that successful projects are characterised by the right people, making the right decisions, at the right time, using the right information. Furthermore, 'competence' ensures there are the right people available to specify the information required to make those decisions, and correctly interpret the received information. It can also be argued that productive work only occurs when competent people specify the work that must be done, which is in turn, carried out by other competent people. The productivity of people comes from effective mediation between competency supply and demand, and enabled by the flow of appropriate and correct information.

Table 1. Key research questions & direction'

	Question *(No.)*	Impact	Maturity (of research)	Lead by
Short-term	How do we design/manage/implement a 'collaborative competence management' approach that enables/facilitates competence as a currency in the labour market? *(Q5.)*	HIGH	NOVEL	JOINT
Short-term	The following subset of questions were determined necessary to answer this wider question: There is a lack of consistency, even understanding, of the terms 'competence', 'competency', and 'competent' within the built environment sector. Without this, any work on 'competence management' and 'competence development' will have a limited effect. What is the best way of redressing this? *(Q1.)*	HIGH	MEDIUM	JOINT
Short-term	Who are the stakeholders that any built environment sector will have to engage with in terms of 'competence management' and 'competence development'? How are they interconnected with other stakeholders? What is their sphere of influence? *(Q2.)*	HIGH	MEDIUM	JOINT
Short-term	The general consensus is that 'competence' is activity based, not role based, yet there is no consistent definition/ontology relating to the activities that contribute to the built environment. Thus, how can we define/develop a flexible/extensible ontology that describes the activities of the built environment? *(Q4.)*	HIGH	MEDIUM	JOINT
Short-term	If productivity is linked to the competence of individuals within the market place and, in particular, the mediation of competence supply and demand, then the low productivity of the built environment sector is likely to be a result of this not happening. Therefore, why is this not happening? And what is required to circumvent the blockers. *(Q10.)*	HIGH	MEDIUM	JOINT
Short-term	How does curriculum development need to adapt to align to a 'collaborative competence management' approach? *(Q7.)*	HIGH	NOVEL	JOINT
Short-term	If the fundamental question is about empowering individuals to consciously evolve from one competence profile to another, then what are the infrastructure, technology and processes required to facilitate/enable this? *(Q11.)*	HIGH	NOVEL	JOINT
Medium-term	'Upskilling' typically implies a 'skills gap', but may also imply a natural result of 'skills decay', where competence is a function of education, skills, experience, and behaviour. The general consensus is that the built environment sector has a 'competence mismatch'. So what is 'upskilling' in the context of an 'individual's competence development' and 'competence management'? *(Q3.)*	HIGH	MEDIUM	JOINT
Medium-term	The UK suffers from poor data relating to its labour force. To address this, what do we need to measure and how do we measure it? How does this relate to competence management/ competence development? *(Q6.)*	MEDIUM	MEDIUM	JOINT
Medium-term	How can we ascertain the likely return on investment of moving from a skills gap philosophy to a conscious competence evolution philosophy? What do we need to measure and how can we measure it? *(Q12.)*	HIGH	NOVEL	JOINT
Medium-term	For a conscious competence evolution to emerge then 'unlearning' is an important part of evolving. In this context, what do we understand as unlearning? And how do we 'teach' this? *(Q13.)*	MEDIUM	NOVEL	ACADEMIA
Medium-term	How does the conscious competence evolution lead to agility and resilience in the workforce? *(Q14.)*	HIGH	NOVEL	ACADEMIA
Medium-term	How does an individual competence development approach allow individuals displaced by the new industrial revolution to thrive and contribute to society? *(Q15.)*	HIGH	NOVEL	JOINT
Long-term	How does education need to adapt to become 'demand led' and 'competence based?' Is our current university education fit for purpose? If all education is demand led, how do we educate the future 'pathfinders'? Should educators be competent in the subjects they are teaching? *(Q8.)*	LOW	NOVEL	JOINT
Long-term	What and how to enable individuals to thrive in the workplace and access opportunities for more meaningful work? What are the means to enable competent and honest individuals to distinguish themselves from incompetent and dishonest individuals? *(Q9.)*	LOW	MEDIUM	JOINT

In order to be successful, we should ensure that we reach out beyond traditional built environment professionals and include experts from other domains and industries with new propositions, e.g. manufacturing, finance, IT, agile business practice, and competence management. The learning from other industries should help in the design and development of the digital competency management ecosystem. This learning has a critical function as it should set out the norms of interactions between individuals and organisations, and organisations with other organisations (many-to-many relationship of competency demand and supply). Process and technology are the means by which these interactions take place. This is even more important when considered with the growing organisational complexity of the Digital Built Environment. This chapter has shown that such complexity leads to high uncertainty, and taken with the central role of collaboration, the proposed ecosystem approach being defined in this chapter becomes more important. A multidisciplinary approach to competency-based management across sectors, disciplines, professions, etc., matters more than targeting productivity improvement through rapidly changing digital technologies.

This chapter has placed competence at the core of any new initiatives to increase productivity in the (digital) built environment. In addition, the need to build competency profiles of individuals to understand competency supply, and to conduct a more intelligent analysis of work and the value of work to understand demand and produce target competency profiles, has been discussed. Competence is not just about knowledge and skills, but is the application of a combination of such dimensions as knowledge, skills, abilities, experience, behaviours and attitude at work, if we are to see productivity gains of the like set by global governments. Ultimately, all people interacting in the built environment need 'competence assurance' (as recommended by Hackitt (2018)), meaning that the sector requires the capability (including technologies and processes) to intelligently manage competence, people, and work. However, one of the key challenges of this is: *Do employers actually know enough about the competence(ies) of their workers?*

In understanding the challenges posed in this chapter, it can be concluded that there are two fundamental questions to be addressed:

1. What does upskilling/reskilling look like through the lens of competence and competency evolution?, and
2. What is the (digital) infrastructure required to support the mediation of competency supply and demand and the fulfilment of competence needs by competence development, planning, migration and assurance?

The focus of the work presented has highlighted the need to create and support a digitally-enabled, agile, competent, and ultimately, productive workforce, and determine the key research questions that need to be addressed to both provide return on investment, and succeed as the catalyst for evolving the manner in which all conceive, plan, design, construct, operate, and interact with the built environment. The vision is to imagine a digital competency management ecosystem where all stakeholders are incentivised to work together in coopetition to create, capture, infer, interpret, specify, integrate, accredit, apply, use, monitor, and evolve competence as a working asset just like any other data asset. Moreover, this needs to be in a consistent, objective, explicit, and scalable manner; with end2end transparency and traceability for all stakeholders that overcome the challenges of competency management. There is an increasing overlap of competencies across sectors, professions, trades, roles, and disciplines. A core element must be an ecosystem organised around digital infrastructure of competency frameworks, competency frameworks in digital operation, and dynamic context; as opposed to the current traditional informational model of competence.

ACKNOWLEDGMENT

We would like to express thanks to the members of the CDBB Pedagogy & Upskilling Network for their commitment, dedication, and valued contribution to the network.

This research was supported by the University of Cambridge Centre for Digital Built Britain (CDBB) [grant number RG96233].

We would also like to acknowledge Dr Gang Zhao, the founder of the concept of a digital competency management ecosystem, for his contributions from his extensive research base and insight into competency and its management for human resource management.

REFERENCES

Aksenova, G., Kiviniemi, A., Kocaturk, T., & Lejeune, A. (2018). From Finnish AEC knowledge ecosystem to business ecosystem: Lessons learned from the national deployment of BIM. *Construction Management and Economics*, (July), 1–19.

All Party Parliamentary Group for Excellence in the Built Environment. (2017). *Building on Brexit. How leaving the EU must drive modernisation and training in the built environment.* London: Author.

Arayici, Y., Coates, P., Koskela, L., Kagioglou, M., Usher, C., & O'Reilly, K. (2011b). Technology adoption in the BIM implementation for lean architectural practice. *Automation in Construction, 20*(2), 189–195. doi:10.1016/j.autcon.2010.09.016

Arayici, Y., Coates, S. P., Koskela, L. J., Kagioglou, M., Usher, C., & O'Reilly, K. (2011a). BIM adoption and implementation for architectural practices. *Structural Survey, 29*(1), 7–25. doi:10.1108/02630801111118377

Autio, E., & Thomas, L. D. W. (2014). Innovation Ecosystems: Implications for Innovation Management. In M. Dodgson, D. M. Gann, & N. Phillips (Eds.), *The Oxford Handbook of Innovation Management* (pp. 204–228). Oxford University Press.

BAF. (2013). *Embedding BIM (Building Information Modelling) within the taught curriculum: Supporting BIM implementation and adoption through the development of learning outcomes within the UK academic context for Built Environment programmes*. Higher Education Academy (HEA).

BAF. (2015). *Current Position and Associated Challenges of BIM Education in UK Higher Education.* BIM Academic Forum.

Baker, M. (2020). Why is primary schooling still based on Victorian policies? *The Guardian*. Available: https://www.theguardian.com/education/2009/sep/15/primary-schools-policy-history-victorian

Ball, M. (2008). *Markets and institutions in real estate and construction*. John Wiley & Sons.

Bartley, T. (2019). Process, Technology and … oh yeah. *People, BIM2050,* •••. https://bim2050.wordpress.com/2015/12/22/process-technology-and-oh-yeah-people/

Bernstein, P. G. (2018). *Architecture, Design and Data: Practice Competency in the era of Computation.* Birkhauser. doi:10.1515/9783035610444

BIM2050. (2014). *A Report on Our Digital Future*. Construction Industry Council.

BIMe Initiative. (2020). *BIM Excellence Initiative*. https://bimexcellence.org/

Bolton, A., Butler, L., Dabson, I., Enzer, M., Evans, M., Fenemore, T., Harradence, F., Keaney, E., Kemp, A., Luck, A., Pawsey, N., Saville, S., Schooling, J., Sharp, M., Smith, T., Tennison, J., Whyte, J., Wilson, A., & Makri, C. (2018). The Gemini Principles: Guiding values for the national digital twin and information management framework. Academic Press.

Brinkley, I., & Crowley, E. (2017). From 'inadequate to 'outstanding': making the UK's skills system world class. Policy report. London: Academic Press.

BSI. (2020). *BSI Flex 8670 v1.0 Built environment – Overarching framework for competence of individuals – Specification*. BSI.

Bush, R., & Robinson, M. (2018). *Developing a BIM Competency Framework: Research & Key Principles*. Scottish Futures Trust.

Carlton, D. (2019). *A competent workforce mitigates risk?* Internal report. Dynamic Knowledge. Unpublished.

CCSIC. (2020). *Flourishing Systems: Re-envisioning infrastructure as a platform for human flourishing*. Cambridge Centre for Digital Built Britain.

CDBB. (2020). *The pathway towards an Information Management Framework – A commons for Digital Built Britain*. Centre for Digital Built Britain.

CDBB PUN. (2018). *Centre for Digital Built Britain: Network - Pedagogy and Upskilling Network (PUN)*. https://www.cdbb.cam.ac.uk/CDBBResearchBridgehead/Networks/2018Network_PUN

Cedefop. (2013). *Quantifying Skill Needs in Europe. Occupational Skills Profiles: Methodology and Application*. The European Centre for the Development of Vocational Training (Cedefop).

CIC. (2019). *Raising the Bar Interim Report: Improving Competence Building a Safer Future. Industry Response Group Steering Group on Competence for Building a Safer Future*. Construction Industry Council UK.

CIC. (2020). *Setting the Bar Final Report: Improving Competence Building a Safer Future. Industry Response Group Steering Group on Competence for Building a Safer Future*. Construction Industry Council.

Clarysse, B., Wright, M., Bruneel, J., & Mahajan, A. (2014). Creating value in ecosystems: Crossing the chasm between knowledge and business ecosystems. *Research Policy*, *43*(7), 1164–1176. doi:10.1016/j.respol.2014.04.014

Client Leadership Council (CLC). (2019, June). *Future Skills Report*. https://www.constructionleadershipcouncil.co.uk/wp-content/uploads/2019/06/CLC-Skills-Workstream_Future-Skills-Report_June-2019_A4-Print-Version.pdf

Client Leadership Council (CLC). (2020, June 1). *Roadmap to Recovery: An Industry Recovery Plan for the UK Construction Sector*. https://cdn.ca.emap.com/wp-content/uploads/sites/8/2020/05/Roadmap-to-Recovery.pptx

Construction Manager. (2020, June 1). *Annual BIM survey: Poor digital skills hold back adoption.* Charted Institute of Building (CIOB). https://www.constructionmanagermagazine.com/annual-bim-survey-poor-digital-skills-hold-back-adoption/

Cook, D., & Chatterjee, P. (2015). Our Changing World: Let's Be Ready. London: Academic Press.

DCOM. (2019, December 1). *A Call to Digitise. Digitisation of Requirements, Regulations, Standards, and Compliance Checking in the Built Environment.* https://www.dcom.org.uk/wp-content/uploads/2019/10/DCOM-Consultation-Note.pdf

DEC. (2020). *Design Engineer Construct!®: Educating the Future of Construction. Inspiring a new generation of Built Environment professionals.* https://designengineerconstruct.com/what-is-dec/

Department for Business. Innovation & Skills (BIS). (2013). UK Construction: An Economic Analysis of the Sector (BIS/13/958). Department for Business, Innovation & Skills.

Designing Buildings Wiki. (2019). *Small and medium-sized enterprises SME.* https://www.designingbuildings.co.uk/wiki/Small_and_medium-sized_enterprises_SME

Egan, J. (1998). *Rethinking Construction – The Egan Report.* Department of Trade and Industry, HMSO.

ESCO. (2020). *European Skills/Competences, Qualifications, and Occupations.* https://ec.europa.eu/esco/portal/home

Farmer, M. (2016). *The Farmer Review of the UK Construction Labour Model.* Construction Leadership Council.

Flinn, S. (2018). Data-to-Learning-to-Action. In Optimizing Data-to-Learning-to-Action (pp. 29-47). Apress.

Hackitt, J. (2018). *Building a Safer Future. Independent Review of Building Regulations and Fire Safety: Final Report.* London: Academic Press.

HBR. (2016 November 3). *Why the Problem with Learning Is Unlearning.* https://hbr.org/2016/11/why-the-problem-with-learning-is-unlearning

Heiskanen, A. (2017 June 2). *Automating Construction Processes by Making BIMs Machine-Readable.* AEC Business. https://aec-business.com/automating-construction-processes-making-bims-machine-readable/

Hughes, W., & Hughes, C. (2013). Professionalism and professional institutions in times of change. *Building Research and Information, 41*(1), 28–38. doi:10.1080/09613218.2013.737096

ICE. (2018). *ICE professional skills: Report by the Skills Review Group, 2018.* Institution of Civil Engineers.

IPPR. (2017). *Skills 2030: Why the adult skills system is failing to build an economy that works for everyone.* London: IPPR.

Irvine, G., White, D., & Diffley, M. (2018). *Measuring Good Work: The Final Report of the Measuring Job Quality Working Group*. RSA Future Work Centre. Carnegie UK Trust. https://www.carnegieuktrust.org.uk/publications/measuring-good-work-the-final-report-of-the-measuring-job-quality-working-group/

Jacobides, M. G., Cennamo, C., & Gawer, A. (2018). Towards a theory of ecosystems. *Strategic Management Journal, 39*(8), 2255–2276. doi:10.1002mj.2904

Langford, D., & Male, S. (2008). *Strategic Management in Construction*. Blackwell Science Ltd.

Moore, J. F. (1997). *The death of competition: leadership and strategy in the age of business ecosystems*. Harper Business.

Morrell, P. (2015 May 19). *Collaboration for Change. The Edge Commission Report on the Future of Professionalism*. https://www.designingbuildings.co.uk/wiki/Collaboration_for_Change,_The_Edge_Commission_Report_on_the_Future_of_Professionalism#:~:text=19%20May%202015-,Collaboration%20for%20Change%2C%20The%20Edge%20Commission%20Report%20on%20the%20Future,to%20design%20and%20problem%20solving

Mosey, D., Bahram, D., Dartnell, R., Hallam, C., Howard, C., Maqbool, A., Murray, K., Rawlinson, S., & Winfield, M. (2016). *Enabling BIM through Procurement and Contracts*. http://alliancecontracting-electroniclawjournal.com/wp-content/uploads/2019/10/Kings-College-London.-2016-Enabling-BIM-through-Procurement-and-Contracts-Centre-of-Construction-Law-and-Dispute-Resolution.pdf

NATSPEC. (2014). *BIM Education – Global: 2014 Update Report*. NATSPEC.

NATSPEC. (2015). *BIM Education – Global: 2015 Update Report*. NATSPEC.

NATSPEC. (2016). *BIM Education – Global: 2016 Update Report*. NATSPEC.

NATSPEC. (2017). *BIM Education – Global: 2017 Update Report*. NATSPEC.

NATSPEC. (2018). *BIM Education – Global: 2018 Update Report*. NATSPEC.

NBS. (2020). *10th Annual BIM Report 2020*. NBS Enterprises Ltd.

NBS. (2020), The Construction Leaders' Summit: The Digital Future, NBS Enterprises Ltd.

NHA. (2016 May 1). *New UKAS Competency Management Standard for Housing Sector*. https://www.nationalhousingacademy.org.uk/news/new-ukas-standard

OECD. (2017a). Better Use of Skills in the Workplace: Why It Matters for Productivity and Local Jobs. Organisation for Economic Co-operation and Development (OECD).

OECD. (2017b). Getting Skills Right: United Kingdom. Organisation for Economic Co-operation and Development (OECD).

Oxford Economics. (2020). *The Digital Opportunity for Small Businesses: How improved digital skills and connectivity boost local areas*. Intuit QuickBooks.

Payne, J. (2008). Skills in context: What can the UK learn from Australia's skill ecosystem projects? *Policy and Politics, 36*(3), 307–323. doi:10.1332/030557308X307685

Plan Grid, Inc. (2019). *The Digital Groundwork: Beyond construction's productivity gap*. Plan Grid Inc.

Quebec Construction 4.0 Initiative. (2020). *Quebec Initiative for Construction 4.0: Improve the performance of the Quebec construction industry through the digital shift*. www.constructionnumerique.ca

Rinkinen, S., & Harmaakorpi, V. (2017). The Business Ecosystem Concept in Innovation Policy Context: Building a Conceptual Framework. *Innovation (Abingdon)*, *31*(3), 333–349. doi:10.1080/13511610.2017.1300089

Royal Academy of Engineers. (2007). *Educating Engineers for 21st Century*. The Royal Academy of Engineering.

Royal Academy of Engineers. (2010). *Engineering Graduates for Industry*. The Royal Academy of Engineering.

Russell, M. G., & Smorodinskaya, N. V. (2018). Leveraging complexity for ecosystemic innovation. *Technological Forecasting and Social Change*, *136*, 114–131. doi:10.1016/j.techfore.2017.11.024

Schouten, C. (2016 January 22). Does construction have a 'productivity problem'? *Construction News*. https://tinyurl.com/ya6ojxnt

SFT. (2018 July). *Developing a BIM Competency Framework: Research & Key Principles*. https://www.scottishfuturestrust.org.uk/storage/uploads/bimcompetencyframeworkresearchjul18.pdf

Simpson, M., & Carlton, D. (2019 January 12). *Upskilling Briefing Paper*. UK BIM Alliance. https://tinyurl.com/ycn7bgcg

Susskind, R., & Susskind, D. (2015). *The Future of the Professions: How Technology will Transform the Work of Human Experts*. Oxford University Press. doi:10.1093/oso/9780198713395.001.0001

Taylor Review. (2017). *Good Work: The Taylor Review of Modern Working Practices*. https://assets.publishing.service.gov.uk/government/uploads/system/uploads/attachment_data/file/627671/good-work-taylor-review-modern-working-practices-rg.pdf

Uff, J. (2016). *UK Engineering 2016: An independent review*. http://www.keatingchambers.com/wp-content/uploads/2017/05/UK-Engineering-2016-An-Independent-review-by-Professor-John-Uff-CBE-QC-FREng.pdf

UK Universities, . (2019). *The Future of Degree Apprenticeships*. Author.

UKBF. (2020). *Learning Outcomes*. https://ukbimframework.org/wp-content/uploads/2020/07/UK-BIM-Framework-Learning-Outcomes_Edition1.pdf

Underwood, J. (2014, November 20). Educating BIM: Changing the Face of Construction. Planning & Control. *BIM Today*. https://www.pbctoday.co.uk/news/bim-news/educating-bim-changing-face-construction/11725/

Ward, D., Whysall, D., Le Grand, Z., Guthrie, W., & Davies, R. (2009). *Never Waste a Good Crisis. A Review of Progress since Rethinking Construction and Thoughts for our Future*. Constructing Excellence.

Wolstenholme, A., Austin, S. A., Bairstow, M., Blumenthal, A., Lorimer, J., McGuckin, S., & Rhys Jones, S. (2009). *Never Waste a Good Crisis: A Review of Progress since Rethinking Construction and Thoughts for our Future*. Loughborough University. https://hdl.handle.net/2134/6040

World Economic Forum. (2018a). *The Future of Jobs Report 2018*. World Economic Forum.

World Economic Forum. (2018b). *Future Scenarios and Implications for the Industry*. Prepared in collaboration with The Boston Consulting Group. World Economic Forum.

Zen Internet. (2019). *Bricks, Mortar and Digital Transformation Report*. Zen Internet.

Zhao, G. (2012). Competence semantics: Engineering and application. *International Journal of Knowledge and Learning*, *8*(1/2), 122–132. doi:10.1504/IJKL.2012.047553

Zhao, G. (2017a). *Competency with Activity Semantics*. Internal report. Intelartes. Unpublished.

Zhao, G. (2017b). *Competency Analytics, Business Context & Approach*. Internal report. Intelartes. Unpublished.

Zhao, G. (2018). *Competency Analytics & Management*. Internal report. Intelartes. Unpublished.

Zhao, G. (2019). *A System Model of Competency Frameworks*. Internal report. Intelartes. Unpublished.

Zhao, G. & Carlton, D. (2015a). *Dynamic, competency-based career planning and CPD: Vision statement and requirements analysis*. Institution of Civil Engineers.

Zhao, G., & Carlton, D. (2015b). Forecast Competency Migration by a Methodology of Competency Analytics. *Open Journal of Social Sciences*, *3*(11), 16–22. doi:10.4236/jss.2015.311003

Zhao, Y. (2014). *Who's Afraid of the Big Bad Dragon?: Why China Has the Best (and Worst) Education System in the World*. Wiley.

KEY TERMS AND DEFINITIONS

Capability: The complex combination of an appropriate set of competences in order to achieve a specific organisational objective(s). It emphasises the key role of strategic management in appropriately adapting, integrating, and in re-configuring the internal and external organisational skills, resources, and functional competences in a changing environment (Teece et al., 1997).

Competence/Competence Frameworks (CF) or Models: Are designed with respect to an organisation, function, or occupational activity. CF are textual descriptions, but they are designed along conceptual dimensions, e.g. knowledge, skills, abilities (KSA) or other attributes such as behaviours, etc. Therefore, a CF can be a collection of competence dimensions or KSAs with relationships, levels, and other attributes such as importance, currency, roll-up rules. Competence Frameworks are typically designed to deliver a breadth of outcomes relating not only to training, but also talent management, professional development/ CPD, appraisal, performance review, organisational design, and development, etc.

Competence(s): 'A capability demonstrated in an activity at work (who can do what, why, how, where, when, with whom, with what'). It is an ability to conduct an activity or perform a task for a professional duty1. Competence has multiple dimensions, expressed in such terms as duties, abilities, skills, experience, expertise, aptitude, knowledge, education, qualifications, behaviour, values, interests and attitudes'. Competence is multi-dimensional, multi-faceted and context dependent and increasingly many competences overlap sectors, disciplines, professions, and roles in the dynamic world of work. But, they all refer to, implicitly or explicitly, to activities at work (Zhao, 2017a, 2017b).

Competency Analytics: A set of technologies and methodologies for competency management. It is a decision support system for bi-directionally mediating competency supply and demand. It enables the analysis of competency state and evolution in order to facilitate the match of competency supply and demand, the fulfilment of competency needs by competency development, competency planning, migration and assurance.

Competency(ies) (in Ecosystem Context): A competence profile; a set of competences associated with a person, team, task, role, project, profession, service, process, practices, courses, publications, and policy (Zhao, 2017b).

Competency Assurance: Set within the context of competency demand and supply mediation and is primarily concerned with understanding people@work risks more dynamically within an organisation or ecosystem where the need, existence, extent, currency, validity, and meaning of competency can be understood and securely shared at a granular level and relevant to the next in line work activity, HRM/people processes (e.g. workforce planning), individual development/career planning, CPD, re-certification/licensing, or in response to a specific competency assurance request (Carlton, 2018).

Competency Management: Competency and its management concerns identifying, acquiring, assessing, matching, foreseeing, and assuring competency at work, in order to mitigate its demands and supplies including minimising risks.

Digital Competency Management Ecosystem: A digital community platform of inter-dependent stakeholders in cooperation within or across sectors and occupations. It is an electronic-market of services facilitating and mediating supply and demand of data, information and knowledge about competences, professions, trades and roles, disciplines, learning opportunities, qualifications, best practices, standards, regulations and policies. It concerns structural innovation.

Reskilling: In terms of re-skilling, the main context is migration – planning the transition and moving to a new activity, role, job, discipline, profession, sector which may require new competences and unlearning. It terms of transition it often concerns both stocks (personal, social and institutional resources) and flows (moving in, moving through, and moving on).

Skill: The functional ability the person or agent has to perform activities and actions.

Skills Gap: Occurs when businesses do not have sufficient staff with the required skills in their existing workforces and willing to move to the next role or job or activity.

Skills Mismatch: Are where individuals are mismatched to their jobs in terms of competences, or qualifications or field of study (discipline). The mismatch can be over-skilled or qualified or under-skilled or qualified for example.

Skills Shortage: Are where businesses have difficulty finding suitably skilled individuals from the potential pool of available recruits or talent pool at the going rate of pay and working conditions. They can be broadly defined in terms of an inadequate supply of workers in high-demand occupations and/or inadequate supply of skills required to perform the tasks associated to such occupations.

Skills Surplus: Are characterised by a relatively high supply and low demand for a given skill (often identified by high unemployment). One can't consider skills gaps or shortages without considering surplus.

Upskilling: Upskilling concerns the progression from one competency profile to another. When considering Professional Upskilling the key questions is 'upskilled to do what to what level, why, where, when, with what, with whom etc.? This may concern learning new competences or ways of working.

ENDNOTES

1. Skills is the current language of the sector and used as a proxy definition or abstraction for competence when in fact skill is a dimension of competence.
2. Up-skilling. re-skilling or de-skilling is the current language of the sector where, for example, upskilling is typically defined as a process or event of learning new skills (i.e. often only application) rather than an evolution (acquisition, application, appreciation & depreciation) from one competency profile to another (dynamic, multi-dimensional, context-dependent).
3. 'Adopting an ecosystem view of business technology' (McKinsey, 2017).

Chapter 18
Smart City Development and Implications

Olusoyi Richard Ashaye
Brunel University, UK & University of Wales Trinity Saint David, UK

Husam Helmi Alharahsheh
https://orcid.org/0000-0002-0194-3700
University of Wales Trinity Saint David, UK

ABSTRACT

Smart cities can be described as the use of modern technology to improve urban space, as well as interacting with citizens to increase quality of life. It covers every aspect of society's and people's livelihoods, from monitoring of public space to public transport services, home automation, cloud storage, and electronic public and business services. The advancement and global practice of information and communication technology (ICT) have led to its fundamental consideration in the development and navigation of cities. This has called for the need to have innovations using technology to improve the quality of life and enhance natural environment. Therefore, this chapter is aiming to provide further understanding and clarity on smart cities with inclusion of several initiatives and examples already applied in practice within the UK and across other countries. The research is mainly including secondary sources from the literature to review smart cities. Key recommendations are provided for policy developers, local authorities, academics, and students interested in the field.

INTRODUCTION AND BACKGROUND

Smart city has become a global trend in the development of advanced cities in the 1st century. The concept is a new concept, which is still in its infancy stage as the idea was first initiated by IMB in the USA in 2008 to study how the functions of a city could be optimised in order to promote development of talented-based economy as well as improve the quality of life.

There is currently no standard definition of smart city, although practitioners and researchers have made attempts to describe the concept. Whilst it has been described as a city smarter than traditional

DOI: 10.4018/978-1-7998-6600-8.ch018

areas and capitalising on new technology and insights to transform and enhance its systems, operations and service delivery; other academicians have classified smart city by a vision that includes five key factors: digital technology, environmental sustainability, civic initiatives, mobility and businesses (IBM 2010; Cavada et al 2015).

Thus, smart city can also be described as the use of modern technology to improve urban space, as well as interacting with citizens to increase quality of life. It covers every aspect of society's and people's livelihood, from monitoring of public space to public transport services, home automation, cloud storage and electronic public and business services.

Cities like Singapore and Copenhagen are believed to be on top of their game in using technology to enable their citizens to enjoy a better quality of life. Whilst Singapore focuses on people using technology – wireless communications and energy efficient appliances or vehicles to help them reduce the need to move cleanly; Copenhagen concentrates on 'green' economy by engaging people, companies, and government top create a green economy and ensuring that urban planning enables the city to develop in a creative way that would improve the quality of life.

Prof Rogers in 'Smart cities better defined by new research', University of Birmingham (2016) stated that, "Singapore and Copenhagen demonstrate that each smart city adopts solutions that fit its own circumstances. It's vital to learn from these global examples and understand how they became smart – understanding these key factors will help smart cities to fulfil their potential. Smartness is a complex and ever-changing concept, but can be expressed as 'talent-green-technology'. It's important to have talented people and a citizenry that is receptive to working towards all of the goals of a smart, sustainable and resilient city.

"This will involve innovative use of new technology, but it must equally support the delivery of smart low carbon dioxide initiatives that both improve the quality of life of citizens and enhance the natural environment."

Smart cities are generally believed to share a common goal: the use of innovative ideas and methods or the application of ICT in various aspects of the city for better synergy as well as more efficient use of resources with a view to improving the management of city and impact on service delivery including the quality of life of the citizens. More so, there is the need to reduce environmental footprint to support innovative development and a low-carbon economy.

Rationale

The advancement and global practice of in information and communication technology (ICT) have led to its fundamental consideration in the development and navigation of cities. This has called for the need to have innovations using technology to improve the quality of life and enhance natural environment. For instance, initiatives to reduce carbon dioxide emissions are a core part of work promoting environmental sustainability.

There is limited research on smart city since it is a new concept, and researchers have not fully incorporated into the smart citizens agenda, necessary for 'smart city' to become truly smart. There is therefore the need for further research on this concept.

Since local authorities/Councils have the unique ability to understand and shape services to local circumstances and needs, and the tremendous ability to reach to the local communities, particularly through housing and social care services; this calls for focus on the public sector organisations to create a business-friendly environment especially in the areas of business regulation, street cleaning, commodity safety and planning. The Councils are therefore well placed to broker the links between job seekers and innovative programmes for the citizens.

The increasing pressure and cuts in the budget to local and central governments justify the call for new approaches to help tackle the costly social challenges arising from the long-term youth unemployment and vulnerable and troubled families who are observed to be living in places that hindering their progress and quality of life. As a result, local authorities have begun to think differently and creative the space for alternative solutions in order to alleviate the risks and maximising expectation and set realistic outcomes for the citizens.

Further, local authorities have realise they are in a trap and there is need for strategies to escape the taps all over the sectors. Hence, the only way is to work differently with public services, communities and users to achieve better outcomes, with radical innovation and creative programmes by the local authorities. There is thus the need for sustainable, programmed approach to partnership and knowledge exchange in order to embrace change, and build partnership.

Smart City Wheel

The smart city wheel was initiated by Boyd Cohen in 2012. It outlines the functions and goals of a smart city with reference to major indicators and rankings; it also listed 6main components, as follows:

1. Smart Economy
2. Smart Environment
3. Smart People
4. Smart Mobility
5. Smart Living, and
6. Smart Government

The smart wheel also consists of 16 sub-domains and 62 assessment indicators.

Examples of Smart City Initiatives and Programmes

The smart city concentrates on the following are:

1. Parking and mobility – enables easy navigation using geo-location and LBS (Location Based Services) functionalities
2. Services – all chargeable services offered by the city in one place i.e. purchase of bus tickets or parking spaces
3. Payments – city e-commerce payments for all services and purchases using a single mobile application
4. Retail – also enables easy navigation using geo-location and LBS functionalities
5. Citizenship – It relates to e-governance and a source of tips for handling cases and appointment booking in local councils, hospitals and other public services
6. Entertainment – Also for handling cases and appointment booking in local councils, hospitals and other public services

Examples of initiatives and programmes to support smart cities in the UK are as follows:

Figure 1.
(Extracted from Central Policy Unit, 2015)

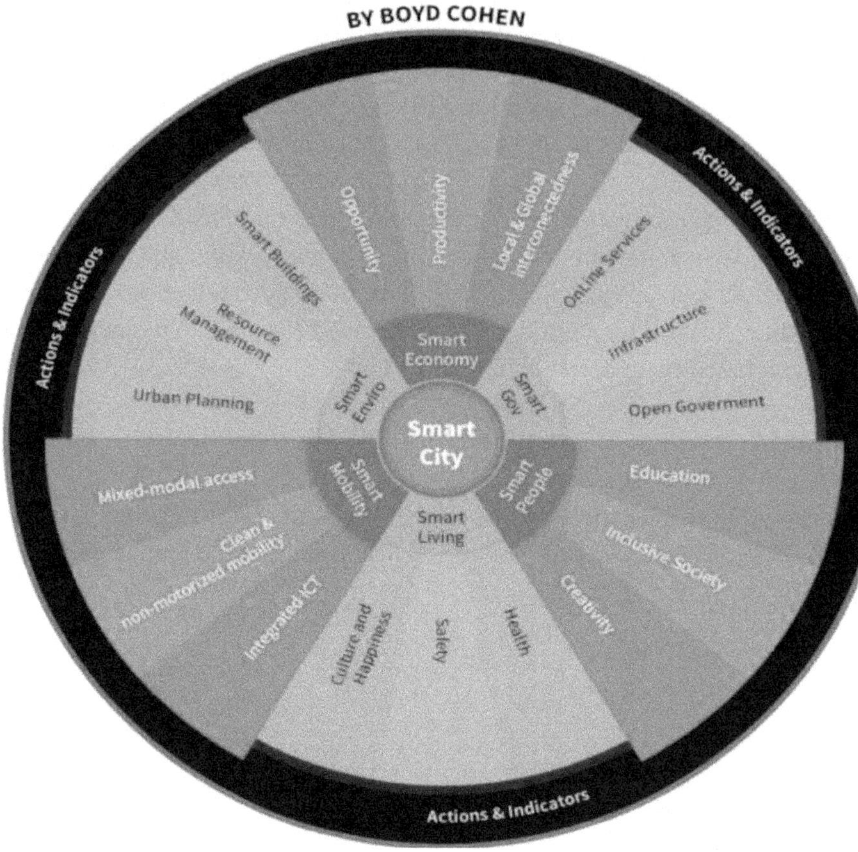

1. Creative Industries (CIs) – Relates to study of work-related learning (WRL) opportunities in the creative industries for young people aged 14-19 in the UK. The CIs embody idea, technology and ways of working that characterised the future economy and world of work, for creative businesses, schools and policy makers. This programme helps to embrace change and to build partnership working.
2. Creative Councils – Based on pioneering work in trialling radical new ways of delivering public services in the UK. It was launched in April 2011and has now moved to the second phase and the programme is being run by a company called Nesta together with the Local Government Authority. There are six creative councils recognised for pioneering innovations, such as:
    a. Cornwall – Implementing 'Shaped by Us', a technology platform and open innovation approach that makes it easier for local communities to come with creative ideas in solving the county's biggest challenges
    b. Derbyshire – Developed 'Uni-fi', a bespoke package of support aimed at developing aspiration among young people in care

c. Monmouthshire – Initiated practical 'Your County Your Way' programme, a cultural transformation within the council to listen and respond more creatively to the needs of its communities through 'Intrepreneurship School' - internal training programme
d. Rotherham – Developing 'Rotherham Ready', council-based social enterprise that works with the teachers of students aged 4-19 to engage them and their schools in the development of an enterprise-based curriculum and ensuring young people have skills relevant for the future
e. Stoke – Developing an efficient 'Great Working City, programme aimed at pushing boundaries of energy supply and reimagining the role of the council as a strategy broker of resources
f. Wigan – Creating a new economic model for social care whereby they meet their services and financial challenged through harnessing of underutilised and unexploited resources within the local community though volunteering and micro-enterprises development.

3. Westminster City Council – Supported the development of 'Hub Westminster', an affordable space in the heart of London for social business to work, collaborate, and access business support and investment. The Hub enables social enterprises to build closer links with local authorities to help grow their ventures. There is also franchising work that supports unemployed people into sustainable work by helping them to access franchising opportunities (Cavada, 2012; Nesta, 2012, 2013)

CONSIDERATIONS FOR EMPLOYABILITY OF WORKERS

Employability skills enhancement within higher education levels should be reflective of the current changes in the labour market (Helmi et al., 2018a, Helmi et al., 2018b; Helmi and Pius, 2018). However, recent research argued that development of employability skills should be more of a collective responsibility to further develop employability skills such as IT skills to be able to reflect key rapid changes, this collective responsibility should include different stakeholders such as universities, employers, policymakers, students, and public sector to further bridge the gap of expectations (Alharahsheh and Pius, 2020). Smart cities certainly need further technology equipped graduates to meet the demands of the labour market

Higher education institutions have increasing pressure to further reflect the current changes at a global level including more dependence on technology and greener use of technology, leading to the development of smart cities that would require different set of skills graduates should have. This was evidenced by research conducted by (Alharahsheh, 2019: Alharahsheh and Pius, 2019) that students lacked application to practice during their reflective assessment, and highlighted the key challenges as lack of ability to apply to particular contexts, lack of work experience and unemployment, lack of self-confidence and awareness, lack of access to training, and lack of skills understanding. Therefore, it is clear that higher education institutions should work in further collaboration to enhance students' understanding of key developments in smart cities, considering that this might significantly affect their ability to be employable in the labour market.

Furthermore, recent research focused on different aspects of applying cloud computing in organisations and digitalization reflected that the idea of using digital resources to enable success of business are becoming imperative, considering that public organisations are investing in developing their digital capabilities to ensure enhancement of the smart cities required infrastructures (Mazumdar and Alharahsheh, 2019). Therefore, it is important for organisations to consider opportunities that can be gained through the digitalization process, and consider any challenges encountered (Mazumdar and Alharahsheh, 2019a; Mazumdar and Alharahsheh, 2019b and Mazumdar and Alharahsheh, 2019c). However, this would require working closely with other stakeholders including customers, government, policy developers and educational institutions, and employees to have an inclusive approach in the development process.

CONCLUSION

The work provided in this chapter reflected several key aspects in relation to smart cities in which can be summarised as the following: Firstly, the rapid advancement of technology in its all forms have certainly contributed to the initiating phase and continues development of smart cities, and technology enabled policy makers at senior levels to start initiatives and programmes to improve key aspects in cities such as quality of life and promotion of environmental sustainability. Secondly, there is limited research on smart cities as a concept. Therefore, there is an increasing need to further understand and explore the concepts with appreciation of its interconnection with other key aspects included in the rational of the review. Thirdly, it was established that local authorises at city of even district level can further develop their public policy and allocation to further utilize the concept of smart cities in key areas such as development of a business-friendly environment, regulations, cleaning and implementation of health and safety measures. Fourthly, Smart city as a concept can lead to cost saving measures as it aims to reduce waste and increase application of technology-oriented solutions bringing benefits for residents and tackling unemployment. However, further research in this area can be conducted to establish further clarity. Fifthly, Local authorities also can work in partnerships with other parties such as educational institutions and the private sector to exchange knowledge and embrace change in developing smarter cities.

REFERENCES

Alharahsheh, H. (2019). Employability skills development through assessment in higher education: Students' voices on reflective assessments. London: University of Greenwich: EDD weekend.

Alharahsheh, H., & Pius, A. (2019). Employability Skills Development through Assessment in Higher Education: Students' Voices on Reflective Assessments. *Global Academic Journal of Economics and Business*, *1*(2), 36–42.

Alharahsheh, H., & Pius, A. (2020). Exploration of Employability Skills in Business Management Studies Within Higher Education Levels. *International Journal of Sustainable Economies Management*, *9*(1), 52–69. doi:10.4018/IJSEM.2020010105

Cavada, N., Hunt, D. V. L., & Rogers, C. D. F. (2015). Do smart cities realise their potential for lower carbon dioxide emissions. *Proceedings of the Institution of Civil Engineers. Engineering Sustainability*, *169*(6), 243–252. doi:10.1680/jensu.15.00032

Central Policy Unit. (2015). Research Report on Smart City. Government of Hong Kong Special Administrative Region, September 2015.

Comarch. (2016). *Smart City: Driven by Innovation.* https://smartcity.comarch.com/?gclid=EAIaIQobChMI2NmLo-K11AIVoxbTCh1kXAcXEAAYASAAEgJnwPD_BwE#about

Helmi, H., & Pius, A. (2018). A Review of International Students' Attraction and Recruitment Strategies: Learning from UK Top Universities. *Scholars Journal of Economics, Business and Management*, *5*(8), 779–785.

Helmi, H., Pius, A., & Guenane, I. (2018a). UK Higher Education Institutions Role in Students' Skills Development for Employment. *Journal of Advances in Education and Philosophy*, *2*(3), 85–88.

Helmi, H., Pius, A., & Guenane, I. (2018b). The Importance of Research Informed Teaching in Higher Education: A Case Study of London College of Contemporary Arts (LCCA). *Journal of Advances in Education and Philosophy*, *2*(3), 89–94.

IBM Institute for Business Value. (2010). *Smaller cities for smarter growth*. Executive Report.

Mazumdar, A., & Alharahsheh, H. (2019a). Insights of Trends and Developments in Cloud Computing. *South Asian Research Journal of Engineering and Technology*, *01*(03), 98–107. doi:10.36346arjet.2019.v01i03.003

Mazumdar, A., & Alharahsheh, H. (2019b). Small & Medium Enterprises (Smes) – Insights of Bangladeshi Smes in Different Contexts of Adopting Cloud Computing. *Cross Current International Journal of Economics, Management and Media Studies*, *01*(05), 130–140.

Mazumdar, A., & Alharahsheh, H. (2019c). Key Challenges and Barriers Facing Small & Medium Enterprises (SMEs) in Bangladesh. *South Asian Research Journal of Business and Management*, *01*(03), 118–120. doi:10.36346arjbm.2019.v01i03.005

Mazumdar, A., & Alharahsheh, H. (2020). Digital Bangladesh – Vision 2021: What is the Digital Bangladesh Concept*? South Asian Research Journal of Engineering and Technology*, *2*(1), 6–9. doi:10.36346arjet.2020.v02i01.002

Nesta. (2012). *How councils can make their economy grow.* Creative Councils. http://www.nesta.org.uk/blog/how-councils-can-make-their-economy-grow#sthash.jzVUQwEr.dpuf

Nesta. (2013). *Six Creative Councils recognised for pioneering innovations*. Creative Councils. http://www.nesta.org.uk/news/six-creative-councils-recognised-pioneering-innovations#sthash.laZFt4jo.EdMELxLf.dpuf

Rutherford, K. (2008). *Creative Opportunities*. https://www.nesta.org.uk/publications/creative-opportunities

Tom Fleming Creative Consultancy & Nesta. (2013). *Six Creative Councils recognised for pioneering innovations*. Creative Councils. http://www.nesta.org.uk/news/six-creative-councils-recognised-pioneering-innovations#sthash.laZFt4jo.EdMELxLf.dpuf

University of Birmingham. (2016). *Smart cities better defined by new research*. https://www.birmingham.ac.uk/news/latest/2016/02/Smart-cities-better-defined-by-new-research-100216.aspx

Compilation of References

Abanda, F. H., Vidalakis, C., Oti, A. H., & Tah, J. H. (2015). A critical analysis of Building Information Modelling systems used in construction projects. *Advances in Engineering Software, 90*, 183–201. doi:10.1016/j.advengsoft.2015.08.009

Abbas, A., Din, Z. U., & Farooqui, R. (2016). Integration of BIM in construction management education: An overview of Pakistani engineering universities. *Procedia Engineering, 145*, 151–157. doi:10.1016/j.proeng.2016.04.034

Abbasnejad, B., Nepal, M., & Drogemuller, R. (2016). Key enablers for effective management of BIM implementation in construction firms. In *Proceedings of the CIB World Building Congress 2016: Volume I-Creating built environments of new opportunities* (Vol. 1, pp. 622-633). TUT–Tampere University of Technology.

Abdel Lattif, A., & Al Shamrani, M. (2013). A comparison of methods for estimating the mean annual runoff coefficient at ungauged stream in southwestern region, Kingdom of Saudi Arabia. *International Journal of Environmental Science and Toxicology Research, 1*(1), 12–20.

Abdel-Basset, M., Manogaran, G., & Mohamed, M. (2018). Internet of Things (IoT) and its impact on supply chain: A framework for building smart, secure and efficient systems. *Future Generation Computer Systems, 86*, 614–628. doi:10.1016/j.future.2018.04.051

Abdirad, H., & Dossick, C. S. (2016). BIM curriculum design in architecture, engineering, and construction education: a systematic review. *Journal of Information Technology in Construction, 21*(17), 250-271.

Abdirad, H. (2017). Metric-based BIM implementation assessment: A review of research and practice. *Architectural Engineering and Design Management, 13*(1), 52–78. doi:10.1080/17452007.2016.1183474

Abed, H. R., Hatem, W., & Jasim, N. (2019). Adopting BIM Technology in Fall Prevention Plans. *Civil Engineering Journal, 5*(10), 227–228.

Abergel, T., Dean, B., & Dulac, J. (2017). Towards a zero-emission, efficient, and resilient buildings and construction sector: Global Status Report 2017. UN Environment and International Energy Agency.

ABET Accreditation Criteria and Supporting Documents. (2020). *Applied and Natural Science Accreditation Commission (ANSAC). 2020-2021 Criteria.* https://www.abet.org/accreditation/accreditation-criteria/

AbouRizk, S. (2010). Role of simulation in construction engineering and management. *Journal of Construction Engineering and Management, 136*(10), 1140–1553. doi:10.1061/(ASCE)CO.1943-7862.0000220

Abubakar, M., Ibrahim, Y. M., Kado, D., & Bala, K. (2014). Contractors perception of the factors affecting Building Information Modeling (BIM) adoption in the Nigerian construction industry. In R. I. Issa & I. Flood (Eds.), *Computing in Civil and Building Engineering* (pp. 167–178). American Society of Civil Engineers. (Original work published 2014) doi:10.1061/9780784413616.022

Adamu, Z. (2016). How universities are teaching BIM: A review and case study from the UK. *Journal of Information Technology in Construction*, *21*(8), 119–139.

Afanasev, M. Y., Fedosov, Y. V., Krylova, A. A., & Shorokhov, S. A. (2018). *An application of blockchain and smart contracts for machine-to-machine communications in cyber-physical production systems. In 2018 IEEE Industrial Cyber-Physical Systems*. ICPS.

Agranoff, R., & McGuire, M. (2003). *Collaborative public management: new strategies for local governments*. Georgetown University Press.

Ahamad, N. R., & Ariffin, M. (2018). Assessment of knowledge, attitude and practice towards sustainable consumption among university students in Selangor, Malaysia. *Sustainable Production and Consumption*, *16*, 88–98. doi:10.1016/j.spc.2018.06.006

Ahankoob, A., Abbasnejad, B., & Wong, P. S. (2020). The Support of Continuous Information Flow Through Building Information Modeling (BIM). In *The 10th International Conference on Engineering, Project, and Production Management* (pp. 125-137). Springer.

Ahmad, M., Zhao, Z.-Y., & Li, H. (2019). Revealing stylized empirical interactions among construction sector, urbanization, energy consumption, economic growth and CO2 emissions in China. *The Science of the Total Environment*, *657*, 1085–1098. doi:10.1016/j.scitotenv.2018.12.112 PMID:30677876

Ahmed, A. L., & Kassem, M. J. (2018). A unified BIM adoption taxonomy: Conceptual development, empirical validation and application. *Automation in Construction*, *96*, 103–127. doi:10.1016/j.autcon.2018.08.017

Ahmed, S. (2019). A review on using opportunities of augmented reality and virtual reality in construction project management. *Organization. Technology and Management in Construction: an International Journal*, *11*(1), 1839–1852. doi:10.2478/otmcj-2018-0012

Ahmed, S., Hossain, M. M., & Hoque, M. I. (2017). A brief discussion on augmented reality and virtual reality in construction industry. *Journal of System and Management Sciences*, *7*(3), 1–33.

AIA-CA. (2012). *BIM education - BIM in practice. A position paper by the Australian Institute of architects and consult Australia*. Consult Australia and Australian Institute of Architects.

Aimi, S. (2019). Enhancing the graduates' employability and career development through Building Information Modelling intensive training. *International Journal of Built Environment and Sustainability*, *6*(1-2), 91–99.

AIQS & NZIQS. (2018). *Australia and New Zealand BIM best practice guidelines*. Retrieved from https://www.nziqs.co.nz/Portals/9/Documents/Miscellaneous/ANZIQS_BIM_Best_Practice_Guidelines_Nov2018.pdf?ver=2018-11-26-140519-660

Ajayi, S. O., Jones, W., & Unuigbe, M. (2019). Occupational stress management for UK construction professionals. *Journal of Engineering, Design and Technology*, 819-832.

Akbarnezhad, A., & Xiao, J. (2017). Estimation and minimization of embodied carbon of buildings: a review. *Buildings, 7*(1), 5.

Åkerlind, G. S. (2012). Variation and commonality in phenomenographic research methods. *Higher Education Research & Development*, *31*(1), 115–127.

Åkerlind, G., McKenzie, J., & Lupton, M. (2014). The Potential of Combining Phenomenography. *Variation Theory and Threshold Concepts to Inform Curriculum Design in Higher Education*, *10*, 227–247.

Akinade, O. O., Oyedele, L. O., Ajayi, S. O., Bilal, M., Alaka, H. A., Owolabi, H. A., & Arawomo, O. O. (2018). Designing out construction waste using BIM technology: Stakeholders' expectations for industry deployment. *Journal of Cleaner Production*, *180*, 375–385. doi:10.1016/j.jclepro.2018.01.022

Akintoye, A., & Main, J. (2007). Collaborative relationships in construction: The UK contractors' perception. *Engineering, Construction, and Architectural Management*, *14*(6), 597–617. doi:10.1108/09699980710829049

Akram, R., Thaheem, M. J., Nasir, A. R., Ali, T. H., & Khan, S. (2019). Exploring the role of building information modeling in Construction safety through science mapping. *Safety Science*, *120*, 456–470. doi:10.1016/j.ssci.2019.07.036

Aksenova, G., Kiviniemi, A., Kocaturk, T., & Lejeune, A. (2018). From Finnish AEC knowledge ecosystem to business ecosystem: Lessons learned from the national deployment of BIM. *Construction Management and Economics*, 1–19.

Al Hasan, A. A. S., & Mattar, E. (2014). Mean runoff coefficient estimation for ungauged streams in the Kingdom of Saudi Arabia. *Arabian Journal of Geosciences*, *7*(5), 2019–2029. doi:10.100712517-013-0892-7

Al Saleem, S. S. (2018). Rainfall-Runoff Analysis for Sustainable Stormwater Drainage for the City of Madinah, Saudi Arabia. *Arabian Journal for Science and Engineering*, *43*(4), 1955–1967. doi:10.100713369-017-2892-5

Alaloul, W. S., Liew, M. S., Zawawi, N. A. W. A., & Kennedy, I. B. (2020). Industrial Revolution 4.0 in the construction industry: Challenges and opportunities for stakeholders. *Ain Shams Engineering Journal*, *11*(1), 225–230. doi:10.1016/j.asej.2019.08.010

AlBahnassi & Hammad. (2012). Near Real-Time Motion Planning and Simulation of Cranes in Construction: Framework and System Architecture. *Journal of Computing in Civil Engineering*, (Jan/Feb), 54–63.

Albert, P. C. C., & Ada, P. L. C. (2004). Key performance indicators for measuring construction success. *Benchmarking*, *11*(2), 203–221. doi:10.1108/14635770410532624

Alcatel Lucent Enterprise (ALE). (2018). *The Internet of Things in Education*. https://www.al-enterprise.com/ko-kr/-/media/assets/internet/documents/iot-for-education-solutionbrief-en.pdf

Alcayaga, A., Wiener, M., & Hansen, E. G. (2019). Towards a framework of smart-circular systems: An integrative literature review. *Journal of Cleaner Production*, *221*, 622–634. doi:10.1016/j.jclepro.2019.02.085

Alharahsheh, H. (2019). Employability skills development through assessment in higher education: Students' voices on reflective assessments. London: University of Greenwich: EDD weekend.

Alharahsheh, H., & Pius, A. (2019). Employability Skills Development through Assessment in Higher Education: Students' Voices on Reflective Assessments. *Global Academic Journal of Economics and Business*, *1*(2), 36–42.

Alharahsheh, H., & Pius, A. (2020). Exploration of Employability Skills in Business Management Studies Within Higher Education Levels. *International Journal of Sustainable Economies Management*, *9*(1), 52–69. doi:10.4018/IJSEM.2020010105

Ali, T., Mert, G., Ferhat, K., Zhanar, B., Aiganym, S., & Gulzat, S. (2019). A Comprehensive Construction and Demolition Waste Management Model using PESTEL and 3R for Construction Companies Operating in Central Asia. *Sustainability*, *11*(6), 1593. doi:10.3390u11061593

Alizadehsalehi, S., Hadavi, A., & Huang, J.C. (2019, June). BIM/MR-Lean Construction Project Delivery Management System. In *2019 IEEE Technology & Engineering Management Conference (TEMSCON)*. IEEE.

Alizadehsalehi, S., Yitmen, I., Celik, T., & Arditi, D. (2018). The effectiveness of an integrated BIM/UAV model in managing safety on construction sites. *International Journal of Occupational Safety and Ergonomics*, 1–16. doi:10.1080/10803548.2018.1504487 PMID:30043680

Al-Jaroodi, J., & Mohamed, N. (2019). Blockchain in Industries: A Survey. *IEEE Access: Practical Innovations, Open Solutions*, *7*, 36500–36515. doi:10.1109/ACCESS.2019.2903554

All Party Parliamentary Group for Excellence in the Built Environment. (2017). *Building on Brexit. How leaving the EU must drive modernisation and training in the built environment.* London: Author.

Allan, J. (1996). Learning outcomes in higher education. *Studies in Higher Education*, *21*(1), 93–108.

Almazroui, M., Islam, M. N., Athar, H., Jones, P. D., & Rahman, M. A. (2012). Recent climate change in the Arabian Peninsula: Annual rainfall and temperature analysis of Saudi Arabia for 1978–2009. *International Journal of Climatology*, *32*(6), 953–966. doi:10.1002/joc.3446

Alonso Díaz, L., & Blázquez Entonado, F. (2009). Are the functions of teachers in e-learning and face-to-face learning environments really different? *Journal of Educational Technology & Society*, *12*(4), 331–343.

American Institutes of Architects. (2007). *Integrated Project Delivery.* A Working Definition, Version 2. https://help.aiacontracts.org/public/wp-content/uploads/2020/03/IPD_Guide.pdf

Amin, K. F., & Abanda, H. F. (2017, April). Developing a Business Case for BIM for a Design and Build Project in Egypt. In *International Conference on Engineering, Applied Sciences and System Modeling* (pp. 149-159). Springer.

An Roinn Oideanchais Agus Scileanna Department of Education and Skills. (2020). *Digital Strategy for Schools 2015-2020 ICT Infrastructure Funding.* https://www.education.ie/en/Schools-Colleges/Services/Grants-and-Additional-Support/ict-infrastructure/digital-strategy-for-schools-2015-2020-ict-infrastructure-funding.html

Ancona, D. G., & Caldwell, D. F. (1992). Bridging the boundary: External activity and performance in organisational teams. *Administrative Science Quarterly*, *37*(4), 634–665. doi:10.2307/2393475

Anderson, T. (2016). Theories for Learning with Emerging Technologies. In G. Veletsianos (Ed.), *Emergence and Innovation in Digital Learning: Foundations and Applications* (pp. 35–50). Athabasca University Press.

Androulaki, E., Barger, A., Bortnikov, V., Cachin, C., Christidis, K., De Caro, A., ... Yellick, J. (2018). Hyperledger Fabric: A Distributed Operating System for Permissioned Blockchains. In *Proceedings of the Thirteenth EuroSys Conference* (pp. 30:1–30:15). New York, NY: ACM. 10.1145/3190508.3190538

Aneziris, O., Papazoglou, I., Mud, M., Damen, M., Kuiper, J., Baksteen, H., & Oh, J. (2008). Towards risk assessment for crane activities. *Safety Science*, *46*(6), 872–884. doi:10.1016/j.ssci.2007.11.012

Anon. (2015a, Jan. 9). How do bitcoin transactions work? How a bitcoin transaction is processed. *The Economist Magazine*.

Anon. (2015b, Oct. 31). The promise of the blockchain. The trust machine. The technology behind bitcoin could transform how the economy works. *The Economist Magazine*.

Anthony, M. (2019). Blockchain in UAE construction. In Construction week. Auckland, NZ: Balasoglou Books.

Antwi-Afari, M. F., Li, H., Pärn, E. A., & Edwards, D. J. (2018). Critical success factors for implementing building information modelling (BIM): A longitudinal review. *Automation in Construction*, *91*, 100–110. doi:10.1016/j.autcon.2018.03.010

Anumba, C. J., & Wang, X. (2012). *Mobile and Pervasive Computing in Construction.* John Wiley & Sons. doi:10.1002/9781118422281

Araiza-Aguilar, J. A., Gutiérrez-Palacios, C., Rojas-Valencia, M. N., Nájera-Aguilar, H. A., Gutiérrez-Hernández, R. F., & Aguilar-Vera, R. A. (2019). Selection of sites for the treatment and the final disposal of construction and demolition waste, using two approaches: An analysis for Mexico City. *Sustainability (Switzerland), 11*(15), 4077. Advance online publication. doi:10.3390u11154077

Araszkiewicz, K. (2017). Digital Technologies in Facility Management - The state of Practice and Research Challenges. *Procedia Engineering*. <> doi:10.1016/j.proeng.2017.08.059

Arayici, Y., Coates, S. P., Koskela, L. J., Kagioglou, M., Usher, C., & O'Reilly, K. (2011a). BIM adoption and implementation for architectural practices. *Structural Survey, 29*(1), 7–25. doi:10.1108/02630801111118377

Arayici, Y., Coates, S. P., Koskela, L. J., Kagioglou, M., Usher, C., & O'Reilly, K. (2011b). Technology adoption in the BIM implementation for lean architectural practice. *Automation in Construction, 20*(2), 189–195. doi:10.1016/j.autcon.2010.09.016

Arayici, Y., Fernando, T., Munoz, V., & Bassanino, M. (2018). Interoperability specification development for integrated BIM use in performance-based design. *Automation in Construction, 85*, 167–181. doi:10.1016/j.autcon.2017.10.018

Arslan, M., Cruz, C., & Ginhac, D. (2019a). Visualizing intrusions in dynamic building environments for worker safety. *Safety Science, 120*(July), 428–446. doi:10.1016/j.ssci.2019.07.020

Arslan, M., Cruz, C., & Ginhac, D. (2019b). Understanding Occupant Behaviors in Dynamic Environments using OBiDE framework. *Building and Environment, 166*(May), 106412. doi:10.1016/j.buildenv.2019.106412

Aslam, M. S., Huang, B. J., & Cui, L. F. (2020). Review of construction and demolition waste management in China and USA. *Journal of Environmental Management, 264*, 110445. Advance online publication. doi:10.1016/j.jenvman.2020.110445 PMID:32217323

Association for Unmanned Vehicle Systems International. (2020). https://www.auvsi.org/our-impact/economic-report

Astour, H., & Franz, V. (2014). BIM-and simulation-based site layout planning. In R. I. Issa & I. Flood (Eds.), *Computing in Civil and Building Engineering* (pp. 291–298). American Society of Civil Engineers. (Original work published 2014) doi:10.1061/9780784413616.037

Attaran, M., & Gunasekaran, A. (2019). Financial Services: The Largest Blockchain Market. In Applications of Blockchain Technology in Business (pp. 21-26). Springer.

Auberger, L., & Kloppmann, M. (2017). *Combine business process management and blockchain*. IBM Tutorial on Blockchain implementation. Available online and accessed on June 15, 2020 at https://www.ibm.com/developerworks/library/mw-1705-auberger-bluemix/1705-auberger.html

Australasian BIM Advisory Board (ABAB). (2020). *Australian BIM Strategic Framework*. https://www.abab.net.au/wp-content/uploads/2020/01/ABAB_Australian_BIM_Strategic_Framework_Feb_2019.pdf

Australian Institute of Building (AIB). (2016). *Standards for the Accreditation of Building & Construction Management Programs- Academic Standards for Course Accreditation*. https://aib.org.au/wp-content/uploads/2014/10/AIB-Accreditation-Academic-Standards-Document-Version-4-5-April-2017-1.pdf

Australian Institute of Quantity Surveyors (AIQS). (2017). *AIQS Course Accreditation Policy*. https://www.aiqs.com.au/sites/default/files/uploaded-content/field_f_content_file/aiqs_course_accreditation_policy_nb_v2.pdf

Australian Safety and Compensation Council. (2006). *Guidance on Occupational health and Safety in Government procurement*. https://www.safeworkaustralia.gov.au/system/files/documents/1702/guidanceonohsingovernementprocurement_2006_pdf.pdf

Autio and Thomas, (2014). The Fifth Facet: the Ecosystem as an Organisational Field. *Proceedings of the DRUID Society Conference 2014.*

Autio, E., & Thomas, L. D. W. (2014). Innovation Ecosystems: Implications for Innovation Management. In M. Dodgson, D. M. Gann, & N. Phillips (Eds.), *The Oxford Handbook of Innovation Management* (pp. 204–228). Oxford University Press.

Azhar, S., Hein, M., & Sketo, B. (2008, April) Building information modeling (BIM): benefits, risks and challenges. In *Proceedings of the 44th ASC Annual Conference* (pp. 2-5). Academic Press.

Azhar, S., Sattineni, A., & Hein, M. (2010). *BIM Undergraduate Capstone Thesis: Student Perceptions and Lessons Learned* [Paper presentation]. The 46th ASC Annual Conference, Boston, MA.

Azhar, S. (2011). Building Information Modeling (BIM): Trends, Benefits, Risks, and Challenges for the AEC Industry. *Leadership and Management in Engineering, 11*(3), 241–252.

Azhar, S. (2011). Building information modeling (BIM): Trends, benefits, risks, and challenges for the AEC industry. *Leadership and Management in Engineering, 11*(3), 241–252. doi:10.1061/(ASCE)LM.1943-5630.0000127

Azhar, S. (2017). Role of visualization technologies in safety planning and management at Construction jobsites. *Procedia Engineering, 171*, 215–226. doi:10.1016/j.proeng.2017.01.329

Azhar, S., & Bahringer, A. (2013). A BIM-based approach for communicating and implementing a construction site safety plan. In *Proceedings of the 49th ASC Annual International Conference*. Associated Schools of Construction. http://ascpro0.ascweb.org/archives/cd/2013/paper/CPRT43002013.pdf

Azhar, S., Khalfan, M., & Maqsood, T. (2012). Building information modelling (BIM): Now and beyond. *Construction Economics and Building, 12*(4), 15–28. doi:10.5130/AJCEB.v12i4.3032

Badcock, K. (2019, February/March). Digitisation of the construction industry. *New Zealand Construction News (NZCN)*, 21.

Badrinath, A.C., Chang, Y., & Hsieh, S. (2016) A review of tertiary BIM education for advanced engineering communication with visualization. *Visualization in Engineering, 4*(9).

Badrinath, A. C., Chang, Y. T., & Hsieh, S. H. (2016). A review of tertiary BIM education for advanced engineering communication with visualization. *Visualization in Engineering, 4*(1), 9. doi:10.118640327-016-0038-6

BAF. (2013). *Embedding BIM (Building Information Modelling) within the taught curriculum: Supporting BIM implementation and adoption through the development of learning outcomes within the UK academic context for Built Environment programmes*. Higher Education Academy (HEA).

BAF. (2015). *Current Position and Associated Challenges of BIM Education in UK Higher Education*. BIM Academic Forum.

Baiden, B. K., Price, A. D. F., & Dainty, A. R. J. (2006). The extent of team integration within construction projects. *International Journal of Project Management, 24*(1), 13–23. doi:10.1016/j.ijproman.2005.05.001

Baillie, C., MacNish, C., Tavner, A., Trevelyan, J., Royle, G., Hesterman, D., Leggoe, J., Guzzomi, A., Oldham, C., Hardin, M., Henry, J., Scott, N., Doherty, J., & Male, S. (2012). Engineering Thresholds: an Approach to Curriculum Renewal. Final Report 2012, Australian Office for Learning and Teaching.

Bakchan, A., Faust, K. M., & Leite, F. (2019). Seven-dimensional automated construction waste quantification and management framework: Integration with project and site planning. *Resources, Conservation and Recycling, 146*, 462–474. doi:10.1016/j.resconrec.2019.02.020

Baker, M. (2020). Why is primary schooling still based on Victorian policies? *The Guardian*. Available: https://www.theguardian.com/education/2009/sep/15/primary-schools-policy-history-victorian

Baldwin, M. (2018). *buildingSMART Professional Certification. PHASE 1: individual qualification. Overview and Sponsorship Plan*. Retrieved from education.buildingsmart.org

Balfour Beatty Construction. (2015). *BIM for Zero Harm*. Author.

Ball, M. (2008). *Markets and institutions in real estate and construction*. John Wiley & Sons.

Banco Central de la Republica Dominicana. (2015). *Informe de la Economia Dominicana*. Enero-Diciembre 2014 Retrieved from https://www.bancentral.gov.do/

Banco Central de la Republica Dominicana. (2016). *Informe de la Economia Dominicana*. Enero-Diciembre 2015 Retrieved from https://www.bancentral.gov.do/

Banco Central de la Republica Dominicana. (2017). *Informe de la Economia Dominicana*. Enero-Diciembre 2016 Retrieved from https://www.bancentral.gov.do/

Banco Central de la Republica Dominicana. (2018). *Informe de la Economia Dominicana*. Enero-Diciembre 2017 Retrieved from https://www.bancentral.gov.do/

Banco Central de la Republica Dominicana. (2019). *Informe de la Economía Dominicana*. Enero-Junio 2019. Retrieved from https://www.bancentral.gov.do/

Banerjee, M., Lee, J., & Choo, K.-K. R. (2018). A blockchain future for Internet of Things security: A position paper. *Digital Communications and Networks*, *4*(3), 149–160. doi:10.1016/j.dcan.2017.10.006

Barbosa, F., Woetzel, J., Mischke, J., Ribeirinho, M. J., Sridhar, M., Parsons, M., . . . Brown, S. (2017). *Reinventing construction through a productivity revolution*. Tech. rep., McKinsey Global Institute.

Barison, M.B., & Santos, E.T. (2011). The Competencies of BIM Specialists: A Comparative Analysis of the Literature Review and Job Ad Descriptions. *Computing in Civil Engineering, ASCE 2011*.

Barison, M. B., & Santos, E. T. (2010). *BIM teaching strategies: An overview of the current approaches* [Paper presentation]. *International Conference on Computing in Civil and Building Engineering*, Nottingham, UK.

Barles, S. (2014). History of waste management and the social and cultural representations of waste. In *The Basic Environmental History* (pp. 199–226). Springer. doi:10.1007/978-3-319-09180-8_7

Barradell, S. (2012). The identification of threshold concepts: A review of theoretical complexities and methodological challenges. *Higher Education*, *65*(2), 265–276.

Barradell, S., & Kennedy-Jones, M. (2015). Threshold concepts, student learning and curriculum: Making connections between theory and practice. *Innovations in Education and Teaching International*, *52*(5), 536–545.

Barratt, M. (2004). Understanding the meaning of collaboration in the supply chain. *Supply Chain Management*, *9*(1), 30–42. doi:10.1108/13598540410517566

Bartley, T. (2019). Process, Technology and . . . oh yeah. *People, BIM2050*, •••. https://bim2050.wordpress.com/2015/12/22/process-technology-and-oh-yeah-people/

Bashabsheh, A. K., Alzoubi, H. H., & Ali, M. Z. (2019). The application of virtual reality technology in architectural pedagogy for building constructions. *Alexandria Engineering Journal*, *58*(2), 713–723. doi:10.1016/j.aej.2019.06.002

BCA. (2011). *The BIM issue*. Retrieved from https://www.bca.gov.sg/publications/BuildSmart/others/buildsmart_11issue9.pdf

Bealtham, S., Anumba, C. J., & Thorpe, T. (2004). KPIs: A critical appraisal of their use in construction. *Benchmarking*, *11*(1), 93–117. doi:10.1108/14635770410520320

Beare, M. E. (2007). The devil made me do it: Business partners in crime. *Journal of Financial Crime*, *14*(1), 34–48. doi:10.1108/13590790710721792

BECA. (2020). *Supporting water treatment for a growing city*. Retrieved from https://www.beca.com/what-we-do/projects/water/pukete-wastewater-treatment-plant-bim-implementatic

Becerik-Gerber, B., Gerber, D. J., & Ku, K. (2011). Pace of Technological Innovation in AEC Education: Integrating Recent Trends into the Curricula. *Journal of Information Technology in Construction*, *16*, 411-432.

Becerik-Gerber, B., Gerber, D. J., & Ku, K. (2011). The pace of technological innovation in architecture, engineering, and construction education: Integrating recent trends into the curricula. *Journal of Information Technology in Construction*, *16*(24), 411–432.

Becerik-Gerber, B., Jazizadeh, F., Li, N., & Calis, G. (2011). Application areas and data requirements for BIM-enabled facilities management. *Journal of Construction Engineering and Management*, *138*(3), 431–442. doi:10.1061/(ASCE)CO.1943-7862.0000433

Bechky, B. A. (2006). Gaffers, gofers, and grips: Role-based coordination in temporary organizations. *Organization Science*, *1*(17), 3–21. doi:10.1287/orsc.1050.0149

Bee Smart City. (2018). *ViewSonic ViewBoard® Interactive Display Solution for Smart Education*. https://www.beesmart.city/solutions/viewsonic-viewboard-r-interactive-display-solution-for-smart-education

Behan, A. (2013). Update on the BIM education of Geomatic Surveyors. *Proceedings of the CITA BIM Gathering Conference*, 163-168.

Behzadi, A. (2016). Using augmented and virtual reality technology in the construction industry. *American Journal of Engineering Research*, *5*(12), 350–353.

Benecke, M. (2002). Magnetic Mountains: Short Communications. *The Skeptical Inquirer*, *26*(3).

Benedetto, L., Cremonesi, P., & Parenti, M. (2019). *A Virtual Teaching Assistant for Personalized Learning*. https://www.researchgate.net/publication/331343430_A_Virtual_Teaching_Assistant_for_Personalized_Learning

Bengtsson, J., Howard, N., & Kneppers, B. (2010). *Weighting of environmental impacts in Australia*. https://edgeenvironment.com/weighting-of-environmental-impacts-in-australia/

Benjaoran, V., & Bhokha, S. (2010). An integrated safety management with construction management using 4D CAD model. *Safety Science*, *48*(3), 395–403. doi:10.1016/j.ssci.2009.09.009

Bercerik-Gerber, B., Gerber, D. J., & Ku, K. (2011). The pace of technological innovation in architecture, engineering, and construction education: Integrating recent trends into the curricula. *Journal of Information Technology in Construction.*, *V16*, 411–432.

BergC.DavidsonS.PottsJ. (2017). Blockchains industrialise trust. Available at SSRN 3074070.

BergC.DavidsonS.PottsJ. (2018). Ledgers. Available at SSRN 3157421.

Bernstein, P. G. (2018). *Architecture, Design and Data: Practice Competency in the era of Computation*. Birkhauser. doi:10.1515/9783035610444

Berwald, S. (2008). From CAD to BIM: The Experience of Architectural Education with Building Information Modelling. *Proceedings of the American Society of Civil Engineers*. 10.1061/41002(328)8

Beyea, S. C., & Nicoll, L. H. (2000). Learn more using focus groups. *AORN Journal, 71*(4), 897-900. doi:10.1016/S0001-2092(06)62276-X

Bianconi, F., Filippucci, M., & Buffi, A. (2019). Automated design and modeling for mass-customised housing. A web-based design space catalog for timber structures. *Automation in Construction, 103*(March), 13–25. doi:10.1016/j.autcon.2019.03.002

Bierema, L. L. (2019). Enhancing Employability through Developing T-Shaped Professionals. *New Directions for Adult and Continuing Education, 2019*(163), 67–81.

Biesta, M., & Priestley, G. (2013). *Reinventing the Curriculum: New Trends in Curriculum Policy and Practice*. Bloomsbury Academic.

Bilal, M., Oyedele, L. O., Akinade, O. O., Ajayi, S. O., Alaka, H. A., Owolabi, H. A., Qadir, J., Pasha, M., & Bello, S. A. (2016). Big data architecture for construction waste analytics (CWA): A conceptual framework. *Journal of Building Engineering, 6*(C), 144–156. doi:10.1016/j.jobe.2016.03.002

Billinghurst & Duenser. (2012). *Augmented Reality in the Classroom*. https://www.researchgate.net/publication/234793015_Augmented_Reality_in_the_Classroom

Biluca, J., de Aguiar, C. R., & Trojan, F. (2020). Sorting of suitable areas for disposal of construction and demolition waste using GIS and ELECTRE TRI. *Waste Management (New York, N.Y.), 114*, 307–320. doi:10.1016/j.wasman.2020.07.007 PMID:32688063

BIM Academic Forum (BAF) UK. (2013). Embedding Building Information Modelling (BIM) within the taught curriculum. York, UK: The Higher Education Academy (HEA).

BIM Academic Forum (BAF). (2015). *BIM Learning Outcomes Framework*. Author.

BIM Academic Forum UK. (2013). *Embedding Building Information Modelling (BIM) within the taught curriculum: Supporting BIM implementation and adoption through the development of learning outcomes within the UK academic context for built environment programmes*. The Higher Education Academy.

BIM Level 2. (2016). *BIM learning outcomes framework*. Retrieved from http://bim-level2.org/globalassets/pdfs/learning-outcomes-framework.pdf

BIM Task Group. (2012). *The BIM task group publishes the initial BIM learning outcomes framework*. Retrieved from www.bimtaskgroup.org

BIM Wiki. (2015). *BIM task group*. Author.

BIM2050. (2014). *A Report on Our Digital Future*. Construction Industry Council.

BIMe Initiative. (2019). *BIMe initiative. 201in Competency Table v2.1*. Retrieved from https://bimexcellence.org/resources/200series/201in/

BIMe Initiative. (2020). *BIM Excellence Initiative*. https://bimexcellence.org/

BIM-E Initiative. (n.d.). *Building Information Modelling (BIM)*. Disponível em: https://bimdictionary.com/en/building-information-modelling/1/

BIMinNZ. (2016). *Benefits of adopting BIM*. Retrieved from https://bim-in-nz.squarespace.com/benefits-of-adopting-bim

BIMinNZ. (2019). *Ara Institute of Canterbury Kahukura block*. Retrieved from https://bim-in-nz.squarespace.com/casestudies/2017/ara-institute-of-canterbury-kahukura-block

Binder, A. K. (2007). *Ward's Automotive Yearbook 2007*. Wards Communications.

Blanco, J. L., Janauskas, M., & Ribeirinho, M. J. (2016). *Beating the low-productivity trap: How to transform construction operations*. https://www.mckinsey.com/industries/capital-projects-and-infrastructure/our-insights/beating-the-low-productivity-trap-how-to-transform-construction-operations#

Blau, I., Weiser, O., & Eshet-Alkalai, Y. (2016). Face-to-face versus one-way and two-way videoconferencing: How medium naturalness and personality traits influence achievement and perceived learning? *11th Iberian Conference on Information Systems and Technologies*. doi:10.1109/CISTI.2016.7521581

Bloom, B. (1956). *Taxonomy of educational objectives: Handbook I: Cognitive domain*. David McKay Company.

Bogue, R. (2018). What are the prospects for robots in the construction industry? *Industrial Robot. International Journal (Toronto, Ont.)*.

Bolpagi, M. (2013). *The Implementation of BIM within the Public Procurement: A Model-Based Approach for the Construction Industry*. VTT Technology.

Bolton, A., Butler, L., Dabson, I., Enzer, M., Evans, M., Fenemore, T., Harradence, F., Keaney, E., Kemp, A., Luck, A., Pawsey, N., Saville, S., Schooling, J., Sharp, M., Smith, T., Tennison, J., Whyte, J., Wilson, A., & Makri, C. (2018). The Gemini Principles: Guiding values for the national digital twin and information management framework. Academic Press.

Bortolini, R., Formoso, C. T., & Viana, D. D. (2019). Site logistics planning and control for engineer-to-order prefabricated building systems using BIM 4D modeling. *Automation in Construction*, *98*, 248–264. doi:10.1016/j.autcon.2018.11.031

Bosch-Sijtsema, P. M., Gluch, P., & Sezer, A. A. (2019). Professional Development of the BIM Actor Role. *Automation in Construction*, *V97*, 44–51. doi:10.1016/j.autcon.2018.10.024

Bosch-Sijtsema, P., Isaksson, A., Lennartsson, M., & Linderoth, H. C. J. (2017). Barriers and facilitators for BIM use among Swedish medium-sized contractors - We wait until someone tells us to use it. *Visualization in Engineering*, *5*(3), 1–12. doi:10.118640327-017-0040-7

Boton, C. (2018). Supporting constructability analysis meetings with Immersive Virtual Reality-based collaborative BIM 4D simulation. *Automation in Construction*, *96*, 1–15. doi:10.1016/j.autcon.2018.08.020

Boton, C., Forgues, D., & Halin, G. (2018). *A framework for Building Information Modelling Implementation in Engineering Education. Canadian Journal of Civil Engineering*.

Bowden, J. A. (2004). Capabilities-driven curriculum design. In C. Baillie & I. Moore (Eds.), *Effective Learning and Teaching in Engineering, Effective learning and teaching in engineering* (pp. 36–47). RoutledgeFalmer.

Bowles, G., & Morgan, J. (2016). An evaluation of the performance of a large scale collaborative procurement initiative in the social housing sector. *Engineering, Construction, and Architectural Management*, *23*(1), 60–74. doi:10.1108/ECAM-03-2014-0036

Bown, M., & Miller, K. (2018). The Use of Unmanned Aerial Vehicles for Sloped Roof Inspections – Considerations and Constraints. *Journal of Facility Management Education and Research*, *1*(2), 12–18. doi:10.22361/jfmer/93832

Bozoglu, J. (2016). Collaboration and coordination learning modules for BIM education. *Journal of Information Technology in Construction.*

Bozoglu, J. (2016). Collaboration and coordination learning modules for BIM education. *Journal of Information Technology in Construction*, *21*(10), 152–163.

BRANZ. (2020). Material use. *Level.* Retrieved from http://www.level.org.nz/material-use/minimising-waste/

Bresnen, M., & Marshall, N. (2000). Partnering in construction: A critical review of issues, problems and dilemmas. *Construction Management and Economics*, *18*(2), 229–237. doi:10.1080/014461900370852

Bressan, P., Garlaschelli, L., & Barracano, M. (2003). Antigravity Hills are Visual Illusions, *American Psychological Society*. *Psychological Science*, *14*(5), 441–449. doi:10.1111/1467-9280.02451 PMID:12930474

Brinkley, I., & Crowley, E. (2017). From 'inadequate to 'outstanding': making the UK's skills system world class. Policy report. London: Academic Press.

British Standards Institution. (2000). *Occupational health and safety management systems - guidelines for the implementation of OHSAS 18001.*

British Standards Institution. (2013). PAS 1192-2: *Specification for information management for the capital/delivery phase of construction projects using building information modelling.*

British Standards Institution. (2015). PAS 1192-5: *Specification for security-minded building information modelling, digital built environments and smart asset management.*

British Standards Institution. (2019). *BS EN ISO 19650: Organisation and digitisation of information about buildings and civil engineering works, including building information modelling - Information management using building information modelling*. BSI.

Brockmann, C., Brezinski, H., & Erbe, A. (2016). Innovation in construction megaprojects. *Journal of Construction Engineering and Management*, *142*(11), 04016059. doi:10.1061/(ASCE)CO.1943-7862.0001168

Brokenshire, J. (2018). *Building a Safer Future: An Implementation Plan.* Presented to Parliament, Secretary of State for Housing, Communities and Local Government, APS Group on behalf of the Controller of Her Majesty's Stationery Office.

Bruno, S., de Fino, M., & Fatiguso, F. (2018). Historic Building Information Modelling: performance assessment for diagnosis-aided information modelling and management. Automation in Construction, 86, 256–276.

Bryde, D., Broquetas, M., & Volm, J. M. (2013). The project benefits of building information modelling (BIM). *International Journal of Project Management*, *31*(7), 971–980. doi:10.1016/j.ijproman.2012.12.001

BSI. (2013). *Design management systems. Part 4: Guide to managing design in construction.* BS 7000-4:2013.

BSI. (2020). *BSI Flex 8670 v1.0 Built environment – Overarching framework for competence of individuals – Specification*. BSI.

Budhedeo. (2016). *Issues And Challenges in Bringing ICT Enabled Education To Rural India.* http://ijsae.in/ijsaeems/index.php/ijsae/article/viewFile/1055/726

Buijs, A., & Silvester, S. (1996). Demonstration projects and sustainable housing: The significance of demonstration projects as a so called second generation steering instrument is examined by the use of case studies. *Building Research and Information*, *24*(4), 195–202. doi:10.1080/09613219608727529

Building and Construction Authority. (2013). *Singapore BIM Guide*. https://www.orenet.gov.sg/media/586132/Singapore-BIM-Guide_V2.pdf

buildingSMART Australasia. (2017). *National Building Information Modelling Initiative*. http://buildingsmart.org.au/wp-content/uploads/2014/03/ NationalBIMIniativeReport_6June2012.pdf

buildingSMART. (2017). *Individual Qualification Scheme Launched*. Retrieved from www.buildingsmart.org

buildingSMART. (2020a). *buildingSMART Professional Certification Program. The global benchmark for openBIM training*. Retrieved from https://education.buildingsmart.org/

buildingSMART. (2020b). *buildingSMART Professional Certification. Frequently Asked Questions*. Retrieved from education.buildingsmart.org

Buis, A. (2019). *A Degree of Concern: Why Global Temperatures Matter*. NASA's Global Climate Change Website.

Bureau of Labor Statistics (BLS). (2019). *National Census of Fatal Occupational Injuries in 2017*. https://www.bls.gov/news.release/pdf/cfoi.pdf

Bureau of Transportation Service. (2017). *Municipal Solid Waste and Construction & Demolition Debris*. United States Department of Transportation. https://www.bts.gov/archive/subject_areas/freight_transportation/faf/faf4/debris

Burgess, G., Jones, M., & Muir, K. (2018). *BIM in the UK house building industry: Opportunities and barriers to adoption*. University of Cambridge.

Burke, M. J., Sarpy, S. A., Smith-Crowe, K., Chan-Serafin, S., Salvador, R. O., & Islam, G. (2006). Relative effectiveness of worker safety and health training methods. *American Journal of Public Health*, *96*(2), 315–324. doi:10.2105/AJPH.2004.059840 PMID:16380566

Burritt, R. L., Herzig, C., Schaltegger, S., & Viere, T. (2019). Diffusion of environmental management accounting for cleaner production: Evidence from some case studies. *Journal of Cleaner Production*, *224*, 479–491. doi:10.1016/j.jclepro.2019.03.227

Burr, K. L. (2009). Creative course design: A study in student-centred course development for a sustainable building/BIM class. In *Proceedings of the 45th Annual Conference by Associated Schools of Construction*. University of Florida.

Bush, R., & Robinson, M. (2018). *Developing a BIM Competency Framework. Research and key principles*. Scottish Future Trust. Retrieved from https://www.scottishfuturestrust.org.uk/

Bush, R., & Robinson, M. (2018). *Developing a BIM Competency Framework: Research & Key Principles*. Scottish Futures Trust.

Büyüközkan, G., & Göçer, F. (2018). Digital Supply Chain: Literature review and a proposed framework for future research. *Computers in Industry*, *97*, 157–177. doi:10.1016/j.compind.2018.02.010

Bygballe, L. E., & Sward, A. (2019). Collaborative project delivery models and the role of routines in institutionalizing partnering. *Project Management Journal*, *50*(2), 161–176. doi:10.1177/8756972818820213

Bynum, P., Issa, R. R. A., & Olbina, S. (2013). Building information modeling in support of sustainable design and construction. *Journal of Construction Engineering and Management*, *139*(1), 24–34.

Byrne, C. (2015). *Building information modelling in Australia, lessons from the UK*. Higher Education and Skills Group.

Cable, V., Fallon, M., & Higgins, D. (2013). Construction 2025. Industrial Strategy: Government and Industry in Partnership. In *Industrial Strategy: Government and Industry in Partnership*. Department for Business, Innovation & Skills.

Cachin, C., & Vukolić, M. (2017). *Blockchain consensus protocols in the wild.* arXiv preprint arXiv:1707.01873.

Cannella, S., Framinan, J. M., Bruccoleri, M., Barbosa-Póvoa, A. P., & Relvas, S. (2015). The effect of inventory record inaccuracy in information exchange supply chains. *European Journal of Operational Research, 243*(1), 120–129. doi:10.1016/j.ejor.2014.11.021

Carlton, D. (2019). *A competent workforce mitigates risk?* Internal report. Dynamic Knowledge. Unpublished.

Carroll, J. M., & Bellotti, V. (2015). Creating Value Together: The Emerging Design Space of Peer-to-Peer Currency and Exchange. In *Proceedings of the 18th ACM Conference on Computer Supported Cooperative Work & Social Computing* (pp. 1500–1510). New York, NY: ACM. 10.1145/2675133.2675270

Carson, B., Romanelli, G., Walsh, P., & Zhumaev, A. (2018). *Blockchain beyond the hype: What is the strategic business value?* McKinsey & Company.

Carson, S. J., Madhok, A., & Wu, T. (2006). Uncertainty, opportunism, and governance: The effects of volatility and ambiguity on formal and relational contracting. *Academy of Management Journal, 49*(5), 1058–1077. doi:10.5465/amj.2006.22798187

Casasayas, O., Hosseini, M. R., Edwards, D. J., Shuchi, S., & Chowdhury, M. (2021). Integrating BIM in Higher Education Programs: Barriers and Remedial Solutions in Australia. *Journal of Architectural Engineering, 27*(1), 05020010.

Casey, M. J. (2008). *Work in Progress: How Building Information Modelling may unify the Civil Engineering Curriculum.* Paper presented to the 38th ASEE/IEEE Frontiers in Education Conference, Saratoga Springs, NY. 10.1109/FIE.2008.4720644

Castello, S. (2020). *UK BIM Framework publishes new resource to support BIM training providers.* Retrieved from www.cdbb.cam.ac.uk

Cavada, N., Hunt, D. V. L., & Rogers, C. D. F. (2015). Do smart cities realise their potential for lower carbon dioxide emissions. *Proceedings of the Institution of Civil Engineers. Engineering Sustainability, 169*(6), 243–252. doi:10.1680/jensu.15.00032

CBA-Media. (2016). *Commonwealth Bank, Wells Fargo and Brighann Cotton pioneer landmark blockchain trade transaction.* Retrieved July 25, 2020, from https://www.commbank.com.au/guidance/newsroom/CBA-Wells-Fargo-blockchain-experiment-201610.html

CCSIC. (2020). *Flourishing Systems: Re-envisioning infrastructure as a platform for human flourishing.* Cambridge Centre for Digital Built Britain.

CDBB PUN. (2018). *Centre for Digital Built Britain: Network - Pedagogy and Upskilling Network (PUN).* https://www.cdbb.cam.ac.uk/CDBBResearchBridgehead/Networks/2018Network_PUN

CDBB. (2020). *The pathway towards an Information Management Framework – A commons for Digital Built Britain.* Centre for Digital Built Britain.

Cedefop. (2013). *Quantifying Skill Needs in Europe. Occupational Skills Profiles: Methodology and Application.* The European Centre for the Development of Vocational Training (Cedefop).

CEDEFOP. (2013). *Quantifying skill needs in Europe: Occupational skills profiles: methodology and application.* Author.

Census and Statistics Department, The Government of the Hong Kong Special Administrative Region. (2020). *Table 036: Gross Domestic Product (GDP) by major economic activity - percentage contribution to GDP at basic prices.* https://www.censtatd.gov.hk/hkstat/sub/sp250.jsp?tableID=036&ID=0&productType=8

Central Policy Unit. (2015). Research Report on Smart City. Government of Hong Kong Special Administrative Region, September 2015.

Central Pollution Control Board. (2017). *Guidelines on Environmental Management of Construction & Demolition (C & D) Wastes*. Ministry of Environment Forests & Climate Change. http://jkspcb.nic.in/WriteReadData/userfiles/file/cand%20D%20guidelines/guidelines%20for%20cd%20waste.pdf

Cerić, A. (2015). *Trust in construction projects*. Routledge. doi:10.4324/9781315747675

Ceschin, F. (2013). Critical factors for implementing and diffusing sustainable product-Service systems: Insights from innovation studies and companies' experiences. *Journal of Cleaner Production*, *45*, 74–88. doi:10.1016/j.jclepro.2012.05.034

Chakkol, M., Selviaridis, K., & Finne, M. (2018). The governance of collaboration in complex projects. *International Journal of Operations & Production Management*, *38*(4), 997–1019. doi:10.1108/IJOPM-11-2017-0717

Chan, L. M., & Yu, S. S. W. (2012). Managing the stress of Hong Kong expatriate construction professionals in mainland China: Focus group study exploring individual coping strategies and organisational support (Author abstract). *Journal of Construction Engineering and Management*, *138*(10), 1150–1160. doi:CO.1943-7862.0000533 doi:10.1061/(ASCE)

Chan, A. P. C., Chan, D. W. M., Chiang, Y. H., Tang, B. S., Chan, E. H. W., & Ho, K. S. K. (2004). Exploring critical success factors for partnering in construction projects. *Journal of Construction Engineering and Management*, *130*(2), 188–198. doi:10.1061/(ASCE)0733-9364(2004)130:2(188)

Chan, C. (2014). Barriers of Implementing BIM in Construction Industry from the Designers' Perspective: A Hong Kong Experience. *Journal of System and Management Sciences*, *4*, 24–40.

Change Agents. (2019). *What is BIM Excellence?* Retrieved from https://changeagents.com.au/bim-excellence.html

Chang, P.-Y., Hwang, M.-S., & Yang, C.-C. (2018). A Blockchain-Based Traceable Certification System. In S.-L. Peng, S.-J. Wang, V. E. Balas, & M. Zhao (Eds.), *Security with Intelligent Computing and Big-data Services* (pp. 363–369). Springer International Publishing. doi:10.1007/978-3-319-76451-1_34

Chang, S. E., Chen, Y. C., & Lu, M. F. (2019). Supply chain re-engineering using blockchain technology: A case of smart contract-based tracking process. *Technological Forecasting and Social Change*, *144*, 1–11. doi:10.1016/j.techfore.2019.03.015

Chapman, L. J. (2013). *Literature review and environmental scan for better translation of research to practice in residential construction*. Retrieved from https://www.cpwr.com/sites/default/files/publications/residential_construction_r2p_literature_review_chapman.pdf

Chapron, G. (2017). The environment needs cryptogovernance. *Nature*, *545*(7655), 403–405. doi:10.1038/545403a PMID:28541332

Chartered Institute of Building (CIOB). (2018). *CIOB Undergraduate Education Framework*. https://www.ciob.org/ciob-education-framework

Chartered Institute of Building (CIOB). (2020). *Professional Review Guidance – Assessment Criteria for Candidates*. Available from https://www.ciob.org/sites/default/files/Professional%20Review%20Guidance%20Notes%20for%20Industry%20Candidates_5.pdf

Chartered Institute of Building. (2018). *Undergraduate Education Framework*. Author.

Chaturvedi, K., Matheus, A., Nguyen, S. H., & Kolbe, T. H. (2019). Securing Spatial Data Infrastructures for Distributed Smart City applications and services. *Future Generation Computer Systems*, *101*, 723–736. doi:10.1016/j.future.2019.07.002

Che Ibrahim, C. K. I., & Belayutham, S. (2020). A knowledge, attitude and practices (KAP) study on prevention through design: A dynamic insight into civil and structural engineers in Malaysia. *Architectural Engineering and Design Management, 16*(2), 131–149. doi:10.1080/17452007.2019.1628001

Chegu Badrinath, A., & Hsieh, S.-H. (2019). Empirical Approach to Identify Operational Critical Success Factors for BIM Projects. *Journal of Construction Engineering and Management, 145*(3), 04018140. doi:10.1061/(ASCE)CO.1943-7862.0001607

Chen, H., & Samarasinghe, D. A. S. (2020). *The factors constraining the adoption of prefabrication in New Zealand residential construction sector.* Paper presented at the New Zealand Built Environment Research Symposium, Auckland, New Zealand.

Chen, B., Liu, A. M., & Hua, Y. (2017). An exploration of the interaction between BIM technology and the business process of a construction organization in BIM implementation. *WIT Transactions on the Built Environment, 169*, 177–189. doi:10.2495/BIM170171

Cheng, A. S., & Sturtevant, V. E. (2012). A framework for assessing collaborative capacity in community-based public forest management. *Environmental Management, 49*(3), 675–689. doi:10.100700267-011-9801-6 PMID:22209779

Cheng, J. C. P., & Lu, Q. (2015). A review of the efforts and roles of the public sector for BIM adoption worldwide. *Journal of Information Technology in Construction, 20*, 442–478.

Cheng, J.-C., Lee, N.-Y., Chi, C., & Chen, Y.-H. (2018). *Blockchain and smart contract for digital certificate. In 2018 IEEE international conference on applied system invention.* ICASI.

Cheng, R., Ryan, N., & Kelly, S. (2012). Exploring the perceived influence of safety management practices on project performance in the construction industry. *Safety Science, 50*(2), 363–369. doi:10.1016/j.ssci.2011.09.016

Chen, H. M., Chang, K. C., & Lin, T. H. (2016). A cloud-based system framework for performing online viewing, storage, and analysis on big data of massive BIMs. *Automation in Construction, 71*, 34–48. doi:10.1016/j.autcon.2016.03.002

Chen, I. J., & Paulraj, A. (2004). Towards a theory of supply chain management: The constructs and measurements. *Journal of Operations Management, 22*(2), 119–150. doi:10.1016/j.jom.2003.12.007

Chen, J., Su, Y., Si, H., & Chen, J. (2018). Managerial Areas of Construction and Demolition Waste: A Scientometric Review. *International Journal of Environmental Research and Public Health, 15*(11), 2350. Advance online publication. doi:10.3390/ijerph15112350 PMID:30356018

Chen, K., & Lu, W. (2021). University–industry collaboration for BIM education: Lessons learned from a case study. *Industry and Higher Education, 34*(6), 401–409.

Chen, P. H., Cui, L., Wan, C., Yang, Q., Ting, S. K., & Tiong, R. L. (2005). Implementation of IFC-based web server for collaborative building design between architects and structural engineers. *Automation in Construction, 14*(1), 115–128. doi:10.1016/j.autcon.2004.08.013

Chen, W. T., Merrett, H. C., Lu, S. T., & Mortis, L. (2019). Analysis of Key Failure Factors in Construction Partnering—A Case Study of Taiwan. *Sustainability, 11*(14), 3994. doi:10.3390u11143994

Chen, X., Liu, C., & Wu, I. (2018). A BIM-based visualisation and warning system for fi re rescue. *Advanced Engineering Informatics, 37*(April), 42–53. doi:10.1016/j.aei.2018.04.015

Chen, Y., Dib, H., & Cox, R. F. (2014). A measurement model of building information modelling maturity. *Construction Innovation, V14*(2), 186–209. doi:10.1108/CI-11-2012-0060

Chcung, F. K., Rihan, J., Tah, J., Duce, D., & Kurul, E. (2012). Early stage multi-level cost estimation for schematic BIM models. *Automation in Construction, 27*, 67–77. doi:10.1016/j.autcon.2012.05.008

Chiang, C., Chu, C., & Chou, C. (2015). BIM-enabled power consumption data management platform for rendering and analysis of energy usage patterns. *Procedia Engineering, 118*, 554–562. doi:10.1016/j.proeng.2015.08.480

Chiang, Y. H., Wong, F. K. W., & Liang, S. (2017). Fatal construction accidents in Hong Kong. *Journal of Construction Engineering and Management, 144*(3), 04017121. doi:10.1061/(ASCE)CO.1943-7862.0001433

Chileshe, N., & Dzisi, E. (2010). Benefits and barriers of construction health and safety management (HSM). *Perceptions of practitioners within design organisations*, (Unit 07), 1–5.

Chileshe, N., Jayasinghe, R. S., & Rameezdeen, R. (2019). Information flow-centric approach for reverse logistics supply chains. *Automation in Construction, 106*(June), 102858. doi:10.1016/j.autcon.2019.102858

Chileshe, N., Rameezdeen, R., Hosseini, M. R., Martek, I., Li, H. X., & Panjehbashi-Aghdam, P. (2018). Factors driving the implementation of reverse logistics: A quantified model for the construction industry. *Waste Management (New York, N.Y.), 79*, 48–57. doi:10.1016/j.wasman.2018.07.013 PMID:30343778

Choi, T. N. Y., Chan, D. W. M., & Chan, A. P. C. (2012). Potential difficulties in applying the Pay for Safety Scheme (PFSS) in construction projects. *Accident; Analysis and Prevention, 48*, 145–155. doi:10.1016/j.aap.2011.04.015 PMID:22664678

Chong, H. Y., Wong, J. S., & Wang, X. (2014). An explanatory case study on cloud computing applications in the built environment. *Automation in Construction, 44*, 152–162. doi:10.1016/j.autcon.2014.04.010

Chou, C., Chiang, C.-T., Wu, P.-Y., Chu, C.-P., & Lin, C.-Y. (2017). Spatiotemporal analysis and visualisation of power consumption data integrated with building information models for energy savings. *Resources, Conservation and Recycling, 123*, 219–229. doi:10.1016/j.resconrec.2016.03.008

Christensen, C., & Raynor, M. (2013). *The innovator's solution: Creating and sustaining successful growth.* Harvard Business Review Press.

Christensen, S., McNamara, J., & O'Shea, K. (2007). Legal and contracting issues in electronic project administration in the construction industry. *Structural Survey, 25*(3/4), 191–203.

Christidis, K., & Devetsikiotis, M. (2016). Blockchains and smart contracts for the Internet of Things. *IEEE Access: Practical Innovations, Open Solutions, 4*, 2292–2303. doi:10.1109/ACCESS.2016.2566339

Chua, D. K. H., Kog, Y. C., & Loh, P. K. (1999). Critical success factors for different project objectives. *Journal of Construction Engineering and Management, 125*(3), 142–150. doi:10.1061/(ASCE)0733-9364(1999)125:3(142)

Chu, M., Matthews, J., & Love, P. E. (2018). Integrating mobile building information modelling and augmented reality systems: An experimental study. *Automation in Construction, 85*, 305–316. doi:10.1016/j.autcon.2017.10.032

CIC. (2013). *BIM planning guide for facility owners. version 1.02.* Retrieved from https://bim.psu.edu/resources/owner/bim_planning_guide_for_facility_owners-version_2.0.pdf

CIC. (2019). *Raising the Bar Interim Report: Improving Competence Building a Safer Future. Industry Response Group Steering Group on Competence for Building a Safer Future*. Construction Industry Council UK.

CIC. (2020). *Setting the Bar Final Report: Improving Competence Building a Safer Future. Industry Response Group Steering Group on Competence for Building a Safer Future*. Construction Industry Council.

CIDB. (2016). *Construction Industry Transformation Programme (CITP) 2016-2020.* Construction Industry Development Board Malaysiafile. doi:10.11113/jt.v78.4056

Ciribini, A. L. C., Pasini, D., Tagliabue, L. C., Manfren, M., Daniotti, B., Rinaldi, S., & De Angelis, E. (2017). Tracking Users' Behaviors through Real-time Information in BIMs: Workflow for Interconnection in the Brescia Smart Campus Demonstrator. *Procedia Engineering*, *180*, 1484–1494. doi:10.1016/j.proeng.2017.04.311

Clarysse, B., Wright, M., Bruneel, J., & Mahajan, A. (2014). Creating value in ecosystems: Crossing the chasm between knowledge and business ecosystems. *Research Policy*, *43*(7), 1164–1176. doi:10.1016/j.respol.2014.04.014

Clevenger, C., & Rush, M. (2011). Collaborating with Industry and Facilities Management to teach BIM. In *EcoBuild America 2011: BIM Academic Forum*. National Institute of Building Sciences.

Client Leadership Council (CLC). (2019, June). *Future Skills Report*. https://www.constructionleadershipcouncil.co.uk/wp-content/uploads/2019/06/CLC-Skills-Workstream_Future-Skills-Report_June-2019_A4-Print-Version.pdf

Client Leadership Council (CLC). (2020, June 1). *Roadmap to Recovery: An Industry Recovery Plan for the UK Construction Sector.* https://cdn.ca.emap.com/wp-content/uploads/sites/8/2020/05/Roadmap-to-Recovery.pptx

Climo, D. (2018). VR and safety training. *BRANZ Build*, 167.

Coates, P., Arayici, Y., Koskela, L., & Usher, C. (2010). The changing perception in the artefacts used in the design practice through BIM adoption. In P. S. Barrett, R. D. G. Amaratunga, R. P. Haigh, K. P. Keraminiyage, & C. Pathirage (Eds.), *2010 CIB World Congress Proceedings*. International Council for Research and Innovation in Building and Construction.

Coates, P., Arayici, Y., Koskela, L., Kagioglou, M., Usher, C., & O'Reilly, K. (2010). The key performance indicators of the BIM implementation process. *Proceedings of the ICCBE Conference*.

Cobbett, S., & Snelgrove-Clarke, E. (2016). Virtual versus face-to-face clinical simulation in relation to student knowledge, anxiety, and self-confidence in maternal-newborn nursing: A randomized controlled trial. *Nurse Education Today*, *45*, 179–184. doi:10.1016/j.nedt.2016.08.004 PMID:27537670

Comarch. (2016). *Smart City: Driven by Innovation*. https://smartcity.comarch.com/?gclid=EAIaIQobChMI2NmLoK11AIVoxbTCh1kXAcXEAAYASAAEgJnwPD_BwE#about

Comiskey, D., Alexander, G., Eadie, R., Hamill, M., McKane, M., & Weatherup, R. (2013). BIM Education - A Multidisciplinary Analysis. *Proceedings of the CITA BIM Gathering Conference*.

Comiskey, D., McKane, M., Eadie, R., & Goldberg, D. (2015). Providing Collaborative Education with an International Dimension. An Ulster University and Pennsylvania State University Case Study. *Proceedings of the CITA BIM Gathering Conference*.

Comiskey, D., McKane, M., O'Shea, E., Hughes, J., McNiff, S. and Eadie, R. (2016), Collaborative and multidiscipline working - From theory to practice in 48 hours: A case study from BIM Region Northern Ireland. *International Journal of 3-D Information Modelling*, *5*(2), 55-71.

Computer Integrated Construction Research Program. (2011). *BIM Project Execution Planning Guide – Version 2.1.* The Pennsylvania State University.

Conceição, P. (2019). *Human Development Report 2019. Beyond income, beyond averages, beyond today: Inequalities in human development in the 21st century.* The United Nations Development Programme. http://hdr.undp.org/en/content/human-development-report-2019

Construction Industry Council Hong Kong. (2014). *Roadmap for BIM strategic implementation in Hong Kong's Construction Industry.* Author.

Construction Industry Council UK. (2013). *Growth through BIM.* Author.

Construction Industry Review Committee Hong Kong Government. (2001). Construction for Excellence Report of Construction Industry Review Committee. *Construction Management and Economics*, *19*(7), 741–751. doi:10.1080/01446190110067037

Construction Manager. (2020, June 1). *Annual BIM survey: Poor digital skills hold back adoption*. Charted Institute of Building (CIOB). https://www.constructionmanagermagazine.com/annual-bim-survey-poor-digital-skills-hold-back-adoption/

Cook, D., & Chatterjee, P. (2015). Our Changing World: Let's Be Ready. London: Academic Press.

Cook, D., & Chatterjee, P. (2015). *RICS Futures: Our Changing World*. RICS.

Cooney, R., & Sohal, A. (2004). Teamwork and total quality management: A durable partnership. *Total Quality Management & Business Excellence*, *15*(8), 1131–1142. doi:10.1080/1478336042000255442

Cortés-Pérez, J. P., Cortés-Pérez, A., & Prieto-Muriel, P. (2020). BIM-integrated management of occupational hazards in building Construction and maintenance. *Automation in Construction*, *113*, 103–115.

Costa, A. A., & Grilo, A. (2015). BIM-Based E-Procurement: An Innovative Approach to Construction E-Procurement. *The Scientific World Journal*. doi:10.1155/2015/905390

Costa, A. A., & Tavares, L. V. (2012). Social e-business and the Satellite Network model: Innovative concepts to improve collaboration in construction. *Automation in Construction*, *22*, 387–397. doi:10.1016/j.autcon.2011.09.017

Costin, A., Adibfar, A., Hu, H., & Chen, S. S. (2018). Building Information Modeling (BIM) for transportation infrastructure – Literature review, applications, challenges, and recommendations. *Automation in Construction*, *94*(July), 257–281. doi:10.1016/j.autcon.2018.07.001

Cousin, G. (2006). An introduction to threshold concepts. *Planet*, *17*, 4–5.

Crawford, R. H., Mathur, D., & Gerritsen, R. (2017). Barriers to Improving the Environmental Performance of Construction Waste Management in Remote Communities. *Procedia Engineering*, *196*, 830–837. doi:10.1016/j.proeng.2017.08.014

Cressie, N. (1990). The Origins of Kriging. *Mathematical Geology*, *22*(3), 239–252. doi:10.1007/BF00889887

Creswell, J. W. (2014). *Research design: qualitative, quantitative, and mixed methods approach* (4th ed.). SAGE Publications.

Creswell, J. W., & Poth, C. N. (2017). *Qualitative inquiry and research design: Choosing among five approaches*. Sage Publications.

Crocker, R., Saint, C., Chen, G., & Tong, Y. (2018). *Unmaking waste in production and consumption: towards the circular economy*. Emerald Publishing Limited. https://find.library.unisa.edu.au/primo-explore/fulldisplay?vid=UNISA&search_scope=All_Resources&docid=UNISA_ALMA21167209160001831

Crumpton, A., & Miller, B. (2008). Building Information Modeling: State of the A&D Industry and BIM integration into Design Education. *Proc., Interior Design Educator's 2008 Annual Conference*.

CSG & NZCIC. (2017). *The NZ construction industry – Briefing to Incoming Government, 2017: How a ministerial champion will benefit all New Zealanders*. Retrieved from https://nzcic.co.nz/wp-content/uploads/2018/10/Construction-Strategy-Brochure-V10.pdf

Dahlbo, H., Bachér, J., Lähtinen, K., Jouttijärvi, T., Suoheimo, P., Mattila, T., Sironen, S., Myllymaa, T., & Saramäki, K. (2015). Construction and demolition waste management – a holistic evaluation of environmental performance. *Journal of Cleaner Production*, *107*, 333–341. doi:10.1016/j.jclepro.2015.02.073

Dallasega, P., Rauch, E., & Linder, C. (2018). Industry 4.0 as an enabler of proximity for construction supply chains: A systematic literature review. *Computers in Industry*, *99*, 205–225. doi:10.1016/j.compind.2018.03.039

Damelio, R. (2011). *The basics of process mapping*. CRC (Taylor & Francis) Press.

Darabseh, M., & Martins, J. P. (2020). Risks and Opportunities for Reforming Construction with Blockchain: Bibliometric Study. *Civil Engineering Journal*, *6*(6), 1204–1217. doi:10.28991/cej-2020-03091541

Darko, A., Chan, A. P., Adabre, M. A., Edwards, D. J., Hosseini, M. R., & Ameyaw, E. E. (2020). Artificial intelligence in the AEC industry: Scientometric analysis and visualization of research activities. *Automation in Construction*, *112*, 103081. doi:10.1016/j.autcon.2020.103081

Das, M., Luo, H., & Cheng, J. C. (2020). Securing interim payments in construction projects through a blockchain-based framework. *Automation in Construction*, *118*, 103211. doi:10.1016/j.autcon.2020.103284

Dave, B. (2018). A framework for integrating BIM and IoT through open standards. Automation in Construction, 95, 35–45. doi:10.1016/j.autcon.2018.07.022

Dave, B., Buda, A., Nurminen, A., & Främling, K. (2018). A framework for integrating BIM and IoT through open standards. *Automation in Construction*, *95*, 35–45.

David, M., Aubry, A., & Derigent, W. (2018). Towards energy efficient buildings: How ICTs can convert advances? *IFAC-PapersOnLine*, *51*(11), 758–763. doi:10.1016/j.ifacol.2018.08.410

Davies, A., & Brady, T. (2000). Organisational capabilities and learning in complex product systems: Towards repeatable solutions. *Research Policy*, *29*(7-8), 931–953. doi:10.1016/S0048-7333(00)00113-X

Davies, A., Dodgson, M., & Gann, D. (2016). Dynamic capabilities in complex projects: The case of London Heathrow Terminal 5. *Project Management Journal*, *47*(2), 26–46. doi:10.1002/pmj.21574

Davies, A., & Hobday, M. (2005). *The business of projects: managing innovation in complex products and systems*. Cambridge University Press. doi:10.1017/CBO9780511493294

Davis, R., Wilen, A., Crane, T., Bryer, L., Ward, D., Pottier, F., & Blackwell, M. (2015). UK Industry Performance Report-2015. *Construction Industry Key Performance Indicators*.

DCOM. (2019, December 1). *A Call to Digitise. Digitisation of Requirements, Regulations, Standards, and Compliance Checking in the Built Environment*. https://www.dcom.org.uk/wp-content/uploads/2019/10/DCOM-Consultation-Note.pdf

de Andrade, A. R., Gomes, M. C., & Duque, J. (2018). New Approach for Optimization of Construction and Demolition Waste Management: Application to the Lisbon Metropolitan Area. In A. I. F. Vaz, J. P. Almeida, J. F. Oliveira, & A. A. Pinto (Eds.), *Operational Research* (Vol. 223, pp. 1–10). doi:10.1007/978-3-319-71583-4_1

De Smith, M. J., Goodchild, M. F., & Longley, P. (2007). *Geospatial Analysis: A Comprehensive Guide to Principles, Techniques and Software Tools*. Troubador Publishing Ltd.

Deacon, C., & Smallwood, J. (2016). The Effect of the Integration of Design, Procurement, and Construction Relative to Health and Safety (H&S) [Paper presentation]. The 2016 International Sustainable Ecological Engineering Design Society (SEEDS) Conference, Leeds Beckett University.

DEC. (2020). *Design Engineer Construct!®: Educating the Future of Construction. Inspiring a new generation of Built Environment professionals*. https://designengineerconstruct.com/what-is-dec/

DeFillippi, R., & Sydow, J. (2016). Project networks: Governance choices and paradoxical tensions. *Project Management Journal*, *47*(5), 6–17. doi:10.1177/875697281604700502

DEFRA. (2020). *Waste Management Plan for England*. Department for Environment Food and Rural Affairs. www.gov.uk/government/publications

Delgado-Mohatar, O., Tolosana, R., Fierrez, J., & Morales, A. (2020). *Blockchain in the Internet of Things: Architectures and Implementation*. arXiv preprint arXiv:2004.05817.

Demian, P., & Walters, D. (2013). The advantages of information management through building information modelling. *Construction Management and Economics*, *32*(12), 1153–1165.

Deng, Y., Gan, V. J., Das, M., Cheng, J. C., & Anumba, C. (2019). Integrating 4D BIM and GIS for Construction Supply Chain Management. *Journal of Construction Engineering and Management*, *145*(4), 04019016. doi:10.1061/(ASCE)CO.1943-7862.0001633

Deniz, G. O. (2018). Emerging CAD and BIM trends in AEC education: An analysis from students' perspective. *Journal of Information Technology in Construction*, *V23*, 138–156.

Denzer, A., & Hedges, K. (2008). From CAD to BIM: Educational Strategies for the Coming Paradigm Shift. *AEI*, *2008*, 1–11. doi:10.1061/41002(328)6

Department for Business. Innovation & Skills (BIS). (2013). UK Construction: An Economic Analysis of the Sector (BIS/13/958). Department for Business, Innovation & Skills.

Department of Education. (2016). *Realizing the potential of technology in education*. https://assets.publishing.service.gov.uk/government/uploads/system/uploads/attachment_data/file/791931/DfE-Education_Technology_Strategy.pdf

Department of State Development, Manufacturing, Infrastructure and Planning (DSDMIP), State of Queensland. (2018). *Digital Enablement for Queensland Infrastructure - Principles for BIM implementation*. http://www.dsdmip.qld.gov.au/resources/guideline/infrastructure/bim-principles.pdf

Designing Buildings Wiki. (2019). *Small and medium-sized enterprises SME*. https://www.designingbuildings.co.uk/wiki/Small_and_medium-sized_enterprises_SME

Dhillon, V., Metcalf, D., & Hooper, M. (2017). The hyperledger project. In *Blockchain enabled applications* (pp. 139–149). Springer. doi:10.1007/978-1-4842-3081-7_10

Dhull & Sakshi. (2017). *Online Learning*. https://www.researchgate.net/publication/332833360_Online_Learning

Di Marco, M. K., Taylor, J. E., & Alin, P. (2010). Emergence and role of cultural boundary spanners in global engineering project networks. *Journal of Management Engineering*, *26*(3), 123–132. doi:10.1061/(ASCE)ME.1943-5479.0000019

Di Serio, Á., Ibáñez, M. B., & Delgado Kloos, C. (2013). Impact of an augmented reality system on students' motivation for a visual art course. *Computers & Education*, *68*, 586–596. doi:10.1016/j.compedu.2012.03.002

Diamond, R. M. (2008). *Designing and Assessing Courses and Curricula. A Practical Guide* (3rd ed.). John Wiley & Sons, Inc.

Dick, W., Carey, L., & Carey, J. O. (2005). *The systematic design of instruction* (6th ed.). Allyn and Bacon.

Digital Engineering (DE). (2018). *Digital Engineering Framework*. https://www.transport.nsw.gov.au/digital-engineering/digital-engineering-framework

Ding, L., Wei, R., & Che, H. (2014). Development of a BIM-based Automated Construction System. *Procedia Engineering*, *85*, 123–131.

Dixit, M. K., Fernández-Solís, J. L., Lavy, S., & Culp, C. H. (2012). Need for an embodied energy measurement protocol for buildings: A review paper. *Renewable & Sustainable Energy Reviews*, *16*(6), 3730–3743. doi:10.1016/j.rser.2012.03.021

Dong, S., Li, H., & Yin, Q. (2018). Building information modeling in combination with real-time location systems and sensors for safety performance enhancement. *Safety Science*, *102*, 226–237. doi:10.1016/j.ssci.2017.10.011

Dowsett, R. M., & Harty, C. F. (2019). Assessing the implementation of BIM–an information systems approach. *Construction Management and Economics*, *37*(10), 551–566. doi:10.1080/01446193.2018.1476728

Drach-Zahavy, A., & Somech, A. (2010). From an intrateam to an interteam perspective of effectiveness: The role of interdependence and boundary activities. *Small Group Research*, *41*(2), 143–174. doi:10.1177/1046496409356479

Drexler, J. A. Jr, & Larson, E. W. (2000). Partnering: Why project owner-contractor relationships change (Statistical Data Included). *Journal of Construction Engineering and Management*, *126*(4), 293–297. doi:10.1061/(ASCE)0733-9364(2000)126:4(293)

Duan, H., Miller, T. R., Liu, G., & Tam, V. W. (2019). Construction debris becomes growing concerns of growing cities. *Waste Management, 83*, 1-5.

Duff, A. R., Robertson, I. T., Phillips, R. A., & Cooper, M. D. (1994). Improving safety by the modification of behaviour. *Construction Management and Economics*, *12*(1), 67–78. doi:10.1080/01446199400000008

Du, J., Zou, Z., Shi, Y., & Zhao, D. (2018). Zero latency: Real-time synchronization of BIM data in virtual reality for collaborative decision-making. *Automation in Construction*, *85*, 51–64. doi:10.1016/j.autcon.2017.10.009

Duke University. (n.d.). *Comparison: online learning environment vs. classroom-based learning environment.* Online teaching guide. https://sites.duke.edu/onlineguide/why-online-learning/comparison/

Dunmore, M. (2019, April/May). Westpac offers prefab choices in New Zealand-first scheme. *New Zealand Construction News (NZCN)*, 17.

Dunphy, P., & Petitcolas, F. A. (2018). A first look at identity management schemes on the blockchain. *IEEE Security and Privacy*, *16*(4), 20–29. doi:10.1109/MSP.2018.3111247

Dünser, A. (2008). Supporting Low-Ability Readers with Interactive Augmented Reality. In *Ann. Rev. CyberTherapy and Telemedicine: Changing the Face of Healthcare* (pp. 41–48). Interactive Media Institute. https://www.researchgate.net/publication/216867629_Supporting_Low_Ability_Readers_With_Interactive_Augmented_Reality

Dzidonu. (2010). *The Role of ICTs in Achieving the MDGs in Education: An Analysis of the Case of African Countries. Accra Ghana.* https://pdfs.semanticscholar.org/957b/2981c7f275f4e8b047f8695173805da0794e.pdf

Eadie, R., Browne, M., Odeyinka, H., McKeown, C., & McNiff, S. (2013). BIM implementation throughout the UK construction project lifecycle: An analysis. *Automation in Construction*, *36*, 145–151.

Eadie, R., Browne, M., Odeyinka, H., McKeown, C., & McNiff, S. (2013). BIM implementation throughout the UK Construction project lifecycle: An analysis. *Automation in Construction*, *36*, 145–151.

Eadie, R., Browne, M., Odeyinka, H., McKeown, C., & McNiff, S. (2015). A survey of current status of and perceived changes required for BIM adoption in the UK. *Built Environment Project and Asset Management*, *5*(1), 4–21. doi:10.1108/BEPAM-07-2013-0023

Eadie, R., Odeyinka, H., Browne, M., McKeown, C., & Yohanis, M. (2013b). An analysis of the drivers for adopting Building Information Modelling. *Journal of Information Technology in Construction*, *18*, 338–352.

Eastman, C. M., Eastman, C., Teicholz, P., Sacks, R., & Liston, K. (2011). *BIM handbook: A guide to building information modeling for owners, managers, designers, engineers and contractors*. John Wiley & Sons.

Eastman, C., Teicholz, P., Sacks, R., & Liston, K. (2008). *BIM handbook: a guide to building information modeling for owners, managers, designers, engineers, and contractors*. John Wiley & Sons. doi:10.1002/9780470261309

Eberhardt, L. C. M., Birgisdottir, H., & Birkved, M. (2019). Potential of Circular Economy in Sustainable Buildings. *IOP Conference Series. Materials Science and Engineering*, *471*, 092051. doi:10.1088/1757-899X/471/9/092051

Ed Tech Blog. (n.d.). *e-Classroom – The Future of Education*. www.edtechnyc.org/e-classroom-the-future-of-education

Edirisinghe, R., & Woo, J. (2020). *BIM-based performance monitoring for smart building management*. Facilities.

Edmondson, V., Cerny, M., Lim, M., Gledson, B., Lockley, S., & Woodward, J. (2018). A smart sewer asset information model to enable an 'Internet of Things' for operational wastewater management. *Automation in Construction*, *91*, 193–205. doi:10.1016/j.autcon.2018.03.003

Edmunds, H. (1999). *The Focus group research handbook*. NTC Business Books.

Egan, J. (1998). *Rethinking Construction – The Egan Report*. Department of Trade and Industry, HMSO.

Egan, J. (1998). *Rethinking Construction (the Egan Report)*. Tech. rep., Department of the Environment, Transport and the Regions, HMSO.

Eisenhardt, K. M. (1989). Building theories from case study research. *Academy of Management Review*, *14*(4), 532–550. doi:10.5465/amr.1989.4308385

Elghaish, F., & Abrishami, S. (2020). Developing a framework to revolutionise the 4D BIM process: IPD-based solution. *Construction Innovation*.

El-Sayegh, S., Romdhane, L., & Manjikian, S. (2020). A critical review of 3D printing in construction: Benefits, challenges, and risks. *Archives of Civil and Mechanical Engineering*, *20*(2), 1–25. doi:10.100743452-020-00038-w

Emmitt, S. (2010). Managing interdisciplinary projects: a primer for architecture, engineering and construction (1st ed.). Padstow: Spon Press, Taylor & Francis Group. doi:10.4324/9780203885338

Emmitt, S. (2016). BIM and the Future of Design Management. In Construction Manager's BIM Handbook. John Wiley & Sons Ltd. doi:10.1002/9781119163404.ch18

Emmitt, S. (2016). The construction design manager–a rapidly evolving innovation. *Architectural Engineering and Design Management, 12*(2), 138-148.

Emmitt, S., & Ruikar, K. (2013). *Collaborative Design Management*. Routledge. doi:10.4324/9780203819128

ENCORD. (2013a). *Construction Waste Measurement Protocol. A guide to measuring and reporting waste from construction activities*. European Network of Construction Companies for Research and Development (ENCORD). http://www.encord.org/wp-content/uploads/2013/06/2013-05-ENCORD-Waste-Measurement-Protocol-Issue-V1-Low-Res.pdf

ENCORD. (2013b). *Construction CO2 e Measurement Protocol. A guide to reporting against the Green House Gas Protocol for construction companies*. European Network of Construction Companies for Research and Development (ENCORD). https://ghgprotocol.org/sites/default/files/ENCORD-Construction-CO2-Measurement-Protocol-Lo-Res_FINAL_0.pdf

Engineers Ireland. (2014a). *Accreditation Criteria for Professional Titles*. Author.

Engineers Ireland. (2014b). *Guidelines for Chartered Engineer*. Author.

Engineers Ireland. (2014c). *Chartered Engineer – 2014 Regulations for the Registered Professional Title*. Author.

Enshassi, A., AbuHamra, L., & Mohamed, S. (2016). Barriers to implementation of building information modelling (BIM) in the Palestinian construction industry. *International Journal of Construction Project Management*, *8*(2), 103–123.

Enshassi, A., Mohamed, S., Mustafa, Z. A., & Mayer, P. E. (2007). Factors affecting labour productivity in building projects in the Gaza Strip. *Journal of Civil Engineering and Management*, *13*(4), 245–254. doi:10.3846/13923730.2007.9636444

Ensor, P. (2004). Contesting discourses in higher education curriculum restructuring in South Africa. *Higher Education*, *48*, 339–359.

Enterprise Ireland & CitA. (2016). *Leadership in Transition Survey Report*. Author.

Entwistle, N. (1997). Introduction: Phenomenography in Higher Education. *Higher Education Research & Development*, *16*(2), 127–134.

Environment and Climate Change Canada. (2018). *Canadian Environmental Sustainability Indicators: Solid waste diversion and disposal*. Environment and Climate Change Canada. https://www.canada.ca/content/dam/eccc/documents/pdf/cesindicators/solid-waste/2018/solid-waste-diversion-disposal-en.pdf

Erboz, G. (2018, July). How to define Industry 4.0. *The Main Pillars of Industry 4.0*.

Erdogan, B., Anumba, C., Bouchlaghem, D., & Nielsen, Y. (2008). Collaboration environments for construction: Implementation case studies. *Journal of Management Engineering*, *24*(4), 234–244. doi:10.1061/(ASCE)0742-597X(2008)24:4(234)

ESCO. (2020). *European Skills/Competences, Qualifications, and Occupations*. https://ec.europa.eu/esco/portal/home

European Commission. (1995). *Green Paper on Innovation*. Author.

European Parliament. (2014). Directive 2014/24/EU of the European parliament and of the council of 26 February 2014 on public procurement and repealing directive 2004/18/EC. *Official Journal of the European Union*. https://eur-lex.europa.eu/legal-content/EN/TXT/?uri=celex%3A32014L0024

European Union. (2010). *Health and safety at work in Europe (1999–2007) A statistical portrait*. Retrieved from http://ec.europa.eu/eurostat/documents/3217494/5718905/KS-31-09-290- EN.PDF/88eef9f7-c229-40deb1cd-43126bc4a946

Eurostat. (2019). *Generation of waste by waste category,hazardousness and NACE Rev. 2 activity*. European Commission. https://ec.europa.eu/eurostat/web/products-datasets/-/env_wasgen

Eurostat. (2020). *Waste Statistics*. https://ec.europa.eu/eurostat/statistics-explained/index.php/Waste_statistics

Evans, N., & Counsell, J. (2009). Web-mediated student peer group assessment of Building Information Modelling performance. Paper presented at VIZ09 – Visualisation in Built and Rural Environments, Barcelona, Spain. doi:10.1109/VIZ.2009.34

Eynon, J. (2013). *The Design Manager's Handbook*. Wiley-Blackwell. doi:10.1002/9781118486184

Fang, E., Palmatier, R. W., Scheer, L. K., & Li, N. (2008). Trust at different organisational levels. *Journal of Marketing*, *72*(2), 80–98. doi:10.1509/jmkg.72.2.80

Fang, Y., Cho, Y. K., Zhang, S., & Perez, E. (2016). Case study of BIM and cloud-enabled real-time RFID indoor localization for construction management applications. *Journal of Construction Engineering and Management*, *142*(7), 05016003.

Fapohunda, J. A., & Stephenson, P. P. (2014). *Knowledge, Attitude, and Perception (KAP) of Construction Participants towards Resource Waste in the Construction Industry*. Academic Press.

Farmer, M. (2016). *The Farmer Review of of the UK Construction Labour Model, Modernise or Die: Time to decide the industry's future.* Construction Leadership Council (CLC). Retrieved Feb 28, 2020, from https://www.cast-consultancy.com/wp-content/uploads/2016/10/Farmer-Review-1.pdf

Farmer, M. (2016). *The Farmer Review of the UK Construction Labour Model.* Construction Leadership Council.

Farzad, K., & Yusuf, A. (2012). Roadmap for implementation of BIM in the UK construction industry. *Engineering, Construction, and Architectural Management, 19*(6), 610–635. doi:10.1108/09699981211277531

Fellows & Liu. (2008). *Research methods for construction* (3rd ed.). Wiley Blackwell.

Fellows, R., & Liu, A. M. (2012). Managing organisational interfaces in engineering construction projects: Addressing fragmentation and boundary issues across multiple interfaces. *Construction Management and Economics, 30*(8), 653–671. doi:10.1080/01446193.2012.668199

Feng, Y. (2013). Effect of safety investments on safety performance of building projects. *Safety Science, 59*, 28–45.

Feng, Y., Zhang, S., & Wu, P. (2015). Factors influencing workplace accident costs of building projects. *Safety Science, 72*, 97–104.

Fernandez Gonzalez, A. (2018). *BIM strategy and education in chile.* Universidad de Chile.

Fialho, B. C., & Codinhoto, R. F. (2019). Trends in BIM and IoT for Reactive Maintenance. *Proceeding of the 36th CIB W78 2019 Conference.*

Finck, M. (2018). Blockchains and data protection in the european union. *Eur. Data Prot. L. Rev., 4*(1), 17–35. doi:10.21552/edpl/2018/1/6

Fiorentino, M., Monno, G., & Uva, A. (2009). Interactive "touch and see" FEM Simulation using Augmented Reality. *International Journal of Engineering Education, 25*, 1124–1128.

Firmin & Genesi. (2013). *History and Implementation of Classroom Technology.* https://core.ac.uk/download/pdf/82211949.pdf

Fitz, T., Theiler, M., & Smarsly, K. (2019). A metamodel for cyber-physical systems. *Advanced Engineering Informatics, 41*(May), 100930. doi:10.1016/j.aei.2019.100930

Flinn, S. (2018). Data-to-Learning-to-Action. In Optimizing Data-to-Learning-to-Action (pp. 29-47). Apress.

Flyvbjerg, B. (2006). Five misunderstandings about case-study research. *Qualitative Inquiry, 12*(2), 219–245. doi:10.1177/1077800405284363

FMI. (2007). *Eighth Annual Survey of Owners.* FMI.

Formoso, C., Isatto, E. L., & Hirota, E. H. (1999). Method for waste control in the building industry. *Proceedings IGLC.* https://www.researchgate.net/publication/228646326_Method_for_Waste_Control_in_the_Building_Industry/citation/download

Fortineau, V., Paviot, T., & Lamouri, S. (2019). Automated business rules and requirements to enrich product-centric information. *Computers in Industry, 104*, 22–33. doi:10.1016/j.compind.2018.10.001

Foton, V. R. (n.d.). *Teach abstruse concepts of science by adding Fun with fotonVR.!!* https://fotonvr.com/?gclid=Cj0KCQjwz4z3BRCgARIsAES_OVfkomaWeMp7Pj59byJAkP1vGl5mI687ZsTzPIcwtXMg908lZUsPHj8aAvCgEALw_wcB

Fox, J., & McEwan, B. (2017). Distinguishing technologies for social interaction: The perceived social affordances of communication channels scale. *Communication Monographs, 84*(3), 298–318. doi:10.1080/03637751.2017.1332418

Fox, S., & Hietanen, J. (2007). Interorganisational use of building information models: Potential for automational, informational and transformational effects. *Construction Management and Economics*, *25*(3), 289–296. doi:10.1080/01446190600892995

Fredrickson, J. W. (1986). The strategic decision process and organisational structure. *Academy of Management Review*, *11*(2), 280–297. doi:10.5465/amr.1986.4283101

Freeman. (2006). 'Best practice' in focus group research: making sense of different views. *Journal of Advanced Nursing*, *56*(5), 491-497. doi:10.1111/j.1365-2648.2006.04043.x

Friedman, H. S. (2019). Introduction to the Special Issue on Theory in Nonverbal Communication. *Journal of Nonverbal Behavior*, *43*(2), 107–109. doi:10.100710919-019-00299-x

Furneaux, C., & Kivvits, R. (2008). *BIM – implications for government.* CRC for Construction Innovation.

Future, M. (2019). *Smart Education, Goodbye to Classroom teaching.* https://www.morningfuture.com/en/article/2018/02/09/smart-education-andrea-cioffi/227/

Gallardo, A., Carlos, M., Peris, M., & Colomer, F. J. (2014). Methodology to design a municipal solid waste generation and composition map: A case study. *Waste Management (New York, N.Y.)*, *34*(11), 1920–1931. doi:10.1016/j.wasman.2014.05.014 PMID:25008298

Gálvez-Martos, J.-L., Styles, D., Schoenberger, H., & Zeschmar-Lahl, B. (2018). Construction and demolition waste best management practice in Europe. *Resources, Conservation and Recycling*, *136*, 166–178. doi:10.1016/j.resconrec.2018.04.016

Gambetti, R. C., Biraghi, S., Martinelli, E. M., & Tunisini, A. (2018). Supply Chain Communication. The International Encyclopedia of Strategic Communication, 1-6.

Ganah, A., & John, G. A. (2015). An overview of the feasibility of achieving level 2 building information modeling by 2016 in the UK. *Journal of Civil Engineering and Architecture*, *9*, 885–894.

Ganah, A., & John, G. A. (2015). Integrating building information modeling and health and safety for onsite construction. *Safety and Health at Work*, *6*(1), 39–45. doi:10.1016/j.shaw.2014.10.002 PMID:25830069

Gao, X., & Pishdad-Bozorgi, P. (2019). BIM-enabled facilities operation and maintenance: A review. Advanced Engineering Informatics, 39, 227–247.

García-Bañuelos, L., Ponomarev, A., Dumas, M., & Weber, I. (2017). Optimized Execution of Business Processes on Blockchain. In J. Carmona, G. Engels, & A. Kumar (Eds.), *Business Process Management* (pp. 130–146). Springer International Publishing. doi:10.1007/978-3-319-65000-5_8

Gardner, J. C., Hosseini, M. R., Rameezdeen, R., & Chileshe, N. (2014). Building information modelling (BIM) education in South Australia: Industry needs. *5th International Conference on Engineering, Project, and Production Management (EPPM 2014)*, 293-302.

Garner-Purkis, Z. (2018). Deaths in Construction rise 27% year on year. *Construction News*. https://www.Construction-news.co.uk/best-practice/health-and-safety/deathsin-Construction-rise-27-year-on-year/10032802.article

Gegenfurtner, A., Quesada-Pallarès, C., & Knog, M. (2014). Digital simulation-based training: A meta-analysis. *British Journal of Educational Technology*, *45*(6), 1097–1114. doi:10.1111/bjet.12188

Gerges, M., Austin, S., Mayouf, M., Ahiakwo, O., Jaeger, M., Saad, A., & Gohary, T. (2017). An investigation into the implementation of building information modelling in the Middle East. *Journal of Information Technology in Construction, 22*, 1–15.

Getuli, V., Capone, P., Bruttini, A., & Isaac, S. (2020). BIM-based immersive Virtual Reality for construction workspace planning: A safety-oriented approach. *Automation in Construction, 114*, 103160. doi:10.1016/j.autcon.2020.103160

Gha, A. (2019). BIM-integrated knowledge-based building management: Inspecting post- construction energy efficiency. Automation in Construction, 97, 13–28.

Ghazifard, A., & Chamchali, M. M. (2019). The use of fuzzy logic spatial modeling via GIS for landfill site selection (case study: Rudbar-Iran). *Environmental Earth Sciences, 78*(10), 1–16. doi:10.100712665-019-8296-y

Ghobakhloo, M. (2019). Industry 4.0, Digitization, and Opportunities for Sustainability. *Journal of Cleaner Production*, 119869.

Gier, D. M. (2008). *What Impact Does Using Building Information Modeling Have on Teaching Estimating to Construction Management Students?* [Paper presentation]. 44th Annual Conference of the Associated Schools of Construction, Auburn, AL.

Giesekam, J. J. (2016). *The contribution to UK climate mitigation targets from reducing embodied carbon in the construction sector* [PhD, University of Leeds]. https://core.ac.uk/download/pdf/77023765.pdf

Gioia, D. A., Corley, K. G., & Hamilton, A. L. (2012). Seeking qualitative rigor in inductive research: Notes on the Gioia methodology. *Organizational Research Methods, 16*(1), 15–31. doi:10.1177/1094428112452151

Gladstein, D. L. (1984). Groups in context: A model of task group effectiveness. *Administrative Science Quarterly, 29*(4), 499–517. doi:10.2307/2392936

Glaser, F. (2017). *Pervasive decentralisation of digital infrastructures: a framework for blockchain enabled system and use case analysis*. Academic Press.

Gledson, B. J. (2016). Hybrid project delivery processes observed in constructor BIM innovation adoption. *Construction Innovation, 16*(2), 229–246. doi:10.1108/CI-04-2015-0020

Gledson, B. J., & Greenwood, D. (2017). The adoption of 4D BIM in the UK construction industry: An innovation diffusion approach. *Engineering, Construction, and Architectural Management, 24*(6), 950–967. doi:10.1108/ECAM-03-2016-0066

Goh, Y. M., & Chua, S. (2015). Knowledge, attitude and practices for design for safety: A study on civil & structural engineers. *Accident; Analysis and Prevention*. Advance online publication. doi:10.1016/j.aap.2015.09.023 PMID:26477455

Goldstein, I. L. (1980). Training in work organizations. *Annual Review of Psychology, 31*(1), 229–272. doi:10.1146/annurev.ps.31.020180.001305

Goodchild, M. (2016). GIS in the Era of Big Data. *Cybergeo: European Journal of Geography*, 1-7. https://journals.openedition.org/cybergeo/27647

Good, N., Ellis, K. A., & Mancarella, P. (2017). Review and classification of barriers and enablers of demand response in the smart grid. *Renewable & Sustainable Energy Reviews, 72*(January), 57–72. doi:10.1016/j.rser.2017.01.043

Gough, D., Oliver, S., & Thomas, J. (2012). *Systematic reviews* (1st ed.). SAGE.

Gould, F. E., & Joyce, N. (2009). *Construction project management* (3rd ed.). Pearson Prentice Hall.

Government, H. M. (2014, July). *New Models of Construction Procurement.* Retrieved May 18, 2020, from https://assets.publishing.service.gov.uk/government/uploads/system/uploads/attachment_data/file/325011/New_Models_of_Construction_Procurement_-_Introduction_to_the_Guidance_-_2_July_2014.pdf

Government, H. M. (2016). *Government Construction Strategy: 2016-2020.* Retrieved March 03, 2020, from https://assets.publishing.service.gov.uk/government/uploads/system/uploads/attachment_data/file/510354/Government_Construction_Strategy_2016-20.pdf

Government, H. M. (2016). *Press release: Launch of Digital Built Britain.* Retrieve from https://www.gov.uk/government/news/launch-of-digital-built-britain

Government, H. M. (2013). *Industrial strategy: Government and industry in partnership.* Retrieved from https://assets.publishing.service.gov.uk/government/uploads/system/uploads/attachment_data/file/210099/bis-13-955-construction-2025-industrial-strategy.pdf

Grant, J. (2002). Learning needs assessment: Assessing the need. *British Medical Journal, 324*(7330), 156–159. doi:10.1136/bmj.324.7330.156 PMID:11799035

Gray, B. (1985). Conditions facilitating inter-organizational collaboration. *Human Relations, 38*(10), 911–936. doi:10.1177/001872678503801001

Grbich, C. (2013). *Qualitative Data Analysis - An Introduction.* SAGE Publications Limited.

Green, B. (2016). *Productivity in construction: creating a framework for the industry to thrive.* Chartered Institute of Building.

Grilo, A., & Jardim-Goncalves, R. (2010). Value proposition on interoperability of BIM and collaborative working environments. *Automation in Construction, 19*(5), 522–530. doi:10.1016/j.autcon.2009.11.003

Gruber, T. R. (1995). Toward principles for the design of ontologies used for knowledge sharing? *International Journal of Human-Computer Studies, 43*(5-6), 907–928. doi:10.1006/ijhc.1995.1081

Gu, N., & London, K. (2010). Understanding and facilitating BIM adoption in the AEC industry. *Automation in Construction, 19*(8), 988–999. doi:10.1016/j.autcon.2010.09.002

Gu, N., Singh, V., Taylor, C., London, K., & Brankovic, L. (2010). BIM adoption: expectations across disciplines. Handbook of research on building information modeling and construction informatics: concepts and technologies, 501-520.

Guedes, A. L. A., Alvarenga, J. C., Goulart, M. S. S., Rodriguez, M. V., & Soares, C. A. P. (2018). Smart Cities: The Main Drivers for Increasing the Intelligence of Cities. *Sustainability, 2018*(10), 3121. doi:10.3390u10093121

Guillen, A., Crespo, A., Gómez, J., González-Prida, V., Kobbacy, K., & Shariff, S. (2016). Building Information Modelling as an Asset Management Tool. *IFAC-PapersOnLine, 49*(28), 191–196. doi:10.1016/j.ifacol.2016.11.033

Gulati, R. (2007). Silo busting: How to execute on the promise of customer focus. *Harvard Business Review, 85*(5), 98–108. PMID:17494254

Gumucio, S. (2011). *Data collection Quantitative methods The KAP survey model.* Knowledge, Attitude & Practices.

Gu, N., & London, K. (2010). Understanding and facilitating BIM adoption in the AEC industry. *Automation in Construction, 19*(8), 988–999.

Gunduz, M., Isikdag, U., & Basaraner, M. (2017). *Integration of BIM, Web maps and IOT for supporting comfort analysis.* ISPRS Annals of the Photogrammetry, Remote Sensing and Spatial Information Sciences. doi:10.5194/isprs-annals-IV-4-W4-221-2017

Guo, L., Li, H., Chan, G., & Skitmore, M. (2012). Using game technologies to improve the safety of construction plant operations. *Accident; Analysis and Prevention*, *48*, 204–213. doi:10.1016/j.aap.2011.06.002 PMID:22664683

H.K.G. Labour Department. (2002). *Code of Practice on Safety Management.* Author.

H.K.G. Labour Department. (2014). *Occupational Safety and Health Statistics, 2013.* Author.

Hackett, M., & Statham, G. (Eds.). (2016). *The aqua group guide to procurement, tendering and contract administration* (2nd ed.). Wiley Blackwell, John Wiley & Sons.

Hackitt, D. J. (2018). *Building a Safer Future. Independent Review of Building Regulations and Fire Safety: Final (Hackitt) Report.* Tech. rep.

Hackitt, J. (2018). *Building a Safer Future Independent Review of Building Regulations and Fire Safety: Final Report.* Retrieved Feb 28, 2020, from https://assets.publishing.service.gov.uk/government/uploads/system/uploads/attachment_data/file/707785/Building_a_Safer_Future_-_web.pdf

Hackitt, J. (2018). *Building a Safer Future. Independent Review of Building Regulations and Fire Safety: Final Report.* London: Academic Press.

Hackney, K. J., & Perrewé, P. L. (2018). A review of abusive behaviors at work: The development of a process model for studying abuse. *Organizational Psychology Review*, *8*(1), 70–92. doi:10.1177/2041386618755724

Haddad, H., Bouyahia, Z., & Chaudhry, S. A. (2019). A Multiagent Geosimulation and IoT-based Framework for Safety Monitoring in Complex Dynamic Spatial Environments. Procedia Computer Science, 151, 527–534. doi:10.1016/j.procs.2019.04.071

Hadikusumo, B., & Rowlinson, S. (2002). Integration of virtually real Construction model and design-for-safety-process database. *Automation in Construction*, *11*(5), 501–509.

Hafsia, M., Monacelli, E., & Martin, H. (2018). Virtual Reality Simulator for Construction workers. In *Proceedings of the Virtual Reality International Conference-Laval Virtual* (pp. 1-7). Academic Press.

Hafsia, M., Monacelli, E., & Martin, H. (2018). Virtual Reality Simulator for Construction Workers. *Proceedings of the virtual Reality Conference – Labal Virtual*. Available from: https://dl.acm.org/citation.cfm?id=3234298

Hager, I., Golonka, A., & Putanowicz, R. (2016). 3D printing of buildings and building components as the future of sustainable construction. *Procedia Engineering*, *151*, 292–299. doi:10.1016/j.proeng.2016.07.357

Halcomb, E. J., Patterson, E., & Davidson, P. M. (2006). Evolution of practice nursing in Australia. *Journal of Advanced Nursing*, *55*(3), 388–390.

Ham, H., Teng, M. A., Wijaya, E., & Wikopratama, R. A. (2018). Integration Citizen' Suggestion System for the Urban Development: Tangerang City Case. *Procedia Computer Science*, *135*, 570–578. doi:10.1016/j.procs.2018.08.210

Hamilton City Council. (2019, Nov 14). *Pukete wastewater treatment plant*. Retrieved from BIM: https://www.biminnz.co.nz/casestudies/2019/hcc-pukete-wwtp

Hamilton, S., & Srivastava, N. (2020). *Member Webinar: DLT Labs & Walmart Improve Margins, Visibility & Transparency in the Supply Chain*. Video recording accessed on July 17, 2020 on https://www.youtube.com/watch?v=XqXL3kPsBj8

Hamma-adama, M., & Kouider, T. (2019). What are the barriers and drivers towards BIM adoption in Nigeria? In M. J. Skibniewski & M. Hajdu, (Eds.), *Proceedings of the 2019 Creative Construction Conference Budapest Hungary*. Budapest: Diamond Congress Ltd.

Hammond, G. P., & Jones, C. I. (2008). Embodied energy and carbon in construction materials *Proceedings of the Institution of Civil Engineers-Energy.*

Han, K. K., & Golparvar-Fard, M. (2017). Potential of big visual data and building information modeling for construction performance analytics: An exploratory study. *Automation in Construction, 73*, 184–198. doi:10.1016/j.autcon.2016.11.004

Hao, J., Yuan, H., & Shen, L. (2010). *Overview of research in construction and demolition waste management*. doi:10.1680/ener.2008.161.2.87

Hapep, A. M., & Al-Bakri, M. (2020). Comparison of Different DEM Generation Methods based on Open Source Datasets. *Journal of Engineering (Stevenage, England), 26*(1), 63–85.

Harðarson, A. (2012). Why the aims of education cannot be settled. *Journal of Philosophy of Education, 46*(2), 223–235.

Hardin, B., & McCool, D. (2015). *BIM and construction management: proven tools, methods, and workflows*. John Wiley & Sons.

Hardin, B., & McCool, D. (2015). *BIM and construction management: Proven tools, methods, and workflows*. John Wiley & Sons.

Hardy, C., Phillips, N., & Lawrence, T. B. (2003). Resources, knowledge and influence: The organizational effects of interorganizational collaboration. *Journal of Management Studies, 40*(2), 321–347. doi:10.1111/1467-6486.00342

Hargaden, V., Papakostas, N., Newell, A., Khavia, A., & Scanlon, A. (2019). The role of blockchain technologies in construction engineering project management. In *2019 IEEE International Conference on Engineering, Technology and Innovation ICE/ITMC*. IEEE. 10.1109/ICE.2019.8792582

Hargadon, A. B. (1998). Firms as knowledge brokers: Lessons in pursuing continuous innovation. *California Management Review, 40*(3), 209–227. doi:10.2307/41165951

Harry, N., Boone, J., & Boone, D. A. (2012). Analyzing Likert Data. *Journal of Extension, 50*(2), 1456–1466. doi:10.100711172-017-1908-3

Harty, C. (2008). Implementing innovation in construction: Contexts, relative boundedness, and actor-network theory. *Construction Management and Economics, 26*(10), 1029–1041. doi:10.1080/01446190802298413

HBR. (2016 November 3). *Why the Problem with Learning Is Unlearning*. https://hbr.org/2016/11/why-the-problem-with-learning-is-unlearning

HBR. (2016). *Why the Problem with Learning Is Unlearning*. Available from https://hbr.org/2016/11/why-the-problem-with-learning-is-unlearning

Health and Safety Executive (HSE). (2018). *Improving Health and Safety Outcomes in Construction Making the Case for Building Information Modelling (BIM).* https://www.hse.gov.uk/construction/lwit/assets/downloads/improving-health-and-safety-outcomes-in-construction.pdf

Health and Safety Executive. (2018). *Improving Health and Safety Outcomes in Construction Making the Case for Building Information Modelling (BIM).* https://www.hse.gov.uk/construction/lwit/assets/downloads/improving-health-and-safety-outcomes-in-construction.pdf

Heaton, J., & Parlikad, A. K. (2019). A conceptual framework for the alignment of infrastructure assets to citizen requirements within a Smart Cities framework. Cities, 90, 32–41. doi:10.1016/j.cities.2019.01.041

Hedges, K. E., & Denzer, A. S. (2008). *How Collaborative Architecture influences Structural Engineering Education.* Paper presented to the Structures Congress 2008, Vancouver, Canada. 10.1061/41016(314)147

Heiskanen, A. (2017 June 2). *Automating Construction Processes by Making BIMs Machine-Readable*. AEC Business. https://aec-business.com/automating-construction-processes-making-bims-machine-readable/

Heiskanen, A. (2017). The technology of trust: How the Internet of Things and blockchain could usher in a new era of construction productivity. *Construction Research and Innovation*, *8*(2), 66–70. doi:10.1080/20450249.2017.1337349

Helmi, H., & Pius, A. (2018). A Review of International Students' Attraction and Recruitment Strategies: Learning from UK Top Universities. *Scholars Journal of Economics, Business and Management*, *5*(8), 779–785.

Helmi, H., Pius, A., & Guenane, I. (2018a). UK Higher Education Institutions Role in Students' Skills Development for Employment. *Journal of Advances in Education and Philosophy*, *2*(3), 85–88.

Helmi, H., Pius, A., & Guenane, I. (2018b). The Importance of Research Informed Teaching in Higher Education: A Case Study of London College of Contemporary Arts (LCCA). *Journal of Advances in Education and Philosophy*, *2*(3), 89–94.

Hengl, T., Bajat, B., Blagojevic´, D., & Reuter, H. I. (2008). Geostatistical Modelling of Topography using Auxiliary Maps. *Computers & Geosciences*, *34*(12), 1886–1899. doi:10.1016/j.cageo.2008.01.005

Hermann, B. G., Kroeze, C., & Jawjit, W. (2007). Assessing environmental performance by combining life cycle assessment, multi-criteria analysis and environmental performance indicators. *Journal of Cleaner Production*, *15*(18), 1787–1796. doi:10.1016/j.jclepro.2006.04.004

Herr, C. M., & Fischer, T. (2019). BIM adoption across the Chinese AEC industries: An extended BIM adoption model. *Journal of Computational Design and Engineering*, *6*(2), 173–178. https://doi.org/10.1016/j.jcde.2018.06.001

Herrington, A., & Herrington, J. (2007). *Authentic mobile learning in higher education*. In AARE 2007 International Educational Research Conference, Fremantle, Australia. https://pdfs.semanticscholar.org/3c71/e83fb19005cf1c340ceb8af02fe22aa5fae0.pdf

Higginsonm, M., Nadeau, M.-C., & Raigopal, K. (2019). *Blockchain's Occam Problem. McKinsey Quarterly*. Insights on Financial Services.

Higgins, S., & Moseley, D. (2011). Teachers' Thinking about ICT and Learning: Believes and Outcomes. *Journal of Teacher Development*, *5*(2), 191–210. doi:10.1080/13664530100200138

Hjelseth, E. (2017). BIM understanding and activities. *WIT Transactions on the Built Environment*, 3–14. doi:10.2495/BIM170011

Hoang, N. H., Ishigaki, T., Kubota, R., Tong, T. K., Nguyen, T. T., Nguyen, H. G., Yamada, M., & Kawamoto, K. (2020). Waste generation, composition, and handling in building-related construction and demolition in Hanoi, Vietnam. *Waste Management (New York, N.Y.)*, *117*, 32–41. doi:10.1016/j.wasman.2020.08.006 PMID:32805599

Hoffman, W. H., & Schlosser, R. (2001). Success factors of strategic alliances in small and medium sized enterprises: An empirical survey. *Long Range Planning*, *34*(3), 357–381. doi:10.1016/S0024-6301(01)00041-3

Hogle, L. (2013). *Industry + academia: The perfect partnership*. Academic Press.

Holland, P. J., Allen, B., & Douglas-Lenders, R. C. (2014). *Building a better workforce: assessing the impact of the experiential and simulation-based learning in the Australian building and construction industry*. Monash University.

Holweg, M., Disney, S., Holmström, J., & Småros, J. (2005). Supply chain collaboration: Making sense of the strategy continuum. *European Management Journal*, *23*(2), 170–181. doi:10.1016/j.emj.2005.02.008

Holzer, R. H. (2015). BIM for procurement - procuring for BIM. In R. H. Crawford and A. Stephan (Eds.), *Living and Learning: Research for a Better Built Environment: 49th International Conference of the Architectural Science Association* (pp. 237-246). The Architectural Science Association and The University of Melbourne.

Hong, Y., Hammad, A., Zhong, X., Wang, B., & Akbarnezhad, A. (2020). Comparative modeling approach to capture the differences in BIM adoption decision-making process in Australia and China. *Journal of Construction Engineering and Management*, *146*(2), 04019099. doi:10.1061/(ASCE)CO. 1943-7862.0001746

Hore. (2017a). *BICP Global BIM Study – Lessons for Ireland's BIM Programme*. Academic Press.

Hore. (2017b). *BIM in Ireland 2017 report*. Academic Press.

Hosseini, M. R., Banihashemi, S., Chileshe, N., Namzadi, M. O., Udaeja, C., Rameezdeen, R., & McCuen, T. (2016). BIM adoption within Australian small and medium-sized enterprises (SMEs): An innovation diffusion model. *Construction Economics and Building*, *16*(3), 71–86.

Hosseini, M. R., Martek, I., Papadonikolaki, E., Sheikhkhoshkar, M., Banihashemi, S., & Arashpour, M. (2018). Viability of the BIM manager enduring as a distinct role: Association rule mining of job advertisements. *Journal of Construction Engineering and Management*, *144*(9), 04018085. doi:10.1061/(ASCE)CO.1943-7862.0001542

Hosseini, M. R., Pärn, E. A., Edwards, D. J., Papadonikolaki, E., & Oraee, M. (2018). Roadmap to mature BIM use in Australian SMEs: Competitive dynamics perspective. *Journal of Management Engineering*, *34*(5), 05018008. doi:10.1061/(ASCE)ME.1943-5479.0000636

Houssin, R., & Coulibaly, A. (2011). An approach to solve contradiction problems for the safety integration in innovative design process. *Computers in Industry*, *62*(4), 398–406. doi:10.1016/j.compind.2010.12.009

Howell, S., Rezgui, Y., & Beach, T. (2017). Integrating building and urban semantics to empower smart water solutions. *Automation in Construction*, *81*, 434–448. doi:10.1016/j.autcon.2017.02.004

HSE. (2018b). *Improving Health and Safety Outcomes in Construction. Making the Case of Building Information Modelling (BIM)*. Health and Safety Executive, RR 235. Available from: <https://www.hse.gov.uk/research/rrpdf/rr235.pdf>

Huang, Y. (2017). American Society for Engineering Education (ASEE). *ASEE Annual Conference*. https://www.asee.org/public/conferences/78/papers/17619/view

Huang, B., Wang, X., Kua, H., Geng, Y., Bleischwitz, R., & Ren, J. (2018). Construction and demolition waste management in China through the 3R principle. *Resources, Conservation and Recycling*, *129*, 36–44. doi:10.1016/j.resconrec.2017.09.029

Huang, X., & Hinze, J. (2006). Owner's Role in Construction Safety. *Journal of Construction Engineering and Management*, *132*(2), 164–173.

Huang, Y. (2018). A Review of Approaches and Challenges of BIM Education in Construction Management. *Journal of Civil Engineering and Architecture*, *12*, 401–407.

Huang, Y. (2018). Developing a Modular Advanced BIM Course in Construction Management. *Journal of Building Construction and Planning Research*, *6*(4), 198–214. doi:10.4236/jbcpr.2018.64014

Huckle, S., Bhattacharya, R., White, M., & Beloff, N. (2016). Internet of things, blockchain and shared economy applications. *Procedia Computer Science*, *98*, 461–466. doi:10.1016/j.procs.2016.09.074

Hughes, W., & Hughes, C. (2013). Professionalism and professional institutions in times of change. *Building Research and Information*, *41*(1), 28–38. doi:10.1080/09613218.2013.737096

Hui, W., Hu, P. J. H., Clark, T. H. K., Tam, K. Y., & Milton, J. (2008). Technology-assisted learning: A longitudinal field study of knowledge category, learning effectiveness and satisfaction in language learning. *Journal of Computer Assisted Learning*, *3*(24), 245–259. doi:10.1111/j.1365-2729.2007.00257.x

Hultgren, M., & Pajala, F. (2018). *Blockchain technology in construction industry: Transparency and traceability in supply chain*. Academic Press.

Hung, R. Y. Y., Yang, B., Lien, B. Y. H., McLean, G. N., & Kuo, Y. M. (2010). Dynamic capability: Impact of process alignment and organisational learning culture on performance. *Journal of World Business*, *45*(3), 285–294. doi:10.1016/j.jwb.2009.09.003

Hu, S., Corry, E., Curry, E., Turner, W. J. N., & O'Donnell, J. (2016). Building performance optimisation: A hybrid architecture for the integration of contextual information and time-series data. *Automation in Construction*, *70*, 51–61. doi:10.1016/j.autcon.2016.05.018

Husain, A. H., Razali, M. N., & Eni, S. (2018). Stakeholders' expectations on building information modelling (BIM) concept in Malaysia. *Property Management*, *36*(4), 400–422. doi:10.1108/PM-02-2017-0013

Hussain & Baggaley. (2017). *Bolton College used IBM Watson to build a virtual assistant that enhances teaching, learning, and information access*. https://www.ibm.com/blogs/watson/2017/08/bolton-college-uses-ibm-watson-ai-to-build-virtual-assistant-that-enhances-teaching-learning-and-assessment/

Hutcheson, P., Pearson, A. W., & Ball, D. F. (1996). Sources of technical innovation in the network of companies providing chemical process plant and equipment. *Research Policy*, *25*(1), 25–41. doi:10.1016/0048-7333(94)00805-1

Hyperledger. (2017). Hyperledger Announces Production-Ready Hyperledger Fabric 1.0. *Hyperledger Announces Production-Ready Hyperledger Fabric 1.0*. Retrieved from https://www.hyperledger.org/news

IBM Institute for Business Value. (2010). *Smaller cities for smarter growth*. Executive Report.

ICE. (2018). *ICE professional skills: Report by the Skills Review Group, 2018*. Institution of Civil Engineers.

Institute of Civil Engineers (ICE). (2018). *P13 Blueprint*. Available from http://www.p13.org.uk/wp-content/uploads/2018/06/P13-Blueprint-Web.pdf

International Data Corporation. (2020). IDC FutureScape: Worldwide IT industry 2020 predictions. *IDC FutureScape: Worldwide IT industry 2020 predictions*. Online and accessed on June 20, 2020 from www.idc.com/getdoc.jsp\?containerId=US45599219

International Labour Office - Occupational Safety and Health Branch. (2014). *A 5 step guide for employers, workers and their representatives on conducting workplace risk assessments*. Author.

International Labour Organisation. (1992). *Safety and health in Construction. An ILO code of practice*. https://www.ilo.org/wcmsp5/groups/public/---ed_protect/---protrav/---safework/documents/normativeinstrument/wcms_107826.pdf

International Labour Organisation. (2020). *World Statistics*. https://www.ilo.org/moscow/areas-of-work/occupational-safety-and-health/WCMS_249278/lang--en/index.htm

International Organization for Standardization. (2009). *ISO 31000 Risk Management - Principles and Guidelines*. Author.

Intsiful, J., Okyere, P. F., & Oase, S. (2003). Use of ICT for education, research, and development in Ghana: Challenges, opportunities, and potentials. *The Journal of Technology Transfer*, *25*, 329–342. http://www.ejds.org/meeting2003/ictp/papers/Intsiful.pdf

IPPR. (2017). *Skills 2030: Why the adult skills system is failing to build an economy that works for everyone*. London: IPPR.

Ireland, R., & Bruce, R. (2000). CPFR. *Supply Chain Management Review, 1*, 80-88.

Irish Government. (2014). *Construction Strategy 2020*. Author.

Irish Government. (2018). *Project Ireland 2040 National Development Plan 2018-2027*. Author.

Irish, BIM. (2017). *Council*. Roadmap to Digital Transition for Ireland's Construction Industry.

Irizarry, J., Karan, E. P., & Jalaei, F. (2013). Integrating BIM and GIS to improve the visual monitoring of construction supply chain management. *Automation in Construction*, *31*, 241–254. doi:10.1016/j.autcon.2012.12.005

Irvine, G., White, D., & Diffley, M. (2018). *Measuring Good Work: The Final Report of the Measuring Job Quality Working Group*. RSA Future Work Centre. Carnegie UK Trust. https://www.carnegieuktrust.org.uk/publications/measuring-good-work-the-final-report-of-the-measuring-job-quality-working-group/

Isaaks, E. H., & Srivastava, R. M. (1989). *An Introduction to Applied Geostatistics*. Oxford University Press.

Ishikawa, T. (2013). Geospatial thinking and spatial ability: An empirical examination of knowledge and reasoning in geographical science. *The Professional Geographer*, *65*(4), 636–646. doi:10.1080/00330124.2012.724350

Ismail, N. A. A., Chiozzi, M., & Drogemuller, R. (2017). An overview of BIM uptake in Asian developing countries. *AIP Conference Proceedings*, *1903*, 080008. https://doi.org/10.1063/1.5011596

ISO 19650, (2019), ISO BIM - Building Information Modelling - ISO 19650

ISO, B. (2018). ISO 19650-1:2018. BSI Standards Publication Organization and digitization of information about buildings and civil engineering works, including building information modelling (BIM)-Information management using building information modelling.

ISO, B. (2018). ISO 19650-2: 2018. Organization and Digitization of Information about Buildings and Civil Engineering Works, Including Building Information Modeling (BIM)-Information Management using Building Information Modeling. Part, 2.

ISO, B. (2020). ISO 19650-5: 2020. Organization and Digitization of Information about Buildings and Civil Engineering Works, Including Building Information Modeling (BIM)-Security-minded approach to information management.

Ivanov, D., Sethi, S., Dolgui, A., & Sokolov, B. (2018). A survey on control theory applications to operational systems, supply chain management, and Industry 4.0. *Annual Reviews in Control*, *46*, 134–147. doi:10.1016/j.arcontrol.2018.10.014

Jackson, S. (2017). Innovation incentives. *Construction Journal*, 25. Large Cities. *International Journal for Housing Science*, *36*(2), 99–107.

Jacobides, M. G., Cennamo, C., & Gawer, A. (2018). Towards a theory of ecosystems. *Strategic Management Journal*, *39*(8), 2255–2276. doi:10.1002mj.2904

Jaf, P. M. (2009). Optical Illusion and the Magnetic Hills in Koya Region (Kurdistan-Iraq), *Journal of Kirkuk University. Scientiae Studia*, *4*(1), 40–47.

Jain, S., Singhal, S., & Jain, N. K. (2019). Construction and demolition waste generation in cities in India: An integrated approach. *International Journal of Sustainable Engineering*, *12*(5), 333–340. doi:10.1080/19397038.2019.1612967

Jamil, R. (2018). Digital Elevation Modeling Analysis for Investigation of Gravity Hill Phenomena. *International Journal of 3-D Information Modeling, 7*(3), 25-38.

Jamil, R. (2020). Hydroelectricity consumption forecast for Pakistan using ARIMA modeling and supply-demand analysis for the year 2030. *Renewable Energy*, *154*, 1–10. doi:10.1016/j.renene.2020.02.117

Janis, I. L. (1972). *Victims of groupthink: A psychological study of foreign-policy decisions and fiascoes*. Academic Press.

Jhurree, V. (2005). Technology Integration in Education in Developing Countries: Guidelines to Policy Makers. *International Education Journal*, *6*(4), 467–483. https://ehlt.flinders.edu.au/education/iej/articles/v6n4/jhurree/paper.pdf

Jia, M. (2019). Adopting Internet of Things for the development of smart buildings: A review of enabling technologies and applications. Automation in Construction, 101, 111–126.

Jikeka, S., Smallwood, J., Emuze, F., & Simpeh, E. (2020, September 11-12). *Waste Management: The Case of Construction and Demolition Waste in Port Elizabeth.* International Conference of Sustainable Ecological Engineering Design for Society (SEEDS) 2019, Ipswich, UK. https://www.springer.com/gp/book/9783030443801

Jin, R., Zou, P. X., Li, B., Piroozfar, P., & Painting, N. (2019). Comparisons of students' perceptions on BIM practice among Australia, China and UK. *Engineering, Construction and Architectural Management.*

Jin, F., Liu, X. H., Chen, W. C., Fan, Z. L., & Wang, H. L. (2019). High initial (1, 3) Beta-D-Glucan concentration may be a predictor of satisfactory response of c aspofungin combined with TMP/SMZ for HIV-negative patients with moderate to severe Pneumocystis jirovecii pneumonia. *International Journal of Infectious Diseases*, *88*, 141–148. doi:10.1016/j.ijid.2019.08.015 PMID:31442630

Jirgensons, M., & Kapenieks, J. (2018). Blockchain and the future of digital learning credential assessment and management. *Journal of Teacher Education for Sustainability*, *20*(1), 145–156. doi:10.2478/jtes-2018-0009

Joannides, M. M., Olbina, S., & Issa, R. R. A. (2012). Implementation of Building Information Modeling into Accredited Programs in Architecture and Construction Education. *International Journal of Construction Education and Research*, *8*(2), 83–100.

Joannides, M. M., Olbina, S., & Issa, R. R. A. (2012). Implementation of Building Information Modelling into Accredited Programmes in Architecture and Construction Education. *International Journal of Construction Education and Research*, *8*(2), 83–100. doi:10.1080/15578771.2011.632809

Ju, C. R. S., & Rowlinson, S. (2014). Institutional determinants of construction safety management strategies of contractors in Hong Kong. *Construction Management and Economics*, *32*(7-8), 725–736. doi:10.1080/01446193.2014.909048

Jung, Y., & Joo, M. (2011). Building information modelling (BIM) framework for practical implementation. *Automation in Construction*, *20*(2), 126–133. doi:10.1016/j.autcon.2010.09.010

Jurva, R., & Matinmikko-Blue, M. (2017). *Stakeholder analysis for digital campus development with 5G micro operators. The ITS Europe Conference.* https://www.econstor.eu/bitstream/10419/205186/1/Jurva-Matinmikko-Blue.pdf

Jurva, R., Matinmikko-Blue, M., Niemeta, V., & Nenonen, S. (2020). *Architecture and Operational Model for Smart Campus Digital Infrastructure.* https://www.researchgate.net/publication/340131385_Architecture_and_Operational_Model_for_Smart_Campus_Digital_Infrastructure

Juszczyk, M. (2017). On the Neural Networks Applications in BIM-Based Macro Level Cost Analyzes in Construction Projects. *DEStech Transactions on Economics, Business and Management.*

Kabirifar, K., Mojtahedi, M., Wang, C. X., & Tam, V. W. Y. (2020). Construction and demolition waste management contributing factors coupled with reduce, reuse, and recycle strategies for effective waste management: A review. *Journal of Cleaner Production*, *263*, 121265. Advance online publication. doi:10.1016/j.jclepro.2020.121265

Kahn, K. B., & Mentzer, J. T. (1996). Logistics and interdepartmental integration. *International Journal of Physical Distribution & Logistics Management*, *26*(8), 6–14. doi:10.1108/09600039610182753

Kamardeen, I. (2011). E-OHS planning system for builders. *Architectural Science Review, 54*(1), 50–64.

Kamble, S. S., Gunasekaran, A., & Gawankar, S. A. (2018). Sustainable Industry 4.0 framework: A systematic literature review identifying the current trends and future perspectives. *Process Safety and Environmental Protection, 117*, 408–425. doi:10.1016/j.psep.2018.05.009

Kam, C., Fischer, M., Rinella, T., Mak, D., & Oldfield, J. (2014). Realising the Promise of BIM in Hong Kong's Construction Industry. *Journal of Hong Kong's Construction Industry*, (May), 29–33.

Kapp, K., & Stahl, B. (2018, January 10). *3 lessons from converting a face-to-face role-play into an online simulation.* Association for Talent Development. https://www.td.org/blog-post/3-lessons-from-converting-a-face-to-face-role-play-into-an-online-simulation

Karjalainen, A., Alha, K., & Jutila, S. (2006). *Give me time to think: Determining student workload in higher education; has been written as part of the project titled" five years, two degrees", funded by the ministry of education, 2004-2006.* Oulu University Press.

Kassem, M., Kelly, G., Dawood, N., Serginson, M., & Lockley, S. (2015). BIM in facilities management applications: a case study of a large university complex. *Built Environment Project and Asset Management, 5*(3), 261-277. Accessed 8 Dec. 2019 at: http://www.emeraldinsight.com/doi/pdfplus /10.1108/BEPAM-02-2014-0011

Kassem, M., Benomran, L., & Teizer, J. (2017). Virtual environments for safety learning in construction and engineering: Seeking evidence and identifying gaps for future research. *Visualization in Engineering, 5*(16), 2–15. doi:10.118640327-017-0054-1

Keengwe, J., Onchwari, G., & Wachira, P. (2011). Computer Technology Integration and Student Learning: Barriers and Promise. *Journal of Science Education and Technology, 17*(6), 560–570. doi:10.100710956-008-9123-5

Kelton, A. P. (2019). Internet of Things: A survey on machine learning-based intrusion detection approaches. *Computer Networks, 151*, 147–157. doi:10.1016/j.comnet.2019.01.023

Kensek, K., & Kahn, W. (2013). *Integration of Environmental Sensors with BIM Seven Case Studies. BESS-SB13 California: Advancing Towards Net Zero.*

Kerin, M., & Pham, D. T. (2019). A review of emerging industry 4. 0 technologies in remanufacturing. *Journal of Cleaner Production, 237*, 117805. doi:10.1016/j.jclepro.2019.117805

Kfir, S., & Fournier, C. (2019). DAML: The contract language of distributed ledgers. *Communications of the ACM, 62*(9), 48–54. doi:10.1145/3343046

Kho, D. D., Lee, S., & Zhong, R. Y. (2018). Big Data Analytics for Processing Time Analysis in an IoT-enabled manufacturing Shop Floor. *Procedia Manufacturing, 26*, 1411–1420. doi:10.1016/j.promfg.2018.07.107

Kidd, P. S., & Parshall, M. B. (2000). Getting the focus and the group: Enhancing analytical rigour in focus group research. *Qualitative Health Research, 10*(3), 293–308. doi:10.1177/104973200129118453 PMID:10947477

Kim, J. (2012). Use of BIM for effective visualisation teaching approach in construction education. *Journal of Professional Issues in Engineering Education and Practice, 138*(3), 214–223. doi:10.1061/(ASCE)EI.1943-5541.0000102

Kim, K., Cho, Y., & Zhang, S. (2016). Integrating work sequences and temporary structures into safety planning: Automated scaffolding-related safety hazard identification and prevention in BIM. *Automation in Construction, 70*, 128–142.

Kim, K., & Teizer, J. (2014). Automatic design and planning of scaffolding systems using building information modeling. *Advanced Engineering Informatics, 28*(1), 66–80.

Kim, M., & Chai, S. (2017). The impact of supplier innovativeness, information sharing and strategic sourcing on improving supply chain agility: Global supply chain perspective. *International Journal of Production Economics, 187*, 42–52. doi:10.1016/j.ijpe.2017.02.007

Kim, N. K., Rahim, N. F. A., Iranmanesh, M., & Foroughi, B. (2019). The role of the safety climate in the successful implementation of safety management systems. *Safety Science*, *118*, 48–56. doi:10.1016/j.ssci.2019.05.008

Kinnaird, C., Geipel, M., & Bew, M. A. (2017). Blockchain technology: how the inventions behind bitcoin are enabling a network of trust for the built environment. London: Arup.

Kinnane, O., & West, R. (2013). BIM introduction into the curriculum of Civil and Structural Engineering students: A project-based active learning approach. *Proceedings of the CITA BIM Gathering Conference*, 175-184.

Kitaoka, A. (2015). Slope Illusion (Magnetic Hill). Art and Its Role in the History: Between Durability and Transient-ISMS, 751-760.

Kiviniemi, M., Sulankivi, K., Kähkönen, K., Mäkelä, T., & Merivirta, M. L. (2011). BIM-based safety management and communication for building construction. *VTT Research Notes, 2597.*

Kline, A. R., Leite, F. L., & Isatto, E. L. (2020). *The perspectives of three university's building information modeling course development* [Paper presentation]. *ASEE Virtual Annual Conference, Virtual, Online.*

Knirsch, F., Unterweger, A., & Engel, D. (2018). Privacy-preserving blockchain-based electric vehicle charging with dynamic tariff decisions. *Computer Science-Research and Development*, *33*(1), 71–79. doi:10.100700450-017-0348-5

Knotten, V., Lædre, O., & Hansen, G. K. (2017). Building design management–key success factors. *Architectural Engineering and Design Management*, *13*(6), 479–493. doi:10.1080/17452007.2017.1345718

Kocaturk, T., & Kiviniemi, A. (2013). Challenges of integrating BIM in architectural education. *eCAADe 2013: Computation and Performance–Proceedings of the 31st International Conference on Education and Research in Computer Aided Architectural Design in Europe.*

Kosba, A., Miller, A., Shi, E., Wen, Z., & Papamanthou, C. (2016). Hawk: The blockchain model of cryptography and privacy-preserving smart contracts. In 2016 IEEE symposium on security and privacy (SP) (pp. 839-858). IEEE.

Koseoglu, O., & Nurtan-Gunes, E. T. (2018). Mobile BIM implementation and lean interaction on construction site. *Engineering, Construction, and Architectural Management*, *25*(10), 1298–1321. doi:10.1108/ECAM-08-2017-0188

Koskela, L. (1992). *Application of the new production philosophy to construction* (Vol. 72). Stanford University.

Kotzab, H., Darkow, I. L., Bäumler, I., & Georgi, C. (2019). Coordination, cooperation and collaboration in logistics and supply chains: A bibliometric analysis. *Production*, *29*, 29. doi:10.1590/0103-6513.20180088

Kugbeadjor, W., Suresh, S., & Renukappa, S. (2015). BIM awareness and readiness of postgraduate built environment students in west midlands universities, UK. *CIB International Conference Proceedings. Going North for Sustainability: Leveraging Knowledge and Innovation for Sustainable Construction and Development*, 23-25.

Kuiper, I., & Holzer, D. (2013). Rethinking the contractual context for Building Information Modelling (BIM) in the Australian built environment industry. *The Australasian Journal of Construction Economics and Building*, *13*(4), 1–17.

Ku, K., & Taiebat, M. (2011). BIM Experiences and Expectations: The Constructors' Perspective. *International Journal of Construction Education and Research*, *7*(3), 175–197. doi:10.1080/15578771.2010.544155

Kukulska-Hulme, A., Sharples, M., Milrad, M., Arnedillo-Sánchez, I., & Vavoula, G. (2009). Innovation in mobile learning: A European perspective. *International Journal of Mobile and Blended Learning*, *1*(1), 13–35. doi:10.4018/jmbl.2009010102

Kumar, B. (2015). *A practical guide to adopting BIM in construction projects*. Whittles Publishing. Retrieved from http://lib.myilibrary.com?ID=883866

Kundu & Nath. (2018). *Barriers to Utilizing ICT in Education in India with a Special Focus on Rural Areas*. https://www.researchgate.net/publication/327671947_Barriers_to_Utilizing_ICT_in_Education_in_India_with_a_Special_Focus_on_Rural_Areas

Kunieda, Y., Codinhoto, R., & Emmitt, S. (2019). Increasing the efficiency and efficacy of demolition through computerised 4D simulation. *Engineering, Construction, and Architectural Management*, *26*(10), 2186–2205. doi:10.1108/ECAM-11-2018-0492

Labour Department Hong Kong Government. (2015). *Occupational Safety and Health Statistics, 2014*. Author.

Labour Department Hong Kong. (2000). *Factories and Industrial Undertakings (Safety Management) Regulation*. Retrieved from http://www.legislation.gov.hk/blis_ind.nsf/WebView?OpenAgent&vwpg=CurAllEngDoc*59.31*0*59.31#59.31

Lahti, M., Shamsuzzoha, A. H. M., & Helo, P. (2009). Developing a maturity model for Supply Chain Management. *International Journal of Logistics Systems and Management*, *5*(6), 654–678. doi:10.1504/IJLSM.2009.024796

Lambe, C. J., Wittmann, C. M., & Spekman, R. E. (2001). Social exchange theory and research on business-to-business relational exchange. *Journal of Business-To-Business Marketing*, *8*(3), 1–36. doi:10.1300/J033v08n03_01

Lam, P. T. I., Yu, A. T. W., Wu, Z., & Poon, C. S. (2019). Methodology for upstream estimation of construction waste for new building projects. *Journal of Cleaner Production*, *230*, 1003–1012. doi:10.1016/j.jclepro.2019.04.183

Lamport, L., Shostak, R., & Pease, M. (1982). The Byzantine generals problem. *ACM Transactions on Programming Languages and Systems*, *4*(3), 382–401. doi:10.1145/357172.357176

Land, R., Rattray, J., & Vivian, P. (2014). Learning in the liminal space: A semiotic approach to threshold concepts. *Higher Education*, *67*(2), 199–217.

Langford, D., & Male, S. (2008). *Strategic Management in Construction*. Blackwell Science Ltd.

Larsson, J., & Holmström, I. (2007). Phenomenographic or phenomenological analysis: Does it matter? Examples from a study on anaesthesiologists' work. *International Journal of Qualitative Studies on Health and Well-being*, *2*(1), 55–64.

Lateef, F. (2010). Simulation-based learning: Just like the real thing. *Journal of Emergencies, Trauma and Shock*, *3*(4), 348–352. doi:10.4103/0974-2700.70743 PMID:21063557

Latham, S. M. (1994). *Constructing the Team.Joint Review of Procurement and Contractual Arrangements in the United Kingdom Construction Industry. Final Report.* Tech. rep., HMSO.

Latiffi, A. A., Mohd, S., & Rakiman, U. S. (2016). *Potential improvement of building information modeling (BIM) implementation in Malaysian construction projects*. In The 12th IFIP International Conference on Product Lifecycle Management, Doha.

Latiffi, A. A., Brahim, J., & Fathi, M. S. (2017). Building information modelling (BIM) after ten years: Malaysian construction players' perception of BIM. *IOP Conference Series. Earth and Environmental Science*, *81*(1), 012147. Advance online publication. doi:10.1088/1755-1315/81/1/012147

Lauer, T., & Franke, K. (2020, February). Behavioral Analysis of Information Exchange Digitalization in the Context of Demand Planning. In *International Conference on Intelligent Human Systems Integration* (pp. 973-979). Springer. 10.1007/978-3-030-39512-4_148

Lavis, J. N., Robertson, D., Woodside, J. M., McLeod, C. B., & Abelson, J.Knowledge Transfer Group. (2003). How can research organizations more effectively transfer research knowledge to decision makers? *The Milbank Quarterly*, *81*(2), 221–248. doi:10.1111/1468-0009.t01-1-00052 PMID:12841049

Lee, N., & Hollar, D. (2013). *Probing BIM education in construction engineering and management programs using industry perceptions*. Paper presented at the 49th ASC Annual Int. Conf. Proc., California Polytechnic State Univ., San Luis Obispo, CA.

Leedy, P. D., & Ormrod, J. E. (2010). Practical research: planning and design (9th ed., international ed.). Upper Saddle River, NJ: Pearson Education International.

Lee, J., Azamfar, M., & Singh, J. (2019). A blockchain enabled Cyber-Physical System architecture for Industry 4.0 manufacturing systems. *Manufacturing Letters*, *20*, 34–39. doi:10.1016/j.mfglet.2019.05.003

Lee, N., & Dossick, C. S. (2012). *Leveraging Building Information Modeling Technology in Construction Engineering and Management Education. American Society for Engineering Education*. ASEE.

Lee, N., Dossick, C. S., & Foley, S. P. (2013). Guideline for Building Information Modeling in Construction Engineering and Management Education. *Journal of Professional Issues in Engineering Education and Practice*, *139*(4), 266–274.

Lee, N., & Hollar, D. A. (2013). *Probing BIM education in construction engineering and management programs using industry perceptions* [Paper presentation]. *49th Annual International Conference Associated Schools of Construction*, Windsor, CA.

Lei, L., Zhou, Y., Luo, H., & Love, P. E. (2019). A CNN-based 3D patch registration approach for integrating sequential models in support of progress monitoring. *Advanced Engineering Informatics*, *41*, 100923. doi:10.1016/j.aei.2019.100923

Le, N. Q., Er, M., Sankaran, S., & Ta, N. B. (2020). Perspectives on BIM Profession of BIM Specialists and non-BIM Specialists: Case Study in Vietnam. In *CIGOS 2019, Innovation for Sustainable Infrastructure* (pp. 1223–1228). Springer. doi:10.1007/978-981-15-0802-8_196

Li, H., & Samarasinghe, D. A. S. (2020). *Analysis of factors affecting design Buildability in New Zealand construction Projects*. Paper presented at the New Zealand Built Environment Research Symposium, Auckland, New Zealand.

Li, J., Greenwood, D., & Kassem, M. (2018). *Blockchain in the built environment: analysing current applications and developing an emergent framework*. Academic Press.

Li, M., & Sond, W. (2020). IEEE standard for data format for blockchain systems. IEEE Std 2418.2-2020, Developed by the Standards Activities Board, IEEE Computer Society.

Li, X., Wu, P., & Yue, T. (2019). Integrating Building Information Modeling and Prefabrication Housing Production. Automation in Construction, 100, 46–60. doi:10.1016/j.autcon.2018.12.024

Liao, L., & Teo, E. A. L. (2019). Managing critical drivers for building information modelling implementation in the Singapore construction industry: An organisational change perspective. *International Journal of Construction Management*, *19*(3), 240–256. doi:10.1080/15623599.2017.1423165

Liao, Y., Deschamps, F., Loures, E. F. R., & Ramos, L. F. P. (2017). Past, present and future of Industry 4.0 - a systematic literature review and research agenda proposal. *International Journal of Production Research*, *55*(12), 3609–3629. doi:10.1080/00207543.2017.1308576

Li, C. Z., Xue, F., Li, X., Hong, J., & Shen, G. Q. (2018). An Internet of Things-enabled BIM platform for on-site assembly services in prefabricated construction. *Automation in Construction*, *89*, 146–161. doi:10.1016/j.autcon.2018.01.001

Li, H., Lu, M., Hsu, S. C., Gray, M., & Huang, T. (2015). Proactive behavior-based safety management for construction safety improvement. *Safety Science*, *75*, 107–117. doi:10.1016/j.ssci.2015.01.013

Li, J., Greenwood, D., & Kassem, M. (2019). Blockchain in the built environment and construction industry: A systematic review, conceptual models and practical use cases. *Automation in Construction*, *102*, 288–307. doi:10.1016/j.autcon.2019.02.005

Liker, J. K. (2004). *The Toyota Way: 14 Management Principles from the World's Greatest Manufacturer*. McGraw–Hill.

Liker, J. K., & Meier, D. (2006). *The Toyota Way Fieldbook: A practical guide for implementing Toyota's 4Ps*. McGraw–Hill.

Li, M., Yu, H., & Liu, P. (2018). An automated safety risk recognition mechanism for underground Construction at the pre-Construction stage based on BIM. *Automation in Construction*, *91*, 284–292.

Lim, S., Buswell, R. A., Le, T. T., Austin, S. A., Gibb, A. G., & Thorpe, T. (2012). Developments in construction-scale additive manufacturing processes. *Automation in Construction*, *21*, 262–268. doi:10.1016/j.autcon.2011.06.010

Lincoln, Y. S., & Guba, E. G. (2016). *The constructivist credo*. Routledge. doi:10.4324/9781315418810

Lindblad, H. (2013). *Study of the implementation process of BIM in construction projects*. http://kth.diva-portal.org/smash/get/diva2:633132/FULLTEXT01

Lines, B. C., Sullivan, K., Smithwick, J., & Mischung, J. (2015). Overcoming resistance to change in engineering and construction: Change management factors for owner organizations. *International Journal of Project Management*, *33*(5), 1170–1179. doi:10.1016/j.ijproman.2015.01.008

Lingard, G., Graham, P., & Smithers, G. (2000). Employee perceptions of the solid waste management system operating in a large Australian contracting organisation: Implications for company policy implementation. *Construction Management and Economics*, *18*(4), 383–393. doi:10.1080/01446190050024806

Lingard, H., Oswald, D., & Le, T. (2019). Embedding occupational health and safety in the procurement and management of infrastructure projects: Institutional logics at play in the context of new public management. *Construction Management and Economics*, *37*(10), 567–583.

Lin, J., & Zhang, J. (2018). Review and Exploratory Text Mining of Building Information Modelling Policies in China [in Chinese]. *Construction Technology*, *47*(6), 73–78.

Liu, D., Chen, J., Hu, D., & Zhang, Z. (2019). Dynamic BIM-augmented UAV safety inspection for water diversion project. *Computers in Industry*, *108*, 163–177. doi:10.1016/j.compind.2019.03.004

Liu, J., Liu, Y., & Wang, X. (2020). An environmental assessment model of construction and demolition waste based on system dynamics: A case study in Guangzhou. *Environmental Science and Pollution Research International*, *27*(30), 37237–37259. doi:10.100711356-019-07107-5 PMID:31893359

Liu, R., & Issa, R. (2012). *3D visualization of sub-surface pipelines in connection with the building utilities: Integrating GIS and BIM for facility management. In Computing in Civil Engineering*. ASCE Publications.

Li, X., Jiang, P., Chen, T., Luo, X., & Wen, Q. (2017). A survey on the security of blockchain systems. *Future Generation Computer Systems*.

Li, X., Shen, G. Q., Wu, P., Xue, F., Chi, H., & Li, C. Z. (2019). Developing a conceptual framework of smart work packaging for constraints management in prefabrication housing production. *Advanced Engineering Informatics*, *42*(June), 100938. doi:10.1016/j.aei.2019.100938

Li, X., Yi, W., Chi, H., Wang, X., & Chan, A. (2018). A Critical Review of Virtual and Augmented Reality (VR/AR) Applications in Construction Safety. *Automation in Construction*, *86*, 150–162. doi:10.1016/j.autcon.2017.11.003

Li, Y., & Guldenmund, F. W. (2018). Safety management systems: A broad overview of the literature. *Safety Science*, *103*, 94–123. doi:10.1016/j.ssci.2017.11.016

Li, Z., Barenji, A. V., & Huang, G. Q. (2018). Toward a blockchain cloud manufacturing system as a peer to peer distributed network platform. *Robotics and Computer-integrated Manufacturing*, *54*, 133–144. doi:10.1016/j.rcim.2018.05.011

Lokshina, V., Greguš, M., & Thomas, W. L. (2019). Application of Integrated Building Information Modeling, IoT and Blockchain of Technologies in System Design of Smart Building. *Procedia Computer Science*, *160*, 497–502. doi:10.1016/j.procs.2019.11.058

London, K. (2015). *Global Passport through Co-Integration of Construction Immersive Environments*. RMIT Learning and Teaching Investment Fund Final Report.

London, K., & Bavinton, N. (2006). A Reflexive Capability Model for Sustainable E-Business Environments in Construction Supply Chain. In K. Brown & K. Hampson (Eds.), Clients Driving Construction Innovation (pp. 313-336). Icon.Net Pty Ltd.

London, K., Bavinton, N., Mentink, J., & Egan, B. (2006). *e-Adoption Profile Report*. Cooperative Research Centre for Construction Innovation, Icon.Net.

London, K., Feng, Y., & Ang, K. (2018). Collaborative Robotics in Construction: A Literature Review for Landcom NSW. Academic Press.

London, K., Mentink, J., Bavinton, N., Egan, G., & Aranda-Mena, G. (2006). *E-Business Adoption in the Construction Industry*. Icon.Net Pty Ltd.

London, K., Zhang, P., Khalfan, M., Maqsood, T., & Siva, J. (2014). *A supply chain management self assessment framework for waste minimisation for the residential sector*. https://find.library.unisa.edu.au/primo-explore/fulldisplay?vid=UNISA&search_scope=All_Resources&docid=UNISA_ALMA51129309780001831

London, K., & Pablo, Z. (2017). An actor–network theory approach to developing an expanded conceptualization of collaboration in industrialized building housing construction. *Construction Management and Economics*, *35*(8-9), 553–577. doi:10.1080/01446193.2017.1339361

London, K., & Singh, V. (2013). Integrated construction supply chain design and delivery solutions. *Architectural Engineering and Design Management*, *9*(3), 135–157.

London, K., & Singh, V. (2013). Integrated Construction supply chain design and delivery solutions. *Architectural Engineering and Design Management*, *9*(3), 135–157.

Lopez del Puerto, C., & Clevenger, C. (2010). *Enhancing safety throughout Construction using BIM/VDC* [Paper presentation]. The ECOBuild 2010 BIM Academic Forum, Washington, DC.

López-Pintado, O., Dumas, M., García-Bañuelos, L., & Weber, I. (2018). *Dynamic Role Binding in Blockchain-Based Collaborative Business Processes*. arXiv preprint arXiv:1812.02909.

Louis, J., & Dunston, P. S. (2018). Integrating IoT into operational work flows for real-time and automated decision-making in repetitive construction operations. *Automation in Construction*, *94*(April), 317–327. doi:10.1016/j.autcon.2018.07.005

Love, P. E. D., Matthews, J., & Gates, B. (2019). The 'how' of benefits management for digital technology: From engineering to asset management. *Automation in Construction*, *107*(July), 102930. doi:10.1016/j.autcon.2019.102930

Love, P. E., Matthews, J., Lockley, S., Kassem, M., Kelly, G., Dawood, N., & Serginson, M. (2015). BIM in facilities management applications: A case study of a large university complex. *Built Environment Project and Asset Management.*

Luo, Y. (2002). Contract, cooperation, and performance in international joint ventures. *Strategic Management Journal*, *23*(10), 903–919. doi:10.1002mj.261

Lu, P., Guo, S., Qian, L., He, P., & Xu, X. (2015). The effectiveness of contractual and relational governances in construction projects in China. *International Journal of Project Management*, *33*(1), 212–222. doi:10.1016/j.ijproman.2014.03.004

Lu, W., Chen, X., Ho, D. C., & Wang, H. (2016). Analysis of the construction waste management performance in Hong Kong: The public and private sectors compared using big data. *Journal of Cleaner Production*, *112*, 521–531. doi:10.1016/j.jclepro.2015.06.106

Lu, Y. (2017). Industry 4.0: A survey on technologies, applications and open research issues. *Journal of Industrial Information Integration*, *6*, 1–10. doi:10.1016/j.jii.2017.04.005

Lyytinen, K., & Newman, M. (2008). Explaining information systems change: A punctuated socio-technical change model. *European Journal of Information Systems*, *17*(6), 589–613. doi:10.1057/ejis.2008.50

Macdonald, J. A. (2011). *BIM – Adding Value by Assisting Collaboration* [Paper presentation]. The LSAA 2011 Conference, Sydney, Australia.

Macdonald, J. A. (2012). A Framework for Collaborative BIM Education across the AEC Disciplines. In *Proceedings of the 37th Annual Conference of Australasian University Building Educators Association (AUBEA) 2012* (pp. 223-230). https://www.academia.edu/download/22716367/Macdonald_AUBEA2012.pdf

Macdonald, J. A., & Mills, J. E. (2011). *The potential of BIM to facilitate collaborative AEC education.* Paper presented at the American Society for Engineering Education.

Macdonald, J. A. (2012). A framework for collaborative BIM education across the AEC disciplines. *37th Annual Conference of Australasian University Building Educators Association (AUBEA)*, *4*, 4-6.

Macdonald, J., & Granroth, M. Multidisciplinary AEC education utilising BIM/PLIM tools and processes. *10th Product Lifecycle Management for Society (PLM)*, 663-674.

Madakan, S., Ramaswamy, S., & Tripathi, S. (2015). Internet of Things (IoT): A Literature Review. *Journal of Computer and Communications*, *3*(05), 164–173. doi:10.4236/jcc.2015.35021

Madi, N., & Srour, I. (2019). Managing emergency construction and demolition waste in Syria using GIS. *Resources, Conservation and Recycling*, *141*, 163–175. doi:10.1016/j.resconrec.2018.10.018

Maghiar, M. (2014). *Planning and scheduling curriculum integration of BIM with industry input* [Paper presentation]. *BIM Academic Symposium*, Washington, DC.

Maghool, S. A. H., Moeini, S. H. I., & Arefazar, Y. (2018). An educational application based on virtual reality technology for learning architectural details: Challenges and benefits. *Archnet-IJAR: International Journal of Architectural Research*, *12*(3), 246. doi:10.26687/archnet-ijar.v12i3.1719

Magiera, J. (2013). *Integrated teaching for integrated engineering practice-BIM on campus*. Paper presented at the 20th International conference on computer methods in mechanics, Poznan, Poland.

Magruk, A. (2015). The most important aspects of uncertainty on the Internet of Things field – context of smart buildings. Procedia Engineering, 122, 220–227.

Maher, A. (2004). Learning Outcomes in Higher Education: Implications for Curriculum Design and Student Learning. *The Journal of Hospitality Leisure Sport and Tourism*, *3*(2), 46–54.

Ma, K., Pal, R., & Gustafsson, E. (2019). What modelling research on supply chain collaboration informs us? Identifying key themes and future directions through a literature review. *International Journal of Production Research*, *57*(7), 2203–2225. doi:10.1080/00207543.2018.1535204

Malekitabar, H., Ardeshir, A., Sebt, M. H., & Stouffs, R. (2016). Construction safety risk drivers: A BIM approach. *Safety Science*, *82*, 445–455. doi:10.1016/j.ssci.2015.11.002

Malik, P., & Lenka, U. (2019). Overcoming destructive deviance: Propositioning an integrated conceptual framework. *The International Journal of Organizational Analysis*, *27*(4), 1148–1167. doi:10.1108/IJOA-06-2018-1441

Manley, K. (2006). The innovation competence of repeat public sector clients in the Australian construction industry. *Construction Management and Economics*, *24*(12), 1295–1304. doi:10.1080/01446190600934953

Manseau, A. (2019). *Building tomorrow: innovation in construction and engineering*. Routledge. doi:10.4324/9781315261133

Manu, E., Ankrah, N., Chinyio, E., & Proverbs, D. (2013). A cognitive approach to trust management in construction. *Proceedings of the Institution of Civil Engineers-Management, Procurement and Law, 166*(5), 232-239. 10.1680/mpal.12.00006

Manu, E., Ankrah, N., Chinyio, E., & Proverbs, D. (2015). Trust influencing factors in main contractor and subcontractor relationships during projects. *International Journal of Project Management*, *33*(7), 1495–1508. doi:10.1016/j.ijproman.2015.06.006

Marcellus, K. A., Spatari, S., & Gallagher, P. M. (2012). Utilizing GIS as a geospatial tool to inventory LEED certified buildings and construction and demolition (C&D) waste flows in the United States. *3rd International Symposium on Life-Cycle Civil Engineering, IALCCE 2012*, Vienna, Austria.

Marcellus-Zamora, K. A., Gallagher, P. M., & Spatari, S. (2020). Can Public Construction and Demolition Data Describe Trends in Building Material Recycling? Observations From Philadelphia. *Frontiers in Built Environment*, *6*, 131. Advance online publication. doi:10.3389/fbuil.2020.00131

Marrero, M., Puerto, M., Rivero-Camacho, C., Freire-Guerrero, A., & Solís-Guzmán, J. (2017). Assessing the economic impact and ecological footprint of construction and demolition waste during the urbanization of rural land. *Resources, Conservation & Recycling, 117*(PB), 160-174. doi:10.1016/j.resconrec.2016.10.020

Marsal-Llacuna, M.-L. (2018). Future living framework: Is blockchain the next enabling network? Technological Forecasting & Social Change, 128, 226–234.

Martien, G., Schut, M., McBreen, J., & Tams, D. (2013). *Smart Infrastucture*. https://www.commscope.com/globalassets/digizuite/1022-1012-center-for-digital-education-smart-infrastructure-report.pdf?utm_source=ruckus&utm_medium=redirect

Martínez-Aires, M. D., López-Alonso, M., & Martínez-Rojas, M. (2018). Building information modeling and safety management: A systematic review. *Safety Science*, *101*, 11–18.

Martinez-Sanchez, V., Levis, J. W., Damgaard, A., Decarolis, J. F., Barlaz, M. A., & Astrup, T. F. (2017). Evaluation of Externality Costs in Life-Cycle Optimization of Municipal Solid Waste Management Systems. *Environmental Science & Technology*, *51*(6), 3119–3127. doi:10.1021/acs.est.6b06125 PMID:28263562

Martin, J. H., Collins, A. R., & Diener, R. G. (1995). A sampling protocol for composting, recycling, and re-use of municipal solid waste. *Journal of the Air & Waste Management Association*, *45*(11), 864–870. doi:10.1080/10473289.1995.10467416

Marton, F. (1986). Phenomenography—A Research Approach to Investigating Different Understandings of Reality. *Journal of Thought*, *21*(3), 28–49.

Matarneh, R., & Hamed, S. (2017). Barriers to the adoption of building information modeling in the Jordanian building industry. *Open Journal of Civil Engineering*, *7*(3), 325–335.

Mathews, M., Robles, D., & Bowe, B. (2017). *BIM+ blockchain: A solution to the trust problem in collaboration?* Academic Press.

Matinmikko-Blue, M., & Latva-aho, M. (2017). Micro operators accelerating 5G deployment. *12th IEEE international conference on industrial and information systems (IEEE ICIIS)*. https://www.researchgate.net/publication/323347676_Micro_operators_accelerating_5G_deployment

Matthews, J., Love, P. E., Mewburn, J., Stobaus, C., & Ramanayaka, C. (2018). Building information modelling in construction: Insights from collaboration and change management perspectives. *Production Planning and Control*, *29*(3), 202–216. doi:10.1080/09537287.2017.1407005

Maylor, H., Turner, N., & Murray-Webster, R. (2015). "It worked for manufacturing…!": Operations strategy in project-based operations. *International Journal of Project Management*, *33*(1), 103–115. doi:10.1016/j.ijproman.2014.03.009

Mayo, G., & Issa, R. R. (2014). Processes and standards for BIM closeout information deliverables for owners. In *Proceedings of International Conference on Computing in Civil and Building Engineering* (pp. 673-680). Orlando, FL: Academic Press.

Ma, Z., & Ren, Y. (2017). Integrated Application of BIM and GIS: An Overview. *Procedia Engineering*, *196*(June), 1072–1079. doi:10.1016/j.proeng.2017.08.064

Mazumdar, A., & Alharahshch, II. (2019a). Insights of Trends and Developments in Cloud Computing. *South Asian Research Journal of Engineering and Technology*, *01*(03), 98–107. doi:10.36346arjet.2019.v01i03.003

Mazumdar, A., & Alharahsheh, H. (2019b). Small & Medium Enterprises (Smes) – Insights of Bangladeshi Smes in Different Contexts of Adopting Cloud Computing. *Cross Current International Journal of Economics, Management and Media Studies*, *01*(05), 130–140.

Mazumdar, A., & Alharahsheh, H. (2019c). Key Challenges and Barriers Facing Small & Medium Enterprises (SMEs) in Bangladesh. *South Asian Research Journal of Business and Management*, *01*(03), 118–120. doi:10.36346arjbm.2019.v01i03.005

Mazumdar, A., & Alharahsheh, H. (2020). Digital Bangladesh – Vision 2021: What is the Digital Bangladesh Concept? *South Asian Research Journal of Engineering and Technology*, *2*(1), 6–9. doi:10.36346arjet.2020.v02i01.002

MBIE. (2019). *National Construction Pipeline Report 2019*. Retrieved from https://www.mbie.govt.nz/assets/national-construction-pipeline-report-2019.pdf

McAllister, M., Levett-Jones, T., Downer, T., Harrison, P., Harvey, T., Reid-Searl, K., Lynch, K., Arthur, C., Layh, J., & Calleja, P. (2013). Snapshots of simulation: Creative strategies used by Australian educators to enhance simulation learning experiences for nursing students. *Nurse Education in Practice*, *13*(6), 567–572. doi:10.1016/j.nepr.2013.04.010 PMID:23707534

McAuley, B., Hore, A., & West, R. (2017). *Building information modelling in Ireland 2017*. BIM Innovation Capability Programme, CitA Ltd.

McCuen, T. (2014). *The Challenges of Advancing BIM in the Curriculum while Addressing Current Accreditation Standards for Construction* [Paper presentation]. *BIM Academic Symposium*, Washington, DC.

McDonnell, P., & West, R. P. (2015). The adoption of Building Information Modelling (BIM) to improve existing teaching methods and support services within a Higher Education Institution in Ireland. *Proceedings of the CITA BIM Gathering Conference*, 271-278.

Mcglinn, K. (2019). Interlinking geospatial and building geometry with existing and developing standards on the web. Automation in Construction, 103, 235–250. doi:10.1016/j.autcon.2018.12.026

McGovern, E., & Behan, A. (2014). *Geomatics and developments in BIM education in Ireland*. Paper presented at the FIG Congress 2014 'Engaging the Challenges - Enhancing the relevance, Kuala Lumpar, Malaysia.

McGovern, E., Behan, V., & Furlong, K. (2014). *Geomatics and Developments in BIM Education in Ireland*. FIG Congress 2014, Engaging the Challenges - Enhancing the Relevance, Kuala Lumpur, Malaysia.

McGraw Hill Construction. (2014). *The Business Value of BIM for Construction in Major Global Markets: How Contractors around the World are driving Innovation with Building Information Modeling*. Author.

McGraw Hill Construction. (2014). *The Business Value of BIM in Major Global Markets. How Contractors Around the World Are Driving Innovation With Building Information Modeling*. McGraw Hill Construction.

McGraw-Hill Construction. (2008). *Building Information Modeling (BIM): Transforming Design and Construction to Achieve Greater Industry Productivity*. McGraw-Hill Construction.

McGraw-Hill Construction. (2009). *The Business Value of BIM: Getting Building Information Modeling to the Bottom Line*. McGraw-Hill Construction.

McGraw-Hill Construction. (2014). *The business value of BIM owners*. Retrieved from https://group.skanska.com/48f2cd/globalassets/about-us/building-information-modeling/business-value-of-bim-for-owners-smr-2014.pdf

McKinsey & Company. (2015). *Building for a better future Vision 2030 for the Hong Kong Construction Industry*. Author.

McKinsey & Company. (2017). Blockchain technology in the insurance sector. Quarterly Meeting of the Federal Advisory Committee on Insurance (FACI). *Blockchain technology in the insurance sector. Quarterly Meeting of the Federal Advisory Committee on Insurance (FACI).*

McKnight, D. H., Choudhury, V., & Kacmar, C. (2002). Developing and validating trust measures for e-commerce: An integrative typology. *Information Systems Research*, *13*(3), 334–359.

McWaters, R. J., Galaski, R., & Chatterjee, S. (2016). The future of financial infrastructure: An ambitious look at how blockchain can reshape financial services. In *World Economic Forum* (Vol. 49). Academic Press.

Mearian, L. (2019). Walmart launches 'world's largest' blockchain-based freight-and-payment network. *Computerworld*.

Mehrbod, S., Staub-French, S., Mahyar, N., & Tory, M. (2019). Characterizing interactions with BIM tools and artifacts in building design coordination meetings. *Automation in Construction*, *98*, 195–213. doi:10.1016/j.autcon.2018.10.025

Meisenzahl, M. (2019). This building in Dubai is the largest 3D-printed structure in the world — and it took just 3 workers and a printer to build it. *Business Insider Australia*. Retrieved from https://www.businessinsider.com.au/dubai-largest-3d-printed-building-apis-cor-photos-2019-12?r=US&IR=T

Melzner, J., Zhang, S., Teizer, J., & Bargstädt, H. J. (2013). A case study on automated safety compliance checking to assist fall protection design and planning in building information models. *Construction Management and Economics*, *31*(6), 661–674.

Members of Congress from United. (2018). *Institute for information industry (III) Attract Multidisciplinary Expert to Make Cities Smarter Through Education.* https://thetechrevolutionist.com/2018/03/institute-for-information-industry-iii.html

Meng-Han, T., Chen, K.-L., & Chang, Y.-L. (2019). Development of a Project-Based Online Course for BIM Learning. *Sustainability*, *11*(5772).

Meng, X. (2020). Proactive management in the context of construction supply chains. *Production Planning and Control*, *31*(7), 527–539. doi:10.1080/09537287.2019.1657977

Merschbrock, C., & Munkvold, B. E. (2015). Effective digital collaboration in the construction industry–A case study of BIM deployment in a hospital construction project. *Computers in Industry*, *73*, 1–7. doi:10.1016/j.compind.2015.07.003

Meszaros, R., Adachi, D., Dharamsi, H., Yetiskin, B., & Thomas, P. (2016). *Blockchain Technology: How banks are building a real-time global payment network*. Accenture Mobility.

Meyer, J. H. F., & Land, R. (2003). Threshold Concepts and Troublesome Knowledge: linkages to ways of thinking and practising within the disciplines. In C. Rust (Ed.), *Improving Student Learning – Ten Years On* (pp. 412–424). OCSLD.

MHURD. (2011). *2011-2015 Outline of Development for Informationalisation in the Construction Industry*. Ministry of Housing and Urban Rural Development of China.

MHURD. (2016). *2016-2020 Outline of Development for Informationalisation in the Construction Industry*. Ministry of Housing and Urban Rural Development of China.

Miettinen, R., & Paavola, S. (2014). Beyond the BIM utopia: Approaches to the development and implementation of building information modelling. *Automation in Construction*, *43*, 84–91. doi:10.1016/j.autcon.2014.03.009

Mihindu, S., & Arayici, Y. (2008). Digital construction through BIM systems will drive the re-engineering of construction business practices. In *Proceedings of the International Conference Visualisation*. IEEE Computer Society. 10.1109/VIS.2008.22

Miller, R. (2018). *Camunda hauls in $28M investment as workflow automation remains hot*. Published online in *Tech Crunch* on 5 December 2018. Retrieved June 25, 2020 from https://techcrunch.com/2018/12/05/camunda-hauls-in-28m-investment-as-workflow-automation-remains-hot/

Miller, G., Sharma, S., Donald, C., & Amor, R. (2013). Developing a building information modelling educational framework for the tertiary sector in New Zealand. *IFIP International Conference on Product Lifecycle Management*, 606-618.

Miller, G., Sharma, S., Donald, C., & Amor, R. (2013). Developing a Building Information Modelling Educational Framework for the Tertiary Sector in New Zealand. In A. Bernard, L. Rivest, & D. Dutta (Eds.), *Product Lifecycle Management for Society. Springer*. doi:10.1007/978-3-642-41501-2_60

Mills, A., Ashford, P., & McLaughlin, P. (2010). *The value of experiential learning for providing a contextual understanding of the construction process* [Paper presentation]. The 31st Australasian University Building Educators Association Conference, Sydney, Australia.

Mills, J., Tran, A., Parks, A., Kelly, S., Smith, E., Jupp, J., Macdonald, J., Williams, A., & Sher, W. (2013). Collaborative building design education using Building Information Modelling (CodeBIM). Final report 2013, Australian Government Office for Learning & Teaching, Canberra, Australia.

Mills, J., Tran, A., Parks, A., Kelly, S., Smith, E., & Jupp, J. (2013). *Collaborative building design education using Building Information Modelling (CodeBIM), Final Report 2013*. Australian Government Office for Learning and Teaching.

Mirarchi, C., Lupica Spagnolo, S., Daniotti, B., & Pavan, A. (2020). Structuring General Information Specifications for Contracts in Accordance with the UNI 11337:2017 Standard. In B. Daniotti, M. Gianinetto, & S. Della Torre (Eds.), Digital Transformation of the Design, Construction and Management Processes of the Built Environment. Research for Development (pp. 103–112). Springer. https://doi.org/10.1007/978-3-030-33570-0_10.

Mirus, A., Patel, Y., & McPherson, P. (2018). *Prefabrication: New Zealand's golden ticket?* Academic Press.

Mitra, S. S. (2018). *Site-Suitability Analysis for the Identification of Potential Sites to Construct Rain Water Harvesting Systems. In Hydrology and Water Resource Management: Breakthroughs in Research and Practice*. IGI Global.

Mixed Reality for Education. (n.d.). *Educational Tools and Scenario*. www.mr4education.com

Miyaoka, M. (2015). *Sustainability Assessment for Strategic Material Flows Between Planned Construction Projects in the Stockholm County* [MSc, Halmstad University]. http://hh.diva-portal.org/smash/get/diva2:849623/FULLTEXT01.pdf

Mohamad, S. F., Abidin, U. F. U. Z., Saari, N., Zakaria, M. H., Ismail, A., & Man, N. A. (2019). Knowledge, attitude and practice towards the implementation of outcome-based education among the academics in Universiti Putra Malaysia. *International Journal of Engineering and Advanced Technology*, *8*(5C), 1044–1053. doi:10.35940/ijeat.E1148.0585C19

Mohammadi, A., Tavakolan, M. & Khosravi, Y. (2018) Factors in Fluencing Safety Performance on Construction Projects: A Review. *Safety Science, 109*(October), 382–397.

Mohammed, N. Z., Ghazi, A., & Mustafa, H. E. (2013). Positional Accuracy Testing of Google Earth. *International Journal of Multidisciplinary Sciences and Engineering*, *4*(6), 6–9.

Molavi, J., & Shapoorian, B. (2013). Implementing an Interactive Program of BIM Applications for Graduating Students [Paper presentation]. ICSDEC 2012: Developing the Frontier of Sustainable Design, Engineering, and Construction, Fort Worth, TX.

Moore, J. F. (1997). *The death of competition: leadership and strategy in the age of business ecosystems*. Harper Business.

Morabito, V. (2017). Blockchain and Enterprise Systems. In *Business Innovation Through Blockchain* (pp. 125–142). Springer. doi:10.1007/978-3-319-48478-5_7

Morandi, M. I. W. M., & Camargo, L. F. R. (2015). Revisão sistemática da literatura. In A. Dresch (Ed.), *Design Science Research - método de pesquisa para avanço da ciência e tecnologia* (pp. 141–172). Bookman.

Mordue, S., & Finch, R. (2014). *BIM for construction health and safety*. RIBA Publishing.

Morgan, D. (1993). *Successful focus groups: advancing the state of the art*. Sage Publications., doi:10.4135/9781483349008

Morgan, D. (2012). Focus groups and social interaction. In J. Gubrium (Ed.), *The Sage handbook of interview research: The complexity of the craft* (pp. 161–176). Sage.

Morrell, P. (2015 May 19). *Collaboration for Change. The Edge Commission Report on the Future of Professionalism*. https://www.designingbuildings.co.uk/wiki/Collaboration_for_Change,_The_Edge_Commission_Report_on_the_Future_of_Professionalism#:~:text=19%20May%202015-,Collaboration%20for%20Change%2C%20The%20Edge%20Commission%20Report%20on%20the%20Future,to%20design%20and%20problem%20solving

Morrell, P. (2015). *Collaboration for Change.* The Edge Commission Report on the Future of Professionalism. Available at http://www.edgedebate.com/?page_id=2829

Mosey, D., Bahram, D., Dartnell, R., Hallam, C., Howard, C., Maqbool, A., Murray, K., Rawlinson, S., & Winfield, M. (2016). *Enabling BIM through Procurement and Contracts.* http://alliancecontractingelectroniclawjournal.com/wp-content/uploads/2019/10/Kings-College-London.-2016-Enabling-BIM-through-Procurement-and-Contracts-Centre-of-Construction-Law-and-Dispute-Resolution.pdf

Motamedi, A., Soltani, M. M., Setayeshgar, S., & Hammad, A. (2016). Extending IFC to incorporate information of RFID tags attached to building elements. *Advanced Engineering Informatics, 30*(1), 39–53. doi:10.1016/j.aei.2015.11.004

Mukuni, J. (2019). *Challenges of educational digital infrastructure in Africa: A tale of hope and disillusionment.* https://www.researchgate.net/publication/334809878_Challenges_of_educational_digital_infrastructure_in_Africa_A_tale_of_hope_and_disillusionment

Müller, R., Pemsel, S., & Shao, J. (2015). organisational enablers for project governance and governmentality in project-based organisations. *International Journal of Project Management, 33*(4), 839–851. doi:10.1016/j.ijproman.2014.07.008

Murphy, M. E. (2014). Implementing innovation: A stakeholder competency-based approach for BIM. *Construction Innovation, 14*(4), 433–452. doi:10.1108/CI-01-2014-0011

Mutai, A., & Guidera, S. (2010). *Building Information Modeling in Construction: Current Practices and their Implications for Construction Engineering Education* [Paper presentation]. 117th American Societyfor Engineering Education (ASEE) *Annual Conference*, Louisville, KY.

Mutis, I., & Paramashivam, A. (2019). Cybersecurity Management Framework for a Cloud-Based BIM Model. Advances in Informatics and Computing in Civil and Construction Engineering, 325-333. doi:10.1007/978-3-030-00220-6_39

Muzafar, M. (2019). Building information modelling to mitigate the health and safety risks associated with the construction industry: A review. *International Journal of Occupational Safety and Ergonomics*, 1-26.

Nakamoto, S. (2008). *Bitcoin: A peer-to-peer electronic cash system.* Academic Press.

Nakamura, S. (2020). Tracking the Product Origins of Waste for Treatment Using the WIO Data Developed by the Japanese Ministry of the Environment. *Environmental Science & Technology, 54*(23), 14862–14867. doi:10.1021/acs.est.0c06015 PMID:33205952

Nambisan, S., Lyytinen, K., Majchrzak, A., & Song, M. (2017). Digital Innovation Management: Reinventing innovation management research in the digital world. *Management Information Systems Quarterly, 41*(1), 223–238. doi:10.25300/MISQ/2017/41:1.03

Nam, C. H., & Tatum, C. B. (1997). Leaders and champions for construction innovation. *Construction Management and Economics, 15*(3), 259–270. doi:10.1080/014461997372999

Natephra, W., & Motamedi, A. (2019) BIM-based live sensor data visualization using virtual reality for monitoring indoor conditions. *Proceedings of the 24th CAADRIA Conference*, 2, 191-200.

National BIM Specification (NBS) Australia. (2019). *Australia and New Zealand BIM Report 2019.* https://www.thenbs.com.au/bim-report

National Building Specification (NBS). (2020a). *BIM Levels Explained.* https://www.thenbs.com/knowledge/bim-levels-explained

National Building Specification (NBS). (2020b). *Digital Built Britain- Building the Future*. https://www.thenbs.com/knowledge/digital-built-britain-building-the-future

National Forum for the Enhancement of Teaching and Learning in Higher Education. (2020) *DELTA Framework*. Author.

National Institute of Building Sciences (NIBS). (2015). *buildingSMART alliance*. National BIM Standard – United States® Version 3. Available at https://www.nationalbimstandard.org/

National, BIM Council. (2017). *Roadmap to Digital Transition for Ireland's Construction Industry*. Author.

NATSPEC. (2011). *NATSPEC National BIM Guide*. https://bim.natspec.org/documents/natspec-national-bim-guide

NATSPEC. (2014). *BIM Education – Global: 2014 Update Report*. NATSPEC.

NATSPEC. (2015). *BIM Education – Global: 2015 Update Report*. NATSPEC.

NATSPEC. (2016). *BIM Education – Global: 2016 Update Report*. NATSPEC.

NATSPEC. (2017). *BIM Education – Global: 2017 Update Report*. NATSPEC.

NATSPEC. (2018). *BIM Education – Global: 2018 Update Report*. NATSPEC.

Navon, R., & Berkovich, O. (2005). Development and on-site evaluation of an automated materials management and control model. *Journal of Construction Engineering and Management*, *131*(12), 1328–1336. doi:10.1061/(ASCE)0733-9364(2005)131:12(1328)

Nawari, N. O., & Ravindran, S. (2019). Blockchain and the built environment: Potentials and limitations. *Journal of Building Engineering Journal, 25*.

Nawari, N. O., & Ravindran, S. (2019). Blockchain and the built environment: Potentials and limitations. *Journal of Building Engineering*, *25*, 100832. doi:10.1016/j.jobe.2019.100832

NBIMS. (2015). *National BIM Standard - United States® Version 3 - Scope*. https://www.nationalbimstandard.org/

NBS, National BIM Report 2019. (2020). *National Skills Council/Expert Group on Future Skills Needs*. NBS.

NBS. (2012). *BIM Legal Roundtable Discussion* [Webcast]. NBS Website. Retrieved from https://www.thenbs.com

NBS. (2020), The Construction Leaders' Summit: The Digital Future, NBS Enterprises Ltd.

NBS. (2020). *10th Annual BIM Report 2020*. NBS Enterprises Ltd.

NBS. (2021). *What is BIM?* NBS Website. Retrieved from https://www.thenbs.com/knowledge/what-is-building-information-modelling-bim

Nelson, A., Reuter, H. I., & Gessler, P. (2009). DEM Production Methods and Sources, *Chapter-3. Developments in Soil Science*, *33*, 65–85. doi:10.1016/S0166-2481(08)00003-2

Nesta. (2012). *How councils can make their economy grow.* Creative Councils. http://www.nesta.org.uk/blog/how-councils-can-make-their-economy-grow#sthash.jzVUQwEr.dpuf

Nesta. (2013). *Six Creative Councils recognised for pioneering innovations*. Creative Councils. http://www.nesta.org.uk/news/six-creative-councils-recognised-pioneering-innovations#sthash.laZFt4jo.EdMELxLf.dpuf

Neti, C. (2018). *How Watson Education, Scholastic, and Edmodo are using AI to close the learning gap.* https://www.ibm.com/blogs/watson/2018/06/using-ai-to-close-learning-gap/

Neuendorf, K. A. (2002). *The content analysis guidebook*. Sage Publications.

New Education Forum (NEF) Brussels. (2017). *Smart Education in Smart Cities and Smart Regions*. http://ciedu.eu/wp-content/uploads/2018/08/NEF_2017_Report.pdf

New York City, U. (2013). *Building Information Modeling Site Safety Submission Guidelines and Standards*. BIM MANUAL.

New Zealand Government. (2019). *Using virtual reality to construct a safer workplace*. Retrieved from https://www.acc.co.nz/newsroom/stories/using-virtual-reality-to-construct-a-safer-workplace/

New, M., Liverman, D., Schroder, H., & Anderson, K. (2011). *Four degrees and beyond: the potential for a global temperature increase of four degrees and its implications*. Academic Press.

Newton, S. (2011). *Building and Construction. Learning and Teaching Academic Standards Statement*. Australian Learning and Teaching Council.

NHA. (2016 May 1). *New UKAS Competency Management Standard for Housing Sector*. https://www.nationalhousingacademy.org.uk/news/new-ukas-standard

Nicholas, J. (1989). Successful Project Management. *J. of Systems Management*, *40*, 24–30.

Nicholson, G. (2013). Investigating the Application of BIM on Small-Scale Construction Projects. *Proceedings of the CITA BIM Gathering Conference*, 49-60.

Nickerson, R. S. (1998). Confirmation Bias: A Ubiquitous Phenomenon in Many Guises. *Review of General Psychology*, *2*(2), 175–220. doi:10.1037/1089-2680.2.2.175

Nigam, A., & Caswell, N. S. (2003). Business artifacts: An approach to operational specification. *IBM Systems Journal*, *42*(3), 428–445. doi:10.1147j.423.0428

Nisar, M. W., Munir, E. U., & Shafqat, A. (2011). Usage and Impact of ICT in Education Sector: A Study of Pakistan. *Australian Journal of Basic and Applied Sciences*, *5*(12), 578–583. https://arxiv.org/ftp/arxiv/papers/1206/1206.5132.pdf

Nisbet, R. I., & Collins, J. M. (1978). Barriers and resistance to innovation. *The Australian Journal of Teacher Education*, *3*(1), 1. doi:10.14221/ajte.1978v3n1.1

Niska, H., & Serkkola, A. (2018). Data analytics approach to create waste generation profiles for waste management and collection. *Waste Management (New York, N.Y.)*, *77*, 477–485. doi:10.1016/j.wasman.2018.04.033 PMID:29724480

Niu, S., Yang, Y., & Pan, W. (2019). Logistics Planning and Visualization of Modular Integrated Construction Projects Based on BIM-GIS Integration and Vehicle Routing Algorithm. *Modular and Offsite Construction (MOC) Summit Proceedings*, 579-586.

Niu, Y., Lu, W., Xue, F., Liu, D., Chen, K., Fang, D., & Anumba, C. (2019). Towards the "third wave": An SCO-enabled occupational health and safety management system for construction. *Safety Science*, *111*, 213–223. doi:10.1016/j.ssci.2018.07.013

Noor, K. B. M. (2008). Case study: A strategic research methodology. *American Journal of Applied Sciences*, *5*(11), 1602–1604. doi:10.3844/ajassp.2008.1602.1604

Nørkjaer Gade, P., Nørkjaer Gade, A., Otrel-Cass, K., & Svidt, K. (2019). A holistic analysis of a BIM-mediated building design process using activity theory. *Construction Management and Economics*, *37*(6), 336–350. doi:10.1080/01446193.2018.1533644

Novais, L., Manuel, J., & Ortiz-Bas, Á. (2019). A systematic literature review of cloud computing use in supply chain integration. *Computers & Industrial Engineering, 129*, 296–314.

Novais, L., Maqueira, J. M., Ortiz, Á., & Bruque, S. (2019). Strategic simulation models as a new methodological approach: An application to information technologies integration, lean/just-in-time and lead-time. *Central European Journal of Operations Research*, 1–34. doi:10.100710100-019-00654-9

NSAI. (2021). *National Standards Authority of Ireland - ISO 19650-2 BIM - Building Information Modelling*. Available from: https://www.nsai.ie/certification/management-systems/bim-building-information-modelling/

Nunes, K. R. A., & Mahler, C. F. (2020). Comparison of construction and demolition waste management between Brazil, European Union and USA. *Waste Management & Research*, *38*(4), 415–422. doi:10.1177/0734242X20902814 PMID:32043421

Nývlt, V. (2018). The role of managing knowledge and information in BIM implementation processes in the Czech Republic. *9th International Scientific Conference Building Defects (Building Defects 2017).*

NZCN. (2019a). RMD changes the way customers look at designes with new visualization tool. *New Zealand Construction News (NZCN)*, 16.

NZCN. (2019b). Virtual reality becomes an educational reality. *New Zealand Construction News (NZCN)*, 7.

NZCN. (2019c, October/November). Fletcher Building opens New Zealand's largest offsite home manufacturing facility. *New Zealand Construction News (NZCN)*, 8.

NZCN. (2019d, October/November). Technology and collaboration star in award-wining buildings. *New Zealand Construction News (NZCN)*, 8.

O'Donnell, J., Truong-Hong, L., Boyl, N., Corry, E., Cao, J., & Laefer, D. (2019). LiDAR Point-Cloud mapping of building facades for building energy performance simulation. *Automation in Construction*, *107*, 102905. Advance online publication. doi:10.1016/j.autcon.2019.102905

O'Donnell, R. (2012). *Profession of Engineering Education: Advancing Teaching, Research and Careers* [Paper presentation]. *23rd Annual Conference of the Australasian Association for Engineering Education*, Melbourne, Australia.

O'Reilly, C. A. III, Caldwell, D. F., & Barnett, W. P. (1989). Work group demography, social integration, and turnover. *Administrative Science Quarterly*, 21–37.

O'Shea, P. (2008). Lessons Learned about Designing Augmented Realities. *Int'l J. Gaming and Computer-Mediated Simulations*, *1*(1), 1-15. https://pdfs.semanticscholar.org/6cde/77becd373273bb096e5f7b40543b6b8b8266.pdf

O'Sullivan, D., & Dooley, L. (2009). *Applying innovation.* SAGE Publications.

OECD. (2017a). Better Use of Skills in the Workplace: Why It Matters for Productivity and Local Jobs. Organisation for Economic Co-operation and Development (OECD).

OECD. (2017b). Getting Skills Right: United Kingdom. Organisation for Economic Co-operation and Development (OECD).

Oesterreich, T. D., & Teuteberg, F. (2016). Understanding the implications of digitisation and automation in the context of Industry 4.0: A triangulation approach and elements of a research agenda for the construction industry. *Computers in Industry*, *83*, 121–139. doi:10.1016/j.compind.2016.09.006

Ogunmakinde, O. E. (2019). A Review of Circular Economy Development Models in China, Germany and Japan. *Recycling, 4*(3), 27.

Oladapo, A. A. (2007). An investigation into the use of ICT in the Nigerian construction industry. *Electronic Journal of Information Technology in Construction*, *12*, 261–277.

Olorunniwo, F. O., & Li, X. (2010). Information sharing and collaboration practices in reverse logistics. *Supply Chain Management*, *15*(6), 454–462. doi:10.1108/13598541011080437

Olowa, T. O. O., Witt, E., & Lill, I. (2019). *BIM for Construction Education: Initial Findings from a Literature Review* [Paper presentation]. The 10th Nordic Conference on Construction Economics and Organization, Tallinn, Estonia.

Olowa, T., Witt, E., & Lill, I. (2020). Conceptualising building information modelling for construction education. *Journal of Civil Engineering and Management*, *26*(6), 551–563. doi:10.3846/jcem.2020.12918

Oluwole, A. (2011). A preliminary review on the legal implications of BIM and model ownership. *Journal of Information Technology in Construction*, *16*, 687–696.

Oraee, M., Hosseini, M. R., Edwards, D. J., Li, H., Papadonikolaki, E., & Cao, D. (2019). Collaboration barriers in BIM-based construction networks: A conceptual model. *International Journal of Project Management*, *37*(6), 839–854. doi:10.1016/j.ijproman.2019.05.004

Orcutt, M. (2018). *The block chain issue*. MIT Technology Review.

Orlikowski, W. J., Yates, J., Okamura, K., & Fujimoto, M. (1995). Shaping electronic communication: The metastructuring of technology in the context of use. *Organisation Science*, *6*(4), 423-444.

Orlikowski, W. J., & Baroudi, J. J. (1991). Studying information technology in organizations: Research approaches and assumptions. *Information Systems Research*, *2*(1), 1–28. doi:10.1287/isre.2.1.1

Osipova, E. (2015). Establishing cooperative relationships and joint risk management in construction projects: Agency theory perspective. *Journal of Management Engineering*, *31*(6), 05014026. doi:10.1061/(ASCE)ME.1943-5479.0000346

Otto, M., Prieur, M., Agethen, P., & Rukzio, E. (2016). Dual reality for production verification workshops: A comprehensive set of virtual methods. *Procedia CIRP*, *44*, 38–43. doi:10.1016/j.procir.2016.02.140

Oxford Economics. (2020). *The Digital Opportunity for Small Businesses: How improved digital skills and connectivity boost local areas*. Intuit QuickBooks.

Ozorhon, B., Oral, K., & Demirkesen, S. (2016). Investigating the components of innovation in construction projects. *Journal of Management Engineering*, *32*(3), 04015052. doi:10.1061/(ASCE)ME.1943-5479.0000419

Pablo, Z., & London, K. (2017, June). Developing collaborative capacities in industrialized building: roadmap for knowledge transfer'. In E. Fidelis and M. Behm (Eds.), *Proceedings of the Joint CIB W099 and TG59 International Safety, Health, and People in Construction Conference: Towards better Safety, Health, Wellbeing, and Life in Construction* (pp. 593-602). Department of Built Environment, Central University of Technology, South Africa.

Pablo, Z., & London, K. (2016, November). A collaborative practices typology for Australian prefabricated housing networks: Convergence, alignment and coordination. In R. Amoêda, & C. Pinheiro (Eds.), *Proceedings of the 2016 International Conference on Sustainable Housing Planning, Management, and Usability* (pp. 725-734). Green Lines Institute for Sustainable Development.

Pablo, Z., & London, K. (2020). Stable relationality and dynamic innovation: Two models of collaboration in SME-driven offsite manufacturing supply chains in housing construction. *Engineering, Construction, and Architectural Management*, *27*(7), 1553–1577. doi:10.1108/ECAM-07-2019-0346

Panarello, A., Tapas, N., Merlino, G., Longo, F., & Puliafito, A. (2018). Blockchain and IoT integration: A systematic survey. *Sensors (Basel)*, *18*(8), 2575. doi:10.339018082575 PMID:30082633

Pan, W., & Goodier, C. (2011). House-building business models and off-site construction take-up. *Journal of Architectural Engineering*, *18*(2), 84–93. doi:10.1061/(ASCE)AE.1943-5568.0000058

Papadonikolaki, E. (2017, June). Aligning BIM Adoption with Implementation In Loosely Coupled construction systems. In *Proceedings of the EPOC-MW Conference*. Engineering Project Organization Society.

Papadonikolaki, E., van Oel, C., & Kagioglou, M. (2019). Organising and Managing boundaries: A structurational view of collaboration with Building Information Modelling (BIM). *International Journal of Project Management*, *37*(3), 378–394. doi:10.1016/j.ijproman.2019.01.010

Papadonikolaki, E., & Wamelink, H. (2017). Inter-and intra-organisational conditions for supply chain integration with BIM. *Building Research and Information*, *45*(6), 649–664. doi:10.1080/09613218.2017.1301718

Park, J., Kim, K., & Cho, Y. K. (2016). Framework of automated construction-safety monitoring using cloud-enabled BIM and BLE mobile tracking sensors. *Journal of Construction Engineering and Management*, *143*(2), 05016019. doi:10.1061/(ASCE)CO.1943-7862.0001223

Park, K. S., & Kim, K. P. (2017). BIM application and adoption in the UK housing sector. In P. Wu, H. Li, & X. Wang (Eds.), *Integrated Building Information Modelling, UAE*. Bentham Science Publishers.

Pärn, E. A., & Edwards, D. J. (2017). Conceptualising the FinDD API plug-in: A study of BIM-FM integration. *Automation in Construction*, *80*, 11–21. doi:10.1016/j.autcon.2017.03.015

PAS 1192-5. (2015). *Specification for security-minded building information modelling, digital built environments and smart asset management*. Tech. rep., BIM Task Group's security working group of the British Standards Institute.

Pasquire, C. L., & Connolly, G. E. (2002). Leaner construction through off-site manufacturing. *Proceedings IGLC*, 263–266.

Patil, G., Chen, D., Mayo, G., Smithwick, J., & Barclay, N. (2019). Implementing BIM and Finite Element Analysis in a Structural Integrity Investigation of Curtain Walls. *Journal of Facility Management Education and Research*, *3*(1), 27–37. doi:10.22361/jfmer/112193

Pauwels, P., Zhang, S., & Lee, Y. (2017). Semantic web technologies in AEC industry: A literature overview. *Automation in Construction*, *73*, 145–165. doi:10.1016/j.autcon.2016.10.003

Payne, J. (2008). Skills in context: What can the UK learn from Australia's skill ecosystem projects? *Policy and Politics*, *36*(3), 307–323. doi:10.1332/030557308X307685

Paz, D. H. F., Lafayette, K. P. V., & Sobral, M. C. (2018). GIS-based planning system for managing the flow of construction and demolition waste in Brazil. *Waste Management & Research*, *36*(6), 541–549. doi:10.1177/0734242X18772096 PMID:29776320

PDST Technology in Education (2020). *Technology Overview in Schools - ICT Infrastructure.* https://www.pdsttechnologyineducation.ie/en/Technology/Technology%20Overview/

Pedagogy & Upskilling Network (PUN). (2019). *Evolve or Die: Transforming the productivity of Built Environment Professionals and Organisations of Digital Built Britain through a new digitally enabled ecosystem underpinned by the mediation between competence supply and demand.* Available from https://ciat.org.uk/uploads/assets/uploaded/101f628c-061d-445d-817368422197b626.pdf

Peeraer, J., & Petergem, P. (2011). ICT in Teacher Education in an Emerging Developing Country: Vietnam's Baseline Situation at the Start of the Year of ICT. *Journal of Computers & Education*, *56*(4), 974–982. doi:10.1016/j.compedu.2010.11.015

Peng, Y., Li, S., & Hu, Z. (2019). A self-learning dynamic path planning method for evacuation in large public buildings based on neural networks. *Neurocomputing*, *365*, 71–85. doi:10.1016/j.neucom.2019.06.099

Penzes, B. (2018). *Blockchain technology in the construction industry*. Institution of Civil Engineers.

Penzes, B. (2018). *Blockchain technology in the construction industry*. Technical report, Institution of Civil Engineers.

Pereira, A. C., & Romero, F. (2017). A review of the meanings and the implications of the Industry 4.0 concept. *Procedia Manufacturing*, *13*, 1206–1214. doi:10.1016/j.promfg.2017.09.032

Perera, S., Nanayakkara, S., Rodrigo, M., Senaratne, S., & Weinand, R. (2020). Blockchain technology: Is it hype or real in the construction industry? *Journal of Industrial Information Integration*, *17*, 100125. doi:10.1016/j.jii.2020.100125

Perović, N., & Young, C. (2016). Rapid and Creative Course Design: As Easy as ABC? *Procedia: Social and Behavioral Sciences*, *228*, 390–395. doi:10.1016/j.sbspro.2016.07.058

Petri, I., Barati, M., Rezgui, Y., & Rana, O. F. (2020). Blockchain for energy sharing and trading in distributed prosumer communities. *Computers in Industry*, *123*, 103282. doi:10.1016/j.compind.2020.103282

Petri, I., Beach, T., Rana, O. F., & Rezgui, Y. (2017). Coordinating multi-site construction projects using federated clouds. *Automation in Construction*, *83*, 273–284. doi:10.1016/j.autcon.2017.08.011

Pezeshki, Z., Soleimani, A., & Darabi, A. (2019). Application of BEM and using BIM database for BEM: A review. *Journal of Building Engineering*, *23*, 1–17.

Pham, T. N. T., Skelton, L., & Samarasinghe, D. A. S. (2020). *BIM implementation barriers in the architecture, engineering and construction (AEC) industry in New Zealand.* Paper presented at the 54th International Conference of the Architectural Science Association (ANZAScA) 2020, Auckland, New Zealand.

Philp, D., & Thompson, N. (2014). *Built Environment 2050: A Report on Our Digital Future*. Construction Industry Council.

Pickin, J., Randell, P., & Latimer, G. (2018). *Improving national waste data and reporting*. D. o. t. E. a. Energy. https://www.environment.gov.au/protection/waste-resource-recovery/publications/improving-national-waste-data-reporting

Pikas, E., Sacks, R., & Hazzan, O. (2013). Building Information Modeling Education for Construction Engineering and Management. II: Procedures and Implementation Case Study. *Journal of Construction Engineering and Management*, *139*(11).

Pikas, E., Sacks, R., & Hazzan, O. (2013). Building information modeling education for construction engineering and management. II: Procedures and implementation case study. *Journal of Construction Engineering and Management*, *139*(11), 05013002.

Pikas, E., Sacks, R., & Hazzan, O. (2013). Building Information Modelling Education for Construction Engineering and Management. *Journal of Construction Engineering and Management*, *139*(11). Advance online publication. doi:10.1061/(ASCE)CO.1943-7862.0000765

Pilkington, M. (2016). Blockchain technology: principles and applications. In Research Handbook on Digital Transformations (vol. 225). Edward Elgar Publishing. doi:10.4337/9781784717766.00019

Pinti, L. (2019). BIM-based Facility Management. *Embedded information management for maintenance: Building Information Modelling and database integration for the Italian Public Administration.*

Piroozfar, P., Farr, E. R., Zadeh, A. H., Inacio, S. T., Kilgallon, S., & Jin, R. (2019). Facilitating Building Information Modelling (BIM) using Integrated Project Delivery (IPD): A UK perspective. *Journal of Building Engineering*, *26*, 100907. doi:10.1016/j.jobe.2019.100907

Pishdad-Bozorgi, P. (2017, September). Future Smart Facilities: State-of-the-Art BIM-Enabled Facility Management. *Journal of Construction Engineering and Management*, *143*(9), 02517006. doi:10.1061/(ASCE)CO.1943-7862.0001376

Plageras, A. P., Psannis, K. E., Stergiou, C., Wang, H., & Gupta, B. B. (2018). Efficient IoT-based sensor BIG Data collection – processing and analysis in smart buildings. *Future Generation Computer Systems*, *82*, 349–357. doi:10.1016/j.future.2017.09.082

Plan BIM. (2017). *Roles BIM and Matriz de Roles BIM*. Comite de Transformacion Digital (CORFO). Retrieved from planbim.cl.

Plan BIM. (2019). *Estandar BIM para proyectos publicos. Intercambio de Informacion entre Solicitante y Proveedores.* Comite de Transformacion Digital (CORFO). Retrieved from planbim.cl.

Plan Grid, Inc. (2019). *The Digital Groundwork: Beyond construction's productivity gap*. Plan Grid Inc.

PlanBIM. (2018). *Plan BIM*. Retrieved from https://planbim.cl/

Poirer, E., Forgues, D., & Staub-French, S. (2016). Collaboration through innovation: Implications for expertise in the AEC sector. *Construction Management and Economics*, *34*(11), 769–789. doi:10.1080/01446193.2016.1206660

Poirier, E. A., Staub-French, S., & Forgues, D. (2015). Measuring the impact of BIM on labor productivity in a small specialty contracting enterprise through action-research. *Automation in Construction*, *58*, 74–84. doi:10.1016/j.autcon.2015.07.002

Polidoro, M. (2007). Spook Hills in the Lab: Notes on a Strange World. *The Skeptical Inquirer*, *31*(6).

Pontell, H. N., & Geis, G. (2007). *International handbook of white-collar and corporate crime*. Springer. doi:10.1007/978-0-387-34111-8

Porwal, A., & Hewage, K. N. (2013). Building Information Modeling (BIM) partnering framework for public construction projects. *Automation in Construction*, *31*, 204–214. doi:10.1016/j.autcon.2012.12.004

Prajogo, D., & Olhager, J. (2012). Supply chain integration and performance: The effects of long-term relationships, information technology and sharing, and logistics integration. *International Journal of Production Economics*, *135*(1), 514–522. doi:10.1016/j.ijpe.2011.09.001

Prasanna, S. R., & Haavisto, I. (2018). Collaboration in humanitarian supply chains: An organisational culture framework. *International Journal of Production Research*, *56*(17), 5611–5625. doi:10.1080/00207543.2018.1475762

Prefab, N. Z. (2019). *How to Prefab: A series of New Zealand offsite construction case studies.* Retrieved from https://www.prefabnz.com/Downloads/Assets/13422/1/PrefabNZ%20HOW%20TO%20Prefab%20Book%202019.pdf

Pro, G. E. (2019). *Google Earth Pro*. https://www.google.com/earth/

ProBIM. (2016). *Equipo ProBIM.do. 1ra reunión de coordinación.* Retrieved from https://www.instagram.com/p/BOAv_cojY4c/

ProBIM. (2017a). *Conferencia Agosto 10*. Retrieved from https://www.instagram.com/p/BXDl_kgDdmu/

ProBIM. (2017b). *ProBIM. promoviendo una construcción más BIM y LEAN*. Retrieved from https://www.linkedin.com/company/probim-do/

ProBIM. (2019a). *¿Qué es BIM?* Retrieved from https://www.youtube.com/watch?v=SBwgYAihh1o

ProBIM. (2019b). *2da Jornada OPEN BIM*. Retrieved from https://www.instagram.com/p/B45vPgEH0dh/

Proffitt, D. R., Creem, S. H., & Zosh, W. D. (2001). Seeing Mountains in Mole Hills: Geographical-Slant Perception. *American Psychological Society. Psychological Science*, *12*(5), 418–423. doi:10.1111/1467-9280.00377 PMID:11554677

Pučko, Z., Šuman, N., & Rebolj, D. (2018). Automated continuous construction progress monitoring using multiple workplace real time 3D scans. *Advanced Engineering Informatics*, *38*, 27–40. doi:10.1016/j.aei.2018.06.001

Pugacheva, N., Kirillova, T., Kirillova, O., Luchinina, A., Korolyuk, I., & Lunev, A. (2020). Digital Paradigm in Educational Management: The Case of Construction Education Based on Emerging Technologies. *International Journal of Emerging Technologies in Learning, 15*(13), 96-115. Retrieved March 10, 2021 from https://www.learntechlib.org/p/217607/

Puolitaival, T., & Forsythe, P. (2016). Practical challenges of BIM education. *Structural Survey*, *34*(4/5), 351–366.

Quality and Qualifications Ireland (QQI). (2016). *Awards Standards for Architectural Technology*. Author.

Quebec Construction 4.0 Initiative. (2020). *Quebec Initiative for Construction 4.0: Improve the performance of the Quebec construction industry through the digital shift*. www.constructionnumerique.ca

Quinn, C. N. (2011). *Designing mLearning: Tapping into the mobile revolution for organizational performance*. John Wiley & Sons. https://quinnovation.com/Designing_mLearningSample.pdf

Rahman, N. A., Taher, S., Alahmadi, F., & Nasir, K. A. M. (2016). Arid Hydrological Modeling at Wadi Alaqiq, Madinah, Saudi Arabia. *Jurnal Teknologi*, *78*(7), 51–58. doi:10.11113/jt.v78.4516

Rahman, R., Suleiman, A., Pingbo, T., & Ayer, S. K. (2016). Comparing Building Information Modelling Skills of Project Managers and BIM Managers based on Social Media Analysis, International Conference on Sustainable Design, Engineering and Construction. *Procedia Engineering*, *V145*, 812–819. doi:10.1016/j.proeng.2016.04.106

Raj, S., & Sharma, A. (2014). Supply chain management in the cloud. *Accenture Global Management Consulting*, 1-12.

Raja Sreedharan, V., & Unnikrishnan, A. (2017). Moving towards industry 4.0: A systematic review. *International Journal of Pure and Applied Mathematics, 117*(20), 929–936.

Rajendran & Clarke. (2011). Building Information Modeling: Safety Benefits & Opportunities. *Professional Safety, 56*(10), 44-51.

Ramachandra, T., & Rotimi, J. O. (2016). The construction contracts act and its implications for payment defaults in New Zealand. *Journal of Legal Affairs and Dispute Resolution in Engineering and Construction*, *8*(1), C4515006. doi:10.1061/(ASCE)LA.1943-4170.0000179

Ramburuth, P., & McCormick, J. (2001). Learning diversity in higher education: A comparative study of Asian international and Australian students. *Higher Education*, *42*, 333–350.

Ramprasad, B. (2018). Leveraging existing sensor networks as IoT devices for smart buildings. *IEEE World Forum on Internet of Things, WF-IoT 2018 - Proceedings*. 10.1109/WF-IoT.2018.8355121

Ranganathan, J., Corbier, L., Bhatia, P., Schmitz, S., Gage, P., & Oren, K. (2004). *The greenhouse gas protocol: a corporate accounting and reporting standard* (revised edition). https://www.wri.org/sites/default/files/pdf/ghg_protocol_2004.pdf

Rashid, K. M., Louis, J., & Fiawoyife, K. K. (2019). Wireless electric appliance control for smart buildings using indoor location tracking and BIM-basсd virtual environments. Automation in Construction, 101, 48–58.

Rathnasinghe, A.P., & Kulatunga, A., (n.d.). *Investigation on comparative roles of major stakeholders for BIM implementation in a traditional construction project environment.* Academic Press.

Ratnasabapathy, S., Alashwal, A., & Perera, S. (2020). Investigation of waste diversion rates in the construction and demolition sector in Australia. *Built Environment Project and Asset Management*. Advance online publication. doi:10.1108/BEPAM-01-2020-0012

Raynsford, N., Briginshaw, J., Burke, A., Cook, R., Curtis-Thomas, C., Davies, H., . . . Smith, N. (2016). *Modern methods of construction: Views from the industry.* Technical report based on primary research. Online publication retrieved on 3 Apr 2019 from: www.nhbcfoundation.org/publications/

Reality, E. (2020). *How VR and AR can Help in the pandemic.* https://eonreality.com/vr-ar-remote-learning-working-coronavirus/

Rebecca, W., & Marshall, S. (2012). *A New Face of Education: Bring Technology into the Classroom in the Developing World.* Global Economy and Development, Brookings. https://www.brookings.edu/research/a-new-face-of-education-bringing-technology-into-the-classroom-in-the-developing-world/

Rebolj, D., Pučko, Z., Babič, N. Č., Bizjak, M., & Mongus, D. (2017). Point cloud quality requirements for Scan-vs-BIM based automated construction progress monitoring. *Automation in Construction*, *84*, 323–334. doi:10.1016/j.autcon.2017.09.021

Redwood, J., Thelning, S., Elmualim, A., & Pullen, S. (2017). The proliferation of ICT and digital technology systems and their influence on the dynamic capabilities of construction firms. *Procedia Engineering*, *180*, 804–811.

Ren, S., Li, X., & Liu, X. (2013). The 3D Visual Research of Improved DEM Data Based on Google Earth and ACIS. *Pervasive Computing and The Networked World, Joint International Conference, ICPCA/SWS 2013, LNCS 8351*, 497-507.

Reuter, H. I., Hengl, T., Gessler, P., & Soille, P. (2009). Preparation of DEMs for Geomorphometric Analysis, *Chapter-4. Developments in Soil Science*, *33*, 87–120. doi:10.1016/S0166-2481(08)00004-4

Reutter & Ford. (1996). Perceptions of public health nursing: views from the field. *Journal of Advanced Nursing, 24*(1), 7-15. doi:10.1046/j.1365-2648.1996.01681.x

Riaz, Z., Arslan, M., & Azhar, S. (2014). CoSMoS: A BIM and wireless sensor based integrated solution for worker safety in confined spaces. *Automation in Construction*, *45*(8), 96–106.

Rice, J., & Martin, N. (2018). Smart infrastructure technologies: Crowdsourcing future development and benefits for Australian communities. *Technological Forecasting and Social Change*, (March), 119256.

Richardon, L. (2018, October/November). Designing in a virtual world. *New Zealand Construction News (NZCN)*, 21.

RICS. (2014). *What is BIM?* Adapted from https://www.rics.org/uk/knowledge/glossary/bim-intro/

Rinkinen, S., & Harmaakorpi, V. (2017). The Business Ecosystem Concept in Innovation Policy Context: Building a Conceptual Framework. *Innovation (Abingdon)*, *31*(3), 333–349. doi:10.1080/13511610.2017.1300089

Ritchie, H., & Roser, M. (2020). CO_2 and Greenhouse Gas Emissions. *OurWorldInData.org*. https://ourworldindata.org/co2-and-other-greenhouse-gas-emissions

Rivest, R. L., Shamir, A., & Adleman, L. (1978). A method for obtaining digital signatures and public-key cryptosystems. *Communications of the ACM*, *21*(2), 120–126. doi:10.1145/359340.359342

Robertson, F.J. (2017). Tri-varsity, interdisciplinary BIM Modelling/Management (BIM) workshop: An action research international example. *International Journal of 3-D Information Modelling, 6*(4), 37-56.

Roehrich, J. K., & Lewis, M. A. (2010). Towards a model of governance in complex (product–service) inter-organisational systems. *Construction Management and Economics*, *28*(11), 1155–1164. doi:10.1080/01446191003762249

Rogage, K., Clear, A., Alwan, Z., Lawrence, T., & Kelly, G. (2019). Assessing building performance in residential buildings using BIM and sensor data. *International Journal of Building Pathology and Adaptation.*

Rogers, E. M. (2010). *Diffusion of innovations*. Simon and Schuster.

Rooney, K. (2017). *BIM education - global - 2017 update report.* Retrieved from https://www.canbim.com/single-post/2017/05/29/NATSPEC-Global-BIM-Education-Report-2017

Rooney, K. (2019). *BIM education – global – 2019 updated report.* ICIS Integrating Construction. Retrieved from https://www.icis.org/wp-content/uploads/2019/05/BIM_Education_Global_2019_Update_Report_V6.0.pdf

Rothenberg, J., Widman, L. E., Loparo, K. A., & Nielsen, N. R. (1989). *The nature of modeling* (Vol. 3027). Rand.

Rowlinson, S. (2004). *Construction safety management systems*. Routledge.

Royal Academy of Engineers. (2007). *Educating Engineers for 21st Century*. The Royal Academy of Engineering.

Royal Academy of Engineers. (2010). *Engineering Graduates for Industry*. The Royal Academy of Engineering.

Royal Institute of Architects Ireland (RIAI). (2016). *Statement on Policy on Education.* Author.

Royal Institute of British Architects Enterprises Limited. (2015). *NBS National BIM Report 2015.* Author.

Royal Institute of British Architects. (2012). *BIM Overlay to the RIBA Outline Plan of Work.* Author.

Royal Institution of Chartered Surveyors (RICS). (2018). *RICS Requirements and Competencies Guide.* Available from https://www.rics.org/globalassets/rics-website/media/qualify/pathway-guides/pathway-guides-requirements-and-competencies-v1.3-oct-2018.pdf

Royal Institution of Chartered Surveyors (RICS). (2019). *RICS Global Accreditation –Policy and Process.* https://www.rics.org/globalassets/rics-website/media/upholding-professional-standards/standards-of-qualification/rics-global-accreditation-policy-and-processes.pdf

Rubio, A. (2018). A common-sense based system for Geo-IoT. *Procedia Computer Science*, *126*, 665–674. doi:10.1016/j.procs.2018.07.301

Rudkovska, A., & Sui, W. (2020, April 29). *Why FaceTime can't replace face-to-face time during social distancing.* The Conversation. https://theconversation.com/why-facetime-cant-replace-face-to-face-time-during-social-distancing-136206

Rusinko, C. (2010). Integrating sustainability in higher education: A generic matrix. *International Journal of Sustainability in Higher Education*, *11*(3), 250–259. doi:10.1108/14676371011058541

Rusli, M., & Majid, M. R. (2012). Digital Elevation Model (DEM) Extraction from Google Earth: a Study in Sungai Muar Watershed. *Proceedings of the Applied Geoinformatics for Society and Environment 2012 (AGSE 2012),* 24-28.

Rusli, M., Majid, M. R., & Din, A. H. M. (2014). Google Earth's Derived Digital Elevation Model: A Comparative Assessment with ASTER and SRTM Data. *8th International Symposium of the Digital Earth (ISDE8). IOP Conference Series. Earth and Environmental Science*, *18*, 012065. doi:10.1088/1755-1315/18/1/012065

Russell, M. G., & Smorodinskaya, N. V. (2018). Leveraging complexity for ecosystemic innovation. *Technological Forecasting and Social Change*, *136*, 114–131. doi:10.1016/j.techfore.2017.11.024

Rüßmann, M., Lorenz, M., Gerbert, P., Waldner, M., Justus, J., Engel, P., & Harnisch, M. (2015). Industry 4.0: The Future of Productivity and Growth in Manufacturing. *The Boston Consulting Group*, *20*. Advance online publication. doi:10.100712599-014-0334-4

Rutherford, K. (2008). *Creative Opportunities*. https://www.nesta.org.uk/publications/creative-opportunities

Sabongi, F. J. (2009). The integration of BIM in the undergraduate curriculum: An analysis of undergraduate courses. *Proceedings of the 45th ASC Annual Conference*, 1-4.

Sacks, P., Perlman, A., & Barak, R. (2013). Construction safety training using immersive virtual reality. *Construction Management and Economics*, *31*(9), 1005–1018. doi:10.1080/01446193.2013.828844

Sacks, R., & Barak, R. (2010). Teaching Building Information Modelling as an integral part of freshmen year civil engineering education. *Journal of Professional Issues in Engineering Education and Practice*, *136*(1), 30–38. doi:10.1061/(ASCE)EI.1943-5541.0000003

Sacks, R., Eastman, C., Lee, G., & Teicholz, P. (2018). *BIM handbook: a guide to building information modeling for owners, designers, engineers, contractors, and facility managers* (3rd ed.). John Wiley & Sons, Inc. doi:10.1002/9781119287568

Sacks, R., & Pikas, E. (2013). Building information modeling education for construction engineering and management. I: Industry requirements, state of the art, and gap analysis (Author abstract). *Journal of Construction Engineering and Management*, *139*(11), 04013016. doi:10.1061/(ASCE)CO.1943-7862.0000759

Sacks, R., & Pikas, E. (2013). Building Information Modeling Education for Construction Engineering and Management. I: Industry Requirements, State of the Art, and Gap Analysis. *Journal of Construction Engineering and Management*, *139*(11), 04013016.

Sáez, P. V., & Santa Cruz Astorqui, J. (2018). Estimation of construction and demolition waste in building energy efficiency retrofitting works of the vertical envelope. *Journal of Cleaner Production*, *172*, 2978–2985. doi:10.1016/j.jclepro.2017.11.113

Safe Work Australia. (2017). *Construction*. https://www.safeworkaustralia.gov.au/Construction

Safe Work Australia. (2020). *Australian Work Health and Safety Strategy 2012-2022*. https://www.safeworkaustralia.gov.au/about-us/australian-work-health-and-safety-strategy-2012-2022#

Sahil, A. Q. (2016). *Adoption of building information modeling in developing countries: a phenomenological perspective* (Master's thesis). Colorado State University.

Sakin, M., & Kiroglu, Y. C. (2017). 3D Printing of Buildings: Construction of the Sustainable Houses of the Future by BIM. *Energy Procedia*, *134*, 702–711. doi:10.1016/j.egypro.2017.09.562

Salas, E., Rosen, M. A., Held, J. D., & Weissmuller, J. J. (2009). Performance measurement in simulation-based training: A review and best practices. *Simulation & Gaming*, *40*(3), 328–376. doi:10.1177/1046878108326734

Samarasinghe, D. A. S. (2020). *Development of framework for building materials-related decisions: A case study*. Paper presented at the Fifth Australasia and South-East Asia Structural Engineering and Construction Conference (ASEA SEC 5), Christchurch, New Zealand.

Samarasinghe, D. A. S., Abd-Latif, S., & Baghaei, N. (2019). *Virtual reality models for promoting learner engagement in construction studies.* Paper presented at the IEEE EDUCON - Global Engineering Education, Dubai, UAE. 10.1109/EDUCON.2019.8725120

Samarasinghe, D. A. S., Baghaei, N., & Stemmet, L. (2020). *Persuasive Virtual Reality: Promoting Earth Buildings in New Zealand.* Paper presented at the 15th International Conference on Persuasive Technologies, Aalborg, Denmark.

Samnani, A. K., & Singh, P. (2016). Workplace bullying: Considering the interaction between individual and work environment. *Journal of Business Ethics*, *139*(3), 537–549. doi:10.100710551-015-2653-x

Sanchez, A. X., Kraatz, J. A., & Hampson, K. D. (2014). *Towards a national strategy. research report 1.* Retrieved from http://www.sbenrc.com.au/wp-content/uploads/2013/10/National-Strategy-research-report.pdf

Sánchez, A., González-Gaya, C., Zulueta, P., & Sampaio, Z. (2019). *A Comparative Study of the Use of Building Information Modelling in Teaching Engineering Projects. IEEE Access Journal.*

Sánchez-Lozano, J. M., & Bernal-Conesa, J. A. (2017). Environmental management of Natura 2000 network areas through the combination of Geographic Information Systems (GIS) with Multi-Criteria Decision Making (MCDM) methods. Case study in south-eastern Spain. *Land Use Policy*, *63*, 86–97. doi:10.1016/j.landusepol.2017.01.021

San, K. M., Choy, C. F., & Fung, W. P. (2019). The Potentials and Impacts of Block chain Technology in Construction Industry: A Literature Review. In *IOP Conference Series: Materials Science and Engineering,*. IOP Publishing.

Santos, A. C., Mendes, P., & Ribau Teixeira, M. (2019). Social life cycle analysis as a tool for sustainable management of illegal waste dumping in municipal services. *Journal of Cleaner Production*, *210*, 1141–1149. doi:10.1016/j.jclepro.2018.11.042

Santos, R., Costa, A. A., & Grilo, A. (2017). Bibliometric analysis and review of Building Information Modelling literature published between 2005 and 2015. *Automation in Construction*, *80*, 118–136. doi:10.1016/j.autcon.2017.03.005

Saunders, L., & Mecca, J., & Schaefer, B. (2013). Analysing the Practice of Life Cycle Assessment Focus on the Building Sector. *Journal of Industrial Ecology*, *17*(5), 777–788. doi:10.1111/jiec.12028

Saviano, M., Polese, F., Caputo, F., & Walletzký, L. (2016). *A T-shaped model for rethinking higher education programs* [Paper presentation]. The 19th Toulon-Verona International Conference Excellence in Services, Huelva, Spain.

Schöttle, A., Haghsheno, S., & Gehbauer, F. (2014, June 25-27). Defining cooperation and collaboration in the context of Lean Construction. In *22nd Annual Conference of the International Group for Lean Construction* (pp. 1269-1280). Oslo, Norway: IGLC.

Schouten, C. (2016 January 22). Does construction have a 'productivity problem'? *Construction News.* https://tinyurl.com/ya6ojxnt

Schweiger, D. M., Sandberg, W. R., & Rechner, P. L. (1989). Experiential effects of dialectical inquiry, devil's advocacy and consensus approaches to strategic decision making. *Academy of Management Journal*, *32*(4), 745–772.

Scott, H., Hofmeister, N., Rogness, N., & Rogers, A. E. (2010). Implementing a fatigue countermeasures program for nurses: A focus group analysis. *The Journal of Nursing Administration*, *40*(5), 233–240. doi:10.1097/NNA.0b013e3181da4078 PMID:20431458

Scottish Futures Trust. (2018). *Developing a BIM Competency Framework: Research and Key Principles report.* Author.

Scott, L. (2015). The Changing Landscape of Construction Higher Education. *International Journal of Construction Education and Research*, *11*(2), 78–78. doi:10.1080/15578771.2015.1022454

Sebastian, R. (2011). Changing roles of the clients, architects and contractors through BIM. *Engineering, Construction and Architectural Management, 18*(2), 176-187.

Sebastian, R. (2011). Changing roles of the Clients, architects and contractors through BIM. *Engineering, Construction, and Architectural Management*, *18*(2), 176–187.

Seebacher, S., & Maleshkova, M. (2018). A model-driven approach for the description of blockchain business networks. *Proceedings of the 51st Hawaii International Conference on System Sciences*. 10.24251/HICSS.2018.442

Senge, P. M. (1990). *The fifth discipline: the art and practice of the learning organization*. Doubleday/Currency.

Senior, B. A., & Rodriguez, T. A. (2012). *Analyzing Barriers to Construction Productivity Improvement in the Dominican Republic. In 20th Annual Conference-International Group for Lean Construction (IGLC-20)*. Montezuma Publishing.

Seror, N., & Portnov, B. A. (2018). Identifying areas under potential risk of illegal construction and demolition waste dumping using GIS tools. *Waste Management (New York, N.Y.)*, *75*, 22–29. doi:10.1016/j.wasman.2018.01.027 PMID:29402616

SFT. (2018 July). *Developing a BIM Competency Framework: Research & Key Principles*. https://www.scottishfuturestrust.org.uk/storage/uploads/bimcompetencyframeworkresearchjul18.pdf

Shaawat, M. E., & Jamil, R. (2014). A Guide to Environmental Building Rating System for Construction of New Buildings in Saudi Arabia. *Emirates Journal for Engineering Research*, *19*(2), 47–56.

Shaawat, M. E., Jamil, R., & Al Enezi, M. M. (2018). Analysis of Challenges in Sustainable Construction Industry by Using Analytic Hierarchy Process: A Case Study of Jubail Industrial City, Saudi Arabia. *International Journal of Sustainable Real Estate and Construction Economics*, *1*(2), 109–122. doi:10.1504/IJSRECE.2018.092254

Shafique, M., & Rafiq, M. (2019). An Overview of Construction Occupational Accidents in Hong Kong: A Recent Trend and Future Perspectives. *Applied Sciences (Basel, Switzerland)*, *9*(10), 2069. doi:10.3390/app9102069

Shahrokni, H., van der Heijde, B., Lazarevic, D., Brandt, N., Höjer, M., Lago, P., & Wangel, J. (2014). *Big Data GIS Analytics towards Efficient Waste Management in Stockholm*. Academic Press.

Sharag-Eldin, A., & Nawari, N. O. (2010). BIM in AEC education. *Structures Congress 2010*, 1676-1688.

Sharif, H. O., Al-Juaidi, F. H., Al-Othman, A., Al-Dousary, I., Fadda, E., Jamal-Uddeen, S., & Elhassan, A. (2014). Flood hazards in an urbanizing watershed in Riyadh, Saudi Arabia. *Geomatics, Natural Hazards & Risk*, *7*(2), 702–720. doi:10.1080/19475705.2014.945101

Shaw, K. (2001). Designing online learning opportunities, orchestrating experiences and managing learning. In J. Stephenson (Ed.), *Teaching and learning online: Pedagogies for new technologies* (pp. 53–66). Stylus.

Shelbourn, M., Macdonald, J., McCuen, T., & Lee, S. (2017). Students' perceptions of BIM education in the higher education sector: A UK and US perspective. *Industry and Higher Education*, *31*(5), 293–304.

Shen, X., & Marks, E. (2016). Near-Miss Information Visualization Tool for BIM. *Journal of Construction Engineering and Management*, *142*(4), 04015100.

Shepherd, G. W., Kahler, R. J., & Cross, J. (2000). Crane fatalities— A taxonomic analysis. *Safety Science*, *36*(2), 83–93. doi:10.1016/S0925-7535(00)00017-5

Sher, D. (2019). Winsun completes world's first, 500 meters long, 3D printed river revetment wall. *3D Printing Media Network*. Retrieved from https://www.3dprintingmedia.network/winsun-completes-worlds-first-3d-printed-river-revetment-wall-in-largest-construction-3d-printing-project-yet

Sherman, J. D., & Keller, R. T. (2011). Suboptimal assessment of interunit task interdependence: Modes of integration and information processing for coordination performance. organisation. *Science*, *22*(1), 245–261.

Shi, Y., Jing, D., Ragan, E., Cho, K., & Ma, S. (2018). Social Influence on Construction Safety Behaviours: A Multi-User Virtual Reality Experiment. Construction Research Congress, 174–183.

Shin, I. (2015). Factors that affect safety of tower crane installation/dismantling in construction industry. *Safety Science*, *72*, 379–390. doi:10.1016/j.ssci.2014.10.010

Shou, W., Hou, L., Wang, J., & Wang, X. (2015). Case studies of BIM-based dynamic scaffolding design and safety prevention. In *Proceedings of 2nd International Conference on Civil and Building Engineering Informatics (ICCBEI 2015)* (pp. 54-56). https://sbenrc.com.au/app/uploads/2013/10/Publication-3-27-2_CASE-STUDIES-OF-BIM-BASED-DYNAMIC-SCAFFOLDING-DESIGN-AND-SAFETY-PREVENTION.pdf

Shrestha, S. S., Biswas, K., & Desjarlais, A. O. (2014). A protocol for lifetime energy and environmental impact assessment of building insulation materials. *Environmental Impact Assessment Review*, *46*, 25–31. doi:10.1016/j.eiar.2014.01.002

Sidani, A., Duarte, J., Baptista, J. S., Martins, J. P., & Soeiro, A. (2018). Improving construction safety using BIM-Based sensor technologies. *2.0 Congresso Português de Building Information Modelling,* (1), 161–169.

Silverio, A. K. (2020). *Implementation of Building Information Modelling in the Dominican Republic construction industry* (Unpublished doctoral dissertation). University of Wolverhampton, Wolverhampton, UK.

Silverio, A. K., Suresh, S., Heesom, D., & Renukappa, S. (2017a) BIM Education Framework for Clients and Professionals of the Construction Industry. *International Journal of 3-D Information Modeling, 6*(2), 57-79.

Silverio, A. K., Suresh, S., Suresh, S., & Heesom, D. (2017) Building information management (BIM) education in the Dominican Republic: An empirical study. In *International Conference on Sustainable Futures*. Applied Science University.

Silverio, A., Suresh, S., Suresh, R., & Heesom, D. (2016). *BIM Education Framework for Clients and Professionals of the Construction Industry*. Glasgow Caledonian University.

Simion, I. M., Ghinea, C., Maxineasa, S. G., Taranu, N., Bonoli, A., & Gavrilescu, M. (2013). Ecological footprint applied in the assessment of construction and demolition waste integrated management. *Environmental Engineering and Management Journal*, *12*(4).

Sim, J. (1998). Collecting and analysing qualitative data: Issues raised by the focus group. *Journal of Advanced Nursing*, *28*(2), 345–352. doi:10.1046/j.1365-2648.1998.00692.x PMID:9725732

Simpson, M., & Carlton, D. (2019 January 12). *Upskilling Briefing Paper*. UK BIM Alliance. https://tinyurl.com/ycn7bgcg

Simpson, M., Underwood, J., Shelbourn, M., & Mollasalehi, S. (2019). *Evolve or Die: Transforming the productivity of Built Environment Professionals and Organisations of Digital Built Britain through a new digitally enabled ecosystem underpinned by the mediation between competence supply and demand.* Pedagogy and Upskilling CDBB Network Report.

Simpson, M., Underwood, J., Shelbourn, M., Carlton, D., Aksenova, G., & Mollasalehi, S. (2019). *Pedagogy and Upskilling Network: Evolve or Die-Transforming the productivity of Built Environment Professionals and Organisations of Digital Built Britain through a new, digitally enabled ecosystem underpinned by the mediation between competence supply and demand.* Academic Press.

Singhal, R., & Rana, R. (2015). Chi-square test and its application in hypothesis testing. *Journal of the Practice of Cardiovascular Sciences*, *1*(1), 69. doi:10.4103/2395-5414.157577

Singh, S. (2000). *The Code Book: The Code Book: The Science of Secrecy from Ancient Egypt to Quantum Cryptograph.* Anchor Books.

Singh, V., Gu, N., & Wang, X. (2011). A theoretical framework of a BIM-based multi-disciplinary collaboration platform. *Automation in Construction*, *20*(2), 134–144. doi:10.1016/j.autcon.2010.09.011

Slaughter, S. (1998). Models of Construction Innovation. *Journal of Construction Engineering and Management*, *124*(3), 226–231. doi:10.1061/(ASCE)0733-9364(1998)124:3(226)

Slaughter, S. (2000). Implementation of construction innovations. *Building Research and Information*, *28*(1), 2–17. doi:10.1080/096132100369055

Smallwood, J. (2019). Clients and Construction health and safety (H&S). In *Proceedings of the Creative Construction Conference 2019* (pp. 656-664). doi:10.3311/CCC2019-090

Smallwood, J. J., & Emuze, F. (2013). Assessing the Contributions of a Public Sector Client to Contractor Health and Safety (H&S) Improvement in South Africa. In *Proceedings of the 2nd International Conference on Infrastructure Development in Africa-ICIDA* (pp. 346-354). https://www.researchgate.net/profile/Joshua_Ayarkwa/publication/312175534_Comparative_residential_benefits/links/5874cd6c08ae329d62201a5a/Comparative-residential-benefits.pdf

Smith, D. K. T. M. (2009). *Building information modeling: a strategic implementation guide for architects, engineers, constructors, and real estate asset managers.* Wiley. doi:10.1002/9780470432846

Smith, P. (2014). BIM implementation - Global strategies. *Procedia Engineering*, *85*, 482–492.

Smith, P. (2016). Project cost management with 5D BIM. *Procedia: Social and Behavioral Sciences*, *226*, 193–200. doi:10.1016/j.sbspro.2016.06.179

Society of Chartered Surveyors Ireland (SCSI). (n.d.). *APC Pathway Guides to Quantity Surveying and Construction, Building Surveying, Building Control and Project Management.* Author.

Söderberg, J. (2015). *Hacking capitalism: The free and open source software movement.* Routledge. doi:10.4324/9780203937853

Soemardi, B., & Erwin, R. G. (2017). Using BIM as a Tool to Teach Construction Safety. *MATEC Web of Conferences*, *138*(4), 05007. doi:10.1051/matecconf/201713805007

Soler, J., Salcedo, N., & Núñez, Y. (2013). *Estudio prospectivo de la formación profesional del sector de la construcción civil.* INFOTEP. Retrieved from https://www.oitcinterfor.org/node/5539

Soosay, C. A., & Hyland, P. (2015). A decade of supply chain collaboration and directions for future research. *Supply Chain Management*, *20*(6), 613–630. doi:10.1108/SCM-06-2015-0217

Spady, W. G. (1994). *Oucome-Based Education: Critical Issues and Answers.* American Association of School Administrators.

Srour, I., AlZaghrini, N., & Srour, F. J. (2019). Using GIS and optimization to manage construction and demolition waste: The case of abandoned quarries in Lebanon. *Waste Management (New York, N.Y.)*, *95*, 139–149. doi:10.1016/j.wasman.2019.06.011 PMID:31351599

Standards Australia / Standards New Zealand. (2001). *Occupational health and safety management systems: general guidelines on principles, systems and supporting techniques* (2nd ed.). Standards Australia Int'l Ltd.

Stasz, C. (2001). Assessing skills for work: Two perspectives. *Oxford Economic Papers*, *53*(3), 385–405. doi:10.1093/oep/53.3.385

Statsbygg. (2013). *Statsbygg BIM Manual Version 1.2.1.* http://www.statsbygg.no/Files/ publikasjoner/manualer/ StatsbyggBIM-manual-ver1-2-1eng-2013-12-17.pdf

Stefan, W. K. (1996) Using DTMs and GIS to Define Input Variables for Hydrological and Geomorphological Analysis. *Proceeding of the Vienna Conference, April 1996, HydroGIS 96: Applications of GIS in Hydrology and Water Resources Management*, 183-190.

Stefanucci, J. K., Proffitt, D. R., Banton, T., & Epstein, W. (2005). Distances Appear Different on Hills. *Perception & Psychophysics*, *67*(6), 1052–1060. doi:10.3758/BF03193631 PMID:16396013

Steinmetz, R., & Wehrle, K. (Eds.). (2005). *Peer-to-peer systems and applications* (Vol. 3485). Springer. doi:10.1007/11530657_2

STF. (2018). *Developing a BIM Competency Framework: Research & Key Principles.* STF.

Strozzi, F., Colicchia, C., Creazza, A., & Noè, C. (2017). Literature review on the 'Smart Factory'concept using bibliometric tools. *International Journal of Production Research*, *55*(22), 6572–6591. doi:10.1080/00207543.2017.1326643

Subyani, A. M., & Al Amri, N. S. (2015). IDF curves and daily rainfall generation for Al-Madinah city, Western Saudi Arabia. *Arabian Journal of Geosciences*, *8*(12), 11107–11119. doi:10.100712517-015-1999-9

Subyani, A. M., Qari, M. H., & Matsah, M. I. (2012). Digital elevation model and multivariate statistical analysis of morphometric parameters of some wadis, western Saudi Arabia. *Arabian Journal of Geosciences*, *5*(1), 147–157. doi:10.100712517-010-0149-7

Succar, B. (2010b). *The Five Components of BIM Performance Measurement.* Paper presented at the CIB World Congress, Salford, UK.

Succar, B. (2014). *Building information modelling: conceptual constructs and performance improvement tools* (Ph.D. thesis). University of Newcastle.

Succar, B., & Sher, W. (n.d.). A competency knowledge-base for BIM learning. *Australasian Journal of Construction Economics and Building-Conference Series, 2*(2) 1-10.

Succar, B., Agar, C., Beazley, S., Berkemenier, P., Choy, R., Rosetta, D. G., Donaghey, S., Linning, C., Macdonald, J., Perey, R., & Plume, J. (2012). *BIM in Practice – BIM Education.* Paper by the Australian Institute of Architects and Consult Australia.

Succar, B., Sher, W., & Aranda-Mena, G. (2007). *A proposal Framework to Investigate Building Information Modelling through Knowledge Elicitation and Visual Models.* Paper presented at the Australian Universities Building Education (AUBEA 2007), Melbourne, Australia.

Succar, B. (2009). Building information modelling framework: A research and delivery foundation for industry stakeholders. *Automation in Construction*, *18*(3), 357–375. doi:10.1016/j.autcon.2008.10.003

Succar, B. (2010a). Building Information Modelling Maturity Matrix. In J. Underwood & U. Iskidag (Eds.), *Handbook of Research on Building Information Modelling and Construction Informatics: Concepts and Technologies, 65-103, Information Science Reference*. IGI Publishing. doi:10.4018/978-1-60566-928-1.ch004

Succar, B., & Kassem, M. (2015). Macro-BIM adoption: Conceptual structures. *Automation in Construction*, *57*, 64–79. doi:10.1016/j.autcon.2015.04.018

Succar, B., & Poirier, E. (2020). Lifecycle information transformation and exchange for delivering and managing digital and physical assets. *Automation in Construction*, *112*, 103090. doi:10.1016/j.autcon.2020.103090

Succar, B., & Sher, W. (2014). A Competency Knowledge Base for BIM Learning. *Australian Journal of Construction Economics and Building Conference Series, V2*(2), 1–10. doi:10.5130/ajceb-cs.v2i2.3883

Succar, B., Sher, W., & Williams, A. (2013). An integrated approach to BIM competency assessment, acquisition and application. *Automation in Construction, 35*, 174–189. doi:10.1016/j.autcon.2013.05.016

Sulankivi, K., Kähkönen, K., Mäkelä, T., & Kiviniemi, M. (2010). 4D-BIM for Construction safety planning. In *Proceedings of W099-Special Track the 18th CIB World Building Congress* (pp. 117-128). https://www.irbnet.de/daten/iconda/CIB_DC24339.pdf

Sulankivi, K., Kähkönen, K., Mäkelä, T., & Kiviniemi, M. (2010, May) 4D-BIM for construction safety planning. In *Proceedings of W099-Special Track 18th CIB World Building Congress* (Vol. 2010, pp. 117-128). Academic Press.

Surfer 13. (2015). Full User's Guide. Golden Software LLC.

Susskind, R., & Susskind, D. (2015). *The Future of the Professions: How Technology will Transform the Work of Human Experts*. Oxford University Press. doi:10.1093/oso/9780198713395.001.0001

Suwal, S., Jäväjä, P., Rahman, M. A., & Gonzalez, V. (2013). Exploring BIM-based education perspectives. *Aubea–2013*.

Suwal, S., & Singh, V. (2018). Assessing students' sentiments towards the use of a building information modelling (BIM) learning platform in a construction project management course. *European Journal of Engineering Education, 43*(4), 492–506.

Swan, M. (2018). Blockchain Economics: "Ripple for ERP". *Eur. Financ. Rev*, 24–27.

Swan, M. (2015). *Blockchain: Blueprint for a new economy*. O'Reilly Media, Inc.

Swan, M. (2018). Blockchain enlightenment and smart city cryptopolis. In *Proceedings of the 1st Workshop on Cryptocurrencies and Blockchains for Distributed Systems* (pp. 48-53). 10.1145/3211933.3211942

Swanson, T. (2015). *Consensus-as-a-service: a brief report on the emergence of permissioned, distributed ledger systems*. Retrieved from http://www.ofnumbers.com/wp-content/uploads/2015/04/Permissioned-distributed-ledgers.pdf

Syed, T. A., Alzahrani, A., Jan, S., Siddiqui, M. S., Nadeem, A., & Alghamdi, T. (2019). A comparative analysis of blockchain architecture and its applications: Problems and recommendations. *IEEE Access: Practical Innovations, Open Solutions, 7*, 176838–176869. doi:10.1109/ACCESS.2019.2957660

Szabo, N. (1997). *The God Protocols*. Satoshi Nakamoto Institute. Retrieved from nakamotoinstitute.org/the-god-protocols/

Tabassi, A. A., & Abu Bakar, A. H. (2009). Training, motivation, and performance: The case of human resource management in construction projects in Mashhad, Iran. *International Journal of Project Management, 27*(5), 471–480. doi:10.1016/j.ijproman.2008.08.002

Tah, J. H. (2012). Virtual planning and knowledge-based decision support. *Construction Innovation and Process Improvement*, 347-362.

Tam, C., & Fung, I. W. (2011). *Tower crane safety in the construction industry: A Hong Kong study*. Academic Press.

Tam, C., & Fung, I. W. IV. (1998). Effectiveness of safety management strategies on safety performance in Hong Kong. *Construction Management and Economics, 16*(1), 49–55. doi:10.1080/014461998372583

Tam, V. W. Y., & Lu, W. (2016). Construction Waste Management Profiles, Practices, and Performance: A Cross-Jurisdictional Analysis in Four Countries. *Sustainability, 8*(2), 190. doi:10.3390u8020190

Tan, S., Hashim, H., Lee, C., Lim, J., & Kanniah, K. (2014). *Optimal waste-to-energy strategy assisted by GIS For sustainable solid waste management*. Academic Press.

Tangpong, C., Hung, K. T., & Ro, Y. K. (2010). The interaction effect of relational norms and agent cooperativeness on opportunism in buyer–supplier relationships. *Journal of Operations Management*, *28*(5), 398–414. doi:10.1016/j.jom.2009.12.001

Tang, S., Shelden, D. R., Eastman, C. M., Pishdad-Bozorgi, P., & Gao, X. (2019). A review of building information modeling (BIM) and the internet of things (IoT) devices integration: Present status and future trends. *Automation in Construction*, *101*(February), 127–139. doi:10.1016/j.autcon.2019.01.020

Tan, T., Chen, K., Xue, F., & Lu, W. (2019). Barriers to Building Information Modeling (BIM) implementation in China's prefabricated Construction: An interpretive structural modeling (ISM) approach. *Journal of Cleaner Production*, *219*, 949–959.

Tapscott, D., & Tapscott, A. (2017). How blockchain will change organizations. *MIT Sloan Management Review*, *58*, 10.

Taylor Review. (2017). *Good Work: The Taylor Review of Modern Working Practices*. https://assets.publishing.service.gov.uk/government/uploads/system/uploads/attachment_data/file/627671/good-work-taylor-review-modern-working-practices-rg.pdf

Taylor, J., & Levitt, R. (2004). Understanding and managing systemic innovation in project-based industries. *Innovations: Project management research*, 83-99.

Tay, Y. W. D., Panda, B., Paul, S. C., Noor Mohamed, N. A., Tan, M. J., & Leong, K. F. (2017). 3D printing trends in building and construction industry: A review. *Virtual and Physical Prototyping*, *12*(3), 261–276. doi:10.1080/17452759.2017.1326724

Teizer, J., Wolf, M., Golovina, O., Perschewski, M., Propach, M., Neges, M., & König, M. (2017). *Internet of Things (IoT) for integrating environmental and localization data in Building Information Modeling (BIM)* [Paper presentation]. The 34th International Symposium on Automation and Robotics in Construction and Mining (ISARC 2017), Taipei, Taiwan.

Teo, M. M. M., & Loosemore, M. (2001). *A theory of waste behaviour in the construction industry*. Academic Press.

Teo, E. A. L., & Ling, F. Y. Y. (2006). Developing a model to measure the effectiveness of safety management systems of construction sites. *Building and Environment*, *41*(11), 1584–1592. doi:10.1016/j.buildenv.2005.06.005

Terroso-Saenz, F., González-Vidal, A., Ramallo-González, A. P., & Skarmeta, A. F. (2019). An open IoT platform for the management and analysis of energy data. *Future Generation Computer Systems*, *92*, 1066–1079. doi:10.1016/j.future.2017.08.046

The British Standard Institute (BSI). (2014). *PAS 180 - Smart cities-Vocabulary*. Disponível em: http://shop.bsigroup.com/upload/PASs/Free-Download/PAS180.pdf

The Construction Industry Council. (2020). *CIC BIM Standards – General*. Construction Industry Council.

Thomas, K., Chisholm, G., Dempsey, B., Graham, B., & Stubbs, R. (2013). Collaborative BIM learning via an academic-industry partnership. *Proceedings of the CITA BIM Gathering Conference*, 201-206.

Thomas. (2014) Collaborative BIM Learning via an Academia-Industry Partnership. *International Journal of 3-D Information Modelling (IJ3DIM)*, *3*(1), 40-48.

Thompson, J. D. (2003). Organisations in action: Social science bases of administrative theory. Transaction Publishers.

Thunberg, M., Rudberg, M., & Gustavsson, T. K. (2017). Categorising on-site problems. *Construction Innovation*, *17*(1), 90-111. doi:http://dx.doi.org.ezproxy.auckland.ac.nz/10.1108/CI-10-2015-0059

Tight, M. (2014). Theory Development and Application in Higher Education Research: The Case of Threshold Concepts. *Theory and Method in Higher Education Research II (International Perspectives on Higher Education Research*, *10*, 249-267.

Tissenbaum & Slotta. (2019). *Developing a smart classroom infrastructure to support real-time student collaboration and inquiry: a 4-year design study.* doi:10.100711251-019-09486-1

Tisserant, A., Pauliuk, S., Merciai, S., Schmidt, J., Fry, J., Wood, R., & Tukker, A. (2017). Solid Waste and the Circular Economy: A Global Analysis of Waste Treatment and Waste Footprints. *Journal of Industrial Ecology*, *21*(3), 628–640. doi:10.1111/jiec.12562

Tobias, M. (2020). Prefabrication: How China built two Coronavirus hospitals in two weeks. *Nearby Engineers*. Retrieved from https://www.ny-engineers.com/blog/prefabrication-how-china-built-two-coronavirus-hospitals-in-two-weeks

Toh, Y. Z., Goh, Y. M., & Guo, B. H. W. (2016). Knowledge, Attitude, and Practice of Design for Safety. Multiple Stakeholders in the Singapore Construction Industry, 1–11. doi:10.1061/(ASCE)CO.1943-7862.0001279

Tom Fleming Creative Consultancy & Nesta. (2013). *Six Creative Councils recognised for pioneering innovations*. Creative Councils. http://www.nesta.org.uk/news/six-creative-councils-recognised-pioneering-innovations#sthash.laZFt4jo.EdMELxLf.dpuf

Torrell, B. (2006). The role of narration in high stakes decision-making: translating knowing into telling. In G. Schreyogg & J. Koch (Eds.), *Knowledge management and narratives: organizational effectiveness through storytelling* (pp. 247–272). Erich Schmidt Verlag GmbH & Cp.

Törrönen, A. (2017). Automating Construction Processes by Making BIMs Machine-Readable. In A. Heiskanen (Ed.), *On the fast KIRA-digi track. AEC-Business.*

Tortorella, G. L., & Fettermann, D. (2018). Implementation of Industry 4.0 and lean production in Brazilian manufacturing companies. *International Journal of Production Research*, *56*(8), 2975–2987. doi:10.1080/00207543.2017.1391420

Tougwa, F. N. (2018). *The Construction Industry and Its Research Rationale in the Construction Industry and Its Research*. Academic Press.

Training Industry Report. (2018). *Training Magazine*. Retrieved from https://trainingmag.com/sites/default/files/trn-2018-industry-report.pdf

Tran, A., Mills, J., Morris, D., & Phillips, M. (2012). *"All hands on deck": Collaborative building design education for architects and engineers.* Paper presented at the Conference Innovation, Practice and Research in Engineering Education, Coventry University, UK.

Traxler, J. (2007). Defining, discussing, and evaluating mobile learning: The moving finger writes and having writ…. *International Review of Research in Open and Distance Learning*, *8*(2), 1–11. doi:10.19173/irrodl.v8i2.346

Truong, N. B., Sun, K., Lee, G. M., & Guo, Y. (2020). GDPR-Compliant Personal Data Management: A Blockchain-Based Solution. *IEEE Transactions on Information Forensics and Security*, *15*, 1746–1761. doi:10.1109/TIFS.2019.2948287

Tsai, W. T., Luo, Y., Deng, E., Zhao, J., Ding, X., Li, J., & Yuan, B. (2020). Blockchain systems for trade clearing. *The Journal of Risk Finance*, *21*(5), 469–492. doi:10.1108/JRF-02-2017-0022

Tsanos, C. S., & Zografos, K. G. (2016). The effects of behavioural supply chain relationship antecedents on integration and performance. *Supply Chain Management, 21*(6), 678–693. doi:10.1108/SCM-06-2016-0211

Tsuinashi, S. (2008). Visual Illusion of a Vertical Gradient: A Pitched Visual Environment and a Pitched Floor. *Japanese Journal of Phycology, 79*(2), 125–133. PMID:18678062

Turk, Ž., & Klinc, R. (2017). Potentials of blockchain technology for construction management. *Procedia Engineering, 196*, 638–645. doi:10.1016/j.proeng.2017.08.052

Tvedt, I. M., & Dyb, K. A. (2019). The Soft Factors in Design Management: A Hidden Success Factor? *10th Nordic Conference on Construction Economics and Organization*, 2, 111-117. 10.1108/S2516-285320190000002032

U.S. Institute of Building Documentation (USIBD). (2016). *C120 (v3) Guide for Level of Accuracy (LOA) Specification for Building Documentation.* http://www.usibd.org/products/level-accuracy-loa-specificationversion

Udomdech, P., Papadonikolaki, E., & Davies, A. (2018, September). An alternative project-based learning model for building information modelling-using teams. Association of Researchers in Construction Management (ARCOM), 34, 57-66.

Uff, J. (2016). *UK Engineering 2016: An independent review.* http://www.keatingchambers.com/wp-content/uploads/2017/05/UK-Engineering-2016-An-Independent-review-by-Professor-John-Uff-CBE-QC-FREng.pdf

Uhm, M., Lee, G., & Jeon, B. (2017). An analysis of BIM jobs and competence based on the use of terms in the industry. *Automation in Construction, 81*, 67–98. doi:10.1016/j.autcon.2017.06.002

UK Government. (2015). *Digital Built Britain. Level 3 Building Information Modelling – Strategic Plan.* UK Government. Available from https://assets.publishing.service.gov.uk/government/uploads/system/uploads/attachment_data/file/410096/bis-15-155-digital-built-britain-level-3-strategy.pdf

UK Universities, . (2019). *The Future of Degree Apprenticeships.* Author.

UKBF. (2020). *Learning Outcomes.* https://ukbimframework.org/wp-content/uploads/2020/07/UK-BIM-Framework-Learning-Outcomes_Edition1.pdf

Ullah, K., Lill, I., & Witt, E. (2019). An Overview of BIM Adoption in the Construction Industry: Benefits and Barriers. In *10th Nordic Conference on Construction Economics and Organization* (Vol. 2), Emerald Publishing Limited.

Underwood, J. (2014, November 20). Educating BIM: Changing the Face of Construction. Planning & Control. *BIM Today.* https://www.pbctoday.co.uk/news/bim-news/educating-bim-changing-face-construction/11725/

Underwood, J., & Ayoade, O. (2015). *Current Position and Associated Challenges of BIM Education in UK Higher Education.* Prepared for the BIM Academic Forum UK.

Underwood, J., Ayoade, O., Khosrowshahi, F., Greenwood, D., Pittard, S., & Garvey, R. (2015). *Current position and associated challenges of BIM education in UK higher education.* BIM Academic Forum.

Underwood, J., Khosrowshahi, F., Pittard, S., Greenwood, D., & Platts, T. (2013). *Embedding building information modelling (BIM) within the taught curriculum: Supporting BIM implementation and adoption through the development of learning outcomes within the UK academic context for built environment programmes.* Academic Press.

UNIBE. (2013). *Especialista de BIM y AUTODESK imparte conferencia sobre implementación de procesos BIM e IPD en proyectos de construcción.* Retrieved from https://www.unibe.edu.do/es/noticias/thu-2013-07-18-1525/especialista-bim-y-autodesk-imparte-conferencia-implementacion-procesos-bim-e-ipd-proyectos-construccion

UNIBE. (2019). *Diplomado de innovación en arquitectura.* Retrieved from https://www.unibe.edu.do/programa/diplomado-de-innovacion-en-arquitectura/

United Nations. (1998). *Kyoto Protocol to the United Nations framework convention on climate change.* United Nations Framework Convention on Climate Change, Kyoto, Japan. https://unfccc.int/resource/docs/convkp/kpeng.pdf

United Nations. (2017). *World Population Prospects The 2017 Revision. Key Findings and Advance Tables.* Department of Economic and Social Affairs. Population Division. https://www.un.org/development/desa/publications/world-population-prospects-the-2017-revision.html

University of Birmingham. (2016). *Smart cities better defined by new research.* https://www.birmingham.ac.uk/news/latest/2016/02/Smart-cities-better-defined-by-new-research-100216.aspx

University of Nottingham. (2019, September 16). Virtual reality training could improve employee safety. *Science Daily.* Retrieved July 8, 2020 from www.sciencedaily.com/releases/2019/09/190916212516.htm

US EPA. (2020). *Advancing Sustainable Material Management 2018 Fact Sheet.* Office of Land and Emergency Management. The United States Environmental Protection Agency. https://www.epa.gov/sites/production/files/2020-11/documents/2018_ff_fact_sheet.pdf

Utama, W. P., Chan, A. P. C., Sesmiwati, S., & Khan, H. Z. A. (2019). Internationalization of construction enterprises: An overview of motivation. *International Journal of Technology, 10*(1), 36–46. doi:10.14716/ijtech.v10i1.1558

Uusitalo, P., Seppänen, O., Peltokorpi, A., & Olivieri, H. (2019). Solving design management problems using lean design management: The role of trust. *Engineering, Construction, and Architectural Management, 26*(7), 1387–1405. doi:10.1108/ECAM-03-2018-0135

Vaidya, S., Ambad, P., & Bhosle, S. (2018). Industry 4.0 - A Glimpse. *Procedia Manufacturing, 20*, 233–238. doi:10.1016/j.promfg.2018.02.034

Valencia, N. (2017). *World's First 3D Printed Bridge Opens in Spain.* Archdaily. Retrieved from https://www.archdaily.com/804596/worlds-first-3d-printed-bridge-opens-in-spain

Valk, J. H., Rashid, A. T., & Elder, L. (2010). Using mobile phones to improve educational outcomes: An analysis of evidence from Asia. *International Review of Research in Open and Distance Learning, 11*(1), 117. doi:10.19173/irrodl.v11i1.794

Van der Zee, D. J., Holkenberg, B., & Robinson, S. (2010). Conceptual modelling for simulation-based serious gaming. *Decision Support Systems, 54*(1), 33–45. doi:10.1016/j.dss.2012.03.006

Vandecasteele, F., Merci, B., & Verstockt, S. (2017). Fireground location understanding by semantic linking of visual objects and building information models. *Fire Safety Journal, 91*, 1026–1034. doi:10.1016/j.firesaf.2017.03.083

Vandis. (2019). *The Importance of Cloud Computing in Education.* https://www.vandis.com/insights/the-importance-of-cloud-computing-in-education/

Vangen, S., Hayes, J. P., & Cornforth, C. (2015). Governing cross-sector, inter-organisational collaborations. *Public Management Review, 17*(9), 1237–1260. doi:10.1080/14719037.2014.903658

Victorian Digital Asset Strategy (VDAS). (2019). *Victorian Digital Asset Strategy.* http://www.opv.vic.gov.au/Victorian-Chief-Engineer/Victorian-Digital-Asset-Strategy

Vigneault, M. A., Boton, C., Chong, H. Y., & Cooper-Cooke, B. (2019). An innovative framework of 5D BIM solutions for construction cost management: A systematic review. *Archives of Computational Methods in Engineering*, 1–18.

Viriyasitavat, W., Da Xu, L., Bi, Z., & Sapsomboon, A. (2018). Blockchain-based business process management (BPM) framework for service composition in industry 4.0. *Journal of Intelligent Manufacturing*, 1–12.

Vizerra, S. A. (2012-2020). *Revizto*. Retrieved May 29, 2020, from Revizto: https://revizto.com/en

Von Ahsen, A. (2014). The integration of quality, environmental and health and safety management by car manufacturers - A long-term empirical study. *Business Strategy and the Environment*, *23*(6), 395–416. doi:10.1002/bse.1791

Vrijhoef, R., & Koskela, L. (2000). The four roles of supply chain management in construction. *European Journal of Purchasing & Supply Management, 6*(3-4), 169-178.

Vukolić, M. (2016). *Hyperledger fabric: towards scalable blockchain for business.* Tech. rep., Tech. rep. Trust in Digital Life 2016. IBM Research. Recovered 1 Apr 2019 from trustindigitallife.eu/wp-content/uploads/2016/07/marko_vukolic.pdf

Vukolić, M. (2017). Rethinking Permissioned Blockchains. *Proceedings of the ACM Workshop on Blockchain, Cryptocurrencies and Contracts*, 3–7. 10.1145/3055518.3055526

Walia, J. S., Hämmäinen, H., & Matinmikko, M. (2017). 5G Micro-operators for the future campus: A techno-economic study. *Conference internet of things business models, users, and networks*. http://jultika.oulu.fi/files/nbnfi-fe201802093294.pdf

Walker, D. H. T. (2002). Enthusiasm, commitment, and project alliancing: An Australian experience. *Construction Innovation*, *2*(1), 15–31. doi:10.1108/14714170210814667

Walker, D., & Walker, B. (2015). *Collaborative project procurement arrangements*. PMI Inc.

Waller, D. L. (2003). *Operations management: a supply chain approach*. International Thomson Business Press.

Walters, A. (2017). *Press Release: Government Announce New Centre at Cambridge*. Retrieved from https://www.cdbb.cam.ac.uk/news/2017NovPressRelease

Wan, A. (2017). *The Application of Building Information Modelling for Safety in the Construction Industry* [PhD Dissertation]. Leeds Beckett University.

Wan, A., Platten, A., & Briggs, T. (2013). Study of Safety Auditors' Views on the Use of BIM for Safety in Hong Kong. *International Journal of 3-D Information Modeling, 2*(1).

Wang, H., Pan, Y., & Luo, X. (2019). Integration of BIM and GIS in sustainable built environment: A review and bibliometric analysis. Automation in Construction, 103, 41–52.

Wang, H., Zhang, J., & Lin, H. (2019). Satellite-based analysis of landfill landslide: the case of the 2015 Shenzhen landslide. *International Journal of Geotechnical Engineering*, 1-8.

Wang, W., Hoang, D. T., Xiong, Z., Niyato, D., Wang, P., Hu, P., & Wen, Y. (2018). *A survey on consensus mechanisms and mining management in blockchain networks.* arXiv preprint arXiv:1805.02707.

Wang, J. Y., Wu, H. Y., Tam, V. W. Y., & Zuo, J. (2019). Considering life-cycle environmental impacts and society's willingness for optimizing construction and demolition waste management fee: An empirical study of China. *Journal of Cleaner Production, 206*, 1004–1014. doi:10.1016/j.jclepro.2018.09.170

Wang, J., Wang, Q., Zhou, N., & Chi, Y. (2017). A novel electricity transaction mode of microgrids based on blockchain and continuous double auction. *Energies*, *10*(12), 1971. doi:10.3390/en10121971

Wang, L., Huang, M., Zhang, X., Jin, R., & Ynag, T. (2020). Review of BIM Adoption in the Higher Education of AEC Disciplines. *Journal of Civil Engineering Education*, *146*(3).

Wang, P., Wu, P., Wang, J., Chi, H. L., & Wang, X. (2018). A critical review of the use of virtual reality in construction engineering education and training. *International Journal of Environmental Research and Public Health*, *15*(6), 1204. doi:10.3390/ijerph15061204 PMID:29890627

Wang, Q., & Kim, M. K. (2019). Applications of 3D point cloud data in the construction industry: A fifteen-year review from 2004 to 2018. *Advanced Engineering Informatics*, *39*, 306–319. doi:10.1016/j.aei.2019.02.007

Wang, S., Wang, W., Wang, K., & Shih, S. (2015). Applying building information modeling to support fire safety management. *Automation in Construction*, *59*, 158–167.

Wang, S., Yuan, Y., Wang, X., Li, J., Qin, R., & Wang, F.-Y. (2018). An overview of smart contract: architecture, applications, and future trends. *2018 IEEE Intelligent Vehicles Symposium (IV)*, 108–113. 10.1109/IVS.2018.8500488

Wang, X. P., Yu, R., Shui, Z. H., Song, Q. L., Liu, Z., Bao, M., Liu, Z. J., & Wu, S. (2019). Optimized treatment of recycled construction and demolition waste in developing sustainable ultra-high performance concrete. *Journal of Cleaner Production*, *221*, 805–816. doi:10.1016/j.jclepro.2019.02.201

Wang, Y., Thangasamy, V. K., Hou, Z., Tiong, R. L., & Zhang, L. (2020). Collaborative relationship discovery in BIM project delivery: A social network analysis approach. *Automation in Construction*, *114*, 103147. doi:10.1016/j.autcon.2020.103147

Wang, Z., Wang, T., Hu, H., Gong, J., Ren, X., & Xiao, Q. (2020). Blockchain-based framework for improving supply chain traceability and information sharing in precast construction. *Automation in Construction*, *111*, 103063. doi:10.1016/j.autcon.2019.103063

Wankmüller, C., & Reiner, G. (2020). Coordination, cooperation, and collaboration in relief supply chain management. *Journal of Business Economics*, *90*(2), 239–276. doi:10.100711573-019-00945-2

Wanzala, W. (2013). Quest for quality and relevant higher education, training, and learning in Kenya: An overview. *Education Journal*, *2*(2), 36–49. doi:10.11648/j.edu.20130202.13

Ward, D., Whysall, D., Le Grand, Z., Guthrie, W., & Davies, R. (2009). *Never Waste a Good Crisis. A Review of Progress since Rethinking Construction and Thoughts for our Future*. Constructing Excellence.

Warne, L. (2016). Crime in the Construction Industry. CIOB Policy Document, Chartered Instute of Building.

Warszawski, A. (2003). *Industrialized and automated building systems: A managerial approach*. Routledge. doi:10.4324/9780203223697

Waterhouse, R., Bew, M., Parkinson, K., Malleson, A., May, I., Matthews, A., . . . Hamil, S. (2017). *NBS National BIM Report 2017*. Academic Press.

Waterhouse, R., Highfield, B., Hamil, S., Malleson, A., Kell, A., Tate, G., . . . Ainscough, N. (2019). *NBS Construction Technology Report 2019*. Academic Press.

Watson, A. (2010). BIM-a driver for change. *Proceedings of the 13th International Conference on Computing in Civil and Building Engineering*. http://www.engineering.nottingham.ac.uk/icccbe/proceedings/pdf/pf69.pdf

Weber, D., & Hedges, K. (2008). *From CAD to BIM: The Engineering Student Perspective. Proceedings from the American Society of Engineers*.

Weber, I., Xu, X., Riveret, R., Governatori, G., Ponomarev, A., & Mendling, J. (2016). Untrusted business process monitoring and execution using blockchain. *International Conference on Business Process Management*, 329–347. 10.1007/978-3-319-45348-4_19

Wells, J., & Hawkins, J. (2010). Increasing 'local content' in infrastructure procurement. Part 2. *Proceedings of the Institution of Civil Engineers – Management, Procurement and Law*, *163*(2), 71–75. doi:10.1680/mpal.2010.163.2.71

Werbach, K. (2018). *The blockchain and the new architecture of trust*. MIT Press. doi:10.7551/mitpress/11449.001.0001

Whitlock, K., & Abanda, H. F. (2017, April). Making a Business Case for BIM Adoption. In *International Conference on Engineering, Applied Sciences and System Modeling* (pp. 231-246). Springer.

Wilkins, J. R. (2011). Construction workers' perceptions of health and safety training programmes. *Construction Management and Economics*, *29*(10), 1017–1026. doi:10.1080/01446193.2011.633538

Wilkinson, S., Noktehdan, M., Shahbazpour, M., & Finch, R. (2015). Learning from innovation. *Build*, (149), 86-87. Retrieved from https://www.buildmagazine.org.nz/assets/PDF/Build-149-86-Innovation-Learning-From-Innovation.pdf

Wilkinson, S., Tookey, J., Potangaroa, R., MacGregor, C., Milicich, M., Ghodrati, N., ... (2017). Transforming the Building Industry. *Urban Research Network*. Retrieved from https://www.buildingbetter.nz/publications/SRA6/SRA6_NSC11_State_of_Nation_knowledge_report_2017.pdf

Winch, G. (1998). Zephyrs of creative destruction: Understanding the management of innovation in construction. *Building Research and Information*, *26*(5), 268–279. doi:10.1080/096132198369751

Witt, E., & Kähkönen, K. (2019). BIM-enabled education: A systematic literature review. *10th Nordic Conference on Construction Economics and Organization*, 261-269.

Wolcott, H. F. (1994). Transforming qualitative data. *Sage (Atlanta, Ga.)*.

Wolstenholme, A., Austin, S. A., Bairstow, M., Blumenthal, A., Lorimer, J., McGuckin, S., & Rhys Jones, S. (2009). *Never Waste a Good Crisis: A Review of Progress since Rethinking Construction and Thoughts for our Future*. Loughborough University. https://hdl.handle.net/2134/6040

Womack, J. P., & Jones, D. T. (2003). *Lean Thinking: Banish Waste and Create Wealth in Your Corporation*. Simon and Schuster.

Womack, J. P., Jones, D. T., & Roos, D. (1990). *Machine that changed the world*. Simon and Schuster.

Wong, J. K. W., Ge, J., & He, S. X. (2018). Digitisation in facilities management: A literature review and future research directions. *Automation in Construction*.

Wong, K. A., Wong, F. K. W., & Nadeem, A. (2011). Building information modelling for tertiary construction education in Hong Kong. *Journal of Information Technology in Construction*, *16*, 467-476.

Wong, K., Wong, F. K., & Nadeem, A. (2011). Building information modelling for tertiary construction education in Hong Kong. *Journal of Information Technology in Construction, 16*, 467–476.

Wong, K., Wong, F. K., & Nadeem, A. (2011). Building information modelling for tertiary construction education in hong kong. *Journal of Information Technology in Construction*.

Wong, A. K. D., Wong, F. K. W., & Nadeem, A. (2010). Attributes of building information modelling implementations in various countries. *Architectural Engineering and Design Management*, *6*(4), 288–302.

Wong, S. Y., & Gray, J. (2019). Barriers to implementing building information modelling (BIM) in the Malaysian construction industry. *Materials Science and Engineering*, *495*(1), 012002.

Won, J., Lee, G., Dossick, C., & Messner, J. (2013). Where to Focus for Successful Adoption of Building Information Modeling within organisation. *Journal of Construction Engineering and Management*, *139*(11), 04013014. doi:10.1061/(ASCE)CO.1943-7862.0000731

Wood, D. J., & Gray, B. (1991). Towards a comprehensive theory of collaboration. *The Journal of Applied Behavioral Science*, *27*(2), 139–162. doi:10.1177/0021886391272001

Woodhead, R., Stephenson, P., & Morrey, D. (2018). Digital construction: From point solutions to IoT ecosystem. *Automation in Construction, 93*(March), 35–46. doi:10.1016/j.autcon.2018.05.004

WorkSafe. (2020). *Data centre*. Retrieved from https://data.worksafe.govt.nz/graph/summary/fatalities

World Economic Forum. (2016). *Shaping the Future of Construction: A Breakthrough in Mindset and Technology*. Prepared in collaboration with the Boston Consulting Group.

World Economic Forum. (2017). *Shaping the Future of Construction: Inspiring innovators to redefine the industry*. Prepared in collaboration with the Boston Consulting Group.

World Economic Forum. (2018). *An Action Plan for solve the Industry's Talent Gap*. Prepared in collaboration with the Boston Consulting Group.

World Economic Forum. (2018a). *The Future of Jobs Report 2018*. World Economic Forum.

World Economic Forum. (2018b). *Future Scenarios and Implications for the Industry*. Prepared in collaboration with The Boston Consulting Group. World Economic Forum.

WRAP. (2010). *Guidelines for measuring and reporting construction, demolition and excavation waste*. The UK Waste Recycling Action Programme. https://www.wrap.org.uk/sites/files/wrap/WRAP%20Waste%20Reporting%20Guidance%20Update%20-%20FINAL1.pdf

Wu, H., Duan, H., Wang, J., Wang, T., & Wang, X. (2015). Quantification of carbon emission of construction waste by using streamlined LCA: a case study of Shenzhen, China. Japan Society of Material Cycles and Waste Management, 17(4), 637-645. doi:10.100710163-015-0404-9

Wu, W., & Issa, R. R. (n.d.). BIM education for new career options: An initial investigation. Paper presented at the BIM Academic Workshop, Washington, DC.

Wu, W., Mayo, G., McCuen, T., Issa, R., & Smith, D. (2018). Building Information Modeling Body of Knowledge I: Background, Framework, and Initial Development. *Journal of Construction Engineering Management*, (144).. doi:10.1061/(ASCE)CO.1943-7862.0001518

Wu, G., Zhao, X., & Zuo, J. (2017). Relationship between project's added value and the trust–conflict interaction among project teams. *Journal of Management Engineering, 33*(4), 04017011. doi:10.1061/(ASCE)ME.1943-5479.0000525

Wu, H., Wang, J., Duan, H., Ouyang, L., Huang, W., & Zuo, J. (2016b). An innovative approach to managing demolition waste via GIS (geographic information system): A case study in Shenzhen city, China. *Journal of Cleaner Production, 112*(P1), 494–503. doi:10.1016/j.jclepro.2015.08.096

Wu, H., Zuo, J., Yuan, H., Zillante, G., & Wang, J. (2019). A review of performance assessment methods for construction and demolition waste management. *Resources, Conservation and Recycling, 150*, 104407. Advance online publication. doi:10.1016/j.resconrec.2019.104407

Wu, H., Zuo, J., Yuan, H., Zillante, G., & Wang, J. (2020). Cross-regional mobility of construction and demolition waste in Australia: An exploratory study. *Resources, Conservation and Recycling, 156*, 104710. doi:10.1016/j.resconrec.2020.104710

Wüst, K., Kostiainen, K., Čapkun, V., & Čapkun, S. (2019). PRCash: Fast, private and regulated transactions for digital currencies. In *International Conference on Financial Cryptography and Data Security* (pp. 158-178). Springer. 10.1007/978-3-030-32101-7_11

Wu, W. H., Jim Wu, Y. C., Chen, C. Y., Kao, H. Y., Lin, C. H., & Huang, S. H. (2012). Review of trends from mobile learning studies: A meta-analysis. *Computers & Education*, *59*(2), 817–827. doi:10.1016/j.compedu.2012.03.016

Wu, W., & Issa, R. R. A. (2014). BIM Education and Recruiting: Survey-Based Comparative Analysis of Issues, Perceptions, and Collaboration Opportunities. *J Prof Iss Eng Ed Pr*, *140*(2), 04013014. doi:10.1061/(ASCE)EI.1943-5541.0000186

Wu, W., & Issa, R. R. A. (2014). BIM Education and Recruiting: Survey-Based Comparative Analysis of Issues, Perceptions, and Collaboration Opportunities. *Journal of Professional Issues in Engineering Education and Practice*, *140*(2), 04013014.

Wu, W., & Issa, R. R. A. (2014). BIM education and recruiting: Survey-based comparative analysis of issues, perceptions, and collaboration opportunities. *Journal of Professional Issues in Engineering Education and Practice*, *140*(2).

Wu, W., Mayo, G., McCuen, T., Issa, R., & Smith, D. (2021). *Developing BIM Talent: A Guide to the BIM Body of Knowledge with Metrics, KSAs, and Learning Outcomes*. Wiley.

Xiong, Z., Zhang, Y., Niyato, D., Wang, P., & Han, Z. (2018). When mobile blockchain meets edge computing. *IEEE Communications Magazine*, *56*(8), 33–39. doi:10.1109/MCOM.2018.1701095

Xu, G. (2018). Cloud asset enabled integrated IoT platform for lean prefabricated construction. Automation in Construction, 93, 123–134. doi:10.1016/j.autcon.2018.05.012

Xu, J. (2019). In the shadow of a negative past: repairing and developing trust in construction projects. *Proceedings of the Institution of Civil Engineers-Management, Procurement and Law*, 1-9.

Xu, Q., Chong, H., & Liao, P. (2019). Collaborative information integration for construction safety monitoring. Automation in Construction, 102, 120–134. doi:10.1016/j.autcon.2019.02.004

Xu, J., Shi, Y., Xie, Y., & Zhao, S. (2019). A BIM-Based construction and demolition waste information management system for greenhouse gas quantification and reduction. *Journal of Cleaner Production*, *229*, 308–324. doi:10.1016/j.jclepro.2019.04.158

Xu, L. D., Xu, E. L., & Li, L. (2018). Industry 4.0: State of the art and future trends. *International Journal of Production Research*, *56*(8), 2941–2962. doi:10.1080/00207543.2018.1444806

Xu, X., Weber, I., & Staples, M. (2019). *Architecture for blockchain applications*. Springer. doi:10.1007/978-3-030-03035-3

Yalcinkaya, M., & Singh, V. (2015). Patterns and trends in Building Information Modelling (BIM) research: A Latent Semantic Analysis. *Automation in Construction*, *V59*, 68–80. doi:10.1016/j.autcon.2015.07.012

Yamamura, S., Fan, L., & Suzuki, Y. (2017). Assessment of urban energy performance through integration of BIM and GIS for smart city planning. *Procedia Engineering*, *180*, 1462–1472. doi:10.1016/j.proeng.2017.04.309

Yan, H., & Damian, P. (2008). Benefits and barriers of Building Information Modelling. *Proceedings of the 12th International Conference on Computing in Civil and Building Engineering (ICCCBE-2008)*. https://pdfs.semanticscholar.org/0594/9fac948a129d821f6c9922d92c8dcbb82408.pdf

Yang, R., Wakeeld, R., Lyu, S., Jayasuriya, S., Han, F., Yi, X., Yang, X., Amarasinghe, G., & Chen, S. (2020). Public and private blockchain in construction business process and information integration. *Automation in Construction*, *118*, 103276. doi:10.1016/j.autcon.2020.103276

Yang, Y., Ng, S. T., Xu, F. J., & Skitmore, M. (2018). Towards sustainable and resilient high-density cities through better integration of infrastructure networks. *Sustainable Cities and Society*, *42*(April), 407–422. doi:10.1016/j.scs.2018.07.013

Yan, W., & Sakairi, T. (2019). Geo CPS: Spatial challenges and opportunities for CPS in the geographic dimension. *Journal of Urban Management*, *8*(3), 331–341. doi:10.1016/j.jum.2019.09.005

Yaqoob, I., Salah, K., Uddin, M., Jayaraman, R., Omar, M., & Imran, M. (2020). Blockchain for digital twins: Recent advances and future research challenges. *IEEE Network*, *34*(5), 290–298. doi:10.1109/MNET.001.1900661

Yazdani, M., Kabirifar, K., Frimpong, B. E., Shariati, M., Mirmozaffari, M., & Boskabadi, A. (2021). Improving construction and demolition waste collection service in an urban area using a simheuristic approach: A case study in Sydney, Australia. *Journal of Cleaner Production*, *280*, 124138. Advance online publication. doi:10.1016/j.jclepro.2020.124138

Ye, Z. (2018). Cup-of-Water Theory: A Review on the Interaction of BIM, IoT and Blockchain During the Whole Building Lifecycle. *Proceedings of the 35th International Symposium on Automation and Robotics in Construction (ISARC).*

Yik, F. W., Lai, J. H., Chan, K. T., & Yiu, E. C. (2006). Problems with specialist subcontracting in the construction industry. *Building Services Engineering Research and Technology*, *27*(3), 183–193. doi:10.1191/0143624406bse160oa

Yin, X. (2019). Building information modelling for o ff -site construction: Review and future directions. Automation in Construction, 101, 72–91.

Yiu, N. S., Sze, N. N., & Chan, D. W. (2018). Implementation of safety management systems in Hong Kong construction industry–A safety practitioner's perspective. *Journal of Safety Research*, *64*, 1–9. doi:10.1016/j.jsr.2017.12.011 PMID:29636156

Young, M., & Muller, J. (2010). Three Educational Scenarios for the Future: Lessons from the sociology of knowledge. *European Journal of Education*, *45*(1), 11–27. doi:10.1111/j.1465-3435.2009.01413.x

You, Z., Wu, C., Zheng, L., & Feng, L. (2020). An Informatization Scheme for Construction and Demolition Waste Supervision and Management in China. *Sustainability*, *12*(4), 1672. doi:10.3390u12041672

Yu, D., Duan, H., Song, Q., Li, X., Zhang, H., Zhang, H., Liu, Y., Shen, W., & Wang, J. (2018). Characterizing the environmental impact of metals in construction and demolition waste. *Environmental Science and Pollution Research International*, *25*(14), 13823–13832. doi:10.100711356-018-1632-z PMID:29508202

Yue, X., Wang, H., Jin, D., Li, M., & Jiang, W. (2016). Healthcare data gateways: Found healthcare intelligence on blockchain with novel privacy risk control. *Journal of Medical Systems*, *40*(10), 218. doi:10.100710916-016-0574-6 PMID:27565509

Yu, S., Shen, Q., Kelly, J., & Hunter, K. (2006). Investigation of critical success factors in construction project briefing by way of content analysis (Author abstract). *Journal of Construction Engineering and Management*, *132*(11), 1178–1186. doi:10.1061/(ASCE)0733-9364(2006)132:11(1178)

Yusuf, M. O. (2005). Information and Communication Education: Analyzing the Nigerian National Policy for Information Technology. *International Education Journal*, *6*(3), 316–321. https://files.eric.ed.gov/fulltext/EJ854985.pdf

Zachary, S. (2019). Developing a theory of an object-oriented city: Building energy for urban problems. *Energy Procedia*, *158*, 4210–4217. doi:10.1016/j.egypro.2019.01.807

Zahrizan, Z., Ali, N. M., Haron, A. T., Marshall-Ponting, A., & Abd Hamid, Z. (2013). Exploring the adoption of Building Information Modelling (BIM) in the Malaysian Construction industry: A qualitative approach. *International Journal of Research in Engineering and Technology*, *2*(8), 384–395.

Zaigham, N. A., Aburizaiza, O. S., Mahar, G. A., Nayyar, Z. A., & Siddique, A. (2015). Watershed Analysis of Rabigh Drainage Basin, Saudi Arabia. *International Journal of Water Resources and Arid Environments*, *4*(2), 138–145.

Zailani, S., Ariffin, H. A. M., Iranmanesh, M. M., Moeinzadeh, S., & Iranmanesh, M. M. (2016). The Moderating Effect of Project Risk Mitigation Strategies on the Relationship between Delay Factors and Construction Project Performance. *Journal of Science and Technology Policy Management*, *7*(3), 346–368. doi:10.1108/JSTPM-12-2015-0041

Zen Internet. (2019). *Bricks, Mortar and Digital Transformation Report*. Zen Internet.

Zhai, Y., Chen, K., Zhou, J. X., Cao, J., Lyu, Z., Jin, X., Shen, G. Q. P., Lu, W., & Huang, G. Q. (2019). An Internet of Things-enabled BIM platform for modular integrated construction: A case study in Hong Kong. *Advanced Engineering Informatics*, *42*(October), 100997. doi:10.1016/j.aei.2019.100997

Zhang, L., Huang, S., & Peng, Y. (2018). Collaboration in Integrated Project Delivery: The effects of trust and formal contracts. *Engineering Management Journal*, *4*(30), 262–273. doi:10.1080/10429247.2018.1498259

Zhang, L., Wu, X., Ding, L., Skibniewski, M., & Lu, Y. (2016). BIM-Based Risk Identification System in tunnel construction. *Journal of Civil Engineering and Management*, *22*(4), 529–539.

Zhang, N., Zheng, L., Duan, H., Yin, F., Li, J., & Niu, Y. (2019). Differences of methods to quantify construction and demolition waste for less-developed but fast-growing countries: China as a case study. *Environmental Science and Pollution Research International*, *26*(25), 25513–25525. doi:10.100711356-019-05841-4 PMID:31264149

Zhang, S., Sulankivi, K., Kiviniemi, M., Romo, I., Eastman, C. M., & Teizer, J. (2015). BIM-based fall hazard identification and prevention in Construction safety planning. *Safety Science*, *72*, 31–45.

Zhang, S., Sulankivi, K., Kiviniemi, M., Romo, I., Eastman, C. M., & Teizer, J. (2015). BIM-based fall hazard identification and prevention in construction safety planning. *Safety Science*, *72*, 31–45. doi:10.1016/j.ssci.2014.08.001

Zhang, S., Teizer, J., Lee, J. K., Eastman, C. M., & Venugopal, M. (2013). Building information modeling (BIM) and safety: Automatic safety checking of Construction models and schedules. *Automation in Construction*, *29*, 183–195.

Zhang, X., Li, M., Lim, J. H., Weng, Y., Tay, Y. W. D., Pham, H., & Pham, Q. C. (2018). Large-scale 3D printing by a team of mobile robots. *Automation in Construction*, *95*, 98–106. doi:10.1016/j.autcon.2018.08.004

Zhao, G. & Carlton, D. (2015a). *Dynamic, competency-based career planning and CPD: Vision statement and requirements analysis.* Institution of Civil Engineers.

Zhao, G. (2017a). *Competency with Activity Semantics*. Internal report. Intelartes. Unpublished.

Zhao, G. (2017b). *Competency Analytics, Business Context & Approach*. Internal report. Intelartes. Unpublished.

Zhao, G. (2018). *Competency Analytics & Management*. Internal report. Intelartes. Unpublished.

Zhao, G. (2019). *A System Model of Competency Frameworks*. Internal report. Intelartes. Unpublished.

Zhao, D., & Lucas, J. (2015). Virtual reality simulation for Construction safety promotion. *International Journal of Injury Control and Safety Promotion*, *22*(1), 57–67.

Zhao, D., McCoy, A. P., Bulbul, T., Fiori, C., & Nikkhoo, P. (2015). Building Collaborative Construction Skills through BIM-integrated Learning Environment. *International Journal of Construction Education and Research*, *11*(2), 97–120.

Zhao, G. (2012). Competence semantics: Engineering and application. *International Journal of Knowledge and Learning*, *8*(1/2), 122–132. doi:10.1504/IJKL.2012.047553

Zhao, G., & Carlton, D. (2015b). Forecast Competency Migration by a Methodology of Competency Analytics. *Open Journal of Social Sciences*, *3*(11), 16–22. doi:10.4236/jss.2015.311003

Zhao, X. (2017). A scientometric review of global BIM research: Analysis and visualisation. *Automation in Construction, 80*, 37–47. doi:10.1016/j.autcon.2017.04.002

Zhao, Y. (2014). *Who's Afraid of the Big Bad Dragon?: Why China Has the Best (and Worst) Education System in the World.* Wiley.

Zhengdao, C. (2018). An Internet of Things-enabled BIM platform for on-site assembly services in prefabricated construction. Automation in Construction, 89, 146–161.

Zhengdao, C. (2016a). SWOT analysis and Internet of Things-enabled platform for prefabrication housing production in Hong Kong. *Habitat International, 57*, 74–87. doi:10.1016/j.habitatint.2016.07.002

Zhengdao, C. (2016b). Schedule risks in prefabrication housing production in Hong Kong: A social network analysis. *Journal of Cleaner Production, 134*, 482–494. doi:10.1016/j.jclepro.2016.02.123

Zhengdao, C. (2017). Integrating RFID and BIM technologies for mitigating risks and improving schedule performance of prefabricated house construction. *Journal of Cleaner Production, 165*, 1048–1062. doi:10.1016/j.jclepro.2017.07.156

Zheng, Z., Xie, S., Dai, H.-N., Chen, X., & Wang, H. (2018). Blockchain challenges and opportunities: A survey. *International Journal of Web and Grid Services, 14*(4), 352–375. doi:10.1504/IJWGS.2018.095647

Zhong, L., Wu, Q., Xie, J., Li, J., & Qin, B. (2019). A secure versatile light payment system based on blockchain. *Future Generation Computer Systems, 93*, 327–337. doi:10.1016/j.future.2018.10.012

Zhong, R. Y. (2015). Towards Physical Prefabricated Housing Construction in Towards Physical Prefabricated Housing Construction in Towards Physical Prefabricated Housing Construction in Towards Physical Internet-enabled Prefabricated Housing Construction in. *IFAC-PapersOnLine, 48*(3), 1079–1086. doi:10.1016/j.ifacol.2015.06.227

Zhong, R. Y., Peng, Y., Xue, F., Fang, J., Zou, W., Luo, H., Thomas Ng, S., Lu, W., Shen, G. Q. P., & Huang, G. Q. (2017). Prefabricated construction enabled by the Internet-of-Things. *Automation in Construction, 76*, 59–70. doi:10.1016/j.autcon.2017.01.006

Zhong, R. Y., Xu, X., Klotz, E., & Newman, S. T. (2017). Intelligent Manufacturing in the Context of Industry 4.0: A Review. *Engineering, 3*(5), 616–630. doi:10.1016/J.ENG.2017.05.015

Zhong, R. Y., Xu, X., & Wang, L. (2017). IoT-enabled Smart Factory Visibility and Traceability using Laser- scanners. *Procedia Manufacturing, 10*, 1–14. doi:10.1016/j.promfg.2017.07.103

Zhou, C. (2019). Cyber-physical-system-based safety monitoring for blind hoisting with the internet of things: A case study. Automation in Construction, 97, 138–150. doi:10.1016/j.autcon.2018.10.017

Zhou, I., Irizarry, J., & Li, Q. (2013). Applying advanced technology to improve safety management in the construction industry: A literature review. *Construction Management and Economics, 31*(6), 606–622. doi:10.1080/01446193.2013.798423

Zhou, X., Zhao, J., Wang, J., Guo, M., Liu, J., & Shi, H. (2020). Towards product-level parallel computing of large-scale building information modeling data using graph theory. *Building and Environment, 169*, 106558. doi:10.1016/j.buildenv.2019.106558

Zhu, Z., Sun, Y., & Riezebos, P. (2016). Introducing the smart education framework: core elements for successful learning in a digital world. *Int. J. Smart Technology and Learning, 1*(1). https://www.researchgate.net/publication/305953715_Introducing_the_smart_education_framework_core_elements_for_successful_learning_in_a_digital_world

Zou, P. X. W., Lun, P., Cipolla, D., & Mohamed, S. (2017). Cloud-based safety information and communication system in infrastructure construction. *Safety Science*, *98*, 50–69.

Zou, P. X., Xu, X., Jin, R., Painting, N., & Li, B. (2019). AEC students' perceptions of BIM practice at Swinburne university of technology. *Journal of Professional Issues in Engineering Education and Practice*, *145*(3).

Zyskind, G., Nathan, O., & Associates. (2015). Decentralizing privacy: Using blockchain to protect personal data. Security and Privacy Workshops (SPW), 2015 IEEE, 180–184.

About the Contributors

Jason Underwood is a Professor of Construction ICT & Digital Built Environments and Programme Director of the MSc. BIM & Digital Built Environments programme in the School of Science, Engineering & Environment at the University of Salford. He has over 25 year's teaching & learning, research, and enterprise experience in the areas of concurrent engineering, integrated and collaborative computing in construction, product and building information modelling, digital built environments, collaborative construction and behaviours, BIM maturity assessment, BIM education and training in the development of BIM capabilities, and organisational e-readiness towards delivering strategic value from ICT/BIM investment. He is actively engaged in the digital transformation of the UK construction industry through affiliations at international, national, and regional levels, and the current Chair of the UK BIM Academic Forum, and Director of Construct IT For Business, an industry-led non-profit making collaborative membership-based network. He is also Editor-in-Chief of the Journal of Digital Innovation in the Built Environments (IJDIBE).

Mark Shelbourn is an Associate Professor and Head of Department for Built Environment at Birmingham City University, UK. He is responsible for the effective strategic leadership and management, promotion, development and enhancement of the teaching, research, enterprise & wider engagement provision in the Dept. He creates a working environment that supports the team and encourages active engagement and a positive contribution from all individuals, with a specific emphasis of engaging industry in all activities of the Dept. Dr Shelbourn is an executive board member of the Council of Heads of Built Environment (CHOBE); the Region 8 Director of the Associated Schools of Construction; Joint Coordinator of the CIB W89 Working Group – Education in the Built Environment; and ongoing work with the Academic Interoperability Coalition in Digital Education in the Built Environment. Dr Shelbourn uses his 25years of research and enterprise activities, industry contacts and case studies to bring a fresh look to Digital construction in the classroom to the future digital leaders of our industry. Research continues in the areas of collaborative working & the use of BIM tools & processes to support industry needs.

* * *

Behzad Abbasnejad currently works as a Lecturer in the School of Property, Construction and Project Management at RMIT University.

Bamiji Adeleye is currently a PhD research candidate at University of Makarere, Uganda with interest in digital spatial planning and land management.

Alireza Ahankoob is a lecturer in Building Information Modelling (BIM) within the School of Property, Construction and Project Management at RMIT University. His research interest lies in the areas of BIM, construction innovation and technologies, knowledge management, and lean practices in the construction industry. He has published several articles and research papers in reputable journals. He currently teaches into the construction management discipline and is looking for any opportunity to collaborate with the construction industry.

Gulnaz Aksenova is a Lecturer in Building Information Modelling and Management at the Bartlett School of Construction and Project Management, UCL. She studies business model innovation, industry evolution, and how digital innovation is changing the business context. Her interest lays within platforms and ecosystems design and governance, and industry specific applications of Building Information Management, Mixed Reality, industrialised construction, competency management, and digital transformation of the design, construction and operation industry. Gulnaz's research is both empirical using qualitative data and conceptual (theory-building) at the intersection of organization science and strategic management. Gulnaz was awarded a PhD (2020) from the University of Liverpool in Strategic Management in Digital Built Environment Sector. She conducted an empirical study on the dark side of ecosystem orchestration around BIM implementation in Finland and California. While at Liverpool, she extensively presented her work at high-profile industry events in Finland, UK, Spain and Canada. Gulnaz holds a Master of Science (2012) in Digital Architectural Design from the University of Salford which she has completed with Distinctions, and a Specialist degree in Architectural Design (2009) from the Ufa State Petroleum Technological University.

Husam Helmi Alharahsheh is an experienced senior academic with an extensive experience of teaching, supervision and research within higher education context. His higher education qualification at the undergraduate and postgraduate levels were focused on the fields of business management and higher education management. He is an active contributor to the development of process of wide range of higher education courses at the undergraduate and postgraduate levels in the fields of business, hospitality and education management. He has excellent experience in academic management and leadership within UK higher education institutions. He is also a visiting academic across various universities as well as other academic institutions based in the UK and abroad. His academic and professional experience is developed through working across a wide number of UK based higher education institutions within the public, private as well as not for profit sectors. He has developed links with external networks within the industry to maintain understanding of the key trends and challenges that can be proactively and reactively researched and implemented within curriculum. His contributions in UK and abroad are giving him the opportunity to have further global exposure and understanding of key trends within various sectors. His key research interest are mainly in employability enhancement, internationalization, Globalization, sustainability, Education and enhancement of curriculum at various levels within higher education. His research is communicated nationally and internationally in the form of academic papers and books/book chapters published through highly reputable international academic journals and publishers. He is also actively contributing to enhancement of research quality published through acting as an executive editor of the Global Academic Journal of Economics and Business, associate editor of East African Journal of

Economics, Business and Management, and Cross Current International Journal of Economics, Management and Media Studies. Furthermore, He has varied contributions to a wide number of international academic journals as a reviewer.

Olusoyi Richard Ashaye is a Freelance lecturer at the Brunel Business School, Brunel University London and University of Wales Trinity St David, London Campus. He holds a PhD degree in Management Studies with specialist interest in e-Government, public sector business process, change management, and professional knowledge in real estate. He is a member of Centre for Digital Governance and Sustainable Operations Management (DG-SOM) research institute and other professional bodies. He has publications from reputable journals.

Popoola Ayobami holds a PhD from University of KwaZulu Natal with interest in infrastructure planning and development.

Sheila Belayutham is currently a senior lecturer at the School of Civil Engineering, College of Engineering, , Universiti Teknologi MARA, Malaysia. Her Ph.D in Civil Engineering was awarded by the University of Auckland, New Zealand, while her B.Eng (Civil) and MSc. (Construction Management) was from Universiti Teknologi Malaysia. Her passion for subjects related to Lean Production, Environmental Sustainability, Cleaner Production, Construction Project Management, Social Innovation and Construction Technological Advancements has been translated into articles published in high impact journals. Currently, she is actively involved in training and consultation works related to construction safety (Prevention through Design), besides having research grant on the subject of Building Information Modelling and Lean Management.

George Blumberg is a senior lecturer in the School of the Built Environment. He was educated at the George Washington University as a mechanical engineer and subsequently worked in the construction industry as a site surveyor. He later became the head engineer of the Marine Seismic Lab at the Massachusetts Institute of Technology (MIT) and was awarded a master degree (SM) based on seismic exploration of the Mid-Atlantic Ridge. After a stint as an environmental consultant, he received a DPhil at the University of Oxford, where his thesis was on the frequency and severity of heat waves in urban centres. His current interests are in automation in the construction process with specialization in the application of Blockchains to construction and engineering.

Udayan Chanda is currently working as Associate Professor in Department of Management, Birla Institute of Technology & Science (BITS) Pilani. Earlier he was associated with Industrial Statistics Lab., Department of Information & Industrial Engineering Yonsei University as Post Doctoral Fellow and Department of Operational Research, University of Delhi as Assistant Professor (Ad-hoc). He received his Ph.D. degree in Marketing Models and Optimization (Operational Research) from University of Delhi, Delhi. He has published numerous papers in the area of Marketing Models, Optimization, Software Reliability and Inventory Management in international journals and conference proceedings. His current research interests include Marketing Models, Inventory Modeling, Software Reliability Growth Modeling, and Dynamic Optimization Techniques.

Che Khairil Izam Che Ibrahim is an Associate Professor at the School of Civil Engineering, College of Engineering, , Universiti Teknologi MARA, Malaysia, where he is currently the Deputy Dean (Academic). He holds a PhD Degree from the University of Auckland, New Zealand. He is currently leading a project on prevention through design in construction funded by National Institute of Occupational Safety and Health (NIOSH), Malaysia and has been involved in several projects funded by public and private organisations. His research mainly concentrates on the relationship-based procurement, team integration, organisation performance, safe design and digital collaboration. More than 25 of his papers were as an article in various scientific and professional construction journals. In 2016, his publication on team integration has achieved the Emerald Literati Network Awards of Excellence, a 'Highly Commended Paper of 2016'. He also received a 'Best Reviewer Award, 2016' in recognition of his contribution in Construction Economics and Building journal. In addition, his recent publication on alliance team received the Emerald Literati Network Awards of Excellence, a 'Highly Commended Paper of 2019'.

Ricardo Codinhoto's expertise is BIM for life-cycle assessment. His research focus on understanding how to anticipate the implications that design decisions have on use, maintenance, recycling and decommissioning of buildings. He led the 2 stage commissioned research project for the Manchester Town Hall Complex Project (first UK facility to operate with BIM support). Regarding large and complex buildings involving vulnerable user groups and wellbeing, he led and participated on research funded by EPSRC, NIHR, MRC and also working closely to the NHS Estates. He contributed to CPD initiatives led by CIOB, RIBA, RICS and CIAT. He also developed commissioned research for the Government Office for Science that supported policy-making. Current research also focus on Design for Disassembly and Recycling.

Mark Costello is Director of Digital Design, BIM & Technical Services with RPS. He has over twenty-five years' experience delivering large infrastructural projects at all stages. Founded in 1970, RPS is a leading global professional services firm. Mark was appointed Adjunct Lecturer in Galway-Mayo Institute of Technology (GMIT) in recognition of his work, expertise, dedication to education and his passion for delivering large scale infrastructural projects in a new collaborative environment. He was also appointed Board Director of the Construction IT Alliance (CitA) in 2018 to further promote its vision of harnessing the potential of ICT in the Irish Construction Industry. Mark was awarded BRE Academy Fellowship in 2017 and is Director responsible for achieving BIM Level 2 Business Systems Certification for RPS in 2015 and the new ISO 19650 (organization and digitization of information) Certification in 2019.

Robert Crocker, DPhil (oxon), is Deputy Director of the China Australia Centre for Sustainable Development in the University of South Australia. With a background in the history of science and philosophy, his research focuses on the relationship between consumption, waste, and design for sustainability. His book Somebody Else's Problem: Consumerism, Design and Sustainability (Greenleaf 2016), won gold in the Axiom Best Business Books for 2017 (sustainability), and he is co-editor of four published collections of essays, including most recently Subverting Consumerism: Reuse in an Accelerated World (Routledge 2018) and Unmaking Waste: Towards a Circular Economy (Emerald 2018). He has convened four international colloquia on various themes in design and consumption and 'unmaking waste' over the last 5 years.

Marcio Fabricio is a professor at the Institute of Architecture and Urbanism, University of São Paulo (IAU USP). Doctor of Engineering from the Polytechnic School of USP (2002), Master in Architecture and Urbanism from EESC USP (1996), Civil Engineer from the Federal University of São Carlos (1993). Chairman of the Graduate IAU USP (November 2013 to October 2015) and Leader of the research group Architecture, Innovation and Technology (Arquitec). Editor-in-chief of the Gestão & Tecnologia de Projetos journal (2006 - 2021). Research areas: Design Management and BIM.

Yingbin Feng is an Associate Professor in the School of Built Environment at Western Sydney University. He obtained a PhD degree at the National University of Singapore (NUS). His main research interests include construction health and safety, sustainable procurement, BIM and robotics in construction, management and organisation in construction. He has over 100 peer reviewed publications demonstrating national and international recognition of his research outcomes. He is the winner of the Premier Award in the Research Paper Category at the prestigious CIOB Innovation and Research Awards 2014.

Beatriz Campos Fialho is an Architect and Urbanist and works as an architect for the Federal University of Sao Carlos (UFSCar), Brazil. She specialises in developing architectural design and guidelines for the university's physical infrastructure development and maintenance. Master's in architecture and Urbanism, she has experience of teaching building design courses at the undergraduate level. Her current PhD research concerns digital transformation at the building lifecycle project stages, with the focus on the application of Building Information Modelling (BIM) and Internet of Things (IoT) solutions for improving the efficiency of building maintenance services.

Angel Gigante-Barrera is interested in data structures associated with project BIM lifecycle processes. Having completed a Construction Management MSc, as well as a M.Arch, he is currently working towards a PhD at the University of Birmingham in the UK. As a supplementary role, he also undertakes research and teaching duties for the University and especially on the MSc Construction Management Programme. Currently, he is also working for Ove Arup & Partners International Limited, at the department of Information Management.

Barry Gledson, MCIOB, MAPM, SFHEA, is a Senior Lecturer in Construction Project Management at Northumbria University. In addition to academia, he also has extensive industry experience in the construction sector from his time working in a project management capacity for major contractors. Barry's full profile can be found on his Northumbria University staff page: https://www.northumbria.ac.uk/about-us/our-staff/g/barry-gledson/. or here: www.linkedin.com/in/drbarrygledson.

David Heesom holds a BEng (Hons) degree in Civil Engineering from Loughborough University and an MSc (distinction) in CAD for Construction from the University of Wolverhampton. He completed his PhD at the University of Wolverhampton in 2004, which developed a suite of software tools to allow more effective planning on construction sites by utilising real time '4D' BIM techniques. He is Chartered member of the Chartered Institute of Architectural Technologists (CIAT) and a Member of the Institute of Knowledge Transfer. His professional experience as an infrastructure engineer working with Atkins involved various projects including a large infrastructure project in Dublin, Ireland and the 3D computer-modelling for a large project in the North of England. David joined the University of Wolverhampton as a Research Fellow undertaking a collaborative EPSRC funded project entitled 'The

Virtual Construction Site' with University College London, UMIST, Carillion, AMEC, Atkins and Teesside University. Since then his research and teaching have focused on the areas of Building Information Modelling, VR and AR for the built environment, HD laser scanning, the Internet of Things and 4D/5D modelling for construction management. He has published this work widely and has been responsible for delivering over £2.5m worth of R&D projects. This has included successfully operated numerous Knowledge Transfer Partnerships (KTP) with each of these achieving high grades following independent assessment. The most recently completed KTP focused on the implementation of BIM and 3D technologies for the nuclear ventilation sector was awarded an 'A-Grade (Outstanding) by an independent review. His expertise in the field of BIM led to collaboration with CITB, which saw him deliver over 25 seminars throughout the UK for various industry bodies and construction companies on the impact of BIM on the construction industry. In 2014, he was awarded the prestigious Industrial Secondment Grant from the Royal Academy of Engineering to spend time back in the industry implementing serious games approaches with BIM data for construction training. David has presented work and run workshops in the field of BIM and 3D technologies at various international events including Germany, Eire, Ukraine, Canada, China, Egypt, USA, and Canada.

Rehan Jamil holds Master's degree in Civil Engineering. Since his Bachelors in 2005, he has served in various international engineering consultancy organizations. He has rich experience of engineering design in plumbing systems for buildings and communities including water supply, drainage and irrigation systems. Currently, he is a faculty member at College of Architecture and Planning, Imam Abdulrahman Bin Faisal University, Dammam, Saudi Arabia. His areas of research are hydraulics, pipe flow, water treatment & supply, hydrology, surface & sub-surface drainage, and irrigation systems. He is actively involved in research having a number of articles to his name, published in high impact factor scientific journals. He is also a part of the editorial committees of various journals and extends his services as a scientific reviewer.

Mark Kelly is a lecturer in the Department of Building and Civil Engineering in the Galway-Mayo Institute of Technology (GMIT). His research interests include Building Information Management (BIM), construction and demolition waste management, construction resource efficiency, the circular economy and education for sustainability.

Farzad Khosrowshahi is the Dean of Engineering & Science at Victoria University. He is Professor of Construction IT and has expertise in financial forecasting, modelling and management. He is previous Director of Construct IT. He is CIOB Fellow and ambassador and has been the assessor/evaluator of many international funding bodies. Farzad has served as the chair of ARCOM; founded BIM Academic Forum; Information Visualisation Society and Uniting Construction Information. He has been committee member of ECCREDI; ECTP; AEEBC; CIOB; CIC; buildingSMART and BS555. Farzad has organised and chaired 3 and co-chaired 10 international conferences and been a member of scientific committee and symposium chair of over 90 international conferences and symposiums. He has presented at over 80 international conferences and 3 keynotes. On 20 occasions, he was visiting professor. Complementary activities include External REF mock examiner; editorial board of 2 international journals and on the editorial team of over 17 international journals. He has been on the review and assessment panels of several funding organisations including EPSRC, Leverhulm, CERG, DTI, TSB and National Research Foundation.

Kerry London is Pro Vice Chancellor at Torrens University having recently been the Professor Built Environment and Urban Transformation at Western Sydney University and Dean School of Built Environment. She has degrees in Architecture and a postgraduate Master Research degree in Building from the University of Newcastle. She was awarded her PhD from University of Melbourne in 2005. She was the first female Chair in construction management in Australia. She has an international research reputation and is a leader in construction supply chain theory and practice. Her work focusses on adoption of innovative technologies and collaborative practice.

Glenda Mayo is an Assistant Professor at UNC Charlotte in the Engineering Technology and Construction Management Department. She received her Ph.D. in Building Construction from the University of Florida in 2014. She specializes in areas of facilities management documentation, BIM, and the management practices in the later phases of a construction project. Prior to academia, her experience includes a position as Assistant Director of Facilities Management.

Tamera McCuen, PhD., LEED AP, Assoc. DBIA, is a Professor of Construction Science at the University of Oklahoma. Tamera specializes and teaches in the areas of construction management, project controls, construction communication, design-build project delivery, and BIM. She is also the Director of the Kenneth Robson BIM+Viz Lab at the OU College of Architecture. Tamera has completed funded research and has more than 70 publications in the areas of BIM and project delivery. She was a co-author of the National BIM Standard – United States™ Version 2 and National BIM Standard – United States® Version 3. Tamera is frequently invited to speak on the topics of BIM, project delivery, and cost engineering at both U.S. and international construction industry events. She has also delivered 23 training workshops and continuing education seminars for AECO professionals. Tamera was awarded the Building Information Modeling Award for Curriculum Development by the American Institute of Architects (AIA) Technology in Architecture Practice in 2008 and has been recognized twice by the Association for the Advancement of Cost Engineering (AACE) for her leadership of the BIM technical committee and work toward the development of information exchange to support BIM use for cost estimating.

Samuel Medayese is currently a PhD research candidate at University of KwaZulu Natal, with interest in development of a new paradigm in Urban Development planning in Africa. Seeking to mark the nexus between Sustainable Urban Development, Inclusive Physical Planning and Urban Liveability at Regional Level in Nigeria, West Africa.

Soroush Moeinzadeh is the last year Ph.D. Candidate in Sustainable Logistics & Supply Chain Management at School of Management, Universiti Sains Malaysia. Scopus Author ID: 56941868800 ORCID Code: https://orcid.org/0000-0001-7199-3001.

Gaurav Nagpal is an Assistant Professor in Management Department of Birla Institute of Technology and Science Pilani. His area of expertise is Operations and Supply Chain Management, Operations Research, Industrial Engineering, Business Statistics and Data Analytics. He received his Bachelor's degree in Mechanical Engineering from YMCA Institute of Engineering in 2005. Post that, he completed his MBA from Indian Institute of Foreign Trade, New Delhi. He has been a GATE topper with 99.83 percentile (AIR-41) in mechanical engineering. He has also done EPMBD from IIM Calcutta, and is CSCP qualified from APICS. He is currently pursuing his PhD from BITS Pilani in the area of inventory management. He has twelve years of industry experience, of which the former seven years were into Operations and supply chain profiles with NTPC, Daimler India Commercial Vehicles and with Escorts Agri-machinery Group. After that he worked for five years in business strategy, planning and analysis profiles. Before joining BITS as Assistant Professor in 2018, he was working with the leading grocery e-Commerce firm "Grofers" as Associate Director- Business Planning and Analysis.

Abd'razack Nelson is a Town Planning with Interest in Urban Ecology and has a PhD in Ecological Footprint Estimation. He has been Teaching in the University for about 15 years.

Gerard Nicholson is a lecturer and researcher in the Department of Building and Civil Engineering at the Galway-Mayo Institute of Technology.

Odunayo Omojola is a PhD research candidate at University of Ibadan with interest in spatial and transportation planning.

Zelinna Pablo is a full-time senior researcher who analyses the complex interactions between people, organisations and technology using qualitative techniques. She completed her PhD in Economics and Commerce with a concentration on organization and management theory at The University of Melbourne, under an Australian Postgraduate Award. She completed a graduate diploma at the University of Amsterdam with distinction, under a full scholarship from the European Union. She was postdoctoral research fellow on an Australian Research Council (ARC) Linkage Project on collaborative practice in construction, and Nominated Senior Research Fellow on an ARC Discovery on digital collaboration in housing. Prior to doing full-time research, Zelinna was an international lecturer at The University of Melbourne (Australia) and at De La Salle University (Philippines), where she developed and delivered courses and established new programs for large (1000+ students) and small cohorts, resulting in multiple awards for teaching excellence.

Lidia Pinti is a research fellow at Department of Architecture, Built Environment and Construction Engineering of Politecnico di Milano, Italy, and she is a co-director of the "BIM Manager" Master of Politecnico di Milano. She receives the M.Sc. degree in Architecture at Politecnico di Milano in 2013 and thereafter obtains a PhD degree with honors from Department of Architecture, Built Environment and Construction Engineering of Politecnico di Milano, Italy, in 2019. She performs research on advanced methods and models for project documentation development and projects analysis, systems validation, risk factors, production planning, cost management and consistency analysis results. She is also assistant professor in Projects Economic Evaluation courses at the Politecnico di Milano. In addition, Lidia contributes in the role of co-supervisor to several master thesis on the use of BIM in different fields. Her current areas of research are the use of BIM in Facility Management phase, embedded Information Management for Maintenance, Building Information Modelling and database integration for the Italian Public Administration.

Payam Rahnamayiezekavat is the Associate Dean (International) and Director of Postgraduate Construction Management programs in the School of Built Environment. Payam teaches and researches in the area of Building Compliance. He offers extensive experience in development and delivery of educational materials for a range of Programs including Construction Management, Building Surveying,

Fire Safety Engineering, Project Management, and Civil Engineering. The recent 2020 students' survey results recorded 93% satisfaction with the learning materials Payam had developed in his Interpreting Building Regulations unit. Besides being an overwhelmingly high satisfaction rate, it is important to note that an absolute majority of the students in his class work in the building certification sector. Dr Zekavat's research has evolved around performance of construction industry. His research particularly emphasises the consequences of enhanced construction processes in terms of development of more live-able and resilient built environment. He is currently investigating the prevalent issue of non-compliant buildings and is navigating his research to assess the consequence of non-compliance in the context of disaster management. Payam supervises eight HDR students at the moment.

Kay Rogage is a Senior Lecturer at Northumbria University. Kay's profile and CV can be found on her Northumbria University staff page: https://www.northumbria.ac.uk/about-us/our-staff/r/kay-rogage/.

Darshan Ruikar works with Ove Arup and Partners.

Kirti Ruikar is a senior lecturer at Loughborough University.

Swapan Saha received his Master of Engineering degree in Construction in 1989 from the Asian Institute of Technology in Bangkok, Thailand and obtained his PhD in 2003 from the University of Western Sydney. He has been working at the Western Sydney for the last 25 years in the various roles and lead the Construction Management discipline for last 15 years. He is currently working as the Deputy Dean of the School of Built Environment. As a Deputy Dean, Swapan is primarily responsible for assisting the Dean in the provision of leadership within the School, and oversight of School planning, performance, and quality management. Swapan is also responsible for the supervision of the School's submissions for relevant quality processes such as TEQSA audits, internal and external course or discipline reviews, in collaboration with the Quality and Reviews Unit. In 2018, Swapan received prestigious Frank E Crowle award established by the Australian Institute of Building for his outstanding leadership and sustained contribution in teaching and scholarship for the Construction Management discipline at WSU. Swapan's teaching and research are mainly in the areas of construction management and simulation, construction and fire safety, Work Health and Safety, energy efficiency and sustainability, Building Information Modelling (BIM), construction information systems and construction technology. He worked in the construction industry in Australia and Thailand before joining as an academic member of staff with UWS/WSU. Swapan has worked on landmark projects including the Sydney Harbour and Tunnel Project Tomago Aluminium Smelter Construction Project during 1990 and 1993, plus the Bangkok elevated expressway project in 1989. He has published score of book chapters, journal and conferences articles. He has strong links with industry and professional organisations both in Australian and overseas. He maintains strong links with Tiers 1 and Ties 2 construction companies in Australia through the School of Built Environment Industry Advisory committees which are represented by high profile industry executives. Swapan keeps regular contacts with WSU Alumni through various social networking medium including LinkedIn, He is a member of Engineers Australia and the Fellow of the Australian Institute of Building, actively collaborating with professional institutes such as Master Builders Association (MBA), Chartered Institute of Building (CIOB) and Australian Institute of Quantity Surveyors (AIQS) and the Australian Institute of Building Surveyors (AIBS). He also collaborates with Academic communities from Universities in Australian and overseas. He has worked on number of projects with RMIT, Deakin,

UTS, UNSW and Sydney University, CSIRO, University of Hawaii, Asian institute of Technology, Anna University (India) and Bangladesh University of Engineering and Technology and Chongqing University China, City University in Hong Kong and Hong Kong Polytechnic University. Swapan has strong track record in research. He is currently working on an ARC grant as the Principal investigator.

Don Samarasinghe is a highly enthusiastic research and education professional with a passion for developing and unleashing the potential of tertiary students. His key strengths include conducting construction industry based research, developing programmes of learning and delivering virtual and face to face Degree level classes on key construction management subjects. His qualifications include a PhD in Engineering (Construction Management), a First Class Honours Bachelor of Civil Engineering and a Graduate Diploma in Tertiary Education (GDTE). He has extensive experience in curriculum design, delivery, assessment and moderation in the New Zealand tertiary education sector. This includes developing qualifications such as a Bachelor of Construction, a Graduate Diploma and a Masters of Professional Practice in Construction/Engineering Management.

Himanshu Seth is a research scholar from department of management, BITS Pilani. His research area includes working capital management in Indian manufacturing sector. Further he is also working on data envelopment analysis. He has published several research papers in journal of national and international repute.

Soroosh Sharifi is a Lecturer in Water Engineering at the University of Birmingham (UoB) with over 15 years of expertise in numerical modelling, data modelling, calibration and uncertainty analysis of hydro-environmental systems. Sharifi has extensive experience in river hydraulics, hydrologic modeling, including climate downscaling and rainfall-runoff modeling, with particular focus on deterministic and probabilistic calibration and uncertainty analysis of such models. He is familiar with and in command of various mechanistic modelling and data-mining tools, which he uses to explore and study sets of large experimental data.

Ana Karina Silverio Rodriguez holds a BArch (Hons) degree in Architecture from Universidad Autonoma de Santo Domingo (UASD) in Santo Domingo, Dominican Republic and an MSc in Building Information Modelling for Integrated Design from the University of Wolverhampton. She completed her PhD at the University of Wolverhampton in 2020 entitled "Implementation of Building Information Modelling in the Dominican Republic construction industry". She has almost ten years of experience in the Dominican Republic and UK, starting in the industry and now focused in Academia. She is currently a lecturer in Architecture and Architecture Technology at the University of Wolverhampton in which she teaches at both UG and PG levels and also undertakes research. She is also an associate member of the Chartered Institute of Architectural Technologists (CIAT).

Pooja Sukhdeve received her Bachelor's degree in Computer Science and Engineering from the University of Swami Ramanand Teerth Marathwada, Nanded, Maharashtra, India in 2016. She is currently a web developer at Universe Exports, India, and a Volunteer at Youth for Unity and Volunteer Action, India. She worked as an Intern and Software Analyst at SSIT, Pvt. Ltd., India from 2016 to 2017. In 2015-2016, she researched face identification using Artificial Intelligence. Also, in 2018, she worked and wrote on a project on the environmental issues of the Nag river, Nagpur, India. She writes and presents widely on the issues and implementation of "Digital-enabled Infrastructure." Currently, she is researching the implementation of Augmented Reality in immersive learning education. Her research interests include Virtual Reality, Augmented Reality, Mixed Reality, Extended Reality, and Artificial Intelligence.

Renukappa Suresh currently serves as a Senior Lecturer in the Faculty of Science and Engineering at the University of Wolverhampton. He holds a PhD in managing change and knowledge associated with sustainability initiatives for improved competitiveness. He has over 20 years of research, consultancy, project management and teaching experience in a wide range of business and management areas across industrial sectors in both developed and emerging economies. His research interests cover, but not limited to, sustainability strategies for competitive advantage; carbon reduction strategies; corporate social responsibility; smart cities development; leading change towards sustainability; knowledge management; public-private partnerships; cloud computing; infrastructure asset management; and sustainable infrastructure investment and development. He has successfully executed more than 30 large projects and authored over 80 papers which have been published in journals, book chapters and conference proceedings.

Subashini Suresh has nearly 20 years' experience in research, teaching and practice in the area of Project Management. Her work in the Engineering and Construction sector has informed 13 countries (For example: USA, UK, Italy, Netherlands, Dominican Republic, Morocco, Ghana, Nigeria, Oman, UAE, Saudi Arabia, Qatar and India). Subashini has worked on European Union and The Engineering and Physical Sciences Research Council (EPSRC) along with other projects. The project's cost range from €1,200 to £850,000 (approx. €1 million). She has successfully executed many research projects in various capacities and has been the Project lead in 10 projects. Under her supervision, 17 PhD students have successfully graduated with PhD's. She has published widely in academic publications, which include journal papers, conference papers, articles, book chapters, reports, and books.

Emma Tallet is a Knowledge Transfer Project (KTP) Associate working on a Northumbria University research project funded by Innovate UK and Turner & Townsend. She holds a first degree in Architecture BA (Hons) and completed an MSc Computer Science in 2020. She has a range of Architectural, Engineering and Construction (AEC) experience, developing particular expertise in Building Information Modelling (BIM) roles.

Anna Thompson is Director of BIM at Turner & Townsend. Anna's profile and CV can be found on her LinkedIn page: https://uk.linkedin.com/in/anna-thompson-23713129.

Angelica Vårhammar joins the Centre for Work Health and Safety from the University sector where she has held several senior research support roles. She has a Master of Science in Environmental Biology and has extensive experience in the management of large research projects. This includes coordinating data collection, managing and interpreting data, supervising, and managing associated staff and facilities, as well as communicating research outcomes to stakeholders.

Gabrielle Wallace joined Western Sydney University in August 2018 as Adjunct Professor. Prior to this she held senior executive roles in New South Wales Government as Director Building Professionals Board (BPB) and Director WHS Roadmap 2022 Coordination and was a member of Board of the BPB. Prof Wallace is a member of Australian Institute of Company Directors (AICD), Western

Sydney University's Postgraduate Construction Management Committee and has sat on various NSW government committees associated with her roles such as NSW Building Regulators' Committee and the Building Regulation Advisory Committee. She has an extensive operational network across government and industry also holds roles with the Housing Industry Association, including as a judge of state housing awards and national professional builder business awards. In addition to being a regulatory and policy specialist, Prof Wallace has led statutory review, strategic and operational policy development and implementation, governance and risk reform, organisational transformation, customer service improvement, and fostered stakeholder engagement, leadership development and collaboration at senior levels. She has addressed various industry conferences and holds a sound understanding of government processes at state and local levels. Prior to entering state government, Prof Wallace completed her PhD in political economy and held roles in building surveying and as a university lecturer.

Allen Wan, after studying or working in China, Hong Kong, the UK, the US & Canada, qualified as a chartered HSEQ professional with over 20 years of experience, ranging from chemical testing & commercial laboratory, environmental management & auditing, and occupational safety & health enforcement & teaching in manufacturing and construction. Dr. Wan's current role is a HSEQ manager for a specialized contractor in Hong Kong and Macau, with research interests in integrated management systems, life cycle assessment, green building and Building Information Modeling.

Drew Wiggett is an Associate Director at Turner & Townsend. Drew's profile and CV can be found on his LinkedIn page: https://uk.linkedin.com/in/drewwiggett.

Nicola Willand, PhD, is a Lecturer in the School of Property, Construction and Project Management at RMIT University, Melbourne, Australia. As an architect by background, her research interests focus on holistic strategies to sustainability in the built environment. Her interdisciplinary research focuses on housing, health, and equity as a socio-technical system, shaped by the interaction of material, social, structural, and cultural contexts.

Peter S. P. Wong is the Associate Dean (Construction Management) of the School of Property Construction and Project Management, RMIT University, Australia. His research into virtual reality in application to construction, offsite construction, carbon reduction, organizational learning, organizational culture, and dispute resolution has been recognized through his publications. As at 2020, his research endeavours have led to more than 90 articles accepted or published in a number of top ranked refereed journals and conference proceedings. He was awarded multiple research grants amounting to A$1.7 million dollars. This includes the Australia Research Council Linkage project entitled "Efficient construction: analysis of adoption of integrated supply chains for innovative off site housing manufacturing systems". The project helps develop an integrated supply chain practice model that include measures of collaboration that will guide effectiveness of offsite manufacturing.

Emma Wood is an experienced broadcast journalist, writer and communications specialist, with a career spanning 20 years in New Zealand and the United Kingdom.

Nadia Safura Zabidin is currently a Ph.D candidate at the School of Civil Engineering, College of Engineering, , Universiti Teknologi MARA, Malaysia. She has graduated with a Master's Degree in

Construction Engineering with her dissertation entitled "Steering towards BIM-ready Civil Engineering graduates in Malaysian Higher Institution". Her passion in combining technological advancements and education in the construction sector has driven her to pursue her Ph.D in developing an educational model for construction engineering curriculum, which is aligned with Industry 4.0, called Construction Engineering Education 4.0 (CEE 4.0). Her recent works have been translated into high impact journal articles, as well as proceeding.

Peng Zhang, PhD, is an Associate Lecturer at the School of Built Environment, Western Sydney University, Australia. His research focuses on productivity, BIM adoption, off site manufacturing, competitiveness, and sustainability.

Gang Zhao graduated in MSc in computer science and PhD in artificial intelligence from the University of Manchester. He has been involved in R&D in intelligent system applications in both academic and industrial contexts since late 1980s. He taught computer science at British and Beglian universities and has been consulting in multi-national European R&D projects. He has architected and implemented AI technologies such as speech recognition applications, natural language understanding, human-machine dialogue system, machine translation, knowledge based reasoning, data and knowledge mining, algorithm trading systems and automatic system optimisation. His recent research in information technology focuses on human resources ecosystems, competency analytics and management, knowledge systems for digital compliance and multimedia information extraction.

Sambo Zulu is a Reader in Quantity Surveying & Construction Management in the School of the Built Environment, Engineering and Computing at Leeds Beckett University. His research interests span across issues in quantity surveying and construction management. His research focuses on construction and project management issues in the Architecture, Engineering and Construction (AEC) industry including sustainability, organisational & management issues, project management, information & communications technology, Building Information Modelling and built environment education.

Index

D

E

F

G

H

I

K

L

M

N

O

P

Q

R

S

T

U

V

W

www.ingramcontent.com/pod-product-compliance
Lightning Source LLC
LaVergne TN
LVHW080618300926
849162LV00005B/1206
* 9 7 8 1 7 9 9 8 6 6 0 0 8 *